École de Physique des Houches

Special Issue, 28 May–15 June 2012

Advanced Data Assimilation for Geosciences

Edited by

Éric Blayo, Marc Bocquet, Emmanuel Cosme, Leticia F. Cugliandolo

OXFORD
UNIVERSITY PRESS

OXFORD
UNIVERSITY PRESS

Great Clarendon Street, Oxford, OX2 6DP,
United Kingdom

Oxford University Press is a department of the University of Oxford.
It furthers the University's objective of excellence in research, scholarship,
and education by publishing worldwide. Oxford is a registered trade mark of
Oxford University Press in the UK and in certain other countries

First Edition published in 2015

Published in the United States of America by Oxford University Press
198 Madison Avenue, New York, NY 10016, United States of America

British Library Cataloguing in Publication Data
Data available

Library of Congress Control Number: 2014945196

ISBN 978–0–19–872384–4

École de Physique des Houches

Service inter-universitaire commun
à l'Université Joseph Fourier de Grenoble
et à l'Institut National Polytechnique de Grenoble

Subventionné par l'Université Joseph Fourier de Grenoble,
le Centre National de la Recherche Scientifique,
le Commissariat à l'Énergie Atomique

Directeur:
Leticia F. Cugliandolo, Sorbonne Universités, Université Pierre et Marie Curie, Laboratoire de Physique Théorique et Hautes Energies, Paris, France

Directeurs scientifiques de la session:
Éric Blayo, LJK, Université Grenoble Alpes, CNRS, INRIA, Grenoble, France
Marc Bocquet, Université Paris-Est and INRIA, CEREA, joint lab École des Ponts ParisTech and EdF R&D, Marne-la-Vallée, France
Emmanuel Cosme, LGGE, Université Grenoble Alpes, CNRS, Grenoble, France
Leticia F. Cugliandolo, Sorbonne Universités, Université Pierre et Marie Curie, Laboratoire de Physique Théorique et Hautes Energies, Paris, France

Previous sessions

I	1951	Quantum mechanics. Quantum field theory
II	1952	Quantum mechanics. Statistical mechanics. Nuclear physics
III	1953	Quantum mechanics. Solid state physics. Statistical mechanics. Elementary particle physics
IV	1954	Quantum mechanics. Collision theory. Nucleon-nucleon interaction. Quantum electrodynamics
V	1955	Quantum mechanics. Non equilibrium phenomena. Nuclear reactions. Interaction of a nucleus with atomic and molecular fields
VI	1956	Quantum perturbation theory. Low temperature physics. Quantum theory of solids. Ferromagnetism
VII	1957	Scattering theory. Recent developments in field theory. Nuclear and strong interactions. Experiments in high energy physics
VIII	1958	The many body problem
IX	1959	The theory of neutral and ionized gases
X	1960	Elementary particles and dispersion relations
XI	1961	Low temperature physics
XII	1962	Geophysics; the earths environment
XIII	1963	Relativity groups and topology
XIV	1964	Quantum optics and electronics
XV	1965	High energy physics
XVI	1966	High energy astrophysics
XVII	1967	Many body physics
XVIII	1968	Nuclear physics
XIX	1969	Physical problems in biological systems
XX	1970	Statistical mechanics and quantum field theory
XXI	1971	Particle physics
XXII	1972	Plasma physics
XXIII	1972	Black holes
XXIV	1973	Fluids dynamics
XXV	1973	Molecular fluids
XXVI	1974	Atomic and molecular physics and the interstellar matter
XXVII	1975	Frontiers in laser spectroscopy
XXVIII	1975	Methods in field theory
XXIX	1976	Weak and electromagnetic interactions at high energy
XXX	1977	Nuclear physics with heavy ions and mesons
XXXI	1978	Ill condensed matter
XXXII	1979	Membranes and intercellular communication
XXXIII	1979	Physical cosmology

XXXIV	1980	Laser plasma interaction
XXXV	1980	Physics of defects
XXXVI	1981	Chaotic behavior of deterministic systems
XXXVII	1981	Gauge theories in high energy physics
XXXVIII	1982	New trends in atomic physics
XXXIX	1982	Recent advances in field theory and statistical mechanics
XL	1983	Relativity, groups and topology
XLI	1983	Birth and infancy of stars
XLII	1984	Cellular and molecular aspects of developmental biology
XLIII	1984	Critical phenomena, random systems, gauge theories
XLIV	1985	Architecture of fundamental interactions at short distances
XLV	1985	Signal processing
XLVI	1986	Chance and matter
XLVII	1986	Astrophysical fluid dynamics
XLVIII	1988	Liquids at interfaces
XLIX	1988	Fields, strings and critical phenomena
L	1988	Oceanographic and geophysical tomography
LI	1989	Liquids, freezing and glass transition
LII	1989	Chaos and quantum physics
LIII	1990	Fundamental systems in quantum optics
LIV	1990	Supernovae
LV	1991	Particles in the nineties
LVI	1991	Strongly interacting fermions and high Tc superconductivity
LVII	1992	Gravitation and quantizations
LVIII	1992	Progress in picture processing
LIX	1993	Computational fluid dynamics
LX	1993	Cosmology and large scale structure
LXI	1994	Mesoscopic quantum physics
LXII	1994	Fluctuating geometries in statistical mechanics and quantum field theory
LXIII	1995	Quantum fluctuations
LXIV	1995	Quantum symmetries
LXV	1996	From cell to brain
LXVI	1996	Trends in nuclear physics, 100 years later
LXVII	1997	Modeling the earths climate and its variability
LXVIII	1997	Probing the Standard Model of particle interactions
LXIX	1998	Topological aspects of low dimensional systems
LXX	1998	Infrared space astronomy, today and tomorrow
LXXI	1999	The primordial universe
LXXII	1999	Coherent atomic matter waves
LXXIII	2000	Atomic clusters and nanoparticles
LXXIV	2000	New trends in turbulence
LXXV	2001	Physics of bio-molecules and cells
LXXVI	2001	Unity from duality: Gravity, gauge theory and strings

LXXVII	2002	Slow relaxations and nonequilibrium dynamics in condensed matter
LXXVIII	2002	Accretion discs, jets and high energy phenomena in astrophysics
LXXIX	2003	Quantum entanglement and information processing
LXXX	2003	Methods and models in neurophysics
LXXXI	2004	Nanophysics: Coherence and transport
LXXXII	2004	Multiple aspects of DNA and RNA
LXXXIII	2005	Mathematical statistical physics
LXXXIV	2005	Particle physics beyond the Standard Model
LXXXV	2006	Complex systems
LXXXVI	2006	Particle physics and cosmology: the fabric of spacetime
LXXXVII	2007	String theory and the real world: from particle physics to astrophysics
LXXXVIII	2007	Dynamos
LXXXIX	2008	Exact methods in low-dimensional statistical physics and quantum computing
XC	2008	Long-range interacting systems
XCI	2009	Ultracold gases and quantum information
XCII	2009	New trends in the physics and mechanics of biological systems
XCIII	2009	Modern perspectives in lattice QCD: quantum field theory and high performance computing
XCIV	2010	Many-body physics with ultra-cold gases
XCV	2010	Quantum theory from small to large scales
XCVI	2011	Quantum machines: measurement control of engineered quantum systems
XCVII	2011	Theoretical physics to face the challenge of LHC
Special Issue	2012	Advanced data assimilation for geosciences

Publishers

- Session VIII: Dunod, Wiley, Methuen
- Sessions IX and X: Herman, Wiley
- Session XI: Gordon and Breach, Presses Universitaires
- Sessions XII–XXV: Gordon and Breach
- Sessions XXVI–LXVIII: North Holland
- Session LXIX–LXXVIII: EDP Sciences, Springer
- Session LXXIX–LXXXVIII: Elsevier
- Session LXXXIX– : Oxford University Press

Preface

Data assimilation aims at determining as accurately as possible the state of a dynamical system by combining heterogeneous sources of information in an optimal way. Generally speaking, the mathematical methods of data assimilation comprise algorithms for forming optimal combinations of observations of a system, a numerical model that describes its evolution, and appropriate prior information. Data assimilation has a long history of application to high-dimensional geophysical systems dating back to the 1960s, with application to the estimation of initial conditions for weather forecasts. It has now become an intensive field of research, with numerous applications in oceanography and atmospheric chemistry and with extensions to other geophysical sciences. Data assimilation is now a key issue in most numerical forecasting systems in geophysics. Because of the high dimensionality of these systems (in meteorological applications, up to 10^9 variables at present, and 10^7 scalar observations per 24-hour period), the mathematics employed should be relatively simple. However, it has been proven in operational meteorology that the use of advanced methods such as optimal control theory could improve data assimilation systems significantly. Since then, the data assimilation community has concurrently contributed to high-dimensional and operational applications and to the development of the fundamental theory of data assimilation.

In this context, the three-week school of which these are the proceedings was focused on the theory of data assimilation. However, because applications motivate and often dictate the use of specific methodologies, a large number of lectures and seminars were given on applications during the session. This Les Houches session gathered 24 lecturers, experts in theoretical and applied data assimilation, and 52 selected students or young researchers of the field.

The theoretical landscape of data assimilation is shaped by two broad classes of methods. First, sequential methods, and in particular the ensemble Kalman filter, a filter that scales well in high-dimensional systems, are introduced by Chris Snyder (Chapter 3). Second, variational methods inspired by optimal control theory are introduced by Olivier Talagrand (Chapter 1), and Andrew Lorenc (Chapter 2). However good the mathematical methods are, the performance of geophysical data assimilation systems can only be appreciated when using real observations. That is why Carla Cardinali presents the Earth global observational system, the information content of its observations, and the sensitivity of the data assimilation systems to those observations (Chapters 5 and 6).

Further on the road set out by Chris Snyder, Emmanuel Cosme presents the theory and implementations of smoothers (extensions of the Kalman filter for retrospective analysis) (Chapter 4), and Peter Jan van Leeuwen describes the promising use of particle filters with high-dimensional systems (Chapter 13). In spite of its elegance, the ensemble Kalman filter requires the use of parameterizations and efficient tricks, such as localization, which are the subject of Peter Houtekamer's chapter (Chapter 12).

Following on from Olivier Talagrand's and Éric Blayo's contributions, it is clear that the power of variational methods is also mitigated by the quality of error covariance modelling and of the a posteriori diagnostic tools used to tune those covariances. Those topics are introduced by Loïk Berre (Chapter 7), Gérald Desroziers (Chapter 8), and Olivier Talagrand (Chapter 9). Four-dimensional variational methods require the use of the adjoint of the model, and Laurent Hascoët showed how to build one, using advanced automatic software (Chapter 15).

Data assimilation in atmospheric and oceanic sciences has been developed in parallel with the inverse modelling techniques of solid-Earth physics. The inverse modelling aspects of data assimilation are more prominent in atmospheric chemistry and in new fields where many of the model parameters are unknown. This is Marc Bocquet's topic in Chapter 20. It is emphasized even more strongly in Chapter 21 by Frédéric Chevallier on the inverse modelling of greenhouse gases fluxes. In Chapter 25, Maëlle Nodet et al. show the promises of data assimilation and inverse modelling in glaciology.

The strongest motivation for the community to develop new data assimilation methods is the operational use of these algorithms in the forecasting geophysical systems. In Chapter 19, Florence Rabier and Michael Fisher describe the application of data assimilation methods to meteorology and introduce some promising new approaches. Ichiro Fukumori (Chapter 23) and Javier Zavala-Garay et al. (Chapter 24) describe the application of data assimilation to large-scale and coastal oceanography. Hendrik Elbern et al. (Chapter 22) describe an example of the use of data assimilation methods in atmospheric chemistry and air quality.

In the desire to apply data assimilation to more and more complex systems, data assimilation methodologies need better applicability to multiscale systems, since the atmosphere and the ocean are fundamentally multiscale systems. This is the subject of the chapters by Marc Bocquet et al. (Chapter 18) and Laurent Debreu et al. (Chapter 17).

New observations and the finer extraction of information from these observations are a constant endeavour of data assimilation. In this respect, Arthur Vidard (Chapter 16) presents recent advances in the assimilation of images and Lagrangian data.

Numerical applications are essential in data assimilation to validate and convince, and two ambitious training sessions were organized by Laurent Debreu, Bénédicte Lemieux-Dudon, Pierre-Antoine Bouttier, and Emmanuel Cosme, and were dedicated to the implementation of the data assimilation methods and tools described here in Chapters 1–3, 7, and 10.

Acknowledgements

This Les Houches session was made possible by substantial financial support from

- EdF R&D, and very strong support from Damien Bilbault and Luc Musson-Genon;

- Centre National d'Étude Spatial (CNES);
- Centre National de la Recherche Scientifique (CNRS);
- Institut Pierre-Simon Laplace (IPSL), with crucial support from Catherine Ottlé and Slimane Bekki;
- the French MEDDTL Ministry, and support from the Mission Changement Global Observation de la Terre (GMES), especially from Isabelle Benezeth and Vincent Pircher;
- Institut National de Recherche en Informatique et en Automatique (INRIA);
- École des Ponts ParisTech, and support and advice from Philippe Courtier, Serge Piperno, and Christian Seigneur;
- Maison de la modélisation et de la simulation (MaiMoSiNE in Grenoble),
- Météo France, and support from Philippe Bougeault;
- Région Rhônes-Alpes.

This financial support also enabled the funding of ten students.

The organization of this school at the École de Physique des Houches arose from the enthusiastic proposal of Jacques Verron and the LEFE-ASSIM scientific committee of INSU. The organizers wish to express their gratitude to the advisory scientific committee composed of Chris Snyder, Olivier Talagrand, and Jacques Verron, and to the administrative staff of the Les Houches School of Physics—Murielle Gardette, Isabelle Lelièvre, and Brigitte Rousset—for their support and help before, during, and after the session, as well as to the staff of the restaurant, who contributed to the success of the school.

É. Blayo
M. Bocquet
E. Cosme
L. F. Cugliandolo

Paris, Grenoble, June 2013

Contents

List of participants

Organizers

BLAYO Éric
Laboratoire Jean Kuntzmann, Université Grenoble Alpes, INRIA, CNRS, 51 rue des Mathématiques, BP 53, 38041 Grenoble Cedex 09, France.

BOCQUET Marc
École des Ponts ParisTech, CEREA, 6–8 avenue Blaise Pascal, Cité Descartes Champs-sur-Marne, 77455 Marne la Vallée Cedex 2, France.

COSME Emmanuel
Laboratoire de Glaciologie et Géophysique de l'Environnement, Université Grenoble Alpes, CNRS, BP 53, 38041 Grenoble Cedex 9, France.

Lecturers

BERRE Loïk
Météo France, 42 avenue Gustave Coriolis, 31057 Toulouse Cedex 1, France.

BRASSEUR Pierre
Laboratoire de Glaciologie et Géophysique de l'Environnement, Université Grenoble Alpes, CNRS, BP53, 38041 Grenoble Cedex 9, France.

CARDINALI Carla
ECMWF, Shinfield Park, Reading, Berks hire RG2 9AX, UK.

CHEVALLIER Frédéric
Laboratoire des Sciences du Climat et de l'Environnement, L'Orme des Merisiers, Bat 701, Point courrier 129, 91191 Gif-sur-Yvette Cedex, France.

DEBREU Laurent
Laboratoire Jean Kuntzmann, Université Grenoble Alpes, INRIA, CNRS, 51 rue des Mathématiques, BP 53, 38041 Grenoble Cedex 09, France.

DESROZIERS Gérald
Météo France, 42 avenue Gustave Coriolis, 31057 Toulouse Cedex 1, France.

ELBERN Hendrik
Rheinisches Institut für Umweltforschung, Institut für Geophysik und Meteorologie–EURAD-Projekt, Universität zu Köln, Aachener Strasse 209, 50931 Köln, Germany.

FISHER Mike
ECMWF, Shinfield Park, Reading, Berkshire RG2 9AX, UK.

FUKUMORI Ichiro
Jet Propulsion Laboratory, M/S 300-323, 4800 Oak Grove Drive, Pasadena, CA 91109, USA.

HANEA Remus
TU Delft, Mekelweg 4, 2628 CD Delft, The Netherlands.

HASCOËT Laurent
INRIA Sophia-Antipolis, 2004 route des Lucioles, BP 93, 06902 Valbonne, France.

HOUTEKAMER Peter
Meteorological Research Division, 2121 Transcanadian Highway, Dorval, Quebec, Canada H9P 1J3.

LE DIMET François-Xavier
Laboratoire Jean Kuntzmann, Université Grenoble Alpes, INRIA, CNRS, 51 rue des Mathématiques, BP 53, 38041 Grenoble Cedex 09, France.

LEMIEUX Bénédicte
Laboratoire Jean Kuntzmann, Université Grenoble Alpes, INRIA, CNRS, 51 rue des Mathématiques, BP 53, 38041 Grenoble Cedex 09, France.

LORENC Andrew C.
Met Office, FitzRoy Road, Exeter EX1 3PB, UK.

NODET Maëlle
Laboratoire Jean Kuntzmann, Université Grenoble Alpes, INRIA, CNRS, 51 rue des Mathématiques, BP 53, 38041 Grenoble Cedex 09, France.

RABIER Florence
Météo France, 42 avenue Gustave Coriolis, 31057 Toulouse Cedex 1, France.

SNYDER Chris
NCAR, 3450 Mitchell Lane, Boulder, CO 80307-3000, USA.

TALAGRAND Olivier
Laboratoire de Météorologie Dynamique, École Normale Supérieure, 24 rue Lhomond, 75231 Paris Cedex 05, France.

VAN LEEUWEN Peter Jan
Department of Meteorology, University of Reading, Earley Gate, PO Box 243, Reading RG6 6BB, UK.

VERRON Jacques
Laboratoire de Glaciologie et Géophysique de l'Environnement, Université Grenoble Alpes, CNRS, BP 53, 38041 Grenoble Cedex 9, France.

VIDARD Arthur
Laboratoire Jean Kuntzmann, Université Grenoble Alpes, INRIA, CNRS, 51 rue des Mathématiques, BP 53, 38041 Grenoble Cedex 09, France.

ZAVALA-GARAY Javier
Institute of Marine and Coastal Sciences, Rutgers University, 71 Dudley Road, New Brunswick, NJ 08901, USA.

Students and Auditors

AUTRET Emmanuelle
IFREMER, Laboratoire d'Océanographie Spatiale, Technopôle Brest Iroise, BP 70, 29280 Plouzané, France.

BARTHÉLEMY Antoine
Georges Lemaître Centre for Earth and Climate Research, Earth and Life Institute, Université Catholique de Louvain, chemin du Cyclotron 2, L7.01.11, B-1348 Louvain-la-Neuve, Belgium.

BELL Simon
Aston University, Aston Triangle, Birminghan B4 7ET, UK.

BELLSKY Tom
Arizona State University, School of Mathematical and Statistical Sciences, Physical Sciences, A-Wing, PO Box 871804, Tempe, AZ 85287-1804, USA.

BERCHET Antoine
Laboratoire des Sciences du Climat et de l'Environnement, CNRS, CEA, UVSQ, CEA-Orme des Merisiers, 91191 Gif-sur-Yvette Cedex, France.

BIANCHI Blandine
Ecole Polytechnique Fédérale de Lausanne, route Cantonale, 1015 Lausanne, Switzerland.

BONAN Bertrand
Laboratoire Jean Kuntzmann, Université Grenoble 1, INRIA, CNRS, 51 rue des Mathématiques, BP 53, 38041 Grenoble Cedex 09, France.

BOUTTIER Pierre-Antoine
Laboratoire Jean Kuntzmann, Université Grenoble 1, INRIA, CNRS, 51 rue des Mathématiques, BP 53, 38041 Grenoble Cedex 09, France.

BUCANEK Antonin
Czech hydrometeorological institute, Na Šabatce 2050/17, 143 06 Praha 4, Komořany, Czech Republic.

BUCHARD-MARCHANT Virginie
NASA/USRA GSFC, Building 33, Greenbelt, MD 20771, USA.

CHABOT Vincent
Laboratoire Jean Kuntzmann, Université Grenoble 1, INRIA, CNRS, 51 rue des Mathématiques, BP 53, 38041 Grenoble Cedex 09, France.

CHATTERJEE Abhishek
Carnegie Institution for Science, 260 Panama Street, Stanford, CA 94305, USA.

DINIZ Fabio
CPTEC/INPE, Rod. Presidente Dutra km 40, Cachoeira Paulista/SP, ZIP: 12.630-000, Brazil.

DREANO Denis
King Abdullah University of Science and Technology, Thuwal 23955-6900, Saudi Arabia.

EL GHARAMTI Mohamad
King Abdullah University of Science and Technology, Thuwal 23955-6900, Saudi Arabia.

EYYAMMADICHI Divya
Indian Institute of Tropical Meteorology, Dr Homi Babha Road, Pashan, Pune-411008, India.

FEKRI Majid
McGill University, Department of Atmospheric and Oceanic Sciences, Burnside Hall, 805 Sherbrooke Street West, Montreal, Quebec, Canada H3A 2K6.

GAUBERT Benjamin
Laboratoire Inter-universitaire des Systèmes Atmosphériques, Universités Paris-Est Créteil et Paris Diderot, CNRS, 61 avenue du Général de Gaulle, 94000 Créteil.

GEPPERT Gernot
Max Planck Institute for Meteorology, Bundesstrasse 53, 20146 Hamburg, Germany.

HASLEHNER Mylene
Hans Ertel Center for Data Assimilation, Meteorological Institute, LMU Munich, Theresienstrasse. 37, 80333 Munich, Germany.

HU Huiquin
Department of Atmospheric and Oceanic Sciences, School of Physics, Peking University, 5 Yiheyuan Road Haidian District, Beijing, PR China100871.

HUANG Ling
Department of Atmospheric and Oceanic Sciences, School of Physics, Peking University, 5 Yiheyuan Road Haidian District, Beijing, PR China100871.

IGRI Moudi Pascal
University of Yaounde 1, Department of Physics, PO Box 812, Yaounde, Cameroon.

ITO Kosuke
Department of Atmospheric Sciences, National Taiwan University No. 1, Section 4, Roosevelt Road, Taipei 10673, Taiwan.

KAY Jun Kyung
Yonsei University, 134 Sinchon-Dong, Seodaemun-Gu, Seoul 120-749, South Korea.

KIZHAKKENIYIL Manoj
University of Northern British Columbia, Department of Atmospheric Science, Environmental Science and Engineering Programs, 3333 University Way, Prince George, Canada V2N 4Z9.

LEE Ite
High Altitude Observatory, National Center for Atmospheric Research, 3080 Center Green, Boulder, CO 80301, USA.

LEE Jin
Centre for Australian Weather and Climate Research, Australian Bureau of Meteorology, GPO Box 1289, Melbourne, Victoria 3001, Australia.

MASSMANN Silvia
Federal Maritime and Hydrographic Agency (BSH), Bernhard-Nocht-Strasse 78, 20359 Hamburg, Germany.

MECHRI Rihab
Laboratoire des Sciences du Climat et de l'Environnement, L'Orme des Merisiers, Bâtiment 701, Point courrier 129, 91191 Gif-sur-Yvette Cedex, France.

MONGIN Mathieu
CSIRO Division of Marine and Atmospheric Research, GPO Box 1538, Hobart, Tasmania 7001, Australia.

MOSTAMANDI Suleyman
Russian State Hydrometeorological University, Malookhtinskii 98, St Petersburg, Russia.

NAVARRO Thomas
Laboratoire de Météorologie Dynamique, 4 place Jussieu, 75252 Paris Cedex 05, France.

OSYCHNY Vladimir
National Centers for Environmental Prediction, 5200 Auth Road, Camp Springs, MD 20746, USA.

PENTAKOTA Sreenivas
Indian Institute of Tropical Meteorology, Dr Homi Babha Road, Pashan, Pune-411008, India.

RONG Xinyao
Chinese Academy of Meteorological Sciences, 46 ZhongGuanCun South Street, Beijing, 100081, PR China.

RUGGIERO Giovanni
LJAD, UMR6621, CNRS, Université de Nice-Sophia Antipolis, 06108 Nice Cedex 02, France.

SAUCEDO Marcos
CIMA/CONICET-UBA, UMI IFAECI/CNRS, Buenos Aires, Argentina.

SHLYAEVA Anna
Hydrometeorological Research Centre of Russia, 11–13B. Predtechensky sidest., Moscow, Russia 123242.

SIRISUP Sirod
National Institute of Informatics, 2-1-2 Hitotsubashi, Chiyoda-ku, Tokyo 101-8430, Japan.

SLIVINSKI Laura
Brown University, 182 George Street, Box F, Providence, RI 02912, USA.

SOMMER Matthias
Hans Ertel Center for Data Assimilation, Meteorological Institute, LMU Munich, Theresienstrasse 37, 80333 Munich, Germany.

SPERREVIK Ann Kristin
Norwegian Meteorological Institute, Postboks 43 Blindern, 0313 Oslo, Norway.

TOQUE Nathalie
CMCC, Viale Aldo Moro 44, 40127 Bologna, Italy.

VRETTAS Michail
Aston University, Aston Triangle, Birminghan B4 7ET, UK.

WANG Yiguo
École des Ponts ParisTech, CEREA, 6–8 avenue Blaise Pascal, Cité Descartes Champs-sur-Marne, 77455 Marne la Vallée Cedex 2, France.

WECHT Kevin
Harvard University, Department of Earth and Planetary Sciences, 29 Oxford Street, Cambridge, MA 02138, USA.

WILSON Gregory
Oregon State University, 104 COAS Admin Building, Corvallis, OR 97331-5503, USA.

WINIAREK Victor
École des Ponts ParisTech, CEREA, 6-8 avenue Blaise Pascal, Cité Descartes Champs-sur-Marne, 77455 Marne la Vallée Cedex 2, France.

WU Shu
Center for Climatic Research, 1225 W. Dayton Street, Madison, Wisconsin 53706-1695, USA.

YUE Jian
Department of Atmospheric and Oceanic Sciences, School of Physics, Peking University, Beijing, PR China.

ZHAO Shunliu
Carleton University, Department of Civil and Environmental Engineering, 1125 Colonel By Drive, Ottawa, Ontario, Canada K1S 5B6.

Part I

Key lectures

1
4D-VAR: four-dimensional variational assimilation

O. Talagrand

Laboratoire de Météorologie Dynamique
Ecole Normale Supérieure
Paris, France

Advanced Data Assimilation for Geosciences. First Edition.
Edited by É. Blayo, M. Bocquet, E. Cosme, and L. F. Cugliandolo.

Chapter Contents

1.1 Introduction

The expression *variational assimilation* designates a class of assimilation algorithms in which the fields to be estimated are explicitly determined as minimizers of a scalar function, called the objective function, that measures the misfit to the available data. In particular, *four-dimensional variational assimilation*, usually abbreviated as *4D-Var*, minimizes the misfit between a temporal sequence of model states and the observations that are available over a given assimilation window. As such, and contrary to the standard Kalman filter and, more generally, to sequential algorithms for assimilation, it propagates the information contained in the data both forward and backward in time.

From a numerical point of view, variational algorithms require the minimization of a scalar function defined over a large-dimensional space. That is possible in practice through the systematic use of the *adjoint* of the assimilating model.

We first describe variational assimilation in the context of statistical linear estimation, which also underlies the theory of the Kalman filter (Section 1.2). This leads to the definition of a general form for the objective function to be minimized. Minimization methods and the adjoint approach for computing gradients, are then succinctly described (Section 1.3), as well as practical implementation of variational assimilation (Section 1.4). A number of problems, associated in particular with the strong nonlinearity of the governing equations, are discussed (Section 1.5). The adjoint approach is further discussed, concerning in particular uses other than variational assimilation (Section 1.6). Conclusions follow in Section 1.7. A large part of what follows is derived in the framework of Bayesian and statistical estimation. $E[.]$ will denote statistical expectation and $\mathcal{N}(a, C)$ the Gaussian probability distribution (either scalar or vector) with expectation a and covariance C. Superscript T will denote transposition.

1.2 Variational assimilation in the context of statistical linear estimation

For an elementary introduction, consider the following situation. One wants to determine an unknown scalar quantity x^t (i.e. true state) from two observations of the form

$$z_1 = x^t + \varepsilon_1, \tag{1.1a}$$

$$z_2 = x^t + \varepsilon_2. \tag{1.1b}$$

In these expressions, ε_1 and ε_2 are observational errors, whose exact values are unknown but whose statistical properties are known. More precisely, it is assumed that these errors are centred ($E[\varepsilon_1] = E[\varepsilon_2] = 0$), are mutually uncorrelated ($E[\varepsilon_1 \varepsilon_2] = 0$), and have respective variances $E[\varepsilon_1^2] = s_1$ and $E[\varepsilon_2^2] = s_2$. We look for an estimate of x^t, of the form $x^a = \alpha_1 z_1 + \alpha_2 z_2$, with $\alpha_1 + \alpha_2 = 1$, and α_1 and α_2 chosen to minimize the statistical quadratic estimation error $s = E[(x^a - x^t)^2]$. The answer is

$$x^a = \frac{s_2 z_1 + s_1 z_2}{s_1 + s_2}; \tag{1.2}$$

i.e. each of the two measurements is weighted in inverse proportion to the variance of the error on that measurement. The corresponding quadratic estimation error, which minimizes s, and which we denote by s_a, is given by

$$\frac{1}{s_a} = \frac{1}{s_1} + \frac{1}{s_2}. \tag{1.3}$$

The same estimate x^a would be obtained by considering z_1 as a 'background' estimate for x, and z_2 as an 'observation' (or the reverse), and then applying the standard formulas for the Kalman filter.

The same estimate can also be obtained as the minimizer of the function

$$x \longrightarrow J(x) = \frac{1}{2}\left[\frac{(x - z_1)^2}{s_1} + \frac{(x - z_2)^2}{s_2}\right]. \tag{1.4}$$

The meaning of this expression is clear. The squared deviation of x from either one of the two observations is weighted in inverse proportion to the variance of the error on that observation. Minimization of $J(x)$ therefore imposes that x must fit either observation to within its own accuracy. This leads to the estimate given by (1.2).

Variational assimilation, as it is implemented at present in meteorological and oceanographical applications, minimizes a function that generalizes (1.4). In particular, in the linear case, and as in the elementary example above, it minimizes the statistical quadratic estimation error (on any component of the estimated fields individually), and is actually another algorithm for solving the same problem as the Kalman filter.

Consider the following more general estimation problem. Estimate an unknown vector $\mathbf{x}^t$ (with components $x_i^t, i = 1, \ldots, n$) belonging to *state space* $\mathcal{S}$, with dimension n, from a known *data vector* $\mathbf{z}$ (with components $z_j, j = 1, \ldots, m$) belonging to *data space* $\mathcal{D}$, with dimension m, of the form

$$\mathbf{z} = \mathbf{\Gamma}\mathbf{x}^t + \boldsymbol{\varepsilon}. \tag{1.5}$$

In (1.5), $\mathbf{\Gamma}$ is a known linear operator from $\mathcal{S}$ into $\mathcal{D}$, called the data operator, and represented by an $m \times n$ matrix. $\boldsymbol{\varepsilon}$ is a random vector in $\mathcal{D}$, called the error vector. The problem is therefore to invert the operator $\mathbf{\Gamma}$, taking into account, as far as possible, the statistical properties of the error $\boldsymbol{\varepsilon}$. The estimate of $\mathbf{x}^t$ is sought in the form of a linear (and a priori non-homogeneous) function of $\mathbf{z}$, namely

$$\mathbf{x}^a = \mathbf{a} + \mathbf{A}\mathbf{z}, \tag{1.6}$$

where $\mathbf{a}$ is a vector of $\mathcal{S}$ and $\mathbf{A}$ is a linear operator from $\mathcal{D}$ into $\mathcal{S}$. $\mathbf{a}$ and $\mathbf{A}$ are to be determined under the following two conditions:

(i) The estimate $\mathbf{x}^a$ is invariant under a change of origin in state space (for instance, if the unknown $\mathbf{x}^t$ contains temperatures, the result must be independent of whether those temperatures are expressed in degrees Celsius or in kelvins);
(ii) For any component x_i^t of $\mathbf{x}^t$, the statistical expectation of the square of the corresponding estimation error $x_i^a - x_i^t$ is minimized.

The solution to this problem is given by

$$\mathbf{x}^a = \left(\mathbf{\Gamma}^{\mathrm{T}}\mathbf{S}^{-1}\mathbf{\Gamma}\right)^{-1}\mathbf{\Gamma}^{\mathrm{T}}\mathbf{S}^{-1}\left(\mathbf{z}-\boldsymbol{\mu}\right), \tag{1.7}$$

i.e. $\mathbf{A} = \mathbf{\Gamma}^{\mathrm{T}}\mathbf{S}^{-1}\mathbf{\Gamma})^{-1}\mathbf{\Gamma}^{\mathrm{T}}\mathbf{S}^{-1}$ and $\mathbf{a} = -\mathbf{A}\boldsymbol{\mu}$, where $\boldsymbol{\mu} = E[\boldsymbol{\varepsilon}]$ and $\mathbf{S} = E[(\boldsymbol{\varepsilon}-\boldsymbol{\mu})(\boldsymbol{\varepsilon}-\boldsymbol{\mu})^{\mathrm{T}}]$ are respectively the expectation and covariance matrix of the error $\boldsymbol{\varepsilon}$. It is seen that $\mathbf{A}$ is a left-inverse of $\mathbf{\Gamma}$ (i.e. $\mathbf{A}\mathbf{\Gamma} = \mathbf{I}_n$, where $\mathbf{I}_n$ is the unit matrix of order n), with the consequence that the estimate $\mathbf{x}^a$ is unbiased ($E[\mathbf{x}^a - \mathbf{x}^t] = 0$), and that the corresponding estimation error has covariance

$$\mathbf{P}^a = E\left[\left(\mathbf{x}^a - \mathbf{x}^t\right)\left(\mathbf{x}^a - \mathbf{x}^t\right)^{\mathrm{T}}\right] = \left(\mathbf{\Gamma}^{\mathrm{T}}\mathbf{S}^{-1}\mathbf{\Gamma}\right)^{-1}. \tag{1.8}$$

Condition (ii) means that the trace of $\mathbf{P}^a$ is the minimum trace that can be obtained among all possible linear estimates of $\mathbf{x}^t$. Equations (1.7) and (1.8) generalize (1.2) and (1.3). The estimate $\mathbf{x}^a$ is called the *best linear unbiased estimate (BLUE)* of $\mathbf{x}^t$ from $\mathbf{z}$ (the term *best linear unbiased estimator* is also used). Its explicit determination requires knowledge of (at most) the expectation $\boldsymbol{\mu}$ and the covariance matrix $\mathbf{S}$ of the data error $\boldsymbol{\varepsilon}$.

Taking (1.7) at face value, the unambiguous definition of the *BLUE* requires the matrix $\mathbf{S}$, and then the matrix $\mathbf{\Gamma}^{\mathrm{T}}\mathbf{S}^{-1}\mathbf{\Gamma}$, to be invertible. The need for invertibility of $\mathbf{S}$ is only apparent (without going into full details, $\mathbf{S}$ is singular when some components of $\mathbf{x}$ are exactly observed; it then suffices to restrict the estimation to those components that are not exactly observed). The condition for invertibility of $\mathbf{\Gamma}^{\mathrm{T}}\mathbf{S}^{-1}\mathbf{\Gamma}$, once $\mathbf{S}$ is invertible, is, on the other hand, real. It is equivalent to the condition that the null space of the data operator $\mathbf{\Gamma}$ is restricted to the 0-vector:

$$\mathbf{\Gamma}\mathbf{x} = \mathbf{0} \iff \mathbf{x} = \mathbf{0}, \tag{1.9}$$

or equivalently that $\mathbf{\Gamma}$ has rank equal to the dimension n of $\mathbf{x}^t$. This means that the data vector $\mathbf{z}$ contains information, either directly or indirectly, on every component of $\mathbf{x}^t$. In other terms, the problem of determining $\mathbf{x}^t$ from $\mathbf{z}$ is overdetermined. This requires that $m \geq n$. There must be at least as many scalar data in $\mathbf{z}$ as there are scalar parameters to be determined. We will set $m = n + p$. The condition given by (1.9) will be called the *determinacy* condition.

The *BLUE* possesses a number of important properties:

- As already mentioned, the operator $\mathbf{A}$ is a left-inverse of $\mathbf{\Gamma}$. This means that if the data are exact ($\boldsymbol{\varepsilon} = 0$ in (1.9)), then so is the estimate $\mathbf{x}^a$.
- The *BLUE* is invariant under a change of origin in either data or state space. It is also invariant under any invertible linear change of coordinates in either space. This means, for instance, that a profile of observed temperatures can be transformed, through the hydrostatic equation, into a profile of geopotential values without altering the estimated fields. It also means that the horizontal wind can be estimated in terms of geometrical coordinates, or in terms of its divergence and vorticity. The result will be the same. This condition of invariance also means

that the *BLUE* is independent of the choice of a scalar product, either in state or data space. For instance, for any symmetric positive-definite matrix $\mathbf{C}$, the quantity $(\mathbf{x}^a - \mathbf{x}^t)^{\mathrm{T}}\mathbf{C}(\mathbf{x}^a - \mathbf{x}^t)$, which is one (among infinitely many) measure of the magnitude of the estimation error $(\mathbf{x}^a - \mathbf{x}^t)$, is minimized by the *BLUE*. The invariance of the *BLUE* in any invertible change of linear coordinates can also be expressed by saying that (1.7) and (1.8) are more than vector–matrix equations. They are *tensor equations*, valid in any system of linear coordinates;

- When the data error $\boldsymbol{\varepsilon}$ is Gaussian, $\boldsymbol{\varepsilon} \equiv \mathcal{N}(\boldsymbol{\mu}, \mathbf{S})$, the *BLUE* achieves Bayesian estimation, in the sense that the conditional probability distribution for the state vector $\mathbf{x}$, given the data vector $\mathbf{z}$, is the Gaussian distribution with expectation $\mathbf{x}^a$ and covariance matrix $\mathbf{P}^a$, as given by (1.7) and (1.8). In condensed notation, $P(\mathbf{x}|\mathbf{z}) = \mathcal{N}(\mathbf{x}^a, \mathbf{P}^a)$.

It is easily verified that the *BLUE* $\mathbf{x}^a$ can be obtained as the minimizer of the following scalar function, defined over state space:

$$\mathbf{x} \longrightarrow J(\mathbf{x}) = \frac{1}{2}\left[\boldsymbol{\Gamma}\mathbf{x} - (\mathbf{z} - \boldsymbol{\mu})\right]^{\mathrm{T}} \mathbf{S}^{-1}\left[\boldsymbol{\Gamma}\mathbf{x} - (\mathbf{z} - \boldsymbol{\mu})\right]. \tag{1.10}$$

This expression generalizes (1.4). Its significance is clear. For any vector $\mathbf{x}$ in state space, $\boldsymbol{\Gamma}\mathbf{x}$ is what the data operator $\boldsymbol{\Gamma}$ would produce if it were applied on $\mathbf{x}$. $J(\mathbf{x})$ is then a measure of the magnitude of the discrepancy between $\boldsymbol{\Gamma}\mathbf{x}$ and the unbiased data vector $\mathbf{z} - \boldsymbol{\mu}$. Through the inverse covariance matrix $\mathbf{S}^{-1}$, that measure possesses two notable properties. First, it weights the data according to their accuracy. Second, it is physically non-dimensional, making it possible to combine in a consistent way data of a different physical nature.

Variational assimilation, as it exists at present in meteorology and oceanography, minimizes objective functions of the form of (1.10), with the only difference, to be discussed later, that moderately nonlinear operators $\boldsymbol{\Gamma}$ are used. What follows is a more detailed description of how variational assimilation is implemented in practice and of the main results it produces.

The first step in the minimization of a function such as that given by (1.10) is to remove the bias in the data by subtracting the error expectation $\boldsymbol{\mu}$ from the data vector. Unless specified otherwise, it will be assumed below that this has been done, and the expectation $\boldsymbol{\mu}$ will no longer appear explicitly in the equations. But it must be kept in mind that implementation of variational assimilation requires the prior knowledge, and subtraction from the data, of the error expectation, or bias. Failure to properly remove the bias in the data will, in general, result in the presence of residual biases in the estimated fields.

When the determinacy condition (1.9) satisfied, the data vector $\mathbf{z}$ can always be transformed, through linear invertible operations, into two components of the following forms: first, an explicit estimate of the true state vector $\mathbf{x}$, of the form

$$\mathbf{x}^b = \mathbf{x}^t + \boldsymbol{\varepsilon}^b, \tag{1.11}$$

where $\boldsymbol{\varepsilon}^b$ is an error; second, an additional set of data, of the form

$$\mathbf{y} = \mathbf{H}\mathbf{x}^t + \boldsymbol{\varepsilon}^o, \tag{1.12}$$

with dimension $p = m - n$. In this equation, $\mathbf{H}$ is a linear operator, represented by a $p \times n$ matrix, and $\boldsymbol{\varepsilon}^o$ is an error. In addition, the transformations that lead to (1.11) and (1.12) can always be defined in such a way that the errors $\boldsymbol{\varepsilon}^b$ and $\boldsymbol{\varepsilon}^o$ are uncorrelated:

$$E\left[\boldsymbol{\varepsilon}^b(\boldsymbol{\varepsilon}^o)^{\mathrm{T}}\right] = 0 \tag{1.13}$$

It is in the form of (1.11) and (1.12) that data are most often available in meteorological and oceanographical applications. The component $\mathbf{x}^b$ is a prior, or background estimate, of the unknown state vector $\mathbf{x}^t$ at a given time k (usually a recent forecast or a climatological estimate). As for the additional vector $\mathbf{y}$, it consists of observations depending on the state vector through the observation operator $\mathbf{H}$. The uncorrelation hypothesis (1.13), although certainly disputable, is often (if not always) made. Equations (1.11) and (1.12), together with (1.13), are also assumed in the standard Kalman filter. We stress here that (1.11)–(1.13) are no more restrictive than, but exactly equivalent to, (1.5) together with the determinacy condition (1.9).

Introducing the covariance matrices of the errors $\boldsymbol{\varepsilon}^b$ and $\boldsymbol{\varepsilon}^o$,

$$\mathbf{P}^b = E\left[\boldsymbol{\varepsilon}^b(\boldsymbol{\varepsilon}^b)^{\mathrm{T}}\right], \qquad \mathbf{R} = E\left[\boldsymbol{\varepsilon}^o(\boldsymbol{\varepsilon}^o)^{\mathrm{T}}\right], \tag{1.14}$$

(1.7) and (1.8) take the following form, used in particular in the Kalman filter:

$$\mathbf{x}^a = \mathbf{x}^b + \mathbf{P}^b\mathbf{H}^{\mathrm{T}}\left(\mathbf{H}\mathbf{P}^b\mathbf{H}^{\mathrm{T}} + \mathbf{R}\right)^{-1}(\mathbf{y} - \mathbf{H}\mathbf{x}^b), \tag{1.15a}$$

$$\mathbf{P}^a = \mathbf{P}^b - \mathbf{P}^b\mathbf{H}^{\mathrm{T}}\left(\mathbf{H}\mathbf{P}^b\mathbf{H}^{\mathrm{T}} + \mathbf{R}\right)^{-1}\mathbf{H}\mathbf{P}^b. \tag{1.15b}$$

We recall that the vector

$$\mathbf{d} = \mathbf{y} - \mathbf{H}\mathbf{x}^b \tag{1.16}$$

is called the *innovation vector* and that the matrix $\mathbf{H}\mathbf{P}^b\mathbf{H}^{\mathrm{T}} + \mathbf{R}$, the inverse of which appears in (1.15), is the covariance matrix of $\mathbf{d}$:

$$\mathbf{H}\mathbf{P}^b\mathbf{H}^{\mathrm{T}} + \mathbf{R} = E\left[\mathbf{d}\mathbf{d}^{\mathrm{T}}\right]. \tag{1.17}$$

As for the objective function (1.10), it takes the following form under the decomposition (1.11) and (1.12):

$$J(\mathbf{x}) = \frac{1}{2}(\mathbf{x} - \mathbf{x}^b)^{\mathrm{T}}(\mathbf{P}^b)^{-1}(\mathbf{x} - \mathbf{x}^b) + \frac{1}{2}(\mathbf{H}\mathbf{x} - \mathbf{y})^{\mathrm{T}}\mathbf{R}^{-1}(\mathbf{H}\mathbf{x} - \mathbf{y}). \tag{1.18}$$

The meaning of this expression is clear. The first term on the right-hand side of (1.18) is a measure of the deviation of $\mathbf{x}$ from the background, while the second term is a measure of the deviation from the observation.

Several situations are encountered in the practice of meteorology and oceanography, which we are going to describe in some detail, giving more explicit expressions for the general form (1.18) of the objective function.

The simplest situation is when a background $\mathbf{x}^b$, of the form given by (1.11), is available at some time k, together with observations, of the form given by (1.12), that have been performed at the same time (or over a period of time short enough that the flow can be considered stationary). Minimization of the objective function (1.18) will produce an estimate of the state of the flow at time t. One then speaks in that case of three-dimensional variational analysis, often abbreviated as 3D-Var.

A different, more complex, situation is encountered when one wants to assimilate observations that are distributed over a period of time over which the evolution of the flow cannot be neglected. Let us assume that observations are available at successive times $k = 0, 1, \ldots, K$, of the form

$$\mathbf{y}_k = \mathbf{H}_k \mathbf{x}_k^t + \boldsymbol{\varepsilon}_k^o, \tag{1.19}$$

where $\mathbf{x}_k^t$ is the exact true state of the flow at time k, $\mathbf{H}_k$ is a linear observation operator, and $\boldsymbol{\varepsilon}_k^o$ is an observational error with covariance matrix $\mathbf{R}_k$. The observational errors are assumed to be uncorrelated in time. It is assumed in addition that the temporal evolution of the flow is described by the equation

$$\mathbf{x}_{k+1}^t = \mathbf{M}_k \mathbf{x}_k^t + \boldsymbol{\eta}_k, \tag{1.20}$$

with known model linear operator $\mathbf{M}_k$ and random model error $\boldsymbol{\eta}_k$.

Assume in addition that a background $\mathbf{x}_0^b$, with error covariance matrix $\mathbf{P}_0^b$ and error uncorrelated with the observational errors in (1.19), is available at time $k = 0$.

If the model error is ignored, any initial condition $\mathbf{x}_0$ at time $k = 0$ defines a model solution

$$\mathbf{x}_{k+1} = \mathbf{M}_k \mathbf{x}_k, \quad k = 0, \ldots, K-1. \tag{1.21}$$

The objective function

$$J(\mathbf{x}_0) = \frac{1}{2}\left(\mathbf{x}_0 - \mathbf{x}_0^b\right)^{\mathrm{T}}\left(\mathbf{P}_0^b\right)^{-1}\left(\mathbf{x}_0 - \mathbf{x}_0^b\right) + \frac{1}{2}\sum_{k=0}^{K}\left(\mathbf{H}_k\mathbf{x}_k - \mathbf{y}_k\right)^{\mathrm{T}}\mathbf{R}_k^{-1}\left(\mathbf{H}_k\mathbf{x}_k - \mathbf{y}_k\right), \tag{1.22}$$

which is of the general form given by (1.10), measures the distance between the model solution (1.21) and the data. Minimization of $J(\mathbf{x}_0)$ will define the initial condition of the model solution that fits the data most closely. Following a terminology first introduced by Sasaki (1970a–c), this is called *strong-constraint four-dimensional variational assimilation*, often abbreviated as strong-constraint 4D-Var. The words 'strong-constraint' stress the fact that the model identified by (1.21) must be exactly satisfied by the sequence of estimated state vectors.

If the model error is taken into account, (1.20) defines an additional set of 'noisy' data. We assume the model error $\boldsymbol{\eta}_k$ in (1.20) to have covariance matrix $\mathbf{Q}_k$, to be uncorrelated in time, and to be uncorrelated with observation and background errors. Equation (1.10) then gives the following expression for the objective function defining the *BLUE* of the sequence of states $\{\mathbf{x}_k, k = 0, \ldots, K\}$:

$$
\begin{aligned}
J(\mathbf{x}_0, \mathbf{x}_1, \ldots, \mathbf{x}_K) = & \frac{1}{2}\left(\mathbf{x}_0 - \mathbf{x}_0^b\right)^{\mathrm{T}}\left(\mathbf{P}_0^b\right)^{-1}\left(\mathbf{x}_0 - \mathbf{x}_0^b\right) \\
& + \frac{1}{2}\sum_{k=0}^{K}\left(\mathbf{H}_k\mathbf{x}_k - \mathbf{y}_k\right)^{\mathrm{T}}\mathbf{R}_k^{-1}\left(\mathbf{H}_k\mathbf{x}_k - \mathbf{y}_k\right) \\
& + \frac{1}{2}\sum_{k=0}^{K-1}\left(\mathbf{x}_{k+1} - \mathbf{M}_k\mathbf{x}_k\right)^{\mathrm{T}}\mathbf{Q}_k^{-1}\left(\mathbf{x}_{k+1} - \mathbf{M}_k\mathbf{x}_k\right). \qquad (1.23)
\end{aligned}
$$

The objective function is now a function of the whole sequence of states $\{\mathbf{x}_k, k = 0, \ldots, K\}$. Minimization of an objective function of the form given by (1.23), where the model equations are present as noisy data to be fitted by the analysed fields like any other data, is called, again according to the terminology introduced by Sasaki (1970a–c), *weak constraint four-dimensional variational assimilation*, abbreviated as *weak constraint 4D-Var.*

Equations (1.22) and (1.23), with appropriate redefinition of the state and observation spaces, are particular cases of (1.10). Another type of variational algorithm can be defined from (1.15a), which can be written as

$$\mathbf{x}^a = \mathbf{x}^b + \mathbf{P}^b\mathbf{H}^{\mathrm{T}}\mathbf{w}, \qquad (1.24)$$

where the vector $\mathbf{w} = [\mathbf{H}\mathbf{P}^b\mathbf{H}^{\mathrm{T}} + \mathbf{R}]^{-1}\mathbf{d}$ minimizes the objective function

$$K(\boldsymbol{\nu}) = \frac{1}{2}\boldsymbol{\nu}^{\mathrm{T}}\left(\mathbf{H}\mathbf{P}^b\mathbf{H}^{\mathrm{T}} + \mathbf{R}\right)\boldsymbol{\nu} - \mathbf{d}^{\mathrm{T}}\boldsymbol{\nu}. \qquad (1.25)$$

This function is defined on the dual of the observation space, which has dimension p. Minimization of (1.25) corresponds to the dual approach to variational assimilation, in opposition to the primal approach given by (1.18). The dual approach is also known as defining the *physical space assimilation system* (*PSAS*, pronounced 'pizzazz'; the word *physical* is historical). Just as (1.18), (1.22), and (1.23) are particular forms of (1.10), the dual approach can be used in any of the situations corresponding to those three equations. Depending on the conditions of the problem, and especially on the relative dimension of the state and observation spaces, it may be more advantageous to use the primal or the dual approach. A significant difference is that the dual approach uses the error covariance matrices $\mathbf{P}^b$ and $\mathbf{R}$ in their direct forms, while the primal approach requires their inverses. Another difference is that the dual approach requires an explicit background $\mathbf{x}^b$, while the primal approach can be implemented, in the general form given by (1.10), without an explicit background (it only requires the determinacy condition (1.9)).

All forms of variational assimilation given by (1.18), (1.22), (1.23), and (1.25) have been used, or at least extensively studied, for assimilation of meteorological and oceanographical observations. The theory of the *BLUE* requires the data operators ($\boldsymbol{\Gamma}$, $\mathbf{H}$, and $\mathbf{M}_k$ in the above notation) to be linear. In practice, this condition is rarely satisfied. In particular, variational assimilation of form given by (1.22) or (1.23) is almost always implemented with a nonlinear model. From a heuristic point of view, it is clear that, if the nonlinearity is in a sense sufficiently small, variational assimilation, even if it does not solve a clearly identified estimation problem, is likely to produce useful results (this point will be further discussed in Section 1.5). The dual approach, on the other hand, explicitly uses the transpose observation operator $\mathbf{H}^{\mathrm{T}}$, and requires exact linearity.

1.3 Minimization methods. The adjoint approach

1.3.1 Gradient methods for minimization

Variational assimilation aims at minimizing an objective function of one of the forms defined in Section 1.2. The objective functions that we will consider can be exactly quadratic or not. We will make a slight change of notation and systematically denote by $\mathbf{x}$ and call *control variable* the argument of the function to be minimized; in (1.23), the control variable is the whole sequence $\mathbf{x}_0, \mathbf{x}_1, \ldots, \mathbf{x}_K$, while it is $\boldsymbol{\nu}$ in (1.25). The control variable belongs to the *control space*, whose dimension will be denoted by N. We will denote by $\partial J/\partial \mathbf{x}$ the gradient of J with respect to $\mathbf{x}$, i.e. the N-vector whose components are the partial derivatives of J with respect to the components x_i of $\mathbf{x}$, namely

$$\frac{\partial J}{\partial \mathbf{x}} = \left(\frac{\partial J}{\partial x_i}\right)_{i=1,\ldots,N}. \tag{1.26}$$

The gradient is equal to 0 at the minimum of the objective function. One way to determine the minimum could conceivably be (as is actually often done in simple small-dimension problems) to determine analytical expressions for the components of the gradient and then to solve a system of N scalar equations for the minimizing components of $\mathbf{x}$. In meteorological and oceanographic applications, the complexity of the computations defining the objective function (in 4D-Var, these calculations include the temporal integration of a numerical dynamical model of the flow over the assimilation window) makes it totally inconceivable even to obtain analytical expressions for the gradient. Another way to proceed is to implement an iterative minimization algorithm, which determines a sequence of successive approximations $\mathbf{x}^{(l)}$ of the minimizing value of $\mathbf{x}$, namely

$$\mathbf{x}^{(l+1)} = \mathbf{x}^{(l)} - \mathbf{D}^{(l)}, \tag{1.27}$$

where $\mathbf{D}^{(l)}$ is at every iteration an appropriately chosen vector in control space. One possibility is to choose $\mathbf{D}^{(l)}$ along the direction of the local gradient $\partial J/\partial \mathbf{x}$. Algorithms that are based on that choice, called *steepest-descent* algorithms, turn out, however, not to be numerically very efficient. In other algorithms, the vector $\mathbf{D}^{(l)}$ is determined as a combination of the local gradient and a number of gradients computed at previous steps of the iteration (1.27) (see e.g. Bonnans et al., 2003). All minimization methods that are efficient for large dimensions are of the form given by (1.27), and require the explicit determination at each iteration step of the local gradient $\partial J/\partial \mathbf{x}$. They are called *gradient methods*. Since one cannot hope to obtain an analytical expression for the gradient, it must be determined numerically. One possibility could be to determine it by finite differences, by imposing in turn a perturbation Δx_i on all components x_i of the control vector and approximating the partial derivative $\partial J/\partial x_i$ by the difference quotient

$$\frac{\partial J}{\partial x_i} \simeq \frac{J(\mathbf{x} - \Delta x_i) - J(\mathbf{x})}{\Delta x_i}. \tag{1.28}$$

This, however, would require N explicit computations of the objective function, i.e., in the case of four-dimensional assimilation, N integrations of the assimilating model. Although that has actually been done for variational assimilation of meteorological observations, in an experimental setting, and with a relatively small dimension model (Hoffman, 1986), it would clearly be impossible in any practical application.

1.3.2 The adjoint method

The adjoint method allows numerical computation of the gradient of a scalar function at a cost that is at most a few times the cost of the direct computation of that function. Adjoint equations are an extremely powerful mathematical and numerical tool. They are central to the theory of optimal control, i.e. the theory of how the behaviour of a physical system can be controlled by acting on some of its components (see e.g. Lions, 1971). Adjoint equations can also be used for solving mathematical problems in their own right. The use of adjoint equations in meteorological and oceanographical applications was advocated by the Russian school of mathematics at an early stage of development of numerical modelling of the atmosphere and ocean (see e.g. Marchuk, 1974). We are going to demonstrate the method of adjoint equations in the special case of strong-constraint 4D-Var (1.22), in the most general case where the model and observation operators can be nonlinear.

To stress the possible nonlinearity of the model and observation operators, we now introduce the nonlinear model operator M_k and the nonlinear observation operator H_k. The notation for operators used hitherto in this chapter, $\mathbf{M}_k$ and $\mathbf{H}_k$ (denoting linear model and observation operators, respectively), will be reserved henceforth for the Jacobians (matrices of partial derivatives) of M_k and H_k, respectively. We rewrite (1.21) and (1.22) with nonlinear operators as

$$J(\mathbf{x}_0) = \frac{1}{2}(\mathbf{x}_0 - \mathbf{x}_0^b)^{\mathrm{T}}(\mathbf{P}_0^b)^{-1}(\mathbf{x}_0 - \mathbf{x}_0^b) + \frac{1}{2}\sum_{k=0}^{K}[H_k(\mathbf{x}_k) - \mathbf{y}_k]^{\mathrm{T}}\mathbf{R}_k^{-1}[H_k(\mathbf{x}_k) - \mathbf{y}_k], \tag{1.29a}$$

with

$$\mathbf{x}_{k+1} = M_k(\mathbf{x}_k), \quad k = 0, \ldots, K-1. \tag{1.29b}$$

Our purpose is to determine the gradient $\partial J/\partial \mathbf{x}_0$ of J with respect to $\mathbf{x}_0$. That gradient is characterized by the property that, for any perturbation $\boldsymbol{\delta}\mathbf{x}_0$ of $\mathbf{x}_0$, the corresponding variation of J is, to first order with respect to $\boldsymbol{\delta}\mathbf{x}_0$, equal to

$$\boldsymbol{\delta} J = \left(\frac{\partial J}{\partial \mathbf{x}_0}\right)^{\mathrm{T}} \boldsymbol{\delta}\mathbf{x}_0. \tag{1.30}$$

The perturbation $\boldsymbol{\delta}\mathbf{x}_0$ results at later times in perturbations that, through differentiation of (1.29b), are given to first order by

$$\boldsymbol{\delta}\mathbf{x}_{k+1} = \mathbf{M}_k\, \boldsymbol{\delta}\mathbf{x}_k, \quad k = 0, \ldots, K-1, \tag{1.31}$$

where, as said, $\mathbf{M}_k$ is the Jacobian of $\mathbf{x}_{k+1}$ with respect to $\mathbf{x}_k$. Equation (1.31) is called the tangent linear equation of (1.29b). Although the dependence is not explicit in (1.31), it must be kept in mind that the Jacobian will in general depend in the nonlinear case on the local value of $\mathbf{x}_k$.

As for the first-order variation of the objective function J, it is given by differentiation of (1.29a), namely

$$\boldsymbol{\delta} J = (\mathbf{x}_0 - \mathbf{x}_0^b)^{\mathrm{T}} (\mathbf{P}_0^b)^{-1} \boldsymbol{\delta}\mathbf{x}_0 + \sum_{k=0}^{K} [H_k(\mathbf{x}_k) - \mathbf{y}_k]^{\mathrm{T}} \mathbf{R}_k^{-1} \mathbf{H}_k\, \boldsymbol{\delta}\mathbf{x}_k, \tag{1.32}$$

where $\mathbf{H}_k$ is the local Jacobian of H_k and where the $\boldsymbol{\delta}\mathbf{x}_k$'s are given by (1.31).

$\boldsymbol{\delta} J$ is a compound function of $\boldsymbol{\delta}\mathbf{x}_0$ through the $\boldsymbol{\delta}\mathbf{x}_k$'s. Our purpose is to 'skip' the intermediate $\boldsymbol{\delta}\mathbf{x}_k$'s and to obtain a direct dependence of $\boldsymbol{\delta} J$ with respect to $\boldsymbol{\delta}\mathbf{x}_0$ of the form given by (1.30). To that end, we introduce at each time $k = 1, \ldots, K$ a vector $\boldsymbol{\lambda}_k$ belonging to the dual of state space (and therefore with dimension n) and to be defined more precisely later. We form the products $\boldsymbol{\lambda}_k^{\mathrm{T}}(\boldsymbol{\delta}\mathbf{x}_k - \mathbf{M}_{k-1}\, \boldsymbol{\delta}\mathbf{x}_{k-1})$, which, according to (1.31), are equal to 0. Subtracting those products from the right-hand side of (1.32) yields

$$\begin{aligned} \boldsymbol{\delta} J =& (\mathbf{x}_0 - \mathbf{x}_0^b)^{\mathrm{T}} (\mathbf{P}_0^b)^{-1} \boldsymbol{\delta}\mathbf{x}_0 + \sum_{k=0}^{K} [H_k(\mathbf{x}_k) - \mathbf{y}_k]^{\mathrm{T}} \mathbf{R}_k^{-1} \mathbf{H}_k\, \boldsymbol{\delta}\mathbf{x}_k \\ & - \sum_{k=0}^{K} \boldsymbol{\lambda}_k^{\mathrm{T}} (\boldsymbol{\delta}\mathbf{x}_k - \mathbf{M}_{k-1}\, \boldsymbol{\delta}\mathbf{x}_{k-1}) \end{aligned} \tag{1.33}$$

(subtracting rather than adding the products is of course arbitrary, but convenient). We now transform (1.33) by first using the fact that the transpose of a matrix product is the product of the corresponding transposes, taken in reverse order. For instance, the product $[H_k(\mathbf{x}_k) - \mathbf{y}_k]^{\mathrm{T}} \mathbf{R}_k^{-1} \mathbf{H}_k$ is equal to $\{\mathbf{H}_k^{\mathrm{T}} \mathbf{R}_k^{-1} [H_k(\mathbf{x}_k) - \mathbf{y}_k]\}^{\mathrm{T}}$ (where use has been made of the fact that the covariance matrix $\mathbf{R}_k$ is symmetric), thus transforming the (scalar) quantity $[H_k(\mathbf{x}_k) - \mathbf{y}_k]^{\mathrm{T}} \mathbf{R}_k^{-1} \mathbf{H}_k\, \boldsymbol{\delta}\mathbf{x}_k$ into the scalar product of the two n-vectors $\mathbf{H}_k^{\mathrm{T}} \mathbf{R}_k^{-1} [H_k(\mathbf{x}_k) - \mathbf{y}_k]^{\mathrm{T}}$ and $\boldsymbol{\delta}\mathbf{x}_k$. Performing that operation on all terms in (1.33) and gathering all terms with common factor $\boldsymbol{\delta}\mathbf{x}_k$ yields

$$\begin{aligned} \boldsymbol{\delta} J =& \left\{ (\mathbf{P}_0^b)^{-1} (\mathbf{x}_0 - \mathbf{x}_0^b) + \mathbf{H}_0^{\mathrm{T}} \mathbf{R}_0^{-1} [H_0(\mathbf{x}_0) - \mathbf{y}_0] + \mathbf{M}_0^{\mathrm{T}} \boldsymbol{\lambda}_1 \right\}^{\mathrm{T}} \boldsymbol{\delta}\mathbf{x}_0 \\ & + \sum_{k=1}^{K-1} \left\{ \mathbf{H}_k^{\mathrm{T}} \mathbf{R}_k^{-1} [H_k(\mathbf{x}_k) - \mathbf{y}_k] - \boldsymbol{\lambda}_k + \mathbf{M}_k^{\mathrm{T}} \boldsymbol{\lambda}_{k+1} \right\}^{\mathrm{T}} \boldsymbol{\delta}\mathbf{x}_k \\ & + \left\{ \mathbf{H}_K^{\mathrm{T}} \mathbf{R}_K^{-1} [H_k(\mathbf{x}_K) - \mathbf{y}_K] - \boldsymbol{\lambda}_K \right\}^{\mathrm{T}} \boldsymbol{\delta}\mathbf{x}_K. \end{aligned} \tag{1.34}$$

This expression is valid for any choice of the $\boldsymbol{\lambda}_k$'s. It is seen that choosing recursively

$$\boldsymbol{\lambda}_K = \mathbf{H}_K^{\mathrm{T}}\mathbf{R}_K^{-1}[H_k(\mathbf{x}_K) - \mathbf{y}_K], \tag{1.35a}$$
$$\boldsymbol{\lambda}_k = \mathbf{M}_k^{\mathrm{T}}\boldsymbol{\lambda}_{k+1} + \mathbf{H}_k^{\mathrm{T}}\mathbf{R}_k^{-1}[H_k(\mathbf{x}_k) - \mathbf{y}_k] \quad \text{for } k = K-1, \ldots, 1, \tag{1.35b}$$
$$\boldsymbol{\lambda}_0 = \mathbf{M}_0^{\mathrm{T}}\boldsymbol{\lambda}_1 + (\mathbf{P}_0^b)^{-1}(\mathbf{x}_0 - \mathbf{x}_0^b) + \mathbf{H}_0^{\mathrm{T}}\mathbf{R}_0^{-1}[H_0(\mathbf{x}_0) - \mathbf{y}_0] \tag{1.35c}$$

eliminates all $\boldsymbol{\delta}\mathbf{x}_k$ terms in (1.34), except the $\boldsymbol{\delta}\mathbf{x}_0$ term. There remains

$$\boldsymbol{\delta} J = \boldsymbol{\lambda}_0^{\mathrm{T}}\boldsymbol{\delta}\mathbf{x}_0, \tag{1.36}$$

which shows that $\boldsymbol{\lambda}_0$ is the required gradient of the objective function with respect to the initial condition $\mathbf{x}_0$ (see (1.30)).

Equations (1.35a–c) make up the *adjoint* of the tangent linear equation (1.31). The word 'adjoint' comes from the fact that (1.35a–c) are built on the transpose matrices $\mathbf{H}_k^{\mathrm{T}}$ and $\mathbf{M}_k^{\mathrm{T}}$, which are particular cases of the more general notion of adjoint operators. The adjoint equation is defined for the particular solution $\mathbf{x}_k$ of the basic equation (1.29b) for which the gradient is to be determined. It depends on that solution through the terms $[H_k(\mathbf{x}_k) - \mathbf{y}_k]$ and, in the case of either a nonlinear model operator M_k or a nonlinear observation operator H_k, through the transpose Jacobians $\mathbf{M}_k^{\mathrm{T}}$ and/or $\mathbf{H}_k^{\mathrm{T}}$. It is often said for convenience that (1.35a–c) define the adjoint of the basic model given by (1.29b), but it must be kept in mind that the adjoint equation is defined for a particular solution of that model.

The computations to be performed for determining the gradient $\partial J/\partial\mathbf{x}_0$ for a given initial condition $\mathbf{x}_0$ are now clearly defined:

(1) Starting from $\mathbf{x}_0$, integrate the basic equation (1.29b). Store the corresponding solution $\mathbf{x}_k$ in memory.
(2) Starting from the 'final' condition (1.35a) at time K, integrate the adjoint equations (1.35b, c) backward in time. The required gradient is $\boldsymbol{\lambda}_0$. The direct solution $\mathbf{x}_k$ is necessary for computing the terms $\mathbf{H}_k^{\mathrm{T}}\mathbf{R}_k^{-1}[H_k(\mathbf{x}_k) - \mathbf{y}_k]$ and, in case the basic model (1.29b) is nonlinear, for determining the transpose Jacobian $\mathbf{M}_k^{\mathrm{T}}$.

The determination of the gradient therefore requires one forward integration of the basic model (1.29b), followed by one backward integration of the adjoint model (1.35a–c). The latter is a modified form of the direct model, and the corresponding cost must be of similar magnitude to the cost of integrating the direct model. It can be rigorously shown that, in terms of the number of arithmetic operations to be performed, the cost of one adjoint computation of the gradient $\partial J/\partial\mathbf{x}_0$ is at most four times the cost of one computation of the objective function J. In meteorological and oceanographical applications, the cost of one adjoint integration (in terms of elapsed computer time) is typically twice the cost of one direct integration. This ratio is basically independent of the dimension N of the control variable, and it makes the adjoint computation of a gradient much more economical than the N direct model integrations that would be required if the gradient were to be computed by explicit perturbations. It is this fact that made variational assimilation possible at all in the first place.

Not surprisingly, there is a price to be paid for this major reduction in computing time. The price, as seen above, is the necessity to store in memory the direct solution $\mathbf{x}_k$. More precisely, what has to be kept in memory (or else to be recomputed in the course of the adjoint integration) are all quantities that are arguments of nonlinear operations in the direct integration. Relaxing the storage constraint, for instance by using a more economical approximate adjoint, is difficult. Experience shows that minimization algorithms, especially efficient ones, are very sensitive to even slight misspecification of the gradient. The question of how the cost of variational assimilation can be reduced will be discussed in the Section 1.4.

The description that has just been given of the adjoint method is fundamentally sufficient for 4D-Var. It obviously covers the case of 3D-Var (minimization of an objective function of form given by (1.18)), which does not involve a dynamical model of the flow. In that case, of course, only the transpose Jacobian $\mathbf{H}^T$ of the observation operator is needed.

The first attempt at using the adjoint approach for variational assimilation of meteorological observations was made by Penenko and Obraztsov (1976), on a simple one-level linear atmospheric model and with synthetic data. Later attempts were made by Lewis and Derber (1985), Le Dimet and Talagrand (1986), and Talagrand and Courtier (1987). Courtier and Talagrand (1987) first used real data, while Thacker and Long (1988) made the first attempt at using adjoint equations for variational assimilation of oceanographical observations. Thépaut and Courtier (1991) first used a full primitive equation meteorological model. These early works showed that variational assimilation of meteorological or oceanographical observations was numerically feasible at an acceptable cost and produced physically realistic results. Variational assimilation was progressively applied to more and more complex numerical models. It was introduced in 1997 in operational prediction, in the strong-constraint formulation, at the European Centre for Medium-Range Weather Forecasts (ECMWF) (Klinker et al. 2000) and in 2000 at the French Meteorological Service (Météo-France). In both places, operational implementation of variational assimilation has resulted in significant improvements of the ensuing forecasts. Some of these improvements were due to side effects not directly linked to the variational character of the assimilation, but others, especially in a number of specific meteorological situations, were due to better consistency between the assimilated states and the dynamics of the atmospheric flow. Since then, other meteorological services, such as the Japan Meteorological Agency, the Meteorological Office (United Kingdom), the Meteorological Service of Canada, and the China Meteorological Administration, have introduced variational assimilation in their operational prediction system. Except at ECMWF, all these schemes are of the strong-constraint form, and use a 6-hour assimilation window. ECMWF now uses a 12-hour window, and has introduced a weak-constraint component in the assimilation for the stratosphere. In addition, ECMWF, after having produced several sets of reanalysed past observations, all based on sequential assimilation algorithms, is now running a new reanalysis project (the *ERA-Interim* project, ⟨http://www.ecmwf.int/research/era/do/get/era-interim⟩) based on variational assimilation. A specific advantage of variational assimilation in the case of reanalysis of past data is that it propagates information both forwards and backwards in time, thus allowing the use of observations that have been performed after estimation time.

Similar developments have taken place in oceanography, and variational assimilation using the adjoint of oceanographic circulation models is now commonly used for many diverse applications (although not so far for operational oceanographic prediction). Those applications include determination of the initial conditions of the flow, as described above (see e.g. Weaver and Anderson, 1997; Vialard et al., 2003; Ricci et al., 2005), but also identification of 'parameters', such as wind stress at the surface of the ocean (Vossepoel et al., 2004). Egbert et al. (1994) and Louvel (2001) used the dual approach through minimization in dual observation space of an objective function of the form given by (1.25). In that approach, each iteration of the minimization process requires first a backward integration of the adjoint model, followed by a forward integration of the tangent linear model. Variational assimilation has also extended to other fields of geophysics and environmental sciences, such as atmospheric chemistry (Fisher and Lary, 1995; Errera and Fonteyn, 2001; Elbern et al., 2007; Lahoz et al., 2010) or surface hydrology (Reichle, 2000; Walker and Houser, 2001). Other extensions of the variational methodology, which have largely benefited from the experience in meteorology, have been to terrestrial and solar magnetism (Fournier et al., 2007; Sun et al., 2007, Jouve et al., 2011) and seismology (Tromp et al., 2005).

1.4 Practical implementation

If the principle of variational assimilation and of the adjoint method is conceptually perfectly clear and rigorous, practical implementation raises a number of serious problems. We will discuss below the specific problems associated with the development and validation of a code for performing the adjoint computations defined by (1.35) and are going to consider first a number of purely numerical problems.

1.4.1 The incremental approach

The developments of Section 1.3 seem to require that it is the adjoint of the complete model (1.29b) that has to be used for the computation of the gradient of the objective function. A numerical weather prediction (NWP) model is an extremely complex and lengthy code, and the ensuing 'all-or-nothing' choice (take the complete adjoint of the model or else do nothing) seems particularly impractical. Simplifying the adjoint equation as such, without modification of the direct model or of the objective function, is not an appropriate solution. That would lead to an approximate gradient of the objective function, and, as has already been said, experience shows that minimization algorithms, especially efficient ones, are very sensitive to even slight misspecification of the gradient. A convenient and versatile solution, known as the *incremental approach* to variational assimilation, was introduced by Courtier et al. (1994). Several variants of that approach exist. We are going to describe the one that is conceptually the simplest.

The basic idea is to simplify the dynamical model (1.29b) to a form that is both more economical and more manageable, in particular as concerns the adjoint. However, that is not done on the model (1.29b) itself, but rather on the tangent linear model (1.31). A *reference solution* $\mathbf{x}_k^{(0)}$ of the basic equation (1.29b) having been determined (emanating for instance from the background $\mathbf{x}_0^b = \mathbf{x}_0^{(0)}$), the corresponding tangent linear model ((1.31) is modified to

$$\boldsymbol{\delta}\mathbf{x}_{k+1} = \mathbf{L}_k\, \boldsymbol{\delta}\mathbf{x}_k, \quad k = 0, \ldots, K-1, \tag{1.37}$$

where $\mathbf{L}_k$ is, at any time k, an appropriately chosen 'simpler' operator than the Jacobian $\mathbf{M}_k$. Consistency then requires that the basic model (1.29b) be modified in such a way that the tangent linear equation corresponding to the solution $\mathbf{x}_k^{(0)}$ is (1.37). This is achieved by making the initial condition $\mathbf{x}_0 = \mathbf{x}_0^{(0)} + \boldsymbol{\delta}\mathbf{x}_0$ evolve into $\mathbf{x}_k = \mathbf{x}_k^{(0)} + \boldsymbol{\delta}\mathbf{x}_k$, where $\boldsymbol{\delta}\mathbf{x}_k$ itself evolves according to (1.37). That makes the basic dynamics linear.

As for the objective function (1.29a), several possibilities exist, at least when the observation operators are nonlinear. One possibility is to linearize those operators just as the model operator M_k has been linearized. This leads to replacing the quantity $H_k(\mathbf{x}_k)$ by $H_k(\mathbf{x}_k^{(0)}) + \mathbf{N}_k\boldsymbol{\delta}\mathbf{x}_k$, where $\mathbf{N}_k$ is an appropriate simplified linear operator (possibly, but not necessarily, the Jacobian of H_k at point $\mathbf{x}_k$). The objective function (1.29a) is then replaced by

$$\begin{aligned} J_1(\boldsymbol{\delta}\mathbf{x}_0) = {} & \frac{1}{2}\left(\boldsymbol{\delta}\mathbf{x}_0 + \mathbf{x}_0^{(0)} - \mathbf{x}_0^b\right)^{\mathrm{T}}\left(\mathbf{P}_0^b\right)^{-1}\left(\boldsymbol{\delta}\mathbf{x}_0 + \mathbf{x}_0^{(0)} - \mathbf{x}_0^b\right) \\ & + \frac{1}{2}\sum_{k=0}^{K}\left(\mathbf{N}_k\boldsymbol{\delta}\mathbf{x}_k - \mathbf{d}_k\right)^{\mathrm{T}}\mathbf{R}_k^{-1}\left(\mathbf{N}_k\boldsymbol{\delta}\mathbf{x}_k - \mathbf{d}_k\right), \end{aligned} \tag{1.38}$$

where the $\boldsymbol{\delta}\mathbf{x}_k$'s are subject to (1.37), and where $\mathbf{d}_k = \mathbf{y}_k - H_k(\mathbf{x}_k^{(0)})$ is the innovation at time k. The function given by (1.38) is a quadratic function of the initial perturbation $\boldsymbol{\delta}\mathbf{x}_0$. The minimizing perturbation $\boldsymbol{\delta}\mathbf{x}_{0,m}$ defines a new initial state $\mathbf{x}_0^{(1)} = \mathbf{x}_0^{(0)} + \boldsymbol{\delta}\mathbf{x}_{0,m}$, from which a new solution $\mathbf{x}_k^{(1)}$ of the basic equation (1.29b) is computed. The process is then repeated for the solution $\mathbf{x}_k^{(1)}$.

This defines a system of two-level nested loops for minimization of the original objective function (1.29a). The fundamental advantage of the incremental approach is that it allows one to define at will the simplified linearized operators $\mathbf{L}_k$ and $\mathbf{N}_k$. Many degrees of freedom are available for ensuring an appropriate trade-off between practical implementability and meteorological accuracy and usefulness. The simplified dynamics in (1.37) can itself be modified in the course of the minimization, by progressively introducing more and more complex dynamics or 'physics' in the successive outer loops.

It is the incremental method that, after the adjoint method, makes variational assimilation feasible. It is implemented, either in the form that has just been described or in slightly different variants, in most (if not all) operational NWP systems that use variational assimilation. At ECMWF, it is implemented with two outer loops, the approximations introduced in the linearized dynamics (1.37) consisting first of a reduced spatial resolution (from triangular spectral truncation T799 to T255 for the second outer loop) and second of a simplified 'physical' package.

An obvious question is whether the nested-loop process of the incremental process converges and, if it does, to what it converges. In the case where the linearized operators $\mathbf{L}_k$ and $\mathbf{N}_k$ vary from one outer loop to the next, the possible convergence of the process can depend on the way those operators vary. In particular, convergence to the minimum of the original objective function (1.29a) is possible only if the linear operators $\mathbf{L}_k$ and $\mathbf{N}_k$ converge to the corresponding Jacobians $\mathbf{M}_k$ and $\mathbf{H}_k$ at that minimum. The question of the convergence of the incremental process has been studied in some detail by Trémolet (2007) on the ECMWF 4D-Var system. Numerical tests show that the process does not converge asymptotically, at least in the conditions in which it is implemented at ECMWF. The way the incremental approach is implemented, at ECMWF and elsewhere, is largely based on empirical tuning.

1.4.2 First-guess-at-the-right-time 3D-Var

An extreme case of the incremental approach is what is called *first-guess-at-the-right-time 3D-Var*, or *FGAT 3D-Var*. It can be described as a process of the form of (1.37) and (1.38) in which the simplified linear operator $\mathbf{L}_k$ is taken as the identity operator. This process is four-dimensional in that the observations distributed over the assimilation window are compared with their analogues in a time-evolving reference integration of the assimilating model. But it is three-dimensional in that the minimization of the objective function (1.38) does not use any explicit dynamics other than the trivial dynamics expressed by the unit operator and that the numerical implementation is in effect three-dimensional. The FGAT 3D-Var approach, which is implemented through a unique minimization (no nested loops), has been shown to improve the quality of the assimilated fields, simply through the fact that is effectively uses a more exact innovation vector than does standard 3D-Var, in which all observations over the assimilation window are compared with the same first-guess field.

1.5 Further considerations on variational assimilation

Independently of its numerical and algorithmic properties, the major advantage of variational assimilation is that it takes into account, through the adjoint equation, the temporal evolution of the uncertainty in the state of the flow, at least over the assimilation window. Although (contrary to the Kalman filter) it does not explicitly compute the evolution of the uncertainty as such (and, in particular, does not produce an explicit estimate of the uncertainty in the estimated fields), it determines an approximation of the minimizing solution of the objective function (1.29a), which depends on the dynamics of the flow, and of the temporal evolution of the uncertainty. This was shown in full detail by Thépaut et al. (1993), who compared the impact of individual observations in a 3D-Var process, which ignores the temporal evolution of the uncertainty, and a 4D-Var process. The impact was significantly different, and strongly dependent on the dynamical state of the flow, in the latter case.

Significant impact does not of course mean positive impact. All operational implementations of 4D-Var have been preceded by the development and implementation of a 3D-Var system. This is very convenient in that it allows progressive introduction of the various components of the full 4D-Var system. But it also provides the opportunity for systematic comparison of 3D-Var and 4D-Var. The comparison has always shown the superiority of 4D-Var, in particular in terms of the quality of the ensuing forecasts. Similar comparisons have also been performed, with the same conclusions, on other, non-operational assimilation systems. See also Lorenc and Rawlins (2005) for a detailed discussion of 3D-Var and 4D-Var.

All operational implementations of 4D-Var have so far been of the strong-constraint form. In spite of the constant improvement of NWP models, the hypothesis of a perfect model is of course highly disputable. Weak-constraint assimilation, which corresponds to minimization of an objective function of the form given by (1.23), would certainly be desirable. It, however, requires a quantitative estimate, in the form of the covariance matrix $\mathbf{Q}_k$, of the model error. A reliable estimate may be difficult to obtain. Derber (1989) has suggested identifying a possible systematic bias in the model by introducing that bias in the control variable. Other authors (Zupanski, 1997; Trémolet, 2006) have studied algorithms of the general form given by (1.23). There is some indication (M. Fisher, personal communication) that weak-constraint variational assimilation could be useful over longer assimilation windows (24 hours or more) than used in strong-constraint assimilation. That is easily understandable in view of the fact that the perfect model hypothesis becomes less and less valid as the length of the assimilation window increases.

The primal weak-constraint objective function (1.23) becomes singular in the limit of a perfect model ($\mathbf{Q}_k = 0$). As already said, the dual approach uses the data error covariance matrices in their direct form, so that the dual objective function (1.25), as defined for weak-constraint variational assimilation, is regular for $\mathbf{Q}_k = 0$. This means that the same dual algorithm can be used for both strong- and weak-constraint variational assimilation. This is an attractive feature of the dual approach.

Courtier (1997) has shown that, subject to an appropriate preconditioning of the dual variable $\boldsymbol{\nu}$ in (1.25), the numerical conditioning (and therefore the numerical cost) of the dual algorithm is the same as that of the primal approach. In variational assimilation, it is actually the repeated numerical integrations of the direct and adjoint models that take the major part of the computations, and the numerical cost of strong- and weak-constraint variational assimilation is fundamentally the same. This point is discussed in more detail in Louvel (2001).

The dual approach requires strict linearity of the operator $\mathbf{H}$ in (1.25), which, in the case of variational assimilation, means strict linearity of the model and observation operators. Auroux and Blum (2002, 2004) have introduced a double-loop algorithm (which has some similarity with the incremental approach described above) in which successive linear problems of form given by (1.25) are solved, each being based on a linearization about the result of the previous one.

More generally, and independently of the particular numerical algorithm that is used, the validity of the linear approach defined by (1.7) and (1.10) is questionable in meteorological and oceanographical applications. It has already been said that, from a purely heuristic point of view, the linear approach must be valid if the nonlinearities are in a sense small enough. A more accurate description of the real situation that is encountered in meteorology and oceanography is given, rather than by (1.11) and (1.12), by

$$\mathbf{x}^b = \mathbf{x}^t + \varepsilon^b, \tag{1.39}$$

$$\mathbf{y} = H^*\left(\mathbf{x}^t\right) + \varepsilon, \tag{1.40}$$

where H^* denotes a nonlinear observation operator. In the case of 3D-Var, H^* is the observation operator at estimation time. In the case of 4D-Var, the vector $\mathbf{y}$ denotes the complete temporal sequence of observations, and the operator H^* includes the (nonlinear) dynamical model. The knowledge of the data (1.39) and (1.40) is equivalent to the knowledge of (1.39) together with what can be called the nonlinear innovation vector

$$\mathbf{d} = \mathbf{y} - H^*(\mathbf{x}^b) = H^*(\mathbf{x}^t) - H^*(\mathbf{x}^b) + \varepsilon. \tag{1.41}$$

If the background $\mathbf{x}^b$ is close enough to the real unknown state $\mathbf{x}^t$, then $\mathbf{d}$ can be approximated by

$$\mathbf{d} \simeq \mathbf{H}\left(\mathbf{x}^t - \mathbf{x}^b\right) + \varepsilon, \tag{1.42}$$

where $\mathbf{H}$ is here the Jacobian of the full operator H^* at point $\mathbf{x}^b$. If the so-called *tangent linear approximation* defined by (1.42) is valid, then (1.39)–(1.42) define an estimation problem that is linear with respect to the deviation $\mathbf{x} - \mathbf{x}^b$ of the real state with respect to the background $\mathbf{x}^b$. Equations (1.15) and (1.18) are then valid, $\mathbf{H}$ being the Jacobian of H^*. In the case of 4D-Var, this leads to minimization of an objective function of the incremental form given by (1.37) and (1.38), where the operators $\mathbf{L}_k$ and $\mathbf{N}_k$ replace the exact Jacobians $\mathbf{M}_k$ and $\mathbf{H}_k$ along the (full nonlinear) reference model solution.

Both direct (see e.g. Lacarra and Talagrand 1988) and indirect evidence shows that the tangent linear approximation is valid for large-scale geostrophic atmospheric flow (scales larger than 200 km) up to about 24–48 hours. This limit, however, rapidly decreases with decreasing spatial scales, to be of the order of a few hours for convective scales. For oceanic geostrophic flow (scales larger than a few tens of kilometres), the limit is a few weeks.

The developments of this chapter are therefore fully valid within those limits. It is to be stressed, however, that in circumstances where the tangent linear approximation is known or hypothesized to be valid, the linearization in (1.42) is rarely performed explicitly. Either fully nonlinear operators are kept in the objective function to be minimized, or (as is actually the case in the incremental approach described above) approximations that go further than (1.42) are implemented. The only case where the linearization given by (1.42) seems to have explicitly been implemented is in the above-mentioned works of Auroux and Blum (2002, 2004) relative to the dual approach, which requires exactly linear operators.

But the question arises as to what is to be done in circumstances when the tangent linear approximation is not valid. In the context of 4D-Var, there are actually two different questions, depending on the strength of the nonlinearities. If these are weak, the minimization of an objective function of the general form given by (1.29) remains numerically feasible, but may not be justified on the basis of estimation theory. If the nonlinearities are strong, then even the numerical minimization of the objective function, owing for instance to the presence of distinct minima, can raise difficulties.

These questions have not been discussed so far in much depth. One can mention the work of Pires et al. (1996), who studied variational assimilation for a strongly chaotic nonlinear system (specifically, the celebrated three-parameter system of Lorenz, 1963). These authors have shown that the objective function given by (1.29a) possesses an increasing number of local minima with increasing length of the assimilation window. This can be easily understood in view of the repeated folding in state space that is associated with chaos. They have defined a procedure, called *quasi-static variational assimilation (QSVA)*, in which the length of the assimilation window, starting from a value for which the objective function (1.29a) possesses a unique minimum, is progressively increased. Each new minimization is started from the result of the previous one. This allows one to keep track of the absolute minimum of the objective function, at least if the temporal density of observations is in a sense high enough. QSVA has been implemented on a quasi-geostrophic atmospheric model by Swanson et al. (1998), who have been able to usefully extend variational assimilation (in the hypothesis of a perfect model) to assimilation windows as long as five days. This is largely beyond the limit of validity of the tangent linear approximation. QSVA, or a similar algorithm, could possibly be implemented in operational practice, for instance by using successive overlapping assimilation windows.

Other developments have taken place recently at the research level. Carrassi et al. (2008) have defined a 3D-Var system in which the control variable, instead of consisting of the whole state vector, is restricted to the deviations from the background along the (relatively few) unstable modes of the system. This approach is now being extended to 4D-Var (Trevisan, personal communication). Somewhat similar work has been performed by Liu et al. (2008), who have developed a low-order incremental 4D-Var system. The background error covariance matrix $\mathbf{P}_0^b$, (1.38), is defined, not on the basis of an a priori statistical model, but on the basis of the dispersion of an ensemble of background forecasts. As in Carrassi et al. (2008), the control space is not the entire state space, but the state spanned by the background forecasts. Taking advantage of the relatively small dimension of the control space, and of the linearity associated with the incremental character of the procedure, it is not necessary to use an adjoint code for computing the gradient of the objective function. That can be achieved through simple transposition of an appropriate matrix. The results obtained are competitive with a fully fledged 4D-Var. The 'ensemble' feature of those works give them similarity with the ensemble Kalman filter.

Both these works suggest that it could be possible to achieve substantial numerical gain, without significant degradation of the final results (and even maybe without the use of an adjoint), by restricting the control variable to an appropriate subspace of the whole state space.

All the algorithms that have been described above are based on the minimization of an objective function of the general form given by (1.10), (1.18), or (1.29a), which is quadratic in terms of the data-minus-unknown differences, with weights equal to the inverse of the covariance matrices of the corresponding errors. Equations (1.10) and (1.18) correspond to least-variance statistical linear estimation, while (1.29a) corresponds to an extension to weakly nonlinear situations. Other forms for the objective function have also been considered. In particular, Fletcher and Zupanski (2006) and Fletcher (2007), following a general Bayesian approach, propose to maximize the conditional probability density function for the state of the flow, given the data. In the case of linear data operators and Gaussian errors, this leads to minimization of an objective function of form given by (1.10). Those authors consider the case of lognormal distributions, which are more appropriate for bounded variables such as humidity. This leads to a significantly different form for the objective function.

1.6 More on the adjoint method

The adjoint method has been demonstrated above in the particular case of the objective function given by (1.29a). It is actually very general, and defines a systematic approach for computing the (exact) gradient of a differentiable scalar function with respect to its arguments. Although this may not be obvious from the above developments, the adjoint method consists in a systematic use of the chain rule for differentiation of a compound function. Proceeding backwards through the original sequence of computations, it recursively computes the partial derivatives of the scalar function under consideration with respect to the variables in those computations (see e.g. Talagrand, 2003). As such, the adjoint method can be used not only for optimization purposes, as in variational assimilation, but (actually more simply) for determination of gradients as such, and for sensitivity studies.

The advantages and disadvantages of variational assimilation will be further discussed in Section 1.7. But its major disadvantage (at least for variational assimilation as it exists at present) is probably the need to develop the adjoint code that performs computations in (1.38). Not only must the adjoint code be developed, but it must be carefully validated, since experience shows that even minor errors in the computed gradient can significantly degrade the efficiency of the minimization (if not totally inhibit it). In addition, NWP models are constantly modified, and the corresponding modifications must be made on the adjoint code. Writing the adjoint of a code at the same time as the direct code involves only a rather small amount of additional work (10% or 20%). But developing the adjoint of an already-existing code can require a substantial amount of work, and can be a very tedious and time-consuming task. On the other hand, the fact that adjoint computation is in essence a systematic use of the chain rule for differentiation leads to perfectly defined 'adjoint' coding rules, which make the development of an adjoint code, if lengthy and tedious, at least totally straightforward. These rules are described in, for example, Talagrand (1991), Giering and Kaminski (1998), and Kalnay (2002).

Those same rules are at the basis of 'adjoint compilers', i.e., software pieces that are designed to automatically develop the adjoint of a given code (see e.g. ⟨http://www.fastopt.de⟩; Hascoët and Pascual, 2004). The adjoint of a particular piece of code is independent of the rest of the code, and automating the derivation of the adjoint instructions for a sequence of coding instructions, which is a purely local operation, is relatively easy. Other aspects, such as the choice and management of nonlinear variables to be kept in memory from the direct integration, or to be recomputed in the course of the adjoint integration, require a global view of the code, and are more difficult to automate. For that reason, the use of these software pieces still requires experience of adjoint coding as well as some preparatory work, but they are nevertheless extremely useful, and very substantially reduce the amount of time and work necessary for developing the adjoint of an atmospheric or oceanic circulation model.

The adjoint approach is used in assimilation of meteorological and oceanographical observations for numerically solving, through an iterative minimization process, an optimization problem. Now, as said above, what the adjoint equations really do is simply compute the gradient of one scalar output of a numerical process with respect to (potentially all) the input parameters of that process. As such, the adjoint approach can be used for sensitivity studies of outputs with respect to inputs, independently of any optimization or minimization. It will be useful to use the adjoint approach when the number of output parameters whose sensitivity is sought is smaller than the number of input parameters with respect to which the sensitivity is sought (in the inverse case, direct perturbation of the input parameters will be more economical).

Actually, the first proponents of the use of the adjoint approach in meteorology and oceanography had primarily sensitivity studies in mind (Marchuk, 1974; Hall et al., 1982). Adjoint models have been used to perform sensitivity studies of many different kinds: sensitivity of the atmospheric flow with respect to initial or lateral boundary conditions (Errico and Vukisevic, 1992; Rabier et al., 1992; Gustafsson et al., 1998); sensitivity of the global oceanic circulation to parameters (Marotzke et al., 1999); sensitivity of biogeochemical processes (Waelbroeck and Louis, 1995); and sensitivity of atmospheric chemical processes (Zhang et al., 1998). See also the special issue of *Meteorologische Zeitschrift* (Ehrendorfer and Errico, 2007) devoted to Adjoint Applications in Dynamic Meteorology. Two specific types of applications are worthy of particular mention. The first has to do with the identification, for a particular situation, of the unstable components of the flow. In its simplest form, this amounts to determining the so-called *singular vectors* of the flow, i.e. the perturbations that amplify most rapidly, over a period of time, in the tangent linear approximation (Lacarra and Talagrand, 1988; Farrell, 1989; Urban, 1993). This has been extended by Mu and colleagues (Mu, 2000; Mu et al., 2003) to *nonlinear singular vectors (NLSVs)*, i.e. perturbations that amplify most rapidly in the full nonlinear evolution. A condition must then be imposed on the initial amplitude of the perturbation, which leads to a (technically more difficult to solve) constrained optimization problem. Both linear and nonlinear singular vectors allow accurate diagnostic and analysis of instability (Moore and Farrell, 1993; Mu and Zhang, 2006; Rivière et al., 2008). A related, but more specific, application is the identification of the components of the flow to which a particular feature of the future evolution of the flow (such as, for instance, the deepening of a depression) is most sensitive. This allows one to 'target' observations in order to optimize the prediction of the feature under consideration. This has been implemented successfully on the occasion of specific campaigns (see e.g. Langland et al., 1999; Bergot and Doerenbecher, 2002). Observation targeting through adjoint methods is further discussed in Buizza et al. (2007). Another, potentially very promising, application of the adjoint method is the determination of the sensitivity of analysed and predicted fields to observations. It is then the adjoint of the whole assimilation and prediction process, and not only of the assimilating model, that has to be used (Langland and Baker, 2004). This has led to very useful diagnostics of the value and usefulness of various types of observations (Langland and Cardinali, personal communication).

1.7 Conclusions

Variational assimilation has now become a basic tool of numerical meteorology and oceanography, and a major component of operational NWP in several major meteorological services. Together with the ensemble Kalman filter, it is one of the two most advanced and powerful assimilation methods. The specific advantages of variational assimilation are rather obvious. It is very versatile and flexible, and allows for easy introduction of a new type of observation in an assimilation system. It suffices to specify the corresponding observation operator and the first- and second-order statistical moments of the associated error. It automatically propagates information both forwards and backwards in time, and makes it easy to take into account temporal correlation between errors (either observation or model errors). To the author's knowledge, this last possibility has been used so far on only one occasion, for taking into account temporally correlated errors in high-frequency observations of surface pressure (Järvinen et al., 1999). But it can be extremely useful, especially for the treatment of model error and of the associated temporal correlation (the time will presumably come when this will be necessary).

Variational assimilation is costly in that it requires the development, validation, and maintenance of the adjoint of the assimilating model, as well as of the various observation operators. This is a time-consuming task. However, owing to the gain in experience and expertise, and to the continuous improvement of adjoint compilers, that task progressively becomes easier and easier. And, as discussed in Section 1.6, adjoints, once they are available, can be used for many other applications than assimilation, and in particular to powerful diagnostic studies.

Assimilation of meteorological and oceanographical observations may be at a turning point. It seems that the limits of what can be obtained from statistical linear estimation (i.e. from (1.7) and its various generalizations to weakly nonlinear situations) are being reached. The only exception is likely quasi-static variational assimilation, discussed in Section 1.5, which is based on minimization of objective functions of form given by (1.29), but whose limits have not been identified. Statistical linear estimation is at the basis of variational assimilation and of the 'Kalman' component of the ensemble Kalman filter. It can legitimately be said that the ultimate purpose of assimilation is to achieve Bayesian estimation, i.e. to determine the conditional probability distribution for the state of the atmosphere (or the ocean), given all the relevant available information. In view of the large dimension of the state of the atmosphere, the only possible way to describe the conditional probability distribution seems to be through an ensemble of points in state space, as indeed the ensemble Kalman filter already does. A basic question is then to determine whether it is possible to develop methods for ensemble variational assimilation, which would produce a Bayesian ensemble, while retaining the specific advantages of variational assimilation, namely easy propagation of information both forwards and backwards in time, and possibility to easily take error temporal correlations into account. Some results suggest that this should be possible.

References

Auroux, D. and Blum, J. (2002). A dual data assimilation method for a layered quasi-geostrophic ocean model. *RACSAM*, **96**, 315–320.

Auroux, D. and Blum, J. (2004). Data assimilation methods for an oceanographic problem. In *Multidisciplinary Methods for Analysis, Optimization and Control of Complex Systems, Mathematics in Industry*, Vol. 6 (ed. V. Capasso and J. Periaux), pp. 179–194. Springer-Verlag, Berlin.

Bergot, T. and Doerenbecher, A. (2002). A study on the optimization of the deployment of targeted observations using adjoint-based methods. *Q. J. R. Meteorol. Soc.*, **128**, 1689–1712.

Bonnans, J.-F., Gilbert, J.-C., Lemaréchal, C., and Sagatizabal, C. (2003). *Numerical Optimization—Theoretical and Practical Aspects*. Springer-Verlag, Berlin.

Buizza, R., Cardinali, C., Kelly, G., and Thépaut, J.-N. (2007). The value of observations. II: The value of observations located in singular-vector-based target areas. *Q. J. R. Meteorol. Soc.*, **133**, 1817–1832.

Carrassi, A., Trevisan, A., Descamps, L., Talagrand, O., and Uboldi, F. (2008). Controlling instabilities along a 3DVar analysis cycle by assimilating in the unstable subspace: a comparison with the EnKF. *Nonlinear Process. Geophys.*, **15**, 503–521.

Courtier, P. (1997). Dual formulation of four-dimensional variational assimilation. *Q. J. R. Meteorol. Soc.*, **123**, 2449–2461.

Courtier, P. and Talagrand, O. (1987). Variational assimilation of meteorological observations with the adjoint vorticity equation. II: Numerical results. *Q. J. R. Meteorol. Soc.*, **113**, 1329–1347.

Courtier, P., Thépaut, J.-N., and Hollingsworth, A. (1994). A strategy for operational implementation of 4D-Var, using an incremental approach. *Q. J. R. Meteorol. Soc.*, **120**, 1367–1387.

Derber, J. (1989). A variational continuous assimilation technique. *Mon. Weather Rev.*, **117**, 2437–2446.

Egbert, G. D., Bennett, A. F., and Foreman, M. G. C. (1994). Topex/Poseidon tides estimated using a global inverse model. *J. Geophys. Res.*, **99**, 24821–24852.

Ehrendorfer, M. and Errico, R. M., Eds. (2007). *Meteorol. Z.*, **16**, 591–818.

Elbern, H., Strunk, A., Schmidt, H., and Talagrand, O. (2007). Emission rate and chemical state estimation by 4-dimensional variational inversion. *Atmos. Chem. Phys.*, **7**, 3749–3769.

Errera, Q. and Fonteyn, D. (2001). Four-dimensional variational chemical assimilation of CRISTA stratospheric measurements. *J. Geophys. Res.*, **106**, 12253–12265.

Errico, R. M. and Vukisevic, T. (1992). Sensitivity analysis using an adjoint of the PSU-NCAR mesoscale model. *Mon. Weather Rev.*, **120**, 1644–1660.

Farrell, B. F. (1989). Optimal excitation of baroclinic waves. *J. Atmos. Sci.*, **46**, 1193–1206.

Fisher, M. and Lary, D. J. (1995). Lagrangian four-dimensional variational data assimilation of chemical species. *Q. J. R. Meteorol. Soc.*, **121**, 1681–1704.

Fletcher, S. J. and Zupanski, M. (2006). A hybrid normal and lognormal distribution for data assimilation. *Atmos. Sci. Lett.*, **7**, 43–46.

Fletcher, S. J. (2007). Implications and impacts of transforming lognormal variables into normal variables in VAR. *Meteorol. Z.*, **16**, 755–765.

Fournier, A., Eymin, C., and Alboussière, T. (2007). A case for variational geomagnetic data assimilation: insights from a one-dimensional, nonlinear, and sparsely observed MHD system. *Nonlinear Process. Geophys.*, **14**, 163–180.

Giering, R. and Kaminski, T. (1998). Recipes for adjoint code construction. *Trans. Math. Software*, **24**, 437–474.

Gustafsson, N., Källen, E., and Thorsteinsson, S. (1998). Sensitivity of forecast errors to initial and lateral boundary conditions. *Tellus*, **50A**, 167–185.

Hall, M. C. G., Cacuci, D. G., and Schlesinger, M. E. (1982). Sensitivity analysis of a radiative–convective model by the adjoint method. *J. Atmos. Sci.*, **39**, 2038–2050.

Hascoët, L. and Pascual, V. (2004). TAPENADE 2.1 User's Guide. Available at ⟨http://www.inria.fr/rrrt/rt-0300.html⟩.

Hoffman, R. N. (1986). A four-dimensional analysis exactly satisfying equations of motion. *Mon. Weather Rev.*, **114**, 388–397.

Järvinen, H., Andersson, E., and Bouttier, F. (1999). Variational assimilation of time sequences of surface observations with serially correlated errors. *Tellus*, **51A**, 469–488.

Jouve, L., Brun, A. S., and Talagrand, O. (2011). Assimilating data into an $\alpha\Omega$ dynamo model of the Sun: a variational approach, *Astrophys. J.* **735**, 31.

Kalnay, E. (2002). *Atmospheric Modeling, Data Assimilation and Predictability.* Cambridge University Press, Cambridge.

Klinker, E., Rabier, F., Kelly, G., and Mahfouf, J.-F. (2000). The ECMWF operational implementation of four-dimensional variational assimilation. III: Experimental results and diagnostics with operational configuration. *Q. J. R. Meteorol. Soc.*, **126**, 1191–1215.

Lacarra, J.-F. and Talagrand, O. (1988). Short-range evolution of small perturbations in a barotropic model. *Tellus*, **40A**, 81–95.

Lahoz, W. A., Geer, A. J., Bekki, S., Bormann, N., Ceccherini, S., Elbern, H., Errera, Q., Eskes, H. J., Fonteyn, D., Jackson, D. R., Khattatov, B., Marchand, M., Massart, S., Peuch, V.-H., Rharmili, S., Ridolfi, M., Segers, A., Talagrand, O., Thornton, H. E., Vik, A. F., and von Clarmann, T. (2007). The Assimilation of Envisat data (ASSET) project. *Atmos. Chem. Phys.*, **7**, 1773–1796.

Lahoz, W., Khattatov, B., and Menard, R. (eds.) (2010). *Data Assimilation*, Part IV: Astrospheric Chemistry. Springer-Verlag, Berlin.

Langland, R. H. and Baker, N. L. (2004). Estimation of observation impact using the NRL variational data assimilation adjoint system. *Tellus*, **56A**, 189–201.

Langland, R. H., Gelaro, R., Rohaly, G. D., and Shapiro, M. A. (1999). Targeted observations in FASTEX: adjoint-based targeting procedures and data impact experiments in IOP17 and IOP18. *Q. J. R. Meteorol. Soc.*, **125**, 3241–3270.

Le Dimet, F.-X. and Talagrand, O. (1986). Variational algorithms for analysis and assimilation of meteorological observations: theoretical aspects. *Tellus*, **38A**, 97–110.
Lewis, J. M. and Derber, J. C. (1985). The use of adjoint equations to solve a variational adjustment problem with advective constraints. *Tellus*, **37A**, 309–322.
Lions, J.-L. (1971). *Optimal Control of Systems Governed by Partial Differential Equations* (translated from the French). Springer-Verlag, Berlin.
Liu, C., Xiao, Q., and Wang, B. (2008). An ensemble-based four-dimensional variational data assimilation scheme. Part I: Technical formulation and preliminary test. *Mon. Weather Rev.*, **136**, 3363–3373.
Lorenc, A. C. and Rawlins, F. (2005). Why does 4D-Var beat 3D-Var? *Q. J. R. Meteorol. Soc.*, **131**, 3247–3257.
Lorenz, E. N. (1963). Deterministic nonperiodic flow. *J. Atmos. Sci.*, **20**, 130–141.
Louvel, S. (2001). Implementation of a dual variational algorithm for assimilation of synthetic altimeter data in the oceanic primitive equation model MICOM. *J. Geophys. Res.*, **106**, 9199–9212.
Marchuk, G.I. (1974). *Numerical Solution of the Problems of Dynamics of the Atmosphere and the Ocean* (in Russian). Gidrometeoizdat, Leningrad.
Marotzke, J., Giering, R., Zhang, K. Q., Stammer, D., Hill, C., and Lee, T. (1999). Construction of the adjoint MIT ocean general circulation model and application to Atlantic heat transport sensitivity. *J. Geophys. Res.*, **104**, 29529–29547.
Moore, A. M. and Farrell, B. F. (1993). Rapid perturbation growth on spatially and temporally varying oceanic flows determined using an adjoint method: application to the Gulf Stream. *J. Phys. Oceanogr.*, **23**, 1682–1702.
Mu, M. (2000). Nonlinear singular vectors and nonlinear singular values. *Sci. China, Ser. D*, **43**, 375–385.
Mu, M., Duan W. S., and Wang, B. (2003). Conditional nonlinear optimal perturbation and its applications. *Nonlinear Process. Geophys.*, **10**, 493–501.
Mu, M. and Zhang, Z. (2006). Conditional nonlinear optimal perturbations of a two-dimensional quasigeostrophic model. *J. Atmos. Sci.*, **63**, 1587–1604.
Penenko, V. V. and Obraztsov, N. N. (1976). A variational initialization method for the fields of the meteorological elements. *Sov. Meteorol. Hydrol.*, **11**, 1–11.
Pires, C., Vautard, R., and Talagrand, O. (1996). On extending the limits of variational assimilation in nonlinear chaotic systems. *Tellus*, **48A**, 96–121.
Rabier, F., Courtier P. and Talagrand, O. (1992). An application of adjoint models to sensitivity analysis. *Beitr. Phys. Atmos.*, **65**, 177–192.
Reichle, R. H. (2000). Land surface data assimilation. In *Variational Assimilation of Remote Sensing Data for Land Surface Hydrologic Applications*, Doctoral thesis, Massachusetts Institute of Technology, Cambridge, MA.
Ricci, S., Weaver, A. T., Vialard, J., and Rogel, P. (2005). Incorporating state-dependent temperature–salinity constraints in the background error covariance of variational ocean data assimilation. *Mon. Weather Rev.*, **133**, 317–338.
Rivière, G.., Lapeyre G. and Talagrand O. (2008). Nonlinear generalization of singular vectors: behavior in a baroclinic unstable flow. *J. Atmos. Sci.*, **65**, 1896–1911.
Sasaki, Y. (1970a). Some basic formalisms in numerical variational analysis. *Mon. Weather Rev.*, **98**, 875–883.

Sasaki, Y. (1970b). Numerical variational analysis formulated from the constraints as determined by longwave equations and a low-pass filter. *Mon. Weather Rev.*, **98**, 884–898.

Sasaki, Y. (1970c). Numerical variational analysis formulated with weak constraint and application to surface analysis of severe storm gust. *Mon. Weather Rev.*, **98**, 899–910.

Sun, Z., Tangborn A., and Kuang, W. (2007). Data assimilation in a sparsely observed one-dimensional modeled MHD system. *Nonlinear Process. Geophys.*, **14**, 181–192.

Swanson, K., Vautard R., and Pires, C. (1998). Four-dimensional variational assimilation and predictability in a quasi-geostrophic model. *Tellus*, **50A**, 369–390.

Talagrand, O. (1991). The use of adjoint equations in numerical modeling of the atmospheric circulation. In *Automatic Differentiation of Algorithms: Theory, Implementation, and Application* (ed. A. Griewank and G. F. Corliss). Society for Industrial and Applied Mathematics, Philadelphia.

Talagrand, O. (2003). Variational Assimilation. Adjoint Equations. In *Data Assimilation for the Earth System* (ed. R. Swinbank, V. Shutyaev and W. A. Lahoz). Kluwer, Dordrecht.

Talagrand, O. and Courtier, P. (1987). Variational assimilation of meteorological observations with the adjoint vorticity equation. I: Theory. *Q. J. R. Meteorol. Soc.*, **113**, 1311–1328.

Thacker, W. C. and Long, R. B. (1988). Fitting dynamics to data. *J. Geophys. Res.*, **93**, 1227–1240.

Thépaut, J.-N. and Courtier, P. (1991). Four-dimensional variational data assimilation using the adjoint of a multilevel primitive-equation model. *Q. J. R. Meteorol. Soc.*, **117**, 1225–1254.

Thépaut, J.-N., Hoffman, R. N., and Courtier, P. (1993). Interactions of dynamics and observations in a four-dimensional variational assimilation. *Mon. Weather Rev.*, **121**, 3393–3414.

Trémolet, Y. (2006). Accounting for an imperfect model in 4D-Var. *Q. J. R. Meteorol. Soc.*, **132**, 2483–2504.

Trémolet, Y. (2007). Incremental 4D-Var convergence study. *Tellus*, **59A**, 706–718.

Tromp, J., Tape C., and Liu Q. (2005). Seismic tomography, adjoint methods, time reversal, and banana-donut kernels. *Geophys. J. Int.*, **160**, 195–216.

Urban, B. (1993). A method to determine the theoretical maximum error growth in atmospheric models. *Tellus*, **45A**, 270–280.

Vialard, J., Weaver, A. T., Anderson D. L. T., and Delecluse, P. (2003). Three- and four-dimensional variational assimilation with a general circulation model of the tropical Pacific Ocean. Part II: Physical validation. *Mon. Weather Rev.*, **131**, 1379–1395.

Vossepoel, F., Weaver, A. T., Vialard, J., and Delecluse, P. (2004). Adjustment of near-equatorial wind stress with four-dimensional variational data assimilation in a model of the Pacific Ocean. *Mon. Weather Rev.*, **132**, 2070–2083.

Waelbroeck, C. and Louis, J.-F. (1995). Sensitivity analysis of a model of CO_2 exchange in tundra ecosystems by the adjoint method. *J. Geophys. Res.*, **100**, 2801–2816.

Walker, J. P. and Houser, P. R. (2001). A methodology for initializing soil moisture is a global climate model: assimilation of near surface soil moisture observations. *J. Geophys. Res.*, **106**, 761–774.

Weaver, A. T. and Anderson, D. L. T. (1997). Variational assimilation of altimeter data in a multilayer model of the tropical Pacific Ocean. *J. Phys. Oceanogr.*, **27**, 664–682.

Zhang, Y., Bischof, C. H., Easter, R. C., and Wu, P.-T. (1998). Sensitivity analysis of a mixed-phase chemical mechanism using automatic differentiation. *J. Geophys. Res.*, **103**, 18953–18979.

Zupanski, D. (1997). A general weak constraint applicable to operational 4DVAR data assimilation systems. *Mon. Weather Rev.*, **125**, 2274–2292.

2
Four-dimensional variational data assimilation

A. C. Lorenc

Met Office, Exeter
EX1 3PB, United Kingdom

Advanced Data Assimilation for Geosciences. First Edition.
Edited by É. Blayo, M. Bocquet, E. Cosme, and L. F. Cugliandolo.

Chapter Contents

2.1 4D-Var: background and motivation

2.1.1 The data assimilation cycle

Although data assimilation (DA) is conceptually a single process of fitting a model to observations, it is useful to split it into steps. This helps us manage the complexity of a modern system such as numerical weather prediction (NWP); each step can be developed independently, often by different members of a team.

1. Using the best available forecast model, make a forecast of the prior estimate (the *background* state) for the next batch of observations.[1]
2. Also predict the likely distribution of errors in this background state.[2]
3. Do a Bayesian combination of the prior from 1 and 2 with the batch of observations (the *analysis* step), to give a new best estimate. Repeat from 1.
4. Perform a statistical verification of the forecasts to measure the properties of the actual forecast error, and at intervals adapt the algorithm used in 2 (and also in 1 and 3).

This whole cycle will be covered in a more mathematical way in Chapter 3 on the Kalman filter. Here we will start by focusing on the analysis (step 3)—we will see that the background error distribution from step 2 also has to be considered.

For NWP, the best available forecast model is based mainly on physical understanding of the equations governing atmospheric behaviour; it needs a powerful computer.

Predicting the error distribution requires physical insight into the likely structure and evolution of forecast errors, combined with statistics on past forecast errors from step 4. The error description has to be put in a form usable for step 3—usually this is a *covariance model*. Recently, ensemble NWP systems have helped provide insight into error structure and evolution.

The whole NWP cycle is soundly based on *the Scientific Method*; we use our scientific understanding to build models for use in steps 1 and 2, test them against new information in step 4, and eventually improve them.

Because it is run and verified operationally every day, NWP has a much stronger evidence-base for the background distribution from step 2 than most other applications of DA, and hence a better justification for the Bayesian methods in step 3. Other applications of DA frequently use the expedient, frowned on by Bayesian purists, of adjusting the assumptions about the prior distribution to give the best results from the so-called Bayesian scheme.

2.1.2 Bayes' theorem

Bayes' theorem for discrete events

The Bayesian approach is to use probabilities to describe the accuracy of our knowledge about past events. We then have a formalism for modifying the probabilities in the

[1] And, if desired, on into the future to provide a useful forecast.

[2] And, if desired, on into the future to provide confidence estimates for the forecast

light of new knowledge—exactly what we need to do in sequential data assimilation. We introduce this with a discrete example for events A and B. $P(A)$ is the probability of A occurring (this is the usual use of probabilities), or a measure of our certainty that A occurred in the past (this is the Bayesian use of probabilities). Then $P(A \cap B)$ is the probability that A and B both occurred, and $P(A|B)$ is the conditional probability of A given B has occurred. We have two ways of expressing $P(A \cap B)$:

$$P(A \cap B) = P(A|B)P(B) = P(B|A)P(A). \tag{2.1}$$

This leads directly to Bayes' theorem:

$$P(A|B) = \frac{P(B|A)P(A)}{P(B)}. \tag{2.2}$$

So we start with a prior probability of A, and add the information that B has occurred, to give us the posterior probability of A given B. To evaluate this, we can calculate $P(B)$ from

$$P(B) = P(B|A)P(A) + P(B|\bar{A})P(\bar{A}), \tag{2.3}$$

where $\bar{A}$ means not A.

Single-variable Bayesian analysis with Gaussian pdfs

For continuous variables, we use probability distribution functions (pdfs). For probability densities, we use a lower-case p:

$$p(x)dx = P(x \le x^t < x + dx), \tag{2.4}$$

where x^t is the true value. Bayes' theorem becomes

$$p(x|y^o) = \frac{p(y^o|x)p(x)}{p(y^o)}. \tag{2.5}$$

Let us start thinking of x as the model state, which accumulates our knowledge. $p(x)$ is the prior distribution, our knowledge from previous observations. $p(x|y^o)$ is the posterior distribution, after adding the information from the observation y^o. $p(y^o|x)$ is the probability density of getting the observation y^o, given our previous knowledge. Note that this is a density in y-space. Regarded as a function of x, $p(y^o|x)$ is not a probability density (its integral is not necessarily 1); it is called the likelihood function for x.

Let us assume that our prior knowledge is that x is near x^b and that the variance of its deviation from x^b is V_b. The usual way of modelling such a distribution is as an unbiassed Gaussian:

$$p(x) = N(x; x^b, V^b) = (2\pi V^b)^{-1/2} \exp\left[-\frac{1}{2}\frac{(x - x^b)^2}{V^b}\right]. \tag{2.6}$$

If our observation directly measures the variable x, with observational error variance V_o, then the probability of observed value y^o, given that the true value is x, can also be modelled by an unbiased Gaussian pdf:

$$p(y^o|x) = N(y^o; x, V^o) = (2\pi V^o)^{-1/2} \exp\left[-\frac{1}{2}\frac{(x - y^o)^2}{V^o}\right]. \tag{2.7}$$

We can get $p(y^o)$ by integrating over all x:

$$p(y^o) = \int p(y^o|x)p(x)\,\mathrm{d}x. \tag{2.8}$$

For Gaussians (2.6) and (2.7), this gives

$$p(y^o) = N(y^o; x^b, V^o + V^b). \tag{2.9}$$

Substituting in (2.5) gives

$$p(x|y^o) = N(x; x^a, V^a), \tag{2.10}$$

where

$$\frac{x^a}{V^a} = \frac{y^o}{V^o} + \frac{x^b}{V^b}, \tag{2.11}$$

$$\frac{1}{V^a} = \frac{1}{V^o} + \frac{1}{V^b}. \tag{2.12}$$

This is the standard formula for the combination of observations with error, known since the work of Gauss in the 1820s.

Figure 2.1 shows the prior distribution $p(x)$, the posterior distribution $p(x|y^o)$, and the likelihood function $p(y^o|x)$ for four different values for y^o. A unique property of the Gaussian pdfs can be seen in them, and in the above equations: the shape of the posterior distribution, and its position relative to x^b and y^o, are both independent of the observed value. They depend only on the error variances.

The reason for this becomes apparent if we take minus the logarithms of both sides of (2.5). The Bayesian analysis equation becomes

$$-\ln[p(x|y^o)] = -\ln[p(x)] - \ln[p(y^o|x)] + \text{constant}. \tag{2.13}$$

Gaussians become quadratics, which add up to give another quadratic using (2.11) and (2.12):

$$-\ln[p(x|y^o)] = \frac{1}{2}\frac{(x - x^b)^2}{V^b} + \frac{1}{2}\frac{(x - y^o)^2}{V^o} + \text{constant}. \tag{2.14}$$

The Gaussian curves of Fig. 2.1 become the quadratics of Fig. 2.2, with the most probable values for x being at the minimum of the total curve. In preparation for variational methods, we call (2.14) the penalty function $J(x)$; it is the sum of the background (prior) penalty $J_b(x)$ and the observational penalty $J_o(x)$. This simple example shows that the Bayesian combination of Gaussian pdfs gives the same 'best' analysis as a weighted least-squares best fit to the data.

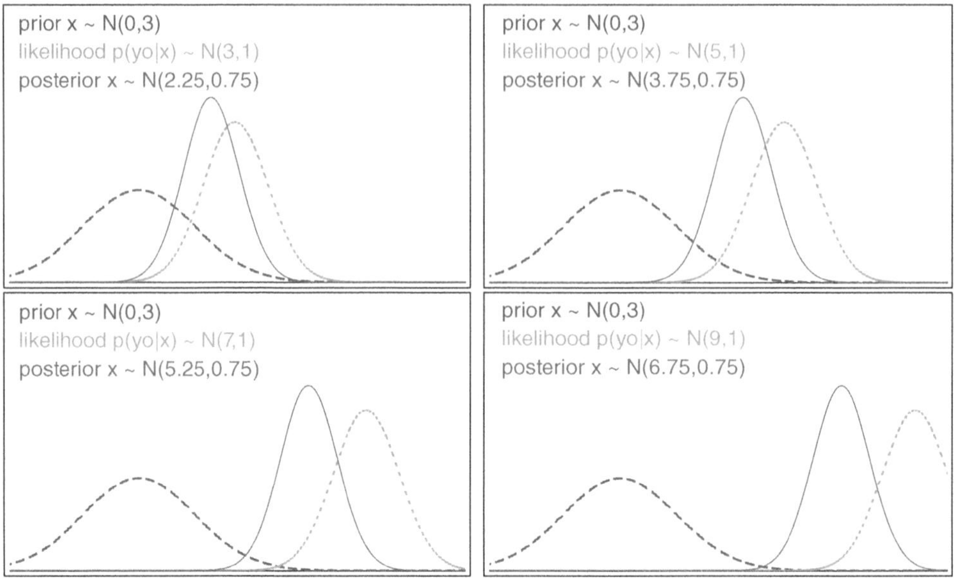

Fig. 2.1 Prior pdf $p(x)$ (dashed line), posterior pdf $p(x|y^o)$ (solid line), and Gaussian likelihood of observation $p(y^o|x)$ (dotted line), plotted against x for various values of y^o. (Adapted from Lorenc and Hammon (1988). Copyright © 1988 Royal Meteorological Society.)

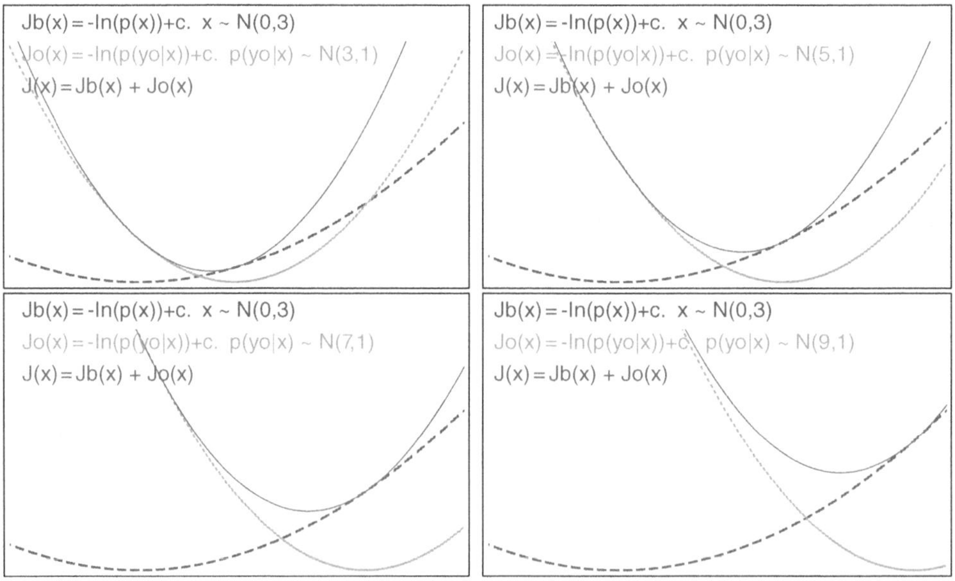

Fig. 2.2 As Fig. 2.1 except that minus logarithms of the curves are shown. The solid line is the penalty function $J(x)$, the dashed line is the prior or background penalty $J_b(x)$, and the dotted line is the observational penalty $J_o(x)$. They are related by $J(x) = J_b(x) + J_o(x) + \text{constant}$, and the most likely x minimizes $J(x)$.

2.1.3 Bayesian analysis equation

Simplest possible example

To extend the above to data assimilation of observations into a model, we need to introduce several new ideas. We do this in the context of the simplest relevant example:

- a model that represents the 'world' with two grid points: $\mathbf{x} = \begin{pmatrix} x_1 \\ x_2 \end{pmatrix}$;
- an instrument that gives one observed value y^o midway between the points; we can interpolate a model estimate y of the observed value:

$$\begin{aligned} \mathbf{y} = H(\mathbf{x}) &= \frac{1}{2}x_1 + \frac{1}{2}x_2 \\ &= \mathbf{Hx} = \begin{pmatrix} \frac{1}{2} & \frac{1}{2} \end{pmatrix} \begin{pmatrix} x_1 \\ x_2 \end{pmatrix}. \end{aligned} \tag{2.15}$$

We need a probability distribution (the background pdf) to describe what we know about $\mathbf{x}$, before getting the observation. Let us suppose we have prior estimates x_1^b and x_2^b with error variance V_b. As in (2.6), we assume a Gaussian:

$$p(x_1) = N(x_1; x_1^b, V^b) = (2\pi V^b)^{-1/2} \exp\left[-\frac{1}{2}\frac{(x_1 - x_1^b)^2}{V^b}\right], \tag{2.16}$$

with a similar equation for $p(x_2)$. Because $\mathbf{x}^b$ is produced by a forecast, errors in x_1^b and x_2^b are correlated:

$$\left\langle \left(x_1^b - x_1^t\right)\left(x_2^b - x_2^t\right)\right\rangle = \mu V^b, \tag{2.17}$$

where μ is the correlation coefficient.[3] The corresponding joint pdf can be modelled by a multidimensional Gaussian:

$$\begin{aligned} p(x_1 \cap x_2) = p(\mathbf{x}) &= N(\mathbf{x}; \mathbf{x}^b, \mathbf{B}) \\ &= \left[(2\pi)^2 |\mathbf{B}|\right]^{-1/2} \exp\left[-\frac{1}{2}\left(\mathbf{x} - \mathbf{x}^b\right)^{\mathrm{T}} \mathbf{B}^{-1} \left(\mathbf{x} - \mathbf{x}^b\right)\right], \end{aligned} \tag{2.18}$$

where $\mathbf{B}$ is the covariance matrix:

$$\mathbf{B} = V^b \begin{pmatrix} 1 & \mu \\ \mu & 1 \end{pmatrix}. \tag{2.19}$$

$p(\mathbf{x})$ is assumed to be the combination of all our prior knowledge about the model state. A schematic contour plot, for our simple two-parameter model, is shown by the long-dashed ellipses in Fig. 2.3. The positive correlation μ means that x_1 and x_2 are more likely to be similar, leading to the major axis being along $x_1 = x_2$.

We now consider the new information provided by the observation. A perfect instrument would measure a 'true' value $\mathbf{y}^t$. Real instruments are not perfect; they

[3] Note that for DA we usually define so-called correlation coefficients by subtracting x^t, not by subtracting $\bar{x}$. They are the same if x is unbiased, but if x is biased then the DA usage allows us to correct for the bias (suboptimally).

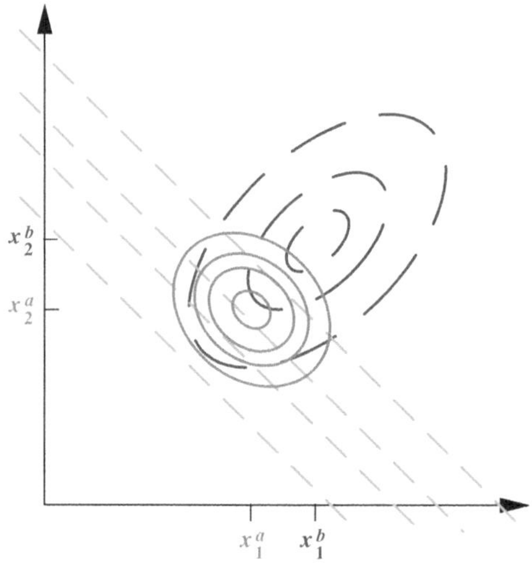

Fig. 2.3 Contour plots for the simple example of Section 2.1.3: long dashes show the prior pdf $p(x_1, x_2)$, a two-dimensional Gaussian with centre (x_1^b, x_2^b); short dashes show the likelihood function $p(y^o|x_1, x_2)$, a Gaussian ridge about the line $\frac{1}{2}(x_1 + x_2) = y^o$; thick lines show the posterior pdf $p(x_1, x_2|y^o)$, a two-dimensional Gaussian with centre (x_1^a, x_2^a).

have instrumental error, which we model by a Gaussian (in our example, this is one-dimensional, but we use a more general notation which can be applied for n observations):

$$\begin{aligned} p(\mathbf{y}^o|\mathbf{y}^t) &= N(\mathbf{y}^o; \mathbf{y}^t, \mathbf{E}) \\ &= [(2\pi)^n|\mathbf{E}|]^{-1/2} \exp\left[-\frac{1}{2}(\mathbf{y}^o - \mathbf{y}^t)^{\mathrm{T}}\mathbf{E}^{-1}(\mathbf{y}^o - \mathbf{y}^t)\right]. \end{aligned} \tag{2.20}$$

We cannot use this directly in Bayes' theorem, because it is a function of $\mathbf{y}^t$, rather than $\mathbf{x}$. We need to specify also the probability of a perfect instrument observing $\mathbf{y}^t$, given that the true model state is $\mathbf{x}$. Because we have to interpolate from $\mathbf{x}$ to $\mathbf{y}$, we cannot know $\mathbf{y}^t$ exactly; the *representativeness error* can be thought of as the error in the interpolation operator H in (2.15):

$$\begin{aligned} p(\mathbf{y}^t|\mathbf{x}) &= N(\mathbf{y}^t; H(\mathbf{x}), \mathbf{F}) \\ &= [(2\pi)^n|\mathbf{F}|]^{-1/2} \exp\left\{-\frac{1}{2}[\mathbf{y}^t - H(\mathbf{x})]^{\mathrm{T}}\mathbf{F}^{-1}[\mathbf{y}^t - H(\mathbf{x})]\right\}. \end{aligned} \tag{2.21}$$

If we assume that instrumental and representativeness errors are independent, we can convolve them to get a combined *observational error*:

$$\begin{aligned} p(\mathbf{y}^o|\mathbf{x}) &= \int p(\mathbf{y}^o|\mathbf{y}^t)p(\mathbf{y}^t|\mathbf{x})\,\mathrm{d}\mathbf{y}^t \\ &= N(\mathbf{y}^o; H(\mathbf{x}), \mathbf{E} + \mathbf{F}) \\ &= [(2\pi)^n|\mathbf{E} + \mathbf{F}|]^{-1/2} \exp\left\{-\frac{1}{2}[\mathbf{y}^o - H(\mathbf{x})]^{\mathrm{T}}(\mathbf{E} + \mathbf{F})^{-1}[\mathbf{y}^o - H(\mathbf{x})]\right\}. \end{aligned} \tag{2.22}$$

The sum of the instrumental and representativeness error covariances, $\mathbf{E} + \mathbf{F}$, is often written as a single observational error covariance $\mathbf{R}$. The above derivation shows that it is composed of two parts: $\mathbf{E}$, which is a function of the instrument characteristics, and $\mathbf{F}$, which is a function of the model resolution. For instance, for a wind observation from a radiosonde, the errors in tracking the balloon might lead to an instrumental error of about $1\,\mathrm{m\,s^{-1}}$. The error of representativeness due to trying to predict such a wind from a model with horizontal grid-length 200 km would be about $3\,\mathrm{m\,s^{-1}}$, and a grid-length on 20 km would reduce the error of representativeness to about $1\,\mathrm{m\,s^{-1}}$.

The likelihood $p(\mathbf{y}^o|x_1, x_2)$ is plotted as a function of (x_1, x_2) as the short-dashed lines in Fig. 2.3. Note that the ridge extends to infinity; all values on the ridge line are, as far as the observed information is concerned, equally likely. To get a unique 'best estimate', we need to combine this with the prior information.

We can substitute (2.18) and (2.22) into the Bayesian analysis equation (2.5), giving

$$p(\mathbf{x}|\mathbf{y}^o) \propto N\left(\mathbf{y}^o; H(\mathbf{x}), \mathbf{R}\right) N\left(\mathbf{x}; \mathbf{x}^b, \mathbf{B}\right). \tag{2.23}$$

It is a property of Gaussians that, if H is linearizable,

$$N\left(\mathbf{y}^o; H(\mathbf{x}), \mathbf{R}\right) N\left(\mathbf{x}; \mathbf{x}^b, \mathbf{B}\right) = N\left(\mathbf{y}^o; H(\mathbf{x}^b), \mathbf{R} + \mathbf{HBH}^{\mathrm{T}}\right) N\left(\mathbf{x}; \mathbf{x}^a, \mathbf{A}\right), \tag{2.24}$$

where $\mathbf{x}^a$ and $\mathbf{A}$ are defined by

$$\begin{aligned} \mathbf{A} &= \mathbf{B} - \mathbf{BH}^{\mathrm{T}}(\mathbf{R} + \mathbf{HBH}^{\mathrm{T}})^{-1}\mathbf{HB}, \\ \mathbf{x}^a &= \mathbf{x}^b + \mathbf{BH}^{\mathrm{T}}(\mathbf{R} + \mathbf{HBH}^{\mathrm{T}})^{-1}\left[\mathbf{y}^o - H(\mathbf{x}^b)\right]. \end{aligned} \tag{2.25}$$

Cancelling the constant of proportionality gives

$$p(\mathbf{x}|\mathbf{y}^o) = N\left(\mathbf{x}; \mathbf{x}^a, \mathbf{A}\right). \tag{2.26}$$

So the posterior probability, after adding the observational information, is a Gaussian with mean $\mathbf{x}^a$ and variance $\mathbf{A}$. As in Fig. 2.1, the posterior pdf in Fig. 2.3 (solid lines) is narrower and taller than the prior. Adding information from the observation increases our confidence, and reduces the error variance from $\mathbf{B}$ to $\mathbf{A}$.

For our simple example, the algebra is easily done by hand, giving

$$\mathbf{x}^a = \begin{pmatrix} x_1^a \\ x_2^a \end{pmatrix} = \begin{pmatrix} x_1^b \\ x_2^b \end{pmatrix} + \frac{\left[V^b\left(\dfrac{1+\mu}{2}\right)\right]^2}{\mathbf{R} + V^b\left(\dfrac{1+\mu}{2}\right)}\left(y^o - \frac{x_1^b + x_2^b}{2}\right)\begin{pmatrix} 1 \\ 1 \end{pmatrix}. \tag{2.27}$$

2.1.4 Practical implementation of analysis step

Section 2.1.3 made the problem seem easy; we can write down the analysis equation (2.25) and solve it directly to give (2.27). However, to give good forecasts, we find that the current NWP models used in Section 2.1.1 step 1 need about a billion degrees of freedom, and we need to use well over a million observed data in a batch. While our powerful supercomputers can store and process vectors of this size, we cannot store full matrices, and matrix operations such as inversion are unaffordable. Approximations are essential. It is the choice of these approximations, and the way they interact in a DA scheme, that makes the subject interesting—they run throughout any course on DA. There are three interacting choices:

- modelling, representing, and estimating prior background error covariances $\mathbf{B}$;
- expressing the analysis equations in a form amenable to solution;
- computing the solution.

We introduce each in turn.

Prior or background error covariances B

The background error covariance can be defined as an expected value $\mathbf{B} = \langle(\mathbf{x}^b - \mathbf{x}^t)(\mathbf{x}^b - \mathbf{x}^t)^{\mathrm{T}}\rangle$ and in principle can be calculated from a large sample of prior estimates $\mathbf{x}^b$ of the true state $\mathbf{x}^t$. In practice, $\mathbf{B}$ is huge and can vary in space and time; we could never sample it (even if we knew $\mathbf{x}^t$, which we do not).

Early approximations, used in *optimal interpolation* (e.g. Lorenc 1981), concentrated on the geopotential height field and assumed that its $\mathbf{B}$ is *stationary* (constant in time), *homogeneous* (the same everywhere), and *isotropic* (the same in all directions). This gives a continuous two-dimensional covariance function of two positions that, because of the assumptions, simplifies to a function of the distance between the positions. Assuming wind errors are geostrophically related to height errors then enables covariances involving winds to be derived, as in Fig. 2.4. Lorenc (1981) extended this 2D covariance model to 3D by assuming that covariances were *separable* into a vertical term that is independent of horizontal position and horizontal terms as just discussed. The model was also extended to global application by having similar covariance functions for height and streamfunction, with a correlation between them that varied with latitude. In modern applications, $\mathbf{B}$ is much too large to calculate each element individually using formulas like these. However, it is still useful to compare implied covariance functions (obtained for instance as the analysis increment due to a single observation) against the patterns in Fig. 2.4.

Parrish and Derber (1992), inspired by Phillips (1985), realized that although we cannot handle $\mathbf{B}$ directly, we do have in NWP models a wide range of operators for global fields such as $\mathbf{x}$ that enable us to calculate dynamical modes. If we define error fields $\mathbf{x}' = \mathbf{x}^b - \mathbf{x}^t$ then a sequence of such operations, including some with empirical coefficients, can construct an operator $\mathbf{U}$ such that $\mathbf{x}' = \mathbf{U}\mathbf{v}$ where $\mathbf{v}$ has a much simpler error structure. With enough steps (an example is given in Section 2.1.5), we can get to a $\mathbf{v}$ with no internal error correlations, i.e. $\langle\mathbf{v}\mathbf{v}^{\mathrm{T}}\rangle = \mathbf{I}$. This gives

$$\mathbf{B} = \langle(\mathbf{x}^b - \mathbf{x}^t)(\mathbf{x}^b - \mathbf{x}^t)^{\mathrm{T}}\rangle = \left\langle\mathbf{U}\mathbf{v}\mathbf{v}^{\mathrm{T}}\mathbf{U}^{\mathrm{T}}\right\rangle = \mathbf{U}\mathbf{U}^{\mathrm{T}}. \tag{2.28}$$

This approach has been at the heart of many 3D-Var schemes (e.g. Courtier et al., 1998; Rabier et al., 1998; Andersson et al., 1998; Gauthier et al., 1999; Lorenc et al., 2000) and hence of the 4D-Var schemes built from them.

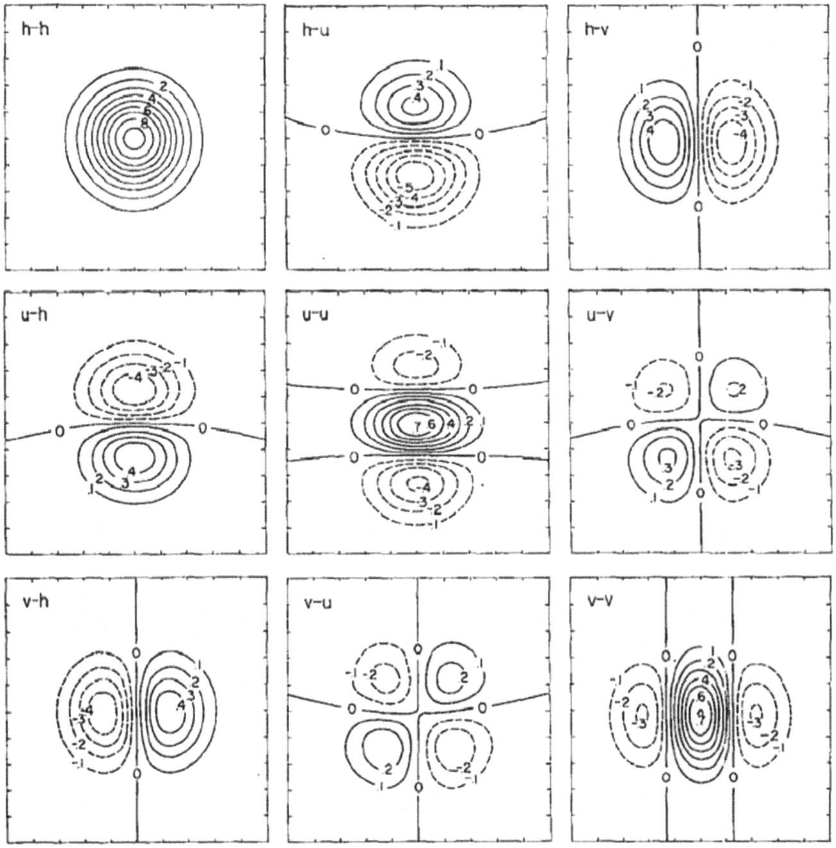

Fig. 2.4 Typical geostrophic multivariate background error covariance functions. (From Schlatter (1975).)

All the 3D-Var schemes referenced above used spectral transforms in the modelling of horizontal correlations—if the correlations are homogeneous and isotropic then the coefficients of a spectral representation are uncorrelated. In practice, however, error correlations are not homogeneous and isotropic, particularly in ocean models near coasts and in the atmosphere near fronts. To be able to represent these local effects in correlations, some schemes model them using diffusion operators (Weaver and Courtier, 2001). Others use recursive filters (Hayden and Purser, 1995; Purser et al., 2013*a, b*).

Early attempts at estimating background error covariances (when forecasts were less accurate) mostly used forecast minus observation difference statistics. It is impossible to separate these statistics into background and observation errors without further assumptions, but by making plausible assumptions, progress can be made (Hollingsworth and Lonnberg, 1986; Desroziers et al., 2005). This has become more difficult as background errors have become relatively smaller, moreover we want covariance estimates where observations are sparse. So, building on the increasing success of ensemble data assimilation methods, many centres now base their covariance estimates on ensembles of forecasts (Fisher, 2003). These ensembles are designed so that the ensemble spread is equal to the error in the ensemble mean. If the ensemble forecast members are $\mathbf{x}_i^f$ for $i = 1, \ldots, K$, with mean $\bar{\mathbf{x}}^f$, then it is convenient to define a rectangular matrix whose columns are the scaled perturbations from the mean. This then can be used to estimate the error covariance—we use the $\approx$ symbol because, with small K, it is a noisy estimate:

$$\begin{aligned}\mathbf{X}^f &= \frac{1}{\sqrt{K-1}}(\mathbf{x}_1^f - \bar{\mathbf{x}}^f, \ldots, \mathbf{x}_i^f - \bar{\mathbf{x}}^f, \ldots, \mathbf{x}_K^f - \bar{\mathbf{x}}^f) \\ \mathbf{B} &\approx \mathbf{X}^f \mathbf{X}^{f^{\mathrm{T}}}.\end{aligned} \tag{2.29}$$

Note that I have followed convention (Ide et al., 1997) and used $\mathbf{B}$ to represent all these modelled forms of the background error covariance. In the Kalman filter literature, it is usual to use $\mathbf{P}$. I reserve that for purely prognostic estimates, as in the Kalman filter equation for the prediction of error covariances:

$$\mathbf{P}^f = \mathbf{M}\mathbf{P}^a\mathbf{M}^{\mathrm{T}} + \mathbf{Q}. \tag{2.30}$$

Analysis equations

The analysis equations (2.25) derived for our simple example are used in many equivalent forms, some of which are not obvious. I list them here for reference, without full derivations.

The Kalman filter defines a Kalman gain $\mathbf{K}$, which means that (2.25) can be written as

$$\begin{aligned}\mathbf{A} &= (\mathbf{I} - \mathbf{K}\mathbf{H})\mathbf{B}, \\ \mathbf{x}^a &= \mathbf{x}^b + \mathbf{K}\left[\mathbf{y}^o - H(\mathbf{x}^b)\right],\end{aligned} \tag{2.31}$$

with $\mathbf{K}$ given by the *observation-space* form used in (2.25) or an equivalent *model-space* form.[4] The *space* denotes the dimension of the matrix to be inverted:

$$\mathbf{K} = \mathbf{B}\mathbf{H}^{\mathrm{T}}\left(\mathbf{R} + \mathbf{H}\mathbf{B}\mathbf{H}^{\mathrm{T}}\right)^{-1}, \tag{2.32}$$

$$\mathbf{K} = \left(\mathbf{H}^{\mathrm{T}}\mathbf{R}^{-1}\mathbf{H} + \mathbf{B}^{-1}\right)^{-1}\mathbf{H}^{\mathrm{T}}\mathbf{R}^{-1}. \tag{2.33}$$

If we use the ensemble form (2.29) to represent $\mathbf{B}$ then it is also possible to derive an *ensemble-space* form for $\mathbf{A}$ (Bishop et al., 2001; Tippett et al., 2003):

$$\begin{aligned}\mathbf{B} &= \mathbf{X}^f \mathbf{X}^{f^{\mathrm{T}}}, \\ \mathbf{X}^a &= \mathbf{X}^f \mathbf{T}, \\ \mathbf{A} &= \mathbf{X}^a \mathbf{X}^{a^{\mathrm{T}}},\end{aligned} \tag{2.34}$$

where the dimension of $\mathbf{T}$ is the ensemble size.

[4] Demonstrated to be equivalent using the Sherman–Morrison–Woodbury formula.

All of the above require an matrix inversion (or similar computation) to derive both a best estimate and its error covariance. They rely on the Gaussian assumption and linear H to make the equations quadratic. We started to introduce an alternative approach in the derivation of equation (2.14) and Fig. 2.2. We can find the most likely estimate by searching for the minimum of a penalty function J equal to minus the logarithm of the posterior pdf—if the pdf is near-Gaussian then the penalty function is near quadratic and minimisation can be made relatively inexpensive. Taking minus the logarithm of (2.23) gives

$$J(\mathbf{x}) = \frac{1}{2}\left(\mathbf{x}-\mathbf{x}^b\right)^{\mathrm{T}}\mathbf{B}^{-1}\left(\mathbf{x}-\mathbf{x}^b\right) + \frac{1}{2}\left[H(\mathbf{x})-\mathbf{y}^o\right]^{\mathrm{T}}\mathbf{R}^{-1}\left[H(\mathbf{x})-\mathbf{y}^o\right]. \qquad (2.35)$$

The most likely $\mathbf{x}^a$ minimizes this—if H is linear, this is identical to that given by (2.31). This minimization does not normally give the error covariance $\mathbf{A}$, but in principle it can be obtained from

$$\mathbf{A}^{-1} = \frac{\partial^2 J}{\partial x^2} = \mathbf{B}^{-1} + \mathbf{H}^{\mathrm{T}}\mathbf{R}^{-1}\mathbf{H}, \qquad (2.36)$$

which again for linear $H = \mathbf{H}$ is identical to (2.31). The form of (2.36) matches the simpler (2.12).

Computing the solution

There are various more technical choices that, when multiplied by the number of different equations above, give a large variety of DA schemes:

1. Incremental or full-field?
2. Global or local?
3. Iterative or explicit?

Most operational DA schemes are incremental, starting from a guess $\mathbf{x}^g$ and seeking to determine $\delta\mathbf{x}$ such that $\mathbf{x}^a = \mathbf{x}^g + \delta\mathbf{x}$. This allows more accurate linearization and, with an accurate $\mathbf{x}^g$, allows approximations in the processing of increments that would be unacceptable for full fields. On the other hand, an approximated incremental approach is more work to develop, especially if an incremental forecast model is needed.

Early DA methods were all local, selecting which observations to use for each variable analysed. This make them cheaper, the main motivation, but also implicitly builds in *covariance localization* that compensates for deficiencies in covariance models and is therefore still widely used in ensemble DA. This effect was seen by Lev Gandin,[5] who suggested a limit of about 25 observations—his conclusions were based on experience using climatology as a background and broad, simplified estimates of background error correlations. Current global variational schemes use several million observations at once!

Explicit inversion of the matrices in equations such as (2.32) and (2.33) is conceptually simpler, allowing direct estimation of the analysis error covariance, useful for

[5] Lev wrote an influential book (Gandin, 1965) about optimum interpolation. In correspondence discussing Lorenc (1981), which used up to 191 data at once, he suggested I was using too many.

ensemble DA methods and observational quality control (e.g. Lorenc, 1981). However, it is much too expensive for use with millions of observations in (2.32) or billions of model variables in (2.33). For such problems, we must use either local methods or else ensemble approaches such as (2.34). Variational methods use instead an iterative approach to minimize an equation such as (2.35), which has the additional advantage of allowing more complicated equations not amenable to explicit solution.

2.1.5 3D-Var

Variational approach

Once we have a practical method for calculating J for any $\mathbf{x}$, using (2.35), we need to find the $\mathbf{x}^a$ that minimizes it. In modern NWP, a state $\mathbf{x}$ has a billion degrees of freedom, so we cannot just try a sample; we need an efficient algorithm. Because of the quadratic form of (2.35) (which comes from the Gaussian pdfs) and because operators H (and those used to model $\mathbf{B}$) are nearly linear, then, rather than searching for the minimum of J, it is better to search for $J'(\mathbf{x}) = 0$ using the iterative Newton method; if in the nth iteration $\mathbf{x}_n$ is an estimate, a better estimate is given by

$$\mathbf{x}_{n+1} = \mathbf{x}_n - \left[J''(\mathbf{x}_n)\right]^{-1} J'(\mathbf{x}_n). \tag{2.37}$$

It is not possible to calculate $\left[J''(\mathbf{x}_n)\right]^{-1}$, so we approximate it in a descent algorithm and do more iterations. However, we keep this method in mind when designing the transforms and preconditioning below—we transform such that approximating $(J'')^{-1}$ is easier.

I like to use matrix and vector notation, with $\delta\mathbf{x}$ a column vector of analysis increments.[6] It is conventional that $\partial J/\partial\mathbf{x}$ is a row vector—we need to transpose it. Bold symbols denote linearizations, e.g. $\partial H(\mathbf{x})/\partial\mathbf{x} = \mathbf{H}$. Applying this and differentiating (2.35) gives

$$\delta\mathbf{x} = \mathbf{x} - \mathbf{x}^g, \tag{2.38}$$

$$\left[\frac{\partial J}{\partial(\delta\mathbf{x})}\right]^{\mathrm{T}} = \mathbf{B}^{-1}\left(\delta\mathbf{x} + \mathbf{x}^g - \mathbf{x}^b\right) + \mathbf{H}^{\mathrm{T}}\mathbf{R}^{-1}\left[H(\delta\mathbf{x} + \mathbf{x}^g) - \mathbf{y}^o\right]. \tag{2.39}$$

Note that the matrix notation automatically gives us the transpose $\mathbf{H}^{\mathrm{T}}$, which is the adjoint of the operator H. In principle, we use a descent algorithm to search for the zero of (2.39); in practice, we need to modify it first.

Change of control variables

The next essential stage is to avoid the large matrix $\mathbf{B}$ and at the same time improve the conditioning. We do this by devising an operator $\mathbf{U}$, which we can implement in a computer program, such that

$$\mathbf{B}_{(x)} = \mathbf{U}\mathbf{U}^{\mathrm{T}}. \tag{2.40}$$

[6] In preparation for the outer loop (Section 2.2.3), I differentiate the latest guess $\mathbf{x}^g$ from the background $\mathbf{x}^b$. Normally in 3D-Var, we set $\mathbf{x}^g = \mathbf{x}^b$.

Note I am following the suggested notation of Ide et al. (1997) and using a subscript (x) to indicated what space is used for $\mathbf{B}$, since we are about to transform to another space. We then transform control variables to $\mathbf{v}$ so we can eliminate $\mathbf{B}_{(x)}$:

$$\delta\mathbf{x} = \mathbf{U}\mathbf{v}, \tag{2.41}$$

$$\mathbf{B}_{(v)} = \mathbf{I}. \tag{2.42}$$

If the guess differs from the background[7] then we define $\mathbf{v}^g$ using

$$\mathbf{x}^g - \mathbf{x}^b = \mathbf{U}\mathbf{v}^g, \tag{2.43}$$

$$J(\mathbf{v}) = \frac{1}{2}\left(\mathbf{v}+\mathbf{v}^g\right)^{\mathrm{T}}\left(\mathbf{v}+\mathbf{v}^g\right) + \frac{1}{2}\left[H(\mathbf{U}\mathbf{v}+\mathbf{x}^g) - \mathbf{y}^o\right]^{\mathrm{T}}\mathbf{R}^{-1}\left[H(\mathbf{U}\mathbf{v}+\mathbf{x}^g) - \mathbf{y}^o\right]. \tag{2.44}$$

It is common (but not essential, see Section 2.2.3) to linearize H for the duration of the minimization of (2.44). We can then precalculate an innovation vector $\mathbf{d}$ using the full nonlinear H, giving a quadratic penalty in (2.46):

$$\mathbf{d} = \mathbf{y}^o - H(\mathbf{x}^g), \tag{2.45}$$

$$J(\mathbf{v}) = \frac{1}{2}\left(\mathbf{v}+\mathbf{v}^g\right)^{\mathrm{T}}\left(\mathbf{v}+\mathbf{v}^g\right) + \frac{1}{2}\left(\mathbf{H}\mathbf{U}\mathbf{v} - \mathbf{d}\right)^{\mathrm{T}}\mathbf{R}^{-1}\left(\mathbf{H}\mathbf{U}\mathbf{v} - \mathbf{d}\right). \tag{2.46}$$

As before, we search for the zero of the gradient rather than the minimum, then apply (2.43) as an additional final step. The equivalents of (2.39) are

$$\left(\frac{\partial J}{\partial \mathbf{v}}\right)^{\mathrm{T}} = \left(\mathbf{v}+\mathbf{v}^g\right) + \mathbf{H}^{\mathrm{T}}\mathbf{R}^{-1}\left[H(\mathbf{U}\mathbf{v}+\mathbf{x}^g) - \mathbf{y}^o\right], \tag{2.47}$$

$$\left(\frac{\partial J}{\partial \mathbf{v}}\right)^{\mathrm{T}} = \left(\mathbf{v}+\mathbf{v}^g\right) + \mathbf{H}^{\mathrm{T}}\mathbf{R}^{-1}\left(\mathbf{H}\mathbf{U}\mathbf{v} - \mathbf{d}\right). \tag{2.48}$$

In (2.47), $\mathbf{H}^{\mathrm{T}}$ is linearized about $\mathbf{U}\mathbf{v} + \mathbf{x}^g$, while in (2.48), $\mathbf{H}$ and $\mathbf{H}^{\mathrm{T}}$ are linearized about $\mathbf{x}^g$. Equation (2.48) comes from the quadratic (2.46) and allows a more efficient descent algorithm; (2.47) allows better treatment of nonlinearities in H and non-Gaussian pdfs, while saving costs by reducing the need for 4D-Var outer loops (Sections 2.2.3, 2.3.2, 2.3.3, and 2.3.4).

A desired side effect of this transformation of control variables is that the Hessian no longer contains $\mathbf{B}^{-1}$; it is now given by

$$\frac{\partial^2 J}{\partial \mathbf{v}^2} = \mathbf{I} + \mathbf{U}^{\mathrm{T}}\mathbf{H}^{\mathrm{T}}\mathbf{R}^{-1}\mathbf{H}\mathbf{U}. \tag{2.49}$$

This allows NWP to use high-resolution NWP models, for which $\mathbf{B}$ becomes increasingly ill-conditioned. The conditioning of (2.49) can be made worse by the use of dense high-accuracy observations. Thinning or 'super-obbing' can be used to avoid this.

[7] If not then $\mathbf{v}^g = 0$,

Control variable transforms

It is common before the transform $\mathbf{U}$ introduced above, especially in 4D-Var, to use a simplifying transform (normally linear) $\delta\mathbf{w} = \mathbf{S}(\delta\mathbf{x})$ (Courtier et al., 1994; Ide et al., 1997; Lorenc et al., 2000; Lorenc, 2003*a*). This can reduce the resolution, saving computation time, especially in 4D-Var. It can also reduce the complexity; for instance, the Met Office system simplifies water vapour and cloud to a total moisture variable, as well as simplifying a non-hydrostatic model to hydrostatic. The simplification removes degrees of freedom from $\delta\mathbf{x}$, so $\mathbf{S}$ has no mathematical inverse, but we can design an approximation using physical knowledge. For moisture, it has to be nonlinear, to allow cloud-free air to make cloud when there is a large moisture increment:

$$\delta\mathbf{x} = S^{-I}(\delta\mathbf{w}, \mathbf{x}). \tag{2.50}$$

We can still define a linearized operator $\mathbf{S}^{-I}$. The generalized inverse S^{-I} introduces strong constraints between the increments to the simplified parameters. For instance, the increments are constructed to be hydrostatic, with cloud and moisture increasing together. Full covariances of errors in the unsimplified model are defined from the simplified covariances, with some state dependence because of the nonlinearity in S^{-I}:

$$\mathbf{B}_{(x)} = \left(\mathbf{S}^{-I}\right)^{\mathrm{T}} \mathbf{B}_{(w)} \mathbf{S}^{-I}. \tag{2.51}$$

Note that $\mathbf{B}_{(x)}$ is singular because $\mathbf{S}^{-I}$ is rectangular. There is no power in non-hydrostatic modes, and there is a diagnostic relationship between cloud and moisture. Its existance is implicit; we never need it. The VAR equations use $\mathbf{B} = \mathbf{B}_{(w)}$, but we avoided the need to manipulate even this by introducing a transform $\mathbf{U}$ that satisfies (2.40). The physically based ideas used to define the transform act as a model for the form of the covariances that make up $\mathbf{B}$. The idea that different types of motion had different (importance of) errors was suggested by Phillips (1985) and used by Parrish and Derber (1992) and many others; I will present a description of the transforms used in Lorenc et al. (2000). Although it is the transform $\mathbf{U}$ and its transpose that are required in the minimization, it is easier first to understand the generalized inverse transform $\mathbf{T} = \mathbf{U}^{-I}$. This is constructed in stages:

- using simple physical ideas to transform parameters ($\mathbf{T}_p$);
- using zonal and seasonal-average statistics to transform into empirical modes in the vertical ($\mathbf{T}_v$);
- filtering, to allow for different scales in the horizontal ($\mathbf{T}_h$).

In our 3D-Var system, the simplified model increments $\delta\mathbf{w}$ are horizontal and vertical wind components, potential temperature, density, pressure, and total humidity. This is consistent with a non-hydrostatic, height coordinate model. We know physical relationships between variables, such as the closeness to balance, and non-divergence, which imply that elements of $\delta\mathbf{w}$ that are different physical parameters (e.g. temperature and wind) are correlated.[8] Following the ideas of Parrish and Derber (1992), we

[8] Strictly, the expected errors are correlated.

use these relationships to design a parameter transform $\mathbf{T}_p$ to separate $\delta\mathbf{w}$ into three-dimensional fields of variables that are uncorrelated with each other: stream function, velocity potential, the unbalanced part of the hydrostatic pressure, and total relative humidity. For a complete representation of our non-hydrostatic model, we would need two additional variables to represent possible deviations of pressure and vertical velocity fields from those diagnosed in a hydrostatic model. We have omitted these from the output of $\mathbf{T}_p$, and in $\mathbf{T}_p^{-1}$ the corresponding inputs are assumed to be zero.

Within each three-dimensional field, there are still correlations between points close in space. Making some assumptions, we can design $\mathbf{T}_v$ to project onto vertical modes and separate each three-dimensional field into two-dimensional fields of coefficients. These coefficients are normalized by the square root of the expected variance of the relevant vertical mode at that location, allowing some horizontal variation of variances and vertical correlations.

Finally, we design $\mathbf{T}_h$ to act on each two-dimensional field, allowing for horizontal correlations. After trying a digital filter[9] and spectral transform for this (Lorenc, 1997), we chose the latter, followed by a scaling based on the correlation power spectrum.

Having designed the transform from the (physically meaningful) $\delta\mathbf{w}$ into the control variable $\mathbf{v}$, it is relatively straightforward to derive its pseudo-inverse $\mathbf{U}$, and then the transpose (or adjoint) $\mathbf{U}^{\mathrm{T}}$, which are used in the variational minimization.

2.2 4D-Var: derivation

2.2.1 Deterministic – fitting a model evolution to observations

I first summarize[10] the traditional way of deriving 4D-Var. The notation follows Ide et al. (1997) with an extension that, to avoid explicit summations over time, underlined variables include the time dimension and underlined operators produce underlined variables. For the time being, I assume a perfect forecast model, so that knowledge of initial conditions $\mathbf{x}$ defines a four-dimensional trajectory $\underline{\mathbf{x}} = \underline{M}(\mathbf{x})$. We want to find the best fit of this trajectory to observations distributed in time ($\underline{\mathbf{y}}^o$) and a prior estimate $\mathbf{x}^b$. We could simply define best to be a minimum-variance solution, but to link with later, I prefer a Bayesian derivation. An expression for the pdf is

$$p\left(\mathbf{x}|\underline{\mathbf{y}}^o\right) \propto p\left(\mathbf{x}\right) p\left(\underline{\mathbf{y}}^o|\mathbf{x}\right). \tag{2.52}$$

We assume the prior pdf for model state $\mathbf{x}$ is a Gaussian with mean $\mathbf{x}^b$ and covariance $\mathbf{B}$:

$$p\left(\mathbf{x}\right) \propto \exp\left[-\frac{1}{2}\left(\mathbf{x}-\mathbf{x}^b\right)^{\mathrm{T}}\mathbf{B}^{-1}\left(\mathbf{x}-\mathbf{x}^b\right)\right]. \tag{2.53}$$

The observations in the time window, $\underline{\mathbf{y}}^o$, are usually assumed to have Gaussian errors in observation space, with covariance $\underline{\mathbf{R}}$, uncorrelated with background errors:

$$p\left(\underline{\mathbf{y}}^o|\mathbf{x}\right) = p\left(\underline{\mathbf{y}}^o|\underline{\mathbf{y}}\right) \propto \exp\left[-\frac{1}{2}\left(\underline{\mathbf{y}}-\underline{\mathbf{y}}^o\right)^{\mathrm{T}}\underline{\mathbf{R}}^{-1}\left(\underline{\mathbf{y}}-\underline{\mathbf{y}}^o\right)\right], \tag{2.54}$$

[9] Suggested by Purser and McQuigg (1982) and further developed by Purser et al. (2003*a*).

[10] Based on a fuller presentation in Lorenc and Payne (2007).

where $\underline{\mathbf{y}}$ represents the estimate of the observations calculated from $\underline{\mathbf{x}}$ using the observation operator $\underline{H}$:

$$\underline{\mathbf{y}} = \underline{H}\left(\underline{M}\left(\mathbf{x}\right)\right). \tag{2.55}$$

$\underline{M}$ and $\underline{H}$ are nonlinear, so the pdf obtained by substituting (2.53)–(2.55) into (2.52) is not Gaussian. It is not practicable to evaluate this full pdf for an NWP system, so we have to assume that the desired estimate is the $\mathbf{x}$ that maximizes $P\left(\mathbf{x}|\underline{\mathbf{y}}^o\right)$, or equivalently minimizes

$$J\left(\mathbf{x}\right) = \frac{1}{2}\left(\mathbf{x} - \mathbf{x}^b\right)^{\mathrm{T}} \mathbf{B}^{-1}\left(\mathbf{x} - \mathbf{x}^b\right) + \frac{1}{2}\left(\underline{\mathbf{y}} - \underline{\mathbf{y}}^o\right)^{\mathrm{T}} \underline{\mathbf{R}}^{-1}\left(\underline{\mathbf{y}} - \underline{\mathbf{y}}^o\right) \tag{2.56}$$

subject to (2.55). This is commonly called the 4D-Var penalty function. The gradient of J with respect to $\mathbf{x}$ is given by

$$\nabla_{\mathbf{x}} J\left(\mathbf{x}\right) = \mathbf{B}^{-1}\left(\mathbf{x} - \mathbf{x}^b\right) + \underline{\mathbf{M}}^*\underline{\mathbf{H}}^*\mathbf{R}^{-1}\left(\underline{\mathbf{y}} - \underline{\mathbf{y}}^o\right), \tag{2.57}$$

where $\underline{\mathbf{M}}^*$ and $\underline{\mathbf{H}}^*$ are the adjoints of the Jacobians of $\underline{M}$ and $\underline{H}$, taken at point $\mathbf{x}$. Equations (2.56) and (2.57) are used in an iterative descent algorithm illustrated schematically in Fig. 2.5. This is affordable, since only a small part of the full pdf is explored.

In practice, this method cannot be applied to current NWP models. The issues are more fundamental than the use of 'IF' tests in software for NWP models' $\underline{M}$; they are due to the physical processes being represented and the principle that each

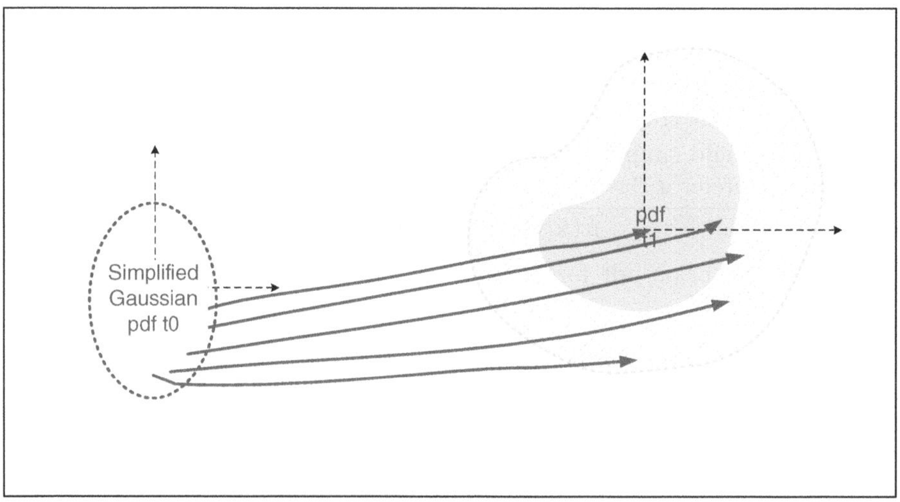

Fig. 2.5 Deterministic 4D-Var. The initial pdf is approximated by a Gaussian. The descent algorithm only explores a small part of the pdf, on the way to a local minimum. The 4D analysis is a trajectory of the full model, optionally augmented by a model error correction term.

component of the complex NWP system should be based on physical understanding. The atmosphere has many processes that often do not give a gradient (2.57) pointing towards the minimum of (2.56):

- *thermostats:* fast processes that are modulated to maintain a longer-timescale 'balance' (e.g. boundary-layer fluxes);
- *limits to growth:* fast processes that in a nonlinear model are limited by some available resource (e.g. evaporation of raindrops);
- *butterflies:* fast processes that are not predictable over a long 4D-Var time window (e.g. eddies with short space- and timescales);
- *observations of intermittent phenomena:* if something (e.g. a cloud or rain) is missing from a state then the gradient does not say what to do to make it appear.

These are fundamental atmospheric processes—it is impossible to write a good NWP model, following the principle that each component is based on a physical understanding, without representing them.

2.2.2 Statistical 4D-Var: the extended Kalman filter

The problem discussed in Section 2.2.1 can be avoided if we seek a more appropriate 'best' estimate. To define what is really best, it is necessary to specify and minimize a cost function measuring the additional costs incurred by issuing an imperfect forecast. The simplest is a quadratic cost: the expected root-mean-square error (RMSE) is minimized by the mean of the pdf:

$$\mathbf{x}^a = \int \mathbf{x}\, P\big(\mathbf{x}|\underline{\mathbf{y}}^o\big)\ \mathrm{d}\mathbf{x}. \tag{2.58}$$

This integral requires evaluation of the whole pdf, rather than the relatively few function and derivative evaluations needed in the descent algorithm of deterministic 4D-Var. Equation (2.58) is not amenable to calculation for pdfs as complicated as (2.52) for a full NWP model. However, rather than minimizing the wrong equation (2.56), it may be better to find an approximate solution to (2.58). Lorenc (2003*a*) suggested one way to do it in the context of 4D-Var, based on the fact that the mode of a pdf is also its mean, as long as the distribution is Gaussian. It is the nonlinear evolution of the prior pdf $P(\mathbf{x})$ by $\underline{M}$ that makes (2.52) non-Gaussian and (2.56) non-quadratic. So let us instead approximate that evolution by $\underline{\bar{M}}$, which predicts the evolution of the mean, and a perturbation forecast (PF) model $\underline{\tilde{\mathbf{M}}}$ that gives linear best estimates of the evolution of finite perturbations about this mean (Fig. 2.6). If necessary, a similar strategy can be applied to H, $\bar{H}$, and $\tilde{\mathbf{H}}$. Note that we can easily extend the approach to models with errors, by augmenting the model variables to contain also parameters describing the errors. I do not show these terms here—they are in Lorenc (2003*a*). The whole pdf (2.52) (rather than its behaviour near its mode) is then approximated using an incremental approach about a guess $\mathbf{x}^g$ of the ensemble mean (Courtier et al., 1994):

$$\mathbf{x} = \mathbf{x}^g + \delta\mathbf{x}, \tag{2.59}$$

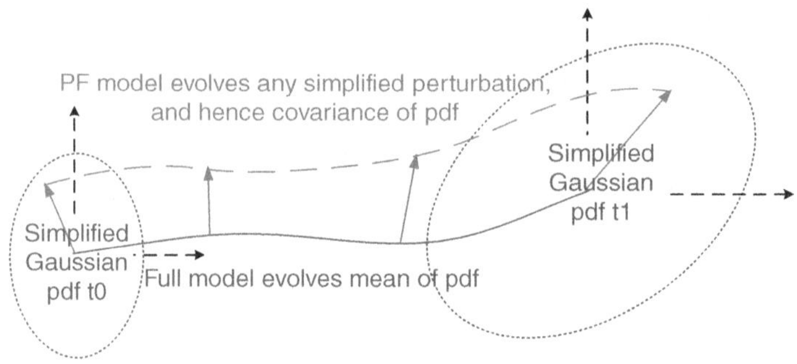

Fig. 2.6 Statistical, incremental, 4D-Var approximates entire pdf by a Gaussian. The 4D analysis increment is a trajectory of the PF model, optionally augmented by a model error correction term.

$$P\left(\delta\mathbf{x}|\mathbf{y}^o\right) \propto \exp\left\{-\frac{1}{2}\left[\delta\mathbf{x}-\left(\mathbf{x}^b-\mathbf{x}^g\right)\right]^{\mathrm{T}}\mathbf{B}^{-1}\left[\delta\mathbf{x}-\left(\mathbf{x}^b-\mathbf{x}^g\right)\right]\right\}$$
$$\times\exp\left[-\frac{1}{2}\left(\underline{\mathbf{y}}-\underline{\mathbf{y}}^o\right)^{\mathrm{T}}\underline{\mathbf{R}}^{-1}\left(\underline{\mathbf{y}}-\underline{\mathbf{y}}^o\right)\right], \tag{2.60}$$

$$\underline{y}=\underline{\tilde{\mathbf{H}}\tilde{\mathbf{M}}}\delta\mathbf{x}+\underline{H}\left(\underline{\bar{M}}\left(\mathbf{x}^g\right)\right). \tag{2.61}$$

Since (2.61) is linear in $\delta\mathbf{x}$, (2.60) is Gaussian. So, if we use the same descent algorithm approach to efficiently approximate the minimum of

$$J\left(\delta\mathbf{x}\right)=\frac{1}{2}\left[\delta\mathbf{x}-\left(\mathbf{x}^b-\mathbf{x}^g\right)\right]^{\mathrm{T}}\mathbf{B}^{-1}\left[\delta\mathbf{x}-\left(\mathbf{x}^b-\mathbf{x}^g\right)\right]$$
$$+\frac{1}{2}\left(\underline{\mathbf{y}}-\underline{\mathbf{y}}^o\right)^{\mathrm{T}}\underline{\mathbf{R}}^{-1}\left(\underline{\mathbf{y}}-\underline{\mathbf{y}}^o\right) \tag{2.62}$$

subject to (2.61), we are also approximating the mean of (2.60). By construction, (2.62) is quadratic, so we avoid the minimization problems that plague (2.56).

2.2.3 The outer loop

Unfortunately this approach is not fully satisfactory either. It has been found to be beneficial to use the best available, high-resolution, model to carry information forward, so that we retain information from past observations as accurately as possible. Components of NWP models are developed based on physical principles. We do not have an accurate model $\underline{\bar{M}}$ to predict the evolution of the mean of the PDF (a non-physical field), as required by (2.61).[11] So in practice we use the normal NWP model

[11] Increased scale-dependent diffusion as the forecast progressed was tried by Jakimow et al. (1992), but the approach was later discarded because the model gave 'unphysical' forecasts that did not look like the real atmosphere.

$\underline{\bar{M}}$. To reduce the effect of linearization errors, (2.62) can be iterated a few times in an outer loop, updating $\mathbf{x}^g$. This gives a compromise between the 'deterministic' and 'statistical' approaches; we are trying to fit the full model to the observations, but the method used to do this considers an approximation to the full pdf rather than just seeking a local mode.

Nonlinear inner loop

The usual approach to incremental VAR (Courtier et al., 1994) is to split (2.61) and subtract the 'guess' part at an early stage to give innovations $\mathbf{d}$:

$$\mathbf{d} = \underline{\mathbf{y}}^o - \underline{H}\left(\underline{\bar{M}}\left(\mathbf{x}^g\right)\right). \tag{2.63}$$

Evaluation of (2.61) can then be made fully incremental—full-field values are only needed in the definition of linear operators $\underline{\tilde{\mathbf{H}}}$ and $\underline{\tilde{\mathbf{M}}}$. The main benefit of this approach is cost—we can use an accurate and expensive model for $\underline{\bar{M}}$, while a cheaper model can be used for the multiple runs of $\underline{\tilde{\mathbf{M}}}$. As discussed above and in Section 2.3.5, there is also a benefit from the regularization of $\underline{\tilde{\mathbf{M}}}$. Because of the fully incremental approach, $\underline{\tilde{\mathbf{H}}}$ has to be linearized too. The Met Office scheme (Lorenc et al., 2000; Rawlins et al., 2007), as described in Section 2.1.5, is, I believe, unique in using instead

$$\mathrm{y} = \underline{H}\left(\underline{\bar{M}}\left(\mathbf{x}^g\right) + \underline{\tilde{\mathbf{M}}}\delta\mathbf{x}\right). \tag{2.64}$$

This allows us to have some nonlinearities in the inner-loop minimization of (2.62), which is no longer strictly quadratic.[12] This allow the treatment of nonlinear operators and non-Gaussian errors (Sections 2.3.2–2.3.4) without the expense of rerunning the full $\underline{\bar{M}}$.

Approximations in (tangent-)linear PF model

It is worth highlighting how large in practice are the approximations in (2.61). We can measure this using the linearization error $\epsilon_{\text{lin}} = \mathbf{M}\delta\mathbf{x} - \left[M\left(\mathbf{x}+\delta\mathbf{x}\right) - M\left(\mathbf{x}\right)\right]$, calculated for typical perturbations $\delta\mathbf{x}$ of similar size to the analysis increment. Then the relative error is a norm $||.||$ given by

$$R = \frac{||\epsilon_{\text{lin}}||}{||M\left(\mathbf{x}+\delta\mathbf{x}\right) - M\left(\mathbf{x}\right)||}. \tag{2.65}$$

If $R > 1$ for any norm relevant to observations, it is likely that the analysed correction $\delta\mathbf{x}$ will not improve the fit of the full model integration to those observations. For some variables (e.g. humidity) and forecast lengths of 6 hours or more, R is much closer to one than zero (Radnóti et al., 2005).

[12] The best minimization method depends on the degree of nonlinearity. In the global model, linear observations dominate, so we split the minimization into a double inner loop. The innermost loop solves a series of quadratic approximations to (2.62) that use $\underline{\tilde{\mathbf{H}}}$ linearized about the latest values from the second loop. In the regional model, nonlinear observations dominate (Section 2.3.4), so we use a more general algorithm such as M1QN3 (Gilbert and Lemaréchal, 1989).

2.3 4D-Var: advanced aspects

2.3.1 Developments to covariance modelling

Flow dependence

An important part of the improvement in forecasts in the last few decades was due to the correct (Bayesian) combination of information from the current observations with the forecast prior, which summarizes the information from previous observations (Simmons and Hollingsworth, 2002). Much important and detailed work is needed to correctly characterize the error pdf of each type of observation—the variance, the bias, and the probability of gross error. An NWP model is very large, and the errors for different variables are strongly related, so the best we can normally attempt is to model the error covariance **B**. The first operational 3D multivariate statistical analysis method (Lorenc, 1981) made the following assumptions about the **B** characterizing background errors, all of which are wrong!

- *stationary*: time- and flow-invariant;
- *balanced*: predefined multivariate relationships exist;
- *homogeneous*: same everywhere;
- *isotropic*: same in all directions;
- *3D separable*: horizontal correlation independent of vertical levels or structure and vice versa.

Since then, many valiant attempts have been made to address them individually, but with limited success because of the errors remaining in the others.

The most attractive ways of addressing them all at once are long-window 4D-Var or hybrid ensemble-Var. Thépaut et al. (1996) showed that a constant 'climatological' covariance evolved into plausible flow-dependent patterns in a 24-hour 4D-Var window, Zhang et al. (2007) showed that even random structures grow to similar patterns in 24–36 hours, while Fisher et al. (2005) showed that there is no advantage to going beyond five days. Fisher (2011) discusses how such a long window might be affordable.[13] Alternatively, ensemble Kalman filters allow a sample of error patterns to grow over a long window. There are many flavours of ensemble Kalman filter for generating them, and two alternative methods of using these *errors of the day* in a variational covariance model. First, one can estimate parameters of the covariance model (as embodied in the variable transforms described in Section 2.1.5) from the ensemble. This approach retains the proven benefits of the existing variational method, but it is difficult to address all the weaknesses listed above (Section 2.4.1). The second approach is to use the ensemble perturbations, after localization, directly to augment (or even replace) the traditional covariance model (Section 2.4.2). The Met Office has just implemented such a hybrid ensemble-4D-Var scheme giving a considerable benefit (Clayton et al., 2013). There is potential for this approach to even replace the linearized perturbation model within 4D-Var; I say more about this in Section 2.4.3.

[13] As Mike Fisher is an author of another chapter is this volume (Chapter 19), I am concentrating here on the alternative ensemble approach.

2.3.2 Non-Gaussian background errors: humidity

It is not possible to represent many aspects of a pdf for as many variables as an NWP model; we usually only consider mean errors and covariances. But in some cases it has been found advantageous to change to variables with more Gaussian errors. ECMWF, HIRLAM, and the Met Office have implemented a nonlinear humidity transform to compensate for the non-Gaussian errors of humidity forecasts (Hólm, 2003; Gustafsson et al., 2011; Ingleby et al., 2013). The largest cause of the non-Gaussianity is the physical limit to humidity—it must always be positive and seldom goes very super-saturated.

Let us assume that our forecast model is unbiased, in that its distribution of model humidity values is the same as that of the atmosphere mapped into model space. This 'truth' state is the goal of our assimilation process, and background errors are in principle measured from it. As we do not know it, we have to study background errors using a proxy—in this section, for illustrative purposes, I use a large set of radiosonde observations mapped to model levels. Figure 2.7, adapted from Lorenc (2007), shows the joint distribution of background and observed (proxy true) relative humidity (RH) values from the Met Office global assimilation. It is close to symmetric about the diagonal, showing that our assumption is reasonable. Yet the distribution of true values conditional on any particular background value is biased, with mean value given by

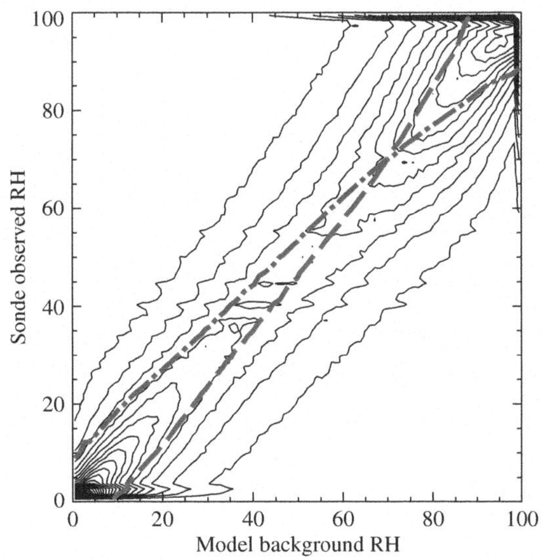

Fig. 2.7 Joint pdf of collocated background relative humidity (RH) from the Met Office global six-hour forecast and observed RH mapped to model levels, for 740 328 radiosonde soundings from December 2005 to July 2007 (Lorenc, 2007). The dash–dot line shows the mean sonde RH for each background RH bin, and the dashed line shows the mean background RH for each sonde RH bin.

the dash–dot line. So, without considering observations, the minimum-variance best estimate of the true relative humidity (RH) would be obtained by bias correcting the background to this line. The resulting overall moisture and cloud distribution would not be correct—no RH would be greater than 90%, so there would be insufficient cloud and precipitation in the subsequent forecast. The method suggested by Hólm (2003) is to use probability distributions conditional on (RHb+RHa)/2. We can illustrate the effect by plotting the joint pdf of the difference in RH and the mean RH (not shown, since it is equivalent to Fig. 2.7 rotated by 45°). This is unbiased. However, having the assumed probability distribution dependent on the analysed value makes the problem implicit, requiring an iterative solution method. ECMWF and HIRLAM put this iteration in their outer loops; the Met Office can solve it more accurately, in their non-quadratic inner loop.

This method performs well: there is no spurious bias; the background error standard deviation (which is a factor in the equation for the analysis increment) can be small when the background RH is near zero and the increment is negative, making negative analysed values unlikely, while for positive increments the standard deviation can be larger, making it possible to change near-zero background RH to any positive value. It is interesting to consider precisely which assumptions about the prior distribution we need to make for this to be the correct Bayesian method:

- The distribution of values in the background, generated by the model, is close to correct—we have the right cloud cover on average.
- It is important to us to retain this correct distribution—more so than to reduce the expected RMSE at each point.

The Hólm transform constructs a (skewed) prior whose mode is the background. We rely on a minimization that finds this mode—*not the mean*—and hence returns the model background unaltered in the absence of observations. I say more about this desire to rely on the model in Section 2.3.5.

2.3.3 Non-Gaussian observation errors: variational quality control

Why quality control?

The quality control we do in data assimilation has two reasons:

1. We have physical reasons for believing certain events may occur that affect the observed value. We wish to detect these events.
2. The distribution of errors associated with a datum is such that there is a non-negligible probability of errors that would be unacceptably large *for the use we are making of the datum.*

Note that reason 2 depends on our use of the observation. If we are using an analysis method based on a quadratic penalty function, it is linear in the observed values. A single large error can then be disastrous (Fig. 2.8, solid line). However, if instead we minimize the mean absolute deviation (this is the correct norm if the observational error probability distribution function is proportional to an exponential of the absolute

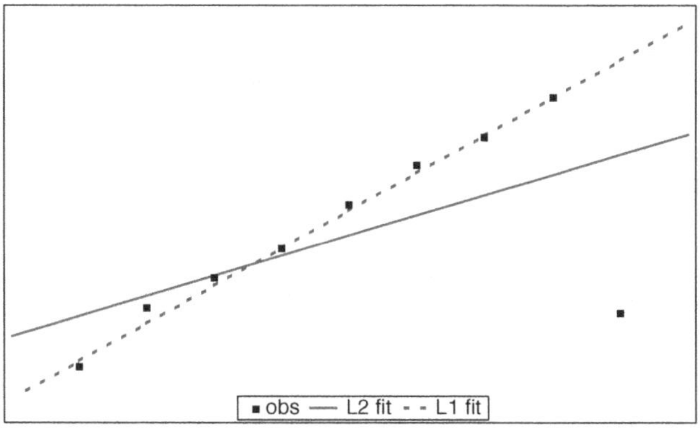

Fig. 2.8 Best-fit straight lines to data including a gross error: the solid line uses a quadratic (L2) norm and the dotted line uses a mean absolute (L1) norm. (Based on Tarantola (1987).)

deviation), the bad datum is ignored (Fig. 2.8, dotted line). Analysis methods designed to ignore such outliers are also considered to be quality control methods.

I do not have time to review all quality control methods – I limit myself to the Bayesian theory[14] which builds on section 2.1.2 and its use within VAR.

Discrete Bayes theorem applied to gross observational errors

The simplest QC is just a decision as to whether an observation is 'bad' or not, as in Table 2.1. The information that I have thrown a six has added to my knowledge, so that the posterior probability that the chosen die is weighted has increased. If I were to throw again, and get another six, the probability would increase again.

Non-Gaussian model for observational errors

The simplest model that allows for the observed fact that observational errors are not in practice Gaussian is to assume that a small fraction of the observations are corrupted and hence worthless. The others have Gaussian errors. For each observation, we have

$$p(y^o|x) = p(y^o|G \cap x)P(G) + p(y^o|\bar{G} \cap x)P(\bar{G}), \tag{2.66}$$

where G is the event 'there is a gross error' and $\bar{G}$ means *not* G. For $p(y^o|G \cap x)$, we can use a Gaussian as in (2.21).

To avoid singularities and give a pdf that integrates to 1, we assume that 'worthless' observations contain no useful information over a range of plausible values (with corrupted observations giving implausible values already rejected):

$$p(y^o|\bar{G} \cap x) = \begin{cases} k & \text{over the range of plausible values,} \\ 0 & \text{elsewhere.} \end{cases} \tag{2.67}$$

[14] The Bayesian approach was suggested to me by Purser (1984) and included in Lorenc (1984).

Table 2.1 Example: Bayesian dice

I have two dice. One is weighted towards throwing sixes. I have performed some experiments with them, and have the prior statistics that:

for the weighted (W) die	$P(6\|W)$	$= \dfrac{58}{60}$
for the good (G) die	$P(6\|G)$	$= \dfrac{10}{60}$
I choose one at random	$P(W) = P(G)$	$= \dfrac{1}{2}$
I throw this die—it shows 6	$P(6) =$	
	$P(6\|W)P(W) + P(6\|G)P(G)$	$= \dfrac{58}{60} \times \dfrac{1}{2} + \dfrac{10}{60} \times \dfrac{1}{2}$
		$= \dfrac{34}{60}$
Applying Bayes' theorem:	$P(G\|6) = \dfrac{P(6\|G)P(G)}{P(6)}$	$= \dfrac{\frac{10}{60} \times \frac{1}{2}}{\frac{34}{60}}$
		$= \dfrac{5}{34}$
	$P(W\|6) = \dfrac{P(6\|W)P(W)}{P(6)}$	$= \dfrac{\frac{58}{60} \times \frac{1}{2}}{\frac{34}{60}}$
		$= \dfrac{29}{34}$

Analysis, allowing for gross errors

One way of dealing with non-Gaussian, long-tailed distributions such as (2.66) is to detect and reject probable gross errors in a quality control step prior to the analysis, and then to assume that the errors of the remaining observations come from a Gaussian distribution. The quality control decision can either use pragmatic criteria or apply the discrete Bayes' theorem to the event G 'there is a gross error' (Lorenc and Hammon, 1988):

$$P(G|y^o) = \frac{P(y^o|G)P(G)}{P(y^o)}. \tag{2.68}$$

In variational quality control we instead use (2.66) directly. We can derive the formula for the posterior pdf $p(x|y^o)$ either directly from the continuous Bayes' theorem, or else from

$$p(x|y^o) = p(x|y^o \cap G)P(G|y^o) + p(x|y^o \cap \bar{G})P(\bar{G}|y^o). \tag{2.69}$$

The posterior pdf is the weighted sum of two Gaussians, corresponding to accepting or rejecting the observation. The weights given to each are the posterior probabilities of G and $\bar{G}$. When the peaks are distinct, these correspond to the areas under each. The results of allowing for gross errors in this way can be quite dramatic, even if $P(G)$ is small. Figure 2.9 shows the equivalent of Fig. 2.1, with errors appropriate for pressure observations from ships, which used to have about 5% gross errors. When the observation and the background agree, there is little difference from Fig. 2.1. But when they disagree, the posterior distribution becomes bimodal.

Figure 2.10 shows the same examples in the log(probability) form of Fig. 2.9. The observational penalty is not quadratic; it has plateaux away from the observed value. Adding this to a quadratic background penalty can give multiple minima: a practical problem in a variational method—we have to decide which is the 'best' minimum, then we have to devise a method for finding it (Dharssi et al., 1992; Andersson and Jarvinen, 1999). This is difficult; moreover, even if done well, the algorithm cannot be robust—an infinitesimal change to the inputs can cause a large change to the outputs. The only way to make it robust is to have a penalty function without multiple minima, and the way to guarantee this is to have the component penalty functions never convex. This can be achieved using the Huber norm shown in Fig. 2.11 (Huber, 1973; Guitton and Symes, 2003)—now the operational method at ECMWF. It has the effect of still giving

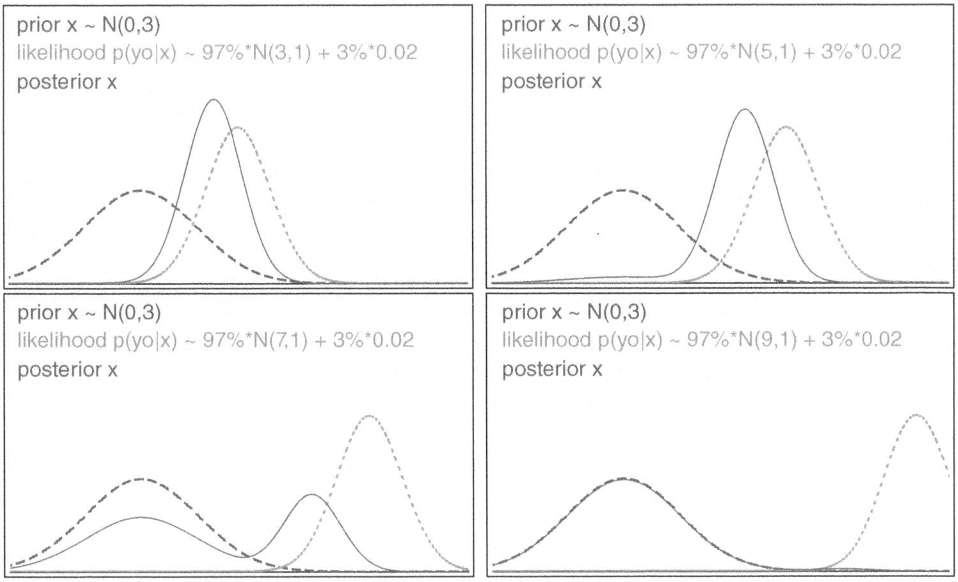

Fig. 2.9 Prior pdf $p(x)$ (dashed line), posterior pdf $p(x|y^o)$ (solid line), and non-Gaussian likelihood of observation $p(y^o|x)$ (dotted line), plotted against x for various values of y^o. For comparison with Fig. 2.1. (Adapted from Lorenc and Hammon (1988).)

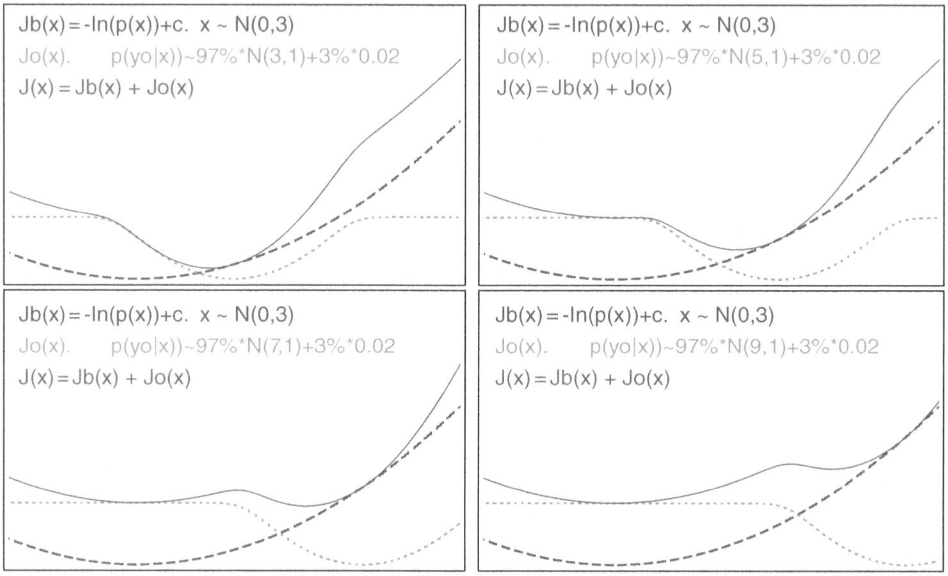

Fig. 2.10 Minus the logarithm of the curves in Fig. 2.9. The solid line is the penalty function $J(x)$, the dashed line is the prior or background penalty $J_b(x)$, and the dotted line is the observational penalty $J_o(x)$. For comparison with Fig. 2.2. (Adapted from Lorenc and Hammon (1988), Copyright © 1988, Royal Meteorological Society.)

'rejected' observations some weight; the impact of an observation (i.e. the innovation times the weight) is limited to the value at the 'rejection' limit—as the innovation increases, the weight decreases accordingly.

2.3.4 Nonlinear observation operators

It is possible to extended the variational QC penalty function shown in Fig. 2.10 to scatterometer observations, where the preprocessing give several possible alias winds, each with a prior probability of being correct. This gives a highly non-quadratic J_o, with a potential minimum for each alias. There are also satellite observation of sea-surface wind speed, which is nonlinearly related to the model's u, v.

Apart from these, the main nonlinearities come from observations affected by the condensation process. The forecast precipitation is determined by the tangent-linear (PF) model's parametrisations, so assimilation of precipitation observations is determined by the realism of these, and relies on the outer loop to handle nonlinearities. Lopez (2011) found that in the ECMWF system only relatively long accumulations (six-hours) could be handled accurately.

Cloud–moisture relationships are more diagnostic, so observations related to these can be handled without an outer loop, or even in 3D-Var, as long as (2.47) is used. Such observations have been used for many years in UK regional models, where short-period forecasts of cloud and visibility are important. For instance Renshaw and Francis

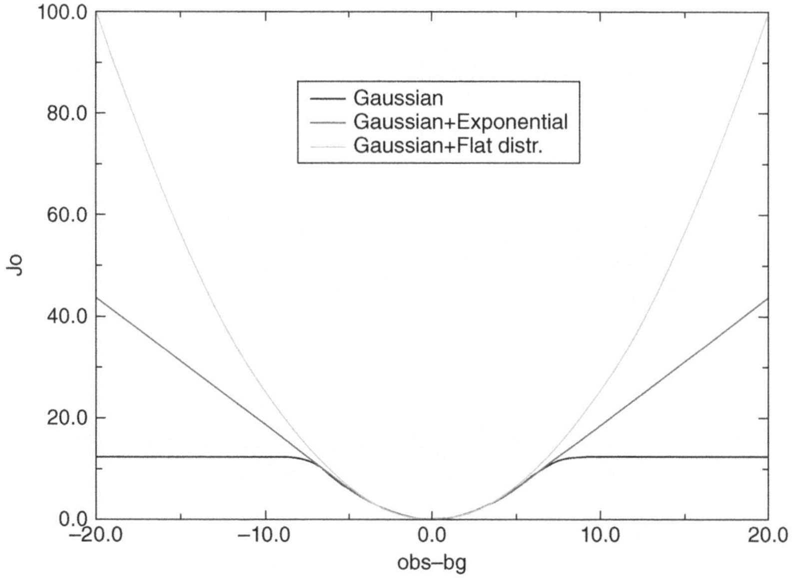

Fig. 2.11 The Huber norm—a compromise between the L2 and L1 norms. (ECMWF DA/SAT Training Course, May 2010. Erik Andersson and Lars Isaksen.)

(2011) show how satellite cloud imagery can be assimilated, while Clark et al. (2008) show the benefit of assimilating visibility observations. The latter have a very nonlinear relationship to humidity, as shown in Fig. 2.12.

2.3.5 Coping with butterflies

Predictability and chaos

Lorenz (1969) pointed out that the atmosphere has many scales of motion and that errors in small scales will quickly grow and affect larger scales. Revised in detail, this is now the accepted picture of the growth of errors in the spectral domain (Tribbia and Baumhefner, 2004), commonly known as the butterfly effect (Palmer, 2005). Lorenc and Payne (2007) showed that because of the butterfly effect, conventional deterministic 4D-Var will not work as model resolutions increase towards the unfiltered continuous limit. They suggested that a solution is to use statistical incremental 4D-Var, with a perturbation forecast model that is filtered to prevent the rapid growth of scales that would otherwise grow excessively over the time window. This idea has been demonstrated in ocean 4D-Var by Hoteit et al. (2005); it is implicit in the designs of most operational NWP 4D-Var systems, which are forced for computational reasons to use a low-resolution linear model in the inner loop.

The butterfly effect is to do with the multiple scales; it means that we cannot necessarily expect 4D-Var algorithms to continue working as we move to higher resolutions on more powerful computers. At any given resolution, an NWP forecast model

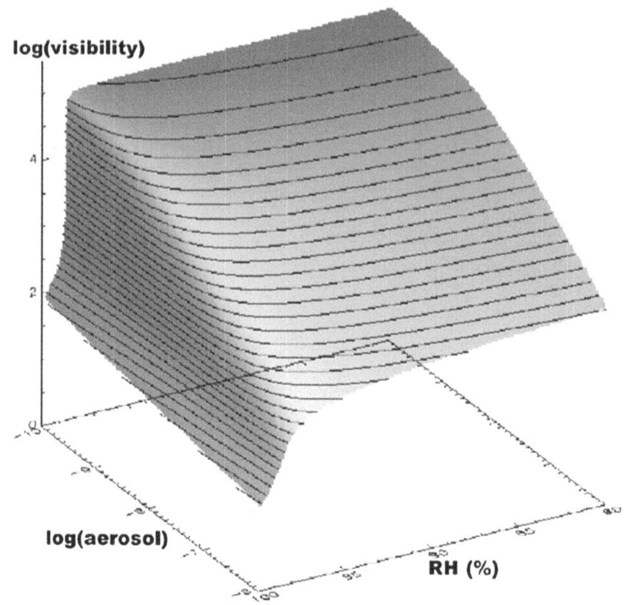

Fig. 2.12 Graph showing the $\log_{10}$(visibility) surface as a function of relative humidity and $\log_{10}$(aerosol), where aerosol is defined as the mass mixing ratio in kg/kg. Temperature is held constant. (From Clark et al. (2008), © Crown Copyright 2008. Reproduced with the permission of the Controller of HMSO. Published by John Wiley & Sons Ltd.)

is chaotic—chaos is different from the butterfly effect in that it can be exhibited in toy models with low resolution (Lorenz, 1963). Abarbanel et al. (2010), approaching data assimilation as synchronized chaos, say that there must be enough (observational) controls to move the positive conditional Lyapunov exponents on the synchronization manifold to negative values. This is normally quite easy to achieve in a low-resolution system with few chaotic Lyapunov vectors, but in modern practical NWP, with varying observational networks over the globe, it is much harder to ensure that everything that grows is sufficiently observed. It is quite easy to test if a data assimilation system (rather than the model it uses) is chaotic. We just run the entire system with identical inputs from initial conditions that differ by a very small perturbation, as in the original (Lorenz, 1963) demonstration of chaos. Figure 2.13 shows the result of such an experiment with a Met Office system. The initially small perturbation grows for several days before saturating on an attractor with RMS differences of order 0.5 m/s for wind components. (An exception is the top of the model, where differences drift to become increasing large owing to the difficulty of controlling model errors and biases at these levels (Polavarapu, 2011).) These differences are of course very much smaller than those between two free-model runs in a similar experiment, but they still represent an irreducible uncertainty in the analyses produced by this DA system. Deterministic 4D-Var, using the exact tangent-linear model to the NWP model used, would probably

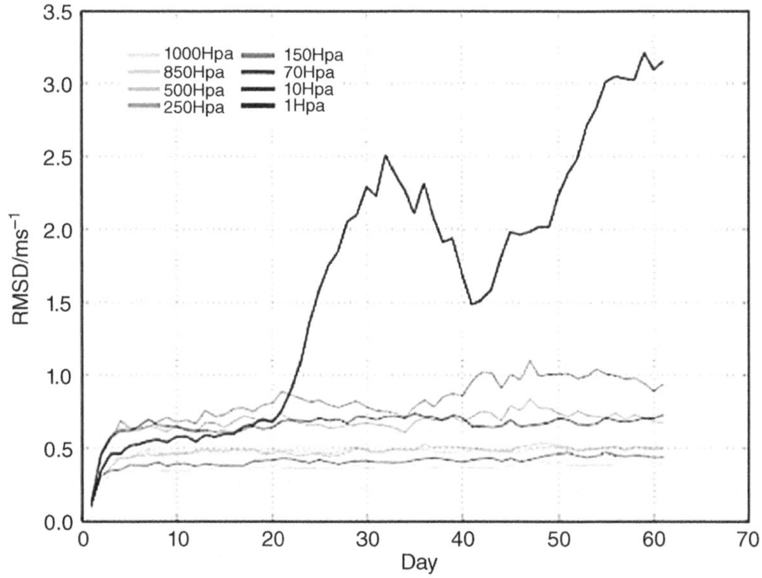

Fig. 2.13 Global RMS differences between u-components in identical NWP assimilations, due to small initial perturbations in the background at day 0. (Peter Jermey, personal communication.)

not work. The uncertainty is another reason why trialling of DA system changes is difficult; identical DA systems can still give apparently random different signals when verified against independent observations, necessitating a longer trial to get significant results about a real change.

Benefiting from the attractor

The problems with long-window 4D-Var for a chaotic model are due to the continuing exponential growth of some infinitesimal perturbations,[15] and hence of similar perturbations of any amplitude in the tangent linear model. In the full nonlinear model, the growth slows as the amplitude increases, leading to saturation often at quite small amplitudes as seen in Fig. 2.13, so an ensemble Kalman filter using the nonlinear model does not have the same problem. Because their predictions remain bounded, chaotic models have an attractor of states that they might pass through that has much lower dimension than the space that could be represented if all the model variables were independent. Meteorologists have long understood this behaviour and developed rules to describe plausible states: both in terms of balance, modes, and power spectra, and also synoptically with conceptual models of fronts, cyclones, cloud-capped inversions, etc. This is important prior information that we should use in data assimilation. For

[15] Measured by the Lyapunov exponent.

instance, a human meteorologist could use a conceptual model of a front to fit scattered observations and draw an analysis that could be used in an accurate forecast of the weather at a point ahead of the front. In NWP data assimilation, we do not usually use this prior knowledge; we compensate by having a good background forecast, but it would be better to use both. Methods for developing flow-dependent correlations are making a start, but the persistent structures are usually maintained by nonlinear processes and do not have Gaussian PDFs. For instance, Lorenc (2007) showed that vertical covariances could not describe the errors associated with a cloud-capped inversion. In practice, the best way we have found of describing the attractor is as a *slow manifold* of 'states that the model likes'. Data assimilation algorithms have regularly used the model in this way, in diabatic nonlinear normal mode initialization, spin-up periods, and other approaches that often seemed at the time like simple trial-and-error engineering and tuning.

Modern incremental VAR methods use it by having a spun-up model state as background, and only altering it (in a smooth way) when there is clear observational evidence to do so. We have used this concept already in Section 2.3.2.

One approach to achieving the goal of smooth increments is to add weak constraints (J_c terms) to the VAR penalty function that penalize the presence of unbalanced structures. For 4D-Var, one of the most promising ideas is to use digital filtering J_c terms. The idea here is to apply low-pass digital filters to the trajectory that is compared with the observations, and penalize the resulting filter increments. This technique has been demonstrated by Gustafsson (1992), Polavarapu et al. (2000), Gauthier and Thépaut (2001), and Wee and Kuo (2004). It was implemented in the Met Office 4D-Var by Adam Clayton, from whose documentation I take this description. The basic idea behind digital filtering J_c terms is to penalize the presence of high-frequency motions within the model trajectory that spans the 4D-Var data window, as revealed by the increments obtained on application of suitably designed low-pass digital filters. Here, one can choose to consider either the full model trajectory $\{\mathbf{x}(t_0), \mathbf{x}(t_1), \ldots, \mathbf{x}(t_n)\}$ or the increment trajectory $\{\mathbf{x}(t_0) - \mathbf{x}^b(t_0), \mathbf{x}(t_1) - \mathbf{x}^b(t_1), \ldots, \mathbf{x}(t_n) - \mathbf{x}^b(t_n)\}$ defined relative to the background trajectory. In general, it is better to work with the latter. There are two reasons for this. First, the background trajectory is normally well balanced in any case, so we only need consider noise within the increments relative to the background. More importantly though, filtering a full model trajectory generally affects legitimate but relatively high-frequency motions (such as the semi-diurnal tides, which have a period of 12 hours) that are best left alone. From now on, then, we will only consider the application of J_c terms to the increment trajectory.

Digital filtering J_c terms are based on non-recursive digital filters of the form

$$\delta\mathbf{x}\left(t_{(l+m)/2}\right)^{\mathrm{DF}} = \sum_{k=l}^{m} \alpha_k \delta\mathbf{x}(t_k), \tag{2.70}$$

where $0 \leq l \leq m \leq n$ and α_k is a (real) filter coefficient. Normally, $l = 0$ and $m = n$, so that the filter spans the whole 4D-Var time window. The high-frequency motions are represented by the filter increments, which can be obtained by replacing α_k with γ_k, where

$$\gamma_k = \begin{cases} \alpha_k, & k \neq (l+m)/2, \\ \alpha_k - 1, & k = (l+m)/2. \end{cases} \tag{2.71}$$

The general form for a digital filtering J_c term is then

$$J_{\mathrm{DF}} = \lambda_{\mathrm{DF}} \left\| \sum_{k=l}^{m} \gamma_k \left[\mathbf{x}(t_k) - \mathbf{x}^b(t_k) \right] \right\|, \tag{2.72}$$

where λ_{DF} is a penalty parameter that determines the weight given to the term, and $\|\cdot\|$ is some kind of norm used to measure the size of the filter increments. With a single digital filtering J_c term included in (2.62), the full 4D-Var cost function is

$$\begin{aligned} J(\delta\mathbf{x}) &= \frac{1}{2}\left[\delta\mathbf{x} - \left(\mathbf{x}^b - \mathbf{x}^g\right)\right]^{\mathrm{T}} \mathbf{B}^{-1} \left[\delta\mathbf{x} - \left(\mathbf{x}^b - \mathbf{x}^g\right)\right] \\ &+ \frac{1}{2}\left(\underline{\mathbf{y}} - \underline{\mathbf{y}}^o\right)^{\mathrm{T}} \underline{\mathbf{R}}^{-1} \left(\underline{\mathbf{y}} - \underline{\mathbf{y}}^o\right) \\ &+ \lambda_{\mathrm{DF}} \left\| \sum_{k=l}^{m} \gamma_k \left[\mathbf{x}(t_k) - \mathbf{x}^b(t_k)\right] \right\|. \end{aligned} \tag{2.73}$$

As we move to higher resolutions, with models that are representing the complex (and poorly observed) structures of convection, the reliance on the model will grow—other simpler concepts of balance do not hold at the convective scale, but there are still clear recognizable structures for convection (and many more structures that are unlikely ever to occur). So, planning for the future, we should select methods that allow the model to spin up and evolve states on the attractor. The particle filter is the extreme way of ensuring this (e.g. van Leeuwen, 2011). Another argument for an outer loop and a long window is that together they seek a spun-up model trajectory that fits the observations. On the other hand, the 4D control variable approach to 4D-Var (Fisher, 2011) deliberately avoids long model runs in the data assimilation, so may have problems of spin up. Some EnKF methods recentre the ensemble each cycle about the ensemble mean analysis. This may be undesirable, because the ensemble mean is not on the attractor—a forecast from it would be expected to give a poor short-period forecast of 'weather' such as cloud and precipitation.

2.4 4D-Var: coupling with ensembles

2.4.1 Use of EDA (co)variances in 4D-Var

Most work so far to use real-time ensembles in the VAR covariance model has concentrated on the variances (e.g. Raynaud et al., 2011; Bonavita et al., 2012). There are plans to extend this; one idea is to implement a flow-dependent extension of the wavelet diagonal correlation model (Fisher, 2003); another is to estimate and use the correlation scale.

These approaches can apply an optimal filtering of the parameters estimated, reducing sampling noise from a small ensemble. But it is difficult for them to address all the flow-dependent properties of covariances. For instance, Montmerle and Berre

(2010) demonstrated a situation dependence to the intervariable correlations, and many studies have shown non-isotropic flow-dependent correlations—neither is easy to parameterize, although attempts can be made along the lines of Ingleby et al. (2013) and Purser et al. (2003*b*).

2.4.2 Hybrid-4D-Var

Another approach is to use the ensemble perturbations directly to replace $\mathbf{B}$, not using the control variable transforms of Section 2.1.5 but instead weighting each perturbation by a new control variable field $\boldsymbol{\alpha}$ as suggested by Lorenc (2003*b*). This is now used operationally at the Met Office; the following description is adapted from Clayton et al. (2013). At the start of each 4D-Var window, the necessary ensemble forecast fields are interpolated onto the analysis grid used by 4D-Var using the simplification operator S, producing states $\mathbf{w}_k$, where k is the member index. It is convenient to introduce a rectangular matrix $\mathbf{W}$ whose columns are scaled differences between the ensemble forecasts and the ensemble mean:

$$\begin{aligned} \mathbf{W} &= \frac{1}{\sqrt{K-1}}(\mathbf{w}_1 - \bar{\mathbf{w}}, \mathbf{w}_2 - \bar{\mathbf{w}}, \ldots, \mathbf{w}_K - \bar{\mathbf{w}}) \\ &= (\, \mathbf{w}'_1, \, \mathbf{w}'_2, \ldots, \, \mathbf{w}'_K \,), \end{aligned} \tag{2.74}$$

where K is the number of perturbed members. The ensemble covariance $\mathbf{P}_e^f$ is then given by

$$\mathbf{P}_e^f = \mathbf{W}\mathbf{W}^{\mathrm{T}}. \tag{2.75}$$

The main problem with this raw ensemble covariance is its low rank (K) and the presence of sampling error. To remedy both problems, Houtekamer and Mitchell (2001) suggested replacing (2.75) by

$$\mathbf{B}_e = \mathbf{P}_e^f \circ \mathbf{C}, \tag{2.76}$$

where $\mathbf{C}$ is a 'localization' covariance (normally correlation) matrix, and the operator $\circ$ denotes the element-by-element product of two same-sized matrices known as the Schur product.[16] The aim of $\mathbf{C}$ is to downweight small ensemble correlations that are likely to be dominated by sampling noise, while leaving larger more robust correlations relatively unaffected. The simplest designs of $\mathbf{C}$ are localizations in a literal sense; removing ensemble correlations beyond a distance where ensemble correlations are likely to be small. More sophisticated schemes (e.g. Anderson, 2007; Bishop and Hodyss, 2011) work more directly on the correlations.

[16] We note that localization is being performed in model space. For non-local observations such as satellite radiances, Campbell et al. (2010) show that model-space localization is superior to the observation-space localisation schemes typically used with the ensemble Kalman filter, such as that of Houtekamer and Mitchell (2001).

Hybrid covariances

The hybrid 4D-Var system seeks to implement a background error covariance $\mathbf{B}$ that is a linear combination of the climatological and ensemble covariances described above, i.e. a covariance of the form

$$\mathbf{B} = \beta_c^2 \mathbf{B}_c + \beta_e^2 \mathbf{B}_e, \tag{2.77}$$

where β_c^2 and β_e^2 are scalar weights.[17] To accomplish this, we use the extended control variable method described in Section 5 of Lorenc (2003*b*). In this formulation, the climatological contribution $\mathbf{Uv}$ to $\delta\mathbf{w}$ is multiplied by β_c, and the introduction of each ensemble error mode $\mathbf{w}'_k$ is controlled via its Schur product with a three-dimensional scalar field $\boldsymbol{\alpha}_k$:

$$\delta\mathbf{w} = \beta_c \mathbf{U}_p \mathbf{U}_v \mathbf{U}_h \mathbf{v} + \beta_e \sum_{k=1}^{K} \mathbf{w}'_k \circ \boldsymbol{\alpha}_k. \tag{2.78}$$

The VAR cost function is then modified to

$$J(\mathbf{v}, \boldsymbol{\alpha}_1, \boldsymbol{\alpha}_2, \ldots, \boldsymbol{\alpha}_K) = \frac{1}{2}\mathbf{v}^{\mathrm{T}}\mathbf{v} + \sum_{k=1}^{K} \boldsymbol{\alpha}_k^{\mathrm{T}} \mathbf{C}^{-1} \boldsymbol{\alpha}_k + J_o + J_c. \tag{2.79}$$

To improve the conditioning of the minimization problem, (2.79) is rewritten in terms of an 'alpha' control vector $\mathbf{v}^{\alpha}$ that is the concatenation of the K vectors $\mathbf{v}_k^{\alpha}$ defined by

$$\boldsymbol{\alpha}_k = \mathbf{U}^{\alpha} \mathbf{v}_k^{\alpha}, \tag{2.80}$$

where $\mathbf{U}^{\alpha} = \mathbf{C}^{1/2}$. Substituting for $\boldsymbol{\alpha}_k$ in (2.79), we then have

$$J(\mathbf{v}, \mathbf{v}^{\alpha}) = \frac{1}{2}\mathbf{v}^{\mathrm{T}}\mathbf{v} + \frac{1}{2}(\mathbf{v}^{\alpha})^{\mathrm{T}}(\mathbf{v}^{\alpha}) + J_o + J_c. \tag{2.81}$$

The proof that (2.78), (2.80) and (2.81) implement the hybrid covariance given by (2.77) is given by Wang et al. (2007).

A similar system for a limited-area model was described by Zhang and Zhang (2012). While they used a similar α, they (and others) define and use the βs differently—they take them out of (2.78) and put them in (2.79) instead. The resulting control variables turn out to be scaled, giving identical results. They use

$$\mathbf{B} = (1-\beta)\mathbf{B}_c + \beta\mathbf{B}_e,$$

$$\delta\mathbf{w} = \mathbf{U}_p \mathbf{U}_v \mathbf{U}_h \mathbf{v} + \sum_{k=1}^{K} \mathbf{w}'_k \circ \boldsymbol{\alpha}_k,$$

$$J(\mathbf{v}, \mathbf{v}^{\alpha}) = \frac{1}{1-\beta}\frac{1}{2}\mathbf{v}^{\mathrm{T}}\mathbf{v} + \frac{1}{\beta}\frac{1}{2}(\mathbf{v}^{\alpha})^{\mathrm{T}}(\mathbf{v}^{\alpha}) + J_o + J_c.$$

[17] It would be natural to set $\beta_c^2 + \beta_e^2 = 1$. At the moment, as explained by Clayton et al. (2013), we increase the βs slightly to preserve the closeness of fit to observations.

2.4.3 4D-Ensemble-Var

It is expected that the computers available for NWP over the next decade will continue to get more powerful, but in terms of the number of processors rather than the speed of each (Isaksen, 2011). Using this power will require a more parallel DA algorithm; the current bottleneck in 4D-Var is the PF model. Fisher (2011) discusses one approach to making 4D-Var more parallel by changing the control variable so that time segments of each PF and adjoint integration can be run in parallel. Here I outline a more radical approach, doing away with the PF model completely.

The main idea is to extend in time the use of the ensemble perturbations, currently used in the Met Office operation hybrid ensemble-4D-Var described above, so they are used to fit the observations in a time window as 4D-Var does, but without the cost of iterating a PF and adjoint model. The potential of ensemble Kalman filters to do this has been recognized for some time (Lorenc, 2003*b*). Hunt et al. (2004) demonstrated it with an ensemble square-root filter for the Lorenz96 model, Fertig et al. (2007) compared a 4D-LETKF with 4D-Var for the same model, and Harlim and Hunt (2007) applied 4D-LETKF to the SPEEDY model. The explicit documentation and testing in a VAR environment has been published by Liu et al. (2008, 2009) and Buehner et al. (2010*a*,*b*). Liu called the technique En4DVAR and Buehner En-4D-Var.[18] I prefer the name 4DEnVar (or, when speaking, 4D-Ensemble-Var) since the key feature is the four-dimensional use of the ensemble; it also is more consistent with the 4DEnKF terminology of Hunt et al. (2004). I reserve En-4D-Var to describe a component of the approach in our current hybrid-4D-Var described above: using the ensemble to estimate the background error covariance **B** at the beginning of the time window,[19] with the fitting of observations distributed in time done as in 4D-Var using a PF and adjoint model. Buehner et al. (2010*b*) presented results from a near-operational-quality Canadian NWP system showing that 4DEnVar is competitive with traditional 4D-Var and with En-4D-Var.

Basic 4DEnVar equations

I consider a four-dimensional best fit to all the observations in an assimilation window from start time t_s to end time t_e. With the addition of the underline notation to denote the extra time dimension, and the replacement of the climatological **B** by the predicted four-dimensional $\underline{\mathbf{P}}$, this has identical form to incremental 3D-Var. We seek to minimize:

$$J\left(\delta\underline{\mathbf{x}}\right) = \frac{1}{2}\delta\underline{\mathbf{x}}^{\mathrm{T}}\underline{\mathbf{P}}^{-1}\delta\underline{\mathbf{x}} + \frac{1}{2}\left[\underline{H}\left(\underline{\mathbf{x}}^b + \delta\underline{\mathbf{x}}\right) - \mathbf{y}^o\right]^{\mathrm{T}}\underline{\mathbf{R}}^{-1}\left[\underline{H}\left(\underline{\mathbf{x}}^b + \delta\underline{\mathbf{x}}\right) - \mathbf{y}^o\right]. \tag{2.83}$$

As usual in NWP DA algorithms, we cannot actually handle (2.83); $\delta\underline{\mathbf{x}}$ is big and $\underline{\mathbf{P}}^{-1}$ is much too big to manipulate! So we seek a representation of $\delta\underline{\mathbf{x}}$ in terms of a reduced set of control variables (see Fig. 2.14). The basic idea of 4DEnVar is that $\delta\underline{\mathbf{x}}$ is made up as a locally weighted linear combination of perturbation trajectories $\underline{\mathbf{x}}'_i$

[18] More recently Mark Buehner says he prefers to call it simply EnVar.
[19] Buehner et al. (2010*b*) called this 4D-Var-Benkf.

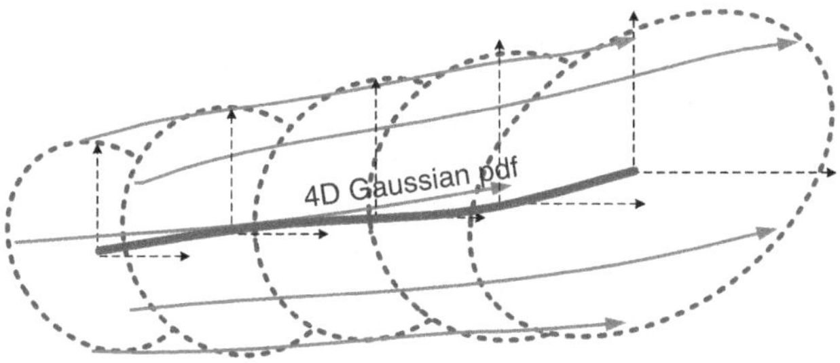

Fig. 2.14 A schematic diagram of 4DEnVar, for comparison with Fig. 2.6. The 4D analysis is a localized linear combination of model trajectories—it is not itself a model trajectory.

that are scaled (and perhaps transformed) differences between ensemble members and the ensemble mean:

$$\delta\underline{\mathbf{x}}_e = \sum_i \underline{\alpha}_i \circ \underline{\mathbf{x}}'_i \tag{2.84}$$

We assume the perturbations are independent, so that we can define each $\underline{\alpha}_i$ independently. To this we can add additional terms, designed to provide scope to correct model errors that are not sampled by the trajectories, and generally to allow the use of hybrid methods to compensate for a small ensemble. I just show a time constant and one time-varying term for each:

$$\delta\underline{\mathbf{x}} = \beta_{e0}\delta\underline{\mathbf{x}}_{e0} + \beta_{e1}\delta\underline{\mathbf{x}}_{e1} + \beta_{c0}\delta\underline{\mathbf{x}}_{c0} + \beta_{c1}\delta\underline{\mathbf{x}}_{c1}. \tag{2.85}$$

The Met Office's existing hybrid ensemble-4D-Var only has hybrid weights β_e and β_c because it does not allow for model error. Note that the climatological term $\delta\underline{\mathbf{x}}_{c0}$ is constant over the time window, since we are not using the PF model; in this aspect, the new method is 3D-Var rather than 4D-Var. Probably, this means that we will want to make more use of the $\delta\underline{\mathbf{x}}_{e0}$ term, which does allow for time evolution, by making the ensemble larger and β_{c0} smaller than in hybrid ensemble-4D-Var. Weak constraint terms allowing for model error are included for completeness: the $\delta\underline{\mathbf{x}}_{c1}$ term allows for a constant model error tendency error and the ensemble term $\delta\underline{\mathbf{x}}_{e1}$ allows for the weights $\underline{\alpha}_i$ to vary in time.

We make the key error modelling assumption that the terms are independent from each other. We can then go on to define independent transforms ($\underline{\mathbf{U}}^{e0}$, $\underline{\mathbf{U}}^{e1}$, $\underline{\mathbf{U}}^{c0}$, $\underline{\mathbf{U}}^{c1}$) to diagonalize the control variables. Actually, to allow for non-Gaussian errors, we currently use a nonlinear parameter transform (Section 2.3.2). To be correct, this has to transform the total increment. We also want to do 'balance aware' localization. So the Met Office's initial design uses

$$\delta\underline{\mathbf{x}} = U_p\left(\beta_{e0}\delta\underline{\mathbf{x}}_{e0} + \beta_{e1}\delta\underline{\mathbf{x}}_{e1} + \beta_{c0}\delta\underline{\mathbf{x}}_{c0} + \beta_{c1}\delta\underline{\mathbf{x}}_{c1}\right), \tag{2.86}$$

where the $\delta\underline{\mathbf{x}}_{e0}$ etc. terms are in transformed parameter space.

These transforms are constructed in the normal VAR way, so the transformed control variables are independent with unit variance. They are combined into a single control vector

$$\mathbf{v} = \left(\mathbf{v}^{c0}, \left(\mathbf{v}_i^{\alpha 0}\right)_{i=1,K}, \mathbf{v}^{c1}, \left(\mathbf{v}_i^{\alpha 1}\right)_{i=1,K}\right). \tag{2.87}$$

This gives us a new penalty function to replace (2.83):

$$J(\mathbf{v}) = \frac{1}{2}\mathbf{v}^{\mathrm{T}}\mathbf{v} + \frac{1}{2}\left[\underline{H}\left(\underline{\mathbf{x}}^b + \delta\underline{\mathbf{x}}\right) - \mathbf{y}^o\right]^{\mathrm{T}} \underline{\mathbf{R}}^{-1} \left[\underline{H}\left(\underline{\mathbf{x}}^b + \delta\underline{\mathbf{x}}\right) - \mathbf{y}^o\right]. \tag{2.88}$$

Prospects

Above, we described a smoother, giving a four-dimensional $\delta\underline{\mathbf{x}}$ over the time window. Work is needed to think more carefully about how to add this to the full model and start the next forecast, i.e. how to convert the smoother solution to an ongoing filter, taking account also of the considerations discussed in Section 2.3.5. Plans are also being developed to apply the same 4DEnVar algorithm (using the same ensemble perturbations) to each ensemble member, replacing the localized ETKF in the current MOGREPS system.

The 4D-Ensemble-Var approach replaces the costly, sequential, PF, and adjoint model integrations by the use of precalculated perturbation trajectories, which have to be input and stored. Current indications are that, even allowing for this, it can be made faster than 4D-Var on our current computer (IBM power6), while results such as those of Buehner et al. (2010*b*) give the expectation that it will be comparable in quality. The new algorithm has the advantage of being highly scalable, so it is more likely to work efficiently at higher resolution on future computers. It also removes the need for an adjoint model. This is otherwise a worry, since independent developments of efficient NWP models for future computers may well lead to a radical restructuring and different grid—recoding of an adjoint for such a model is not an attractive prospect.

References

Abarbanel, H. D. I., Kostuk, M., and Whartenby, W. (2010). Data assimilation with regularized nonlinear instabilities. *Q. J. R. Meteorol. Soc.*, **136**, 769–783.

Anderson, J. L. (2007). Exploring the need for localization in ensemble data assimilation using a hierarchical ensemble filter. *Physica D*, **230**, 99–111.

Andersson, E. and Jarvinen, H. (1999). Variational quality control. *Q. J. R. Meteorol. Soc.*, **125**, 697–722.

Andersson, E., Haseler, J., Undén, P., Courtier, P., Kelly, G., Vasiljevic, D., Brankovic, C., Cardinali, C., Gaffard, C., Hollingsworth, A., Jakob, C., Janssen, P., Klinker, E., Lanzinger, A., Miller, M., Rabier, F., Simmons, A., Strauss, B., Thépaut, J.-N., and Viterbo, P. (1998). The ECMWF implementation of three-dimensional variational assimilation (3D-Var). III: Experimental results. *Q. J. R. Meteorol. Soc.*, **124**, 1831–1860.

Bishop, C. H. and Hodyss, D. (2011). Adaptive ensemble covariance localization in ensemble 4D-Var state estimation. *Mon. Weather Rev.*, **139**, 1241–1255.

Bishop, C. H., Etherton, B. J., and Majumdar, S. J. (2001). Adaptive sampling with the ensemble transform Kalman filter. Part I: Theoretical aspects. *Mon. Weather Rev.*, **129**, 420–436.

Bonavita, M., Isaksen, L., and Holm, E. (2012). On the use of EDA background error variances in the ECMWF 4D-Var. *Q. J. R. Meteorol. Soc.*, **138**, 1540–1559.

Buehner, M., Houtekamer, P. L., Charette, C., Mitchell, H. L., and He, B. (2010*a*). Intercomparison of variational data assimilation and the ensemble Kalman filter for global deterministic NWP. Part I: Description and single-observation experiments. *Mon. Weather Rev.*, **138**, 1550–1566.

Buehner, M., Houtekamer, P. L., Charette, C., Mitchell, H. L., and He, B. (2010*b*). Intercomparison of variational data assimilation and the ensemble Kalman filter for global deterministic NWP. Part II: One-month experiments with real observations. *Mon. Weather Rev.*, **138**, 1567–1586.

Campbell, W. F., Bishop, C. H. and Hodyss, D. (2010). Vertical covariance localization for satellite radiances in ensemble Kalman filters. *Mon. Weather Rev.*, **138**(1), 282–290.

Clark, P. A., Harcourt, S. A., Macpherson, B., Mathison, C. T., Cusack, S., and Naylor, M. (2008). Prediction of visibility and aerosol within the operational Met Office Unified Model. I: Model formulation and variational assimilation. *Q. J. R. Meteorol. Soc.*, **134**, 1801–1816.

Clayton, A. M., Lorenc, A. C., and Barker, D. M. (2013). Operational implementation of a hybrid ensemble/4D-Var global data assimilation system at the Met Office. *Q. J. R. Meteorol. Soc.*, **139**, 1445–1461.

Courtier, P., Andersson, E., Heckley, W., Pailleux, J., Vasiljevic, D., Hamrud, M., Hollingsworth, A, Rabier, F., and Fisher., M. (1998). The ECMWF implementation of three-dimensional variational assimilation (3D-Var). I: Formulation. *Q. J. R. Meteorol. Soc.*, **124**, 1783–1808.

Courtier, P., Thépaut, J.-N., and Hollingsworth, A. (1994). A strategy for operational implementation of 4D-Var, using an incremental approach. *Q. J. R. Meteorol. Soc.*, **120**, 1367–1387.

Desroziers, G., Berre, L., Chapnik, B., and Poli, P. (2005). Diagnosis of observation, background and analysis error statistics in observation space. *Q. J. R. Meteorol. Soc.*, **131**, 3385–3396.

Dharssi, I., Lorenc, A. C., and Ingleby, N. B. (1992). Treatment of gross errors using maximum probability theory. *Q. J. R. Meteorol. Soc.*, **118**, 1017–1036.

Fertig, E. J., Harlim, J., and Hunt, B. R. (2007). A comparative study of 4D-Var and a 4D ensemble Kalman filter: perfect model simulations with Lorenz-96. *Tellus A*, **59**, 96–100.

Fisher, M. (2003). Background error covariance modelling. In *Proceedings of Seminar on Recent Developments in Data Assimilation for Atmosphere and Ocean, 8–12 September 2003*, Shinfield Park, Reading, pp. 45–64. ECMWF.

Fisher, M. (2011). Long window weak-constraint 4D-Var. In *Proceedings of ECMWF Seminar on Data Assimilation for Atmosphere and Ocean, 6–9 September 2011*, Shinfield Park, Reading.

Fisher, M., Leutbecher, M., and Kelly, G. A. (2005). On the equivalence between Kalman smoothing and weak-constraint four-dimensional variational data assimilation. *Q. J. R. Meteorol. Soc.*, **131**, 3235–3246.

Gandin, L. (1965). *Objective Analysis of Meteorological Fields.* Israel Program for Scientific Translations.

Gauthier, P., Charette, C., Fillion, L., Koclas, P., and Laroche, S. (1999). Implementation of a 3D variational data assimilation system at the Canadian Meteorological Centre. Part I: The global analysis. *Atmosphere–Ocean*, **37**, 103–156.

Gauthier, P. and Thépaut, J.-N. (2001). Impact of the digital filter as a weak constraint in the preoperational 4DVAR assimilation system of Météo-France. *Mon. Weather Rev.*, **129**, 2089–2102.

Gilbert, J.-C. and Lemaréchal, C. (1989). Some numerical experiments with variable-storage quasi-Newton algorithms. *Math. Program.*, **45**, 407–435.

Guitton, A. and Symes, W. W. (2003). Robust inversion of seismic data using the Huber norm. *Geophysics*, **68**, 1310–1319.

Gustafsson, N. (1992). Use of a digital filter as a weak constraint in variational data assimilation. In *Proceedings of ECMWF Workshop on Variational Assimilation, with Special Emphasis on Three-dimensional Aspects*, pp. 327–338.

Gustafsson, N., Thorsteinsson, S., Stengel, M., and Hólm, E. (2011). Use of a nonlinear pseudo-relative humidity variable in a multivariate formulation of moisture analysis. *Q. J. R. Meteorol. Soc.*, **137**, 1004–1018.

Harlim, J. and Hunt, B. R. (2007). Four-dimensional local ensemble transform Kalman filter: numerical experiments with a global circulation model. *Tellus A*, **59**, 731–748.

Hayden, C. M. and Purser, R. J. (1995). Recursive filter objective analysis of meteorological fields: applications to NESDIS operational processing. *J. Appl. Meteorol.*, **34**, 3–15.

Hollingsworth, A. and Lonnberg, P. (1986). The statistical structure of short-range forecast error as determined from radiosonde data. Part I: The wind field. *Tellus A*, **38A**, 111–136.

Hólm, E.V. (2003). Revision of the ECMWF humidity analysis: construction of a Gaussian control variable. In *Proceedings of ECMWF/GEWEX Workshop on Humidity Analysis.*

Hoteit, I., Cornuella, B., Kohl, A., and Stammer, D. (2005). Treating strong adjoint sensitivities in tropical eddy-permitting variational data assimilation. *Q. J. R. Meteorol. Soc.*, **131**, 3659–3682.

Houtekamer, P. L. and Mitchell, H. L. (2001). A sequential ensemble Kalman filter for atmospheric data assimilation. *Mon. Weather Rev.*, **129**, 123–137.

Huber, P. J. (1973). Robust regression: asymptotics, conjectures, and Monte Carlo. *Ann. Statist.*, **1**, 799–821.

Hunt, B. R., Kalnay, E, Kostelich, E. J., Ott, E., Patil, D. J., Sauer, T., Szunyogh, I., Yorke, J. A., and Zimin, A. V. (2004). Four-dimensional ensemble Kalman filtering. *Tellus A*, **56**, 273–277.

Ide, K., Courtier, P., Ghil, M., and Lorenc, A. C. (1997). Unified notation for data assimilation: Operational, sequential and variational. *J. Met. Soc. Japan*, **75**, 181–189.

Ingleby, N. B., Lorenc, A. C., Ngan, K., Rawlins, F. R., and Jackson, D. R. (2013). Improved variational analyses using a nonlinear humidity control variable. *Q. J. R. Meteorol. Soc.*, **139**, 1875–1887.

Isaksen, Lars (2011). Data assimilation on future computer architectures. In *Proceedings of ECMWF Seminar on Data Assimilation for Atmosphere and Ocean, 6–9 September 2011*, Shinfield Park, Reading. ECMWF.

Jakimow, G., Yakimiw, E., and Robert, A. (1992). An implicit formulation for horizontal diffusion in gridpoint models. *Mon. Weather Rev.*, **120**, 124–130.

Liu, C., Xiao, Q., and Wang, B. (2008). An ensemble-based four-dimensional variational data assimilation scheme. Part I: Technical formulation and preliminary test. *Mon. Weather Rev.*, **136**, 3363–3373.

Liu, C., Xiao, Q., and Wang, B. (2009). An ensemble-based four-dimensional variational data assimilation scheme. Part II: Observing system simulation experiments with advanced research WRF (ARW). *Mon. Weather Rev.*, **137**, 1687–1704.

Lopez, P. (2011). Direct 4D-Var assimilation of NCEP Stage IV radar and gauge precipitation data at ECMWF. *Mon. Weather Rev.*, **139**, 2098–2116.

Lorenc, A. C. (1984). Analysis methods for the quality control of observations. In *Proceedings of ECMWF Workshop on the Use and Quality Control of Meteorological Observations for Numerical Weather Prediction*, Shinfield Park, Reading.

Lorenc, A. C. (1981). A global three-dimensional multivariate statistical analysis scheme. *Mon. Weather Rev.*, **109**, 701–721.

Lorenc, A. C. (1997). Development of an operational variational assimilation scheme. *J. Met. Soc. Japan*, **75**, 339–346.

Lorenc, A. C. (2003*a*). Modelling of error covariances by four-dimensional variational data assimilation. *Q. J. R. Meteorol. Soc.*, **129**, 3167–3182.

Lorenc, A. C. (2003*b*). The potential of the ensemble Kalman filter for NWP - a comparison with 4D-Var. *Q. J. R. Meteorol. Soc.*, **129**, 3183–3203.

Lorenc, A. C. (2007). A study of o-b monitoring statistics from radiosondes, composited for low-level cloud layers. Met Office Forecasting Research Technical Report 504, Met Office.

Lorenc, A. C. and Hammon, O. (1988). Objective quality control of observations using Bayesian methods. theory, and a practical implementation. *Q. J. R. Meteorol. Soc.*, **114**, 515–543.

Lorenc, A. C. and Payne, T. (2007). 4D-Var and the butterfly effect: statistical four-dimensional data assimilation for a wide range of scales. *Q. J. R. Meteorol. Soc.*, **133**, 607–614.

Lorenc, A. C., Ballard, S. P., Bell, R. S., Ingleby, N. B., Andrews, P. L. F., Barker, D. M., Bray, J. R., Clayton, A. M., Dalby, T., Li, D., Payne, T. J., and Saunders, F. W. (2000). The Met. Office global three-dimensional variational data assimilation scheme. *Q. J. R. Meteorol. Soc.*, **126**, 2991–3012.

Lorenz, E. N. (1963). Deterministic nonperiodic flow. *J. Atmos. Sci.*, **20**, 130–141.

Lorenz, E. N. (1969). The predictability of a flow that possesses many scales of motion. *Tellus A*, **21**, 289–307.

Montmerle, T. and Berre, L. (2010). Diagnosis and formulation of heterogeneous background-error covariances at the mesoscale. *Q. J. R. Meteorol. Soc.*, **136**, 1408–1420.

Palmer, T. N. (2005). Quantum reality, complex numbers, and the meteorological butterfly effect. *Bull. Am. Meteorol. Soc.*, **86**, 519–530.

Parrish, D. F. and Derber, J. C. (1992). The National Meteorological Center's spectral statistical-interpolation analysis system. *Mon. Weather Rev.*, **120**, 1747–1763.

Phillips, N. A. (1985). The spatial statistics of random geostrophic modes and first-guess errors. *Tellus A*, **38A**, 314–332.

Polavarapu, S. (2011). Stratospheric and mesospheric data assimilation. In *Proceedings of ECMWF Seminar on Data Assimilation for Atmosphere and Ocean, 6–9 September 2011.*

Polavarapu, S., Tanguay, M., and Fillion, L. (2000). Four-dimensional variational data assimilation with digital filter initialization. *Mon. Weather Rev.*, **128**, 2491–2510.

Purser, R. J. (1984). A new approach to the optimal assimilation of meteorological data by iterative Bayesian analysis. In *Preprints, 10th Conference on Weather Forecasting and Analysis*, pp. 102–105. American Meteorological Society.

Purser, R. J. and McQuigg, R. (1982). A successive correction analysis scheme using recursive numerical filters. MetO11 Technical Note 154.

Purser, R. J., Wu, W.S., Parrish, D. F., and Roberts, N. M. (2003*a*). Numerical aspects of the application of recursive filters to variational statistical analysis. Part I: Spatially homogeneous and isotropic Gaussian covariances. *Mon. Weather Rev.*, **131**, 1524–1535.

Purser, R. J., Wu, W.S., Parrish, D. F., and Roberts, N. M. (2003*b*). Numerical aspects of the application of recursive filters to variational statistical analysis. Part II: Spatially inhomogeneous and anisotropic general covariances. *Mon. Weather Rev.*, **131**, 1536–1548.

Rabier, F., McNally, A., Andersson, E., Courtier, P., Undén, P., Eyre, J., Hollingsworth, A., and Bouttier., F. (1998). The ECMWF implementation of three-dimensional variational assimilation (3D-Var). II: Structure functions. *Q. J. R. Meteorol. Soc.*, **124**, 1809–1830.

Radnóti, G., Trémolet, Y., Andersson, E., Isaksen, L., Hólm, E. and Janisková, M. (2005). Diagnostics of linear and incremental approximations in 4D-Var revisited for higher resolution analysis. Technical Memorandum 467, ECMWF.

Rawlins, F., Ballard, S. P., Bovis, K. J., Clayton, A. M., Li, DingMin, Inverarity, G. W., Lorenc, A. C., and Payne, T. J. (2007). The Met Office global 4-dimensional data assimilation system. *Q. J. R. Meteorol. Soc.*, **133**, 347–362.

Raynaud, L., Berre, L., and Desroziers, G. (2011). An extended specification of flow-dependent background-error variances in the Meteo-France global 4D-Var system. *Q. J. R. Meteorol. Soc.*, **137**, 607–619.

Renshaw, R. and Francis, P. N. (2011). Variational assimilation of cloud fraction in the operational Met Office unified model. *Q. J. R. Meteorol. Soc.*, **137**, 1963–1974.

Schlatter, T. W. (1975). Some experiments with a multivariate statistical objective analysis scheme. *Mon. Weather Rev.*, **103**, 246–257.

Simmons, A. J. and Hollingsworth, A. (2002). Some aspects of the improvement in skill of numerical weather prediction. *Q. J. R. Meteorol. Soc.*, **128**, 647–677.

Tarantola, A. (1987). *Inverse Problem Theory: Methods for Data Fitting and Model Parameter Estimation.* Elsevier.

Thépaut, J.-N., Courtier, P., Belaud, G., and Lemaître, G (1996). Dynamical structure functions in a four-dimensional variational assimilation: a case study. *Q. J. R. Meteorol. Soc.*, **122**, 535–561.

Tippett, M. K., Anderson, J. L., Bishop, C. H., Hamill, T. M., and Whitaker, J. S. (2003). Ensemble square-root filters. *Mon. Weather Rev.*, **131**, 1485–1490.

Tribbia, J. J. and Baumhefner, D. P. (2004). Scale interactions and atmospheric predictability: an updated perspective. *Mon. Weather Rev.*, **132**, 703–713.

van Leeuwen, P. J. (2011). Nonlinear large-dimensional data assimilation: the potential of particle filters. In *Proceedings of ECMWF Seminar on Data Assimilation for Atmosphere and Ocean, 6–9 September 2011.*

Wang, X., Snyder, C., and Hamill, T. M. (2007). On the theoretical equivalence of differently proposed ensemble–3DVAR hybrid analysis schemes. *Mon. Weather Rev.*, **135**, 222–227.

Weaver, A. and Courtier, P. (2001). Correlation modelling on the sphere using a generalized diffusion equation. *Q. J. R. Meteorol. Soc.*, **127**, 1815–1846.

Wee, T. K. and Kuo, Y. H. (2004). Impact of a digital filter as a weak constraint in MM5 4DVAR: an observing system simulation experiment. *Mon. Weather Rev.*, **132**, 543–559.

Zhang, F., Bei, N., Rotunno, R., Snyder, C., and Epifanio, C. C. (2007). Mesoscale predictability of moist baroclinic waves: convection-permitting experiments and multistage error growth dynamics. *J. Atmos. Sci.*, **64**, 3579–3594.

Zhang, M. and Zhang, F. (2012). E4DVar: coupling an ensemble Kalman filter with four-dimensional variational data assimilation in a limited-area weather prediction model. *Mon. Weather Rev.*, **140**, 587–600.

3
Introduction to the Kalman filter

C. SNYDER

NCAR
Boulder, Colorado, USA

Advanced Data Assimilation for Geosciences. First Edition.
Edited by É. Blayo, M. Bocquet, E. Cosme, and L. F. Cugliandolo.

Chapter Contents

The literature related to the Kalman filter is large, diverse and extends back to the late 1950s and early 1960s (e.g. Kalman, 1960). For atmospheric and oceanic applications, interest was limited by the difficulty of implementing the Kalman filter for systems of the size typically encountered in those applications, together with the rapid development of four-dimensional variational techniques, which treat a very similar problem and avoid at least some of the hurdles to implementation. During the 1990s, several approximate extended Kalman filters were proposed for oceanographic applications (Fukumori and Malanotte-Rizzoli, 1995; Pham *et al.*, 1998; Lermusiaux and Robinson, 1999). A related technique is the ensemble Kalman filter (EnKF) (Evensen, 1994; Houtekamer and Mitchell, 1998), a Monte Carlo approach in which the covariances required in the Kalman filter are estimated from a finite sample, or ensemble, of forecasts. Interest in the Kalman filter has surged again, as the EnKF has proven to be effective in numerous applications. With the return of interest in the Kalman filter, many results from the original literature of the 1960s and 1970s are being utilized or rediscovered.

This chapter is meant mainly as an exposition of my personal perspective on the Kalman filter. I make no attempt to be comprehensive in reviewing the original literature. My main interest is in how the Kalman filter fits within the broader contexts of Bayesian state estimation, which I see as the mathematical foundation for data assimilation, and linear estimation theory, which sheds light both on the Kalman filter's behavior in the typical situation that it is suboptimal and on its relation to the ensemble Kalman filter. I also present some basic results for the original 'low-rank' EnKF that follow naturally by taking the ensemble estimate of the covariances as correct and applying the corresponding Kalman-filter results. Although I see it as perhaps the key element of successful EnKFs, I do not cover covariance localization; that topic appears in Chapter 10 of this volume, by Peter Houtekamer.

There are classic papers from atmospheric science that present concise introductions, especially Cohn (1997) and Lorenc (1986). Good reference books include Anderson and Moore (1979), Gelb (1974), and Jazwinski (1970). A clear omission in what follows is discussion of the continuous-time filter, but the discrete-time filter illustrates most essential points.

3.1 A Bayesian view of data assimilation

As a prelude to examining the Kalman filter in detail, this section outlines the Bayesian view of data assimilation that will form the background for the more detailed discussion. There are important aspects of the filtering problem that become more transparent when viewed from a Bayesian perspective. This section also presents other preliminaries, such as notation.

3.1.1 Statement of the problem

Consider the system

$$\mathbf{x}_k = f(\mathbf{x}_{k-1}) + \eta_k, \tag{3.1}$$

where $\mathbf{x}_k$, the system state at time t_k, is a vector of dimension N_x and η_k, also of dimension N_x, is the system noise or model error. Suppose we observe or measure the variables

$$\mathbf{y}_k = h(\mathbf{x}_k) + \epsilon_k, \tag{3.2}$$

where $\mathbf{y}_k$ has dimension N_y and ϵ_k is the observation error. We will take the system noise η_k and the observation error ϵ_k to be random variables as well, but make no other assumptions about them at this point.

We wish to estimate $\mathbf{x}_k$ given a specific realization of the observations $\mathbf{y}_k = \mathbf{y}_k^o$, or, more generally, given a sequence of observations over time, $\{\mathbf{y}_0^o, \mathbf{y}_1^o, \ldots, \mathbf{y}_k^o\}$. Because (3.1) and (3.2) contain noise in both the evolution and observation equations, the state $\mathbf{x}_k$ is naturally considered a random variable, and we cannot expect to obtain $\mathbf{x}_k$ precisely. Instead, the most that can be known about $\mathbf{x}_k$ at time t_k is its probability distribution function (pdf) $p(\mathbf{x}_k|\mathbf{y}_0^o, \ldots, \mathbf{y}_k^o)$ conditioned on the available observations. Our goal should therefore be to calculate $p(\mathbf{x}_k|\mathbf{y}_0^o, \ldots, \mathbf{y}_k^o)$.

3.1.2 Model error and observation error

In order to understand more clearly the meaning of (3.1) and (3.2), recall that our aim is to predict the evolution of the atmosphere or ocean, which vary continuously in space and time.

Let $\mathbf{s}(x, y, z, t)$ be the continuous state of the fluid, with the components of $\mathbf{s}$ corresponding to different variables (pressure, density, the three components of velocity, etc.) and (x, y, z) representing a spatial location in this subsection only, and suppose that the evolution of $\mathbf{s}$ over discrete times t_k is given by

$$\mathbf{s}(x, y, z, t_k) = g(\mathbf{s}(x, y, z, t_{k-1})).$$

We have a numerical model, which is discretized in space and time with state $\mathbf{x}_k$ at time t_k, and we have observations of the continuous fluid that typically represent averages over some volume and time interval and are subject to measurement errors, at least a portion of which are random.

Assume that the relation between the discretized state $\mathbf{x}_k$ of the numerical model and the continuous atmospheric (or oceanic) field $\mathbf{s}(x, y, z)$ is given by a projection operator that returns the discretized model state corresponding to a given, continuous atmospheric state.

$$\mathbf{x}_k = \Phi(\mathbf{s}(t_k)).$$

For many models, unfortunately, this projection operator is not precisely defined.

Finally, let $f(\mathbf{x}_{k-1})$ be the model prediction for $\mathbf{x}_k$ given the discrete state at time t_{k-1} and define η_k to be the difference between $\mathbf{x}_k$ and the model prediction of $\mathbf{x}_k$ from t_{k-1}; that is,

$$\eta_k = \Phi(\mathbf{s}(t_k)) - f(\mathbf{x}_{k-1}) = \Phi(g(\mathbf{s}(t_{k-1}))) - f(\Phi(\mathbf{s}(t_{k-1}))). \tag{3.3}$$

The discrete system (3.1) then follows directly from these definitions. In particular, η_k is the model error and will generally depend on the continuous state $\mathbf{s}(t_k)$.

A similar approach also clarifies the nature of the observation error in (3.2). The observations are measurements of the continuous fluid:

$$\mathbf{y}_k = m(\mathbf{s}(x, y, z, t_k)) + \mu, \tag{3.4}$$

where μ is random error associated with the instruments used in the measurement. Equation (3.4) may be rewritten as

$$\mathbf{y}_k = h(\mathbf{x}_k) + \mu + \beta,$$

with the additional error β defined by

$$\beta = m(\mathbf{s}(t_k)) - h(\mathbf{x}_k) = m(\mathbf{s}(t_k)) - h(\Phi(\mathbf{s}(t_k))).$$

Analogously to η, β arises from the difference between the correct measurement of the continuous fluid and the discrete observation operator applied to the projection of the continuous state into the discrete space.

The error β may be further decomposed as $\beta = \gamma + \delta$, with

$$\gamma = m(\mathbf{s}(t_k)) - m(\Phi(\mathbf{s}(t_k))), \qquad \delta = m(\Phi(\mathbf{s}(t_k))) - h(\Phi(\mathbf{s}(t_k))).$$

The error δ arises from differences in m and h applied to the same state, that is, from errors in h as an approximation to m, which may arise from imperfectly known physics of the measurement itself. The error γ is the representativeness error, which is associated with the truncation by Φ of scales in the continuous state that affect the observed quantities. For example, point measurements of velocity include contributions from small-scale, turbulent motions that are not retained (or *represented*) in numerical models. Defining $\epsilon_k = \mu + \gamma + \delta$ then yields (3.2).

In principle, both η_k and ϵ_k will depend on $\mathbf{x}_k$ and possibly also $\mathbf{x}_{k-1}$, although such dependence will be ignored here.

3.1.3 Calculating the conditional pdf

There are two results that facilitate calculating $p(\mathbf{x}_k|\mathbf{y}_0^o, \ldots, \mathbf{y}_k^o)$. To make the conditioning statements more compact, let $Y_k^o = \{\mathbf{y}_0^o, \ldots, \mathbf{y}_k^o\}$.

The first result is the 'propagation' rule:

$$p(\mathbf{x}_k|Y_{k-1}^o) = \int_{-\infty}^{\infty} p(\mathbf{x}_k|\mathbf{x}_{k-1}, Y_{k-1}^o)p(\mathbf{x}_{k-1}|Y_{k-1}^o)\,\mathrm{d}\mathbf{x}_{k-1}. \tag{3.5}$$

If we know $p(\mathbf{x}_{k-1}|Y_{k-1}^o)$, then (3.5) gives $p(\mathbf{x}_k|Y_{k-1}^o)$, thereby propagating the information in observations Y_{k-1}^o from earlier times to time t_k.

The system dynamics (3.1) is implicit in (3.5) through $p(\mathbf{x}_k|\mathbf{x}_{k-1}, Y_{k-1}^o)$. Under the typical assumption that η_k is independent of η_j if $k \neq j$ and independent of ϵ_j for all j, then $p(\mathbf{x}_k|\mathbf{x}_{k-1}, Y_{k-1}^o) = p(\mathbf{x}_k|\mathbf{x}_{k-1})$, which is known from (3.1) if the distribution of η_k is known. For example, if $\eta_k \sim N(0, \mathbf{Q})$, then (3.1) implies that $\mathbf{x}_k|\mathbf{x}_{k-1} \sim N(f(\mathbf{x}_{k-1}), \mathbf{Q})$ (Section 3.2.1 gives a review of Gaussian distributions).

The second useful result is Bayes' rule:

$$p\left(\mathbf{x}_k|\mathbf{y}_k^o, Y_{k-1}^o\right) = \frac{p\left(\mathbf{y}_k^o|\mathbf{x}_k, Y_{k-1}^o\right) p\left(\mathbf{x}_k|Y_{k-1}^o\right)}{p\left(\mathbf{y}_k^o|Y_{k-1}^o\right)}. \tag{3.6}$$

This specifies how to correct, or *update*, the probability $p(\mathbf{x}_k|Y_{k-1}^o)$ using the new observation $\mathbf{y}_k^o$ and its likelihood as a function of the state and the previous observations, $p(\mathbf{y}_k^o|\mathbf{x}_k, Y_{k-1}^o)$.

The likelihood function follows from the observation equation (3.2). Again assuming that ϵ_k is independent of ϵ_j for $j \neq k$ and of η_j for all j, the likelihood becomes $p(\mathbf{y}_k^o|\mathbf{x}_k, Y_{k-1}^o) = p(\mathbf{y}_k^o|\mathbf{x}_k)$, which is known from (3.2) if the distribution of ϵ_k is known. For example, if $\epsilon \sim N(0, \mathbf{R})$, then (3.2) implies that

$$p(\mathbf{y}_k^o|\mathbf{x}_k) \propto \exp\left\{\left[\mathbf{y}_k^o - h(\mathbf{x}_k)\right]^{\mathrm{T}} \mathbf{R}^{-1} \left[\mathbf{y}_k^o - h(\mathbf{x}_k)\right]\right\}. \tag{3.7}$$

Together, the propagation rule and Bayes' rule form a recursive, sequential-in-time algorithm to calculate $p(\mathbf{x}_k|Y_k^o)$: starting from the conditional pdf for the state at t_{k-1} given observations up to and including t_{k-1}, (i) propagate from t_{k-1} to t_k using the propagation rule (3.5), then (ii) update given the observations at t_k using Bayes' rule (3.6). The algorithm requires (3.1), (3.2), and knowledge of the (joint) pdfs of η_k, ϵ_k, and the initial state $\mathbf{x}_0$. The Kalman filter can be cast in this form.

An alternative approach is to use the system dynamics to propagate the initial information $p(\mathbf{x}_0)$ from t_0 to time t_k, by calculating the unconditional joint pdf

$$\begin{aligned} p(\mathbf{x}_0, \mathbf{x}_1, \ldots, \mathbf{x}_k) &= p(\mathbf{x}_0)p(\mathbf{x}_1|\mathbf{x}_0)p(\mathbf{x}_2|\mathbf{x}_1, \mathbf{x}_0)\ldots p(\mathbf{x}_k|\mathbf{x}_{k-1}, \mathbf{x}_{k-2}, \ldots, \mathbf{x}_0) \\ &= p(\mathbf{x}_0)p(\mathbf{x}_1|\mathbf{x}_0)p(\mathbf{x}_2|\mathbf{x}_1)\ldots p(\mathbf{x}_k|\mathbf{x}_{k-1}), \end{aligned}$$

where the second equality follows from the independence assumptions on η_k given above. Bayes' rule gives

$$p(\mathbf{x}_0, \ldots, \mathbf{x}_k|Y_k^o) = \frac{p(Y_k^o|\mathbf{x}_0, \ldots, \mathbf{x}_k)p(\mathbf{x}_0, \ldots, \mathbf{x}_k)}{p(Y_k^o)},$$

and the desired $p(\mathbf{x}_k|Y_k^o)$ can be calculated as the marginal of $p(\mathbf{x}_0, \ldots, \mathbf{x}_k|Y_k^o)$ by integrating over $\mathbf{x}_0, \ldots, \mathbf{x}_{k-1}$. In this approach, the update is not sequential, since it treats all the observations at once regardless of their time. The sequential-in-time algorithm, while intuitive, is therefore not fundamental to the update—it is simply convenient because of the Markov form of (3.1), which specifies how to generate $\mathbf{x}_k$ if $\mathbf{x}_{k-1}$ is known.

3.1.4 Smoothing, filtering, and prediction

Other marginals of $p(\mathbf{x}_0, \ldots, \mathbf{x}_k|Y_k^o)$ may be of interest, for example $p(\mathbf{x}_j|Y_k^o)$ with $j < k$. Conditioning on observations valid both before and after t_j provides additional information and a sharper pdf for the state.

Calculating the pdf of the state at a certain time conditioned on observations up to and including that time, i.e. $p(\mathbf{x}_k|Y_k^o)$, is known as the filtering problem and the algorithm that performs the calculation is known as a filter. Conditioning on observations valid at subsequent times is known as the smoothing problem (see Chapter 4). There are various forms of smoothing; for example, calculating $p(\mathbf{x}_{k-n}|Y_k^o)$, for some fixed n, is the 'fixed-lag' smoothing problem. Finally, the prediction problem is concerned with calculating $p(\mathbf{x}_j|Y_k^o)$ for $j > k$. This is a (probabilistic) forecast of the state at times beyond which observations are available.

3.1.5 Simple illustrations

The discussion to this point has been mainly mathematical. This subsection provides simple illustrations of the Bayesian update and of the recursive, propagate-then-update algorithm for computing $p(\mathbf{x}_k|Y_k^o)$.

Figure 3.1 illustrates the update step, via Bayes' rule (3.6), for a one-dimensional (scalar) problem. Figure 3.1(a) shows $p(\mathbf{x}_k|Y_{k-1}^o)$ (gray lines), the pdf for the state at time t_k given observations up to time t_{k-1}. In this case, the pdf is bimodal and non-Gaussian.

Suppose we have a noisy measurement of $\mathbf{x}_k$; that is, $\mathbf{y}_k = \mathbf{x}_k + \epsilon_k$. Figure 3.1(b) shows in addition the likelihood $p(\mathbf{y}_k^o|\mathbf{x}_k)$ (thin black line) for $\mathbf{y}_k^o = 0.8$. Note that $p(\mathbf{y}_k^o|\mathbf{x}_k)$ is considered as a function of $\mathbf{x}_k$ with $\mathbf{y}_k^o$ known and is given by $p(\mathbf{y}_k^o|\mathbf{x}_k) = \rho(\mathbf{y}_k^o - \mathbf{x}_k)$, where $\rho(\epsilon)$ is the observation-error pdf. Here, the observation-error distribution is skewed, with the most likely error being slightly positive and with large negative errors being more likely than large positive errors, which means that the likelihood peaks for $\mathbf{x}_k$ slightly less than $\mathbf{y}_k^o$ and has a long tail for $\mathbf{x}_k > \mathbf{y}_k^o$.

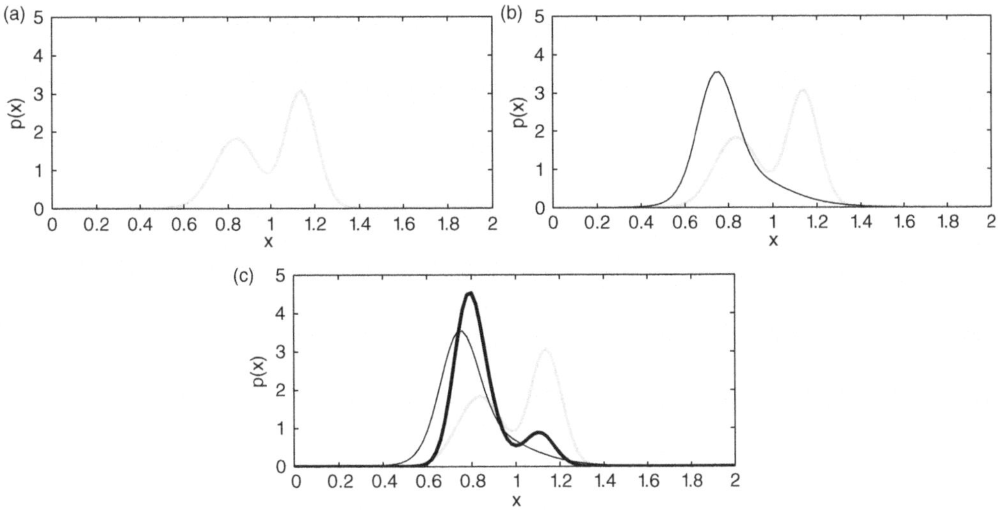

Fig. 3.1 An xample of the update step for a one dimensional-problem: $p(\mathbf{x}_k|Y_{k-1}^o)$ (gray lines, all panels), $p(\mathbf{y}_k^o|\mathbf{x}_k)$ (thin black lines, in (b) and (c)) for $\mathbf{y}_k^o = 0.8$, and $p(\mathbf{x}_k|Y_k^o)$ (thick black line in (c)) given by (3.6).

Figure 3.1(c) shows that the actual update: $p(\mathbf{x}_k|Y^o_{k-1})$ (gray line) and $p(\mathbf{y}_k|\mathbf{x}_k)$ (thin black line) are multiplied (and the product normalized to have unit integral over all $\mathbf{x}_k$) to yield, by (3.6), $p(\mathbf{x}_k|Y^o_k)$ (thick black line). After the update, the most likely states are those that had significant probability under $p(\mathbf{x}_k|Y^o_{k-1})$ and are also consistent with the observation $\mathbf{y}^o_k = 0.8$.

Next, consider a simple, two-dimensional example that illustrates the recursive algorithm of Section 3.1.3. Let $\mathbf{x}_k = [a_k, b_k]$ satisfy

$$a_k = 0.9a_{k-1} - 0.4b_{k-1} + \eta_k, \quad b_k = 0.9b_{k-1} + 0.4a_{k-1}, \tag{3.8}$$

where the noise η_k is Gaussian with zero mean and standard deviation 0.1 and is independent across times. These equations for a_k and b_k are the counterpart of (3.1).

The observations are assumed available every third time step (i.e. at $k = 3n + 1$, $n = 1, 2, \ldots$) and are given by

$$y_k = a_k + \epsilon_k, \tag{3.9}$$

which is the counterpart of (3.2). The observation error ϵ_k is Gaussian, with zero mean and standard deviation 0.1, and is independent of η_k and across times. We define Y^o_k as before to be the set of all observed values over times less than or equal to t_k. Since observations are only available every third step, $Y^o_k = Y^o_{k+1} = Y^o_{k+2}$ for $k = 3n + 1$.

We now perform an experiment in which we compute a solution of (3.8). This is the state whose pdf is to be estimated given observations $\{y^o_k,\ k = 4, 7, \ldots, 28\}$ that are constructed according to (3.9); that is, they are obtained by adding a realization of ϵ_k to a_k. After assuming an initial distribution $p(a_1, b_1)$ (which we take to be Gaussian with standard deviation 0.1 for each component), the conditional pdf $p(a_k, b_k|Y^o_k)$ can be computed as in Section 3.1.3. As will be shown in Section 3.2, the conditional pdf is also Gaussian, since the dynamics and observation operator are linear and the system noise and observation error are Gaussian, but our interest here is simply to illustrate the recursive algorithm for computing $p(a_k, b_k|Y^o_k)$.

One realization of the experiment is shown in Fig. 3.2. The 'true' solution for a_m, $m = 1, \ldots, 30$, appears as the grey line in Fig. 3.2(a) and the observations y^o_k generated from it are shown as black circles centred on the observation value with radius equal to the observation-error standard deviation.

In any actual problem, we may have observations but would not know the true state; thus, we seek to compute $p(a_k|Y^o_k)$, the filtering distribution for the state at t_k conditioned on all observations up to the same time. Consider how one cycle of the algorithm works, starting from $p(a_{k-3}|Y^o_{k-3})$ and yielding $p(a_k|Y^o_k)$. For concreteness, suppose $k = 10$. (Although it is not shown here, the algorithm in general will involve the pdf of the full state, i.e. $p(a_k, b_k|Y^o_k)$ for this example. In particular, the propagation step requires use of the joint pdf, since the dynamics (3.8) depends on both a_k and b_k.)

The mean of $a_{k-3}|Y^o_{k-3}$ is shown by the black dot in Fig. 3.2(b) and the range of $a_{k-3}|Y^o_{k-3}$ between plus or minus one standard deviation is indicated by the grey shading. The pdf at t_{k-2}, t_{k-1}, t_k (still conditioned on Y^o_{k-3}) is then calculated by applying the propagation rule (3.5) three times in succession. The black line in Fig. 3.2(b) shows how the mean evolves from t_{k-3} to t_k. Propagation of the pdf forward in time

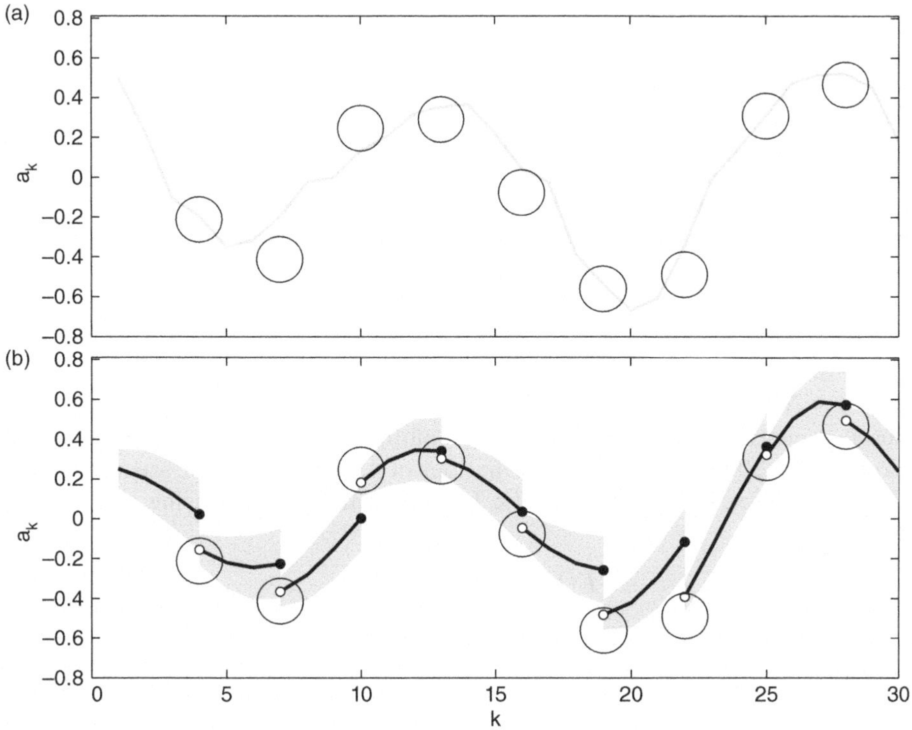

Fig. 3.2 An illustration of the recursive, propagate-then-update algorithm to compute $p(a_k|Y_k^o)$ for the system (3.8), (3.9). (a) The 'true' state (grey line) and the observations $\{y_k^o,\, k = 4, 7, \ldots, 28\}$ (black circles centred on observation values with radius equal to the observation-error standard deviation). (b) Mean (black lines with white initial points at t_k and black final points at t_{k+3}) and standard deviation (plus or minus relative to the mean, indicated by extent of grey shaded area) of $a_m|Y_k^o$, with $m = k, k+1, \ldots, k+3$. Observations are as in (a).

also leads to more uncertainty, as illustrated by the consistent increase of the standard deviation about the mean over the same period.

At $k = 10$, the update step starts from the predictive or forecast distribution $p(a_k|Y_{k-3}^o)$ and uses Bayes' rule to condition on y_k^o and obtain the filtering distribution $p(a_k|Y_k^o)$. After the update, the mean of $a_k|Y_k^o$, which is indicated by a black dot, shifts upwards towards the new observation y_k^o, so that the updated mean lies between y_k^o and the forecast mean (i.e. the mean of $a_k|Y_{k-3}^o$, shown by the black dot). The width of the grey-shaded area also changes discontinuously at $k = 10$, reflecting the decrease of uncertainty from assimilation of the new observation.

In this example, the dynamics and observation operator are linear, and the system noise and observation errors are Gaussian. For such systems, one can obtain analytic expressions for the updated mean and covariance, and for the propagation of the mean and covariance. Section 3.2 derives those expressions, which together form the Kalman filter.

3.2 The Kalman filter (I)

This section derives the Kalman-filter equations by applying the Bayesian framework of Section 3.1 to the case of a linear, Gaussian system.

3.2.1 Facts about the Gaussian distribution

A Gaussian (or normal) distribution is completely determined by the mean and covariance. If a random vector $\mathbf{x}$ is Gaussian, then it has probability density

$$p(\mathbf{x}) = (2\pi)^{-N_x/2}|\mathbf{P}|^{-1/2}\exp\left[-\tfrac{1}{2}(\mathbf{x}-\bar{\mathbf{x}})^{\mathrm{T}}\mathbf{P}^{-1}(\mathbf{x}-\bar{\mathbf{x}})\right], \tag{3.10}$$

where $\bar{\mathbf{x}} = E(\mathbf{x})$ is the mean of $\mathbf{x}$ and $\mathbf{P} = \mathrm{cov}(\mathbf{x}) = E\left((\mathbf{x}-\bar{\mathbf{x}})(\mathbf{x}-\bar{\mathbf{x}})^{\mathrm{T}}\right)$ is the covariance matrix of $\mathbf{x}$. To denote that $\mathbf{x}$ is Gaussian with mean $\bar{\mathbf{x}}$ and covariance $\mathbf{P}$, we write $\mathbf{x} \sim N(\bar{\mathbf{x}}, \mathbf{P})$.

A basic property is that arbitrary linear combinations of Gaussian random variables are also Gaussian. Thus, linear transformations of a Gaussian vector are also Gaussian. In fact, $\mathbf{L}\mathbf{x} \sim N(\mathbf{L}\bar{\mathbf{x}}, \mathbf{L}\mathbf{P}\mathbf{L}^{\mathrm{T}})$ if $\mathbf{x} \sim N(\bar{\mathbf{x}}, \mathbf{P})$. This can be shown using the linearity of the expectation:

$$E(\mathbf{L}\mathbf{x}) = \mathbf{L}E(\mathbf{x}) = \mathbf{L}\bar{\mathbf{x}},$$

and

$$\mathrm{cov}(\mathbf{L}\mathbf{x}) = E\left((\mathbf{L}\mathbf{x}-\mathbf{L}\bar{\mathbf{x}})(\mathbf{L}\mathbf{x}-\mathbf{L}\bar{\mathbf{x}})^{\mathrm{T}}\right) = \mathbf{L}E\left((\mathbf{x}-\bar{\mathbf{x}})(\mathbf{x}-\bar{\mathbf{x}})^{\mathrm{T}}\right)\mathbf{L}^{\mathrm{T}} = \mathbf{L}\mathbf{P}\mathbf{L}^{\mathrm{T}}.$$

An additional property is that the marginal distribution for any subset of the components of $\mathbf{x}$ is also Gaussian. The mean and covariance are simply the corresponding subset of entries from $\bar{\mathbf{x}}$ and $\mathbf{P}$. This can be derived by choosing the linear transformation L that simply selects the desired components of $\mathbf{x}$.

3.2.2 The linear, Gaussian system

Assume (i) that the deterministic dynamics and the observation operator are linear, $f(\mathbf{x}_{k-1}) = \mathbf{M}\mathbf{x}_{k-1}$ in (3.1) and $h(\mathbf{x}_k) = \mathbf{H}\mathbf{x}_k$ in (3.2), and (ii) that the system noise η_k and observation errors ϵ_k are Gaussian. The system dynamics and observation equation then become

$$\mathbf{x}_k = \mathbf{M}\mathbf{x}_{k-1} + \eta_k, \qquad \mathbf{y}_k = \mathbf{H}\mathbf{x}_k + \epsilon_k. \tag{3.11}$$

Time dependence of $\mathbf{M}$ and $\mathbf{H}$ can be handled easily, but is omitted in order to simplify notation.

The system noise and observation errors are taken to have zero mean and known covariances:

$$\eta_k \sim N(0, \mathbf{Q}), \qquad \epsilon_k \sim N(0, \mathbf{R}). \tag{3.12}$$

For the purposes of this section, we will further assume that η_k and ϵ_k are independent and white in time, $\mathrm{cov}(\eta_k, \eta_l) = \mathrm{cov}(\epsilon_k, \epsilon_l) = 0$ if $k \neq l$. Relaxing these assumptions on η_k and ϵ_k will be discussed in Section 3.5.

3.2.3 The importance of covariances

The covariance matrix of the state figures prominently in the Kalman-filter equations that are derived in Section 3.2.4. Here we present a simple example illustrating why covariances are important: the update step can spread information from the observations to unobserved state variables when there is covariance between observed and unobserved variables.

We return to the two-dimensional example of Section 3.1.5, in which (a_k, b_k) is the state at t_k and the observations are given by (3.9); that is, a_k is observed with additive, Gaussian observation error. Suppose that the observation $y_k^o = 1.4$ and that the prior pdf $p(a_k, b_k)$ is known and Gaussian (where we omit for simplicity any conditioning on previous observations).

Figure 3.3(a) shows the prior pdf $p(a_k, b_k)$ (grey contours). Since it is Gaussian, the pdf is constant on ellipses centred at the mean, $a_k = 1$ and $b_k = 1$. The covariance of a_k and b_k is positive; that is, values of a_k larger than the mean are associated with values of b_k that are also larger than the mean, and the ellipses on which the pdf is constant are elongated from the lower left to the upper right of the panel. The marginal distributions for a_k and b_k separately are shown schematically by the thin grey lines along each axis.

Figure 3.3(b) displays in addition the likelihood $p(y_k^o|a_k)$ (dotted contours). Because the observation is independent of b_k, the contours of the likelihood in the (a_k, b_k) plane are straight lines parallel to the b_k axis. The likelihood as a function only of a_k is shown as a thin dotted line along the a_k axis.

The conditional pdf $p(a_k, b_k|y_k^o)$ may be computed, according to Bayes' rule (3.6), by pointwise multiplication of the prior pdf (grey contours) and the likelihood considered as a function of a_k and b_k (dotted contours), followed by normalization by an appropriate constant. The result is shown by the black contours in Fig. 3.3(c). The contours are again ellipses—$p(a_k, b_k|y_k^o)$ is also Gaussian.

It is instructive to examine how the update changes the marginal distributions. The marginals of the conditional distribution, $p(a_k|y_k^o)$ and $p(b_k|y_k^o)$, are indicated in Fig. 3.3(c) by the thin black lines along each coordinate axis. The conditional pdf for

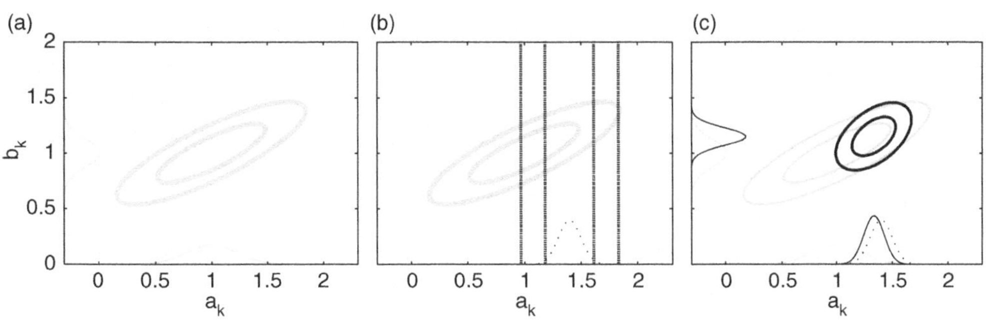

Fig. 3.3 The effect of the update step on an unobserved variable. See text for details.

a_k is very close to the observation likelihood (dotted line), since the likelihood is much narrower than the prior pdf for a_k. The updated mean of a_k shifts close to y_k^o.

The conditional pdf for b_k also differs from the prior pdf. The conditional pdf is narrower, and shifted to larger values of b_k, than the prior. The shift of the pdf to larger values of b_k clearly follows from the positive covariance between b_k and the observed variable (i.e. the positive tilt of the ellipses on which the prior pdf is constant), together with the fact that y_k^o lies to the right of the prior pdf for a_k. Thus, in this case in which $\text{cov}(a_k, b_k) \neq 0$, the update yields information about b_k, even though the observations are independent of b_k.

3.2.4 The conditional pdf for linear, Gaussian systems

Here we derive the conditional pdf $p(\mathbf{x}_k|Y_k^o)$ for linear, Gaussian systems. This involves specializing Bayes' rule (3.6) and the propagation rule (3.5) to the linear, Gaussian case, which shows that $\mathbf{x}_k|Y_k^o$ is Gaussian and yields expressions to update and propagate its mean and covariance. Together, these expressions comprise the Kalman filter.

In what follows, we adopt the notation of Ide et al. (1997), using superscripts a (for 'analysis') to indicate conditioning on observations through the same time as the state and f (for 'forecast') to indicate conditioning only on observations up to the previous time. In particular, the mean and covariance of $\mathbf{x}_k|Y_k^o$ are

$$\mathbf{x}_k^a \equiv E(\mathbf{x}_k|Y_k^o), \qquad \mathbf{P}_k^a \equiv \text{cov}(\mathbf{x}_k|Y_k^o)$$

and the mean and covariance of $\mathbf{x}_k|Y_{k-1}^o$ are

$$\mathbf{x}_k^f \equiv E(\mathbf{x}_k|Y_{k-1}^o), \qquad \mathbf{P}_k^f \equiv \text{cov}(\mathbf{x}_k|Y_{k-1}^o).$$

Quantities with superscripts will be deterministic, rather than random, variables.

We proceed by induction, assuming $\mathbf{x}_0 \sim N(\mathbf{x}_0^f, \mathbf{P}_0^f)$.

First, consider the update step at t_k given new observations $\mathbf{y}_k^o$. Suppose $\mathbf{x}_k|Y_{k-1}^o \sim N(\mathbf{x}_k^f, \mathbf{P}_k^f)$, so that, by (3.10),

$$p(\mathbf{x}_k|Y_{k-1}^o) = a_1 \exp\left[-\left(\mathbf{x}_k - \mathbf{x}_k^f\right)^{\mathrm{T}} \left(\mathbf{P}_k^f\right)^{-1} \left(\mathbf{x}_k - \mathbf{x}_k^f\right)/2\right],$$

where a_1 is a constant that does not depend on $\mathbf{x}_k$. Bayes' rule (3.6) also requires $p(\mathbf{y}_k^o|\mathbf{x}_k)$, which is given by

$$p(\mathbf{y}_k^o|\mathbf{x}_k) = a_2 \exp\left[-\left(\mathbf{y}_k^o - \mathbf{H}\mathbf{x}_k\right)^{\mathrm{T}} \mathbf{R}^{-1} \left(\mathbf{y}_k^o - \mathbf{H}\mathbf{x}_k\right)/2\right],$$

with a_2 another constant that does not depend on $\mathbf{x}_k$. Applying Bayes' rule then gives

$$p(\mathbf{x}_k|Y_k^o) = a_3 \exp\left[-J(\mathbf{x}_k)/2\right], \tag{3.13}$$

where a_3 is yet another constant that does not depend on $\mathbf{x}_k$ and

$$J(\mathbf{x}_k) = \left(\mathbf{x}_k - \mathbf{x}_k^f\right)^{\mathrm{T}} \left(\mathbf{P}_k^f\right)^{-1} \left(\mathbf{x}_k - \mathbf{x}_k^f\right) + \left(\mathbf{y}_k^o - \mathbf{H}\mathbf{x}_k\right)^{\mathrm{T}} \mathbf{R}^{-1} \left(\mathbf{y}_k^o - \mathbf{H}\mathbf{x}_k\right). \tag{3.14}$$

It is immediately clear that $J(\mathbf{x}_k)$ is a quadratic form in $\mathbf{x}_k$ (as long as the observation operator $\mathbf{H}$ is linear) and thus that $p(\mathbf{x}_k|Y_k^o)$ given by (3.13) is a Gaussian density.

To find the mean and covariance of $\mathbf{x}_k|Y_k^o$, we expand the products in (3.14) and group like terms together. This results in

$$J(\mathbf{x}_k) = \mathbf{x}_k^{\mathrm{T}}\left[\left(\mathbf{P}_k^f\right)^{-1} + \mathbf{H}^{\mathrm{T}}\mathbf{R}^{-1}\mathbf{H}\right]\mathbf{x}_k - 2\left[\mathbf{H}^{\mathrm{T}}\mathbf{R}^{-1}\mathbf{y}_k^o + \left(\mathbf{P}_k^f\right)^{-1}\mathbf{x}_k^f\right]^{\mathrm{T}}\mathbf{x}_k + c_1,$$

where c_1 does not depend on $\mathbf{x}_k$. A little algebra then gives

$$J(\mathbf{x}_k) = (\mathbf{x}_k - \mathbf{x}_k^a)^{\mathrm{T}}(\mathbf{P}_k^a)^{-1}(\mathbf{x}_k - \mathbf{x}_k^a) + c_2, \tag{3.15}$$

with c_2 a constant not dependent on $\mathbf{x}_k$ and

$$\mathbf{x}_k^a = \mathbf{P}_k^a\left[\mathbf{H}^{\mathrm{T}}\mathbf{R}^{-1}\mathbf{y}_k^o + \left(\mathbf{P}_k^f\right)^{-1}\mathbf{x}_k^f\right], \qquad \mathbf{P}_k^a = \left[\left(\mathbf{P}_k^f\right)^{-1} + \mathbf{H}^{\mathrm{T}}\mathbf{R}^{-1}\mathbf{H}\right]^{-1}. \tag{3.16}$$

Substituting (3.16) and (3.15) into the expression (3.13) for $p(\mathbf{x}_k|Y_k^o)$ and comparing with the definition (3.10) of the Gaussian density shows that $\mathbf{x}_k^a$ and $\mathbf{P}_k^a$ are the mean and covariance, respectively, of $\mathbf{x}_k|Y_k^o$. Equation (3.16) specifies how to update the forecast mean $\mathbf{x}_k^f$ and forecast covariance $\mathbf{P}_k^f$ given the observations $\mathbf{y}_k^o$.

The updates for both the mean and covariance can be written in numerous forms. One alternative form for the update of the update for the mean is

$$\mathbf{x}_k^a = \mathbf{x}_k^f + \mathbf{P}_k^a\mathbf{H}^{\mathrm{T}}\mathbf{R}^{-1}\left(\mathbf{y}_k^o - \mathbf{H}\mathbf{x}_k^f\right). \tag{3.17}$$

This may be obtained from (3.16) by adding and subtracting $\mathbf{H}\mathbf{x}_k^f$ from $\mathbf{y}_k^o$. The matrix multiplying the observation–forecast difference is known as the *gain* matrix and will be denoted by $\mathbf{K}$ hereinafter:

$$\mathbf{K} = \mathbf{P}_k^a\mathbf{H}^{\mathrm{T}}\mathbf{R}^{-1}. \tag{3.18}$$

As shown in Section 3.8, the gain may also be written as

$$\mathbf{K} = \mathbf{P}_k^f\mathbf{H}^{\mathrm{T}}\left(\mathbf{H}\mathbf{P}_k^f\mathbf{H}^{\mathrm{T}} + \mathbf{R}\right)^{-1}. \tag{3.19}$$

In addition, we will call the observation–forecast difference $\mathbf{y}_k^o - \mathbf{H}\mathbf{x}_k^f$ the *innovation*. An alternative form for the update of the covariance is

$$\mathbf{P}_k^a = \mathbf{P}_k^f - \mathbf{P}_k^f\mathbf{H}^{\mathrm{T}}\left(\mathbf{H}\mathbf{P}_k^f\mathbf{H}^{\mathrm{T}} + \mathbf{R}\right)^{-1}\mathbf{H}\mathbf{P}_k^f, \tag{3.20}$$

which is derived in Section 3.8 and also follows from the Sherman–Morrison–Woodbury formula applied to the form of $\mathbf{P}_k^a$ given in (3.16).

Next, consider the propagation (or forecast) step from t_{k-1} to t_k. (Note that simpler derivations exist for the propagation step of the Kalman filter and will be discussed in subsequent sections.) Suppose $\mathbf{x}_{k-1}|Y_{k-1}^o \sim N(\mathbf{x}_{k-1}^a, \mathbf{P}_{k-1}^a)$. The propagation rule

(3.5) also requires $p(\mathbf{x}_k|\mathbf{x}_{k-1}, Y^o_{k-1})$. Since the system noise and observation errors are assumed independent for each k, the conditioning on Y^o_{k-1} can be ignored. Equation (3.11) then implies that

$$\mathbf{x}_k|\mathbf{x}_{k-1} \sim N(\mathbf{M}\mathbf{x}_{k-1}, \mathbf{Q}),$$

since $\eta_k \sim N(0, \mathbf{Q})$.

The integrand in the propagation rule (3.5) is therefore the product of two Gaussian densities, and the exponent of that product is the sum of the exponents of the original densities. Using the definition of the Gaussian density gives

$$p(\mathbf{x}_k|\mathbf{x}_{k-1})p(\mathbf{x}_{k-1}|Y^o_{k-1}) = c\exp\left[-J(\mathbf{x}_{k-1}, \mathbf{x}_k)/2\right],$$

where c does not depend on $\mathbf{x}_{k-1}$ or $\mathbf{x}_k$ and

$$\begin{aligned} J(\mathbf{x}_{k-1}, \mathbf{x}_k) = {} & (\mathbf{x}_k - \mathbf{M}\mathbf{x}_{k-1})^{\mathrm{T}}\mathbf{Q}^{-1}(\mathbf{x}_k - \mathbf{M}\mathbf{x}_{k-1}) \\ & + (\mathbf{x}_{k-1} - \mathbf{x}^a_{k-1})^{\mathrm{T}}(\mathbf{P}^a_{k-1})^{-1}(\mathbf{x}_{k-1} - \mathbf{x}^a_{k-1}). \end{aligned}$$

Replacing $\mathbf{x}_k - \mathbf{M}\mathbf{x}_{k-1}$ by $(\mathbf{x}_k - \mathbf{M}\mathbf{x}^a_{k-1}) - \mathbf{M}(\mathbf{x}_{k-1} - \mathbf{x}^a_{k-1})$, J can be rewritten as

$$J(\mathbf{x}_{k-1}, \mathbf{x}_k) = \begin{pmatrix} \mathbf{x}_{k-1} - \mathbf{x}^a_{k-1} \\ \mathbf{x}_k - \mathbf{M}\mathbf{x}^a_{k-1} \end{pmatrix}^{\mathrm{T}} \mathbf{S} \begin{pmatrix} \mathbf{x}_{k-1} - \mathbf{x}^a_{k-1} \\ \mathbf{x}_k - \mathbf{M}\mathbf{x}^a_{k-1} \end{pmatrix},$$

where

$$\mathbf{S} = \begin{pmatrix} (\mathbf{P}^a_{k-1})^{-1} + \mathbf{M}^{\mathrm{T}}\mathbf{Q}^{-1}\mathbf{M} & -\mathbf{M}^{\mathrm{T}}\mathbf{Q}^{-1} \\ -\mathbf{Q}^{-1}\mathbf{M} & \mathbf{Q}^{-1} \end{pmatrix}.$$

We now have

$$\mathbf{S}^{-1} = \begin{pmatrix} \mathbf{P}^a_{k-1} & \mathbf{P}^a_{k-1}\mathbf{M}^{\mathrm{T}} \\ \mathbf{M}\mathbf{P}^a_{k-1} & \mathbf{M}\mathbf{P}^a_{k-1}\mathbf{M}^{\mathrm{T}} + \mathbf{Q} \end{pmatrix},$$

which may be verified directly. Thus, the joint distribution of $(\mathbf{x}_{k-1}, \mathbf{x}_k)|Y^o_{k-1}$ is Gaussian, with mean $(\mathbf{x}^a_{k-1}, \mathbf{M}\mathbf{x}^a_{k-1})$ and covariance $\mathbf{S}^{-1}$.

The propagation rule (3.5) integrates over $\mathbf{x}_{k-1}$ to obtain the marginal distribution of $\mathbf{x}_k|Y^o_{k-1}$. Rather than performing the integration, it is sufficient to note that the mean and covariance for the marginal are given by the entries in the full mean and covariance corresponding to $\mathbf{x}_k$. Therefore,

$$\mathbf{x}_k|Y^o_{k-1} \sim N(\mathbf{M}\mathbf{x}^a_{k-1}, \mathbf{M}\mathbf{P}^a_{k-1}\mathbf{M}^{\mathrm{T}} + \mathbf{Q}),$$

and, after defining $\mathbf{x}^f_k = E(\mathbf{x}_k|Y^o_{k-1})$ and $\mathbf{P}^f_k = \mathrm{cov}(\mathbf{x}_k|Y^o_{k-1})$,

$$\mathbf{x}^f_k = \mathbf{M}\mathbf{x}^a_{k-1}, \qquad \mathbf{P}^f_k = \mathbf{M}\mathbf{P}^a_{k-1}\mathbf{M}^{\mathrm{T}} + \mathbf{Q}.$$

This is the propagation or forecast step of the Kalman filter.

3.2.5 Summary

Specializing the Bayesian algorithm of Section 3.1 to the linear, Gaussian system (3.11) shows that both the one-step-ahead prediction distribution $p(\mathbf{x}_k|Y^o_{k-1})$ and the filtering distribution $p(\mathbf{x}_k|Y^o_k)$ are Gaussian. These distributions can then be specified by their mean and covariances, which are denoted by $\mathbf{x}^f_k$ and $\mathbf{P}^f_k$ for the prediction distribution and $\mathbf{x}^a_k$ and $\mathbf{P}^a_k$ for the filtering distribution. The Kalman filter consists of the following recursive equations for propagating and updating the mean and covariance:

- **Propagation (or forecast) step**

$$\mathbf{x}^f_k = \mathbf{M}\mathbf{x}^a_{k-1}, \qquad \mathbf{P}^f_k = \mathbf{M}\mathbf{P}^a_{k-1}\mathbf{M}^{\mathrm{T}} + \mathbf{Q}. \tag{3.21}$$

- **Update (or analysis) step**

$$\mathbf{x}^a_k = \mathbf{P}^a_k\left[\mathbf{H}^{\mathrm{T}}\mathbf{R}^{-1}\mathbf{y}^o_k + \left(\mathbf{P}^f_k\right)^{-1}\mathbf{x}^f_k\right], \qquad \mathbf{P}^a_k = \left[\left(\mathbf{P}^f_k\right)^{-1} + \mathbf{H}^{\mathrm{T}}\mathbf{R}^{-1}\mathbf{H}\right]^{-1}, \tag{3.22}$$

or, equivalently,

$$\mathbf{x}^a_k = \mathbf{x}^f_k + \mathbf{K}\left(\mathbf{y}^o_k - \mathbf{H}\mathbf{x}^f_k\right), \qquad \mathbf{P}^a_k = (\mathbf{I} - \mathbf{K}\mathbf{H})\,\mathbf{P}^f_k, \tag{3.23}$$

where

$$\mathbf{K} = \mathbf{P}^a_k\mathbf{H}^{\mathrm{T}}\mathbf{R}^{-1} = \mathbf{P}^f_k\mathbf{H}^{\mathrm{T}}\left(\mathbf{H}\mathbf{P}^f_k\mathbf{H}^{\mathrm{T}} + \mathbf{R}\right)^{-1}. \tag{3.24}$$

Since it yields the mean and covariance of the desired conditional pdfs, and since those pdfs are Gaussian, the Kalman filter is optimal for linear, Gaussian systems.

3.3 A closer look at the forecast and update steps

3.3.1 A 40-dimensional example

Lorenz (1995) proposed a continuous-time dynamical system of moderate dimensionality for studies of atmospheric predictability. The state of the system can be thought of as representing quantities along a latitude circle, and its ith element will be denoted by $x_i(t)$. (In this subsection, subscripts on scalar variables indicate location on the latitude circle rather than discrete times.) The system evolves according to

$$\frac{\mathrm{d}x_i}{\mathrm{d}t} = (x_{i+1} - x_{i-2})x_{i-1} - x_i + F, \tag{3.25}$$

where $i = 1, \ldots, N_x$, F is a constant, and cyclic conditions are imposed on the indices,

$$x_{i+N_x} = x_{i-N_x} = x_i.$$

Now define a discrete-time system of the form (3.1), to be denoted by L95, with $f(\mathbf{x}_k)$ obtained by a chosen approximation to time integration of the continuous-time system over a time step Δt and whose state vector is $\mathbf{x}_k = [x_1(t_k), \ldots, x_{N_x}(t_k)]^{\mathrm{T}}$, with $t_k = k\Delta t$. We will use a third-order Runge–Kutta scheme with $\Delta t = 0.025$ to specify the deterministic dynamics and will include additive system noise $\eta_k \sim N(0, \mathbf{Q})$ in (3.1).

The applicability of the Kalman filter to L95 is not obvious, since the dynamics (and potentially the observation operator) are nonlinear. Linearization of the dynamics and observation operator about a given sequence of states is one possible strategy. We follow this approach in this section, in order to illustrate some basic properties of the forecast and update steps of the Kalman filter, but reserve a more complete discussion for later sections.

Let $\tilde{\mathbf{x}}_k$ satisfy the deterministic part of (3.1); that is, $\tilde{\mathbf{x}}_k = f(\tilde{\mathbf{x}}_{k-1})$. We are interested in how small deviations about $\tilde{\mathbf{x}}_k$ evolve and how they affect the observed variables. Substituting $\mathbf{x}_k = \tilde{\mathbf{x}}_k + \delta\mathbf{x}_k$ into (3.1) and retaining only terms that are linear in $\delta\mathbf{x}_k$ gives

$$\delta\mathbf{x}_k = \left.\frac{\partial f}{\partial \mathbf{x}_{k-1}}\right|_{\tilde{\mathbf{x}}_{k-1}} \delta\mathbf{x}_{k-1} + \eta_k, \qquad \delta\mathbf{y}_k = \left.\frac{\partial h}{\partial \mathbf{x}_k}\right|_{\tilde{\mathbf{x}}_k} \delta\mathbf{x}_k + \epsilon_k, \tag{3.26}$$

where $\delta\mathbf{y}_k = \mathbf{y}_k^o - h(\tilde{\mathbf{x}}_k)$. The Jacobians $\partial f/\partial\mathbf{x}_{k-1}|_{\tilde{\mathbf{x}}_{k-1}}$ and $\partial h/\partial\mathbf{x}_k|_{\tilde{\mathbf{x}}_k}$ at each t_k are simply matrices, which vary in time as $\tilde{\mathbf{x}}_k$ evolves. We denote these matrices by $\mathbf{M}_k = \mathbf{M}_k(\tilde{\mathbf{x}}_{k-1})$ and $\mathbf{H}_k = \mathbf{H}_k(\tilde{\mathbf{x}}_k)$, and (3.26) is then an equations for $\delta\mathbf{x}_k$ of the form (3.11), but with explicit dependence of the propagation matrix $\mathbf{M}_k$ and observation matrix $\mathbf{H}_k$ on time.

3.3.2 The forecast step

Covariance evolution in the forecast step follows (3.21). Typically, the variance of the state, as measured by the trace of the covariance, increases during the forecast step and the covariance structure is also altered.

There are two effects that contribute to the increase of variance. The first is the system noise and the second is the system dynamics as embodied in $\mathbf{M}$. Both effects are easiest to understand if we suppose for the sake of argument that there are no observations. The update step is then trivial: $\mathbf{x}_k^a = \mathbf{x}_k^f$ and $\mathbf{P}_k^a = \mathbf{P}_k^f$.

In the absence of dynamics (i.e. $\mathbf{M} = \mathbf{I}$), the covariance evolves as

$$\mathbf{P}_k^f = \mathbf{P}_{k-1}^f + \mathbf{Q} = \mathbf{P}_0^f + k\mathbf{Q},$$

and thus $\mathrm{tr}\left(\mathbf{P}_k^f\right) = \mathrm{tr}(\mathbf{P}_0^f) + k\,\mathrm{tr}(\mathbf{Q})$. The system noise in isolation therefore leads to a linear growth of variance with time, and the forecast covariance matrix increasingly reflects the structure of $\mathbf{Q}$.

Next consider the effects of the dynamics in the absence of system noise (i.e. $\mathbf{Q} = 0$). To begin, we ignore the time dependence of the dynamics and take $\mathbf{M}_k = \mathbf{M}$. The deviation from the mean evolves as

$$\mathbf{x}_k - \mathbf{x}_k^f = \mathbf{M}(\mathbf{x}_{k-1} - \mathbf{x}_{k-1}^f) = \mathbf{M}^k(\mathbf{x}_0 - \mathbf{x}_0^f).$$

Let the eigenvalues of $\mathbf{M}$ be $\omega_1, \ldots, \omega_{N_x}$ with $|\omega_1| \geq |\omega_2| \geq \ldots \geq |\omega_{N_x}|$. The eigenvalues of $\mathbf{M}^k$ are then $\omega_1^k, \ldots, \omega_{N_x}^k$. Speaking loosely, $\mathbf{M}^k(\mathbf{x}_0 - \mathbf{x}_0^f)$ will be increasingly dominated as k increases by the leading eigenvectors corresponding to the largest few eigenvalues. More precisely, $\mathbf{M}^k(\mathbf{x}_0 - \mathbf{x}_0^f)$ will converge as $k \to \infty$ to the subspace spanned by the eigenvectors corresponding to eigenvalues with magnitude equal to $|\omega_1|$, and the rate of convergence depends on the separation between the magnitude of those eigenvalues and the magnitude of the next largest eigenvalue. This is illustrated in Fig. 3.4 using a matrix $\mathbf{M}$ obtained from the linearized dynamics of the L95 system over a single time step, where the linearization is performed about a reference solution chosen randomly from the system attractor.

The eigenvectors of $\mathbf{M}$ are not, in general, orthogonal. This means that the projection of $\mathbf{x}_0 - \mathbf{x}_0^f$ onto the eigenvectors does not determine the qualitative character of the evolution, except at long times (k large)—initial conditions projecting only onto eigenvectors with $|\omega_i| < 1$ may grow over finite intervals, for example.

Our interest for the Kalman filter is how the covariance matrix $\mathbf{P}_{k-1}^a$ evolves over the interval from t_{k-1} to t_k. Over finite intervals, the singular vectors and singular values of $\mathbf{M}(\mathbf{P}_{k-1}^a)^{1/2}$ are relevant. The decomposition is

$$\mathbf{M}(\mathbf{P}_{k-1}^a)^{1/2} = \mathbf{U}\Lambda\mathbf{V}^{\mathrm{T}},$$

where Λ is a diagonal matrix of real, non-negative singular values λ_i and $\mathbf{U}$ and $\mathbf{V}$ are orthonormal matrices whose columns are the left and right singular vectors, respectively. The left singular vectors $\mathbf{u}_i$ are the eigenvectors of the evolved covariance matrix and the singular values λ_i are the square roots of its eigenvalues:

$$\mathbf{M}\mathbf{P}_{k-1}^a\mathbf{M}^{\mathrm{T}} = \mathbf{U}\Lambda^2\mathbf{U}^{\mathrm{T}}.$$

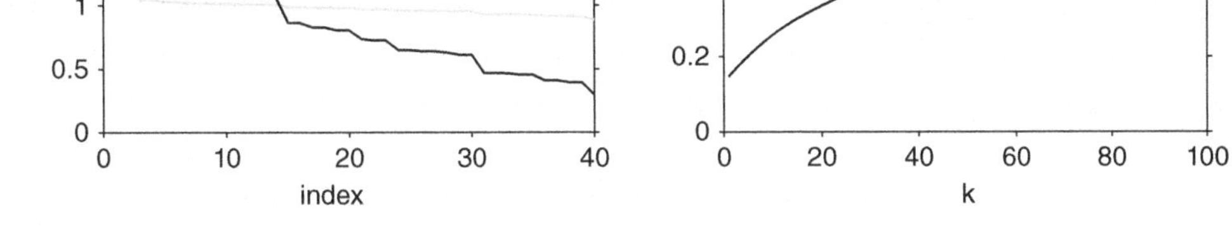

Fig. 3.4 (a) Magnitude of eigenvalues of $\mathbf{M}$ (grey line) and $\mathbf{M}^{10}$ (black line). (b) Expected cosine of the angle θ between $\mathbf{M}^k(\mathbf{x}_0 - \mathbf{x}_0^f)$, where $(\mathbf{x}_0 - \mathbf{x}_0^f)$ is drawn from $N(0, \mathbf{I})$, and the leading eigenvector of $\mathbf{M}$, as a function of the 'time' k. The expectation is estimated by averaging over 10^3 realizations of $\mathbf{x}_0 - \mathbf{x}_0^f$.

As orthogonal eigenvectors of its covariance matrix, the vectors $\mathbf{u}_i$ partition the variance of $\mathbf{x}_k - \mathbf{x}_k^f \sim N(0, \mathbf{M}\mathbf{P}_{k-1}^a\mathbf{M}^\mathrm{T})$, in the sense that the total variance is the sum of the variances of the projections onto the $\mathbf{u}_i$:

$$\mathrm{tr}\Big(\mathrm{cov}\left(\mathbf{x}_k - \mathbf{x}_k^f\right)\Big) = \sum \mathrm{var}\Big(\mathbf{u}_i^\mathrm{T}\left(\mathbf{x}_k - \mathbf{x}_k^f\right)\Big) = \sum \lambda_i^2.$$

Each of the $\mathbf{u}_i$ accounts for a certain portion, given by λ_i^2, of the variance of $\mathbf{x}_k - \mathbf{x}_k^f$. The $\mathbf{u}_i$ are also the directions of the principal axes of a hyper-ellipsoid on which the Gaussian density $N(0, \mathbf{M}\mathbf{P}_{k-1}^a\mathbf{M}^\mathrm{T})$ is constant. The right singular vectors $\mathbf{v}_i$ pertain to time t_{k-1} and can be understood by considering the 'pre-whitened' variable $\mathbf{w} = (\mathbf{P}_{k-1}^a)^{-1/2}(\mathbf{x}_{k-1} - \mathbf{x}_{k-1}^a)$, which satisfies $\mathrm{cov}(\mathbf{w}) = \mathbf{I}$. The $\mathbf{v}_i$ are vectors in the space of $\mathbf{w}$ that evolve into the $\mathbf{u}_i$ at time t_k after scaling by λ_i:

$$\mathbf{M}(\mathbf{P}_{k-1}^a)^{1/2}\mathbf{v}_i = \lambda_i \mathbf{u}_i.$$

The dynamics linearized about a nonlinear solution $\tilde{\mathbf{x}}_k$ will in general vary with time: $\mathbf{M} = \mathbf{M}_k$. The singular value decomposition remains relevant over finite intervals in this case. However, since the directions of growth and decay over successive finite time intervals will vary over time, the cumulative effect, over very long intervals, is not obvious. In fact, fundamental results from dynamical systems theory show that similar qualitative behavior holds (see Legras and Vautard, 1995). All perturbations evolved from the distant past to a specific time exhibit exponential growth with the same average growth rate and converge to a single direction, the leading Lyapunov vector, similar to the way in which all perturbations evolved under $\mathbf{M}^k$ grow exponentially and converge to the leading eigenvector of $\mathbf{M}$ as $k \to \infty$. Unlike the case with $\mathbf{M}$ independent of k, the leading Lyapunov vector is not a fixed direction, but depends on $\tilde{\mathbf{x}}_k$.

The overall picture, illustrated in Fig. 3.5, is that the dynamics tends to steepen the eigenvalue spectrum of the covariance matrix, meaning that fewer directions account for more of the total variance. If we begin from initial conditions distributed isotropically over a sphere in the perturbation space, the system dynamics leads to growth in certain preferred directions and decay in others, so that the forecast perturbations will lie on an ellipse, with the ellipse typically becoming more eccentric with time. The eventual effect of linearized dynamics over long times is to compress that ellipse nearly into a reduced subspace. In high-dimensional systems, the unstable space is often of much lower dimension and the degree of structure imposed on the covariances can be very significant.

3.3.3 The update step

The update step corrects the mean and covariance based on new observational information. Several basic results are analytically accessible in the linear, Gaussian case; we discuss these, in order to build intuition for the nature of those corrections.

One form of the update for the mean is (3.22):

$$\mathbf{x}_k^a = \mathbf{P}_k^a\left[\mathbf{H}^\mathrm{T}\mathbf{R}^{-1}\mathbf{y}_k^o + \left(\mathbf{P}_k^f\right)^{-1}\mathbf{x}_k^f\right].$$

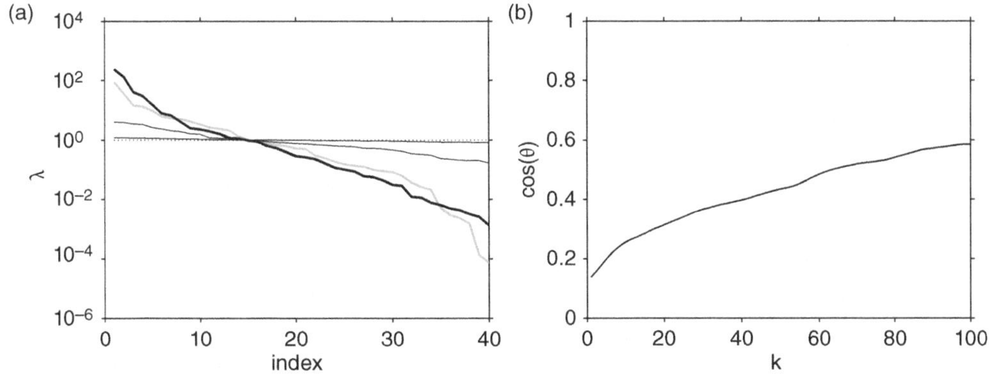

Fig. 3.5 (a) As in Fig. 3.4, but showing singular values of $\mathbf{M}$ and $\mathbf{M}^{10}$ (thin black lines), $\mathbf{M}^{50}$ (thick black lines), and $\prod_{k=1}^{50} \mathbf{M}_k$ (thick grey lines). For this example, we assume the simplest case, $\mathbf{P}^a_{k-1} = \mathbf{I}$; its eigenvalues are shown by the dotted line. (b) Cosine of the angle θ between $\mathbf{M}^k(\mathbf{x}_0 - \mathbf{x}^f_0)$, where $\mathbf{x}_0 - \mathbf{x}^f_0$ is drawn from $N(0, \mathbf{I})$, and the leading Lyapunov vector as a function of the 'time' k, averaged over 400 realizations of the initial perturbations.

The analysis mean depends linearly on both the observations and the forecast mean. Defining a state $\mathbf{x}^o_k$ that maps exactly to the observations, $\mathbf{H}\mathbf{x}^o_k = \mathbf{y}^o_k$, the update of the mean becomes

$$\mathbf{x}^a_k = \mathbf{P}^a_k \left[\mathbf{H}^{\mathrm{T}} \mathbf{R}^{-1} \mathbf{H} \mathbf{x}^o_k + \left(\mathbf{P}^f_k\right)^{-1} \mathbf{x}^f_k \right].$$

Thus, $\mathbf{x}^a_k$ is the weighted sum of two prior estimates in the state space. The weight given to each is proportional to the inverse of a covariance matrix, so that the more accurate estimate, the greater the weight it receives. The similarity of the update to the scalar case is clear, where $x^a = [\sigma_o^2 \sigma_f^2 / (\sigma_o^2 + \sigma_f^2)](x^o/\sigma_o^2 + x^f/\sigma_f^2)$.

In addition, the weights sum to $\mathbf{I}$, since (3.22) shows that $\mathbf{H}^{\mathrm{T}}\mathbf{R}^{-1}\mathbf{H} + (\mathbf{P}^f_k)^{-1} = (\mathbf{P}^a_k)^{-1}$. The fact that the weights sum to $\mathbf{I}$ immediately yields the second form (3.23) of the update using the gain $\mathbf{K}$ and the innovation $\mathbf{y}^o_k - \mathbf{H}\mathbf{x}^f_k$.

Next consider the update of the covariance. The update decreases the total variance, as expected since the observations bring additional information. To see this, begin from (3.23) and use the definition of $\mathbf{K}$ to write

$$\mathbf{P}^a_k = \mathbf{P}^f_k - \mathbf{P}^f_k \mathbf{H}^{\mathrm{T}} \left(\mathbf{H}\mathbf{P}^f_k \mathbf{H}^{\mathrm{T}} + \mathbf{R} \right)^{-1} \mathbf{H}\mathbf{P}^f_k. \tag{3.27}$$

The second term on the right-hand side is the symmetric product of symmetric, positive-definite matrices, and is therefore also symmetric and positive-definite. Thus, it has positive trace and $\mathrm{tr}(\mathbf{P}^a_k) < \mathrm{tr}(\mathbf{P}^f_k)$.

The updated covariance does not depend on the specific realization of the observations. If we know which observation will be available, that is, if we know $\mathbf{H}$ and $\mathbf{R}$, then $\mathbf{P}^a_k$ can be computed before we know $\mathbf{y}^o_k$. This is useful for 'targeted' observations, in which additional observations are included in a network in order to improve specific aspects of the analysis or subsequent forecasts.

3.4 The Kalman filter (II)

This section presents an alternative derivation of the Kalman-filter equations. Rather than making assumptions about the pdfs of the noise and the initial state (i.e. that they are Gaussian), we will consider the class of linear estimators and seek the one that minimizes the expected mean squared error.

3.4.1 Estimators of random variables

While the conditional pdf $p(\mathbf{x}_k|Y_k^o)$ is the most that can be known about $\mathbf{x}_k$ given noisy observations and an imperfect forecast model, it is natural and often necessary to rely on simpler and less costly methods. The simplest approach is to compute a single estimate $(\hat{\mathbf{x}}_k)$ that depends on the available observations Y_k^o.

There are several potential choices. An obvious one is the mode of $p(\mathbf{x}_k|Y_k^o)$, that is, the $\hat{\mathbf{x}}_k$ that maximizes $p(\mathbf{x}_k|Y_k^o)$. This is known as the maximum a posteriori (MAP) estimate. Another possibility is the conditional mean:

$$\hat{\mathbf{x}}_k = E(\mathbf{x}_k|Y_k^o).$$

The conditional mean is also referred to as the minimum variance estimate, because $E\left(|\mathbf{x}_k - \hat{\mathbf{x}}_k|^2\right) = \mathrm{tr}(\mathrm{cov}(\mathbf{x}_k)) + |\hat{\mathbf{x}}_k - E(\mathbf{x}_k)|^2$ is a minimum when $\hat{\mathbf{x}}_k = E(\mathbf{x}_k)$ and the second term is zero. Thus, the conditional mean provides the minimum expected mean squared error. Note that these choices do not arise in the Gaussian case, where the conditional mean and conditional mode are the same.

We wish to examine the properties of an estimate over all possible values of the observations. To do so, we introduce the notion of an estimator, a function that relates the observations, considered as random variables, to the state. An estimator is a random variable (since it is a function of a random variable) and produces an estimate when evaluated for a specific realization of the observations. Thus, the conditional mean estimator is $E(\mathbf{x}_k|Y_k)$, where the expectation is taken over $p(\mathbf{x}_k|Y_k)$ (i.e. with Y_k fixed) but the estimator itself is considered as a function of Y_k.

3.4.2 The best linear unbiased estimator

Consider estimators of the form

$$\mathbf{e}(\mathbf{y}) = \mathbf{A}\mathbf{y} + \mathbf{b}. \tag{3.28}$$

Strictly, $\mathbf{e}$ is an affine function of $\mathbf{y}$, though it is usually termed a *linear* estimator. The best linear unbiased estimator (BLUE), or linear minimum-variance estimator, minimizes the expected squared error over all linear estimators $\mathbf{e}(\mathbf{y})$. Since we are seeking the best estimator regardless of $\mathbf{y}$, the expectation of the squared error should be taken over the joint distribution of $\mathbf{x}$ and $\mathbf{y}$.

Calculation of $\mathbf{A}$ and $\mathbf{b}$ for the BLUE is straightforward. The squared error of an estimator is $\delta^2 = |\mathbf{x} - \mathbf{e}|^2$, which may be expanded as

$$\delta^2 = \mathbf{x}^{\mathrm{T}}\mathbf{x} + \mathbf{y}^{\mathrm{T}}\mathbf{A}^{\mathrm{T}}\mathbf{A}\mathbf{y} + \mathbf{b}^{\mathrm{T}}\mathbf{b} + 2\Big(\mathbf{b}^{\mathrm{T}}\mathbf{A}\mathbf{y} - \mathbf{x}^{\mathrm{T}}\mathbf{A}\mathbf{y} - \mathbf{x}^{\mathrm{T}}\mathbf{b}\Big).$$

Now note that a derivative with respect to a deterministic variable commutes with an expectation, that is,

$$\frac{\partial}{\partial \mathbf{A}} E(g(\mathbf{x}, \mathbf{y}, \mathbf{A})) = E\left(\frac{\partial g}{\partial \mathbf{A}}\right),$$

and use results on derivatives with respect to matrices, as given in Section 3.8, to derive

$$\frac{\partial}{\partial \mathbf{A}} E(\delta^2) = 2\mathbf{A}E(\mathbf{y}\mathbf{y}^{\mathrm{T}}) + 2\mathbf{b}E(\mathbf{y})^{\mathrm{T}} - 2E(\mathbf{x}\mathbf{y}^{\mathrm{T}})$$

and

$$\frac{\partial}{\partial \mathbf{b}} E(\delta^2) = 2\mathbf{b} + 2\mathbf{A}E(\mathbf{y}) - 2E(\mathbf{x}).$$

Setting both partial derivatives equal to zero gives

$$\mathbf{A} = \mathbf{P}_{xy}\mathbf{P}_{yy}^{-1}, \qquad \mathbf{b} = E(\mathbf{x}) - \mathbf{A}E(\mathbf{y}),$$

where $\mathbf{P}_{yy} = E(\mathbf{y}\mathbf{y}^{\mathrm{T}}) - E(\mathbf{y})E(\mathbf{y}^{\mathrm{T}}) = \mathrm{cov}(\mathbf{y})$ and $\mathbf{P}_{xy} = E(\mathbf{x}\mathbf{y}^{\mathrm{T}}) - E(\mathbf{x})E(\mathbf{y}^{\mathrm{T}}) = \mathrm{cov}(\mathbf{x}, \mathbf{y})$. The general form of the BLUE, which we denote by $\tilde{\mathbf{e}}(\mathbf{y})$, is then

$$\tilde{\mathbf{e}}(\mathbf{y}) = E(\mathbf{x}) + \mathbf{P}_{xy}\mathbf{P}_{yy}^{-1}\left[\mathbf{y} - E(\mathbf{y})\right]. \tag{3.29}$$

Clearly, the BLUE is unbiased, $E(\mathbf{x} - \tilde{\mathbf{e}}(\mathbf{y})) = 0$.

It is important to emphasize that the expectations are over $p(\mathbf{x}, \mathbf{y})$. Crucially, no assumptions on $p(\mathbf{x}, \mathbf{y})$ are needed, except knowledge of the mean and covariances. Nor is any assumption necessary on the relation between $\mathbf{x}$ and $\mathbf{y}$.

The expected error $E(\delta^2)$ can be calculated by noting that $\delta^2 = \mathrm{tr}((\mathbf{x}-\mathbf{e})(\mathbf{x}-\mathbf{e})^{\mathrm{T}})$ and using (3.29), which yields

$$(\mathbf{x} - \mathbf{e})(\mathbf{x} - \mathbf{e})^{\mathrm{T}} = \mathbf{x}'\mathbf{x}'^{\mathrm{T}} - \mathbf{x}'\mathbf{y}'^{\mathrm{T}}\mathbf{A}^{\mathrm{T}} - \mathbf{A}\mathbf{y}'\mathbf{x}'^{\mathrm{T}} + \mathbf{A}\mathbf{y}'\mathbf{y}'^{\mathrm{T}}\mathbf{A}^{\mathrm{T}}, \tag{3.30}$$

where primes denote deviations of a variable from its mean. Setting $\mathbf{A} = \mathbf{P}_{xy}\mathbf{P}_{yy}^{-1}$ and taking an expectation over $p(\mathbf{x}, \mathbf{y})$ shows the error covariance matrix to be

$$E\left((\mathbf{x} - \tilde{\mathbf{e}})(\mathbf{x} - \tilde{\mathbf{e}})^{\mathrm{T}}\right) = \mathbf{P}_{xx} - \mathbf{P}_{xy}\mathbf{P}_{yy}^{-1}\mathbf{P}_{xy}^{\mathrm{T}}. \tag{3.31}$$

The error covariance is independent of $\mathbf{y}$.

Next, we specialize to the linear, Gaussian case (3.11). To make direct comparison with the Kalman-filter equations of Section 3.2, suppose that we know $\mathbf{x}_k \sim N(\mathbf{x}_k^f, \mathbf{P}_k^f)$. The necessary means and covariances are

$$E(\mathbf{x}_k) = \mathbf{x}_k^f, \qquad E(\mathbf{y}_k) = \mathbf{H}\mathbf{x}_k^f, \qquad \mathbf{P}_{xy} = \mathbf{P}_k^f\mathbf{H}^{\mathrm{T}}, \qquad \mathbf{P}_{yy} = \mathbf{H}\mathbf{P}_k^f\mathbf{H}^{\mathrm{T}} + \mathbf{R}.$$

If the expectation were only over $p(\mathbf{x}_k)$, rather than over the joint distribution of $\mathbf{x}_k$ and $\mathbf{y}_k = \mathbf{H}\mathbf{x}_k + \epsilon$, then the covariance $\mathbf{P}_{yy}$ would not have included contributions from both the state covariance and the observation-error covariance.

Substituting the means and covariances into (3.29), the BLUE is therefore

$$\tilde{\mathbf{e}}(\mathbf{y}) = \mathbf{x}_k^f + \mathbf{P}_k^f\mathbf{H}^{\mathrm{T}}\left(\mathbf{H}\mathbf{P}_k^f\mathbf{H}^{\mathrm{T}} + \mathbf{R}\right)^{-1}(\mathbf{y} - \mathbf{H}\mathbf{x}_k^f).$$

Similarly, its error covariance becomes

$$E\left((\mathbf{x} - \tilde{\mathbf{e}})(\mathbf{x} - \tilde{\mathbf{e}})^{\mathrm{T}}\right) = \mathbf{P}_k^f - \mathbf{P}_k^f\mathbf{H}^{\mathrm{T}}(\mathbf{H}\mathbf{P}_k^f\mathbf{H}^{\mathrm{T}} + \mathbf{R})^{-1}\mathbf{H}\mathbf{P}_k^f. \tag{3.32}$$

These expressions agree with the Kalman-filter update—the BLUE is the updated, conditional mean $\mathbf{x}_k^a$, its error covariance is the updated covariance $\mathbf{P}_k^a$, and the matrix $\mathbf{A}$ appearing in (3.28) is the Kalman gain $\mathbf{K}$. Therefore, the BLUE and the Kalman-filter update are equivalent in the linear Gaussian case.

Because its derivation requires no assumptions about the form of $p(\mathbf{x}, \mathbf{y})$, the BLUE still provides the minimum expected squared error over all linear estimators, and its error covariance is still given by (3.31), when $p(\mathbf{x}, \mathbf{y})$ is non-Gaussian. In the non-Gaussian case, however, the BLUE generally differs from the conditional mean, and its error covariance, which is independent of $\mathbf{y}$, differs from the conditional covariance, which depends on the specific realization $\mathbf{y}_k^o$ of the observations. Both the BLUE and the conditional mean minimize expected squared error, but in different senses: the BLUE minimizes expected error for linear estimators and over all observations, whereas the conditional mean minimizes expected error for a specific $\mathbf{y}_k^o$.

Another case of interest is when $h(\mathbf{x})$ is nonlinear and the observations have state-independent, additive noise that has covariance $\mathbf{R}$ and is potentially non-Gaussian. Then

$$\mathbf{P}_{xy} = \mathrm{cov}(\mathbf{x}, h(\mathbf{x})), \qquad \mathbf{P}_{yy} = \mathrm{cov}(h(\mathbf{x})) + \mathbf{R}.$$

These expressions are useful in understanding the ensemble Kalman filter.

3.5 Assorted topics

3.5.1 Linear transformations

Starting from the linear, Gaussian system (3.11), linear transformations of either the state or observation variables do not change the BLUE (or the Kalman-filter analysis). Making a change of variables and then performing the update gives the same result as performing the update in the original variables and then making the transformation.

This result is easy to see, though the notation becomes complicated. Let $\mathbf{z} = \mathbf{L}_x\mathbf{x}$ and $\mathbf{w} = \mathbf{L}_y\mathbf{y}$. For now, we assume that $\mathbf{L}_x$ and $\mathbf{L}_y$ are invertible.

If $\mathbf{x} \sim N(\mathbf{x}^f, \mathbf{P}^f)$ and $\mathbf{y} = \mathbf{H}\mathbf{x} + \epsilon$ with $\epsilon \sim N(0, \mathbf{R})$, then the results of Section 3.2.1 show that $\mathbf{z} \sim N(\mathbf{z}^f, \mathbf{P}^f_{zz})$, with

$$\mathbf{z}^f = \mathbf{L}_x\mathbf{x}^f, \quad \mathbf{P}^f_{zz} = \mathbf{L}_x\mathbf{P}^f\mathbf{L}_x^{\mathrm{T}},$$

and $\mathbf{w} = \tilde{\mathbf{H}}\mathbf{z} + \tilde{\epsilon}$, with

$$\tilde{\mathbf{H}} = \mathbf{L}_y\mathbf{H}\mathbf{L}_x^{-1}, \quad \tilde{\epsilon} \sim N(0, \mathbf{L}_y\mathbf{R}\mathbf{L}_y^{\mathrm{T}}).$$

Given observations $\mathbf{w}^o = \mathbf{L}_y\mathbf{y}^o$, applying the Kalman filter (3.23) to the variables $\mathbf{z}$ and $\mathbf{w}$ implies

$$\begin{aligned}
\mathbf{z}^a &= \mathbf{z}^f + \tilde{\mathbf{K}}(\mathbf{w}^o - \tilde{\mathbf{H}}\mathbf{z}^f), \\
\tilde{\mathbf{K}} &= \mathbf{P}^f_{zz}\tilde{\mathbf{H}}^{\mathrm{T}}(\tilde{\mathbf{H}}\mathbf{P}^f_{zz}\tilde{\mathbf{H}}^{\mathrm{T}} + \mathbf{L}_y\mathbf{R}\mathbf{L}_y^{\mathrm{T}})^{-1}, \\
\mathbf{P}^a_{zz} &= (\mathbf{I} - \tilde{\mathbf{K}}\tilde{\mathbf{H}})\mathbf{P}^f_{zz}.
\end{aligned}$$

It is then straightforward to substitute for $\mathbf{z}^f$, $\mathbf{P}^f_{zz}$, and $\tilde{\mathbf{H}}$ and show that the same equations would be obtained if one first updated $\mathbf{x}$ and then transformed the results to $\mathbf{z}$. For example, the gain $\tilde{\mathbf{K}}$ becomes

$$\begin{aligned}
\tilde{\mathbf{K}} &= \mathbf{L}_x\mathbf{P}^f\mathbf{L}_x^{\mathrm{T}}\mathbf{L}_x^{-\mathrm{T}}\mathbf{H}^{\mathrm{T}}\mathbf{L}_y^{\mathrm{T}}\left(\mathbf{L}_y\mathbf{H}\mathbf{L}_x^{-1}\mathbf{L}_x\mathbf{P}^f\mathbf{L}_x^{\mathrm{T}}\mathbf{L}_x^{-\mathrm{T}}\mathbf{H}^{\mathrm{T}}\mathbf{L}_y^{\mathrm{T}} + \mathbf{L}_y\mathbf{R}\mathbf{L}_y^{\mathrm{T}}\right)^{-1} \\
&= \mathbf{L}_x\mathbf{P}^f\mathbf{H}^{\mathrm{T}}(\mathbf{H}\mathbf{P}^f\mathbf{H}^{\mathrm{T}} + \mathbf{R})^{-1}\mathbf{L}_y^{-1} \\
&= \mathbf{L}_x\mathbf{K}\mathbf{L}_y^{-1}.
\end{aligned}$$

The update of the mean is therefore

$$\begin{aligned}
\mathbf{z}^a &= \mathbf{L}_x\mathbf{x}^f + \tilde{\mathbf{K}}(\mathbf{L}_y\mathbf{y}^o - \mathbf{L}_y\mathbf{H}\mathbf{x}^f) \\
&= \mathbf{L}_x\left[\mathbf{x}^f + \mathbf{K}(\mathbf{y}^o - \mathbf{H}\mathbf{x}^f)\right] \\
&= \mathbf{L}_x\mathbf{x}^a.
\end{aligned}$$

3.5.2 Extending the state

The discussion to this point has assumed, at least implicitly, that the state used in the Kalman filter is identical to the model state. In practice, it can be useful algorithmically to extend the state vector to include variables other than those explicitly present in the numerical model. This is accomplished by simply appending the additional variables to the state vector, a technique known as extending the state vector.

There are many applications of this technique. Extending the state vector provides a framework for parameter estimation. Section 3.5.3 will illustrate how temporal correlations in the model error or observation errors can be handled via an extended state vector.

As an example to fix ideas, consider the extended state vector $\mathbf{z}$ of dimension $N_x + N_y$ obtained by appending the observation variables to the state:

$$\mathbf{z} = \begin{pmatrix} \mathbf{x} \\ \mathbf{Hx} \end{pmatrix} = \begin{pmatrix} \mathbf{I} \\ \mathbf{H} \end{pmatrix} \mathbf{x}.$$

This is helpful in parallelizing the serial EnKF (Anderson and Collins, 2007).

In terms of $\mathbf{z}$, the observation equation becomes

$$\mathbf{y} = \tilde{\mathbf{H}}\mathbf{z} + \epsilon, \qquad \tilde{\mathbf{H}} = (\mathbf{0} \;\; \mathbf{I}) .$$

It is also easy to see that if $\mathbf{x}_k \sim N(\mathbf{x}_k^f, \mathbf{P}_k^f)$, then $\mathbf{z}_k \sim N(\mathbf{z}_k^f, \mathbf{Z}_k^f)$, where

$$\mathbf{z}_k^f = \begin{pmatrix} \mathbf{I} \\ \mathbf{H} \end{pmatrix} \mathbf{x}_k^f, \quad \mathbf{Z}_k^f = \begin{pmatrix} \mathbf{I} \\ \mathbf{H} \end{pmatrix} \mathbf{P}_k^f \left(\mathbf{I} \;\; \mathbf{H}^{\mathrm{T}}\right) = \begin{pmatrix} \mathbf{P}_k^f & \mathbf{P}_k^f\mathbf{H}^{\mathrm{T}} \\ \mathbf{HP}_k^f & \mathbf{HP}_k^f\mathbf{H}^{\mathrm{T}} \end{pmatrix}.$$

The update step follows by applying (3.23) with $\mathbf{H}$ replaced by $\tilde{\mathbf{H}}$, and $\mathbf{P}_k^f$ replaced by $\mathbf{Z}_k^f$. The updated mean is

$$\mathbf{z}_k^a = \mathbf{z}_k^f + \tilde{\mathbf{K}}(\mathbf{y}_k^o - \tilde{\mathbf{H}}\mathbf{z}_k^f),$$

where

$$\tilde{\mathbf{K}} = \mathbf{Z}_k^f\tilde{\mathbf{H}}^{\mathrm{T}}(\tilde{\mathbf{H}}\mathbf{Z}_k^f\tilde{\mathbf{H}}^{\mathrm{T}} + \mathbf{R})^{-1}.$$

Simplifying, the analysis equation is

$$\mathbf{z}_k^a = \begin{pmatrix} \mathbf{x}_k^f \\ \mathbf{Hx}_k^f \end{pmatrix} + \begin{pmatrix} \mathbf{P}_k^f\mathbf{H}^{\mathrm{T}} \\ \mathbf{HP}_k^f\mathbf{H}^{\mathrm{T}} \end{pmatrix} \left(\mathbf{HP}_k^f\mathbf{H}^{\mathrm{T}} + \mathbf{R}\right)^{-1} \left(\mathbf{y}_k^o - \mathbf{Hx}_k^f\right).$$

This returns the usual result for state update in the upper block of $\mathbf{z}_k^a$, together with the observation-space update (which may be obtained by applying $\mathbf{H}$ to $\mathbf{x}_k^a$) in the lower block of $\mathbf{z}_k^a$.

Thus, by extending the state vector, a single update step has provided additional information, for the additional variable appended to the state, without altering the update of the original state variables.

3.5.3 Temporal correlations in η and ϵ

The derivations of the Kalman-filter equations given in Sections 3.2 and 3.4 assume that both the model error η_k and the observation error ϵ_k are not temporally correlated, that is, that η_k and η_l are independent for $k \neq l$, as are ϵ_k and ϵ_l. Although this is a convenient assumption, it is seldom truly justified: the unresolved scales that contribute to both η_k and, through representativeness error, to ϵ_k will likely be temporally correlated. Moreover, both η_k and ϵ_k contain contributions from approximations to physical processes, such as simplification of radiative transfer, and these are generally state-dependent and thus temporally correlated.

Suppose η_k or ϵ_k or both are temporally correlated. A simple model for such temporal correlations is

$$\begin{aligned}\eta_k &= \mathbf{N}\eta_{k-1} + \eta'_k, \quad & \eta'_k &\sim N(0, \mathbf{Q}),\\ \epsilon_k &= \mathbf{O}\epsilon_{k-1} + \epsilon'_k, \quad & \epsilon'_k &\sim N(0, \mathbf{R}),\end{aligned}$$

where the primed random variables η'_k and ϵ'_k are now uncorrelated in time and are assumed mutually independent. By the same arguments just given for the temporal correlations in η_k and ϵ_k, the latter assumption is likely not justified.

Correlations for either η_k or ϵ_k, or both, can be treated by extending the state. Here, we will look at time correlations in η_k alone and take ϵ_k uncorrelated in time. Since we will think of η as evolving along with $\mathbf{x}$, it is also convenient to modify the notation introduced in Section 3.2.2 slightly and write the system dynamics as $\mathbf{x}_k = \mathbf{M}\mathbf{x}_{k-1} + \eta_{k-1}$, where both terms on the left-hand side now have the same time index.

The extended state is defined by

$$\mathbf{z}_k = \begin{pmatrix}\mathbf{x}_k\\ \eta_k\end{pmatrix}.$$

The extended state evolves according to

$$\mathbf{z}_k = \tilde{\mathbf{M}}\mathbf{z}_{k-1} + \begin{pmatrix}0\\ \mathbf{I}\end{pmatrix}\eta'_k, \quad \tilde{\mathbf{M}} = \begin{pmatrix}\mathbf{M} & \mathbf{I}\\ 0 & \mathbf{N}\end{pmatrix},$$

and is related to the observations by

$$\mathbf{y}_k = \tilde{\mathbf{H}}\mathbf{z}_k + \epsilon_k, \quad \tilde{\mathbf{H}} = (\,\mathbf{H}\;0\,).$$

First consider the forecast step from t_{k-1} to t_k, with $\mathbf{z}_{k-1} \sim N(\mathbf{z}^a_{k-1}, \mathbf{Z}^a_{k-1})$. Let $\mathbf{Z}^a_{k-1}$ be partitioned into blocks as

$$\mathbf{Z}^a_{k-1} = \begin{pmatrix}\mathbf{P}^a_{k-1} & \mathbf{C}^a_{k-1}\\ \left(\mathbf{C}^a_{k-1}\right)^{\mathrm{T}} & \mathbf{Q}^a_{k-1}\end{pmatrix},$$

where $\mathbf{P}^a_{k-1}$ is the covariance of $\mathbf{x}_{k-1}$, $\mathbf{C}^a_{k-1}$ is the cross-covariance of $\mathbf{x}_{k-1}$ and η_{k-1}, and $\mathbf{Q}^a_{k-1}$ is the covariance of η_{k-1}.

Applying the forecast equations for the Kalman filter gives the forecast mean $\mathbf{z}^f_k = \tilde{\mathbf{M}}\mathbf{z}^a_{k-1}$, or, as is also easily seen directly from the original system,

$$\mathbf{x}^f_k = \mathbf{M}\mathbf{x}^a_{k-1} + \eta^a_{k-1}, \quad \eta^f_k = \mathbf{N}\eta^a_{k-1}.$$

The forecast covariances are given by

$$\mathbf{Z}^f_k = \tilde{\mathbf{M}}\mathbf{Z}^a_{k-1}\tilde{\mathbf{M}}^{\mathrm{T}} + \begin{pmatrix}0 & 0\\ 0 & \mathbf{Q}\end{pmatrix}.$$

Multiplying shows that the blocks $\mathbf{P}_k^f$, $\mathbf{C}_k^f$, and $\mathbf{Q}_k^f$ are given by

$$\begin{aligned}\mathbf{P}_k^f &= \mathbf{M}\mathbf{P}_{k-1}^a\mathbf{M}^{\mathrm{T}} + (\mathbf{C}_{k-1}^a)^{\mathrm{T}}\mathbf{M}^{\mathrm{T}} + \mathbf{M}\mathbf{C}_{k-1}^a + \mathbf{Q}_{k-1}^a,\\ \mathbf{C}_k^f &= \mathbf{M}\mathbf{C}_{k-1}^a\mathbf{N}^{\mathrm{T}} + \mathbf{Q}_{k-1}^a\mathbf{N}^{\mathrm{T}},\\ \mathbf{Q}_k^f &= \mathbf{N}\mathbf{Q}_{k-1}^a\mathbf{N}^{\mathrm{T}} + \mathbf{Q}.\end{aligned}$$

Thus, temporal correlation in η_k alters the evolution of $\mathbf{P}_k^f$, and a non-zero cross-covariance $\mathbf{C}_k^f$ is generated during the forecast step, even if $\mathbf{C}_{k-1}^a = 0$.

Next consider the update step at t_k. The calculation of $\tilde{\mathbf{K}}$ involves

$$\mathbf{Z}_k^f\tilde{\mathbf{H}}^{\mathrm{T}} = \begin{pmatrix}\mathbf{P}_k^f & \mathbf{C}_k^f\\ \left(\mathbf{C}_k^f\right)^{\mathrm{T}} & \mathbf{Q}_k^f\end{pmatrix}\begin{pmatrix}\mathbf{H}^{\mathrm{T}}\\ 0\end{pmatrix} = \begin{pmatrix}\mathbf{P}_k^f\mathbf{H}^{\mathrm{T}}\\ \left(\mathbf{C}_k^f\right)^{\mathrm{T}}\mathbf{H}^{\mathrm{T}}\end{pmatrix}.$$

Similarly, $\tilde{\mathbf{H}}\mathbf{Z}_k^f\tilde{\mathbf{H}}^{\mathrm{T}} = \mathbf{H}\mathbf{P}_k^f\mathbf{H}^{\mathrm{T}}$. The update for the mean becomes

$$\mathbf{z}_k^a = \mathbf{z}_k^f + \begin{pmatrix}\mathbf{P}_k^f\mathbf{H}^{\mathrm{T}}\\ \left(\mathbf{C}_k^f\right)^{\mathrm{T}}\mathbf{H}^{\mathrm{T}}\end{pmatrix}(\mathbf{H}\mathbf{P}_k^f\mathbf{H}^{\mathrm{T}} + \mathbf{R})^{-1}(\mathbf{y}_k^o - \mathbf{H}\mathbf{x}_k^f).$$

The update for η_k^f is proportional to $(\mathbf{C}_k^f)^{\mathrm{T}}$, which is generated by the forecast step.

The update for the covariances follows from $\mathbf{Z}_k^a = \mathbf{Z}_k^f - \mathbf{Z}_k^f\tilde{\mathbf{H}}^{\mathrm{T}}(\tilde{\mathbf{H}}\mathbf{Z}_k^f\tilde{\mathbf{H}}^{\mathrm{T}} + \mathbf{R})^{-1}\tilde{\mathbf{H}}\mathbf{Z}_k^f$. Performing the multiplication shows that the blocks of $\mathbf{Z}_k^a$ are

$$\begin{aligned}\mathbf{P}_k^a &= \mathbf{P}_k^f - \mathbf{P}_k^f\mathbf{H}^{\mathrm{T}}\left(\mathbf{H}\mathbf{P}_k^f\mathbf{H}^{\mathrm{T}} + \mathbf{R}\right)^{-1}\mathbf{H}\mathbf{P}_k^f,\\ \mathbf{C}_k^a &= \mathbf{C}_k^f - \mathbf{P}_k^f\mathbf{H}^{\mathrm{T}}\left(\mathbf{H}\mathbf{P}_k^f\mathbf{H}^{\mathrm{T}} + \mathbf{R}\right)^{-1}\mathbf{H}\mathbf{C}_k^f,\\ \mathbf{Q}_k^a &= \mathbf{Q}_k^f - \mathbf{C}_k^f\mathbf{H}^{\mathrm{T}}\left(\mathbf{H}\mathbf{P}_k^f\mathbf{H}^{\mathrm{T}} + \mathbf{R}\right)^{-1}\mathbf{H}\mathbf{C}_k^f.\end{aligned}$$

Thus, the update for $\mathbf{x}_k^f$, $\mathbf{P}_k^f$ is unchanged from the standard case. In addition, $\mathbf{C}_k^f$ is changed by the update when it is non-zero but is not generated by the update.

The derivation and its results are analogous for temporal correlations in ϵ, except that modifications appear in the update step for $\mathbf{x}$, rather than the propagation step. In addition, cross-covariance between $\mathbf{x}_k$ and ϵ_k is generated in the update step rather than the forecast step. Since ϵ_k directly affects the observed variables, the update for ϵ_k^f does not depend solely on the cross-covariance of $\mathbf{x}_k$ and η_k.

3.5.4 Stability and divergence of the Kalman filter

A classical question is the stability of the Kalman-filter solutions. We ask what happens if the Kalman filter runs for a long time and whether the filter is always well behaved.

The Kalman filter is designed to cope with noise gracefully, since it is meant to 'filter' noise from the observations. At the same time, in most applications, the Kalman filter will be subject to effects not considered in its design, such as unknown model and observation errors, nonlinearity, and non-Gaussianity.

Rigorous results are possible for linear systems; see for example Sections 7.5 and 7.6 of Jazwinski (1970). We do not cover those results here, but present qualitative illustrations and discussion.

Consider the L95 system again, which has dimension 40 and is governed by (3.25) discretized in time. We take $F = 8$ in (3.25) and perform several assimilation experiments. In these experiments, the evolution of a reference, or true, state is obtained by integrating (3.25), and then observations are simulated from the true state and assimilated using the Kalman filter.[1] We assume that (i) the evolution of the true state includes zero system noise, (ii) the observations are of each component of $\mathbf{x}$,

$$\mathbf{y} = \mathbf{x} + \epsilon,$$

where $\epsilon \sim N(0, \mathbf{I})$, and (iii) the observations are available at intervals of 0.1 time units, or every fourth time step of the discrete-time system.

To implement the Kalman filter, one must specify the initial state covariance $\mathbf{P}_0^a$, the observation-error covariance $\mathbf{R}$ and the model-error covariance $\mathbf{Q}$. In the first experiment (Fig. 3.6(a)), all these covariances are specified correctly; that is, $\mathbf{R} = \mathbf{I}$, $\mathbf{Q} = 0$, and $\mathbf{P}_0^a$ is consistent with the statistics of the random error in the ensemble mean at t_0. Five realizations of the experiment are shown. The filter behaves as expected: the analysis error (black lines) asymptotes to a stable value over about 10 cycles and the analysis variance computed by the filter (grey lines) is consistent with the analysis error, although the error, being a specific realization of a random process, varies from cycle to cycle more than the variance.

If $\mathbf{P}_0^a$ is specified too small (Fig. 3.6(b)), then the filter performs markedly worse than in the first experiment. Because the forecast variance is also too small, the update step gives too much weight to the erroneous forecast and too little weight to the observations. The analysis error then decreases only very slowly, or often not at all.

The effect of mis-specifying $\mathbf{P}_0^a$ depends on whether that specification is too optimistic (as in this case) or too conservative (not shown). With a conservative estimate of $\mathbf{P}_0^a$, the filter fits the observations too closely initially, but correctly calculates that the analysis error is decreasing; as $\mathbf{P}_k^a$ decreases, the filter performs better and rapidly approaches the behavior shown in the first experiment.

Another difficulty in implementing the Kalman filter is that the forecast model may contain unknown errors. The third experiment illustrates how an incorrect parameter in the forecast model may affect the assimilation system. If the forecast model uses $F = 9$ (Fig. 3.6(c)), rather than the correct $F = 8$, in (3.25), the filter initially behaves similarly to the case in which the model is perfect. After 10–15 cycles, however, the analysis error begins to increase in the case that the forecast model has the incorrect

[1] Since the system dynamics (3.25) are nonlinear, the Kalman filter will be suboptimal in this problem and the formulas (3.21) for the propagation step, which depend on linearity, must be modified. Here we employ the ensemble Kalman filter (see Section 3.7), but with sufficiently large ensemble size that the results are insensitive to further increasing the number of members. In any event, the effects of nonlinearity are of secondary importance in the results we present next.

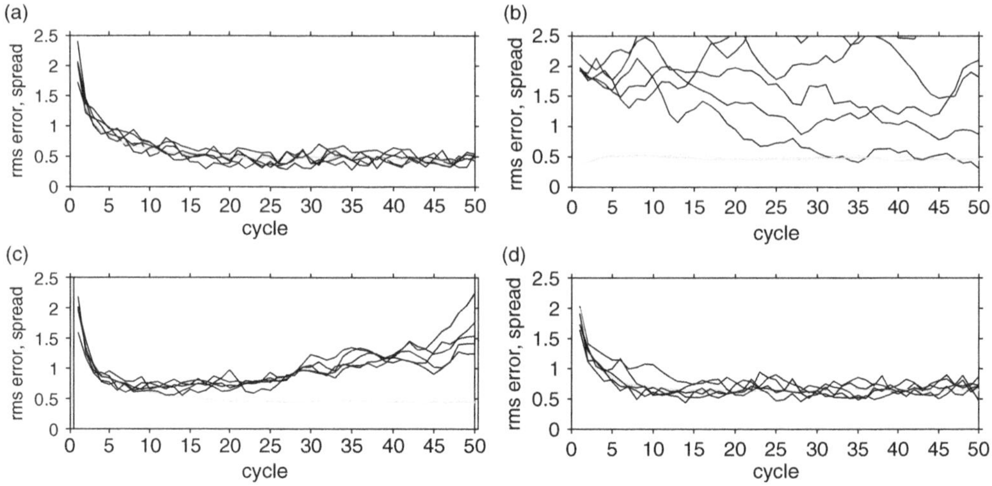

Fig. 3.6 Spread (square root of analysis variance; grey lines) and root-mean-square (rms) analysis error (black lines) as a function of time (in terms of the number of assimilation cycles) for four experiments with the L95 system. In the first experiment (a), the forecast model is perfect and the initial spread and rms error are consistent, though substantially larger than their asymptotic values after many assimilation cycles. The second experiment (b) also uses a perfect forecast model, but the initial spread is much smaller than the initial error. The third experiment (c) is identical to the first, except that the forecast model is imperfect, with $F = 9$ in (3.25). The fourth experiment (d) is identical to the third, except the forecast model also includes additive noise in (3.25) that is Gaussian with zero mean and variance 0.04 and is independent for each variable and each forecast time.

value of F. This continues through the end of the experiment, at which point the root-mean-square (rms) error is roughly a factor of 3 larger than the filter's estimate of the analysis spread.

This behavior is known as filter divergence. The filter systematically underestimates the analysis and forecast variance—in this case because the forecast step does not account for the model error introduced with $F = 9$. In each update step, the filter gives too little weight to the observations and further underestimates the analysis variance. Note that the analysis variance and its evolution in time (grey lines) are nearly identical in Figs. 3.6(a) and (c)—the filter is blind to the error in the forecast model and to its steadily worsening divergence.

Any factor, either in formulation or implementation, that leads to suboptimality of the Kalman filter can contribute to filter divergence. Examples include errors and bias in the forecast model that are not accounted for in the forecast step (as shown in Fig. 3.6), misspecification of observation-error covariances or initial-state covariances, nonlinearity and non-Gaussianity in the system or observations, and approximations in the filter calculations, such as are necessary in the EnKF.

One approach to counter filter divergence is to increase the forecast covariances at each cycle. This may be accomplished by increasing the additive noise in the forecast step (i.e. by increasing $\mathbf{Q}$) or by 'inflating' the forecast covariance after the forecast step (i.e. by replacing $\mathbf{P}_k^f$ with $\alpha^2\mathbf{P}_k^f$ with $\alpha^2 > 1$). In a final experiment (Fig. 3.6(d)), we include an additive noise $\eta \sim N(0, 0.04\mathbf{I})$ in the forecast model, which increases $\mathbf{P}_k^f$ at each cycle by $0.04\mathbf{I}$. This cures the divergence of the filter: the analysis error and analysis spread remain consistent through 50 cycles and the the analysis error at the end of the experiment is less than half of that in the previous experiment. Since the forecast model is degraded by the incorrect value of F, the error at the end of the experiment is larger than when the forecast model is perfect, by roughly 25%.

3.6 Nonlinearity and non-Gaussianity

For linear, Gaussian systems as given in (3.11), the Kalman filter gives the conditional mean and covariance of the state; that is, it calculates $E(\mathbf{x}_k|\mathbf{y}_k^o)$ and $\text{cov}(\mathbf{x}_k|\mathbf{y}_k^o)$. The Kalman filter is optimal for the linear, Gaussian case—it provides the full conditional pdf $p(\mathbf{x}_k|\mathbf{y}_k^o)$, since that pdf is Gaussian for the linear, Gaussian case, and is therefore determined by its mean and covariance.

In most practical applications, the system dynamics (and the forecast model) is nonlinear. The observation-error statistics are also often non-Gaussian, especially for positive-definite variables such as mixing ratios. The conditional pdf will then in general be non-Gaussian and the Kalman filter will be suboptimal, with the updated mean and covariance given by (3.11) differing from the conditional mean and covariance. The effects of non-Gaussianity are discussed in more detail in the Section 3.6.1 and, following that, we examine two approaches to extend the Kalman filter to non-Gaussian situations.

3.6.1 Effects of non-Gaussianity

There are several basic differences between results for the nonlinear, non-Gaussian case and those for the linear, Gaussian case.

The conditional mean $E(\mathbf{x}_k|\mathbf{y}_k^o)$ and conditional mode (the $\mathbf{x}_k$ that maximizes $p(\mathbf{x}_k|\mathbf{y}_k^o)$) differ for non-Gaussian distributions. This means that if we seek a single, 'best' estimate of $\hat{\mathbf{x}}_k$, we now have a choice between the conditional mean and conditional mode or even other possibilities, such as the conditional median.

Since it is not possible to summarize all characteristics of $p(\mathbf{x}_k|\mathbf{y}_k^o)$ in a single estimate $\hat{\mathbf{x}}_k$, a choice for $\hat{\mathbf{x}}_k$ may have advantages or disadvantages depending on the specific situation. Consider the conditional mean. It minimizes the expected squared error: $E\left(\text{tr}((\hat{\mathbf{x}}_k - \mathbf{x}_k|\mathbf{y}_k^o)(\hat{\mathbf{x}}_k - \mathbf{x}_k|\mathbf{y}_k^o)^{\mathrm{T}})\right)$ is smallest when $\hat{\mathbf{x}}_k = E(\mathbf{x}_k|\mathbf{y}_k^o)$. On the other hand, the conditional mean may represent an unlikely state. If, for example, $p(\mathbf{x}_k|\mathbf{y}_k^o)$ is strongly bimodal, then the mean will lie between either peak of the pdf, where the probability is small.

The linear, Gaussian case has other special properties that do not hold in general. For linear, Gaussian systems, the conditional mean $\mathbf{x}_k^a$ and covariance $\mathbf{P}_k^a$ are given by the Kalman filter update, (3.23). The update equations show that $\mathbf{x}_k^a$ depends linearly on $\mathbf{y}_k^o$ and the prior mean $\mathbf{x}_k^f$. In addition, the covariance $\mathbf{P}_k^a$ is independent of the observations, and the analysis variance, $\text{tr}(\mathbf{P}_k^a)$, always decreases relative to $\mathbf{P}_k^f$—see the discussion related to (3.27).

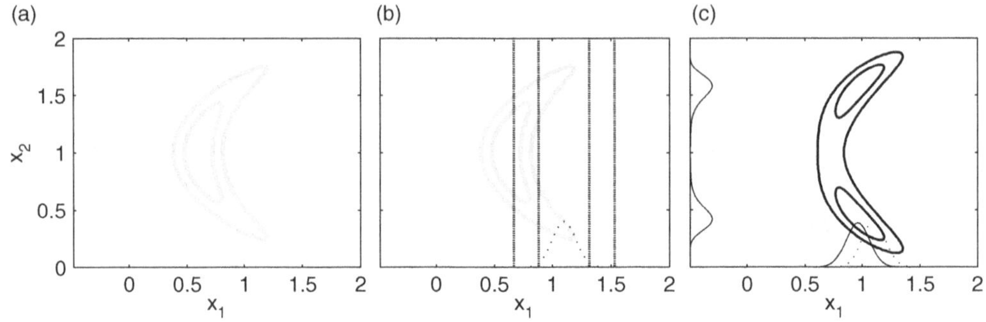

Fig. 3.7 Non-Gaussian effects in a two-dimensional example. The prior distribution $p(x_1, x_2)$ is shown in grey in (a), with marginal distributions $p(x_1)$ and $p(x_2)$ indicated by the thin lines on each axis. Note that $p(x_1)$ is skewed to larger values of x_1. A single observation of x_1 with Gaussian errors, $y = x_1 + \epsilon$, is available and has the value $y^o = 1.1$. The observation likelihood $p(y^o|x_1, x_2)$ is shown by dotted lines in (b), both as a function of x_1 and x_2 (it is independent of x_2) and as function only of x_1 (dotted line on x_1 axis). The conditional distribution $p(x_1, x_2|y^o)$ appears in black in (c) and its marginals $p(x_1|y^o)$ and $p(x_2|y^o)$ are shown on each axis (thin black lines).

The lack of these properties in the general case is illustrated in Fig. 3.7. First, the analysis variance increases relative to the prior variance, both for the observed variable x_1 and, especially, for the unobserved variable x_2. This increase is apparent by comparing the prior and posterior marginal distributions (thin grey and black lines, respectively, in Fig. 3.7(c)). In addition, the degree to which the variance increases, or whether it increases at all, clearly depends on y^o. If y^o were smaller, the range of values of x_2 that are consistent with the observation would decrease, owing to the shape of $p(x_1, x_2)$. Finally, the conditional mean for x_1, $E(x_1|y^o)$, depends nonlinearly on y^o; since the prior distribution is skewed, $E(x_1|y^o)$ draws more closely to the observed value y^o when y^o is less than the prior mode of x_1 than in the case shown, in which y^o is larger than the prior mode of x_1.

3.6.2 The extended Kalman filter

Here and in Section 3.6.3, we consider how the Kalman filter might be generalized to systems (3.1), (3.2) with f and h nonlinear.

Here, we suppose that η_k in (3.1) and ϵ_k in (3.2), as well as $\mathbf{x}_{k-1}$, are approximately Gaussian, where $\mathbf{x}_{k-1} \sim N(\mathbf{x}^a_{k-1}, \mathbf{P}^a_{k-1})$. A crucial assumption for what follows is that $\mathbf{P}^a_{k-1}$ is small, in a sense to be clarified later.

Define $\mathbf{M}_k = (\partial f/\partial \mathbf{x})|_{\mathbf{x}^a_{k-1}}$, the linearization of f at $\mathbf{x}^a_{k-1}$. The state $\mathbf{x}_k$ at the next time can be approximated by expanding $f(\mathbf{x}_{k-1})$ in a Taylor series, using $\mathbf{M}_k$:

$$f(\mathbf{x}_{k-1}) = f(\mathbf{x}^a_{k-1}) + \mathbf{M}_k\,(\mathbf{x}_{k-1} - \mathbf{x}^a_{k-1}) + O(|\mathbf{x}_{k-1} - \mathbf{x}^a_{k-1}|^2). \tag{3.33}$$

Inserting the leading terms from (3.33) into (3.1) yields

$$\mathbf{x}_k \approx f(\mathbf{x}^a_{k-1}) + \mathbf{M}_k(\mathbf{x}_{k-1} - \mathbf{x}^a_{k-1}) + \eta_k. \tag{3.34}$$

Under the assumption that $\mathbf{x}_{k-1} \sim N(\mathbf{x}^a_{k-1}, \mathbf{P}^a_{k-1})$, (3.34) implies that $\mathbf{x}_k \sim N(\mathbf{x}^f_k, \mathbf{P}^f_k)$, with

$$\mathbf{x}^f_k = f(\mathbf{x}^a_{k-1}), \quad \mathbf{P}^f_k = \mathbf{M}_k\mathbf{P}^a_{k-1}\mathbf{M}^{\mathrm{T}}_k + \mathbf{Q}. \tag{3.35}$$

This suggests a forecast step like (3.21) for the Kalman filter, except that the mean is propagated under the nonlinear dynamics f and the covariances are propagated using linear dynamics that depend on the previous analysis $\mathbf{x}^a_{k-1}$.

A similar treatment is possible for $h(\mathbf{x}_k)$ and the update step. Define $\mathbf{H}_k = (\partial h/\partial \mathbf{x})|_{\mathbf{x}^f_k}$, the linearization of h at $\mathbf{x}^f_k$. Then $\mathbf{y}_k$ may be approximated by expanding $h(\mathbf{x}_k)$ in a Taylor series,

$$h(\mathbf{x}_k) = h\left(\mathbf{x}^f_k\right) + \mathbf{H}_k(\mathbf{x}_k - \mathbf{x}^f_k) + O(|\mathbf{x}_k - \mathbf{x}^f_k|^2), \tag{3.36}$$

and inserting the result into (3.2):

$$\mathbf{y}_k \approx h\left(\mathbf{x}^f_k\right) + \mathbf{H}_k(\mathbf{x}_k - \mathbf{x}^f_k) + \epsilon_k. \tag{3.37}$$

Writing $\mathbf{y}_k - h(\mathbf{x}^f_k) = \mathbf{H}_k(\mathbf{x}_k - \mathbf{x}^f_k) + \epsilon_k$, we apply the Kalman-filter update (3.23) for $\mathbf{x}'_k = (\mathbf{x}_k - \mathbf{x}^f_k) \sim N(0, \mathbf{P}^f_k)$, given observations $\mathbf{y}^o_k$, and obtain

$$\mathbf{x}^a_k = \mathbf{x}^f_k + \mathbf{K}\left(\mathbf{y}^o_k - h(\mathbf{x}^f_k)\right), \quad \mathbf{P}^a_k = (\mathbf{I} - \mathbf{K}\mathbf{H}_k)\,\mathbf{P}^f_k, \tag{3.38}$$

where the gain is given by

$$\mathbf{K} = \mathbf{P}^f_k\mathbf{H}_k\left(\mathbf{H}_k\mathbf{P}^f_k\mathbf{H}^{\mathrm{T}}_k + \mathbf{R}\right)^{-1}. \tag{3.39}$$

This suggests an update step as in the Kalman filter, but with the innovation calculated with nonlinear h.

In the linearization (3.33), the neglected terms are $O(|\mathbf{x}_{k-1} - \mathbf{x}^a_{k-1}|^2)$. Their precise magnitude will depend both on $E(|\mathbf{x}_{k-1} - \mathbf{x}^a_{k-1}|^2) = \mathrm{tr}(\mathbf{P}^a_{k-1})$ and on the second derivatives of f, $\partial^2 f/\partial \mathbf{x}^2$. Similarly, the neglected terms in (3.36) will depend on $E(|\mathbf{x}_k - \mathbf{x}^f_k|^2) = \mathrm{tr}\left(\mathbf{P}^f_k\right)$ and, through (3.35), on $\mathbf{P}^a_{k-1}$. Thus, $\mathbf{P}^a_{k-1}$ must be sufficiently small that the Taylor series expansions of f and h are accurate.

The resulting algorithm is known as the extended Kalman filter. Its propagation step (3.35) differs from the Kalman filter in that it evolves the mean using the nonlinear dynamics and propagates the covariance using $\mathbf{M}_k$, the dynamics linearized about the prior mean $\mathbf{x}^a_{k-1}$. Its update step (3.38) differs from the Kalman filter in that it calculates the innovation as $\mathbf{y}^o_k - h(\mathbf{x}^f_k)$, using the nonlinear observation operator h applied to the forecast mean, and employs the linearization $\mathbf{H}_k$ of h in the gain and the covariance update.

The extended Kalman filter provides a generalization to nonlinear systems. It is optimal, however, only in the case of linear, Gaussian systems, where it reduces to Kalman filter, and it lacks a clear probabilistic interpretation except in that case.

3.6.3 The BLUE for nonlinear, non-Gaussian systems

The BLUE is a second generalization of the Kalman filter to nonlinear, non-Gaussian systems. Recall from Section 3.4.2 that the properties of the BLUE do not require any assumptions on the joint pdf $p(\mathbf{x}_k, \mathbf{y}_k)$. In particular, the system dynamics f and the observation operator h may be nonlinear without changing the property that the BLUE gives the minimum expected squared error over all linear estimators, where the expectation is relative to $p(\mathbf{x}_k, \mathbf{y}_k)$.

Comparing the BLUE (3.29) with the update (3.38) for $\mathbf{x}_k^a$ from the extended Kalman filter shows that both are linear estimators of the form

$$\tilde{\mathbf{x}}_k = \mathbf{x}_k^b + \mathbf{A}(\mathbf{y}_k^o - \mathbf{y}_k^b),$$

where $\tilde{\mathbf{x}}_k$ is an updated estimate of $\mathbf{x}_k$, $\mathbf{x}_k^b$ is the estimate before the update, and $\mathbf{y}_k^b$, which is used to compute the 'innovation', is an estimate of the observed variables before the update. The BLUE differs from the extended Kalman filter in its definition of $\mathbf{x}_k^b$, however. The BLUE uses

$$\mathbf{x}_k^b = E(\mathbf{x}_k) = E(f(\mathbf{x}_{k-1})), \qquad \mathbf{y}_k^b = E(\mathbf{y}_k) = E(h(\mathbf{x}_k)), \tag{3.40}$$

whereas the extended Kalman filter uses

$$\mathbf{x}_k^b = f(\mathbf{x}_{k-1}^a) = f(E(\mathbf{x}_{k-1})), \qquad \mathbf{y}_k^b = h\left(\mathbf{x}_k^f\right) = h(E(\mathbf{x}_k)). \tag{3.41}$$

The BLUE takes expectations over the non-Gaussian distributions produced by nonlinear f and h, whereas the extended Kalman filter performs expectations first and then applies f or h.

In general, it is not clear which of these approaches will yield the best results. Their relative effectiveness likely depends on the details of the nonlinearity.

3.7 Basics of the ensemble Kalman filter

The EnKF is a scheme whose update step approximates the Kalman filter for linear, Gaussian systems and approximates the BLUE in the general case. Algorithms to implement the EnKF have two basic forms, either 'stochastic', in which the algorithm involves random sampling and an attendant generation of random numbers, or deterministic. In both types of algorithms, the update step provides an ensemble of analyses, while the forecast step consists of an ensemble of forecasts using the analysis ensemble as initial conditions.

3.7.1 Why approximation is necessary

The earlier sections of this chapter have described recursive schemes for general Bayesian filtering, the Kalman filter and the BLUE. Implementing any of these schemes computationally is, however, far from straightforward for the high-dimensional systems of interest in geophysical problems.

Consider first the problem of general, Bayesian filtering. The update and forecast steps, described by (3.6) and (3.5), involve pdfs for $\mathbf{x}$. These pdfs are continuous functions of $N_x = \dim(\mathbf{x})$ variables. A naive computational scheme would discretize those functions over, say, 10 points in each of the dimensions, giving a numerical discretization on a N_x-dimensional grid with a total of 10^{N_x} grid points. Such a scheme is clearly impossible for the foreseeable future.

The computational requirements are reduced for the Kalman filter, but direct, unapproximated calculations are still prohibitive. The covariance matrices that must be propagated and updated (i.e. in (3.21) and (3.22)) are $N_x \times N_x$ or $N_y \times N_y$. For geophysical applications, where N_x is generally greater than 10^6 and N_y may be nearly as large, the manipulation of such matrices is not feasible on existing computers; these matrices are too large even to store in memory.

3.7.2 Random samples and sample statistics

To fix terminology, let $\mathbf{x}$ be a random variable with density $p(\mathbf{x})$. We say that $\{\mathbf{x}^i,\ i = 1, \ldots, N_e\}$ is a random sample of $\mathbf{x}$ (or 'from $p(\mathbf{x})$') if each $\mathbf{x}^i$ is drawn randomly and independently according to

$$P(\mathbf{x} \in V) = \int_V p(\mathbf{x})\, \mathrm{d}\mathbf{x}.$$

For our purposes, a random sample can be considered equivalent to an ensemble, although ensembles in numerical weather prediction are not always designed to be truly random samples from a specified distribution. Integer superscripts will be used to index the elements (or members) of a random sample, despite the potential for confusion with the superscripts f and a that indicate forecast or analysis quantities from the Kalman filter.

Algorithms that generate and manipulate random samples are known as Monte Carlo algorithms.

Basic but important quantities derived from a random sample are the sample mean

$$\hat{\mathbf{x}} = (N_e)^{-1} \sum_{i=1}^{N_e} \mathbf{x}^i$$

and the sample covariance

$$\hat{\mathbf{P}} = (N_e - 1)^{-1} \sum_{i=1}^{N_e} (\mathbf{x}^i - \hat{\mathbf{x}})(\mathbf{x}^i - \hat{\mathbf{x}})^{\mathrm{T}}.$$

The caret (or 'hat') indicates a sample estimate; $\hat{\mathbf{x}}$ and $\hat{\mathbf{P}}$ are unbiased estimates that approach $E(\mathbf{x})$ and $\mathbf{P} = \mathrm{cov}(\mathbf{x})$, respectively, as $N_e \to \infty$.

Monte Carlo estimates typically have errors that are $O(N_e^{-1/2})$. For example, the expected squared error of the sample mean is

$$E\left(|\hat{\mathbf{x}} - E(\mathbf{x})|^2\right) = \mathrm{tr}(\mathbf{P})/N_e. \tag{3.42}$$

This can be derived by writing $\hat{\mathbf{x}} - E(\mathbf{x}) = N_e^{-1} \sum_{i=1}^{N_e} \mathbf{x}^i - E(\mathbf{x})$, so that

$$N_e^2 |\hat{\mathbf{x}} - E(\mathbf{x})|^2 = \sum_{i=1}^{N_e} |\mathbf{x}^i - E(\mathbf{x})|^2 + \sum_{i,j} [\mathbf{x}^i - E(\mathbf{x})]^{\mathrm{T}} [\mathbf{x}^j - E(\mathbf{x})].$$

Since $\mathbf{x}^i$ is drawn from the distribution of $\mathbf{x}$, $E(|\mathbf{x}^i - E(\mathbf{x})|^2) = \mathrm{tr}(\mathbf{P})$, and since $\mathbf{x}^i$ and $\mathbf{x}^j$ are drawn independently, the expectation of the cross term $[\mathbf{x}^i - E(\mathbf{x})]^{\mathrm{T}}[\mathbf{x}^j - E(\mathbf{x})]$ is zero. The expression (3.42) then follows directly. Errors in the elements of $\hat{\mathbf{P}}$ are also $O(N_e^{-1})$, but depend on both the second and fourth moments of $\mathbf{x}$.

The sample covariance $\hat{\mathbf{P}}$ can also easily be expressed in terms of a square root. Let $\mathbf{X}$ be the $N_x \times N_e$ matrix whose columns are the ensemble perturbations scaled by $(N_e - 1)^{-1/2}$; that is, its columns are $(N_e - 1)^{-1/2}(\mathbf{x}^i - \hat{\mathbf{x}})$. Then

$$\hat{\mathbf{P}} = \mathbf{X}\mathbf{X}^{\mathrm{T}}. \tag{3.43}$$

To see this, note that the jk element of the right-hand side is $\sum_{i=1}^{N_e}(\mathbf{X})_{ji}(\mathbf{X})_{ki} = \sum_{i=1}^{N_e}(\mathbf{x}^i - \hat{\mathbf{x}})_j(\mathbf{x}^i - \hat{\mathbf{x}})_k$, where $(\mathbf{x}^i - \hat{\mathbf{x}})_j$ denotes the jth element of $\mathbf{x}^i - \hat{\mathbf{x}}$. The form (3.43) will be used frequently in the EnKF.

3.7.3 The forecast step for ensemble data assimilation

The idea in ensemble data assimilation, illustrated in Fig. 3.8, is to avoid the direct use of the continuous pdf or of the full covariance matrices, and work instead with random samples from the appropriate distributions. This approximation allows tractable computational algorithms for both recursive, Bayesian filtering and the Kalman filter. An ensemble approximation to the former is the particle filter, which will be discussed elsewhere in this volume. The EnKF approximates the Kalman filter (or, more generally, the BLUE) and is discussed below.

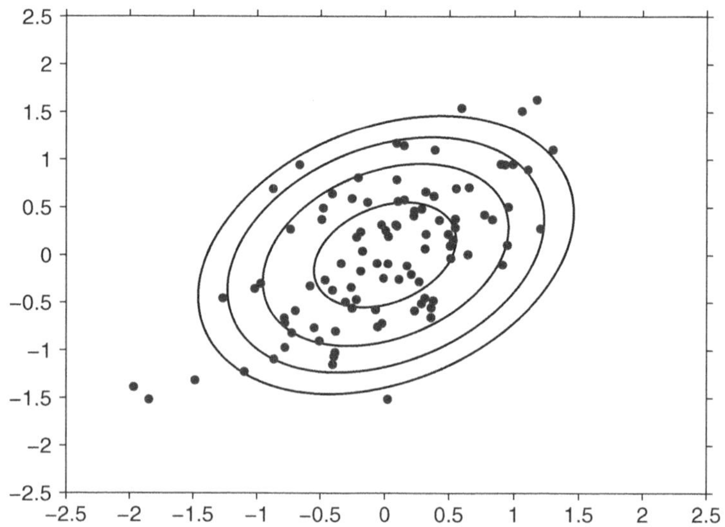

Fig. 3.8 A Gaussian probability density $p(\mathbf{x})$ (black contours) for a 2-vector $\mathbf{x}$ and a sample of 50 realizations of $\mathbf{x}$. Ensemble data assimilation seeks to work with samples instead of $p(\mathbf{x})$ or the covariance matrix for $\mathbf{x}$.

Before turning to the specifics of the EnKF, we examine a general template for sequential ensemble filters, focusing on the forecast step.

The algorithm is recursive, beginning from a sample from $p(\mathbf{x}_k|Y_{k-1}^o)$. Given $\mathbf{y}_k^o$, that sample is updated to approximate one drawn from the distribution $p(\mathbf{x}_k|Y_k^o)$ conditioned also on $\mathbf{y}_k^o$. This update step is non-trivial and may be tackled via several different approaches.

Starting with the updated sample drawn from $p(\mathbf{x}_k|Y_k^o)$, we next wish to generate a sample that approximates a random draw from the forecast distribution $p(\mathbf{x}_{k+1}|Y_k^o)$. Recall that

$$p(\mathbf{x}_{k+1}, \mathbf{x}_k|Y_k^o) = p(\mathbf{x}_{k+1}|\mathbf{x}_k)p(\mathbf{x}_k|Y_k^o), \tag{3.44}$$

which follows from the definition of the conditional distribution and the assumption that η_{k+1} is independent of η_m for $m \neq k+1$ and of ϵ_j for all j. According to (3.44), a sample $\{(\mathbf{x}_k^i, \mathbf{x}_{k+1}^i), i = 1, \ldots, N_e\}$ from the joint distribution $p(\mathbf{x}_{k+1}, \mathbf{x}_k|Y_k^o)$ can be obtained starting from the sample $\{\mathbf{x}_k^i, i = 1, \ldots, N_e\}$ from $p(\mathbf{x}_k|Y_k^o)$ by simply drawing, for each $\mathbf{x}_k^i$, a corresponding $\mathbf{x}_{k+1}^i$ from $p(\mathbf{x}_{k+1}|\mathbf{x}_k^i)$. Restricting the resulting sample to t_{k+1} (i.e. by discarding $\mathbf{x}_k^i$ from each pair) then yields the desired sample from $p(\mathbf{x}_{k+1}|Y_k^o)$. (Note that the propagation rule (3.5) achieves the marginal density for $\mathbf{x}_{k+1}$ by an integral over the joint density.)

If we have the forecast model, drawing from $p(\mathbf{x}_{k+1}|\mathbf{x}_k^i)$ is easy. For a system of the form (3.1), it is only necessary to compute

$$\mathbf{x}_{k+1}^i = f(\mathbf{x}_k^i) + \eta_{k+1}^i,$$

where η_{k+1}^i is drawn randomly from the assumed distribution for η. As a result, a forecast from $\mathbf{x}_k^i$ produces a random draw from $p(\mathbf{x}_{k+1}|\mathbf{x}_k^i)$, and the forecast step is nothing more than an ensemble forecast.

This approach involves no approximations; if $\mathbf{x}_k^i$ is drawn from $p(\mathbf{x}_k|Y_k^o)$, then $\mathbf{x}_{k+1}^i$ generated by this procedure is a random draw from $p(\mathbf{x}_{k+1}|Y_k^o)$. Ensemble forecasts are thus a general technique for the propagation step in a recursive ensemble filter.

3.7.4 The stochastic EnKF

Here we derive the 'stochastic' form of the update step for the EnKF. In essence, this is a Monte Carlo approximation to the BLUE: beginning from the forecast ensemble, each member is updated given the latest observations $\mathbf{y}_k^o$ in such a way that the mean of the updated ensemble approaches the BLUE as $N_e \to \infty$, while sample covariances from the updated ensemble approach the error covariances for the BLUE in the same limit.

In the linear, Gaussian case, the EnKF therefore approaches the Kalman filter as the ensemble size grows. The linear, Gaussian case also allows a stronger probabilistic result: the stochastic EnKF produces an analysis ensemble that approximates a random draw from the correct conditional distribution, $p(\mathbf{x}_k|Y_k^o)$.

We begin with the linear, Gaussian case, where the system and observations are given by (3.11) and assumptions on the errors are as in Section 3.2.2. Suppose that the prior at the present time is $\mathbf{x} \sim N(\mathbf{x}^f, \mathbf{P}^f)$. (More concretely, suppose that the time is t_k, that we know the prior distribution $\mathbf{x}_k|Y^o_{k-1} \sim N(\mathbf{x}^f_k, \mathbf{P}^f_k)$, and that the observations to be used in the update are $\mathbf{y}^o_k$. I will omit subscripts indicating time and conditioning on Y^o_{k-1} in what follows to simplify the exposition.)

Now consider the random variable

$$\xi = \mathbf{x} + \mathbf{K}\left[\mathbf{y}^o - (\mathbf{H}\mathbf{x} + \epsilon)\right], \tag{3.45}$$

with the gain $\mathbf{K} = \mathbf{P}^f\mathbf{H}^{\mathrm{T}}\left(\mathbf{H}\mathbf{P}^f\mathbf{H}^{\mathrm{T}} + \mathbf{R}\right)^{-1}$. Since it is a linear function of the Gaussian random variables $\mathbf{x}$ and ϵ, ξ is also Gaussian. Its mean is easily seen to be

$$E(\xi) = \mathbf{x}^a = \mathbf{x}^f + \mathbf{K}\left(\mathbf{y}^o - \mathbf{H}\mathbf{x}^f\right),$$

and thus agrees with the updated mean from the Kalman filter for $\mathbf{x}|\mathbf{y}^o$. The covariance of ξ is calculated in much the same way as the error covariance of the BLUE (cf. (3.30)). Denoting deviations from the mean with a prime and using the fact that ϵ and $\mathbf{x}$ are independent, we have

$$\begin{aligned}
\mathrm{cov}(\xi) &= E\left(\xi'\xi'^{\mathrm{T}}\right) \\
&= E\left(\left[\mathbf{x}' - \mathbf{K}(\mathbf{H}\mathbf{x}' + \epsilon)\right]\left[\mathbf{x}' - \mathbf{K}(\mathbf{H}\mathbf{x}' + \epsilon)\right]^{\mathrm{T}}\right) \\
&= \mathbf{P}^f - \mathbf{P}^f\mathbf{H}^{\mathrm{T}}\mathbf{K}^{\mathrm{T}} - \mathbf{K}\mathbf{H}\mathbf{P}^f + \mathbf{K}\left(\mathbf{H}\mathbf{P}^f\mathbf{H}^{\mathrm{T}} + \mathbf{R}\right)\mathbf{K}^{\mathrm{T}} \\
&= \mathbf{P}^f - \mathbf{P}^f\mathbf{H}^{\mathrm{T}}\left(\mathbf{H}\mathbf{P}^f\mathbf{H}^{\mathrm{T}} + \mathbf{R}\right)^{-1}\mathbf{H}\mathbf{P}^f \\
&= \mathbf{P}^a.
\end{aligned} \tag{3.46}$$

Thus, $\xi \sim N(\mathbf{x}^a, \mathbf{P}^a)$, with $\mathbf{x}^a$ and $\mathbf{P}^a$ given by the Kalman-filter update equations. This means that ξ has the same distribution as $\mathbf{x}|\mathbf{y}^o$.

A random draw of ξ is therefore a draw from the desired conditional distribution. This makes it easy to draw from the analysis distribution: starting from $\mathbf{x}^i$ drawn from the forecast distribution $\mathbf{N}(\mathbf{x}^f, \mathbf{P}^f)$ and ϵ^i from $N(0, \mathbf{R})$, update each $\mathbf{x}^i$ according to

$$\xi^i = \mathbf{x}^i + \mathbf{K}\left[\mathbf{y}^o - (\mathbf{H}\mathbf{x}^i + \epsilon^i)\right].$$

The relation (3.45) between the random variables ξ and $\mathbf{x}$ forms the basis for sampling from $\mathbf{x}|\mathbf{y}^o$ starting from $\mathbf{x}^i$ drawn from the distribution of $\mathbf{x}$, in the same way that the system dynamics provides a mechanism to sample from $\mathbf{x}_k|\mathbf{x}_{k-1}$, starting from a draw from the distribution of $\mathbf{x}_{k-1}$.

Now that we have an algorithm to sample from $\mathbf{x}_k|\mathbf{y}^o$, it remains to approximate the update equations, whose solution is required in the sampling, so that they are computationally feasible. The natural approximation is to replace the covariances in $\mathbf{K}$ by sample estimates based on the forecast ensemble $\{\mathbf{x}^i,\, i = 1, \ldots, N_e\}$.

Let the sample mean be $\hat{\mathbf{x}}^f = N_e^{-1}\sum \mathbf{x}^i$ and define

$$\begin{aligned}
\mathbf{X} &= \text{matrix with columns } (N_e - 1)^{-1/2}\left(\mathbf{x}^i - \hat{\mathbf{x}}^f\right), \\
\mathbf{Y} &= \text{matrix with columns } (N_e - 1)^{-1/2}\left(\mathbf{H}\mathbf{x}^i + \epsilon^i - \mathbf{H}\hat{\mathbf{x}}^f - \hat{\epsilon}\right).
\end{aligned}$$

Using the result (3.43), the sample estimate for $\mathbf{P}^f\mathbf{H}^{\mathrm{T}}$ is $\mathbf{X}\mathbf{Y}^{\mathrm{T}}$ and that for $\mathbf{H}\mathbf{P}^f\mathbf{H}^{\mathrm{T}}+\mathbf{R}$ is $\mathbf{Y}\mathbf{Y}^{\mathrm{T}}$. Inserting these into (3.19), the approximate gain becomes

$$\hat{\mathbf{K}} = \mathbf{X}\mathbf{Y}^{\mathrm{T}}\left(\mathbf{Y}\mathbf{Y}^{\mathrm{T}}\right)^{-1}. \tag{3.47}$$

The basic, stochastic form of the EnKF is then

$$\xi^i = \mathbf{x}^i + \hat{\mathbf{K}}\left[\mathbf{y}^o - (\mathbf{H}\mathbf{x}^i + \epsilon^i)\right]. \tag{3.48}$$

This update step involves performing a separate analysis for each member, using the approximate gain $\hat{\mathbf{K}}$ and including a realization of the observation error ϵ when mapping each member to the observed variables. The result of the update step is an ensemble of analyses, which approximates a random sample from $p(\mathbf{x}|\mathbf{y}^o)$.

The analysis ensemble $\{\xi^i,\ i=1,\ldots,N_e\}$ satisfies appropriate versions of the Kalman-filter update equations. The sample mean is

$$\hat{\mathbf{x}}^a \equiv \hat{\xi} = \hat{\mathbf{x}}^f + \hat{\mathbf{K}}\left[\mathbf{y}^o - (\mathbf{H}\hat{\mathbf{x}}^f + \hat{\epsilon})\right],$$

which agrees with the Kalman-filter update of the mean, (3.23), when expected values are replaced with sample means and the gain is approximated using sample covariances. Defining $\mathbf{X}^a$ to be the matrix with columns $(N_e-1)^{-1/2}(\xi^i-\hat{\mathbf{x}}^a)$, $i=1,\ldots,N_e$, the deviations from the sample mean update according to

$$\mathbf{X}^a = \mathbf{X} - \hat{\mathbf{K}}\mathbf{Y},$$

so that the updated sample covariance is

$$\begin{aligned}\hat{\mathbf{P}}^a &= \mathbf{X}^a\mathbf{X}^{a\mathrm{T}}\\ &= \mathbf{X}\mathbf{X}^{\mathrm{T}} - \mathbf{X}\mathbf{Y}^{\mathrm{T}}\hat{\mathbf{K}}^{\mathrm{T}} - \hat{\mathbf{K}}\mathbf{Y}\mathbf{X}^{\mathrm{T}} + \hat{\mathbf{K}}\mathbf{Y}\mathbf{Y}^{\mathrm{T}}\hat{\mathbf{K}}^{\mathrm{T}}\\ &= \mathbf{X}\mathbf{X}^{\mathrm{T}} - \mathbf{X}\mathbf{Y}^{\mathrm{T}}\left(\mathbf{Y}\mathbf{Y}^{\mathrm{T}}\right)^{-1}\mathbf{Y}\mathbf{X}^{\mathrm{T}}.\end{aligned}$$

This expression for $\hat{\mathbf{P}}^a$ agrees with the Kalman-filter result (3.20), when covariance matrices are replaced with sample estimates from the ensemble.

As $N_e \to \infty$, the sample estimates will converge to their correct (population) values and the ensemble produced by the stochastic EnKF (3.48) will have mean and covariance that converge to those given by the Kalman-filter update. In this sense, the stochastic EnKF converges to the Kalman filter for large ensembles.

Our motivation for exploring these ensemble approximations was to reduce the computational requirements relative to the exact Kalman filter. Some computational benefits are already apparent. In particular, the full covariance matrices $\mathbf{P}^f$, $\mathbf{P}^f\mathbf{H}^{\mathrm{T}}$, and $\mathbf{H}\mathbf{P}^f\mathbf{H}^{\mathrm{T}}$ are not explicitly calculated or stored. Instead, the algorithm employs the square roots $\mathbf{X}$ and $\mathbf{Y}$, which is equivalent to storing the N_e ensemble perturbations and the N_e predicted observation vectors based on those perturbations.

The gain $\hat{\mathbf{K}}$ still involves the inverse of $\mathbf{Y}\mathbf{Y}^{\mathrm{T}}$, which is $N_y \times N_y$. This inverse is expensive to compute explicitly when N_y is much larger than, say, 10^3, but its effect on $\mathbf{y}^o - (\mathbf{H}\mathbf{x}^i + \epsilon^i)$ can be calculated by solving $\mathbf{Y}\mathbf{Y}^{\mathrm{T}}\mathbf{b} = \mathbf{y}^o - (\mathbf{H}\mathbf{x}^i + \epsilon^i)$ for $\mathbf{b}$, for instance by borrowing techniques from observation-space implementations of variational assimilation schemes. Two other ways of avoiding this matrix inversion—serial assimilation of observations and deterministic EnKF algorithms—are discussed later in this section.

A few additional details can easily reduce the sampling error in the algorithm. First, $\hat{\epsilon}$ appears as a non-zero term in $\hat{\mathbf{x}}^a$ only because N_e is finite. Since $E(\epsilon) = 0$ by assumption, it is better to subtract $\hat{\epsilon}$ from each ϵ^i, effectively setting $\hat{\epsilon} = 0$. In addition, there will be cross terms such as $(\mathbf{x}^i - \hat{\mathbf{x}}^f)\epsilon^i$ in both $\mathbf{YY}^{\mathrm{T}}$ and $\mathbf{XY}^{\mathrm{T}}$, which appear in the expression for $\hat{\mathbf{P}}^a$. Those cross terms, which would approach zero for a large ensemble since ϵ is assumed independent of $\mathbf{x}$, can be avoided by replacing $\mathbf{YY}^{\mathrm{T}}$ with $(\mathbf{HX})(\mathbf{HX})^{\mathrm{T}} + \mathbf{R}$ and $\mathbf{XY}^{\mathrm{T}}$ with $\mathbf{X}(\mathbf{HX})^{\mathrm{T}}$.

An important subtlety is the inclusion of a realization ϵ^i of the observation error in the forward operator when updating each member via (3.48). The need for this is most easily seen by returning to the random variable ξ given by (3.45). Excluding ϵ from (3.45) leads to $\mathrm{cov}(\xi) = \mathbf{P}^a - \mathbf{KRK}^{\mathrm{T}}$, as can easily be seen following the steps of (3.46) with ϵ omitted. Drawing $\mathbf{x}^i$ with ϵ^i omitted therefore produces an analysis ensemble whose sample covariance converges to $\mathbf{P}^a - \mathbf{KRK}^{\mathrm{T}}$ as N_e increases, rather than to $\mathbf{P}^a$.

Rather than using (3.48), stochastic EnKFs in the existing literature update members according to

$$\xi^i = \mathbf{x}^i + \hat{\mathbf{K}}\left(\mathbf{y}^i - \mathbf{H}\mathbf{x}^i\right),$$

where $\mathbf{y}^i = \mathbf{y}^o + \epsilon^i$ is a 'perturbed' observation for the ith member. Of course, this form can be obtained from (3.48) by replacing ϵ^i with $-\epsilon^i$. The two forms are therefore equivalent when the observation-error distribution is symmetric. Moreover, since the updated (sample) mean and covariances depend only on the first and second moments of ϵ^i, both forms yield precisely the same updated mean and covariance (starting from the same ensembles of $\mathbf{x}^i$ and ϵ^i). Differences will appear, however, in higher moments of the analysis ensemble; for example, if ϵ has a skewed distribution, the use of (3.48) will give an analysis ensemble with a skewness of the same sign as that of the correct conditional distribution, while the perturbed-observation approach will give the opposite skewness.

3.7.5 The update as a linear combination of ensemble perturbations

Using (3.48), the analysis increment for each member is

$$\begin{aligned}\xi^i - \mathbf{x}^i &= \hat{\mathbf{K}}\left[\mathbf{y}^o - (\mathbf{H}\mathbf{x}^i + \epsilon^i)\right]\\ &= \mathbf{XY}^{\mathrm{T}}(\mathbf{YY}^{\mathrm{T}})^{-1}\left[\mathbf{y}^o - (\mathbf{H}\mathbf{x}^i + \epsilon^i)\right]\\ &= \mathbf{Xa}.\end{aligned}$$

The vector $\mathbf{a} = \mathbf{Y}^{\mathrm{T}}(\mathbf{YY}^{\mathrm{T}})^{-1}\left[\mathbf{y}^o - (\mathbf{H}\mathbf{x}^i + \epsilon^i)\right]$ has length N_e. The analysis increment therefore lies in the subspace spanned by the columns of $\mathbf{X}$, which are the perturbations of the forecast ensemble about the ensemble mean. Stated differently, $\xi^i - \mathbf{x}^i$ is a linear combination of the ensemble perturbations $\mathbf{x}^i - \hat{\mathbf{x}}^f$, with coefficients specified by the elements of $\mathbf{a}$.

This fact that the analysis increment for each member is simply a linear combination of the forecast ensemble perturbations is central to understanding several aspects of the EnKF, including (i) the computational savings afforded by the EnKF, (ii) the main obstacles in the use of the EnKF with small to moderate ensemble sizes in high-dimensional systems, and (iii) the principal technique to overcome those obstacles, namely localization of the sample covariances.

The potential computational benefits of working in the ensemble subspace are evident already. The update for the sample mean $\hat{\mathbf{x}}^a$ requires estimation of only N_e parameters, rather than N_x as in the exact Kalman filter. More important, the update of the sample covariance only requires estimating $N_e(N_e - 1)$ parameters, as opposed to $O(N_x^2)$ for the Kalman filter. Section 3.7.8 will show that the covariance update can be achieved by computing the eigenvector decomposition of a $N_e \times N_e$ matrix in the deterministic EnKF algorithm.

Difficulties for the EnKF in high-dimensional systems will be explored in Section 3.7.9, while covariance localization is discussed in Chapter 10 of this volume.

3.7.6 The EnKF as an approximation to the BLUE

The cases in which $\mathbf{y} = h(\mathbf{x}) + \epsilon$ or $\mathbf{x}$ is non-Gaussian, or both, can be handled easily. The update equations that emerge naturally in these cases are Monte Carlo approximations to the BLUE and its error covariance.

Following the linear case, let $\hat{h} = (N_e)^{-1} \sum h(\mathbf{x}^i)$ and define the square roots $\mathbf{X}$ and $\mathbf{Y}$ of the sample covariances, where $\mathbf{X}$ is as in Section 3.7.4 and

$$\mathbf{Y} = \text{matrix with columns } (N_e - 1)^{-1/2} \left[h(\mathbf{x}^i) + \epsilon^i - \hat{h} - \hat{\epsilon} \right],$$

The EnKF update follows (3.48), but uses the nonlinear observation operator $h(\mathbf{x}^i)$ in calculating the innovation for each member,

$$\xi^i = \mathbf{x}^i + \hat{\mathbf{K}} \left[\mathbf{y}^o - (h(\mathbf{x}^i) + \epsilon^i) \right]. \tag{3.49}$$

The approximate gain $\hat{\mathbf{K}}$ is again given by (3.47).

The underlying relationship to the BLUE is now obvious. For the approximate gain, $\mathbf{X}\mathbf{Y}^{\mathrm{T}}$ is a sample estimate of $\mathbf{P}_{xy} = \mathrm{cov}(\mathbf{x}, \mathbf{y})$, $\mathbf{Y}\mathbf{Y}^{\mathrm{T}}$ is a sample estimate of $\mathbf{P}_{yy} = \mathrm{cov}(\mathbf{y})$, and, for large N_e,

$$\hat{\mathbf{K}} \to \mathbf{P}_{xy} \mathbf{P}_{yy}^{-1}.$$

Thus, the sample mean converges to the BLUE:

$$\hat{\mathbf{x}}^a = \hat{\mathbf{x}}^f + \hat{\mathbf{K}}[\mathbf{y}^o - (\hat{h} + \hat{\epsilon})] \to E(\mathbf{x}) + \mathbf{P}_{xy}\mathbf{P}_{yy}^{-1}[\mathbf{y}^o - E(\mathbf{y})].$$

The stochastic EnKF with nonlinear observation operators used in estimating $\hat{\mathbf{K}}$ and in the innovation therefore provides a Monte Carlo approximation to the BLUE.

The innovation in the BLUE, which has the form $\mathbf{y}^o - E(\mathbf{y})$, involves an expectation over the joint distribution of $\mathbf{x}$ and $\mathbf{y}$ as emphasized in Section 3.4. That joint distribution reflects the effect of the observation error, via the observation equation in (3.11). This makes clear the need to include the realization of observation error as part of the predicted observation in (3.49), rather than as a perturbation to $\mathbf{y}^o$.

3.7.7 The serial EnKF algorithm

A particularly simple approach to the EnKF update is to assimilate the components of $\mathbf{y}^o$ as scalars, one at a time, and use the result from previous observations as the prior for the assimilation of the next observation. This approach may be applied in either the stochastic or deterministic forms of the EnKF. We will treat the stochastic form here.

When observation errors are componentwise-independent, calculating the conditional pdf serially is equivalent to conditioning on all components of $\mathbf{y}^o$ at once. Applying first Bayes' rule and then using the assumption that each component of the observation-error vector ϵ is independent of other components shows

$$\begin{aligned} p(\mathbf{x}|y_1^o, \ldots, y_n^o) &\propto p(y_n^o|\mathbf{x}, y_1^o, \ldots, y_{n-1}^o)p(\mathbf{x}|y_1^o, \ldots, y_{n-1}^o) \\ &= p(y_n^o|\mathbf{x})p(\mathbf{x}|y_1^o, \ldots, y_{n-1}^o), \end{aligned}$$

where the proportionality hides a normalizing factor that is independent of $\mathbf{x}$. The formula states that the pdf for $\mathbf{x}$ conditioned on the first n components of $\mathbf{y}^o$ can be obtained from the pdf $\mathbf{x}$ conditioned on the first $n-1$ components by multiplying by the likelihood $p(y_n^o|\mathbf{x})$. This is precisely the procedure to assimilate y_n^o with a prior pdf $p(\mathbf{x}|y_1^o, \ldots, y_{n-1}^o)$. Performing such assimilations serially starting from y_1^o and ending with $y_{N_y}^o$ therefore yields $p(\mathbf{x}|\mathbf{y}^o)$, demonstrating the equivalence of the serial update to conditioning on all components of $\mathbf{y}^o$ at once.

Since the Kalman filter implements Bayes' rule for linear, Gaussian systems, serial updating with the Kalman filter is also equivalent to solving the update equations once with the vector $\mathbf{y}^o$, as long as the observation operator is linear. Serial updating with the Kalman filter means applying the update equations in turn for each scalar component y_n^o of $\mathbf{y}^o$, with the forecast covariance $\mathbf{P}^f$ replaced in $\mathbf{K}$ by the updated covariance $\mathbf{P}^a$ from the previous observation component y_{n-1}^o. Finally, the serial update for the EnKF is also identical to updating with all components of $\mathbf{y}^o$ at once, since the EnKF update equations are identical to those of the Kalman filter, but with $\mathbf{P}^f$ replaced by $\hat{\mathbf{P}}^f$. This identity holds only when $\mathbf{H}$ is linear.

The arguments above would still hold if components of $\mathbf{y}^o$ were rearranged in a different order. A corollary is therefore that the ordering of the observation components in the serial update is irrelevant, again if $\mathbf{H}$ is linear.

One appeal of the serial algorithm is its simplicity. It can be concisely summarized as follows:

For each element y_m^o of $\mathbf{y}^o$:

1. Compute $h_m(\mathbf{x}^i)$ and generate ϵ_m^i, $i = 1, \ldots, N_e$
2. Compute $\hat{\mathbf{c}} = (N_e - 1)^{-1} \sum_{i=1}^{N_e} (\mathbf{x}^i - \hat{\mathbf{x}})(h_m(\mathbf{x}^i) - \hat{h}_m)$
3. Compute $\hat{d} = r + (N_e - 1)^{-1} \sum_{i=1}^{N_e} (h_m(\mathbf{x}^i) - \hat{h}_m)^2$, where $r = \mathrm{var}(\epsilon_m)$.
4. For each member $\mathbf{x}^i$:

$$\mathbf{x}^i = \mathbf{x}^i + (\hat{\mathbf{c}}/\hat{d})\left(y_m^o - h_m(\mathbf{x}^i) - \epsilon_m^i\right)$$

end

end

3.7.8 Deterministic EnKFs

Deterministic EnKFs do not attempt to generate an analysis ensemble that approximates a random sample from $p(\mathbf{x}|\mathbf{y}^o)$. Instead, they produce an analysis ensemble with the 'correct' sample mean and covariance, in the sense that the updated sample mean and covariance are consistent with the Kalman-filter update based on $\mathbf{y}^o$ and the mean and covariance from the forecast ensemble. Because of the consistency, deterministic EnKFs also converge to the Kalman filter as N_e becomes large, if the system is linear and Gaussian. These algorithms are termed deterministic because they are a deterministic function of their inputs and do not involve pseudorandom numbers or sampling. A single version of the deterministic EnKF is described below, but there are several variants that are discussed in more detail in Tippett et al. (2003).

Suppose we have a forecast ensemble $\{\mathbf{x}^i, i = 1, \ldots, N_e\}$. The first step for the deterministic EnKF is to write the Kalman-filter update for the mean, using $\mathbf{x}^f = \hat{\mathbf{x}}$ and $\mathbf{P}^f = \mathbf{X}\mathbf{X}^{\mathrm{T}}$, where $\mathbf{X}$ is given in Section 3.7.4. The update of the sample mean then agrees with the Kalman filter by construction and is given by

$$\hat{\mathbf{x}}^a = \hat{\mathbf{x}} + \hat{\mathbf{K}}(\mathbf{y}^o - \mathbf{H}\hat{\mathbf{x}}). \tag{3.50}$$

We write the updated sample covariance as

$$(N_e - 1)^{-1} \sum_{i=1}^{N_e} (\xi^i - \hat{\mathbf{x}}^a)(\xi^i - \hat{\mathbf{x}}^a)^{\mathrm{T}} = \mathbf{X}^a \mathbf{X}^{a\mathrm{T}},$$

where ξ^i are the members of the analysis ensemble and $\mathbf{X}^a$ has columns equal to the deviation of ξ^i from the ensemble mean $\hat{\mathbf{x}}^a$, scaled by $(N_e - 1)^{-1}$. Equating the updated sample covariance with the analysis covariance from the Kalman filter yields

$$\begin{aligned}
\mathbf{X}^a\mathbf{X}^{a\mathrm{T}} &= \mathbf{P}^f - \mathbf{P}^f\mathbf{H}^{\mathrm{T}}\left(\mathbf{H}\mathbf{P}^f\mathbf{H}^{\mathrm{T}} + \mathbf{R}\right)^{-1}\mathbf{H}\mathbf{P}^f \\
&= \mathbf{X}\mathbf{X}^{\mathrm{T}} - \mathbf{X}\mathbf{X}^{\mathrm{T}}\mathbf{H}^{\mathrm{T}}\left(\mathbf{H}\mathbf{X}\mathbf{X}^{\mathrm{T}}\mathbf{H}^{\mathrm{T}} + \mathbf{R}\right)^{-1}\mathbf{H}\mathbf{X}\mathbf{X}^{\mathrm{T}} \\
&= \mathbf{X}\left[\mathbf{I} - \mathbf{X}^{\mathrm{T}}\mathbf{H}^{\mathrm{T}}\left(\mathbf{H}\mathbf{X}\mathbf{X}^{\mathrm{T}}\mathbf{H}^{\mathrm{T}} + \mathbf{R}\right)^{-1}\mathbf{H}\mathbf{X}\right]\mathbf{X}^{\mathrm{T}}.
\end{aligned}$$

That relation is satisfied if $\mathbf{X}^a = \mathbf{X}\mathbf{F}$, with

$$\mathbf{F}\mathbf{F}^{\mathrm{T}} = \mathbf{I} - \mathbf{X}^{\mathrm{T}}\mathbf{H}^{\mathrm{T}}\left(\mathbf{H}\mathbf{X}\mathbf{X}^{\mathrm{T}}\mathbf{H}^{\mathrm{T}} + \mathbf{R}\right)^{-1}\mathbf{H}\mathbf{X}. \tag{3.51}$$

The covariance update therefore involves calculating the $N_e \times N_e$ matrix $\mathbf{F}$ as the square root of the right-hand side above. The columns of $\mathbf{F}$ are again coefficients for linear combinations of the forecast ensemble perturbations. They specify the linear combinations of forecast perturbations that yield each analysis perturbation, in the same way that the analysis increment in Section 3.7.5 was written in terms of a linear combination of the forecast ensemble perturbations.

Note also that $\mathbf{F}$ is determined by (3.51) only up to multiplication by an orthogonal matrix. That is, replacing $\mathbf{F}$ with $\mathbf{F}\mathbf{V}$ leaves $\mathbf{X}^a\mathbf{X}^{a\mathrm{T}}$ unchanged when $\mathbf{V}\mathbf{V}^{\mathrm{T}} = \mathbf{I}$. Thus,

$\mathbf{F}$ is not unique, and it may be chosen in many ways while retaining the consistency of the EnKF update and the Kalman-filter update.

The right-hand side of (3.51) involves the inverse of $\mathbf{HXX}^{\mathrm{T}}\mathbf{H}^{\mathrm{T}} + \mathbf{R}$. Since that matrix is $N_y \times N_y$, the inverse will be computationally demanding if there are many observations. Intuitively, however, the entire problem should reduce to $N_e \times N_e$ matrices. To avoid the use of $\left(\mathbf{HXX}^{\mathrm{T}}\mathbf{H}^{\mathrm{T}} + \mathbf{R}\right)^{-1}$, note that

$$\mathbf{I} - \mathbf{X}^{\mathrm{T}}\mathbf{H}^{\mathrm{T}}\left(\mathbf{HXX}^{\mathrm{T}}\mathbf{H}^{\mathrm{T}} + \mathbf{R}\right)^{-1}\mathbf{HX} = \left(\mathbf{I} + \mathbf{X}^{\mathrm{T}}\mathbf{H}^{\mathrm{T}}\mathbf{R}^{-1}\mathbf{HX}\right)^{-1}. \tag{3.52}$$

The latter matrix is $N_e \times N_e$ and can be inverted easily. The derivation is analogous to that given in Section 3.8 showing the equivalence of two forms for $\mathbf{P}^a$ from the Kalman filter.

3.7.9 Obstacles for the EnKF

The fundamental approximation in the EnKF is to replace the mean and covariances appearing in the Kalman filter with sample estimates. Because N_e is finite, the sample estimates used in the EnKF contain sampling error, which will affect the results of the algorithm in multiple ways.

As discussed in Section 3.7.2, the error in each element of a sample covariance matrix such as $\hat{\mathbf{P}}^f = \mathbf{XX}^{\mathrm{T}}$ is $O(N_e^{-1})$. When N_e is as large as 100, the error in an individual element of a sample covariance matrix is not huge, and indeed probably no larger than the typical error in the nearly homogeneous, isotropic and stationary covariance models employed in other assimilation schemes.

Other properties of the sample covariance suffer larger errors. The rank of $\mathbf{XX}^{\mathrm{T}}$ is at most $N_e - 1$, and thus, in a high-dimensional system with moderate N_e, almost all the eigenvalues of $\hat{\mathbf{P}}^f$ are zero. Crucially, the sampling error is correlated across entries of the sample covariance matrix. Global properties of a covariance matrix, such as its eigenvalues or eigenvectors, can therefore be estimated very poorly with moderate N_e, despite the reasonable accuracy of each entry of $\hat{\mathbf{P}}^f$.

There is little theory to quantify how sampling error affects the EnKF. Some qualitative intuition can, however, be obtained from simple examples.

As a first example, consider the EnKF update for a single, scalar observation that is unrelated to state, that is, an observation $y = h(\mathbf{x}) + \epsilon$ such that $\mathrm{cov}(\mathbf{x}, y) = 0$. (This may be achieved with $h(\mathbf{x}) = 0$ and ϵ independent of $\mathbf{x}$.) The update for the ith member is

$$\xi^i = \mathbf{x}^i + (\hat{\mathbf{c}}/\hat{d})\left[y^o - h(\mathbf{x}^i) - \epsilon^i\right],$$

where $\hat{c}$ is the sample estimate of $\mathrm{cov}(\mathbf{x}, y) = \mathrm{cov}(\mathbf{x}, h(\mathbf{x}))$ and $\hat{d}$ is the sample estimate of $\mathrm{cov}(y) = \mathrm{cov}(h(\mathbf{x}) + \epsilon)$.

The sample covariance $\hat{\mathbf{c}}$ will be non-zero because of sampling error. This will lead to noise in ξ^i, but the noise will be zero-mean and will not accumulate too rapidly as other observations are assimilated. The updated covariance is more problematic. Applying the usual Kalman-filter update, we have

$$\hat{\mathbf{P}}^a = \hat{\mathbf{P}}^f - \hat{\mathbf{c}}\hat{\mathbf{c}}^{\mathrm{T}}/\hat{d}.$$

The updated covariances decrease, despite the uninformative observation. Although the decrease in variance for any element of the state is small (decreasing as N_e^{-1}), assimilating the single observation y will systematically decrease the variance of *every* element of the state.

In real assimilation problems, $\mathrm{cov}(\mathbf{x}, y)$ will not be exactly zero. But non-zero observation-state covariances will typically be restricted to a subset of the elements of $\mathbf{x}$, often of a size much smaller than N_x. The EnKF update will therefore clearly underestimate $\hat{\mathbf{P}}^a$ systematically, and this problem will be compounded when assimilating N_y observations instead of one.

A second example provides another view of the problem. Suppose that

$$\mathbf{x} \sim N(0, \mathbf{I}), \quad \epsilon \sim N(0, \mathbf{I}), \quad \mathbf{H} = \mathbf{I}, \quad \mathbf{y} = \mathbf{x} + \epsilon.$$

Suppose also that we have an ensemble of two members, with $\delta\mathbf{x}$ the deviation of member 1 from the mean. (The deviation of the second member from the sample mean is $-\delta\mathbf{x}$.)

How will $\mathrm{tr}(\mathbf{P}^a)$ from the KF differ from $\mathrm{tr}(\hat{\mathbf{P}}^a)$ estimated by the EnKF? In order to make a fair comparison, we normalize $\delta\mathbf{x}$ so that the sample forecast covariance has the same trace as the actual forecast covariance:

$$\mathrm{tr}(\hat{\mathbf{P}}^f) = \mathrm{tr}(\delta\mathbf{x}\delta\mathbf{x}^{\mathrm{T}}) = \delta\mathbf{x}^{\mathrm{T}}\delta\mathbf{x} = \mathrm{tr}(\mathbf{P}^f) = N_x.$$

For the Kalman filter,

$$\begin{aligned}\mathbf{P}^a &= \mathbf{P}^f - \mathbf{P}^f\mathbf{H}^{\mathrm{T}}\left(\mathbf{H}\mathbf{P}^f\mathbf{H}^{\mathrm{T}} + \mathbf{R}\right)^{-1}\mathbf{H}\mathbf{P}^f \\ &= \mathbf{I} - (\mathbf{I} + \mathbf{I})^{-1} \\ &= \tfrac{1}{2}\mathbf{I},\end{aligned}$$

and therefore $\mathrm{tr}(\mathbf{P}^a) = N_x/2$. For the EnKF, the updated covariance is $\hat{\mathbf{P}}^a = \mathbf{X}^a\mathbf{X}^{a\mathrm{T}}$, with $\mathbf{X}^a = \mathbf{X}\mathbf{F}$ and $\mathbf{F}$ satisfying (3.51). In this example, $\mathbf{X} = \delta\mathbf{x}$ and $\mathbf{F} = a$, a scalar. From (3.51) and (3.52),

$$a^2 = (1 + \delta\mathbf{x}^{\mathrm{T}}\delta\mathbf{x})^{-1} = (N_x + 1)^{-1}.$$

Therefore, $\hat{\mathbf{P}}^a = (N_x + 1)^{-1}\delta\mathbf{x}\delta\mathbf{x}^{\mathrm{T}}$ and $\mathrm{tr}(\hat{\mathbf{P}}^a) = N_x/(N_x + 1)$.

Comparing results shows that the analysis variance estimated by the EnKF is a factor of $2/(N_x + 1)$ smaller than the correct analysis variance. This is of course a substantial underestimation if N_x is large.

The severe rank deficiency of $\hat{\mathbf{P}}^f$ is clearly the root cause of this underestimation. The EnKF sees all the forecast variance, equal to N_x, confined to a subspace of dimension 1 (i.e., along $\delta\mathbf{x}$). At the same time, the observations may be subjected to a rotation of coordinates to align one of the observations with $\delta\mathbf{x}$ without changing the result of the update; the observations orthogonal to $\delta\mathbf{x}$ are ignored since there is no forecast variance in those directions, while the observation along $\delta\mathbf{x}$ reduces the variance to (nearly) the observation-error variance of 1, thereby removing almost all the variance from the ensemble.

Because of these problems arising from sampling error in the covariances estimated from the ensemble, 'localization' of sample covariances is crucial to effective, high-dimensional EnKF. In essence, this technique sets the sample covariances to zero at sufficiently large separation distance (i.e. separating two state variables, or a state variable and an observation variable), thus incorporating the a priori knowledge that covariances typically decrease as spatial separation increases. The two examples in this section have illustrated two ways that the sampling error can be manifested in the EnKF covariance update, both of which lead to substantial underestimation of $\mathbf{P}^a$ by $\hat{\mathbf{P}}^a$. Localization addresses both these manifestations of sampling error, since it directly removes many small and potentially noisy entries from the sample covariance matrices and also (greatly) increases the rank of the covariance matrices implicitly used in $\hat{\mathbf{K}}$, although those matrices are never explicitly constructed. A more complete discussion of this crucial topic appears in Chapter 10.

3.8 Assorted derivations and identities

- Two equivalent forms of the gain matrix $\mathbf{K}$:

$$\begin{aligned}
\mathbf{P}_k^a\mathbf{H}^{\mathrm{T}}\mathbf{R}^{-1} &= \mathbf{P}_k^a\mathbf{H}^{\mathrm{T}}\mathbf{R}^{-1}(\mathbf{H}\mathbf{P}_k^f\mathbf{H}^{\mathrm{T}}+\mathbf{R})(\mathbf{H}\mathbf{P}_k^f\mathbf{H}^{\mathrm{T}}+\mathbf{R})^{-1}\\
&= \mathbf{P}_k^a(\mathbf{H}^{\mathrm{T}}\mathbf{R}^{-1}\mathbf{H}\mathbf{P}_k^f\mathbf{H}^{\mathrm{T}}+\mathbf{H}^{\mathrm{T}})(\mathbf{H}\mathbf{P}_k^f\mathbf{H}^{\mathrm{T}}+\mathbf{R})^{-1}\\
&= \mathbf{P}_k^a[\mathbf{H}^{\mathrm{T}}\mathbf{R}^{-1}\mathbf{H}+\left(\mathbf{P}_k^f\right)^{-1}]\,\mathbf{P}_k^f\mathbf{H}^{\mathrm{T}}(\mathbf{H}\mathbf{P}_k^f\mathbf{H}^{\mathrm{T}}+\mathbf{R})^{-1}\\
&= \mathbf{P}_k^f\mathbf{H}^{\mathrm{T}}(\mathbf{H}\mathbf{P}_k^f\mathbf{H}^{\mathrm{T}}+\mathbf{R})^{-1}.
\end{aligned}$$

- Two equivalent forms for $\mathbf{P}_k^a$, without appealing to Sherman–Morrison–Woodbury:

$$\begin{aligned}
\left[\left(\mathbf{P}_k^f\right)^{-1}+\mathbf{H}^{\mathrm{T}}\mathbf{R}^{-1}\mathbf{H}\right]^{-1} &= \left[\left(\mathbf{P}_k^f\right)^{-1}+\mathbf{H}^{\mathrm{T}}\mathbf{R}^{-1}\mathbf{H}\right]^{-1}\\
&\quad\left(\mathbf{I}+\mathbf{H}^{\mathrm{T}}\mathbf{R}^{-1}\mathbf{H}\mathbf{P}_k^f-\mathbf{H}^{\mathrm{T}}\mathbf{R}^{-1}\mathbf{H}\mathbf{P}_k^f\right)\\
&= \left[\left(\mathbf{P}_k^f\right)^{-1}+\mathbf{H}^{\mathrm{T}}\mathbf{R}^{-1}\mathbf{H}\right]^{-1}\\
&\quad\left\{\left[\left(\mathbf{P}_k^f\right)^{-1}+\mathbf{H}^{\mathrm{T}}\mathbf{R}^{-1}\mathbf{H}\right]\mathbf{P}_k^f-\mathbf{H}^{\mathrm{T}}\mathbf{R}^{-1}\mathbf{H}\mathbf{P}_k^f\right\}\\
&= \mathbf{P}_k^a\left[(\mathbf{P}_k^a)^{-1}\mathbf{P}_k^f-\mathbf{H}^{\mathrm{T}}\mathbf{R}^{-1}\mathbf{H}\mathbf{P}_k^f\right]\\
&= \mathbf{P}_k^f-\mathbf{P}_k^a\mathbf{H}^{\mathrm{T}}\mathbf{R}^{-1}\mathbf{H}\mathbf{P}_k^f\\
&= (\mathbf{I}-\mathbf{K}\mathbf{H})\,\mathbf{P}_k^f\\
&= \mathbf{P}_k^f-\mathbf{P}_k^f\mathbf{H}^{\mathrm{T}}\left(\mathbf{H}\mathbf{P}_k^f\mathbf{H}^{\mathrm{T}}+\mathbf{R}\right)^{-1}\mathbf{H}\mathbf{P}_k^f.
\end{aligned}$$

- Helpful identities: $\mathbf{x}$ and $\mathbf{y}$ are random vectors, and $\mathbf{A}$ and $\mathbf{b}$ are deterministic.
- The derivative with respect to a deterministic variable commutes with the expectation:

$$\frac{\partial}{\partial \mathbf{A}} E(g(\mathbf{x}, \mathbf{y}, \mathbf{A})) = E\left(\frac{\partial g}{\partial \mathbf{A}}\right)$$

- Derivatives of products and traces:

$$\begin{aligned}
\frac{\partial \mathbf{x}^{\mathrm{T}}\mathbf{b}}{\partial \mathbf{b}} = \frac{\partial \mathbf{b}^{\mathrm{T}}\mathbf{x}}{\partial \mathbf{b}} &= \mathbf{x}, \\
\frac{\partial \mathbf{x}^{\mathrm{T}}\mathbf{A}\mathbf{y}}{\partial \mathbf{A}} &= \mathbf{x}\mathbf{y}^{\mathrm{T}}, \\
\frac{\partial \mathbf{y}^{\mathrm{T}}\mathbf{A}^{\mathrm{T}}\mathbf{A}\mathbf{y}}{\partial \mathbf{A}} &= 2\mathbf{A}\mathbf{y}\mathbf{y}^{\mathrm{T}}, \\
\frac{\partial}{\partial \mathbf{A}}\left[\mathrm{tr}\left(\mathbf{A}\mathbf{y}\mathbf{y}^{\mathrm{T}}\mathbf{A}^{\mathrm{T}}\right)\right] &= 2\mathbf{A}\mathbf{y}\mathbf{y}^{\mathrm{T}}.
\end{aligned}$$

References

Anderson, B. D. O. and Moore, J. B. (1979). *Optimal Filtering.* Prentice-Hall, Englewood Cliffs, NJ.

Anderson, J. L. and Collins, N. (2007). Scalable implementations of ensemble filter algorithms for data assimilation. *J. Atmos. Ocean. Tech.*, **24**, 1452–1463.

Cohn, S. E. (1997). An introduction to estimation theory. *J. Meteor. Soc. Japan*, **75**, 257–288.

Evensen, G. (1994). Sequential data assimilation with a nonlinear quasigeostrophic model using monte carlo methods to forecast error statistics. *J. Geophys. Res.*, **99**, 10143–10162.

Fukumori, I. and Malanotte-Rizzoli, P. (1995). An approximate Kalman filter for ocean data assimilation: an example with an idealized Gulf stream model. *J. Geophys. Res.*, **100**, 6777–6793.

Gelb, A. (1974). *Applied Optimal Estimation.* MIT Press, Cambridge, MA.

Houtekamer, P. L. and Mitchell, H. L. (1998). Data assimilation using an ensemble Kalman filter technique. *Mon. Weather Rev.*, **126**, 796–811.

Ide, K., Courtier, P., Ghil, M., and Lorenc, A. C. (1997). Unified notation for data assimilation: Operational, sequential and variational. *J. Meteor. Soc. Japan*, **75**, 181–189.

Jazwinski, A. H. (1970). *Stochastic Processes and Filtering Theory.* Academic Press, New York.

Kalman, R. E. (1960). A new approach to linear filter and prediction problems. *J. Basic. Eng.*, **82**, 35–45.

Legras, B. and Vautard, R. (1995). A guide to liapunov vectors. In *Proceedings of Seminar on Predictability*, ECMWF, Reading, Berkshire, UK, Vol. 1, pp. 143–156.

Lermusiaux, P. F. J. and Robinson, A. R. (1999). Data assimilation via error subspace statistical estimation. Part I: Theory and schemes. *Mon. Weather Rev.*, **127**, 1385–1407.

Lorenc, A. C. (1986). Analysis methods for numerical weather prediction. *Q. J. R. Meteorol. Soc.*, **112**, 1177–1194.

Lorenz, E. (1995). Predictability: a problem partly solved. In *Proceedings of Seminar on Predictability*, ECMWF, Reading, Berkshire, UK, Vol. 1, pp. 1–18.

Pham, D. T., Verron, J., and Roubaud, M. C. (1998). A singular evolutive extended Kalman filter for data assimilation in oceanography. *J. Marine. Syst.*, **16**, 323–340.

Tippett, M. K., Anderson, J. L., Bishop, C. H., Hamill, T. M., and Whitaker, J. S. (2003). Ensemble square root filters. *Mon. Weather Rev.*, **131**, 1485–1490.

4
Smoothers

E. Cosme

Laboratoire de Glaciologie et Géophysique de l'Environnement
Université Grenoble Alpes, CNRS
Grenoble, France

Advanced Data Assimilation for Geosciences. First Edition.
Edited by É. Blayo, M. Bocquet, E. Cosme, and L. F. Cugliandolo.

Chapter Contents

4.1 Introduction

As seen in Chapter 3, the *filtering* problem consists in estimating the system state at one single time, based on present and past observations. The probability density we aim to estimate reads

$$p(\mathbf{x}_k|\mathbf{y}_{1:k}), \tag{4.1}$$

where $\mathbf{y}_{1:k}$ denotes the time series of observations $\{\mathbf{y}_1, \ldots, \mathbf{y}_k\}$. When the state, or the sequence of states, we intend to estimate is posterior to the observations, it is referred to as the *forecasting* problem. Looking for

$$p(\mathbf{x}_{k+k'}|\mathbf{y}_{1:k}) \qquad \text{or} \qquad p(\mathbf{x}_{k+1:k+k'}|\mathbf{y}_{1:k}), \tag{4.2}$$

with $k' \geq 1$, are the main forms of the forecasting problem. Any problem other than filtering or forecasting involves the estimation of a state or a sequence of states based on future observations, and is called a *smoothing* problem. The most commonly encountered formulations, referred to simply as the *smoothing problems* later in the chapter, are the following:

- $p(\mathbf{x}_{k'}|\mathbf{y}_{1:k})$, with $0 \leq k' < k$: the *marginal* or *fixed-point* smoothing problem. Only one state is estimated, based on past, present, and future observations. When $k' = 0$, this is the initial state estimation problem, where all observations are posterior to the estimated state.
- $p(\mathbf{x}_{0:k}|\mathbf{y}_{1:k})$: the *joint* smoothing problem of *fixed-interval* type. All the states within a time interval are estimated based on the observations contained in the same interval.
- $p(\mathbf{x}_{k-L:k}|\mathbf{y}_{1:k})$: the *joint* smoothing problem of *fixed-lag* type. A limited number of states, including the last one, are estimated based on the observations of the time interval. L is called the *lag*.

Each formulation responds to a specific question. The fixed-lag smoother, for instance, is standard for reanalysis issues. The fixed-interval smoother can be invoked to estimate unknown parameters based on observations intensively collected over a limited time range. Further details are given in Section 4.5.

In Section 4.2, Bayesian solutions of the previously introduced smoother formulations are derived. The focus is on those applicable to moderate- to high-dimensional geophysical problems, in which a non-invertible dynamical model is generally involved. In this case, the filtering and forecasting problems have a self-evident solution consisting of model integrations interrupted with observational updates of the state estimate. This is how Kalman, ensemble, and particle filters proceed. To produce a forecast, the model is integrated further after the last observational update. No such evident process exists for the smoothing problems (not to mention the fact that there are several smoothing problem formulations), which has led to a wide variety of algorithms (Cosme et al., 2012).

The linear Gaussian and ensemble forms of smoothers are described in Sections 4.3 and 4.4, respectively. Again, the discussion is limited to algorithms applicable to high-dimensional problems: the *sequential* smoother, the *two-step* smoother, and the *forward–backward* smoother, also known as the RTS (Rauch–Tung–Striebel) smoother.

4.2 Smoothing algorithms in a Bayesian framework

A dynamical model yields an estimate of the state at time k, $\mathbf{x}_k$, starting from an estimate at an earlier time $k-1$, $\mathbf{x}_{k-1}$. In statistical terminology, a dynamical model provides the conditional density $p(\mathbf{x}_k|\mathbf{x}_{k-1})$, sometimes called the *transition* density. If the model is 'good', then this density is peaked; if the model f is perfect, then $p(\mathbf{x}_k|\mathbf{x}_{k-1}) = \delta\left(\mathbf{x}_k - f(\mathbf{x}_{k-1})\right)$. Considering the general sparseness of geophysical observations, the use of the model can rarely be circumvented in data assimilation, and the Bayesian solutions of the smoothing problems must involve the transition densities. Not surprisingly, the observation densities, of the form $p(\mathbf{y}_k|\mathbf{x}_k)$, must be involved as well, since they are the additional source of reliable information. A prior density for the initial state, $p(\mathbf{x}_0)$, forms the third piece of information.

4.2.1 Joint smoothing problems

Applying Bayes' rule and using the hidden Markov chain assumptions (Wikle and Berliner, 2007; Section 3.1.3 in Chapter 3 in this volume), the probability density of the fixed-interval joint smoothing problem can be decomposed as the product of the initial ingredients of the estimation problem, namely the background, transition, and observation densities:

$$p(\mathbf{x}_{0:k}|\mathbf{y}_{1:k}) \propto p(\mathbf{x}_0) \prod_{k'=1}^{k} p(\mathbf{x}_{k'}|\mathbf{x}_{k'-1}) \prod_{k'=1}^{k} p(\mathbf{y}_{k'}|\mathbf{x}_{k'}). \tag{4.3}$$

The fixed-interval joint smoothers (sequential and two-step) differ only in the order in which the multiplications of densities are performed. Two methods prevail. The first alternates multiplications of transition and observation densities, starting from the background. The density sought at time k is written as a function of its counterpart at time $k-1$:

$$p(\mathbf{x}_{0:k}|\mathbf{y}_{1:k}) \propto p(\mathbf{x}_{0:k-1}|\mathbf{y}_{1:k-1}) p(\mathbf{x}_k|\mathbf{x}_{k-1}) p(\mathbf{y}_k|\mathbf{x}_k). \tag{4.4}$$

In practical problems, model integrations alternate with observational updates, as with the filter. It is thus referred to as the *sequential* smoother. The operations of the sequential smoother are schematically reproduced in Fig. 4.1.

The second method first multiplies the transition densities with the background and then multiplies the observation densities. After the former step and before the latter, the intermediate formulation of the problem is

$$p(\mathbf{x}_{0:k}|\mathbf{y}_{1:k}) \propto p(\mathbf{x}_{0:k}) \prod_{k'=1}^{k} p(\mathbf{y}_{k'}|\mathbf{x}_{k'}). \tag{4.5}$$

In practical problems, this is implemented by a model integration over the whole time interval, followed by an observational update using all observations. These can be

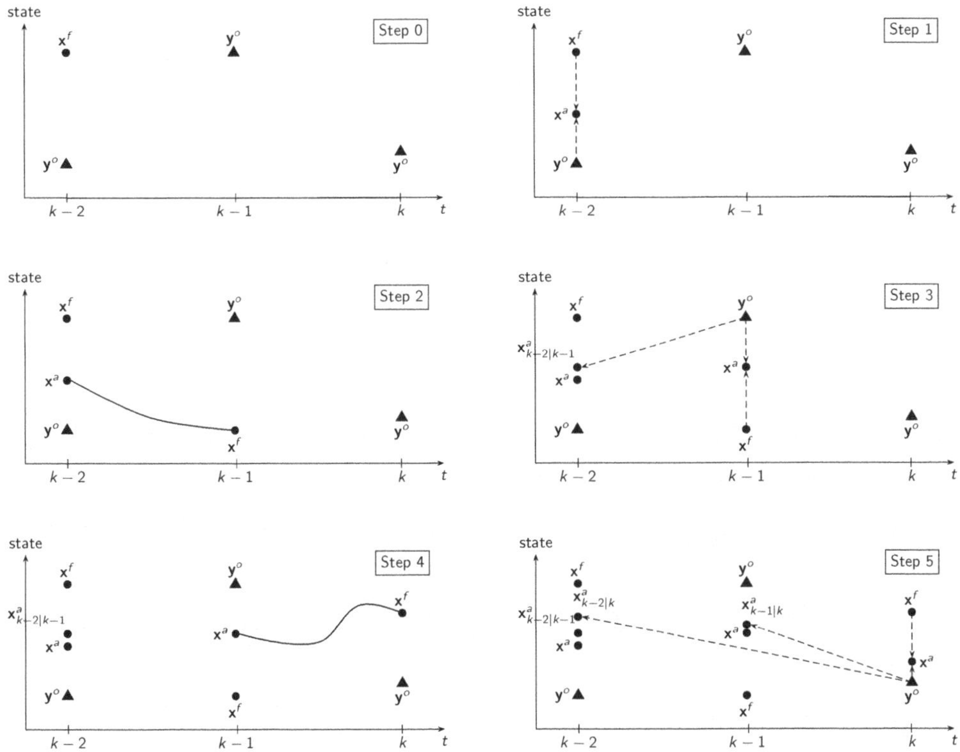

Fig. 4.1 Modus operandi of the sequential smoother. Step 0: the forecast at time $k-2$ is available; Step 1: Filter analysis at time $k-2$; Step 2: Forecast to time $k-1$; Step 3: Filter analysis at time $k-1$ and smoother analysis at time $k-2$; Step 4: Forecast to time k; Step 5: Filter analysis at time k and smoother analyses at times $k-2$ and $k-1$.

processed one at a time or globally. We call this algorithm the *two-step smoother*, the operations of which are depicted in Fig. 4.2.

Equation (4.4) differs from the filter (equations (3.5) and (3.6) in Chapter 3) only by the absence of marginalization (integration) over $\mathbf{x}_{k-1}$, the effect of which would remove $\mathbf{x}_{k-1}$ from the augmented state $\mathbf{x}_{0:k}$ to be estimated. As this marginalization is omitted from the start ($k-1=0$), all the states prior to k remain in the augmented state. If this marginalization operation is *lagged*, meaning it occurs for $k-L-1$ for all k, the *fixed-lag* smoother is obtained:

$$p(\mathbf{x}_{k-L:k}|\mathbf{y}_{1:k}) = p(\mathbf{x}_k|\mathbf{x}_{k-1}) \int p(\mathbf{x}_{k-L-1:k-1}|\mathbf{y}_{1:k-1})\,\mathrm{d}\mathbf{x}_{k-L-1} p(\mathbf{y}_k|\mathbf{x}_k). \tag{4.6}$$

The lagged formulation appears as a smoother localized in time. Each state estimate is affected by future observations up to L time steps.

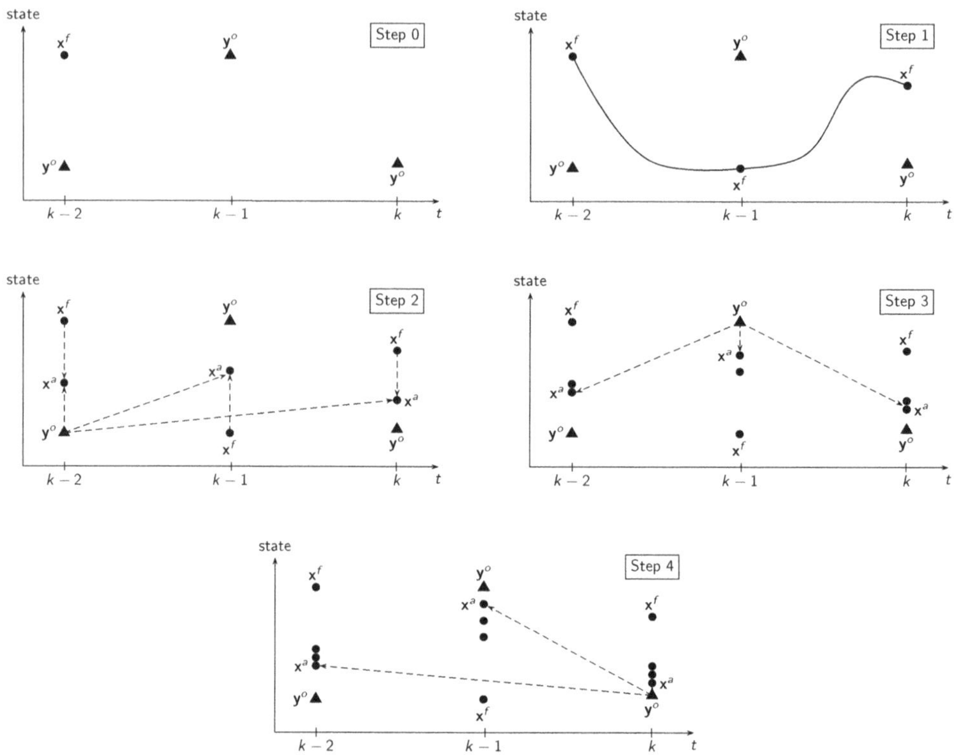

Fig. 4.2 Modus operandi of the two-step smoother. Step 0: the forecast at time $k-2$ is available; Step 1: Forecast from time $k-2$ to times $k-1$ and k; Step 2: Smoother analysis of all states using observations from time $k-2$; Step 3: Smoother analysis of all states using observations from time $k-1$; Step 4: Smoother analysis of all states using observations from time k.

4.2.2 Marginal smoothing problem

The marginal smoothing problem ($p(\mathbf{x}_{k'}|\mathbf{y}_{1:k})$, with $0 \leq k' < k$) focuses on estimating one single state, like the filter. Thus, it is tempting to search for a filter-like sequential solution algorithm. Owing to the presence of future observations in the conditional statement, the state at time $k'+1$ is introduced to estimate the state at k':

$$p(\mathbf{x}_{k'}|\mathbf{y}_{1:k}) = \int p(\mathbf{x}_{k'}|\mathbf{x}_{k'+1}, \mathbf{y}_{1:k})p(\mathbf{x}_{k'+1}|\mathbf{y}_{1:k})\,\mathrm{d}\mathbf{x}_{k'+1}. \tag{4.7}$$

The second term in the integral is the smoothing solution at $k'+1$. However, the first term is not trivial. In particular, it is not the exact *backward* transition density, $p(\mathbf{x}'_k|\mathbf{x}_{k'+1})$, which it would be possible to use if the dynamical model satisfied certain conditions (reversibility at first). This term must be further decomposed to exhibit

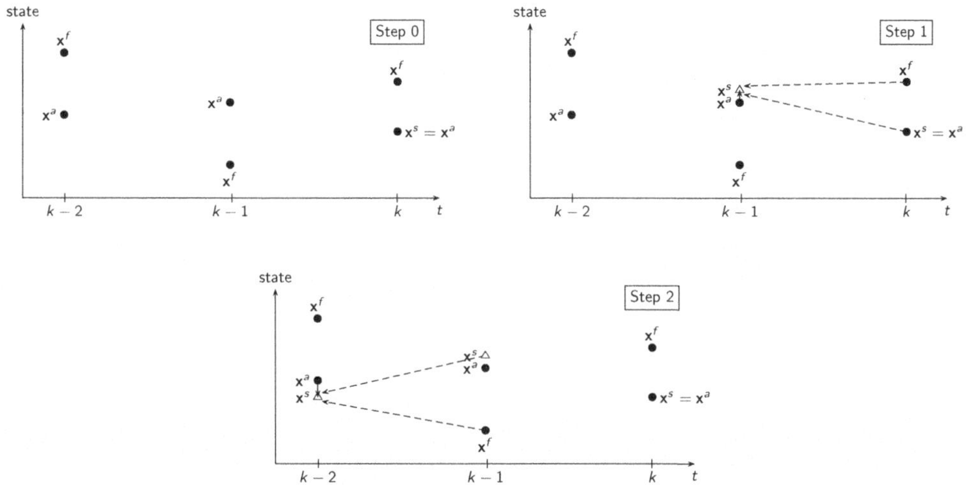

Fig. 4.3 Modus operandi of the forward–backward smoother. Step 0: the filter forecasts and analyses are available at all times after a filter pass. The smoother analysis at time k is initialized with the filter analysis; Step 1: The smoother analysis at time $k-1$ is calculated; Step 2: The smoother analysis at time $k-2$ is calculated.

known densities. This leads to

$$p(\mathbf{x}_{k'}|\mathbf{y}_{1:k}) = p(\mathbf{x}_{k'}|\mathbf{y}_{1:k'}) \int \frac{p(\mathbf{x}_{k'+1}|\mathbf{x}_{k'})}{p(\mathbf{x}_{k'+1}|\mathbf{y}_{1:k'})} p(\mathbf{x}_{k'+1}|\mathbf{y}_{1:k})\, \mathrm{d}\mathbf{x}_{k'+1}, \tag{4.8}$$

in which one recognizes, in order of appearance, the analysis filter estimate at k', a transition density (numerator), the forecast filter estimate at $k'+1$ (denominator), and the smoother estimate at $k'+1$. The filter has to be run first on the interval $[0, k]$; then the smoother estimate is obtained by a backward pass with (4.8). This algorithm is the *forward-backward* smoother. The operations of this smoother are reproduced schematically in Fig. 4.3.

Another possible decomposition for the marginal smoothing density is

$$p(\mathbf{x}_{k'}|y_{1:k}) \propto p(\mathbf{x}_{k'}|\mathbf{y}_{1:k'}) p(\mathbf{y}_{k'+1:k}|\mathbf{x}_{k'}, \mathbf{y}_{1:k'}). \tag{4.9}$$

However, this formulation is much more difficult to implement in a realistic data assimilation setting than the others, and suffers from the lack of an ensemble form. It has not so far been used for high-dimensional geophysical problems. For these reasons, it is not further developed here. Details can be found in Cosme et al. (2012).

4.3 Linear Gaussian smoothers

4.3.1 Notation

With the smoothers, in contrast to the filters, an analysis at a time i is not necessarily conditioned to observations $\mathbf{y}_{1:i}$ only, but can be conditioned to various sets of observations. It is thus necessary to explicitly mention the indices of the observations that have been taken into account. The smoother analysis mean at time i that contains the information from observations up to $k > i$ is written $\mathbf{x}^a_{i|k}$, with the corresponding covariance matrix being $\mathbf{P}^a_{i|k}$. The cross-covariance matrix (defined in (4.10) below) between the smoother analysis at time i and the filter forecast at time k, both conditioned to observations $\mathbf{y}_{1:k-1}$, is written $\mathbf{P}^{af}_{i,k|k-1}$.

4.3.2 Cross-covariance matrices

Perhaps the predominant difference between filters and smoothers lies in the use of *cross-time* covariance matrices (often shortened to *cross-covariance* matrices). Consider $\mathbf{x}_i$ and $\mathbf{x}_k$, states of the dynamical system at two different times indexed by i and k, and form the augmented state vector $(\mathbf{x}_i^{\mathrm{T}}, \mathbf{x}_k^{\mathrm{T}})^{\mathrm{T}}$. Note that $(\mathbf{x}_i^{b\,\mathrm{T}}, \mathbf{x}_k^{b\,\mathrm{T}})^{\mathrm{T}}$ is an estimate of this vector and $\mathbf{e}^b = (\mathbf{e}_i^{b\,\mathrm{T}}, \mathbf{e}_k^{b\,\mathrm{T}})^{\mathrm{T}}$ is the error. The covariance matrix is written as

$$\mathbf{P}^b = E[\mathbf{e}^b \mathbf{e}^{b\,\mathrm{T}}] = \begin{pmatrix} E\left[\mathbf{e}_i^b \mathbf{e}_i^{b\,\mathrm{T}}\right] & E\left[\mathbf{e}_i^b \mathbf{e}_k^{b\,\mathrm{T}}\right] \\ E\left[\mathbf{e}_k^b \mathbf{e}_i^{b\,\mathrm{T}}\right] & E\left[\mathbf{e}_k^b \mathbf{e}_k^{b\,\mathrm{T}}\right] \end{pmatrix} = \begin{pmatrix} \mathbf{P}_i^b & \mathbf{P}_{i,k}^b \\ \mathbf{P}_{k,i}^b & \mathbf{P}_k^b \end{pmatrix}, \tag{4.10}$$

where $\mathbf{P}_i^b$ and $\mathbf{P}_k^b$ are the covariance matrices for the estimates at times i and k, respectively, and $\mathbf{P}_{k,i}^b$ and $\mathbf{P}_{i,k}^b$ are the cross-covariance matrices, which satisfy $\mathbf{P}_{k,i}^b = \mathbf{P}_{i,k}^{b\;\mathrm{T}}$. In contrast to covariance matrices, cross-covariances matrices are generally not symmetric. If $\mathbf{x}_k$ is observed with mean $\mathbf{y}_k^o$, covariance $\mathbf{R}_k$, and observation operator $\mathbf{H}_k$, then a straightforward Kalman filter analysis (Chapter 3, Section 3.2.5) with the augmented state vector updates the estimate $\mathbf{x}_i^b$ as

$$\mathbf{x}^a_{i|k} = \mathbf{x}_i^b + \mathbf{P}_{i,k}^b \mathbf{H}_k \left(\mathbf{H}_k \mathbf{P}_k^b \mathbf{H}_k^{\mathrm{T}} + \mathbf{R}_k\right)^{-1} (\mathbf{y}_k^o - \mathbf{H}_k \mathbf{x}_k^b). \tag{4.11}$$

Equation (4.11) illustrates how the cross-covariance matrix is used to perform a retrospective analysis, that is, an updated estimation of a state based on future observations (assuming $i < k$). This equation is actually used for the sequential smoothers, but all the smoother algorithms described in the following rely on the use of cross-covariance matrices, in different ways.

4.3.3 Joint or marginal? The Gaussian exception

The Gaussian density exhibits many specific and often nice properties. One is the following: if the random, augmented state vector $(\mathbf{x}_i^{\mathrm{T}}, \mathbf{x}_k^{\mathrm{T}})^{\mathrm{T}}$ is Gaussian-distributed with mean $(\mathbf{x}_i^{b\,\mathrm{T}}, \mathbf{x}_k^{b\,\mathrm{T}})^{\mathrm{T}}$ and covariance matrix given by (4.10), then the vector $\mathbf{x}_i$ alone is also Gaussian-distributed, with mean $\mathbf{x}_i^b$ and covariance matrix $\mathbf{P}_i^b$. In other words, it is straightforward to obtain marginal smoother estimates from a joint smoother estimate. Reciprocally, it is possible to determine a joint smoothing solution from a series of marginal smoothing solutions, if a way exists to compute the cross-covariance matrices (typically, if the estimates are represented by samples). This property is implicitly invoked to refer to the *equivalence* of the different smoothers in the linear Gaussian framework.

4.3.4 Sequential smoothers

Sequential smoothers implement (4.4) and (4.6). However, considering the Gaussian property mentioned previously, they can also be used for fixed-point smoothing. The algorithm is the same for both. Sequential smoothers run along with a filter. When the filter processes the observations at time k to update the state estimate at time k, the smoother uses the same observations to update the state estimates at times $i < k$ with (4.11).

In the standard formulation, the sequential smoother equations are those of the Kalman filter, replaced/completed with the following:

Forecast:

$$\mathbf{P}^{fa}_{k,i|k-1} = \mathbf{M}_k \mathbf{P}^{aa}_{k-1,i|k-1}, \quad i \in \Sigma_k, \tag{4.12a}$$

$$\mathbf{P}^{f}_{k|k-1} = \mathbf{M}_k (\mathbf{P}^{fa}_{k,k-1|k-1})^{\mathrm{T}} + \mathbf{Q}_k; \tag{4.12b}$$

smoother analysis:

$$\mathbf{K}_{i|k} = \left(\mathbf{H}_k \mathbf{P}^{fa}_{k,i|k-1}\right)^{\mathrm{T}} \left(\mathbf{H}_k \mathbf{P}^{f}_{k|k} \mathbf{H}_k^{\mathrm{T}} + \mathbf{R}_k\right)^{-1}, \quad i \in \Sigma_k, \tag{4.13a}$$

$$\mathbf{x}^a_{i|k} = \mathbf{x}^a_{i|k-1} + \mathbf{K}_{i|k} \left(\mathbf{y}^o_k - \mathbf{H}_k \mathbf{x}^f_{k|k-1}\right), \quad i \in \Sigma_k, \tag{4.13b}$$

$$\mathbf{P}^{aa}_{k,i|k} = \left(\mathbf{I} - \mathbf{K}_{k|k} \mathbf{H}_k\right) \mathbf{P}^{fa}_{k,i|k-1}, \quad i \in \Sigma_k, \tag{4.13c}$$

$$\mathbf{P}^a_{i|k} = \mathbf{P}^a_{i|k-1} - \mathbf{K}_{i|k} \mathbf{H}_k \mathbf{P}^{fa}_{k,i|k-1}, \quad i \in \Sigma_k. \tag{4.13d}$$

The ensemble Σ_k determines the type of smoother:

- Σ_k is a singleton: *fixed-point* type, for example, $\Sigma_k = \{0\}$;
- $\Sigma_k = \{0, 1, \ldots\ M-1, M\}$, with M fixed: *fixed-interval* type;
- $\Sigma_k = \{k-L, \ldots\ k-1\}$, with L fixed: *fixed-lag* type.

Note that in the full-rank formulation, the number of propagations of cross-covariance matrices, (4.12a), grows linearly with the cardinal of Σ_k (the number of smoother analyses to perform), which leads to a dramatic increase in computational complexity with respect to the filter, as illustrated in Table 4.1.

4.3.5 Two-step smoother

This smoother implements (4.5). After forming a background augmented state vector $\hat{\mathbf{x}}^b = (\mathbf{x}_0^{b\,\mathrm{T}}, \ldots, \mathbf{x}_k^{b\,\mathrm{T}})^{\mathrm{T}}$ with an associated covariance matrix, an analysis is reformed with all observations available in the time interval. This can be done globally, applying the best linear unbiased estimate (BLUE) equations with $\hat{\mathbf{x}}^b$ and the augmented observation vector, or sequentially, performing a BLUE analysis with observations $\mathbf{y}_1$

Table 4.1 Approximate computational complexities of the filter and the linear Gaussian smoothers

METHOD	FORECAST	ANALYSIS	STORAGE
Filter	$(2N_x + 1)kC_m$	$N_x kC_h + {N_x}^2 N_y k$	$2k(N_x + 1)S$
Sequential, fixed-point	$F + N_x(k-1)C_m$	$2F$	$\frac{3}{2}F$
Sequential, fixed-lag	$F + (L-1)kN_x C_m$	LF	$\left(\frac{L}{2}+1\right)F$
Sequential, fixed-interval	$F + (k-1)\frac{k}{2}N_x C_m$	$\frac{k+3}{2}F$	$\frac{k+5}{4}F$
Two-step	$F + (k-1)\frac{k}{2}N_x C_m$	kF	$\frac{k+1}{2}F$
Forward–backward	F	$F + 4{N_x}^3$	$2F$

In each column, F refers to the filter computational complexity seen on the first line. The rest of the notation is N_x, number of variables; N_y, number of observations, $\ll N_x$, constant through time; C_m, model cost ($= {N_x}^2$ in the linear case); C_h, observation operator cost ($= N_x N_y$ in the linear case); S, storage volume for one state vector; k, number of analysis cycles in the time interval; L: Lag, $\ll k$.

to $\mathbf{y}_k$ one by one. As with the sequential smoothers, forming all the cross-covariance matrices is computationally very expensive in the full-rank formulation.

4.3.6 Forward–backward smoother

This smoother was first introduced in the linear Gaussian framework by Rauch, Tung, and Striebel (1965), and is often called the RTS smoother. The process begins with a Kalman-filter pass over the time interval. Then, the smoothed estimates are recursively calculated with a backward pass, using the following equations, initialized with $\mathbf{x}^s_{k|k} = \mathbf{x}^a_{k|k}$ and $\mathbf{P}^s_{k|k} = \mathbf{P}^a_{k|k}$:

$$\mathbf{K}^s_{k'} = \mathbf{P}^a_{k'|1:k'}\mathbf{M}^{\mathrm{T}}_{k',k'+1}\left(\mathbf{P}^f_{k'+1|1:k'}\right)^{-1}, \tag{4.14a}$$

$$\mathbf{x}^s_{k'|1:k} = \mathbf{x}^a_{k'|1:k'} + \mathbf{K}^s_{k'}\left(\mathbf{x}^s_{k'+1|1:k} - \mathbf{x}^f_{k'+1|1:k'}\right), \tag{4.14b}$$

$$\mathbf{P}^s_{k'|1:k} = \mathbf{P}^a_{k'|1:k'} - \mathbf{K}^s_{k'}\left(\mathbf{P}^f_{k'+1|1:k'} - \mathbf{P}^s_{k'+1|1:k}\right)\mathbf{K}^{s\mathrm{T}}_{k'}. \tag{4.14c}$$

Compared with the two previous algorithms, for this algorithm, no cross-covariance matrix needs to be propagated. However, the inverses of the filter forecast matrices are necessary, and obtaining them quickly becomes unaffordable for larger dimensions of the state vector.

4.4 Ensemble smoothers

We adopt the stochastic EnKF viewpoint (see Chapter 3, Section 3.7.4) to describe the ensemble smoothers. The other formulations are briefly discussed at the end of this section.

4.4.1 Cross-covariance matrices

In ensemble methods, the covariance matrices are calculated from samples of anomalies (or perturbations; see Chapter 3, Section 3.7.5). As suggested by (4.10), a cross-covariance matrix can also be calculated using samples of error, represented by anomalies. If $\mathbf{X}^f_{k|k-1}$ represents the forecast ensemble anomalies (where the subscript explicitly denotes the conditioning) and $\mathbf{X}^a_{i|k-1}$ represents the smoother analysis ensemble anomalies at time i, then the corresponding cross-covariance matrix can be calculated as

$$\mathbf{P}^{fa}_{k,i|k-1} = \mathbf{X}^f_{k|k-1}\mathbf{X}^a_{i|k-1}{}^{\mathrm{T}}. \tag{4.15}$$

This ensemble formulation leads to considerable benefits in terms of computational complexity. Where a large number of covariance matrices are propagated in the full-rank case, (4.12a), instead, with the ensemble formulation, no additional matrix propagation is required with respect to the filter.

4.4.2 Sequential smoothers

The EnKF analysis equation for member m is written as

$$\mathbf{x}^{a,m}_{k|k} = \mathbf{x}^{f,m}_{k|k-1} + \mathbf{X}^f_{k|k-1}\gamma^m_{k|k-1}, \tag{4.16}$$

where $\gamma^m_{k|k-1}$ has been defined in Section 3.7.5 in Chapter 3, where it was denoted for short by $\mathbf{a}$. Note that only information from time k (filter forecast and observations) is used to compute these terms. Rewriting the full-rank sequential smoother equations in the ensemble formalism leads to the simple smoother analysis equation, for member m:

$$\mathbf{x}^{a,m}_{i|1:k} = \mathbf{x}^{a,m}_{i|1:k-1} + \mathbf{X}^a_{i|1:k-1}\gamma^m_{k|k-1}. \tag{4.17}$$

The smoother solution for time i conditioned to observations $\mathbf{y}_{1:k}$ appears as a simple update of the previous smoother solution at the same time i, conditioned to $\mathbf{y}_{1:k-1}$, involving the corresponding ensemble anomalies and the same low-dimensional vector $\gamma^m_{k|k-1}$ as the filter. The sequential smoothers thus require a very limited number of additional operations with respect to the filter.

4.4.3 Two-step smoother

Since only ensemble anomalies at each time step are required to obtain the cross-covariance matrices, the first step of the ensemble two-step smoother, introduced by van Leeuwen and Evensen (1996), consists in propagating the ensemble over the whole time interval. The second step is the analysis. A strategy exists to perform the analysis of all observations at the same time (van Leeuwen and Evensen, 1996; Cosme et al., 2012). Another approach, in which the observations are processed sequentially, is possible, and it is probably preferable because it basically uses the equations of the sequential smoothers, which are straightforward to implement. Denoting by $\mathbf{x}_i^{b,m}$ the mth member at time i of the initial background ensemble forecast, observations at time 1 are first used to provide the smoother estimates $\mathbf{x}_{i|1}^{a,m}$. Observations are then recursively introduced to provide the smoother solutions $\mathbf{x}_{i|1:k'}^{a,m}$, until $k' = k$, with

$$\mathbf{x}_{i|1:k'}^{a,m} = \mathbf{x}_{i|1:k'-1}^{a,m} + \mathbf{X}_{i|1:k'-1}^{a} \gamma_{k'|k'-1}^{m}, \tag{4.18}$$

with i running from 0 to the end of the interval k, and

$$\gamma_{k'|k'-1}^{m} = \mathbf{D}^{\mathrm{T}} [\mathbf{D}\mathbf{D}^{\mathrm{T}} + \mathbf{R}_{k'}]^{-1} \left(\mathbf{y}_{k'}^{o,m} - \mathbf{H}_{k'} \mathbf{x}_{k'|1:k'-1}^{a,m} \right), \tag{4.19}$$

with $\mathbf{D} = \mathbf{H}_{k'} \mathbf{X}_{k'|1:k'-1}^{a}$.

4.4.4 Forward–backward smoother

As the first step of the full-rank forward–backward smoother is a filter pass over the time interval, the first step of its ensemble version is an EnKF pass. In order to invert $\mathbf{P}_{k'+1|1:k'}^{f}$, (4.14a), which is actually singular in a reduced-rank approach, the ensemble formulation provides a nice framework (Lermusiaux and Robinson, 1999). The matrix of ensemble anomalies $\mathbf{X}_{k'+1|k'}^{f}$ can be decomposed into singular values as

$$\mathbf{X}_{k'+1|k'}^{f} = \mathbf{U} \Omega \mathbf{V}^{\mathrm{T}}, \tag{4.20}$$

with Ω being a diagonal matrix of dimension $N_e \times N_e$. Then, noting that $\mathbf{P}_{k'|1:k'}^{a} \mathbf{M}_{k',k'+1}^{\mathrm{T}}$ can be decomposed into $\mathbf{X}_{k'|1:k'}^{a} \mathbf{X}_{k'+1|1:k'}^{f}{}^{\mathrm{T}}$, the smoother gain, (4.14a), can be simply written as:

$$\hat{\mathbf{K}}_{k'}^{s} = \mathbf{X}_{k'|1:k'}^{a} \mathbf{V} \Omega^{-1} \mathbf{U}^{\mathrm{T}}. \tag{4.21}$$

An alternative formulation is

$$\mathbf{K}_{k'}^{s} = \mathbf{X}_{k'|1:k'}^{a} \left[\left(\mathbf{X}_{k'+1|1:k'}^{f} \right)^{\mathrm{T}} \mathbf{X}_{k'+1|k'}^{f} \right]^{-1} \mathbf{X}_{k'+1|k'}^{f}{}^{\mathrm{T}}, \tag{4.22}$$

where the above matrix to be inverted is also of dimensions $N_e \times N_e$. As Table 4.2 illustrates, the ensemble formulation saves a large amount of computational complexity. Once the smoother gain has been computed, the smoother update equations (4.14b, c) can be summarized by an update equation carried out for each member of the ensemble:

$$\mathbf{x}_{k'|1:k}^{s,m} = \mathbf{x}_{k'|1:k'}^{a,m} + \hat{\mathbf{K}}_{k'}^{s} \left(\mathbf{x}_{k'+1|1:k}^{s,m} - \mathbf{x}_{k'+1|1:k'}^{f,m} \right). \tag{4.23}$$

Table 4.2 Approximate computational complexities of the filter and the ensemble smoothers

METHOD	FORECAST	ANALYSIS	STORAGE
Filter	$N_e k C_m$	$N_e k C_h + {N_e}^2 N_x K$	$2kN_e S$
Sequential, fixed-point	F	$F + {N_e}^2 N_x k$	$\frac{3}{2}F$
Sequential, fixed-lag	F	$F + {N_e}^2 N_x L k$	$\left(\frac{L}{2}+1\right)F$
Sequential, fixed-interval	F	$F + \frac{k(k+1)}{2} N_x {N_e}^2$	$\frac{k+5}{4}F$
Two-step	F	$F + {N_e}^2 N_x k^2$	$\frac{k+1}{2}F$
Forward–backward	F	$F + {N_e}^2 N_x k$	$\frac{3}{2}F$

Notation is as in Table 4.1. N_e denotes the ensemble size ($N_e \ll N_y$ or $\mathbf{R}$ diagonal).

4.4.5 Other variants of smoothing algorithms

In a first deterministic version of EnKF, the ensemble transform Kalman filter (ETKF; Bishop et al., 2001) or singular, extended, evolutive Kalman filter (SEEK; Pham et al., 1998), the corrections to the mean and the ensemble anomalies are computed by right-multiplying a matrix of ensemble anomalies, similarly to the stochastic EnKF. All the schemes described previously in the context of the stochastic filter are easily applicable to the deterministic filter. A second, deterministic formulation of EnKF involves a right-multiplication of the forecast ensemble matrix (and not anomalies) to compute the analysis (Evensen, 2003):

$$\mathbf{A}_k^a = \mathbf{A}_k^f \mathbf{T}_k, \tag{4.24}$$

where $\mathbf{A}_k^f = (\mathbf{x}_k^{f,1}, \ldots, \mathbf{x}_k^{f,N_e})$ and $\mathbf{T}_k$ is a transformation matrix. The sequential approach is directly applicable to this formulation, for $i < k$:

$$\mathbf{A}_{i|1:k}^s = \mathbf{A}_{i|1:k-1}^s \mathbf{T}_k. \tag{4.25}$$

Ravela and McLaughlin (2007) introduced algorithms allowing more efficient application of this smoother.

4.5 Advantages, drawbacks, and high-dimensional applications

Various smoothers described above have already been implemented, in their ensemble forms, with realistically sized models for oceanic or atmospheric circulation, though not yet at operational levels. The choice of a smoother must be made carefully based not only on the scientific question posed, but also on the algorithmic and computational characteristics of the smoothers. Such algorithms have been presented in Section 4.3 and 4.4. Their computational complexities (leading behavior, assuming $N_x \gg N_y \gg N_e$) are summarized in Tables 4.1 and 4.2.

The fixed-lag smoother, not limited to a fixed interval and not overly demanding in terms of CPU and storage, is well designed for reanalyzing the dynamics over long time periods. Zhu et al. (2003) performed a simplified implementation with an atmospheric general circulation model (AGCM), relying on the physical-space statistical analysis (PSAS) system (Cohn et al., 1998). They found limited benefits (in terms of global error reduction) in using the smoother in addition to the filter, contrary to Khare et al. (2008), who obtained significant improvements with another AGCM at low resolution. Cosme et al. (2010) also drew significant benefits from the fixed-lag smoother with an oceanic GCM.

The fixed-interval sequential smoother is better suited to short time intervals. This is because an observation contains no information about a model state distant in the past, but also because storage requirements increase quadratically with the interval length. The fixed-interval sequential smoother is indicated, for instance, for the estimation of some variables or parameters of a system based on measurements from an intensive but short field campaign. Barth et al. (2010) made use of this smoother to estimate the tidal boundary conditions for a three-dimensional ocean model.

The two-step smoother, of the fixed-interval type, did not receive significant attention after Evensen and van Leeuwen (2000) found that it was outperformed by the sequential smoother when applied to the Lorenz 96 model. The two-step smoother exhibits several assets though, besides its fairly acceptable computational complexity. It is perfectly indicated for *observing simulation system experiments* (OSSE), since any assimilation of a new observing system does not require additional ensemble integration. Also, the scheme can take into account model and observation errors correlated in time without theoretical approximation. Finally, recent advances in ensemble filtering, in particular localization that can be applied in time as well as space, open the way to significant improvements of the two-step smoother. A few oceanic applications were performed (van Leeuwen, 1999, 2001), but not with a large-size operational model.

The forward–backward smoother is the most common in geophysics (Gaspar and Wunsch, 1989; Fukumori, 2002, 2006), but a very small number of applications have been carried out with GCMs (Lermusiaux et al., 2002). The forward–backward smoother is the only known form of the marginal type that can be implemented in an ensemble form. However, considering the equivalence of joint and marginal smoothers in the linear Gaussian case, and the greater simplicity of implementation for the sequential smoother, the forward–backward smoother seems to be of limited interest. However, a thorough examination of the respective merits of these smoothers for high-dimensional, nonlinear systems is necessary to confirm or discount this assertion.

References

Barth, A., Alvera-Azcarate, A., Gurgel, K.-W., Staneva, J., Port, A., Beckers, J.-M., and Stanev, E. V. (2010). Ensemble perturbation smoother for optimizing tidal boundary conditions by assimilation of high-frequency radar surface currents—application to the German Bight. *Ocean Sci.*, **6**, 161–178.

Bishop, C. H., Etherton, B. J., and Majumdar, S. J. (2001). Adaptive sampling with the ensemble transform kalman filter. Part I: Theoretical aspects. *Mon. Weather Rev.*, **129**, 420–436.

Cohn, S. E., da Silva, A., Guo, J., Sienkiewicz, M., and Lamich, D. (1998). Assessing the effects of data selection with the DAO physical-space statistical analysis system. *Mon. Weather Rev.*, **126**, 2913–2926.

Cosme, E., Brankart, J.M., Verron, J., Brasseur, P., and Krysta, M. (2010). Implementation of a reduced-rank, square-root smoother for ocean data assimilation. *Ocean Model.*, **33**, 87–100.

Cosme, E., Verron, J., Brasseur, P., Blum, J., and Auroux, D. (2012). Smoothing problems in a bayesian framework and their linear Gaussian solutions. *Mon. Weather Rev.*, **140**, 683–695.

Evensen, G. (2003). The ensemble Kalman filter: theoretical formulation and practical implementation. *Ocean Dyn.*, **53**, 343–367.

Evensen, G. and van Leeuwen, P. J. (2000). An ensemble Kalman smoother for nonlinear dynamics. *Mon. Weather Rev.*, **128**, 1852–1867.

Fukumori, I. (2002). A partitioned Kalman filter and smoother. *Mon. Weather Rev.*, **130**, 1370–1383.

Fukumori, I. (2006). What is data assimilation really solving, and how is the calculation actually done? In *Ocean Weather Forecasting: An Integrated View of Oceanography* (ed. E. Chassignet and J. Verron), pp. 271–316. Springer, Berlin.

Gaspar, P. and Wunsch, C. (1989). Estimates from altimeter data of barotropic Rossby waves in Northwestern Atlantic Ocean. *J. Phys. Oceanogr.*, **19**, 1821–1844.

Khare, S. P., Anderson, J. L., Hoar, T. J., and Nychka, D. (2008). An investigation into the application of an ensemble Kalman smoother to high-dimensional geophysical systems. *Tellus A*, **60**, 97–112.

Lermusiaux, P. F. J. and Robinson, A. R. (1999). Data assimilation via error subspace statistical estimation. Part I: Theory and schemes. *Mon. Weather Rev.*, **127**, 1385–1407.

Lermusiaux, P. F. J., Robinson, A. R., Haley, P. J. H., and Leslie, W. G. (2002). Advanced interdisciplinary data assimilation: filtering and smoothing via error subspace statistical estimation. In *Proceedings of OCEANS '02 MTS/IEEE*, pp. 795–802. IEEE, New York.

Pham, D. T., Verron, J., and Roubaud, M. C. (1998). A singular evolutive extended Kalman filter for data assimilation in oceanography. *J. Marine. Syst.*, **16**, 323–340.

Rauch, H. E., Tung, F., and Striebel, C. T. (1965). Maximum likelihood estimates of linear dynamic systems. *AIAA J.*, **3**, 1445–1450.

Ravela, S. and McLaughlin, D. (2007). Fast ensemble smoothing. *Ocean Dyn.*, **57**, 123–134.

van Leeuwen, P. J. (1999). The time mean circulation in the Agulhas region determined with the ensemble smoother. *J. Geophys. Res.*, **104**, 1393–1404.

van Leeuwen, P. J. (2001). An ensemble smoother with error estimates. *Mon. Weather Rev.*, **129**, 709–728.

van Leeuwen, P. J. and Evensen, G. (1996). Data assimilation and inverse methods in terms of a probabilistic formulation. *Mon. Weather Rev.*, **124**, 2898–2913.

Wikle, C. K. and Berliner, L. M. (2007). A Bayesian tutorial for data assimilation. *Physica D*, **230**, 1–16.

Zhu, Y., Todling, R., Guo, J., Cohn, S. E., Navon, I. M., and Yang, Y. (2003). The GEOS-3 retrospective data assimilation system: the 6-hour lag case. *Mon. Weather Rev.*, **131**, 2129–2150.

5
Observation influence diagnostic of a data assimilation system

C. CARDINALI

European Centre for Medium-Range Weather Forecasts (ECMWF)
Reading, United Kingdom

Advanced Data Assimilation for Geosciences. First Edition.
Edited by É. Blayo, M. Bocquet, E. Cosme, and L. F. Cugliandolo.

Chapter Contents

The influence matrix is used in ordinary least-squares applications for monitoring statistical multiple-regression analyses. Concepts related to the influence matrix provide diagnostics on the influence of individual data on the analysis, the change in analysis that would occur by leaving one observation out, and the effective information content (degrees of freedom for signal) in any subset of the analysed data. In this chapter, the corresponding concepts are derived in the context of linear statistical data assimilation in numerical weather prediction. An approximate method to compute the diagonal elements of the influence matrix (the self-sensitivities) has been developed for a large-dimension variational data assimilation system (the 4D-Var system of the European Centre for Medium-Range Weather Forecasts, ECMWF). Results show that, in the ECMWF operational system, 18% of the global influence is due to the assimilated observations, and the complementary 82% is the influence of the prior (background) information, a short-range forecast containing information from earlier assimilated observations. About 20% of the observational information is currently provided by surface-based observing systems and 80% by satellite systems.

A toy model is developed to illustrate how the observation influence depends on the data assimilation covariance matrices. In particular, the role of highly correlated observation errors and highly correlated background errors with respect to uncorrelated ones is presented. Low-influence data points usually occur in data-rich areas, while high-influence data points are in data-sparse areas or in dynamically active regions. Background error correlations also play an important role: High correlation diminishes the observation influence and amplifies the importance of the surrounding real and 'pseudo-' observations (prior information in observation space). To increase the observation influence in the presence of highly correlated background errors, it is necessary to also take the observation error correlation into consideration. However, if the observation error variance is too large with respect to the background error variance the observation influence will not increase. Incorrect specifications of the background and observation error covariance matrices can be identified by the use of the influence matrix.

5.1 Introduction

Over the years, data assimilation (DA) schemes have evolved into very complicated systems, such as the four-dimensional variational system (4D-Var; Rabier et al. 2000) at the European Centre for Medium-Range Weather Forecasts ECMWF. The scheme handles a large variety of both space- and surface-based meteorological observations. It combines the observations with prior (or background) information of the atmospheric state- and uses a comprehensive (linearized) forecast model to ensure that the observations are given a dynamically realistic, as well as statistically likely, response in the analysis.

Effective monitoring of such a complex system, with the order of 10^9 degrees of freedom and more than 10^7 observations per 12-hour assimilation cycle, is a necessity.

The monitoring cannot be restricted to just a few indicators, but a complex set of measures is needed to indicate how different variables and regions influence the DA scheme. Measures of the observational influence are useful for understanding the DA scheme itself: How large is the influence of the latest data on the analysis and how much influence is due to the background? How much would the analysis change if one single influential observation were removed? How much information is extracted from the available data? It is the aim of this chapter to provide a description of such analytical tools.

We turn to the diagnostic methods that have been developed for monitoring statistical multiple regression analyses. In fact, 4D-Var is a special case of the generalized least-squares (GLS) problem (Talagrand, 1997) for weighted regression, thoroughly investigated in the statistical literature.

The structure of many regression data sets makes effective diagnosis and fitting a delicate matter. In robust (resistant) regression, one specific issue is to provide protection against distortion by anomalous data. In fact, a single unusual observation can heavily distort the results of ordinary (non-robust) least-squares (LS) regression (Hoaglin et al., 1982). Unusual or influential data points are not necessarily bad data points: they may contain some of the most useful sample information. For practical data analysis, it helps to judge such effects quantitatively. A convenient diagnostic measures the effect of a (small) change in the observation y_i on the corresponding predicted (estimated) value $\hat{y}_i$. In LS regression, this involves a straightforward calculation: any change in y_i has a proportional impact on $\hat{y}_i$. The desired information is available in the diagonal of the *hat matrix* (Velleman and Welsch, 1981), which gives the estimated values $\hat{y}_i$ as a linear combination of the observed values y_i . The term *hat matrix* was introduced by J. W. Tukey (1972) because the matrix maps the observation vector $\mathbf{y}$ into $\hat{\mathbf{y}}$, but it is also referred to as the *influence matrix* since its elements indicate the data influence on the regression fit of the data. The matrix elements have also been referred to as the *leverage* of the data points: in the case of high *leverage*, a unit y-value will strongly disturb the fit (Hoaglin and Welsch, 1978). Concepts related to the influence matrix also provide diagnostics on the change that would occur by leaving one data point out and the effective information content (degrees of freedom for signal) in the data.

These influence matrix diagnostics are explained in Section 5.2 for ordinary LS regression. In Section 5.3, the corresponding concepts for linear statistical DA schemes is derived. It will be shown that observational influence and background influence complement each other. Thus, for any observation y_i, either very large or very small influence could be the sign of inadequacy in the assimilation scheme and may require further investigation. A practical approximate method that enables calculation of the diagonal elements of the influence matrix for large-dimension variational schemes (such as the ECMWF operational 4D-Var system) is described in Cardinali et al. (2004) and is not discussed here. In Section 5.4, results and selected examples related to data influence diagnostics are presented, including an investigation into the effective information content in several of the main types of observational data. Conclusions are drawn in Section 5.5.

5.2 Classical statistical definitions of influence matrix and self-sensitivity

The ordinary linear regression model can be written as

$$\mathbf{y} = \mathbf{X}\boldsymbol{\beta} + \boldsymbol{\varepsilon}, \tag{5.1}$$

where $\mathbf{y}$ is an $m \times 1$ vector for the response variable (predictand); $\mathbf{X}$ is an $m \times q$ matrix of q predictors; $\boldsymbol{\beta}$ is a $q \times 1$ vector of parameters to be estimated (the regression coefficients), and $\boldsymbol{\varepsilon}$ is an $m \times 1$ vector of errors (or fluctuations) with expectation $E(\boldsymbol{\varepsilon}) = 0$ and covariance $\mathrm{var}(\boldsymbol{\varepsilon}) = \sigma^2 \mathbf{I}_m$ (i.e., uncorrelated observation errors). In fitting the model (5.1) by LS, the number of observations m has to be greater than the number of parameters q in order to have a well-posed problem, and $\mathbf{X}$ is assumed to have full rank q.

The LS method provides the solution of the regression equation as $\boldsymbol{\beta} = (\mathbf{X}^{\mathrm{T}}\mathbf{X})^{-1}\mathbf{X}^{\mathrm{T}}\mathbf{y}$. The fitted (or estimated) response vector is thus

$$\hat{\mathbf{y}} = \mathbf{S}\mathbf{y}, \tag{5.2}$$

where

$$\mathbf{S} = \mathbf{X}(\mathbf{X}^{\mathrm{T}}\mathbf{X})^{-1}\mathbf{X}^{\mathrm{T}} \tag{5.3}$$

is the $m \times m$ *influence matrix* (or hat matrix). It is easily seen that

$$\mathbf{S} = \frac{\delta \hat{\mathbf{y}}}{\delta \mathbf{y}}, \tag{5.4}$$

and that

$$\begin{aligned} S_{ij} &= \frac{\delta \hat{y}_i}{\delta y_j}, \\ S_{ii} &= \frac{\delta \hat{y}_i}{\delta y_i} \end{aligned} \tag{5.5}$$

for the off-diagonal ($i \neq j$) and diagonal ($i = j$) elements, respectively. Thus, S_{ij} is the rate of change of $\hat{y}_i$ with respect to y_j variations. The diagonal element S_{ii} instead measures the rate of change of the regression estimate $\hat{y}_i$ with respect to variations in the corresponding observation y_i. For this reason, the *self-sensitivity* (or self-influence, or leverage) of the ith data point is the ith diagonal element S_{ii}, while an off-diagonal element is a *cross-sensitivity* diagnostic between two data points. Hoaglin and Welsch (1978) discuss some properties of the influence matrix. The diagonal elements satisfy

$$0 \leq S_{ii} \leq 1, \quad i = 1, 2, \ldots, m, \tag{5.6}$$

as $\mathbf{S}$ is a symmetric and idempotent projection matrix ($\mathbf{S}=\mathbf{S}^2$). The covariance of the error in the estimate $\hat{\mathbf{y}}$ and the covariance of the residual $\mathbf{r} = \mathbf{y} - \hat{\mathbf{y}}$ are related to $\mathbf{S}$ by

$$\begin{aligned} \text{var}(\hat{\mathbf{y}}) &= \sigma^2 \mathbf{S}, \\ \text{var}(\mathbf{r}) &= \sigma^2 (\mathbf{I}_m - \mathbf{S}). \end{aligned} \tag{5.7}$$

The trace of the influence matrix is

$$\text{tr}(\mathbf{S}) = \sum_{i=1}^{m} S_{ii} = q = \text{rank}(\mathbf{S}) \tag{5.8}$$

(in fact, $\mathbf{S}$ has m eigenvalues equals to 1 and $m - q$ zeros). Thus, the trace is equal to the number of parameters. The trace can be interpreted as the amount of information extracted from the observations or *degrees of freedom for signal* (Wahba et al., 1995). The complementary trace, $\text{tr}(\mathbf{I} - \mathbf{S}) = m - \text{tr}(\mathbf{S})$, on the other hand, is the degree of freedom for noise, or simply the degree of freedom (df) of the error variance, widely used for model checking (F test).

A zero self-sensitivity $S_{ii} = 0$ indicates that the ith observation has had no influence at all in the fit, while $S_{ii} = 1$ indicates that an entire degree of freedom (effectively one parameter) has been devoted to fitting just that data point. The average self-sensitivity value is q/m, and an individual element S_{ii} is considered 'large' if its value is greater than three times the average (Velleman and Welsch, 1981). By a symmetrical argument, a self-sensitivity value that is less than one-third of the average is considered 'small'.

Furthermore, the change in the estimate that occurs when the ith observation is deleted is

$$\hat{y}_i - \hat{y}_i^{(-i)} = \frac{S_{ii}}{1 - S_{ii}} r_i, \tag{5.9}$$

where $\hat{y}_i^{(-i)}$ is the LS estimate of y_i obtained by leaving-out the ith observation of the vector $\mathbf{y}$ and the ith row of the matrix $\mathbf{X}$. The method is useful to assess the quality of the analysis by using the discarded observation but is impractical for large systems. The formula shows that the impact of deleting $(\mathrm{y}_i, \mathbf{x}_i)$ on $\hat{y}_i$ can be computed by knowing only the residual r_i and the diagonal element S_{ii}—the nearer the self-sensitivity S_{ii} is to 1, the greater in the impact on the estimate $\hat{y}_i$. A related result concerns the so-called cross-validation (CV) score: that is, the LS objective function obtained when each data point is in turn deleted (Wahba, 1990, Theorem 4.2.1):

$$\sum_{i=1}^{m} (y_i - \hat{y}_i^{(-i)})^2 = \sum_{i=1}^{m} \frac{(y_i - \hat{y}_i)^2}{(1 - S_{ii})^2}. \tag{5.10}$$

This theorem shows that the CV score can be computed by relying on the all-data estimate $\hat{\mathbf{y}}$ and the self-sensitivities, without actually performing m separate LS regressions on the leaving-out-one samples. Moreover, (5.9) shows how to compute self-sensitivities by the leaving-out-one experiment.

The definitions of influence matrix, (5.4), and self-sensitivity, (5.5), are rather general and can be applied also to non-LS and non-parametric statistics. In spline regression, for example, the interpretation remains essentially the same as in ordinary linear regression and most of the results, like the CV theorem above, still apply. In this context, Craven and Wahba (1979) proposed the generalized CV score, replacing in (5.10) S_{ii} by the mean $\mathrm{tr}(\mathbf{S})/q$. For further applications of influence diagnostics beyond usual LS regression (and further references), see Ye (1998) and Shen et al. (2002). The notions related to the influence matrix that has been introduced here will in the following section be derived in the context of a statistical analysis scheme used for data assimilation in numerical weather prediction (NWP).

5.3 Observational influence and self-sensitivity for a DA scheme

5.3.1 Linear statistical estimation in NWP

Data assimilation systems for NWP provide estimates of the atmospheric state $\mathbf{x}$ by combining meteorological observations $\mathbf{y}$ with prior (or background) information $\mathbf{x}_b$. A simple Bayesian normal model provides the solution as the posterior expectation for $\mathbf{x}$, given $\mathbf{y}$ and $\mathbf{x}_b$. The same solution can be achieved from a classical *frequentist* approach, based on a statistical linear analysis scheme providing the best linear unbiased estimate (BLUE; Talagrand, 1997) of $\mathbf{x}$, given $\mathbf{y}$ and $\mathbf{x}_b$. The optimal GLS solution to the analysis problem (see Lorenc, 1986) can be written as

$$\mathbf{x}_a = \mathbf{K}\mathbf{y} + (\mathbf{I}_n - \mathbf{K}\mathbf{H})\mathbf{x}_b. \tag{5.11}$$

The vector $\mathbf{x}_a$ is the 'analysis'. The gain matrix $\mathbf{K}$ ($n \times m$) takes into account the respective accuracies of the background vector $\mathbf{x}_b$ and the observation vector $\mathbf{y}$ as defined by the $n \times n$ covariance matrix $\mathbf{B}$ and the $m \times m$ covariance matrix $\mathbf{R}$, with

$$\mathbf{K} = (\mathbf{B}^{-1} + \mathbf{H}^{\mathrm{T}}\mathbf{R}^{-1}\mathbf{H})^{-1}\mathbf{H}^{\mathrm{T}}\mathbf{R}^{-1}. \tag{5.12}$$

Here, $\mathbf{H}$ is an $m \times n$ matrix interpolating the background fields to the observation locations, and transforming the model variables to observed quantities (e.g. radiative transfer calculations transforming a model's temperature, humidity and ozone into brightness temperatures as observed by several satellite instruments). In the 4D-Var context introduced below, $\mathbf{H}$ is defined to include also the propagation in time of the atmospheric state vector to the observation times using a forecast model. Substituting (5.12) into (5.11) and projecting the analysis estimate onto the observation space, the estimate becomes

$$\hat{\mathbf{y}} = \mathbf{H}\mathbf{x}_a = \mathbf{H}\mathbf{K}\mathbf{y} + (\mathbf{I}_m - \mathbf{H}\mathbf{K})\mathbf{H}\mathbf{x}_b. \tag{5.13}$$

It can be seen that the analysis state in observation space ($\mathbf{H}\mathbf{x}_a$) is defined as a sum of the background (in observation space, $\mathbf{H}\mathbf{x}_b$) and the observations $\mathbf{y}$, weighted by the

$m \times m$ square matrices $\mathbf{I}_m - \mathbf{HK}$ and $\mathbf{HK}$, respectively. Equation (5.13) is the analogue of (5.1), except for the last term on the right-hand side. In this case, for each unknown component of $\mathbf{Hx}$, there are two data values: a real and a 'pseudo-' observation. The additional term in (5.13) includes these pseudo-observations, representing prior knowledge provided by the observation-space background $\mathbf{Hx}_b$. From (5.13) and (5.4), the analysis sensitivity with respect to the observations is obtained:

$$\mathbf{S} = \frac{\delta \hat{\mathbf{y}}}{\delta \mathbf{y}} = \mathbf{K}^{\mathrm{T}}\mathbf{H}^{\mathrm{T}}. \tag{5.14}$$

Similarly, the analysis sensitivity with respect to the background (in observation space) is given by

$$\frac{\delta \hat{\mathbf{y}}}{\delta(\mathbf{Hx}_b)} = \mathbf{I}_m - \mathbf{K}^{\mathrm{T}}\mathbf{H}^{\mathrm{T}} = \mathbf{I}_m - \mathbf{S}. \tag{5.15}$$

Let us focus here on the expressions (5.14) and (5.15). The influence matrix for the weighted regression DA scheme is actually more complex (see Appendix 1 of this chapter), but it obscures the dichotomy of the sensitivities between data and model in observation space.

The (projected) background influence is complementary to the observation influence. For example, if the self-sensitivity with respect to the ith observation is S_{ii}, the sensitivity with respect to the background projected at the same variable, location and time will be simply $1 - S_{ii}$. It also follows that the complementary trace, $\mathrm{tr}(\mathbf{I}_m - \mathbf{S}) = m - \mathrm{tr}(\mathbf{S})$, is not the *df* for noise but that for background, instead—that is, the weight given to prior information, to be compared with the observational weight $\mathrm{tr}(\mathbf{S})$. These are the main differences with respect to standard LS regression. Note that the different observations can have different units, so that the units of the cross-sensitivities are the corresponding unit ratios. Self-sensitivities, however, are pure numbers (no units), as in standard regression. Finally, as long as $\mathbf{R}$ is diagonal, (5.6) is assured (see Section 5.3.2), but for more general non-diagonal $\mathbf{R}$ matrices, it is easy to find counter-examples to that property. Inserting (5.12) into (5.14), we obtain

$$\mathbf{S} = \mathbf{R}^{-1}\mathbf{H}(\mathbf{B}^{-1} + \mathbf{H}^{\mathrm{T}}\mathbf{R}^{-1}\mathbf{H})^{-1}\mathbf{H}^{\mathrm{T}}. \tag{5.16}$$

As $(\mathbf{B}^{-1} + \mathbf{H}^{\mathrm{T}}\mathbf{R}^{-1}\mathbf{H})^{-1}$ is equal to the analysis error covariance matrix $\mathbf{A}$, we can also write $\mathbf{S} = \mathbf{R}^{-1}\mathbf{HAH}^{\mathrm{T}}$.

5.3.2 R diagonal

In this section, it is shown that as long as $\mathbf{R}$ is diagonal, (5.6) is satisfied. Equation (5.16) can be written as

$$\begin{aligned} \mathbf{S} &= \mathbf{R}^{-1}\mathbf{H}[\mathbf{B} - \mathbf{BH}^{\mathrm{T}}(\mathbf{HBH}^{\mathrm{T}} + \mathbf{R})^{-1}\mathbf{HB}]\mathbf{H}^{\mathrm{T}} \\ &= \mathbf{R}^{-1}\mathbf{HBH}^{\mathrm{T}} - \mathbf{R}^{-1}\mathbf{HBH}^{\mathrm{T}}(\mathbf{HBH}^{\mathrm{T}} + \mathbf{R})^{-1}\mathbf{HBH}^{\mathrm{T}}. \end{aligned} \tag{5.17}$$

Let us introduce the matrix $\mathbf{V}$=$\mathbf{HBH^T}$. Equation (5.17) then becomes

$$\begin{aligned}
S &= \mathbf{R}^{-1}\mathbf{V} - \mathbf{R}^{-1}\mathbf{V}(\mathbf{V}+\mathbf{R})^{-1}\mathbf{V} \\
&= \mathbf{R}^{-1}\mathbf{V}[\mathbf{I} - (\mathbf{V}+\mathbf{R})^{-1}\mathbf{V}] \\
&= \mathbf{R}^{-1}\mathbf{V}[(\mathbf{V}+\mathbf{R})^{-1}(\mathbf{V}+\mathbf{R}) - (\mathbf{V}+\mathbf{R})^{-1}\mathbf{V}] \\
&= \mathbf{R}^{-1}\mathbf{V}(\mathbf{V}+\mathbf{R})^{-1}\mathbf{R} \\
&= \mathbf{R}^{-1}[(\mathbf{V}+\mathbf{R})(\mathbf{V}+\mathbf{R})^{-1} - \mathbf{R}(\mathbf{V}+\mathbf{R})^{-1}]\mathbf{R} \\
&= \mathbf{R}^{-1}[\mathbf{I} - \mathbf{R}(\mathbf{V}+\mathbf{R})^{-1}]\mathbf{R} \\
&= \mathbf{I} - (\mathbf{V}+\mathbf{R})^{-1}\mathbf{R} \\
&= (\mathbf{V}+\mathbf{R})^{-1}\mathbf{V}. \qquad (5.18)
\end{aligned}$$

Since $\mathbf{V}$ and $\mathbf{R}$ are positive-definite covariance matrices, the matrix $\mathbf{V}+\mathbf{R}$ is also positive-definite. In fact, by definition, for a non-zero vector $\mathbf{z}$ with real entries, the quantity $\mathbf{z}^{\mathrm{T}}(\mathbf{V}+\mathbf{R})\mathbf{z} = \mathbf{z}^{\mathrm{T}}\mathbf{V}\mathbf{z} + \mathbf{z}^{\mathrm{T}}\mathbf{R}\mathbf{z} > 0$.

Let us consider the following theorem: If $\mathbf{D}$ is positive-definite matrix then $\mathbf{D}^{-1}$ is positive-definite, and, defining $\mathbf{D}^{-1}=\{\delta_{ij}\}$ and $\mathbf{D}=\{d_{ij}\}$, we have $\delta_{ii} \geq 1/d_{ii}$, where equality holds if and only if $d_{i1} = \ldots = d_{ii-1} = d_{ii+1} = \ldots = d_{in=0}$.

The diagonal elements of $\mathbf{D}^{-1}=(\mathbf{V}+\mathbf{R})^{-1}=\{\delta_{ij}\}$ are then larger than the diagonal elements of $\mathbf{V}+\mathbf{R}$. Moreover, if $\mathbf{V}=\{v_{ij}\}$ and $\mathbf{R}=\mathrm{diag}(r_i)$, we obtain

$$\delta_{ii} \geq \frac{1}{v_{ii}+r_i}. \qquad (5.19)$$

And, since the i-diagonal element of $(\mathbf{V}+\mathbf{R})^{-1}\mathbf{R}$ is

$$(\delta_{i1}, \ldots, \delta_{in}) \begin{pmatrix} 0 \\ \vdots \\ r_i \\ \vdots \\ 0 \end{pmatrix} = \delta_{ii} r_i,$$

we have

$$\delta_{ii} r_i \geq \frac{r_i}{v_{ii}+r_i}. \qquad (5.20)$$

From (5.18), considering that the product of two positive-definite matrix is still a positive-definite matrix,

$$0 < S_{ii} = 1 - \delta_{ii} r_i \leq 1 - \frac{r_i}{v_{ii}+r_i} = \frac{v_{ii}}{v_{ii}+r_i} < 1. \qquad (5.21)$$

Equation (5.21) proves that the diagonal elements of the influence matrix for the weighted regression DA scheme are bound in the interval (0, 1).

5.3.3 Toy model

Let us assume a simplified model with two observations, each coincident with a point of the background—that is, $\mathbf{H} = \mathbf{I}_2$. Assume the error of the background at the two locations have correlation α—that is,

$$\mathbf{B} = \sigma_b^2 \begin{pmatrix} 1 & \alpha \\ \alpha & 1 \end{pmatrix},$$

with variance σ_b^2—and that, similarly,

$$\mathbf{R} = \sigma_o^2 \begin{pmatrix} 1 & \beta \\ \beta & 1 \end{pmatrix},$$

with variance σ_o^2 and correlation β. For this simple case, $\mathbf{S}$ is obtained from (5.14):

$$S_{11} = S_{22} = \frac{\sigma_b^2\sigma_o^2(1-\alpha\beta) + \sigma_b^4(1-\alpha^2)}{\sigma_b^4(1-\alpha^2) + \sigma_o^4(1-\beta^2) + 2\sigma_b^2\sigma_o^2(1-\alpha\beta)}, \tag{5.22}$$

$$S_{12} = S_{21} = \frac{\sigma_b^2\sigma_o^2(\alpha-\beta)}{\sigma_b^4(1-\alpha^2) + \sigma_o^4(1-\beta^2) + 2\sigma_b^2\sigma_o^2(1-\alpha\beta)}. \tag{5.23}$$

For $\alpha \neq \pm 1$ and $\beta \neq \pm 1$ ($\mathbf{R}$ and $\mathbf{B}$ are full-rank matrices), let us define $r = \sigma_o^2/\sigma_b^2$. Then (5.22) and (5.23) reduce to

$$S_{11} = S_{22} = \frac{r(1-\alpha\beta) + 1 - \alpha^2}{r^2(1-\beta^2) + 1 - \alpha^2 + 2r(1-\alpha\beta)}, \tag{5.24}$$

$$S_{12} = S_{21} = \frac{r(\alpha-\beta)}{r^2(1-\beta^2) + 1 - \alpha^2 + 2r(1-\alpha\beta)}. \tag{5.25}$$

Figure 5.1 shows the diagonal elements of the influence matrix as functions of r, $S_{ii} = S_{ii}(r)$, (5.24). From now on, S_{ii} is also termed the observation influence (OI). In general, the observation influence decreases with increasing r. For highly correlated ($\alpha = 0.9$, $\beta = 0.9$) and diagonal ($\alpha = 0, \beta = 0$) $\mathbf{R}$ and $\mathbf{B}$ matrices, the observation influence as a function of r is the same (solid grey line and dashed thick line, respectively, in Fig. 5.1). Maximum observation influence is achieved when $\mathbf{B}$ is diagonal ($\alpha = 0$) and $\mathbf{R}$ is highly correlated ($\beta = 0.9$) (thin black line). The observation influence will constantly decrease from the 'maximum curve' with decreasing correlation degree in $\mathbf{R}$ ($\mathbf{B}$ still diagonal). And the minimum observation influence curve is achieved when $\mathbf{R}$ is diagonal ($\beta = 0$) and $\mathbf{B}$ is highly correlated ($\alpha = 0.9$, thick solid line). It is worth noting that if the observation error variance is larger than the background error variance ($\sigma_o^2 > \sigma_b^2$), then introducing the observation error correlation will slightly increase the observation influence and, for $\sigma_o^2 \gg \sigma_b^2$, the observations will not be more influential in the analysis despite $\mathbf{R}$ not being diagonal.

For $\mathbf{R}$ diagonal and $\mathbf{B}$ non-diagonal ($\alpha \neq 0, \beta = 0$), (5.24) and (5.25) reduce respectively to

$$S_{11} = S_{22} = \frac{r + 1 - \alpha^2}{r^2 + 1 - \alpha^2 + 2r}, \tag{5.26}$$

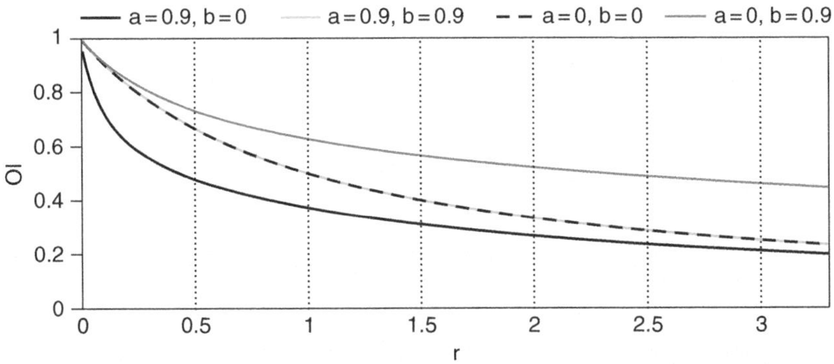

Fig. 5.1 Self-sensitivities or observation influence (OI), as functions of the ratio between the observation error variance and the background error variance. Four different cases are shown: highly correlated **B** and uncorrelated **R** (thick black line); highly correlated **R** and highly correlated **B** (thick grey line); uncorrelated **B** and highly correlated **R** (thin grey line); uncorrelated **R** and uncorrelated **B** (dashed black line).

$$S_{12} = S_{21} = \frac{r\alpha}{r^2 + 1 - \alpha^2 + 2r}. \tag{5.27}$$

It can be seen that if the observations are very close (compared with the scale-length of the background error correlation), i.e. $\alpha \sim 1$ (data-dense area), then

$$S_{11} = S_{22} = S_{12} = S_{21} \cong \frac{1}{r+2}. \tag{5.28}$$

Furthermore, if $\sigma_b = \sigma_o$, that is $r = 1$, we have three pieces of information with equal accuracy and $S_{11} = S_{22} = 1/3$. The background sensitivity at both locations is $1 - S_{11} = 1 - S_{22} = 2/3$. If the observation is much more accurate than the background ($\sigma_b \gg \sigma_o$), that is, $r \sim 0$, then both observations have influence $S_{11} = S_{22} = 1/2$, and the background sensitivities are $1 - S_{11} = 1 - S_{22} = 1/2$.

Let us now turn to the dependence on the background error correlation α, for the case $\sigma_b = \sigma_o$ $(r = 1)$. It is

$$S_{11} = S_{22} = \frac{2 - \alpha^2}{4 - \alpha^2}, \tag{5.29}$$

$$S_{12} = S_{21} = \frac{\alpha}{4 - \alpha^2}, \tag{5.30}$$

If the locations are far apart, such that $\alpha \sim 0$, then $S_{11} = S_{22} = 1/2$, the background sensitivity is also $1/2$ and $S_{12} = S_{21} = 0$. It can be concluded that where observations are sparse, S_{ii} and the background sensitivity are determined by their relative accuracies (r) and the off-diagonal terms are small (indicating that surrounding observations have small influence). Conversely, where observations are dense, S_{ii} tends to be small, the background sensitivities tend to be large, and the off-diagonal terms are also large.

It is also convenient to summarize the case $\sigma_b = \sigma_o$ $(r = 1)$ by showing the projected analysis at location 1:

$$\hat{y_1} = \frac{1}{4-\alpha^2}[(2-\alpha^2)y_1 + 2x_1 - \alpha(x_2 - y_2)]. \tag{5.31}$$

The estimate $\hat{y}_1$ depends on y_1, x_1 and an additional term due to the second observation. It is noticed that, with a diagonal $\mathbf{R}$, the observational contribution is generally devalued with respect to the background, because a group of correlated background values counts more than the single observation $[\alpha \to \pm 1, (2-\alpha^2) \to 1]$. From the expression above, we also see that the contribution from the second observation is increasing with the correlation's absolute value, implying a larger contribution due to the background x_2 and observation y_2 near to observation y_1.

5.4 Results

The diagonal elements of the influence matrix have been computed for the operational 4D-Var assimilation system at T159 spectral truncation 91 model levels for October 2011. For the calculation details, see Cardinali et al. (2004) and Appendix 2 of this chapter. The observation departures $\mathbf{y} - \mathbf{H}\mathbf{x}_b$ were calculated by comparing the observations with a 12-hour forecast integration at T511 resolution. The assimilated observations for each main observation type are given in Table 5.1. A large proportion ($\sim 98\%$) of the used data is provided by satellite systems.

5.4.1 Trace diagnostic: observation influence and DFS

The global average observation influence is defined as

$$\text{OI} = \frac{\text{tr}(\mathbf{S})}{m}, \tag{5.32}$$

where m is the total number of observations. For October 2011, OI = 0.18. Consequently, the average background global influence to the analysis at observation points is equal to 0.82 (see (5.15)). It is clear that in the ECMWF system, the global observation influence is quite low.

In Fig. 5.2, the OI of all the different observation types are plotted. In general, the OI of conventional observations (SYNOP, DRIBU, PROFILER, PILOT, DROP, TEMP, and Aircraft) are larger than that for satellites. The largest OI is provided by DRIBU surface pressure observations, because these are located over the oceans, which are in general very poorly observed (less than continental areas). Moreover, DRIBU and SYNOP observations are very high-quality measurements and the observation error variances are quite small, probably smaller than the background error variance (see the 'toy model' in Section 5.3.3). Similarly, the $OI \sim 0.4$–0.5 of the remaining conventional data are due to their quite small observation error variance. In Section 5.3.3, it has been proved that if $\mathbf{R}$ is diagonal, then the OI is bounded in the range (0, 1),

Table 5.1 Observation type assimilated. The number of data in one assimilation cycle is on average $m \sim 5\,000\,000$.

DATA NAME	DATA KIND	INFORMATION
OZONE (O3)	Backscattered solar UV radiation, retrievals	Ozone, stratosphere
GOES-Radiance	US geostationary satellite infrared sounder radiances	Moisture, mid/upper troposphere
MTSAT-Rad	Japanese geostationary satellite infrared sounder radiances	Moisture, mid/upper troposphere
MET-rad	EUMETSAT geostationary satellite infrared sounder radiances	Moisture, mid/upper troposphere
AMSU-B	Microwave sounder radiances	Moisture, troposphere
MHS	Microwave sounder radiances	Moisture, troposphere
MERIS	Differential reflected solar radiation, retrievals	Total column water vapour
GPS-RO	GPS radio occultation bending angles	Temperature, surface pressure
IASI	Infrared sounder radiances	Temperature, moisture, ozone
AIRS	Infrared sounder radiances	Temperature, moisture, ozone
AMSU-A	Microwave sounder radiances	Temperature
HIRS	Infrared sounder radiances	Temperature, moisture, ozone
ASCAT	Microwave scatterometer backscatter coefficients	Surface wind
MODIS-AMV	US polar atmospheric motion vectors, retrievals	Wind, troposphere
Meteosat-AMV	EUMETSAT geostationary atmospheric motion vectors, retrievals	Wind, troposphere
MTSAT-AMV	Japanese geostationary atmospheric motion vectors, retrievals	Wind, troposphere

(*continued*)

Table 5.1 (*continued*)

DATA NAME	DATA KIND	INFORMATION
GOES-AMV	US geostationary atmospheric motion vectors, retrievals	Wind, troposphere
PROFILER	American, European, and Japanese wind profiles	Wind, troposphere
PILOT	Radiosondes at significant level from land stations	Wind, troposphere
DROP	Dropsondes from Aircraft	Wind, temperature, moisture, pressure, troposphere
TEMP	Radiosondes from land and ships	Wind, temperature, moisture, pressure, troposphere
Aircraft	Aircraft measurements	Wind, temperature, troposphere
DRIBU	Drifting buoys	Surface pressure, temperature, moisture, wind
SYNOP	Surface observations at land stations and on ships	Surface pressure, temperature, moisture, wind

but we can see from Fig. 5.2 that the DRIBU OI is greater than 1. This is due to the approximation of the numerical solution and, in particular, the use in the influence matrix calculation of an estimate of the analysis covariance matrix **A** (for details, see Cardinali et al., 2004). On the contrary, the OI of satellite data is quite small. The largest influence is provided by GPS-RO observations (~0.4), which again are accurate data and probably with uncorrelated observation error (Healy and Thépaut, 2006), followed by AMSU-A measurements (~0.3). All the other observations have an influence of about 0.2. Recently, changes in the assimilation of 'all-sky' observations (TMI and SSMIS) have increased their influence in the analysis (Cardinali and Prates, 2011; Geer and Bauer, 2011).

In Section 5.2, it has been shown that tr(**S**) can be interpreted as a measure of the amount of information extracted from the observations. In fact, in non-parametric statistics, tr(**S**) measures the 'equivalent number of parameters' or *degrees of freedom for signal (DFS)*. Having obtained values of all the diagonal elements of **S** (using (5.16)), we can now obtain reliable estimates of the information content in any subset of the observational data. However, it must be noted that this theoretical measure of information content does not necessarily translate into a value of forecast impact. Figure 5.3 shows the information content for all main observation types. It can be

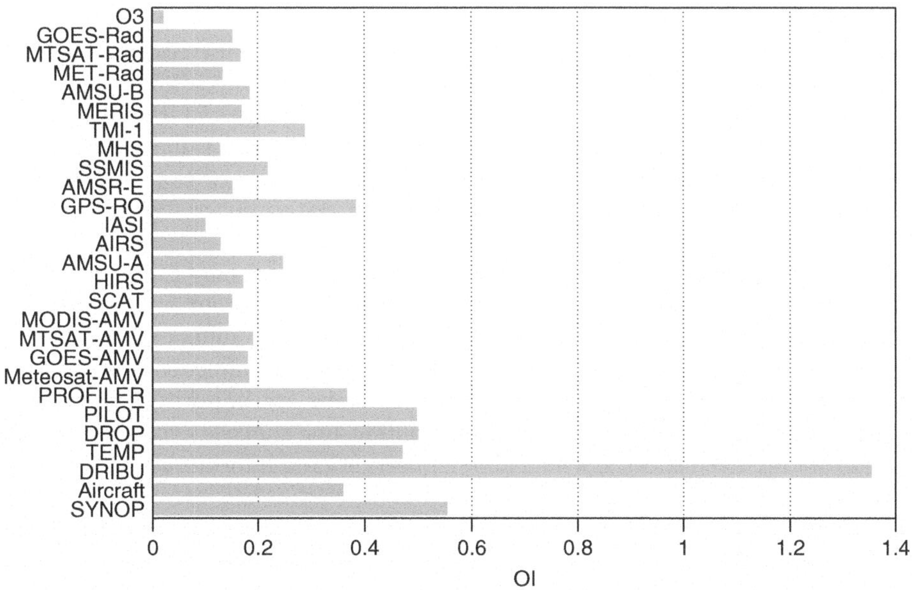

Fig. 5.2 Observation influence (OI) of all assimilated observations in the ECMWF 4D-Var system in October 2011. Observation types are described in Table 5.1.

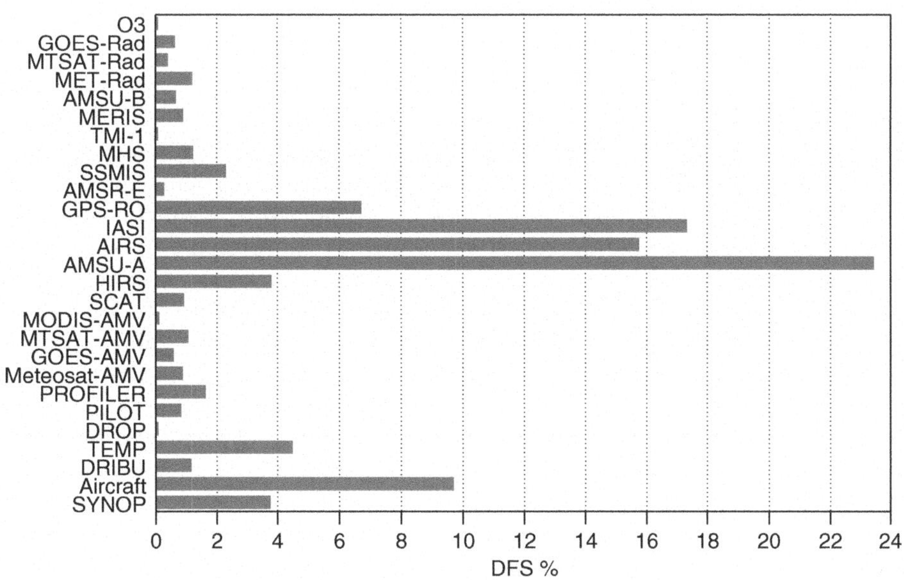

Fig. 5.3 Degree of freedom for signal (DFS) of all observations assimilated in the ECMWF 4D-Var system in October 2011. Observation types are described in Table 5.1.

seen that AMSU-A radiances are the most informative data type, providing 23% of the total observational information, IASI follows with 17% and AIRS with 16%. The information content of Aircraft (10%) is the largest among conventional observations, followed by TEMP and SYNOP (~4%). Noticeable is the 7% of GPS-RO (fourth in the satellite DFS ranking), which combines well with the 0.4 value for the average observation influence. In general, the importance of the observations as defined by, for example, the DFS correlates well with the recent data impact studies by Radnoti et al. (2010).

Similar information content of different observation types may be due to different reasons. For example, DRIBU and OZONE information content is similarly small, but whereas OZONE observations have a very small average influence (Fig. 5.2) and dense data coverage, DRIBU observations have large mean influence but much lower data counts (Fig. 5.2). In any case, the OZONE data are important for the ozone assimilation in spite of their low information content per analysis cycle. In fact, ozone is generally a long-lived species, which allows observational information to be advected by the model over periods of several days.

The difference between OI and DFS comes from the number of observation assimilated. Therefore, despite the generally low OI of satellite measurements, they show quite large DFS because of the large number assimilated. A large discrepancy between OI and DFS points on those observation types where a revision of the assigned covariance matrices **R** and **B** will be beneficial, with more information being extracted from, for example, satellite measurements.

Another index of interest is the partial observation influence for any selected subset of data:

$$\mathrm{OI}_{m_I} = \frac{\sum_{i \subset I} S_{ii}}{m_I}, \tag{5.33}$$

where m_I is the number of data in subset I. The subset I can represent a specific observation type, a specific vertical or horizontal domain, or a particular meteorological variable. In Fig. 5.4, the OI of Aircraft data (I) is plotted as a function of pressure layers and for all observed parameters: temperature (t) and zonal (u) and meridional (v) components of the wind. The largest OI is provided by temperature observations (~0.4), similarly distributed on the different pressure layers. Wind observations have larger influence (0.4) at the top of the atmosphere (above 400 hPa) than at the bottom (0.2), owing to the fact that there are very few wind observations in the troposphere and lower stratosphere, mainly over the oceans. At those levels, temperature information is also provided by different satellite platforms (in terms of brightness temperature or radiance). In Fig. 5.5, the Aircraft DFS with respect to different pressure levels and observed parameters is shown. The largest DFS in the lower troposphere (below 700 hPa) for temperature measurements (~10% with respect to the total Aircraft DFS) with respect to wind measurements is due to the largest temperature influence. For all the other levels, the DFS is quite similar to the OI distribution, with the exception of the layer from 200 to 300 hPa, where the increase to ~50% is due to the increase of number of observations assimilated. Figures 5.6 and 5.7 shows the AMSU-A OI and DFS, respectively, for all the channels assimilated. A large part of the AMSU-A

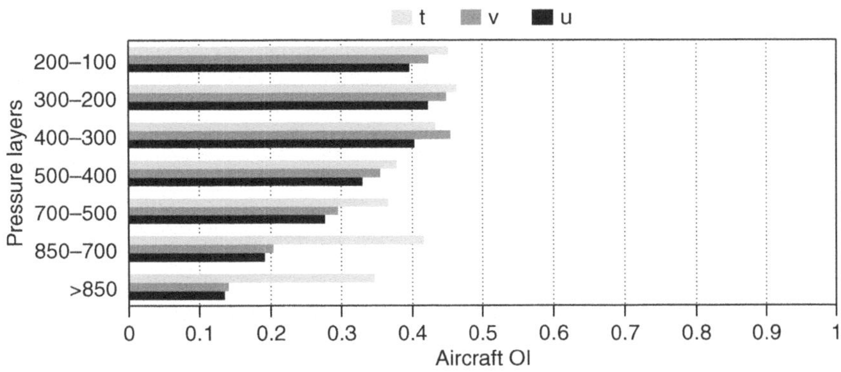

Fig. 5.4 Observation influence (OI) for Aircraft observations and for October 2011 grouped by pressure layer and observed parameter. Parameters are temperature (t, light grey bar), meridional wind (v, dark grey bar), and zonal wind (u, black bar).

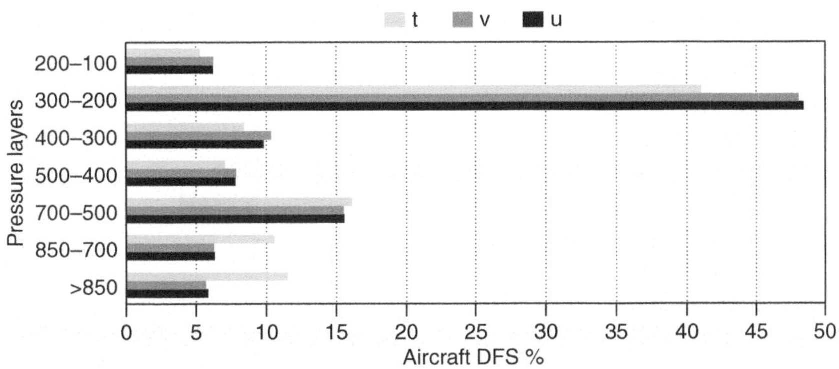

Fig. 5.5 Degree of freedom for signal (DFS) as a percentage for Aircraft observations and for October 2011 grouped by pressure layer and observed parameter. Parameters are temperature (t, light grey bar), meridional wind (v, dark grey bar), and zonal wind (u, black bar). The percentage is relative to the total Aircraft DFS.

information is with respect to stratospheric temperature, and the largest OI at that atmospheric layer is from channels 9 and 10 (~0.4) (Fig. 5.6). Channel 5 (~700 hPa) shows a very large OI (~0.8), the largest influence among all the channels. The reason of this large OI is unclear, and investigations are under way to understand the cause. The channel's OI distribution is similar to the DFS distribution (Fig. 5.7): channels 9 and 10 count for 18% of the AMSU-A DFS and channel 5 for 24%.

5.4.2 Geographical map of OI

The geographical map of OI for SYNOP and DRIBU surface pressure observations is shown in Fig. 5.8. Each box indicates the OI per observation location averaged among

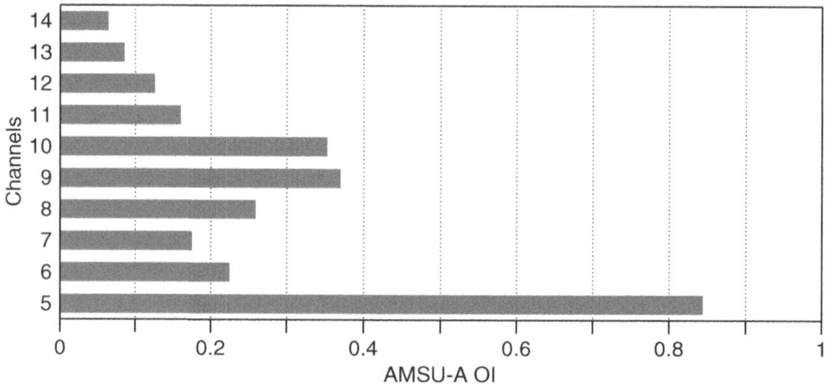

Fig. 5.6 Observation influence (OI) for AMSU-A observations and for October 2011 grouped by channels.

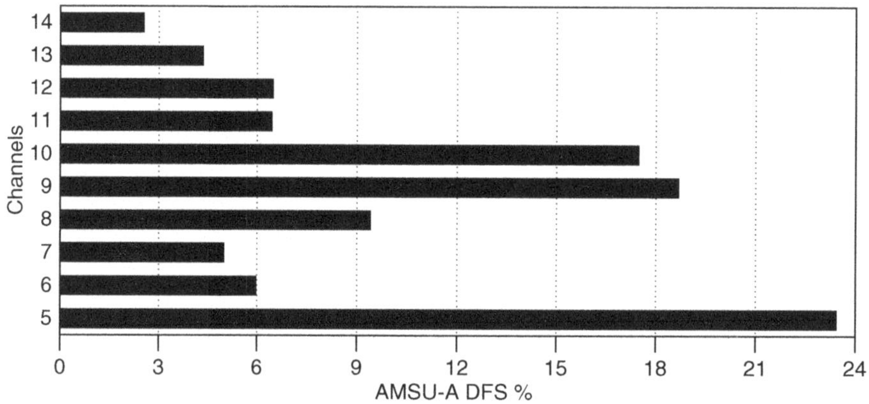

Fig. 5.7 Degree of Freedom for Signal in percentage (DFS) for AMSU-A observations and for October 2011 grouped by channels. The percentage is relative to the total AMSU-A DFS.

all the October 2011 measurements. Data points with influence greater than one are due to the approximation of the computed diagonal elements of the influence matrix (see Cardinali et al. (2004) and Appendix 2 of this chapter).

Low-influence data points have large background influence (see (5.14) and (5.15)), which is the case in data-rich areas such as North America and Europe (observation influence ~0.2) (see also Section 5.3.3). In data-sparse areas, individual observations have larger influence: in the polar regions, where there are only a few isolated observations, the OI is very high (theoretically ~1) and the background has a very small influence on the analysis. In dynamically active areas (Fig. 5.8; e.g. the North Atlantic and North Pacific), several fairly isolated observations have a large influence

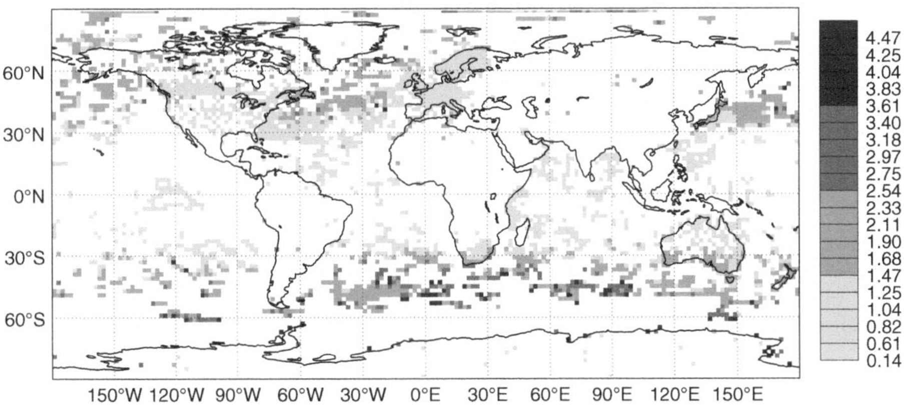

Fig. 5.8 Observation Influence (OI) of SYNOP and DRIBU surface pressure observations for October 2011. High influential points are close to 1 and low influential points are close to 0.

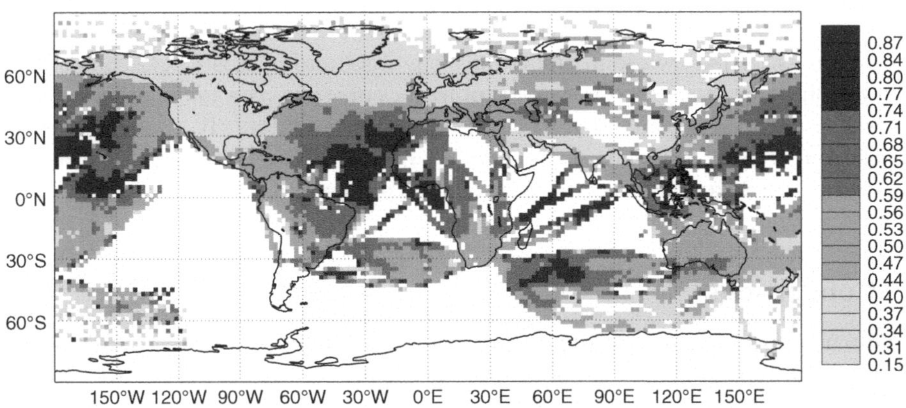

Fig. 5.9 Observation Influence (OI) of Aircraft zonal wind component above 400 hPa for October 2011. High influential points are close to 1 and low influential points are close to 0.

on the analysis. This is also due to the evolution of the background error covariance matrix as propagated by the forecast model in 4D-Var (Thépaut et al., 1993, 1996). As a result, the data assimilation scheme can fit these observations more closely.

Similar features can be seen in Fig. 5.9, which shows the influence of the u-component wind observations for Aircraft data above 400 hPa. Isolated flight tracks over the Atlantic and Pacific Oceans show larger influences than measurements over data-dense areas over America and Europe. The flight tracks over the North Atlantic and North Pacific are also in dynamically active areas where the background error variances are implicitly inflated by the evolution of the background error covariance matrix in the 4D-Var window. Figure 5.10 shows the geographical distribution of AMSU-A

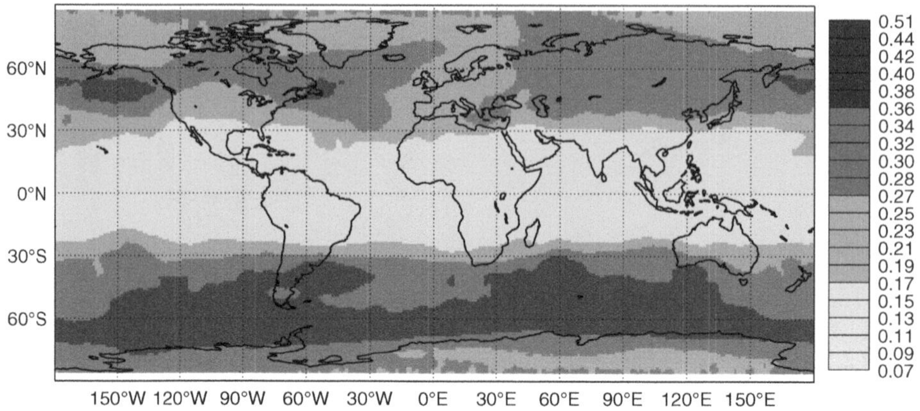

Fig. 5.10 Observation influence (OI) of AMSU-A channel 8 for October 2011. High-influential points are close to 1 and low-influene points are close to 0.

channel 8 observation influence. The largest influence is noticed in the extra-tropical and polar areas (~0.4), while in the tropics the maximum OI is ~0.12. Since channel 8 observation error variances are geographically constant, the main difference in the observed OI pattern is likely due to the **B** covariance matrix. It appears that either the background error correlations are higher or the background error variances are larger in the extra-tropical regions than in the tropics.

5.5 Conclusions

The influence matrix is a well-known concept in multivariate linear regression, where it is used to identify influential data and to predict the impact on the estimates of removing individual data from the regression. In this chapter, the influence matrix in the context of linear statistical analysis schemes has been derived, as used for data assimilation of meteorological observations in numerical weather prediction (Lorenc, 1986). In particular, an approximate method to compute the diagonal elements of the influence matrix (the self-sensitivities or observation influence) in the ECMWF operational data assimilation system (4D-Var) has been derived and implemented. The approach necessarily approximates the solution owing to the large dimension of the estimation problem at hand: the number of estimated parameters is of the order 10^9 and the number of observational data is around 25×10^6.

The self-sensitivity provides a quantitative measure of the observation influence in the analysis. In robust regression, it is expected that the data have similar self-sensitivity (sometimes called leverage)—that is, they exert similar influence in estimating the regression line. Disproportionate data influence on the regression estimate can have different reasons: First, there is the inevitable occurrence of incorrect data. Second, influential data points may be legitimately occurring extreme observations. However, even if such data often contain valuable information, it is constructive to determine the extent to which the estimate depends on these data. Moreover, diagnostics may reveal other patterns, for example that the estimates are based primarily on a specific subset of the data rather than on the majority of the data.

In the context of 4D-Var there are many components, that together determine the influence given to any one particular observation. First, there is the specified observation error covariance **R**, which is usually well known and obtained simply from tabulated values. Second, there is the background error covariance **B**, which is specified in terms of transformed variables that are most suitable to describe a large proportion of the actual background error covariance. The implied covariance in terms of the observable quantities is not immediately available for inspection, but it determines the analysis weight given to the data. Third, the dynamics and the physics of the forecast model propagate the covariance in time, and modify it according to local error growth in the prediction. The influence is further modulated by data density. Examples for surface pressure and aircraft wind observations have been shown, indicating that low-influence data points occur in data-rich areas, while high-influence data points are in data-sparse regions or in dynamically active areas. Background error correlations also play an important role. In fact, very high correlations drastically lessen the observation influence (it is halved in the idealized example presented in Section 5.3.3) in favour of background influence and amplify the influence of the surrounding observations. The observation influence pattern of AMSU-A channel 8 suggests some affectation of the correlation expresses by the **B** covariance matrix.

The global observation influence per assimilation cycle has been found to be 18%, and consequently the background influence is 82%. Thus, on average, the observation influence is low compared with the influence of the background (the prior). However, it must be taken into account that the background contains observation information from the previous analysis cycles. The theoretical information content (the degrees of freedom for signal) for each of the main observation types was also calculated. It was found that AMSU-A radiance data provide the most information to the analysis, followed by IASI, AIRS, Aircraft, GPS-RO, and TEMP. In total, about 20% of the observational information is currently provided by surface-based observing systems, and 80% by satellite systems. It must be stressed that this ranking is not an indication of relative importance of the observing systems for forecast accuracy. Nevertheless, recent studies on the 24-hour observation impact on the forecast with the adjoint methodology have shown similar data ranking (Langland and Baker, 2004; Zhu and Gelaro, 2008; Cardinali, 2009).

If the influence matrix were computed without approximation, then all the self-sensitivities would have been bounded in the interval $(0, 1)$. With the approximate method used, out-of-bound self-sensitivities occur if the Hessian representation based on an eigenvector expansion is truncated, especially when few eigenvectors are used. However, it has been shown that this problem affects only a small percentage of the self-sensitivities computed, and in particular those that are closer to 1.

Self-sensitivities provide an objective diagnostic on the performance of the assimilation system. They could be used in observation quality control to protect against distortion by anomalous data; This aspect has been explored by Junjie et al. (2009) in the context of the ensemble Kalman filter, where **B** is well known and the solution for the diagonal element of the Influence matrix is therefore very accurate. Junjie et al. (2009) have shown that the leaving-out-one observation which is not practical for large system dimension, can be replaced by the self-sensitivities, (5.9), which provide a similar diagnostic without performing separate leastsquares regressions. Self-sensitivities also provide indications on model and observation error specification and tuning. Incorrect specifications can be identified, interpreted, and better understood through observation influence diagnostics, partitioned, for example, by observation types, variables, levels, and regions.

In the near future, more satellite data will be used and will likely be thinned. Thinning has to be performed either to reduce the observation error spatial correlation (Bormann et al., 2003) or to reduce the computational cost of the assimilation. The observation influence provides an objective way of selecting observations dependent on their local influence on the analysis estimate to be used in conjunction with forecast impact assessments. Recently, Bauer et al. (2011) have shown that satellite measurements in sensitive areas as defined by a singular vectors methodology have a larger impact in the forecast than measurements in different regions, and also an impact that is larger than or similar to that of the full amount of data. In this case, a dynamical thinning can be considered that selects, at every assimilation cycle, the most influential measurement partition of a particular remote sensing instrument, from information based on the previous cycle (see also Rabier et al., 2002). Clearly, it can be assumed that components of the observing network remain constant and that the background error variances remain almost unchanged for close assimilation cycles.

Acknowledgements

The author thanks Olivier Talagrand and Sergio Pezzulli for fruitful discussions on the subject. Many thanks also to Mohamed Dahoui and Anne Fouilloux for their precious technical support.

Appendix 1: Influence matrix calculation in weighted regression DA scheme

Under the frequentist approach, the regression equations for observation,

$$\mathbf{y} = \mathbf{H}\boldsymbol{\theta} + \boldsymbol{\varepsilon}_o, \tag{5.34}$$

and for background,

$$\mathbf{x}_b = \boldsymbol{\theta} + \boldsymbol{\varepsilon}_b, \tag{5.35}$$

are assumed to have uncorrelated error vectors $\boldsymbol{\varepsilon}_o$ and $\boldsymbol{\varepsilon}_b$, zero vector means, and variance matrices $\mathbf{R}$ and $\mathbf{B}$, respectively. The $\boldsymbol{\theta}$ parameter is the unknown system state ($\mathbf{x}$) of dimension n. These regression equations are summarized as a weighted regression

$$\mathbf{z} = \mathbf{X}\boldsymbol{\theta} + \boldsymbol{\varepsilon}, \tag{5.36}$$

where $\mathbf{z} = [\mathbf{y}^{\mathrm{T}}, \mathbf{x}_b^{\mathrm{T}}]^{\mathrm{T}}$ is $(m+n) \times 1$; $\mathbf{X} = [\mathbf{H}^{\mathrm{T}}, \mathbf{I}_n]^{\mathrm{T}}$ is $(m+n) \times n$, and $\boldsymbol{\varepsilon} = [\boldsymbol{\varepsilon}_o^{\mathrm{T}}, \boldsymbol{\varepsilon}_b^{\mathrm{T}}]^{\mathrm{T}}$ is $(m+n) \times 1$, with zero mean and variance matrix

$$\boldsymbol{\Omega} = \begin{pmatrix} \mathbf{R} & 0 \\ 0 & \mathbf{B} \end{pmatrix}. \tag{5.37}$$

The generalized LS solution for $\boldsymbol{\theta}$ is the BLUE and is given by

$$\hat{\boldsymbol{\theta}} = (\mathbf{X}^{\mathrm{T}}\boldsymbol{\Omega}^{-1}\mathbf{X})\mathbf{X}^{\mathrm{T}}\boldsymbol{\Omega}^{-1}\mathbf{z} \tag{5.38}$$

(see Talagrand, 1997). After some algebra, this equation is found to be equivalent to (5.11). Thus,

$$\mathbf{z} = \mathbf{X}\hat{\boldsymbol{\theta}} = [\mathbf{H}^{\mathrm{T}}\mathbf{x}_a^{\mathrm{T}}, \mathbf{x}_a^{\mathrm{T}}]^{\mathrm{T}} = \mathbf{X}(\mathbf{X}^{\mathrm{T}}\boldsymbol{\Omega}^{-1}\mathbf{X})^{-1}\mathbf{X}^{\mathrm{T}}\boldsymbol{\Omega}^{-1}\mathbf{z}, \tag{5.39}$$

and, by (5.5), the influence matrix becomes

$$\mathbf{S}_{zz} = \frac{\delta\hat{\mathbf{z}}}{\delta\mathbf{z}} = \frac{\delta\mathbf{X}\hat{\boldsymbol{\theta}}}{\delta\mathbf{z}} = \begin{pmatrix} \mathbf{S}_{yy} & \mathbf{S}_{yb} \\ \mathbf{S}_{by} & \mathbf{S}_{bb} \end{pmatrix} = \begin{pmatrix} \mathbf{R}^{-1}\mathbf{HAH}^{\mathrm{T}} & \mathbf{R}^{-1}\mathbf{HA} \\ \mathbf{B}^{-1}\mathbf{AH}^{\mathrm{T}} & \mathbf{B}^{-1}\mathbf{A} \end{pmatrix}, \tag{5.40}$$

where

$$\mathbf{S}_{yy} = \frac{\delta\mathbf{Hx}_a}{\delta\mathbf{y}}, \; \mathbf{S}_{yb} = \frac{\delta\mathbf{x}_a}{\delta\mathbf{y}}, \; \mathbf{S}_{by} = \frac{\delta\mathbf{Hx}_a}{\delta\mathbf{x}_b}, \; \mathbf{S}_{bb} = \frac{\delta\mathbf{x}_a}{\delta\mathbf{x}_b}.$$

Note that $\mathbf{S}_{yy} = \mathbf{S}$ as defined in (5.14).

Generalized LS regression is different from ordinary LS because the influence matrix is no longer symmetric. For idempotence, using (5.34), it is easy to show that $\mathbf{S}_{zz}\mathbf{S}_{zz} = \mathbf{S}_{zz}$. Finally,

$$\mathbf{S}_{bb} = \mathbf{B}^{-1}\mathbf{A} = \mathbf{I}_n - \mathbf{H}^{\mathrm{T}}\mathbf{R}^{-1}\mathbf{HA}, \tag{5.41}$$

and hence

$$\mathrm{tr}(\mathbf{S}_{bb}) = n - \mathrm{tr}(\mathbf{H}^{\mathrm{T}}\mathbf{R}^{-1}\mathbf{HA}) = n - \mathrm{tr}(\mathbf{S}_{yy}). \tag{5.42}$$

It follows that

$$\mathrm{tr}(\mathbf{S}_{zz}) = \mathrm{tr}(\mathbf{S}_{yy}) + \mathrm{tr}(\mathbf{S}_{bb}) = n. \tag{5.43}$$

The trace of the influence matrix is still equal to the parameter's dimension.

Appendix 2: Approximate calculation of self-sensitivity in a large variational analysis system

In an optimal variational analysis scheme, the analysis error covariance matrix $\mathbf{A}$ is approximately the inverse of the matrix of second derivatives (the Hessian) of the cost function J, i.e. $\mathbf{A} = (J'')^{-1}$ (Rabier and Courtier, 1992). Given the large dimension of the matrices involved, J'' and its inverse cannot be computed explicitly. Following Fisher and Courtier (1995), we use an approximate representation of the Hessian based on a truncated eigenvector expansion with vectors obtained through the Lanczos algorithm. The calculations are performed in terms of a transformed variable $\chi = \mathbf{L}^{-1}(\mathbf{x} - \mathbf{x}_b)$, with $\mathbf{L}$ chosen such that $\mathbf{B}^{-1}=\mathbf{L}^{\mathrm{T}}\mathbf{L}$. The transformation $\mathbf{L}$ thus reduces the covariance of the prior to the identity matrix. In variational assimilation, $\mathbf{L}$ is referred to as the change-of-variable operator (Courtier et al., 1998). Then,

$$J''^{-1} = \mathbf{B} - \sum_{i=1}^{M} \frac{1-\lambda_i}{\lambda_i}(\mathbf{L}\mathbf{v}_i)(\mathbf{L}\mathbf{v}_i)^{\mathrm{T}}. \tag{5.44}$$

The summation in (5.44) approximates the variance reduction $\mathbf{B} - \mathbf{A}$ due to the use of observations in the analysis. $(\lambda_i, \mathbf{v}_i)$ are the eigenpairs of $\mathbf{A}$. The Hessian eigenvectors are also used to precondition the minimization (Fisher and Andersson, 2001). The computed eigenvalues are not used to minimize the cost function, but only to estimate the analysis covariance matrix. It is well known, otherwise, that the minimization algorithm is analogous to the conjugate-gradient algorithm. Because the minimum is found within an iterative method, the operational number of iterations is sufficient to find the solution (with the required accuracy), but does not provide a sufficient number of eigenpairs to estimate the analysis error variances.

The diagonal of the background error covariance matrix $\mathbf{B}$ in (5.44) is also computed approximately, using the randomization method proposed by Fisher and Courtier (1995). From a sample of N random vectors $\mathbf{u}_i$ (in the space of the control vector χ), drawn from a population with zero mean and unit Gaussian variance, a low-rank representation of $\mathbf{B}$ (in terms of the atmospheric state variables x) is obtained by

$$\mathbf{B} = \frac{1}{N}\sum_{i=1}^{N}(\mathbf{L}\mathbf{u}_i)(\mathbf{L}\mathbf{u}_i)^{\mathrm{T}}. \tag{5.45}$$

This approximate representation of $\mathbf{B}$ has previously been used by Andersson et al. (2000) to diagnose background errors in terms of observable quantities, namely $\mathbf{HBH}^{\mathbf{T}}$.

Inserting (5.44) and (5.45) into (5.16), an approximate method for calculating $\mathbf{S}$ is achieved, that is practical for a large dimension variational assimilation (both 3D- and 4D-Var):

$$\mathbf{S} = \mathbf{R}^{-1}\mathbf{H}\left[\frac{1}{N}\sum_{i=1}^{N}(\mathbf{L}\mathbf{u}_i)(\mathbf{L}\mathbf{u}_i)^{\mathrm{T}} - \sum_{i=1}^{M}\frac{1-\lambda_i}{\lambda_i}(\mathbf{L}\mathbf{v}_i)(\mathbf{L}\mathbf{v}_i)^{\mathrm{T}}\right]\mathbf{H}^{\mathrm{T}}. \tag{5.46}$$

Only the diagonal elements of $\mathbf{S}$ are computed and stored—that is, the analysis sensitivities with respect to the observations, or self-sensitivities S_{ii}. The cross-sensitivity S_{ij} for $i \neq j$, which represents the influence of the jth observation on the analysis at the ith location, is not computed. Note that the approximation of the first term is unbiased, whereas the second term is truncated such that variances are underestimated. For small M, the approximate S_{ii} will tend to be overestimates. For the extreme case $M = 0$, (5.46) gives $\mathbf{S} = \mathbf{R}^{-1}\mathbf{HBH}^{\mathrm{T}}$ which in particular can have diagonal elements larger than 1 if elements of $\mathbf{HBH}^{\mathbf{T}}$ are larger than the corresponding elements of $\mathbf{R}$. The effect of the approximation on the calculated S_{ii} values is investigated. Approximations in both of the terms of (5.46) contribute to the problem. For the second term, the degree of overestimation depends on the structure of the covariance reduction matrix $\mathbf{B} - \mathbf{A}$.

For an analysis in which observations lead to strongly localised covariance reduction (such as the surface pressure analysis with its short covariance lengthscales ~180 km, and large observational impacts) a large M is required to approximate $\mathbf{B} - \mathbf{A}$ accurately.

In Fig. 5.11, the proportion of observations for which $S_{ii} > 1$ is plotted versus M, the number of Hessian vectors used. The plot shows a gradual decrease of $S_{ii} > 1$ as M increases, as expected. The curve seems to approach 10 000 observations for which $S_{ii} > 1$ (0.7% in the plot) for M somewhere between 1000 and 2000. However, increasing the number of Hessian vectors slightly increases the number of self-sensitivities less than zero (by 0.5%). This problem can be understood by looking at the approximations introduced through the first term of (5.46). The truncation N of the first term determines the randomization sample size: larger N leads to smaller noise. The noise

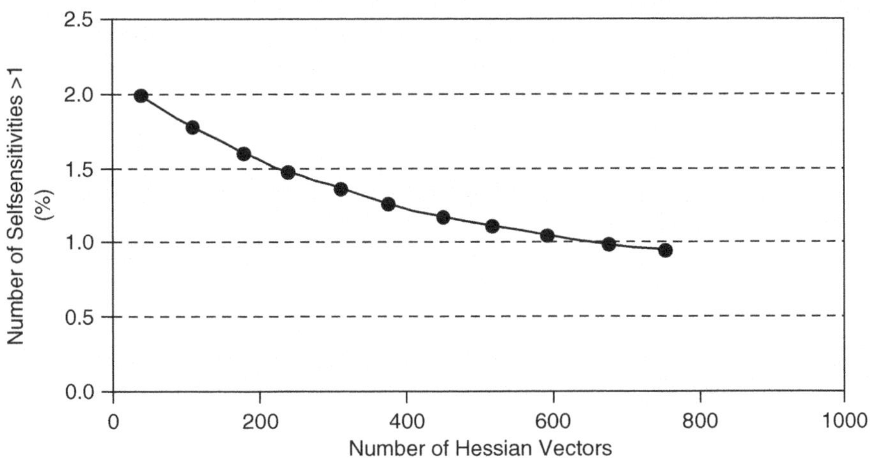

Fig. 5.11 Proportion of self-sensitivity values greater than 1 (as a percentage) versus the number of Hessian vectors used to compute the variances of $\mathbf{B} - \mathbf{A}$, using the approximate method described by (5.46). (From Cardinali et al. (2004). Copyright © 2004 Royal Meteorological Society.)

is unbiased—that is, the term is neither over- nor underestimated on average. The randomization noise in the diagonal elements is in the order 10% with $N = 50$ (Andersson et al., 2000). With $N = 500$, values $S_{ii} < 0$ have all disappeared.

References

Andersson, E., Fisher, M., Munro, R., and McNally, A. (2000). Diagnosis of background errors for radiances and other observable quantities in a variational data assimilation scheme, and the explanation of a case of poor convergence. *Q. J. R. Meteor. Soc.*, **126**, 1455–1472.

Bauer, P., Buizza, R., Cardinali, C., and Thépaut, J.-N. (2011). Impact of singular vector based satellite data thinning on NWP. *Q. J. R. Meteor. Soc.*, **137**, 286–302.

Bormann, N., Saarinen, S., Kelly G., and Thépaut, J.-N. (2003). The spatial structure of observation errors in atmospheric motion vectors from geostationary satellite data. *Mon. Weather Rev.*, **131**, 706–718.

Cardinali, C. (2009). Monitoring the forecast impact on the short-range forecast. *Q. J. R. Meteor. Soc.*, **135**, 239–250.

Cardinali, C. and Prates, F. (2011). Performance measurement with advanced diagnostic tools of all-sky microwave imager radiances in 4D-Var. *Q. J. R. Meteor. Soc.*, **137**, 2038–2046.

Cardinali, C., Pezzulli, S., and Andersson, E. (2004). Influence matrix diagnostics of a data assimilation system. *Q. J. R. Meteor. Soc.*, **130**, 2767–2786.

Courtier, P., Andersson, E., Heckley, W., Vasiljevic, D., Hamrud, M., Hollingsworth, A., Rabier, F., Fisher, M., and Pailleux, J. (1998). The ECMWF implementation of three-dimensional variational assimilation (3D-Var). Part I: Formulation. *Q. J. R. Meteor. Soc.*, **124**, 1783–1807.

Craven, P. and Wahba, G. (1979). Smoothing noisy data with spline functions: estimating the correct degree of smoothing by the method of generalized cross-validation. *Numer. Math.*, **31**, 377–403.

Geer A. J. and Bauer, P. (2011). Observation errors in all-sky data assimilation. *Q. J. R. Meteor. Soc.*, **137**, 2024–2037.

Fisher, M. and Andersson, E. (2001). Developments in 4D-Var and Kalman filtering. ECMWF Tech. Memo. 347.

Fisher, M. and Courtier, P. (1995). Estimating the covariance matrices of analysis and forecast error in variational data assimilation. ECMWF Tech. Memo. **220**.

Junjie, L., Kalnay E., Miyoshi T. and Cardinali C. (2009). Analysis sensitivity calculation within an ensemble Kalman filter. *Q. J. R. Meteor. Soc.*, **135**, 1842–1851.

Healy, S. B. and Thépaut, J.-N. (2006). Assimilation experiments with CHAMP GPS radio occultation measurements. *Q. J. R. Meteor. Soc.*, **132**, 605–623.

Hoaglin, D. C. and Welsch, R. E. (1978). The hat matrix in regression and ANOVA. *Am. Statist.*, **32**, 17–22 [Corrigenda **32**, 146].

Hoaglin, D. C., Mosteller, F., and Tukey, J.W. (1982). *Understanding Robust and Exploratory Data Analysis.* Wiley, New York.

Langland R. and Baker, N.L. (2004). Estimation of observation impact using the NRL atmospheric variational data assimilation adjoint system. *Tellus*, **56A**, 189–201.

Lorenc, A. (1986). Analysis methods for numerical weather prediction. *Q. J. R. Meteor. Soc.*, **112**, 1177–1194.

Rabier, F. and Courtier, P. (1992). Four-dimensional assimilation in the presence of baroclinic instability. *Q. J. R. Meteor. Soc.*, **118**, 649–672.

Rabier, F., Järvinen, H., Klinker, E., Mahfouf, J. F., and Simmons, A. (2000). The ECMWF operational implementation of four-dimensional variational assimilation. Part I: Experimental results with simplified physics. *Q. J. R. Meteor. Soc.*, **126**, 1143–1170.

Rabier, F., Fourrié, N., Chafaï, D., and Prunet, P. (2002). Channel selection methods for infrared atmospheric sounding interferometer radiances. *Q. J. R. Meteor. Soc.*, **128**, 1011–1027.

Radnoti, G., Bauer, P., McNally, A., Cardinali, C., Healy, S., and de Rosnay, P. (2010). ECMWF study on the impact of future developments of the space-based observing system on numerical weather prediction. ECMWF Tech Memo. **638**.

Shen, X., Huang, H., and Cressie, N. (2002). Nonparametric hypothesis testing for a spatial signal. *J. Am. Statist. Assoc.*, **97**, 1122–1140.

Talagrand, O. (1997). Assimilation of observations, an Introduction. *J. Meteorol. Soc. Japan*, **75**, 191–209.

Thépaut, J.-N., Hoffman, R. N., and Courtier, P. (1993). Interactions of dynamics and observations in a four-dimensional variational assimilation. *Mon. Weather Rev.*, **121**, 3393–3414.

Thépaut, J.-N., Courtier, P., Belaud, G., and Lemaître, G. (1996). Dynamical structure functions in four-dimensional variational assimilation: A case study. *Q. J. R. Meteor. Soc.*, **122**, 535–561.

Tukey, J. W. (1972). Data analysis, computational and mathematics. *Q. Appl. Math.*, **30**, 51–65.

Velleman, P. F. and Welsch, R. E. (1981). Efficient computing of regression diagnostics. *Am. Statist.*, **35**, 234–242.

Wahba, G. (1990). *Spline Models for Observational Data.* SIAM, CBMS-NSF, Regional Conference Series in Applied Mathematic, Vol. 59.

Wahba, G., Johnson, D. R., Gao, F., and Gong, J. (1995). Adaptive tuning of numerical weather prediction models: randomized GCV in three- and four-dimensional data assimilation. *Mon. Weather Rev.*, **123**, 3358–3369.

Ye, J. (1998). On measuring and correcting the effect of data mining and model selection. *J. Am. Statist. Assoc.*, **93**, 120–131.

Zhu, Y. and Gelaro, R. (2008). Observation sensitivity calculations using the adjoint of the gridpoint statistical interpolation (GSI) analysis system. *Mon. Weather Rev.*, **136**, 335–351.

6
Observation impact on the short-range forecast

C. Cardinali

European Centre for Medium-Range Weather Forecasts (ECMWF)
Reading, United Kingdom

Advanced Data Assimilation for Geosciences. First Edition.
Edited by É. Blayo, M. Bocquet, E. Cosme, and L. F. Cugliandolo.

Chapter Contents

This chapter illustrates the concept of forecast error sensitivity to observations and its use for diagnostic purposes. The tool presented here computes the contribution of all observations to the forecast error: a positive contribution is associated with forecast error increase and a negative contribution with forecast error decrease. The forecast range investigated is 24 hours. It can be seen that, globally, the assimilated observations decrease the forecast error. Locally however, poor performance can also be found. The forecast deterioration can be related either to the data quality or to the data assimilation and forecast system. The data impact on the forecast is spatially and also temporally variable. It depends on atmospheric regimes, which may or may not be well-represented by the model or by the data. An example of a routine diagnostic assessment of observational impact on the short-range forecast performance is shown. The example also illustrates the tool's flexibility to represent different degrees of detail of forecast improvement or deterioration.

6.1 Introduction

The European Centre for Medium-Range Weather Forecasts (ECMWF) four-dimensional variational system (4D-Var; Rabier et al., 2000) handles a large variety of both space- and surface-based meteorological observations (more than 30 million a day) and combines the observations with the prior (or background) information on the atmospheric state. A comprehensive linearized and nonlinear forecast model is used, counting a number of degrees of freedom of the order of 10^8.

The assessment of the observational contribution to analysis (Cardinali et al., 2004; Chapnik et al., 2004; Lupu et al., 2011) and forecast is among the most challenging diagnostics in data assimilation and numerical weather prediction. For the forecast, the assessment of the forecast performance can be achieved by adjoint-based observation sensitivity techniques that characterize the forecast impact of every measurement (Baker and Daley, 2000; Langland and Baker, 2004; Cardinali and Buizza, 2004; Morneau et al., 2006; Xu et al., 2006; Zhu and Gelaro, 2008; Cardinali, 2009). The technique computes the variation in the forecast error due to the assimilated data. In particular, the forecast error is measured by a scalar function of the model parameters, namely wind, temperature, humidity, and surface pressure, that are more or less directly related to the observable quantities.

In general, the adjoint methodology can be used to estimate the sensitivity measure with respect to any assimilation system parameter of importance. For example, Daescu (2008) derived a sensitivity equation of an unconstrained variational data assimilation system from the first-order necessary condition with respect to the main input parameters: observation, background, and observation and background error covariance matrices.

The *forecast sensitivity to observation* technique (FSO) is complementary to the observing system experiments (OSEs) that have been the traditional tool for estimating data impact in a forecasting system (Bouttier and Kelly, 2001; English et al., 2004; Lord et al., 2004; Kelly, 2007; Radnoti et al., 2010, 2012). Very important is the use of OSEs in complement to FSO to highlight the contribution of, for example, a particular data set and to address the causes of degradation or improvement which FSO measures.

The main differences between adjoint-based and OSE techniques are as follows:

- The adjoint-based observation sensitivity measures the impact of observations when the entire observational dataset is present in the assimilation system, while the observing system is modified in the OSE. In fact, each OSE experiment differs from the others in terms of assimilated observations.
- The adjoint-based technique measures the impact of observations separately at every analysis cycle versus the background, while the OSE measures the total impact of removing data information from both background and analysis.
- The adjoint-based technique measures the response of a single forecast metric to all perturbations of the observing system, while the OSE measures the effect of a single perturbation on all forecast metrics.
- The adjoint-based technique is restricted by the tangent linear assumption and is therefore valid for forecasts up to 2 days, while the OSE can measure the data impact on longer-range forecasts and in nonlinear regimes.

This chapter introduces the mathematical concept and the application of the forecast sensitivity to the observation tool. The general ECMWF system performance in the 24-hour-range forecast is shown as derived by the diagnostic tool. In Section 6.2, the theoretical background of the FSO and the calculation of the forecast error contribution (FEC) from observations are shown. The ECMWF forecast performance is illustrated in Section 6.3 and conclusions are drawn in Section 6.4.

6.2 Observational impact on the forecast

6.2.1 Linear analysis equation

Data assimilation systems for numerical weather prediction provide estimates of the atmospheric state $\mathbf{x}$ by combining meteorological observations $\mathbf{y}$ with prior (or background) information $\mathbf{x}_b$. A simple Bayesian normal model provides the solution as the posterior expectation for $\mathbf{x}$, given $\mathbf{y}$ and $\mathbf{x}_b$. The same solution can be achieved from a classical *frequentist* approach, based on a statistical linear analysis scheme providing the best linear unbiased estimate (BLUE) (Talagrand, 1997) of $\mathbf{x}$, given $\mathbf{y}$ and $\mathbf{x}_b$. The optimal general least-squares solution to the analysis problem (see Lorenc, 1986) can be written as

$$\mathbf{x}_a = \mathbf{K}\mathbf{y} + (\mathbf{I}_n - \mathbf{K}\mathbf{H})\mathbf{x}_b. \tag{6.1}$$

The vector $\mathbf{x}_a$ is called the *analysis*. The gain matrix $\mathbf{K}$ (of dimension $n \times p$, with n being the dimension of the state vector and p that of the observation vector) takes into account the respective accuracies of the background vector $\mathbf{x}_b$ and the observation vector $\mathbf{y}$ as defined by the $(n \times n)$-dimensioned covariance matrix $\mathbf{B}$ and the $(p \times p)$-dimensioned covariance matrix $\mathbf{R}$, with

$$\mathbf{K} = (\mathbf{B}^{-1} + \mathbf{H}^{\mathrm{T}}\mathbf{R}^{-1}\mathbf{H})^{-1}\mathbf{H}^{\mathrm{T}}\mathbf{R}^{-1}. \tag{6.2}$$

$\mathbf{I}_n$ is the $n \times n$ identity matrix. Here, $\mathbf{H}$ is a $(p \times n)$-dimensioned matrix interpolating the background fields to the observation locations, and transforming the model variables to observed quantities (e.g. radiative transfer calculations transforming the model's temperature, humidity, ozone, etc. to brightness temperatures as observed by satellite instruments). In the 4D-Var context introduced above, $\mathbf{H}$ is defined to also include the propagation of the atmospheric state vector by the forecast model to the time at which the observations were recorded.

From (6.1), the sensitivity of the analysis system with respect to the observations can be derived from

$$\frac{\delta \mathbf{x}_a}{\delta \mathbf{y}} = \mathbf{K}^{\mathrm{T}}. \tag{6.3}$$

Equation (6.3) provides the observational influence in the analysis (Cardinali et al., 2004; see also Chapter 4 of this proceedings' book).

6.2.2 Sensitivity equation

Baker and Daley (2000) derived the forecast sensitivity equation with respect to the observations in the context of variational data assimilation. Let us consider a scalar function J of the forecast error. The sensitivity of J with respect to the observations can be obtained using a simple derivative chain as

$$\frac{\delta J}{\delta \mathbf{y}} = \frac{\delta J}{\delta \mathbf{x}_a} \frac{\delta \mathbf{x}_a}{\delta \mathbf{y}}, \tag{6.4}$$

where $\delta J / \delta \mathbf{x}_a$ is the sensitivity of the forecast error to the initial conditions (Rabier et al., 1996; Gelaro et al., 1998). The forecast error is mapped onto the initial conditions by the adjoint of the model, providing, for example, regions that are particularly sensitive to forecast error growth (see Section 6.2.3). By using (6.2) and (6.3), the forecast sensitivity to the observations becomes

$$\frac{\delta J}{\delta \mathbf{y}} = \mathbf{K}^{\mathrm{T}} \frac{\delta J}{\delta \mathbf{x}_a} = \mathbf{R}^{-1}\mathbf{H}(\mathbf{B}^{-1} + \mathbf{H}^{\mathrm{T}}\mathbf{R}^{-1}\mathbf{H})^{-1} \frac{\delta J}{\delta \mathbf{x}_a}, \tag{6.5}$$

where $(\mathbf{B}^{-1} + \mathbf{H}^{\mathrm{T}}\mathbf{R}^{-1}\mathbf{H})^{-1}$ is the analysis error covariance matrix $\mathbf{A}$. In practice, a second-order sensitivity gradient is needed (Langland and Baker, 2004; Errico, 2007) to obtain the information related to the forecast error, because the first-order sensitivity gradient only contains information on the suboptimality of the assimilation system (see Section 6.2.3 and Cardinali, 2009).

The forecast error is defined by $J = (1/2)\langle \mathbf{e}_t, \mathbf{C}\mathbf{e}_t \rangle$, where t stands for the *truth*. $\mathbf{e}$ denotes the forecast error with respect to temperature, vorticity, and divergence, as well as surface pressure. In practice, the forecast error is computed as the difference between the 24-hour forecast and the analysis valid at the same time. This implies that the verifying analysis is considered to be the *truth*:

- The verifying analysis is only a proxy of the truth and thus errors in the analysis can obscure the observation impact in the short-range forecast.

$\mathbf{C}$ is a matrix of weighting coefficients that integrate the elements of the forecast error to a dry energy norm that is a scalar:

- The energy norm is a suitable choice, because it depends directly on the most relevant model parameters also contained in the control vector $\mathbf{x}$ (the vectors used in the minimization process in, e.g., 4D-Var). Nevertheless, alternative functions of model parameters can be used.

Equation (6.5) can be solved (Krylov method; Van der Vorst, 2003) and the forecast error sensitivity to all assimilated observations is then derived. The numerical method used is shown in Section 6.2.4 (see also Cardinali, 2009).

6.2.3 Sensitivity gradient

Let us consider two forecasts of length f starting from $\mathbf{x}_a$ and length g starting from $\mathbf{x}_b$, $\mathbf{x}_b$ being the background field used in the $\mathbf{x}_a$ analysis. Both forecasts verify at time t. Following Langland and Baker (2004) and Errico (2007), the second-order sensitivity gradient is defined as

$$\frac{\delta J}{\delta \mathbf{x}_a} = \frac{\delta J_f}{\delta \mathbf{x}_a} + \frac{\delta J_g}{\delta \mathbf{x}_b}, \tag{6.6}$$

where $J_f = \langle(\mathbf{x}_f - \mathbf{x}_t), \mathbf{C}(\mathbf{x}_f - \mathbf{x}_t)\rangle/2$ and $J_g = \langle(\mathbf{x}_g - \mathbf{x}_t), \mathbf{C}(\mathbf{x}_g - \mathbf{x}_t)\rangle/2$ are quadratic measures of the two forecast errors ($\mathbf{x}_t$ being the verifying analysis) and $\mathbf{C}$ is the matrix of dry energy weighting coefficients. It is clear from (6.4) that the adjoint model maps the sensitivity (with respect to the forecast) of J_f into $\delta J_f/\delta \mathbf{x}_a$ along the trajectory f and the sensitivity of J_g into $\delta J_g/\delta \mathbf{x}_a$ along the trajectory g (for the first-order sensitivity gradient definition and computation, see Rabier et al., 1996; Gelaro et al., 1998). Equation (6.6) is schematically represented in Fig. 6.1. Let us now compare the first-order sensitivity gradient with the second-order one, expressing the variation of the forecast error due to the assimilation of observations, $J(\mathbf{e}_a) - J(\mathbf{e}_b)$, where $\mathbf{e}_a$ and $\mathbf{e}_b$ are the analysis and the background error. Following Langland and Baker (2004), the second-order Taylor series decomposition is used to map such variation:

$$J(\mathbf{e}_b) - J(\mathbf{e}_a) = (\mathbf{e}_b - \mathbf{e}_a)^{\mathrm{T}} J'_{\mathbf{e}_a} + \frac{1}{2}(\mathbf{e}_b - \mathbf{e}_a)^{\mathrm{T}} J''_{\mathbf{e}_a}(\mathbf{e}_b - \mathbf{e}_a). \tag{6.7}$$

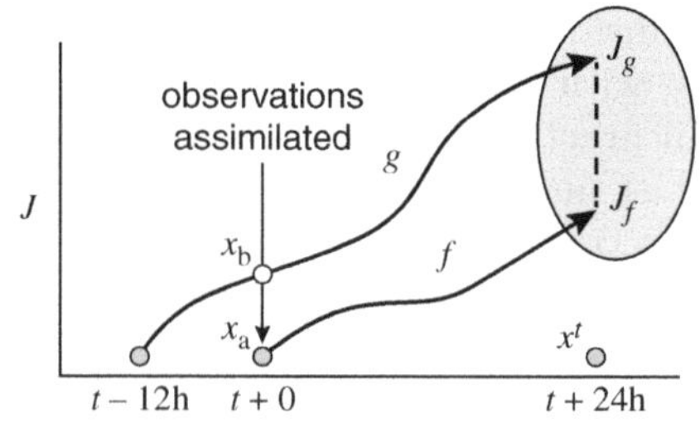

Fig. 6.1 Geometrical representation of the sensitivity gradient calculation expressed in (6.6).

Because the error cost function is quadratic, (6.7) reduces to

$$J(\mathbf{e}_b) - J(\mathbf{e}_a) = 2(\mathbf{e}_b - \mathbf{e}_a)^{\mathrm{T}}\mathbf{e}_a + (\mathbf{e}_b - \mathbf{e}_a)^{\mathrm{T}}(\mathbf{e}_b - \mathbf{e}_a), \tag{6.8}$$

which at first order is

$$J(\mathbf{e}_b) - J(\mathbf{e}_a) = 2\mathbf{d}^{\mathrm{T}}\mathbf{K}^{\mathrm{T}}\mathbf{e}_a. \tag{6.9}$$

In an optimal assimilation system, the right-hand side of this equation is on average zero (Talagrand, 2002), since, statistically, the innovation vector $\mathbf{d} = \mathbf{y} - \mathbf{H}\mathbf{x}_b$ and the analysis error are orthogonal. Therefore, it is clear that the results obtained by using the first-order sensitivity gradient only provide a measure of the suboptimality of the analysis system. It appears that it is necessary to include the second-order term in the FSO calculation.

6.2.4 Numerical solution

In an optimal variational analysis scheme, the analysis error covariance matrix $\mathbf{A}$ is approximately the inverse of the matrix of second derivatives (the Hessian) of the analysis cost function J_a (Rabier et al., 2000), i.e. $\mathbf{A} = (J_a'')^{-1}$ (Rabier and Courtier, 1992). Given the large dimension of the matrices involved, J_a'' and its inverse cannot be computed explicitly. The minimization is performed in terms of a transformed variable $\chi = \mathbf{L}^{-1}(\mathbf{x} - \mathbf{x}_b)$, with $\mathbf{L}$ chosen such that $\mathbf{B} = \mathbf{L}\mathbf{L}^{\mathrm{T}}$. The transformation $\mathbf{L}$ thus reduces the covariance of the prior to the identity matrix. In variational data assimilation, $\mathbf{L}$ is referred to as the change-of-variable operator (Courtier et al., 1998). Let us apply the change of variables in the analysis cost function and write

$$\begin{aligned} J_a(\mathbf{x}) &= \frac{1}{2}(\mathbf{x} - \mathbf{x}_b)^{\mathrm{T}}\mathbf{B}^{-1}(\mathbf{x} - \mathbf{x}_b) + \frac{1}{2}(\mathbf{H}\mathbf{x} - \mathbf{y})^{\mathrm{T}}R^{-1}(\mathbf{H}\mathbf{x} - \mathbf{y}) \\ &= \frac{1}{2}\chi^{\mathrm{T}}\chi + \frac{1}{2}(\mathbf{H}\mathbf{L}\chi - \mathbf{y})^{\mathrm{T}}R^{-1}(\mathbf{H}\mathbf{L}\chi - \mathbf{y}) = J_a(\chi). \end{aligned} \tag{6.10}$$

The Hessian becomes

$$J_a''(\chi) = \mathbf{I} + \mathbf{L}^{\mathrm{T}}\mathbf{H}^{\mathrm{T}}\mathbf{R}^{-1}\mathbf{H}\mathbf{L}. \tag{6.11}$$

By applying the change of variables in (6.7) and using (6.8), the forecast sensitivity to the observations is expressed as

$$\frac{\delta J}{\delta \mathbf{y}} = \mathbf{R}^{-1}\mathbf{H}\mathbf{L}(\mathbf{I} + \mathbf{L}^{\mathrm{T}}\mathbf{H}^{\mathrm{T}}\mathbf{R}^{-1}\mathbf{H}\mathbf{L})^{-1}\mathbf{L}^{\mathrm{T}}\frac{\delta J}{\delta \mathbf{x}_a}. \tag{6.12}$$

Using the conjugate gradient algorithm, first the following equation for $\delta J/\delta\mathbf{y} = \mathbf{R}^{-1}\mathbf{H}\mathbf{z}$ is solved:

$$\begin{aligned} (\mathbf{I} + \mathbf{L}^{\mathrm{T}}\mathbf{H}^{\mathrm{T}}\mathbf{R}^{-1}\mathbf{H}\mathbf{L})\mathbf{z} &= \mathbf{L}\mathbf{z}_a, \\ \mathbf{z}_a &= \frac{\delta J}{\delta \mathbf{x}_a}. \end{aligned} \tag{6.13}$$

The solution $\mathbf{z}$ lies in the Krylov subspace generated by the vector $\mathbf{L}^{\mathrm{T}}\mathbf{z}_a$ and the matrix $\mathbf{I} + \mathbf{L}^{\mathrm{T}}\mathbf{H}^{\mathrm{T}}\mathbf{R}^{-1}\mathbf{HL}$. The Krylov subspace dimension is the degree of the minimal polynomial of $\mathbf{I} + \mathbf{L}^{\mathrm{T}}\mathbf{H}^{\mathrm{T}}\mathbf{R}^{-1}\mathbf{HL}$. Therefore, if the degree is low, the Krylov method searches the solution on a small-dimensional space. The method is very efficient in an iterative solution of a linear system with large and sparse matrices (Van der Vorst, 2003). The forecast sensitivity to observations is then given by interpolating $\mathbf{z}$ (using the $\mathbf{H}$ operator) in the observation space and by normalizing with respect to the observation error covariance matrix $\mathbf{R}$.

6.2.5 Observation impact measure

Once the forecast sensitivity has been computed, the variation δJ of the forecast error expressed by J can be found by rearranging (6.1) and using the adjoint property for the linear operator:

$$
\begin{aligned}
\delta J = \left\langle \frac{\delta J}{\delta \mathbf{x}_a}, \delta \mathbf{x}_a \right\rangle &= \left\langle \frac{\delta J}{\delta \mathbf{x}_a}, \mathbf{K}(\mathbf{y} - \mathbf{H}\mathbf{x}_b) \right\rangle \\
&= \left\langle \mathbf{K}^{\mathrm{T}} \frac{\delta J}{\delta \mathbf{x}_a}, \mathbf{y} - \mathbf{H}\mathbf{x}_b \right\rangle \\
&= \left\langle \mathbf{K}^{\mathrm{T}} \frac{\delta J}{\delta \mathbf{x}_a}, \delta \mathbf{y} \right\rangle = \left\langle \frac{\delta J}{\delta \mathbf{y}}, \delta \mathbf{y} \right\rangle, \qquad (6.14)
\end{aligned}
$$

where $\delta \mathbf{x}_a = \mathbf{x}_a - \mathbf{x}_b$ are the analysis increments and $\delta \mathbf{y} = \mathbf{y} - \mathbf{H}\mathbf{x}_b$ is the innovation vector. δJ is computed across the 12-hour window; the sensitivity gradients $\delta J/\delta \mathbf{x}_a$, valid at the starting time of the 4D-Var window (09 and 21 UTC in the ECMWF system), are distributed by $\mathbf{K}^{\mathrm{T}}$, which incorporates the temporal dimension, over the 12-hour window. From (6.14), a few considerations should be taken into account:

- The forecast impact δJ (hereinafter called the forecast error contribution, FEC) of all observations assimilated depends on the forecast error ($J(\mathbf{e}) \rightarrow \delta J/\delta \mathbf{x}_a$), the assimilation system ($\mathbf{K}^{\mathrm{T}}$), and the difference between the observations and the model ($\mathbf{y} - \mathbf{H}\mathbf{x}_b$).
- Positive forecast error variation $\delta J > 0$ is synonymous of forecast degradation. Negative forecast error variation $\delta J < 0$ is synonymous of forecast improvement.
- The verifying analysis is only a proxy of the truth. Therefore, errors in the analysis can mask the observation impact in the forecast.
- Biases in the model can result in forecast degradation that is erroneously interpreted as an observation-related degradation.
- Since the computation is performed with the linearized model, only errors in the short-range forecast can be diagnosed.
- The forecast error is measured using a dry energy norm that depends on wind, temperature, and surface pressure. Therefore, observables depending on these parameters are rather well assessed. Moreover, the dependence of the forecast error on humidity is represented by the linearized moist process, so that the forecast impact of humidity observations is also fully assessed (Janiskova and Cardinali, in preparation).

- The variation of the forecast error due to a specific measurement can be summed over time and space in different subsets to compute the average contribution of different components of the observing system to the forecast error. For example, the contribution of all AMSU-A satellites (s) and channels (i) over time T will be

$$\delta J_{\text{AMSU-A}} = \sum_{s \subset S} \sum_{i \subset \text{channels}} \sum_{t \subset T} = \delta J^s_{i,t}.$$

  This is one of the most important characteristics of the tool, because it allows any necessary level of analysis granularity for a comprehensive investigation.

Given all the points above, it is clear that a full diagnostic assessment is necessary to establish the causes for a forecast error increase.

6.3 Results

The routinely computed observational impact from the operational ECMWF 4D-Var system (Rabier et al., 2000; Janiskova et al., 2002; Lopez and Moreau, 2005) is shown in Fig. 6.2 for September and October 2011. At ECMWF, the 'observation impact' suite runs one day behind the model suite, in time to recover the actual verifying analysis for the forecast error computation. The 24-hour forecast error contribution (FEC) of all the observing system components is computed and shown in Fig. 6.2(a) for different observation types as defined in Table 5.1 of Chapter 5. For technical reasons, microwave imagers (SSM/I and TMI) have not been considered in this study. The largest contribution to decreasing the forecast error is provided by AMSU-A ($\sim$ 25%), IASI, AIRS, and AIREP (aircraft data and GPS-RO observations account for 10% of the total impact, respectively). TEMP and SYNOP surface pressure observations contribute 5%, followed by AMVs and HIRS ($\sim$4%), then by ASCAT and DRIBU (3%). All other observations contribute less than 3%.

The error of the observation impact measure is also displayed in Fig. 6.2(a); it depends on the standard error and on the number of observation assimilated in that period. If the FEC measured variability is within the error range, the variation is not considered to be significant. In Fig. 6.2(b), the mean impact per individual observation is shown. In this case, the impact is independent of the observation number. The largest mean contribution is provided by DROP and DRIBU (surface pressure) observations, followed by the contribution of a second group of observations comprising MERIS, AMVs, ASCAT, GPS-RO, SYNOP, TEMP, AMSU-B, and AIREP. In contrast to the total forecast impact, which is largely provided by satellite observations, the largest per-observation impact is obtained from conventional observations. The difference between the two impact measures is mainly due to difference in observation accuracy, through which a single conventional observation is on average more influential in the analysis than a single satellite measurement.

The monthly variation of forecast impact is shown in Fig. 6.3 per observation type and for June–October 2011. The only significant temporal variation is observed for

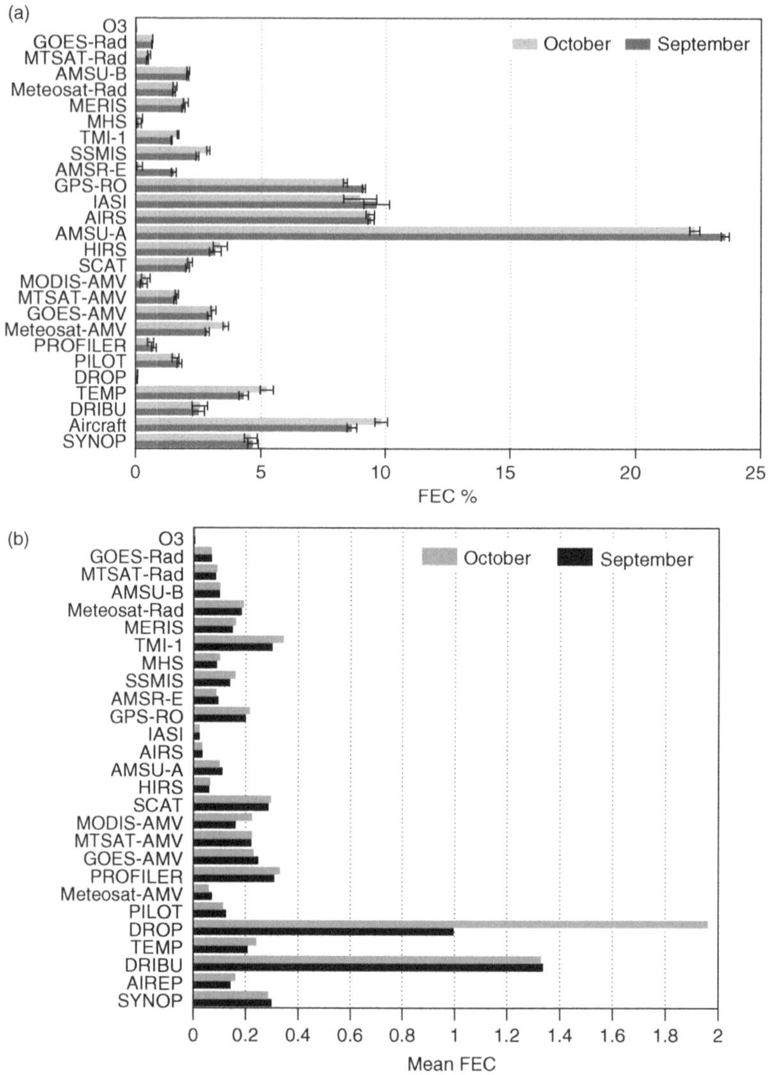

Fig. 6.2 Observation contribution to the global forecast error reduction grouped by observation type as defined in (5.1). The measure is given in percent and for the months of September and October. (a) Total forecast error contribution, where the error bars are computed using the standard error measure. (b) Average forecast error contribution (normalized by the number of observations used; unit J/kg).

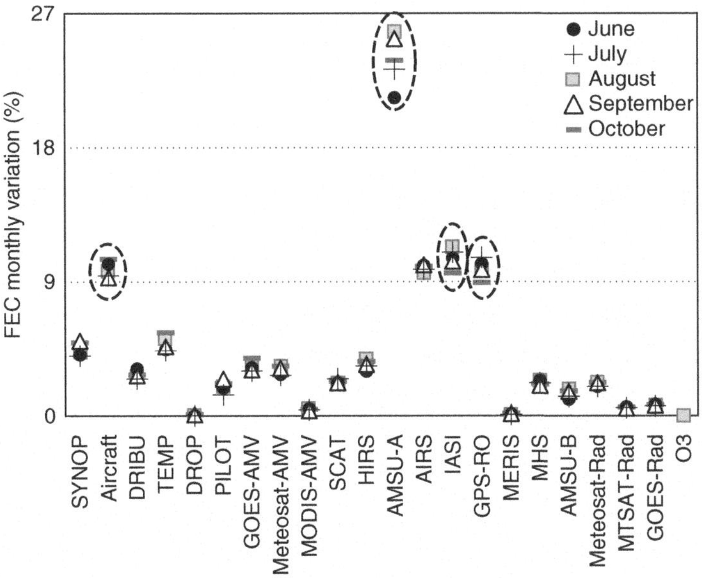

Fig. 6.3 Variation of total forecast error contribution in June, July, August, September, and October 2011 for the different observation types.

AMSU-A, with the largest forecast impact in August and September, and for GPS-RO and IASI in July and August, respectively.

The AMSU-A forecast impact has been analysed in more detail. In Fig. 6.4, the contribution of all channels to the forecast error decrease is shown. Channel 8 has the largest overall impact and the stratospheric channels (11–14) the smallest. There is no significant difference in performance between September and October. The geographical distribution of mean forecast improvement or deterioration from channel 8 is shown in Fig. 6.5 for September–October 2011. The METOP-A AMSU-A performance is compared with that of NOAA-15, since they have a similar satellite orbit. Nevertheless, there is a difference in the measurement time, since METOP-A crosses the equator at around 9:30 and NOAA-15 at 16:30. The overall impact of the instruments on the two satellites is comparable. The geographical location of the improvement instead differs quite substantially, with the exception of the polar and central Southern Hemisphere regions, where both perform similarly well. In the western part of the Southern Hemisphere, METOP-A reduces the forecast error, while NOAA-15 increases it. In contrast, in the eastern part, NOAA-15 shows a large and consistent improvement, whereas METOP-A shows small areas of degradation. A similar impact pattern is observed for the tropics and the Northern Hemisphere.

Once the area of degradation or improvement and the periods of interest have been determined, the addition of OSEs can help to determine the possible causes. For example, it can be necessary to identify the explicit contribution of AMSU-A channel 8

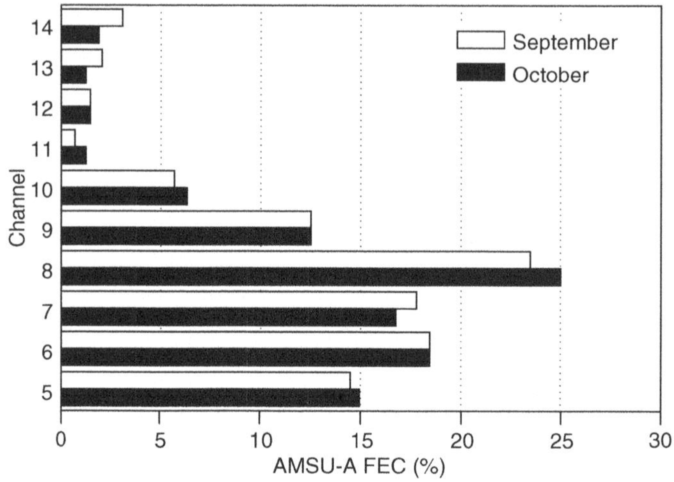

Fig. 6.4 Total forecast error contribution for all AMSU-A instruments as a percentage. The impact is shown for all assimilated channels and for September and October 2011.

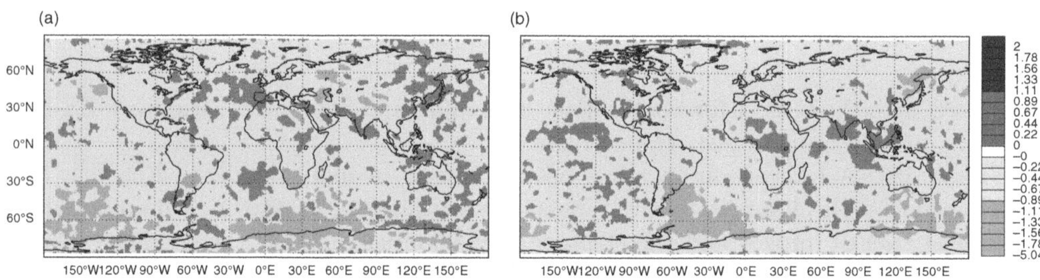

Fig. 6.5 Mean forecast error contribution for AMSU-A channel 8 onboard METOP-A (a) and NOAA-15 (b) for the whole globe. Units are joules.

to the degradation over the Atlantic (METOP-A) or central Africa (NOAA-15). Comparison between the experiment where channel 8 is not assimilated and the control experiment (in which it is assimilated) will add information for the specific case and will help in evaluating suitability of the assimilation procedure for this data.

The variation of forecast impact with time for AMSU-A channel 8 is shown for the North Atlantic region in Fig. 6.6(a,b). Again, METOP-A (a) and NOAA-15 (b) are compared. METOP-A shows much larger temporal variability than NOAA-15 and displays more events of detrimental impact (positive values) than NOAA-15, which, except for a few occasions, performs rather well over the entire period. The observation departures are also different: the departures with respect to the background (black line in Figs. 6.6(c,d)) are smaller for METOP-A (on average 0.05 K) until the beginning of

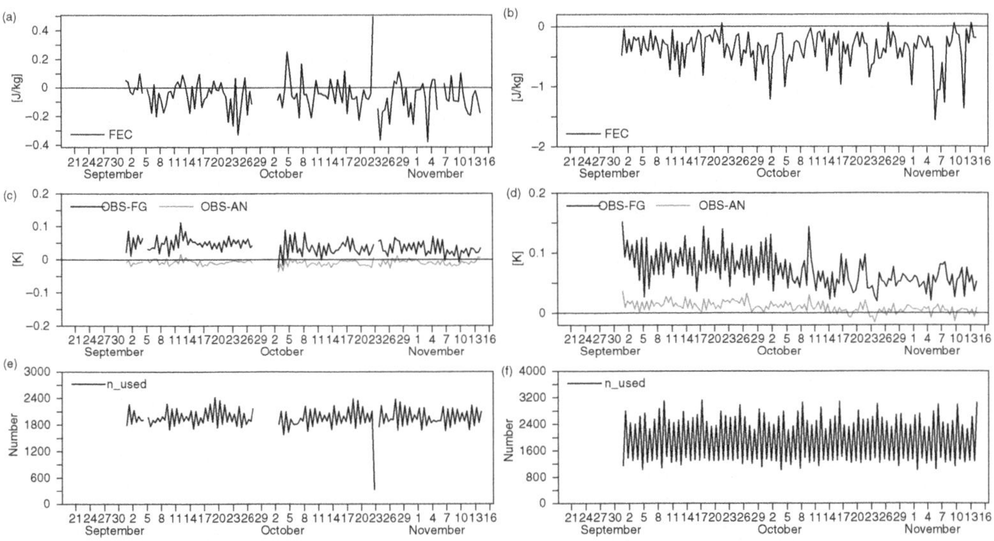

Fig. 6.6 Daily variation of mean FEC (a, b), background (black line) and analysis (grey line) departure (c, d) and observation number (e, f) over the Atlantic region from September to mid-November 2011 for METOP-A (a, c, e) and NOAA-15 (b, d, f) AMSU-A channel 8.

October, when the assimilation of METOP-A restarted after a break of three days due to routine satellite maintenance. After 2 October, METOP-A background departures become smaller, but the largest absolute decrease (0.025 K) is observed instead for NOAA-15. And, from October onwards, the observation departure from the analysis (grey line in Figs. 6.6(c, d)) becomes very similar (close to zero on average), while, before that day, NOAA-15 shows a small positive bias. Interestingly, the forecast reduction also changes: METOP-A shows larger variability than before and, to a lesser extent, so does NOAA-15. However, on average, as shown in Fig. 6.5, the impact of the two satellites is quantitatively similar, though different in terms of location. Over the Pacific, for example, METOP-A and NOAA-15 time series of the forecast performance are more similar, with METOP-A showing also few large improvements (not shown). The number of measurements provided by the two satellites is very similar (Figs. 6.6(e,f)). The larger forecast error reduction of NOAA-15 with respect to METOP-A over the North Atlantic is due to the measurement time (Fig. 6.7). In fact, the NOAA-15 satellite crosses the Atlantic close to 9 UTC, which corresponds to the end of the 12-hour assimilation window in the 4D-Var system used (Fig. 6.7, light grey), while the METOP-A platform is observing the Atlantic at the beginning of the assimilation window (Fig. 6.7, dark grey). Owing to the evolution of the model error covariance matrix **B** across the assimilation window, observations assimilated towards the end of the window are more influential than observations assimilated at the beginning of the window.

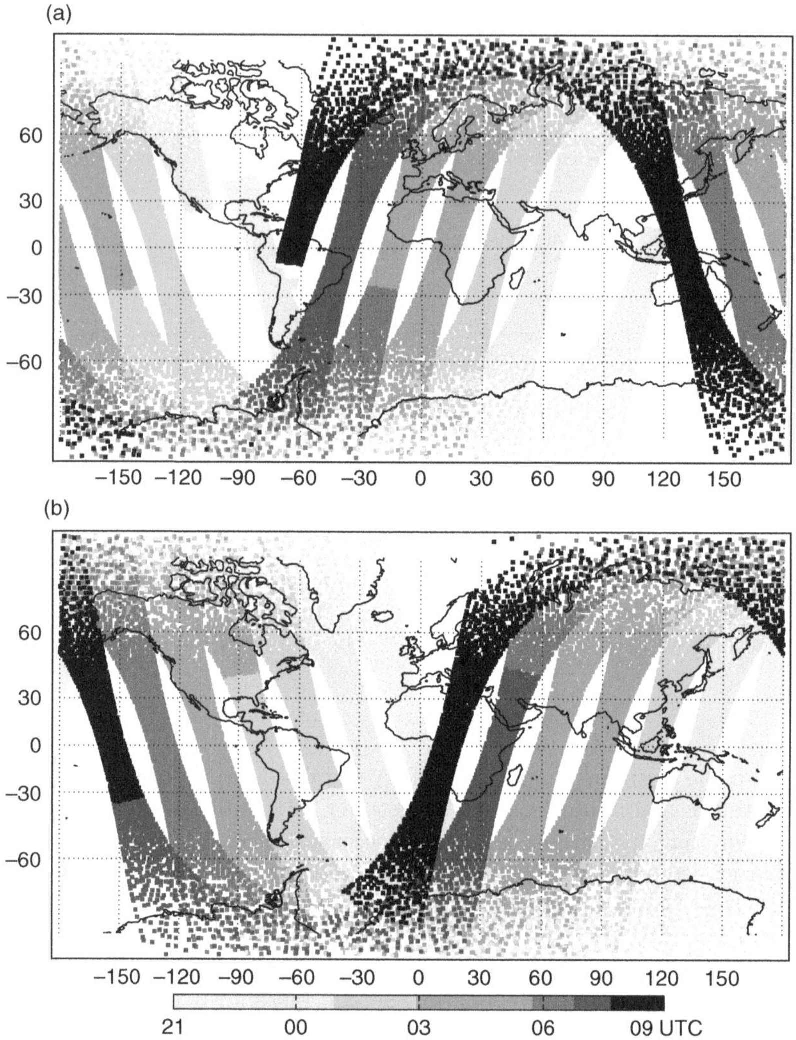

Fig. 6.7 Data coverage for METOP-A (a) and NOAA-15 (b) AMSU-A. The swath shading is related to the measurement time from 21 UTC to 9 UTC.

6.4 Conclusion

Over the last few years, the potential of using derived adjoint-based diagnostic tools has been widely exploited. Recently, a compact derivation of the 4D-Var sensitivity equations using the theoretical framework of the implicit function has been performed (Daescu, 2008). The analytical formulation of the sensitivity equations with respect to an extended set of input parameters has been shown and numerical applications will soon follow. This chapter has introduced the use of the forecast sensitivity with respect to time-distributed observational data, for the first time in a 12-hour 4D-Var assimilation system, as a diagnostic tool to monitor the observation performance in the short-range forecast. The fundamental principles on which the forecast sensitivity diagnostic tool is based have been illustrated and an example of a routine diagnostic has been provided.

The forecast sensitivity to observations can only be used to diagnose the impact on the short-range forecast, namely for periods of 24–48 hours, given the use of the adjoint model and the implied linearity assumption. The tool allows the computation and visualization of the impact for each assimilated measurement, and therefore the diagnostic can be performed from local to global scales and for any period of interest. The use of the second-order sensitivity gradient is necessary to identify the forecast impact of the observations. In fact, the projected first-order sensitivity gradient only contains information on the suboptimality of the assimilation system. The tool use characteristics have been explained: in particular, the dependence of the tool on the verifying analysis used to compute the forecast error and the dependence of the sensitivity tool on the scalar function representing the global forecast error (energy norm). The function of the global forecast error is first mapped onto the initial conditions (using the adjoint operator of the model forecast) and then into the observation space (using the adjoint operator of the analysis system). The forecast error sensitivity of a specific measurement is transformed on forecast error variation via a scalar product with the innovation vector.

The global impact of observations is found to be positive, and the forecast errors decrease for all data type when monthly averaged. In fact, because of the statistical nature of the assimilation procedure, the observation impact must be averaged over a long enough period to be significant.

An example of observation impact monitoring has been shown, and from the global performance assessment the specific performance of one AMSU-A channel has been illustrated for two polar orbiting satellites, namely METOP-A and NOAA-15, covering a similar orbit. The causes of degradation or improvement can be further investigated using observing system experiments.

Given the dependence of some observation types on the meteorological situation, it is suggested that the forecast sensitivity to the observation diagnostic tool be run on an operational basis and in relation to the operational suite error. A constant monitoring of the performance of the model forecast would allow the use of the observation network in an adaptive way in which observations with negative impact can be investigated and potentially denied in real time.

Acknowledgements

The author thanks Mohamed Dahoui, Anne Fouilloux, and Fernando Prates for their continued support in monitoring, displaying, and diagnosing the forecast performance of all observations assimilated at ECMWF.

References

Baker, N. L. and Daley, R. (2000). Observation and background adjoint sensitivity in the adaptive observation targeting problem. *Q. J. R. Meteor. Soc.*, **126**, 1431–1454.

Bouttier, F. and Kelly, G. (2001). Observing system experiments in the ECMWF 4D-Var data assimilation system. *Q. J. R. Meteor. Soc.*, **127**, 1469–1488.

Cardinali, C. (2009). Monitoring the observation impact on the short-range forecast. *Q. J. R. Meteor. Soc.*, **135**, 239–250.

Cardinali, C., Pezzulli, S., and Andersson, E. (2004). Influence matrix diagnostics of a data assimilation system. *Q. J. R. Meteor. Soc.*, **130**, 2767–2786.

Cardinali, C. and Buizza, R. (2004). Observation sensitivity to the analysis and the forecast: a case study during ATreC targeting campaign. In *Proceedings of the First THORPEX International Science Symposium, 6–10 December 2004, Montreal, Canada*, WMO TD 1237 WWRP/THORPEX N. 6.

Chapnik, B., Desroziers, G., Rabier, F., and Talagrand, O. (2006). Diagnosis and tuning of observation error in a quasi-operational data assimilation setting. *Q. J. R. Meteor. Soc.*, **132**, 543–565.

Courtier, P., Andersson, E., Heckley, W., Vasiljevic, D., Hamrud, M., Hollingsworth, A., Rabier, F., Fisher, M., and Pailleux, J. (1998). The ECMWF implementation of three-dimensional variational assimilation (3D-Var). Part I: Formulation. *Q. J. R. Meteor. Soc.*, **124**, 1783–1807.

Daescu, D. N. (2008). On the sensitivity equations of 4D-Var data assimilation. *Mon. Weather Rev.*, **136**, 3050–3065.

English, S., Saunders, R., Candy, B., Forsythe, M., and Collard, A. (2004). Met Office satellite data OSEs. In *Proceedings of Third WMO Workshop on the Impact of Various Observing Systems on Numerical Weather Prediction, Alpbach, Austria.* WMO/TD, 1228, pp. 146–156.

Errico, R. (2007). Interpretation of an adjoint-derived observational impact measure. *Tellus*, **59A**, 273–276.

Gelaro, R., Buizza, R., Palmer, T. N., and Klinker, E. (1998). Sensitivity analysis of forecast errors and the construction of optimal perturbations using singular vectors. *J. Atmos. Sci.*, **55**, 1012–1037.

Janiskova, M., Mahfouf, J.-F., Morcrette, J.-J., and Chevallier, F. (2002). Linearized radiation and cloud schemes in the ECMWF model: development and evaluation. *Q. J. R. Meteorol. Soc.*, **128**, 1505–1527.

Kelly, G. (2007). Evaluation of the impact of the space component of the Global Observation System through observing system experiments. ECMWF Newsletter, Autumn.

Langland, R. and Baker, N. L. (2004). Estimation of observation impact using the NRL atmospheric variational data assimilation adjoint system. *Tellus*, **56A**, 189–201.

Lorenc, A. (1986). Analysis methods for numerical weather prediction. *Q. J. R. Meteorol. Soc.*, **112**, 1177–1194.

Lopez, P. and Moreau, E. (2005). A convection scheme for data assimilation: description and initial tests. *Q. J. R. Meteor. Soc.*, **131**, 409–436.

Lord, S., Zapotocny, T., and Jung, J. (2004). Observing system experiments with NCEP US global forecast system. In *Proceedings of Third WMO Workshop on the Impact of Various Observing Systems on Numerical Weather Prediction, Alpbach, Austria.* WMO/TD 1228, pp. 56–62.

Lupu, C., Gauthier, P., and Laroche, S. (2011). Evaluation of the impact of observations on analyses in 3D- and 4D-Var based on information content. *Mon. Weather Rev.*, **139**, 726–737.

Morneau, J. Pellerin, S., Laroche, S., and Tanguay, M. (2006). Estimation of adjoint sensitivity gradients in observation space using the dual (PSAS) formulation of the Environment Canada Operational 4D-Var. In *Proceedings of Second THORPEX International Science Symposium, 4–8 December 2006, Landshut, Germany,* WMO/TD 1355, WWRP/THORP6X 7, pp. 162–163.

Rabier, F. and Courtier, P. (1992). Four-dimensional assimilation in the presence of baroclinic instability. *Q. J. R. Meteorol. Soc.*, **118**, 649–672.

Rabier, F., Klinker, E., Courtier, P., and Hollingsworth A. (1996). Sensitivity of forecast errors to initial condition. *Q. J. R. Meteorol. Soc.*, **122**, 121–150.

Rabier, F., Järvinen, H., Klinker, E., Mahfouf, J. F., and Simmons, A. (2000). The ECMWF operational implementation of four-dimensional variational assimilation. Part I: experimental results with simplified physics. *Q. J. R. Meteorol. Soc.*, **126**, 1143–1170.

Radnoti, G., Bauer, P., McNally, A., Cardinali, C., Healy, S., and de Rosnay, P. (2010). ECMWF study on the impact of future developments of the space-based observing system on numerical weather prediction. ECMWF Tech. Memo., 638.

Radnoti, G., Bauer, P. , McNally, A., and Horanyi, A. (2012). ECMWF study to quantify the interaction between terrestrial and space-based observing systems on numerical weather prediction skill. ECMWF Project Report.

Talagrand, O. (1997). Assimilation of observations, an Introduction. *J. Meteorol. Soc. Japan*, **75**, 191–209.

Talagrand O. (2002). A posteriori validation of assimilation algorithms. In *Proceedings of NATO Advanced Study Institute on Data Assimilation for the Earth System, Acquafreda, Maratea, Italy.*

Tompkins, A. M. and Janiskova, M. (2004). A cloud scheme for data assimilation: description and initial tests. *Q. J. R. Meteor. Soc.*, **130**, 2495–2517.

Van der Vorst, H. A. (2003). *Iterative Krylov Methods for Large Linear Systems.* Cambridge University Press, Cambridge.

Xu, L., Langland, R., Baber, N., and Rosmond, T. (2006). Development and resting of the adjoint of NAVDAS-AR. In *Proceedings of Seventh latenational Workshop on Adjoint Applications in Dynamic Meterology, 8–13 October 2006, Obergurgl, Austria.*

Zhu, Y. and Gelaro, R. (2008). Observation sensitivity calculations using the adjoint of the gridpoint statistical interpolation (GSI) analysis system. *Mon. Weather Rev.*, **136**, 335–351.

Part II

Specialized lectures

7
Background error covariances: estimation and specification

L. Berre

Météo-France/CNRS
CNRM/GAME
Toulouse, France

Advanced Data Assimilation for Geosciences. First Edition.
Edited by É. Blayo, M. Bocquet, E. Cosme, and L. F. Cugliandolo.

Chapter Contents

This chapter deals with the estimation and specification of background error covariances in data assimilation systems, with some examples for variational data assimilation formulations in meteorology, although issues are relatively general and are often also valid for Kalman-filter formulations and, for example, oceanographic systems.

While the definition of background error is relatively trivial, since it corresponds to the difference between the background state and the true state, there are several scientific challenges to be kept in mind when considering associated covariance modelling aspects for realistic geophysical systems.

A first difficulty is that the true atmospheric state is never exactly known. This means that background errors and associated covariances can, at best, only be estimated.

A second difficulty is that error contributions to the background error are relatively complex, since the background is the result of a complex data assimilation procedure, which involves observations assimilated in the past, the analysis formulation, and the forecast model operator.

A third typical difficulty is that the size of the background error covariance matrix is huge, since it is equal to the square of the number of model gridpoints. This implies that the full covariance matrix cannot be calculated or even stored in memory.

These different challenges have lead to the development of error simulation techniques, the use of observation-based diagnostics, the diagnosis of the different covariance components, and the use of sparse covariance modelling approaches.

The structure of this paper will follow these four types of solutions to the aforementioned challenges.

7.1 Error equations and their simulation

As will be shown in this section, while background error is difficult to estimate as such, one attractive feature is that the fundamental underlying error equations are well known to some extent. This is at the heart of the development and use of error simulation techniques.

In order to make the discussion more concrete, a 6-hour data assimilation cycle will be considered as an example, which means that the background corresponds to a 6-hour forecast in this case.

7.1.1 Analysis and forecast error equations in general

The analysis state equation corresponding to the best linear unbiased estimate (BLUE) formalism (e.g. Talagrand, 1997) can be written as follows:

$$\mathbf{x}_a = (\mathbf{I} - \mathbf{KH})\mathbf{x}_b + \mathbf{K}\mathbf{y}_o, \tag{7.1}$$

where $\mathbf{x}_a$ is the analysis, $\mathbf{x}_b$ is the background, $\mathbf{y}_o$ is the observation vector, $\mathbf{H}$ is the observation operator (assumed here to be linear for the sake of simplicity), and $\mathbf{K}$ is the specified gain matrix (which can be suboptimal). The same kind of equation can be written for the true state equation, which is related to the idempotent property of the BLUE equation:

$$\mathbf{x}_t = (\mathbf{I} - \mathbf{KH})\mathbf{x}_t + \mathbf{KH}\mathbf{x}_t.$$

The difference between these two equations leads to the analysis error equation:

$$\mathbf{e}_a = \mathbf{x}_a - \mathbf{x}_t,$$

i.e.

$$\mathbf{e}_a = (\mathbf{I} - \mathbf{KH})\mathbf{e}_b + \mathbf{K}\mathbf{e}_o. \tag{7.2}$$

It can be seen that the same operators $\mathbf{I} - \mathbf{KH}$ and $\mathbf{K}$ appear in this analysis error equation (7.2) and in the analysis state equation (7.1). Moreover, (7.2) indicates that the analysis error is a relatively simple function of the background error and of the observation error.

Another important aspect to be kept in mind is that this analysis error equation remains valid even in the case where the gain matrix is suboptimal (which is the actual situation in general—e.g. because covariances always have to be estimated and thus approximated). Besides, in the more general context where the analysis formulation is partly nonlinear, as in 4D-Var with nonlinear trajectory updates, it can be shown (Desroziers et al., 2009) that the above analysis error equation remains valid.

The same kind of formalism can be considered for the forecast step. For the sake of simplicity, the forecast model operator will be assumed to be linear. The forecast state equation can thus be written as follows:

$$\mathbf{x}_{f+} = \mathbf{M}\mathbf{x}_a,$$

where $\mathbf{M}$ is the forecast model operator, and $\mathbf{x}_{f+}$ is the forecast state valid at time t_+ (i.e. six hours after the analysis time in the case of six-hour assimilation cycling). The true forecast evolution can be written as follows:

$$\mathbf{x}_{t+} = \mathbf{M}_t\mathbf{x}_t,$$

where $\mathbf{M}_t$ is the exact forecast operator. The forecast error equation can thus be expanded as follows:

$$\begin{aligned}\mathbf{e}_f &= \mathbf{x}_{f+} - \mathbf{x}_{t+} = \mathbf{M}\mathbf{x}_a - \mathbf{M}_t\mathbf{x}_t\\ &= \mathbf{M}\mathbf{x}_a - \mathbf{M}\mathbf{x}_t + \mathbf{M}\mathbf{x}_t - \mathbf{M}_t\mathbf{x}_t\\ &= \mathbf{M}(\mathbf{x}_a - \mathbf{x}_t) + (\mathbf{M} - \mathbf{M}_t)\mathbf{x}_t,\end{aligned}$$

i.e.

$$\mathbf{e}_f = \mathbf{M}\mathbf{e}_a + \mathbf{e}_m, \tag{7.3}$$

where $\mathbf{e}_m = (\mathbf{M} - \mathbf{M}_t)\mathbf{x}_t$ is the model error associated with $\mathbf{M}$. This corresponds to model imperfections (measured by $\mathbf{M} - \mathbf{M}_t$) that are accumulated during the considered forecast integration (i.e. six-hour accumulated model error).

This equation indicates that the forecast error can be seen as resulting from two components. The first, $\mathbf{M}\mathbf{e}_a$, corresponds to the forecast evolution of analysis errors by the (imperfect) model operator $\mathbf{M}$. The second, $\mathbf{e}_m$, corresponds to the model error contribution, associated with specific accumulated effects of imperfections in $\mathbf{M}$.

7.1.2 Analysis and forecast perturbation equations in EnDA

Schematic description of EnDA

The generality and relative simplicity of these two fundamental error equations can be seen as justifying the development and use of ensemble data assimilation techniques. In this chapter, ensemble data assimilation (EnDA) will be described as a method for simulating the two aforementioned error equations, using perturbations whose amplitudes and structures (i.e. variances and correlations) are representative of observation and model errors typically. In practice, this is achieved by running an ensemble of independent data assimilation cycles, which are perturbed by adding representative observation and model perturbations (e.g. Houtekamer et al., 1996; Fisher, 2003; Belo Pereira and Berre, 2006).

The principle of ensemble data assimilation is illustrated in Fig. 7.1 (Ehrendorfer, 2006). The idea is to start from an ensemble of perturbed analyses, with analysis perturbations that are constructed to be consistent with analysis error covariances. These perturbed analyses are evolved by the six-hour forecast model until the next analysis date, and model perturbations should also be added (e.g. at the end of the six-hour forecast) in order to represent six-hour accumulated model errors. During the next analysis step, each perturbed forecast state can then be used as a perturbed background, to be combined with a set of perturbed observations. This provides a new ensemble of perturbed analyses, from which a new ensemble of forecasts can be launched for the next analysis date.

One of the important aspects here is to use observation perturbations and model perturbations that are consistent with associated error covariances. This can be done by drawing random samples of specified observation and model error covariances, whose respective matrices are denoted by $\mathbf{R}$ and $\mathbf{Q}$:

$$\epsilon_o \sim \mathcal{N}(0, \mathbf{R}),$$

$$\epsilon_m \sim \mathcal{N}(0, \mathbf{Q}).$$

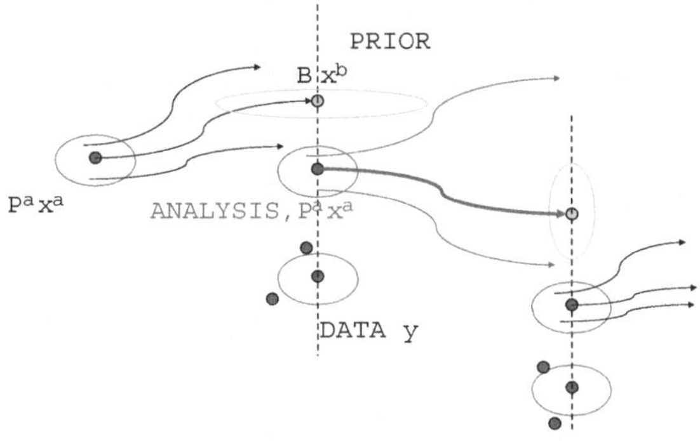

Fig. 7.1 Illustration of the principle of ensemble data assimilation. (From Ehrendorfer (2006).)

Analysis and forecast perturbation equations

It is interesting to show that the equations of the ensemble perturbations are very similar to the error equations that were derived in Section 7.1.1.

The equation for the perturbed analysis state $\mathbf{x}'_a$ (corresponding to a given member of the ensemble) can be written as

$$\mathbf{x}'_a = (\mathbf{I} - \mathbf{KH})\mathbf{x}'_b + \mathbf{K}\mathbf{y}'_o,$$

and the equation for the unperturbed analysis $\mathbf{x}_a$ can be recalled to be

$$\mathbf{x}_a = (\mathbf{I} - \mathbf{KH})\mathbf{x}_b + \mathbf{K}\mathbf{y}_o.$$

This implies that the analysis perturbation equation is simply

$$\epsilon_a = \mathbf{x}'_a - \mathbf{x}_a = (\mathbf{I} - \mathbf{KH})\epsilon_b + \mathbf{K}\epsilon_o, \tag{7.4}$$

where $\epsilon_b = \mathbf{x}'_b - \mathbf{x}_b$ and $\epsilon_o = \mathbf{y}'_o - \mathbf{y}_o$ are background and observation perturbations, respectively.

Similarly, the equation for the perturbed forecast step $\mathbf{x}'_{f+}$ can be written as

$$\mathbf{x}'_{f+} = \mathbf{M}\mathbf{x}'_a + \epsilon_m,$$

while the unperturbed forecast equation corresponds to

$$\mathbf{x}_{f+} = \mathbf{M}\mathbf{x}_a,$$

which leads to the following equation for the forecast perturbation:

$$\epsilon_f = \mathbf{M}\epsilon_a + \epsilon_m. \tag{7.5}$$

It can thus be noticed that the two perturbation equations (7.4) and (7.5) are exactly the same as the two fundamental error equations (7.2) and (7.3). This corresponds to the idea that the analysis and forecast operators can be used with perturbed inputs, in order to simulate the error evolution during the data assimilation cycling.

Some features of EnDA

One advantage of such an EnDA approach is that background error covariances can be calculated in model space from the ensemble of perturbed backgrounds. This can be contrasted with observation-based estimates of background error covariances, which are provided in observation space.

Another attractive feature is that EnDA is based on a good approximation of analysis error covariances, provided that $\mathbf{R}$ and $\mathbf{Q}$ are reasonably well known. This can be shown after noticing that the analysis error covariance matrix can be written as follows, where $E[\ ]$ is the expectation operator, and ^{T} the transpose:

$$\begin{aligned} E[\mathbf{e}_a\mathbf{e}_a^{\mathrm{T}}] &= (\mathbf{I} - \mathbf{KH})\ E[\mathbf{e}_b\mathbf{e}_b^{\mathrm{T}}]\ (\mathbf{I} - \mathbf{KH})^{\mathrm{T}} + \mathbf{K}\ E[\mathbf{e}_o\mathbf{e}_o^{\mathrm{T}}]\ \mathbf{K}^{\mathrm{T}} \\ &= (\mathbf{I} - \mathbf{KH})\mathbf{B}(\mathbf{I} - \mathbf{KH})^{\mathrm{T}} + \mathbf{KRK}^{\mathrm{T}}. \end{aligned}$$

This is to be compared with the analysis perturbation covariance matrix, which corresponds to a similar equation with the same kind of operators:

$$\begin{aligned} E[\epsilon_a \epsilon_a^{\mathrm{T}}] &= (\mathbf{I} - \mathbf{KH})\ E[\epsilon_b \epsilon_b^{\mathrm{T}}]\ (\mathbf{I} - \mathbf{KH})^{\mathrm{T}} + \mathbf{K}\ E[\epsilon_o \epsilon_o^{\mathrm{T}}]\ \mathbf{K}^{\mathrm{T}} \\ &\simeq (\mathbf{I} - \mathbf{KH})\mathbf{B}(\mathbf{I} - \mathbf{KH})^{\mathrm{T}} + \mathbf{KRK}^{\mathrm{T}} \simeq E[\mathbf{e}_a \mathbf{e}_a^{\mathrm{T}}], \end{aligned}$$

provided that $E[\epsilon_b \epsilon_b^{\mathrm{T}}] \simeq \mathbf{B}$ and $E[\epsilon_o \epsilon_o^{\mathrm{T}}] \simeq \mathbf{R}$.

In Section 7.1.3, this attractive feature of EnDA will be compared with what other error simulation techniques can provide, such as the so-called NMC method. It may also be emphasized that this good representation of analysis error covariances is associated in particular with an accurate representation of data density effects.

Another appealing property of EnDA is that forecast uncertainties can thus be simulated for short forecast ranges, such as the 6-hour forecast range, which is of interest for the background state.

7.1.3 Comparison with the NMC method

The so-called NMC method (e.g. Parrish and Derber, 1992; Rabier et al., 1998) used to be a popular method for estimating background error covariances, in particular at early development stages of variational data assimilation schemes. This method is based on differences between forecasts valid at the same time but for different forecast ranges, such as 12 and 36 hours.

In order to understand to which extent such a method may provide reasonable estimates of error covariances, one may notice that differences between 12-hour and 36-hour forecasts originate from the four successive analysis increments (in the case of a 6-hour data assimilation cycle) which are added during the 24-hour interval before launching the 12-hour forecast (Berre et al., 2006):

$$\mathbf{x}_f(36\mathrm{h}) - \mathbf{x}_f(12\mathrm{h}) \simeq \mathbf{M}^2 \sum_{i_t=1,4} \mathbf{M}^{4-i_t}\ \delta\mathbf{x}(i_t),$$

where i_t is the time index for each of the four analysis increments added during the 24-hour interval period ($i_t = 1$ is the oldest analysis time and $i_t = 4$ is the most recent analysis time).

In other words, the analysis increment $\delta\mathbf{x}$ is the fundamental perturbation ingredient that is used in the NMC method. Moreover, it is interesting to compare the equation for the analysis increment $\delta\mathbf{x}$ with that for the analysis error. The equation for $\delta\mathbf{x}$ is

$$\begin{aligned} \delta x &= \mathbf{K}(\mathbf{y}_o - \mathbf{H}\mathbf{x}_b) = \mathbf{K}(\mathbf{y}_o - \mathbf{H}\mathbf{x}_t + \mathbf{H}\mathbf{x}_t - \mathbf{H}\mathbf{x}_b) \\ &= \mathbf{K}(\mathbf{e}_o - \mathbf{H}\mathbf{e}_b) = -\mathbf{K}\mathbf{H}\mathbf{e}_b + \mathbf{K}\mathbf{e}_o, \end{aligned}$$

whereas the analysis error equation is

$$\mathbf{e}_a = (\mathbf{I} - \mathbf{KH})\mathbf{e}_b + \mathbf{K}\mathbf{e}_o.$$

It can thus be noticed that there is some similarity in the two equations, but that the operator $\mathbf{I} - \mathbf{KH}$ applied to $\mathbf{e}_b$ in the analysis error equation is replaced in the analysis increment equation by $\mathbf{KH}$ (or $-\mathbf{KH}$ to be more precise, but the minus sign is not crucial for the remainder of the discussion, because it is to be cancelled in the analysis increment auto-covariance expression). This difference is used in Bouttier (1994) and Berre et al. (2006) to explain in which cases analysis increment covariances may be a possible approximation of analysis error covariances. This discussion can be summarized as follows.

On the one hand, in the case of observations having similar error amplitudes and structures as the background (i.e. $\mathbf{B} \simeq \mathbf{R}$), and with relatively dense observations (i.e. $\mathbf{H} \simeq \mathbf{I}$), it can easily be shown that $\mathbf{I} - \mathbf{KH} \simeq \mathbf{KH} \simeq \mathbf{I}/2$. This is a relatively specific case where analysis increment covariances may be a good approximation of analysis error covariances.

On the other hand, in data-poor areas for instance, analysis increment amplitudes are likely to be relatively small, and thus they are expected to be smaller than analysis error amplitudes. Morever, while $\mathbf{I} - \mathbf{KH}$ tends to be a high-pass filter, $\mathbf{KH}$ tends to be a low-pass filter. This implies that analysis increment correlations are likely to be larger-scale than analysis error correlations. In other words, the NMC method is expected to lead to an overestimation of error correlations, and this has been confirmed experimentally (e.g. Belo Pereira and Berre, 2006).

These limitations of the NMC method are also worsened by the use of relatively long forecast ranges (at least 24 hour typically), whereas the background corresponds to a 6-hour forecast.

For these reasons, the NMC method is less attractive than the EnDA approach, even if it also provides an estimate of covariances in model space.

7.2 Innovation-based estimations

7.2.1 Innovation auto-covariances

Some information on background error covariances can be derived from innovation covariances. The innovation vector $\delta\mathbf{y}$ corresponds to the difference between the observation and the background (projected onto observation space), and it can also be written as the difference between observation errors and background errors:

$$\delta\mathbf{y} = \mathbf{y}_o - \mathbf{H}\mathbf{x}_b = \mathbf{y}_o - \mathbf{H}\mathbf{x}_t + \mathbf{H}\mathbf{x}_t - \mathbf{H}\mathbf{x}_b = \mathbf{e}_o - \mathbf{H}\mathbf{e}_b.$$

This implies that innovation covariances can be written as the sum of observation error covariances and background error covariances, under the assumption that observation errors are uncorrelated with background errors:

$$E[(\mathbf{y}_o - \mathbf{H}\mathbf{x}_b)(\mathbf{y}_o - \mathbf{H}\mathbf{x}_b)^{\mathrm{T}}] = \mathbf{R} + \mathbf{H}\mathbf{B}\mathbf{H}^{\mathrm{T}}.$$

Moreover, observation errors can be assumed to be spatially uncorrelated in some cases, such as for radiosonde observations (whereas this is less valid for satellite radiances, for instance). This property has been used by Hollingsworth and Lönnberg (1986), for instance, in order to derive estimates of $\mathbf{R}$ and of $\mathbf{B}$. The idea is that the spatially correlated component of the innovation vector corresponds to the contribution of background errors, while the residual uncorrelated component is the observation error contribution. Therefore, by extrapolating innovation covariances towards the zero separation distance, it is possible to distinguish $\mathbf{HBH}^{\mathrm{T}}$ and $\mathbf{R}$ contributions to innovation covariances.

It may be noticed that such an innovation-based method provides an estimate of $\mathbf{HBH}^{\mathrm{T}}$, whereas an estimate of $\mathbf{B}$ is required in principle for the specification of the gain matrix. This explains why such a method is sometimes used mostly for validation of an already specified matrix $\mathbf{B}$. Another possible approach is to calibrate well-chosen covariance parameters using this innovation method, such as globally averaged variances and correlation lengthscales.

7.2.2 Covariances of analysis residuals

There are other related innovation-based methods, for obtaining information on background error covariances. One such method has been proposed by Desroziers et al. (2005). It is based on the calculation of covariances between the analysis increment $\mathbf{H}\delta\mathbf{x}$ and the innovation vector $\mathbf{y}_o - \mathbf{H}\mathbf{x}_b$.

It can be shown in fact that these cross-covariances can be seen as an estimate of background error covariances in observation space:

$$\begin{aligned} E[(\mathbf{H}\delta x)(\mathbf{y}_o - \mathbf{H}\mathbf{x}_b)^{\mathrm{T}}] &= \mathbf{HK}\ E[(\mathbf{y}_o - \mathbf{H}\mathbf{x}_b)(\mathbf{y}_o - \mathbf{H}\mathbf{x}_b)^{\mathrm{T}}] \\ &\simeq \mathbf{HK}\ (\mathbf{HB}_t\mathbf{H}^{\mathrm{T}} + \mathbf{R}_t) \\ &\simeq \mathbf{HBH}^{\mathrm{T}}\ (\mathbf{HBH}^{\mathrm{T}} + \mathbf{R})^{-1}\ (\mathbf{HB}_t\mathbf{H}^{\mathrm{T}} + \mathbf{R}_t) \\ &\simeq \mathbf{HB}_t\mathbf{H}^{\mathrm{T}}, \end{aligned}$$

where $\mathbf{B}_t$ and $\mathbf{R}_t$ are the true error covariances.

In practice, this tends to be approximately valid either when the gain matrix is nearly optimal or when focusing on average background error variances, for instance, in the case where background errors and observation errors have very different spatial structures. This corresponds in particular to the case where observation errors are spatially uncorrelated while background errors are spatially correlated.

7.2.3 Practical use of innovation-based estimates

Figure 7.2 shows a comparison between ensemble-based estimates of background error standard deviations (projected onto observation space) and the corresponding estimates provided by covariances of analysis residuals (Berre and Desroziers, 2010) and locally averaged over a 500 km radius. This is done in the space of the satellite channel HIRS-7 for a specific date (28 August 2006 at 00 UTC).

It is striking to notice that similar structures are visible in the two maps, for instance with relatively large standard deviations in the Central Pacific and relatively small standard deviations in the Southern Atlantic. Conversely, there are also areas where differences can be noticed, which could provide some information on the quality of the ensemble-based estimation. This illustrates the idea that innovation-based estimates can be useful to validate ensemble-based estimates for instance.

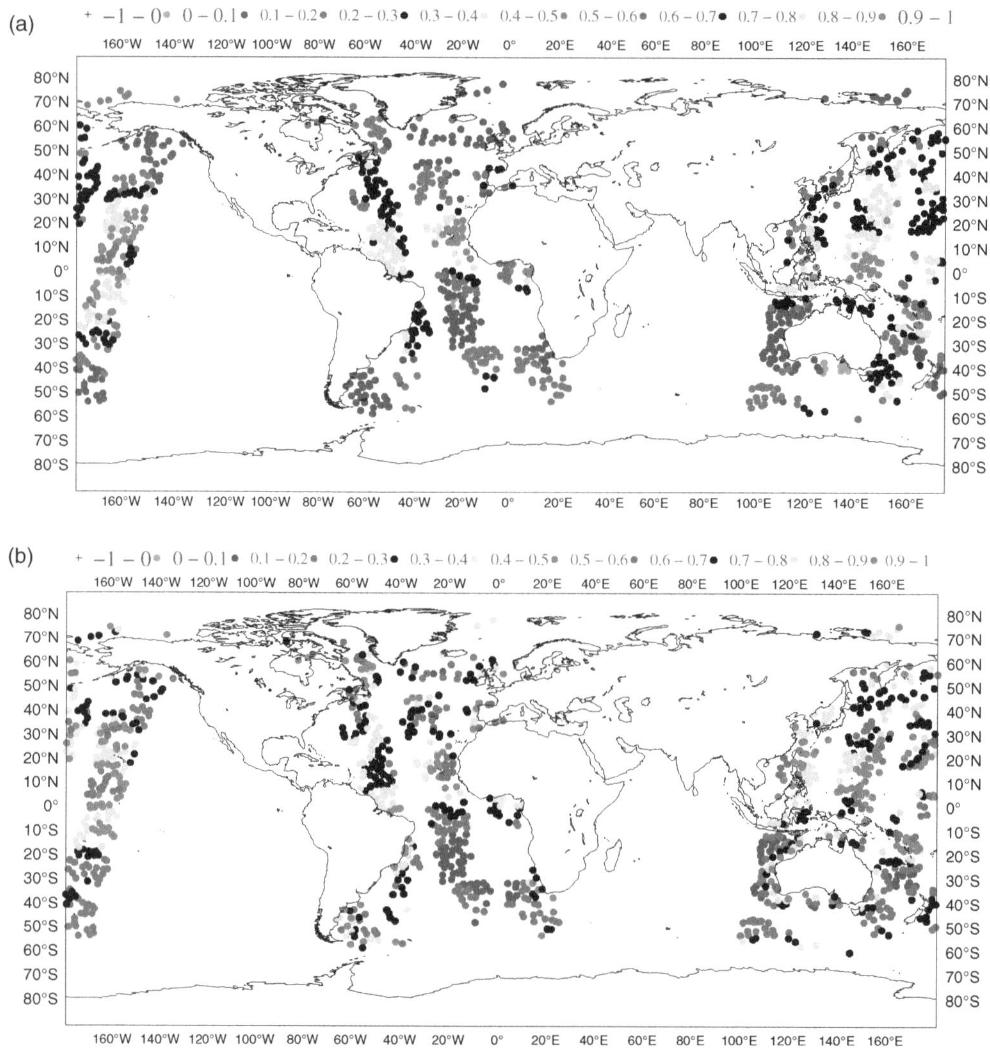

Fig. 7.2 Estimated background error standard deviation fields, in the space of HIRS channel 7, on 28 August 2006 at 00 UTC. (a) smoothed ensemble-based field from a six-member ensemble. (b) smoothed innovation-based field. The smoothing was obtained by locally averaging raw variance values over a 500-km circular radius. Temperature unit is K. ©ECMWF 2007, from Berre et al. (2007). Reproduced by permission of the ECMWF.

This is particularly appealing with respect to model error covariance estimation in particular. This corresponds to the approach proposed by Daley (1992), in particular. In order to explain this, background error covariances may be expressed as follows, according to (7.3), and assuming that model errors are uncorrelated with evolved analysis errors:

$$\mathbf{B} \simeq \mathbf{MAM}^{\mathrm{T}} + \mathbf{Q}.$$

Ensemble data assimilation can be used to estimate $\mathbf{MAM}^{\mathrm{T}}$, by employing perturbed analyses evolved by the imperfect model $\mathbf{M}$. Moreover, innovation-based techniques can be used to estimate $\mathbf{B}$ (or at least $\mathbf{HBH}^{\mathrm{T}}$).

This implies that model error covariances may be estimated from the difference between respective estimates of background error covariances and of evolved analysis error covariances:

$$\mathbf{Q} \simeq \mathbf{B} - \mathbf{MAM}^{\mathrm{T}}.$$

This has been used for instance by Raynaud et al. (2012) in order to estimate model error covariances in the Météo-France ensemble data assimilation system.

7.3 Diagnosis of background error covariances

As stated in the introduction to this chapter, the full background error covariance matrix is huge—far too large to be calculated explicitly or even stored in memory. Beyond these computation and storage issues, this also implies that salient features of the matrix $\mathbf{B}$ have to be diagnosed, so that the content of this matrix can be understood, validated, and monitored. There are several typical components of this covariance matrix that are usually diagnosed (see also e.g. Bannister, 2008a).

7.3.1 Variances and their spatio-temporal variations

For instance, background error variances (which are the diagonal elements of $\mathbf{B}$) correspond to the expected amplitudes of background errors, and they are thus quite important to examine and validate, since they reflect the level of confidence attributed to the background state for different variables and different locations. Figure 7.3 is an example of a background error standard deviation map for wind near 500 hPa, on a given date for the Météo-France global system using an EnDA approach. It can be seen that error standard deviations are relatively large over data-poor and dynamically active areas such as in the Pacific and Atlantic storm tracks, while error standard deviations are relatively small over data rich areas such as North America and Europe.

Figure 7.4 illustrates the day-to-day variability of such error standard deviations, and the connection with the synoptic situation. For this date (24 January 2009), background error standard deviations are relatively large over Europe. This is related to the occurence of a severe winter storm approaching France. The intensity of this meteorological event is associated with large uncertainties in terms of amplitude and position, and this is reflected in the background error standard deviation maps. This is the kind of flow-dependent information that can be provided by EnDA.

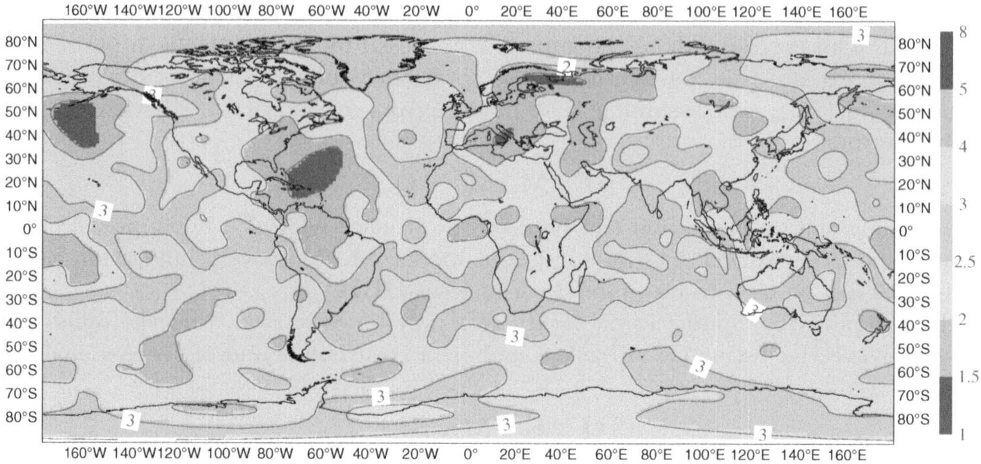

Fig. 7.3 Estimated background error standard deviation fields, for zonal wind (in m/s) near 500 hPa on 21 December 2011, derived from the Météo-France ensemble variational assimilation system.

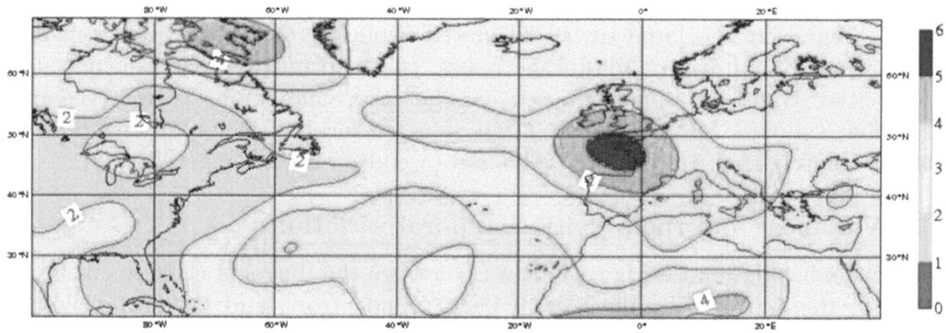

Fig. 7.4 Estimated background error standard deviation fields, for vorticity (unit : $10^{-5}s^{-1}$) near 500 hPa on 24 January 2009, derived from the Météo-France ensemble variational assimilation system.

Calculating time averages of these flow-dependent maps can be interesting to characterize the average contrast of variances between different geographical areas. This allows the average influence of data density contrasts and storm track areas to be highlighted (e.g. Belo Pereira and Berre, 2006).

7.3.2 Horizontal correlations

Spatial correlations of background errors are other quite important components of $\mathbf{B}$. They describe the spatial structure of the errors, i.e. their predominant spatial scales for instance. During the analysis step, they will determine how local observations are spatially filtered and propagated towards neighbouring areas.

In geophysical systems, spatial correlations are three-dimensional, but it is convenient to consider the examination of horizontal and vertical correlations separately as a start, before discussing their possible relations.

It is usual to diagnose horizontal correlation functions, which typically exhibit correlation values that decrease from 1 at the origin (i.e. at zero separation distance) towards 0 at long separation distances. This decrease with distance varies for different variables and different geographical locations. For instance, horizontal correlation functions tend to be sharper near the surface than in the middle of the troposphere. This reflects the larger importance of small-scale error structures near the surface, associated, for instance, with orographic effects. This scale variability of the errors can be diagnosed by examining power spectra of correlation functions, which can be calculated either through a spectral transform of the correlation function or through the calculation of spectral variances, i.e. variances of error spectral coefficients (e.g. Rabier et al., 1998; Berre, 2000).

On the other hand, since there is one horizontal correlation function to be diagnosed for each model gridpoint, computing and examining all correlation functions is impossible. For this reason, it can be useful to synthetize typical features of local horizontal correlation functions, such as their characteristic lengthscale L. This has been defined by Daley (1991) as the inverse value of the Laplacian of the correlation function ρ at its origin (considered here in the one-dimensional case):

$$L = \sqrt{1/(-\Delta\rho)_{s=0}}$$

This can be seen as describing the rate at which the correlation function decreases from the origin. L can also be seen as the distance at which the parabolic approximation of the correlation function at the origin reaches the value of 0.5 (Pannekoucke et al., 2008). It is also possible to estimate this lengthscale from the ratio between the local error variance and the local error gradient variance (Belo Pereira and Berre, 2006) or from a sampled correlation value (Pannekoucke et al., 2008). These latter two approaches are convenient because they do not require the computation of all values of the local correlation function.

It is thus possible to estimate lengthscales for different geographical locations, and to examine geographical maps of such lengthscales. This is illustrated in Fig. 7.5, which corresponds to geographical variations of horizontal correlation lengthscales for wind near 500 hPa, averaged over a 4-day winter period for the Météo-France EnDA (Varella et al., 2011). It can be seen that there is a strong contrast between large lengthscale values over tropical oceans (which are related to large-scale tropical waves) and small lengthscale values in the Northern Atlantic (which are induced by local wind gradients in baroclinic areas).

It must be also noticed that horizontal correlations depend not only on separation distance, but also on separation direction. This corresponds to the anisotropy of correlation functions, which can be elongated along local fronts, for instance. This can be diagnosed by computing eigenvectors of a lengthscale tensor, as illustrated, for example, in Belo Pereira and Berre (2006).

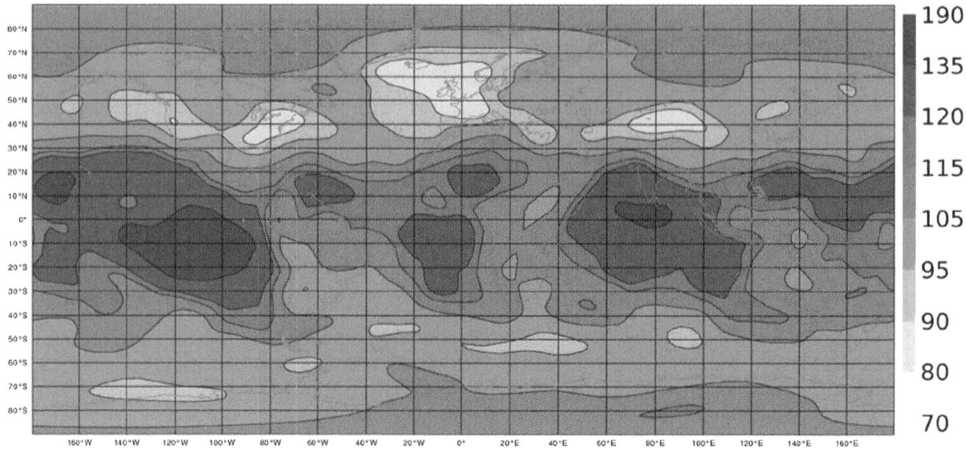

Fig. 7.5 Estimated horizontal lengthscales (in km) of background error correlations, for vorticity near 500 hPa, averaged over a 4-day period (24–27 February 2010), derived from the Météo-France ensemble variational assimilation system.

7.3.3 Vertical correlations

Vertical correlation functions are related to vertical structures of background errors. They are particularly important for the determination of how vertically integrated measurements, such as satellite radiances, are to be interpreted and spread on the different vertical levels of the model.

Vertical correlation functions can be computed and examined in a similar way as for horizontal correlation functions, although the usually irregular vertical discretization makes the diagnosis and comparison for different vertical levels somewhat more complex. Eigenvectors of the vertical correlation matrices can be calculated to diagnose the predominant vertical structures.

Moreover, vertical correlations are often known to be scale-dependent, with sharper vertical correlations for small horizontal scales. This can be diagnosed by computing vertical correlations for different wave vectors in the error spectral coefficients (e.g. Rabier et al., 1998; Berre, 2000). This is a first illustration of the concept of non-separability between horizontal and vertical correlation aspects.

Another example of non-separability is the possible tilting of vertical correlations with height, which can occur, for instance, in baroclinic areas (Thépaut et al., 1996). Moreover, vertical correlations tend to be sharper in the tropics than in the extra-tropical regions (e.g. Derber and Bouttier, 1999; Ingleby, 2001). All this reflects the three-dimensional character of spatial correlations, which makes their diagnosis relatively complex.

7.3.4 Cross-covariances

Another important set of components of **B** corresponds to cross-covariances between the different variables of the model. In the context of numerical weather prediction, this corresponds typically to cross-covariances between temperature, surface pressure, wind, and humidity. These cross-covariances usually reflect physical couplings between the different variables.

A typical example corresponds to cross-covariances between mass and wind background errors, which reflect geostrophic balance to some extent (e.g. Hollingsworth and Lönnberg, 1986; Rabier et al., 1998). This geostrophic balance is visible in cross-covariances, because both the background and true states are in geostrophic balance to some degree and thus the difference between these two states also tends to be in geostrophic balance. The same kind of thinking can be applied to perturbations provided by EnDA. It is possible to diagnose how the degree of geostrophy varies with, for example, latitude, height, and horizontal scale. One way to quantify this is to compute a linear regression between, for example, the mass field errors and the geostrophic wind errors, and to examine the percentage of explained variance as a function of latitude, height, and horizontal scale (e.g. Derber and Bouttier, 1999).

More general couplings can be diagnosed by examining, for example, covariances between vorticity and divergence, which reflect frictional effects in the boundary layer. Another typical example is the examination of cross-covariances between humidity and divergence (Berre, 2000), which exhibit the link between local moistening, convergence below, and divergence above. Similarly, cross-covariances between temperature and humidity reflect the dependence of water content with temperature, and also diabatic effects associated with evaporation (cooling and moistening), for instance.

All these multivariate couplings make their diagnosis interesting from a physical point of view, although interpretations can be relatively complex in some cases. These cross-covariances and couplings are fundamental in data assimilation, in the sense that they allow, for example, a surface pressure observation to produce modifications in the wind field, which are consistent with analysis increments of the mass field. This helps the observed information to be assimilated and preserved during the analysis and forecast steps, instead of simply generating artificial gravity waves (Daley, 1991).

7.4 Modelling and filtering covariances

As stated in the previous sections, several steps are required before modelling and specifying background error covariances in the data assimilation scheme. Error simulation experiments must be carried out (using EnDA typically), the resulting auto- and cross-covariances must be diagnosed, and a comparison with innovation-based diagnostics is also important for validation and possible adjustment of some covariance components (e.g. average variances).

In addition to these tasks, since the full covariance matrix is impossible to handle, a sparse covariance model needs to be developed and calibrated, in order to be able to specify $\mathbf{B}$ in the data assimilation system (see also, e.g., Bannister, 2008b). A related aspect is that raw covariances produced by error simulation techniques are usually affected by sampling noise, because the ensemble size is finite and often relatively small (from 5 to 100 members typically). This issue leads to the development and use of filtering techniques, which can be seen as an important part of the covariance modelling effort.

7.4.1 Modelling $\mathbf{B}^{1/2}$ with sparse operators

It is usual and convenient to model $\mathbf{B}$ through the design of its inverse square root, namely $\mathbf{B}^{-1/2}$. This is related to the fact $\mathbf{B}^{-1/2}$ can itself be conceived as an operator that transforms the background error $\mathbf{e}_b$ into a normalized variable $\mathbf{e}'_b = \mathbf{B}^{-1/2}\mathbf{e}_b$, whose components are uncorrelated and have unit variance (e.g. Gustafsson et al., 2001). In other words, the covariance matrix of $\mathbf{e}'_b$ is the identity matrix:

$$
\begin{aligned}
E\left[\mathbf{e}'_b\mathbf{e}'^{\mathrm{T}}_b\right] &= E\left[\left(\mathbf{B}^{-1/2}\mathbf{e}_b\right)\left(\mathbf{B}^{-1/2}\mathbf{e}_b\right)^{\mathrm{T}}\right] \\
&= \mathbf{B}^{-1/2}\, E\left[\mathbf{e}_b\mathbf{e}_b^{\mathrm{T}}\right]\, \mathbf{B}^{-\mathrm{T}/2} = \mathbf{B}^{-1/2}\, \mathbf{B}\, \mathbf{B}^{-\mathrm{T}/2} = \mathbf{I},
\end{aligned}
$$

where $\mathbf{I}$ is the identity matrix. Note that the diagonality of $\mathbf{I}$ means that cross-covariances between the different components in $\mathbf{e}'_b$ are zero.

In practice, it is usual (e.g. Derber and Bouttier, 1999) to achieve this normalization through the application of several sparse operators, such as

$$
\mathbf{B}^{-1/2} = \mathbf{C}^{-1/2}\,\mathbf{S}^{-1}\,\mathbf{L}^{-1}.
$$

$\mathbf{L}^{-1}$ is an operator that transforms the different variables (e.g. vorticity, divergence, mass, and humidity) into a set of uncorrelated variables, through the use of multivariate linear regressions and possible analytical balance equations. The transformed variables can be, for instance, vorticity and the respective regression residuals for divergence, mass, and humidity. $\mathbf{S}^{-1}$ is a diagonal matrix containing specified background error standard deviations. Owing to the previous application of $\mathbf{L}^{-1}$, standard deviations are specified for transformed variables, i.e. for the first regression predictor (e.g. vorticity) and for regression residuals (which are sometimes called the 'unbalanced variables'). $\mathbf{C}^{-1/2}$ is a normalization by the inverse square root of the spatial correlation matrix, which ensures that the transformed variable is spatially uncorrelated.

It can be noted that the square root of $\mathbf{B}$ can be recovered as $\mathbf{B}^{1/2} = \mathbf{LSC}^{1/2}$. Similarly, even if this is often not needed for practical implementations, the form of the specified matrix $\mathbf{B}$ can be expanded as $\mathbf{B} = \mathbf{B}^{1/2}\mathbf{B}^{\mathrm{T}/2} = \mathbf{LSCSL}^{\mathrm{T}}$.

In the remainder of this section, the different covariance modelling parts of $\mathbf{B}^{1/2}$ will be described.

7.4.2 Cross-covariances and multivariate couplings

The usual approach when modelling $\mathbf{B}^{1/2}$ is to represent these cross-covariances through the use of linear regressions between mass, wind, and moisture. These linear regressions are in fact precisely derived from the corresponding cross-covariances. This can be illustrated for a linear regression between background errors of vorticity ζ (which is here the predictor) and divergence η (which is the predictand):

$$
\eta = \mathcal{M}\zeta + \eta_u,
$$

where $\mathcal{M} = \mathbf{B}_{\eta,\zeta}\mathbf{B}_{\zeta,\zeta}^{-1}$ is the linear regression matrix, $\mathbf{B}_{\eta,\zeta}$ is the cross-covariance matrix between ζ and η, $\mathbf{B}_{\zeta,\zeta}$ is the auto-covariance matrix of ζ, and η_u is the regression residual, which is sometimes called the unbalanced component of divergence (whereas $\mathcal{M}\zeta$ is sometimes called the balanced component, by considering that the linear regression is similar to a balance equation).

An important property is that the regression residual η_u is uncorrelated with the predictor ζ, precisely because of the formula for $\mathcal{M}$:

$$\begin{aligned} E\big[\eta_u\zeta^{\mathrm{T}}\big] &= E\big[(\eta - \mathcal{M}\zeta)\,\zeta^{\mathrm{T}}\big] = E\big[\eta\zeta^{\mathrm{T}}\big] - \mathcal{M}\;E\big[\zeta\zeta^{\mathrm{T}}\big] \\ &= \mathbf{B}_{\eta,\zeta} - \mathbf{B}_{\eta,\zeta}\;\mathbf{B}_{\zeta,\zeta}^{-1}\;\mathbf{B}_{\zeta,\zeta} = \mathbf{B}_{\eta,\zeta} - \mathbf{B}_{\eta,\zeta} = 0. \end{aligned}$$

This illustrates the idea that a linear regression can be used to transform a set of correlated parameters (here vorticity and divergence) into a set of uncorrelated variables (here vorticity and 'unbalanced divergence'). It is this property that explains the use of linear regressions when modelling $\mathbf{B}^{1/2}$, as described previously.

These linear regressions can be calculated in spectral space for instance, which allows the scale dependence of multivariate couplings to be represented. On the other hand, cross-covariances between different wave vectors are usually neglected, which corresponds to an assumption of horizontal homogeneity.

This assumption can nevertheless be relaxed by the use of analytical expressions of mass/wind couplings such as geostrophy (before computing the linear regression coefficients), which account for part of the spatial and temporal variations at play. For instance, the simplest expression of geostrophic balance corresponds to

$$\phi = f\Delta^{-1}\zeta,$$

where ϕ is geopotential height and f is the Coriolis parameter. Using this kind of balance equation before computing linear regressions allows the mass/wind geostrophic coupling to be relaxed as a function of latitude, to reflect the lesser amount of geostrophy in the tropics (Derber and Bouttier, 1999). The nonlinear balance equation and the omega balance equation can also be used to represent flow-dependent effects associated with these couplings (Fisher, 2003).

7.4.3 Sampling noise in ensemble-based variances

Before specifying background error variances, it is useful to diagnose sampling noise in ensemble-based variances and to apply spatial filtering techniques in order to reduce these sampling noise effects. Figure 7.6 corresponds to an ensemble-based variance field $\tilde{\mathbf{v}}$ in an idealized one-dimensional context, using an ensemble with $N = 50$ members (Raynaud et al., 2008):

$$\tilde{v}(i_x) = \frac{1}{N-1}\sum_{i=1}^{N}\epsilon_b^2(i_x, i),$$

where $\epsilon_b(i_x, i)$ is the background perturbation for member i at position i_x.

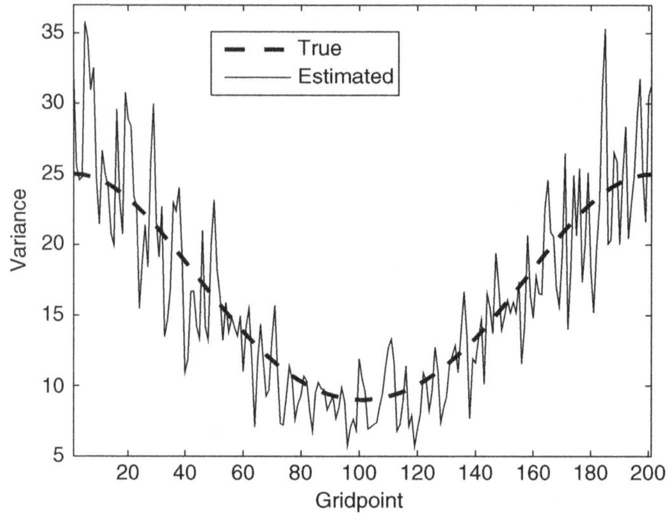

Fig. 7.6 True variance field (thick dashed line) and raw ensemble-based variance field (thin solid line) for a 50-member ensemble in an idealized one-dimensional context. (From Raynaud et al. (2008). Copyright © 2008, Royal Meteorological Society.)

This can be compared with the true variance field $\mathbf{v}_t$, which has been specified to be a simple large-scale sinusoide in this context. Ensemble background perturbations have been produced as random draws from the (specified) true background error covariance matrix. On the one hand, the large-scale signal of interest is clearly visible in the raw ensemble-based variance field of interest. This illustrates the idea that an ensemble of perturbed states contains some relevant information on the error variance. On the other hand, it can also be seen that the large-scale signal is affected by spurious small-scale oscillations. These artificial oscillations correspond to sampling noise effects, which are due to the finite size of the ensemble. The sampling noise affecting the estimated variance field can be defined as follows:

$$\mathbf{v}^e = \tilde{\mathbf{v}} - \mathbf{v}_t.$$

The statistical features of this sampling noise can be studied experimentally by comparing estimates from two independent ensembles (Berre and Desroziers, 2010). Moreover, it is also possible to derive analytical expressions for sampling noise covariances, assuming that background perturbations are Gaussian random variables. This corresponds to the following expression (Raynaud et al., 2009):

$$E[\mathbf{v}^e(\mathbf{v}^e)^{\mathrm{T}}] = \frac{2}{N-1}\mathbf{B}_t \circ \mathbf{B}_t, \tag{7.6}$$

where $\circ$ is the Hadamard product (i.e. the elementwise product) between two matrices. This implies firstly that the standard-deviation of sampling noise (denoted by $\sigma(\mathbf{v}^e)$)

decreases relatively slowly as a function of ensemble size:

$$\sigma(\mathbf{v}^e) = \sqrt{\frac{2}{N-1}}\,\mathbf{v}_t,$$

For instance, the relative average error (i.e. $\sigma(\mathbf{v}^e)/\mathbf{v}_t = \sqrt{2/(N-1)}$) is around 30% for a 25-member ensemble, and it remains around 20% when doubling the ensemble size up to 50 members. This indicates that sampling noise on the variance field is a significant issue to be accounted for.

Moreover, the sampling noise covariance formula, (7.6), can be used to derive an expression for the lengthscale of sampling noise in the variance field:

$$L(\mathbf{v}^e) = \frac{L(\mathbf{e}_b)}{\sqrt{2}}.$$

This indicates that the sampling noise is smaller scale than the background error field. Moreover, it can be shown experimentally that the true background error variance field tends to be larger-scale than the background error field (e.g. Berre and Desroziers, 2010). This difference of scale, between the signal of interest and the noise at play in the estimation, motivates the use of spatial filtering techniques, to be applied to the raw ensemble-based variance field.

7.4.4 Filtering of ensemble-based variances

As discussed in Berre and Desroziers (2010), it is possible to apply an objective filter $\mathbf{F}$ to the raw ensemble-based variance field, which allows sampling noise effects to be reduced, in order to minimize the error on the estimated background error variance field. Assuming that the variance field and sampling noise are nearly homogeneous, it is possible to derive a simple expression of $\mathbf{F}$ in spectral space:

$$\mathbf{F} = \frac{\mathbf{P}(\mathbf{v}_t)}{\mathbf{P}(\tilde{\mathbf{v}})} = \frac{\mathbf{P}(\mathbf{v}_t)}{\mathbf{P}(\mathbf{v}_t) + \mathbf{P}(\mathbf{v}^e)},$$

where $\mathbf{P}(\mathbf{v}_t)$ and $\mathbf{P}(\mathbf{v}^e)$ are the respective power variances of the true variance field and of the sampling noise. The sampling noise power $\mathbf{P}(\mathbf{v}^e)$ can be estimated either from the difference between two independent ensemble estimates or using (7.6). Moreover, the signal power $\mathbf{P}(\mathbf{v}_t)$ can be estimated from the difference between the raw signal power and the sampling noise power: $\mathbf{P}(\mathbf{v}_t) = \mathbf{P}(\tilde{\mathbf{v}}) - \mathbf{P}(\mathbf{v}^e)$.

This filter is the operator that minimizes the residual error on the estimation of the variance field. In this sense, this is very close to the BLUE approach for minimizing the error on the estimated model state, with an objective BLUE filter corresponding to the gain matrix, which can be seen as the ratio between background error covariances and innovation covariances.

Since the sampling noise tends to be smaller-scale than the true signal of interest (as illustrated in Fig. 7.6), $\mathbf{F}$ tends to be a low-pass filter . This reflects the idea that $\mathbf{F}$ will tend to filter out small-scale sampling noise, while retaining the large-scale signal of interest.

Once the background error variance field has been filtered, it can be specified in the form of a diagonal matrix **S** that contains background error standard deviations. If EnDA is run in real time, flow-dependent variances will be specified in this way. Conversely, if EnDA is only run offline (e.g. due to computational resources), a time-averaged estimate of variances will be specified.

Normalizing **B** by these standard deviations allows the remaining modelling effort to focus on spatial correlations.

7.4.5 Modelling of spatial correlations in spectral space

The size of the full correlation matrix **C** is the same as for **B**, and therefore sparse horizontal correlation models are often employed to circumvent this difficulty.

A usual sparse correlation model is the representation of three-dimensional spatial correlations by a block-diagonal matrix in spectral space (e.g. Courtier et al., 1998; Berre, 2000). In a two-dimensional horizontal context, this corresponds to the representation of horizontal correlations by a diagonal matrix in spectral space (e.g. Courtier et al., 1998; Berre, 2000), containing spectral variances, i.e. variances of spectral coefficients of background errors (after normalization by their gridpoint standard deviations). In the case of a spherical global model, the spectral transform corresponds to a spherical harmonic representation. In the context of a limited area model, a bi-Fourier representation is often employed (although a specific procedure is required in this case to make the gridpoint fields periodic). In a three-dimensional context, the off-diagonal elements of each diagonal block correspond to vertical correlations for a given wave vector. This representation thus allows the scale dependence of vertical correlations to be represented.

It must be noted that this sparse block-diagonal representation of correlations in spectral space corresponds to an assumption that three-dimensional correlation functions are horizontally homogeneous. On the one hand, although this assumption tends to be more valid for correlations than for covariances, this hypothesis is a limitation, since it tends to prevent the representation of horizontal heterogeneities in the correlation functions, which have been illustrated in Section 7.3.2. On the other hand, using the homogeneity assumption amounts to estimating correlations through a global spatial average. This allows an increase in the sample size, whose order of magnitude becomes the ensemble size times the number of independent regions over the global or regional domain of interest. This increased sample size allows estimated correlations to be quite robust, and this property is often enhanced by the use of a time average. In other words, sampling noise is negligible in this case, and the main issue is rather the lack of representation of horizontal and temporal variations in the correlations.

7.4.6 Modelling of spatial correlations in wavelet space

This has motivated the development of alternative sparse correlation models. One such model is based on a wavelet diagonal approach (e.g. Fisher, 2003; Deckmyn and Berre, 2005). A wavelet representation of the fields can be seen as intermediate between the gridpoint and spectral approaches. A given wavelet component thus corresponds to a specific scale as in the spectral approach (though at a lower resolution than in spectral space), but it also corresponds to a specific geographical location as in the gridpoint representation (though at a lower resolution than in gridpoint space). Therefore, the error variance of a given wavelet component of background error contains information about the local spatial structures of errors in the corresponding location. It is this property that allows a wavelet diagonal approach to represent horizontal variations of correlation functions. This approach can thus be used to represent heterogeneous correlations, in either a static way (Fisher, 2003) or a flow-dependent way (Varella et al., 2011).

Another attractive feature of this method is that it can also be seen as a spatial filtering technique applied to raw ensemble-based correlations (Berre and Desroziers, 2010). This corresponds to the fact that a wavelet diagonal assumption amounts to computing local spatial averages of the correlation functions (Pannekoucke et al., 2007). This means that the wavelet approach is an attractive compromise between two extreme solutions, which correspond respectively to fully heterogeneous but noisy (unfiltered) local correlations in gridpoint space on one hand, and to robust but homogeneous correlations in spectral space on the other. This corresponds to the idea that, in the wavelet diagonal approach, the use of local spatial averages allows the small-scale sampling noise on correlation heterogeneities to be filtered locally, while preserving the representation of relevant (larger-scale) correlation heterogeneities.

7.4.7 Modelling of spatial correlations in gridpoint space

Other sparse covariance modelling approaches can be considered, for instance in gridpoint space. This corresponds to the use of diffusion-based correlation models (Weaver and Courtier, 1991), in which the application of $\mathbf{B}^{1/2}$ is approximated through a numerical diffusion process. This has been considered for oceanic applications, with the possibility to handle complex coastal boundaries, and also to represent some anisotropies and heterogeneities. Another possible approach is based on the use of recursive spatial filters (Wu et al., 2002), which can be implemented in either homogeneous or heterogeneous contexts. These two approaches require the estimation and specification of correlation lengthscales, which can be made horizontally varying.

Another complementary method is to apply coordinate transformations, in order to make correlations heterogeneous and anisotropic. This can be done by using a semigeostrophic coordinate transformation (Desroziers, 1997), which allows specified correlations to be stretched along the fronts. Another example is the use of potential temperature as a vertical coordinate (Benjamin et al., 1991), which makes vertical correlations vary as function of vertical stability.

7.4.8 Modelling of spatial covariances in ensemble space

Another possibility is to handle a set of ensemble perturbation vectors in the form of a rectangular matrix, denoted by ϵ_b, and to use this as the square root of $\mathbf{B}$ (Lorenc, 2003):

$$\mathbf{B} = \epsilon_b(\epsilon_b)^{\mathrm{T}}/(N-1),$$
$$\mathbf{B}^{1/2} = \epsilon_b/\sqrt{N-1}.$$

However, using such raw ensemble-based covariances leads to noisy correlation values at long separation distances. To reduce this effect, a Schur filter is often employed. This corresponds to a Hadamard multiplication of **B** with a specified correlation matrix, with correlation values that are equal to zero at long distances.

It may be noted that the Schur filter does not enable the error variance field to be filtered. Such a formulation thus requires a relatively large ensemble (of the order of 100) to make the covariance representation sufficiently accurate (with respect to sampling noise).

When the available ensemble size is smaller, a hybrid combination with a static covariance model may be considered, in order to reduce sampling noise effects:

$$\mathbf{B} = \alpha\epsilon_b(\epsilon_b)^{\mathrm{T}}/(N-1) + (1-\alpha)\mathbf{B}_{\mathrm{static}}.$$

7.5 Conclusions

Estimating and specifying background error covariances can be seen as an iterative process, in which efforts for simulating errors are to be combined with covariance diagnosis, modelling, and validation. The research on this topic has evolved from homogeneous and static approaches to heterogeneous and flow-dependent aspects, thanks to the development of ensemble methods and spatial modelling and filtering techniques.

Beyond this progress, error covariance modelling remains a very active area of research, for instance because model errors are largely unknown. The combined use of ensemble data assimilation and innovation-based diagnostics is likely to be helpful on this aspect. Another issue is that the finite sizes of the ensemble and of the innovation sets imply that research efforts on covariance modelling and filtering methods will remain crucial in the future.

Moreover, while background error covariances are usually conceived in a three-dimensional context, new formulations such as the 4D-En-Var (e.g. Buehner, 2010) allow the four-dimensional nature of ensemble trajectories to be exploited in data assimilation, for instance as an alternative to tangent linear and adjoint models. This may be particularly attractive for high-resolution applications, for which nonlinearities are pronounced.

References

Bannister, R. N. (2008a). A review of forecast error covariance statistics in atmospheric variational data assimilation. I: Characteristics and measurements of forecast error covariances. *Q. J. R. Meteor. Soc.*, **134**, 1951–1970.

Bannister, R. N. (2008b). A review of forecast error covariance statistics in atmospheric variational data assimilation. II: Modelling the forecast error covariance statistics. *Q. J. R. Meteor. Soc.*, **134**, 1971–1996.

Belo Pereira, M. and Berre, L. (2006). The use of an ensemble approach to study the background error covariances in a global NWP model. *Mon. Weather Rev.*, **134**, 2466–2489.

Benjamin, S. G., Brewster, K. A., Brummer, R., Jewett, B. F., Schlatter, T. W., Smith, T., and Stamus, P. A. (1991). An isentropic three-hourly data assimilation system using ACARS observations. *Mon. Weather Rev.*, **119**, 888–906.

Berre, L. (2000). Estimation of synoptic and meso scale forecast error covariances in a limited area model. *Mon. Weather Rev.*, **128**, 644–667.

Berre, L. and Desroziers, G. (2010). Filtering of background error variances and correlations by local spatial averaging: a review. *Mon. Weather Rev.*, **138**, 3693–3720.

Berre, L., Ştefănescu, S. E., and Belo Pereira, M. (2006). The representation of the analysis effect in three error simulation techniques. *Tellus*, **58A**, 196–209.

Berre, L., Pannekoucke, O., Desroziers, G., Ştefănescu, S. E., Chapnik, B., and Raynaud, L. (2007). A variational assimilation ensemble and the spatial filtering of its error covariances: increase of sample size by local spatial averaging. In *Proceedings of the ECMWF Workshop on Flow-Dependent Aspects of Data Assimilation, 11–13 June 2007*, pp. 151–168.

Bouttier, F. (1994). Sur la prévision de la qualité des prévisions météorologiques. PhD dissertation, Université Paul Sabatier, Toulouse.

Buehner, M. and Charron, M. (2007). Spectral and spatial localization of background error correlations for data assimilation. *Q. J. R. Meteor. Soc.*, **133**, 615–630.

Buehner, M., Houtekamer, P. L., Charette, C., Mitchell, H. L., and He, B. (2010). Intercomparison of variational data assimilation and the ensemble kalman filter for global deterministic NWP. Part I: Description and single-observation experiments. *Mon. Weather Rev.*, **138**, 1550–1566.

Courtier, P., Andersson, E., Heckley, W., Vasiljevic, D., Hamrud, M., Hollingsworth, A., Rabier, F., Fisher, M. and Pailleux, J. (1998). The ECMWF implementation of three-dimensional variational assimilation (3D-Var). Part I: Formulation. *Q. J. R. Meteor. Soc.*, **124**, 1783–1807.

Daley, R. (1991). *Atmospheric Data Analysis.* Cambridge University Press, Cambridge.

Daley, R. (1992). Estimating model-error covariances for application to atmospheric data assimilation. *Mon. Weather Rev.*, **120**, 1735–1746.

Deckmyn, A. and Berre, L. (2005). A wavelet approach to representing background error covariances in a limited area model. *Mon. Weather Rev.*, **133**, 1279–1294.

Derber, J. and Bouttier, F. (1999). A reformulation of the background error covariance in the ECMWF global data assimilation system. *Tellus*, **51A**, 195–221.

Desroziers, G. (1997). A coordinate change for data assimilation in spherical geometry of frontal structures. *Q. J. R. Meteor. Soc.*, **125**, 3030–3396.

Desroziers, G., Berre, L., Chapnik, B., and Poli, P. (2005). Diagnosis of observation, background and analysis error statistics in observation space. *Q. J. R. Meteor. Soc.*, **131**, 3385–3396.

Desroziers, G., Berre, L., Chabot, V., and Chapnik, B. (2009). A posteriori diagnostics in an ensemble of perturbed analyses. *Mon. Weather Rev.*, **137**, 3420–3436.

Ehrendorfer, M. (2006). Review of issues concerning ensemble-based data assimilation techniques. Oral presentation at the Seventh International Workshop on Adjoint Applications in Dynamic Meteorology, 8–13 October 2006, Obergurgl, Austria.

Fisher, M. (2003). Background error covariance modelling. In *Proceedings of the ECMWF Seminar on Recent Developments in Data Assimilation*, pp. 45–63.

Gustafsson, N., Berre, L., Hörnquist, S., Huang, X.-Y., Lindskog, M., Navasques, B., Mogensen, K. S. and Thorsteinsson, S. (2001). Three-dimensional variational data assimilation for a limited area model. Part I: General formulation and the background error constraint. *Tellus,*, **53A**, 425–446.

Hollingsworth, A. and Lönnberg, P. (1986). The statistical structure of short-range forecast errors as determined from radiosonde data. Part I: The wind field. *Tellus*, **38A**, 111–136.

Houtekamer, P. L., Lefaivre, L., Derome, J., Ritchie, H., and Mitchell, H. L. (1996). A system simulation approach to ensemble prediction. *Mon. Weather Rev.*, **124**, 1225–1242.

Ingleby, N. B. (2001). The statistical structure of forecast errors and its representation in the Met. Office Global 3-D variational data assimilation scheme. *Q. J. R. Meteor. Soc.*, **127**, 209–232.

Lorenc, A. C. (2003). The potential of the ensemble Kalman filter for NWP—a comparison with 4D-Var. *Q. J. R. Meteorol. Soc.*, **129**, 3183–3203.

Pannekoucke, O., Berre, L., and Desroziers, G. (2007). Filtering properties of wavelets for local background error correlations. *Q. J. R. Meteor. Soc.*, **133**, 363–379.

Pannekoucke, O., Berre, L., and Desroziers, G. (2008). Background error correlation length-scales and their sampling statistics. *Q. J. R. Meteor. Soc.*, **133**, 497–508.

Parrish, D. F. and Derber, J. C. (1992). The National Meteorological Center's spectral statistical interpolation analysis system. *Mon. Weather Rev.*, **120**, 1747–1763.

Rabier, F., McNally, A., Andersson, E., Courtier, P., Undén, P., Eyre, J., Hollingsworth, A., and Bouttier, F. (1998). The ECMWF implementation of three-dimensional variational assimilation (3D-Var). II: Structure functions. *Q. J. R. Meteor. Soc.*, **124**, 1809–1829.

Raynaud, L., Berre, L., and Desroziers, G. (2008). Spatial averaging of ensemble-based background-error variances. *Q. J. R. Meteor. Soc.*, **134**, 1003–1014.

Raynaud, L., Berre, L., and Desroziers, G. (2009). Objective filtering of ensemble-based background-error variances. *Q. J. R. Meteor. Soc.*, **135**, 1177–1199.

Raynaud, L., Berre, L., and Desroziers, G. (2012). Accounting for model error in the Météo-France ensemble data assimilation system. *Q. J. R. Meteor. Soc.*, **138**, 249–262.

Talagrand, O. (1997). Assimilation of observations, an introduction. *J. Meteor. Soc. Japan*, **75**, 191–209.

Thépaut, J.-N., Courtier, P., Belaud, G., and Lemaître, G. (1996). Dynamical structure functions in a four-dimensional variational assimilation: a case study. *Q. J. R. Meteor. Soc.*, **122**, 535–561.

Varella, H., Berre, L., and Desroziers, G. (2011). Diagnostic and impact studies of a wavelet formulation of background error correlations in a global model. *Q. J. R. Meteor. Soc.*, **137**, 1369–1379.

Weaver, A. T. and Courtier, P. (2001). Correlation modelling on the sphere using a generalized diffusion equation. *Q. J. R. Meteor. Soc.*, **127**, 1815–1846.

Wu, W.-S., Purser, R. J., and Parrish, D. F. (2002). Three-dimensional variational analysis with spatially inhomogeneous covariances. *Mon. Weather Rev.*, **130**, 2905–2916.

8
Observation error specifications

G. Desroziers

Météo-France
CNRM/GMAP/ALGO
Toulouse, France

Advanced Data Assimilation for Geosciences. First Edition.
Edited by É. Blayo, M. Bocquet, E. Cosme, and L. F. Cugliandolo.

Chapter Contents

The aim of this chapter is to propose an overview of the specifications of observation errors in data assimilation schemes. Different ways of diagnosing the statistics of those errors are in particular presented. The evidence of correlations for a given number of observations and the way they can be represented are also discussed.

8.1 General framework

From a general point of view, the specification of observation errors is related to the Kalman-filter formalism, where one wants to obtain an estimate $\mathbf{x}^a$ of the true state, from two pieces of information: a background $\mathbf{x}^b$ associated with an error covariance matrix $\mathbf{B}$ and observations $\mathbf{y}^o$ associated with an error covariance matrix $\mathbf{R}$. The analysis is given by the following equation:

$$\mathbf{x}^a = \mathbf{x}^b + \delta\mathbf{x} = \mathbf{x}^b + \mathbf{K}\mathbf{d} = \mathbf{x}^b + \mathbf{B}\mathbf{H}^{\mathrm{T}}(\mathbf{H}\mathbf{B}\mathbf{H}^{\mathrm{T}} + \mathbf{R})^{-1}\mathbf{d}. \tag{8.1}$$

It is thus obtained by adding a correction $\delta\mathbf{x}$ to the background, and this correction $\delta\mathbf{x}$ is itself given by the application of the gain matrix $\mathbf{K}$ to the innovation vector $\mathbf{d}$ containing the differences between observations and their equivalents for the background. The gain matrix is a particular expression, where $\mathbf{B}$, the linear or linearized observation operator $\mathbf{H}$, and $\mathbf{R}$ appear. The quality of the analysis will then depend on the correct specification of those three ingredients and thus particularly on the correct specification of matrix $\mathbf{R}$.

It is easy to see that such an expression for $\delta\mathbf{x}$ is also the expression that minimizes the following quadratic cost function:

$$J(\delta\mathbf{x}) = \delta\mathbf{x}^{\mathrm{T}}\mathbf{B}^{-1}\delta\mathbf{x} + (\mathbf{H}\delta\mathbf{x} - \mathbf{d})^{\mathrm{T}}\mathbf{R}^{-1}(\mathbf{H}\delta\mathbf{x} - \mathbf{d}). \tag{8.2}$$

In the case where the observation operator H is nonlinear, the incremental formulation proposes a way to minimize the original non-quadratic cost function, by minimizing a set of successive quadratic cost functions (Courtier et al., 1994):

$$J(\mathbf{x}) = (\mathbf{x} - \mathbf{x}^b)^{\mathrm{T}}\mathbf{B}^{-1}(\mathbf{x} - \mathbf{x}^b) + [H(\mathbf{x}) - \mathbf{y}^o]^{\mathrm{T}}\mathbf{R}^{-1}[H(\mathbf{x}) - \mathbf{y}^o]. \tag{8.3}$$

Even in such a slightly nonlinear problem, one can consider that analysis, background, model and observation errors are linked, at first order, by such a linear relation:

$$\boldsymbol{\epsilon}^a = (\mathbf{I} - \mathbf{K}\mathbf{H})\boldsymbol{\epsilon}^b + \mathbf{K}\boldsymbol{\epsilon}^o, \tag{8.4}$$

with $\boldsymbol{\epsilon}^a = \mathbf{x}^a - \mathbf{x}^t$, $\boldsymbol{\epsilon}^b = \mathbf{x}^b - \mathbf{x}^t$, and $\epsilon^o = \mathbf{y}^o - H(\mathbf{x}^t)$, where $\mathbf{x}^t$ is the unknown true state.

The background error $\boldsymbol{\epsilon}^{b^+}$ for the next assimilation is the result of the application of the tangent-linear model $\mathbf{M}$ to the current analysis error $\boldsymbol{\epsilon}^a$ plus an error $\boldsymbol{\epsilon}^m$ intrinsically related to the model:

$$\boldsymbol{\epsilon}^{b^+} = \mathbf{M}\boldsymbol{\epsilon}^a + \boldsymbol{\epsilon}^m. \tag{8.5}$$

It is easy to check that the exact covariance matrix for the analysis error is given by the expression

$$\mathbf{A}^t = (\mathbf{I} - \mathbf{KH})\mathbf{B}^t(\mathbf{I} - \mathbf{KH})^{\mathrm{T}} + \mathbf{KR}^t\mathbf{K}^{\mathrm{T}}, \tag{8.6}$$

where $\mathbf{B}^t$ and $\mathbf{R}^t$ are the exact covariance matrices and $\mathbf{K}$ is the possibly suboptimal or inexact gain matrix.

A correct representation of the background error covariance matrix $\mathbf{B}^{t^+}$ for the next analysis is then given by the expression

$$\mathbf{B}^{t^+} = \mathbf{MA}^t\mathbf{M}^{\mathrm{T}} + \mathbf{Q}^t, \tag{8.7}$$

where $\mathbf{Q}^t$ is the model error covariance matrix.

From these expressions, it is interesting to note that $\mathbf{R}^t$ is an input in these equations, like $\mathbf{Q}^t$. It is not an output, like the matrix $\mathbf{B}^{t^+}$. That is, it has to be specified, but cannot be documented by an algorithm such as ensemble Kalman filtering.

Actually, what are called observation errors, in assimilation schemes, are the differences between observations and their equivalents for the virtual true model state $\mathbf{x}^t$, obtained by the application of the observation operator H to this true state. These differences can be developed as

$$\begin{aligned}\epsilon^o &= \mathbf{y}^o - H(\mathbf{x}^t) \\ &= \mathbf{y}^o - \mathbf{y}^t + \mathbf{y}^t - H(\mathbf{x}^t) \\ &= \epsilon_i^o - \epsilon_H^o,\end{aligned} \tag{8.8}$$

where $\mathbf{y}^t$ is the true state equivalent of $\mathbf{y}^o$. This expression makes appear the actual instrument observation error ϵ_i^o and an error ϵ_H^o, which is a complex function of

- the type of observation (and in particular, whether they are in situ or integrated observations, like satellite radiances);
- the resolution of the model state, which leads to the appearance of what is called the representativeness error;
- the precision of the observation operator, especially for satellite observations.

8.2 Methods for estimating observation error statistics

8.2.1 Innovation based diagnostics

Observation errors are not explicitly known. However, information on the statistics of those errors can be found in the innovations, that is, in the departures between observations and background. A first approach to the extraction of information on the observation error variance from the innovations is the so-called Hollingsworth and Lönnberg method (Hollingsworth and Lönnberg, 1986). The principle is to calculate a histogram of innovation covariances, stratified against separation. If one assumes that observation errors are spatially uncorrelated and that observation and background errors are uncorrelated, one can fit a model of background covariance to this histogram. The intercept of this covariance model at zero separation is then the background variance, and the observation variance is simply the difference between the variance of the innovations at zero separation and the background error variance.

8.2.2 Optimality criterion

A second diagnostic that was proposed is one based on the so-called $J_{\min}$ diagnostics, that is, the expected value of the above-mentioned cost function at its minimum, for the analysis. It can be shown that the statistical expectation for the total cost function J, including the J^b and J^o terms, should simply be equal to p, the total number of observations in an optimal assimilation (Bennett et al., 1993). More precisely, it can be shown (Talagrand, 1997) that the statistical expectation of any subpart J_i^o of the J^o term of the cost function is given by the expresssion

$$E[J_i^o(\mathbf{x}^a)] = p_i - \mathrm{Tr}(\mathbf{H}_i\mathbf{A}\mathbf{H}_i^{\mathrm{T}}\mathbf{R}_i{}^{-1}), \tag{8.9}$$

where p_i is the number of observations associated with this subpart. $\mathbf{H}_i$ and $\mathbf{R}_i$ are respectively the corresponding observation operator and error covariance matrix. $\mathbf{A}$ is the analysis covariance matrix for the resulting estimation.

If the actual mean value of a subpart of J^o deviates from the previous optimal value, it can be expected that this may be partly due to a bad specification of observation errors for the corresponding observations. Then, a tuning procedure of the observation error variance is to determine a normalization coefficient as the ratio between the observed value of $J_i^o(\mathbf{x}^a)$ and its expected value:

$$\begin{aligned} s_i^{o2} &= J_i^o(\mathbf{x}^a)/E[J_i^o(\mathbf{x}^a)] \\ &= J_i^o(\mathbf{x}^a) \Big/ \Big[p_i - \mathrm{Tr}\left(\mathbf{H}_i\mathbf{A}\mathbf{H}_i^{\mathrm{T}}\mathbf{R}_i{}^{-1}\right)\Big]. \end{aligned} \tag{8.10}$$

This is possible because it happens that this complicated expression can be computed, even in a variational scheme, by a randomization procedure based on a perturbation of observations (Desroziers and Ivanov, 2001; Chapnik et al., 2004).

By the way, the expression $\mathrm{Tr}(\mathbf{H}_i\mathbf{A}\mathbf{H}_i^{\mathrm{T}}\mathbf{R}_i{}^{-1})$ is nothing else than the mean sensitivity of the analysis to the particular subset of observations i.

It happens then that an ensemble variational assimilation, where observations are perturbed provides, as a by-product, a way to tune the error variance of the observations and also a way to measure their mean impact in the analysis (Desroziers et al., 2010).

It can also be shown that such a tuning procedure for the variance of observation errors is equivalent to a maximum-likelihood approach (Dee and da Silva, 1999). Assuming that the innovation vector $\mathbf{d}$ is a Gaussian random vector with mean $\mathbf{0}$ and covariance matrix $\mathbf{D}$ and that $\mathbf{D}$ is a function of the parameter vector $\mathbf{s}$, then the conditional pdf of a certain realization of $\mathbf{d}$ knowing $\mathbf{s}$ is given by the likelihood function

$$f(\mathbf{d}|\mathbf{s}) = \frac{1}{\sqrt{(2\pi)^p \det(\mathbf{D}(\mathbf{s}))}} \exp\left[-\frac{1}{2}\,\mathbf{d}^{\mathrm{T}}\mathbf{D}(\mathbf{s})\mathbf{d}\right]. \tag{8.11}$$

The maximum-likelihood estimate of the true coefficient vector $\mathbf{s}^t$ is the one that mimimizes the log-likelihood function

$$L(\mathbf{s}) = -\log[f(\mathbf{d}|\mathbf{s})]. \tag{8.12}$$

8.2.3 Observation-space diagnostics

Other diagnostics of the consistency of an analysis scheme are available. It can be shown simply (Desroziers et al., 2005) that the covariance between the $\mathbf{d}_b^a$ *analysis-minus-background* differences in observation space and the innovations $\mathbf{d}$ should be equal to

$$E[\mathbf{d}_b^a \mathbf{d}^{\mathrm{T}}] = \mathbf{HBH}^{\mathrm{T}}, \tag{8.13}$$

if the matrix $\mathbf{HK} = \mathbf{HBH}^{\mathrm{T}}(\mathbf{HBH}^{\mathrm{T}} + \mathbf{R})^{-1}$ is in agreement with the true covariances for background and observation errors.

This is a first additional diagnostic to the diagnostic on innovations. It provides a separate consistency check on background error covariances in observation space.

Similarly, the covariance between the $\mathbf{d}_a^o$ *observation-minus-analysis* differences and the innovation $\mathbf{d}$ should correspond to

$$E[\mathbf{d}_a^o \mathbf{d}^{\mathrm{T}}] = \mathbf{R}. \tag{8.14}$$

Finally, the cross-product between the $\mathbf{d}_b^a$ *analysis-minus-background* differences in observation space and the $\mathbf{d}_a^o$ *observation-minus-analysis* differences can also be derived:

$$E[\mathbf{d}_b^a \mathbf{d}_a^{o\mathrm{T}}] = \mathbf{HAH}^{\mathrm{T}}. \tag{8.15}$$

A simple geometrical interpretation of these relations can be found in Fig. 8.1, which symbolizes the simplified case of the construction of the analysis x^a of a single parameter x, using an observation y^o distant of ϵ^o from the truth x^t and a background

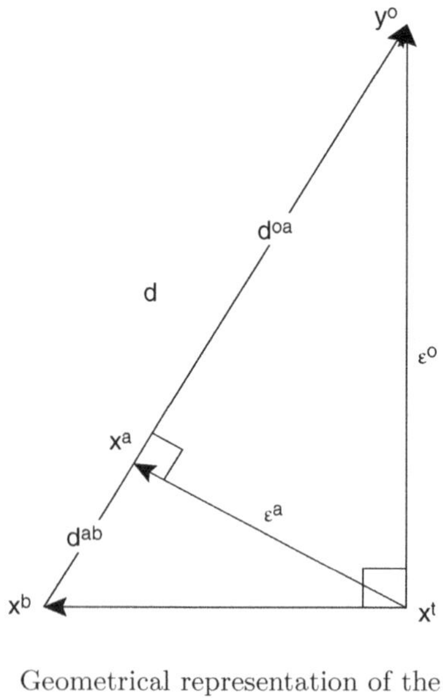

Fig. 8.1 Geometrical representation of the analysis.

x^b distant of ϵ^b from x^t. If the scalar product of two vectors of errors is the statistical covariance, then observation error and background error are orthogonal at x^t, if one assumes that they are statistically uncorrelated. The triangle (x^b, x^t, y^o) is then a right triangle at x^t.

By construction, the analysis is a linear combination of x^b and y^o and is then on the line defined by x^b and y^o. On the other hand, the analysis error ϵ^a must be minimal and then orthogonal to $y^o - x^b$, that is, to the innovation vector. Then, the application of Euclid's theorem $d^{oa}d = \epsilon^{o2}$ leads, in particular, to this relation for **R**.

The expressions (8.13) and (8.14) can be used in turn, as in Desroziers and Ivanov (2001), to tune background or observation error variances, as well as correlations. It has been shown in Desroziers et al. (2005), relying on a toy analysis problem on a circular domain, that the following fixed-point iteration converges towards the exact values of background and observation error variances v^{bt} and v^{ot} (assumed to be homogeneous over the domain), under the condition that background and observation correlations are well specified and sufficiently different:

$$\begin{cases} v^b = \mathrm{Tr}(E[\mathbf{d}_b^a \mathbf{d}^{\mathrm{T}}]), \\ v^o = \mathrm{Tr}(E[\mathbf{d}_a^o \mathbf{d}^{\mathrm{T}}]). \end{cases} \tag{8.16}$$

As in Desroziers et al. (2005), and Ménard et al. (2009), if a uniform data density is assumed over the domain, with one observation at each grid point, a spectral version of the previous iteration can be written:

$$\begin{cases} \boldsymbol{\Lambda}^b = v^b \boldsymbol{\lambda}^b, \\ \boldsymbol{\Lambda}^o = v^o \boldsymbol{\lambda}^o, \\ \boldsymbol{\Lambda}^b = \boldsymbol{\Lambda}^b(\boldsymbol{\Lambda}^{bt} + \boldsymbol{\Lambda}^{ot})/(\boldsymbol{\Lambda}^b + \boldsymbol{\Lambda}^o), \\ \boldsymbol{\Lambda}^o = \boldsymbol{\Lambda}^o(\boldsymbol{\Lambda}^{bt} + \boldsymbol{\Lambda}^{ot})/(\boldsymbol{\Lambda}^b + \boldsymbol{\Lambda}^o), \\ v^b = \sum_{k=1,p} \Lambda_k^b = F(v^b), \\ v^o = \sum_{k=1,p} \Lambda_k^o = G(v^o), \end{cases} \tag{8.17}$$

where $\boldsymbol{\Lambda}^b$ and $\boldsymbol{\Lambda}^o$ are the eigenvalues of the matrices **B** and **R**, respectively, $\boldsymbol{\lambda}^b$ and $\boldsymbol{\lambda}^o$ are their counterparts for the corresponding correlation matrices, and p is the number of eigenvalues. It is easy to check that $\boldsymbol{\Lambda}^b$ and $\boldsymbol{\Lambda}^o$ are such as $\boldsymbol{\Lambda}^b + \boldsymbol{\Lambda}^o = \boldsymbol{\Lambda}^{bt} + \boldsymbol{\Lambda}^{ot}$, where $\boldsymbol{\Lambda}^{bt}$ and $\boldsymbol{\Lambda}^{ot}$ are the exact eigenvalues of **B** and **R**. Observation and background error variances are also linked by $v^b + v^o = F(v^b) + G(v^o) = v^{bt} + v^{ot}$.

Figure. 8.2 shows the convergence of the iteration on v^b and v^o (and accordingly of $\boldsymbol{\Lambda}^b$ and $\boldsymbol{\Lambda}^o$), in the toy analysis problem treated in Desroziers et al. (2005), with a background error lengthscale $L^b = 300$ km and no correlation in observation errors ($L^o = 0$ km). In this case, the fixed-point iteration converges towards the right values. The plot of $G(v^o)$ (Fig. 8.3) explains why the convergence towards the exact values is so fast for this case. Note that there are two undesirable fixed points when applying the iteration, which are also the boundary values of the possible interval for v^o ($v^o = 1$ and $v^o = v^{bt} + v^{ot}$) and which can be easily eliminated.

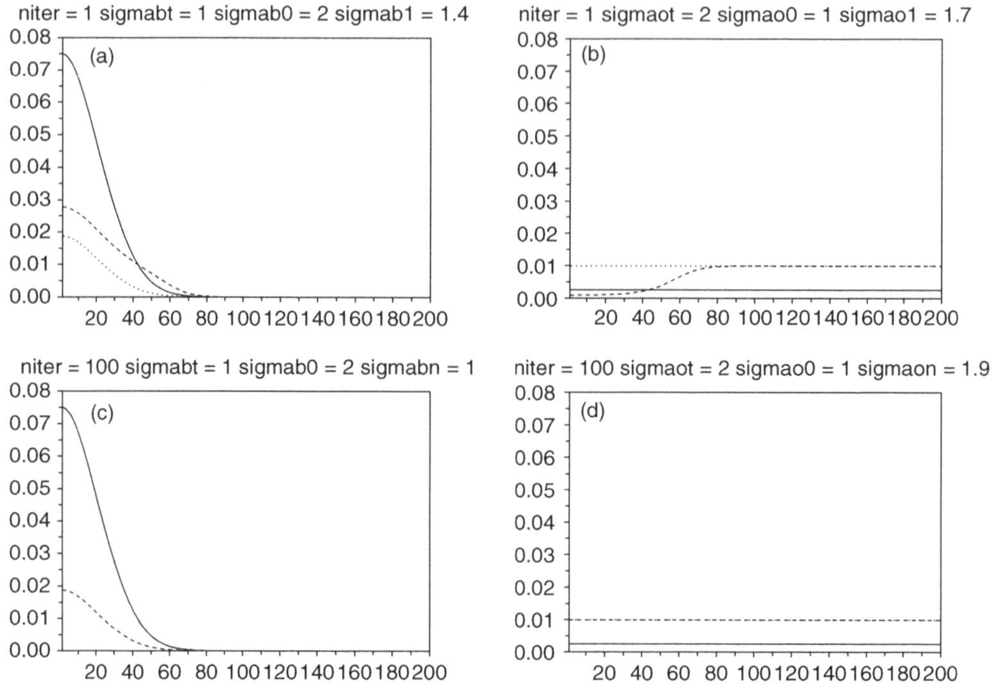

Fig. 8.2 Spectra of background (a, c) and observation errors (b, d). Lengthscales for background and observation errors are $L^b = 300$ km and $L^o = 0$ km, respectively. Dotted lines show exact spectra. (Solid lines) show erroneous spectra specified in the analysis. Dashed lines show retrieved spectra after one iteration (a, b) or convergence (c, d). Exact values of error standard deviations are $\sigma^{bt} = 1$ and $\sigma^{ot} = 2$. Originally specified values in the analysis are $\sigma^{b0} = 2$ and $\sigma^{a0} = 1$. Retrieved values of σ^{bn}/σ^{on} after one iteration or convergence are displayed above each panel.

The convergence of the algorithm is still guaranteed but slower if the lengthscales of background and observation errors become closer (Figs. 8.4 and 8.5; See also Chapnik (2009) for a discussion of the convergence of the algorithm). It could even fail if the two correlation lengthscales are too close, and in this case the sum of background and observation variances is equal to the right innovation variance ($v^{bt} + v^{ot}$), but the ratio between v^b and v^o will stay equal to the ratio specified at the beginning of the iteration. This case is equivalent to the scalar case mentioned in Ménard et al. (2009), where no scale separation allows background error variance to be distinguished from observation error variance.

Figure 8.6 shows that even if the lengthscale of the observation error is not perfectly represented in the matrix **R** (the exact value is $L^{ot} = 100$ km, but the specified value is $L^o = 0$ km), the algorithm converges towards a reasonable value of v^o. However, it has to be noted that the retrieved value underestimates the exact value, whereas a larger value of v^o would be, on the contrary, required to compensate for the lack of representation of error correlation in **R**.

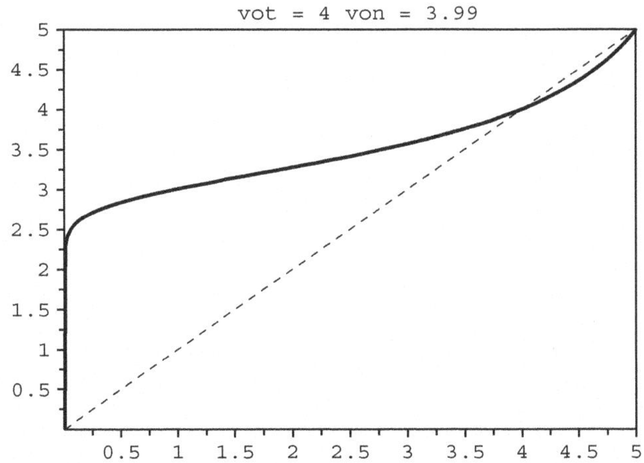

Fig. 8.3 Function $G(v^o)$ (solid line), with $L^b = 300$ km and $L^o = 0$ km, and true value of observation error variance $v^o = 4$.

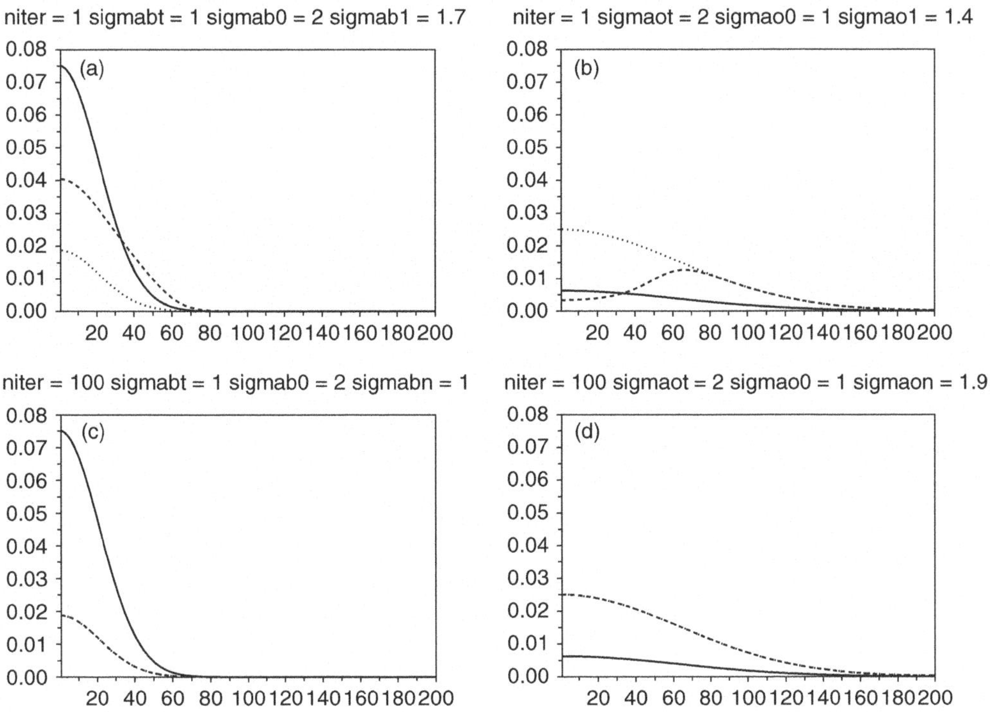

Fig. 8.4 Same as Fig. 8.2, but with $L^b = 300$ km and $L^o = 100$ km.

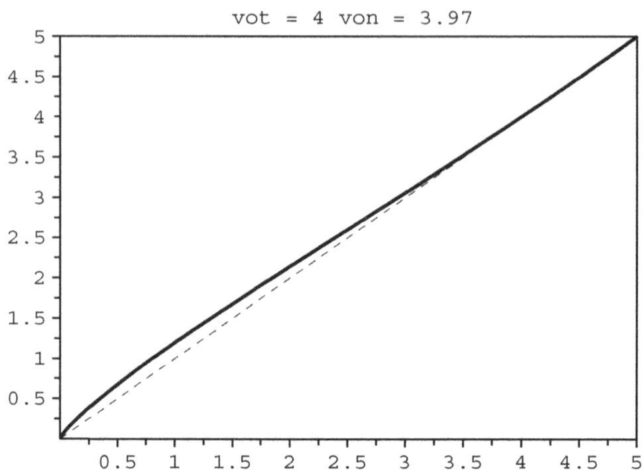

Fig. 8.5 Same as Fig. 8.3, but with $L^b = 300$ km and $L^o = 200$ km.

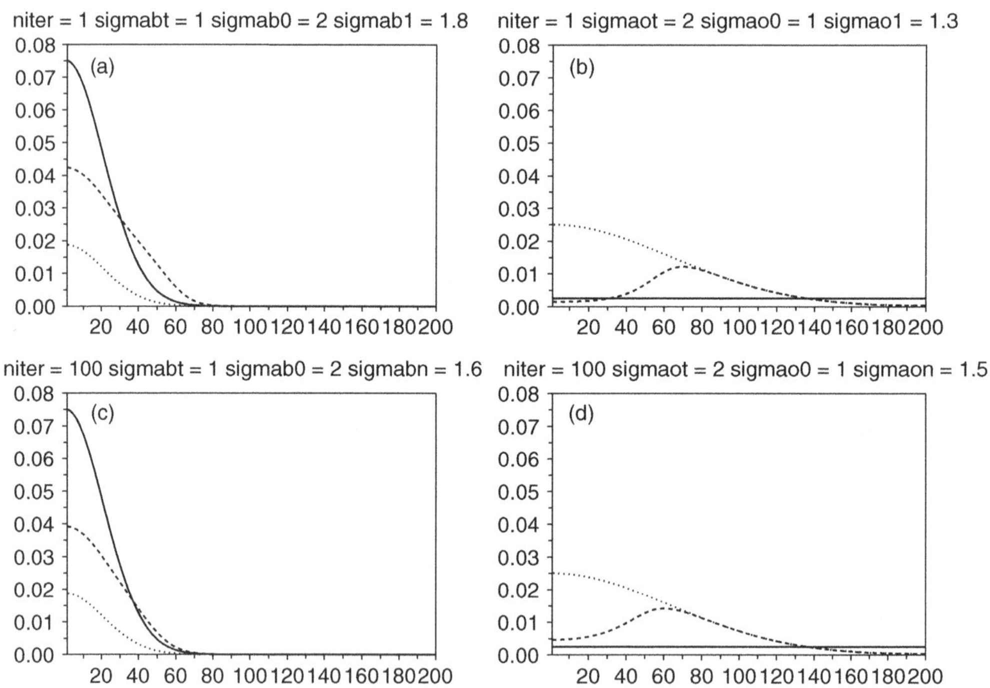

Fig. 8.6 Same as Fig. 8.4 ($L^b = 300$ km and $L^o = 100$ km), but with a mis-specified correlation in matrix $\mathbf{R}$ ($L^o = 0$ km in $\mathbf{R}$).

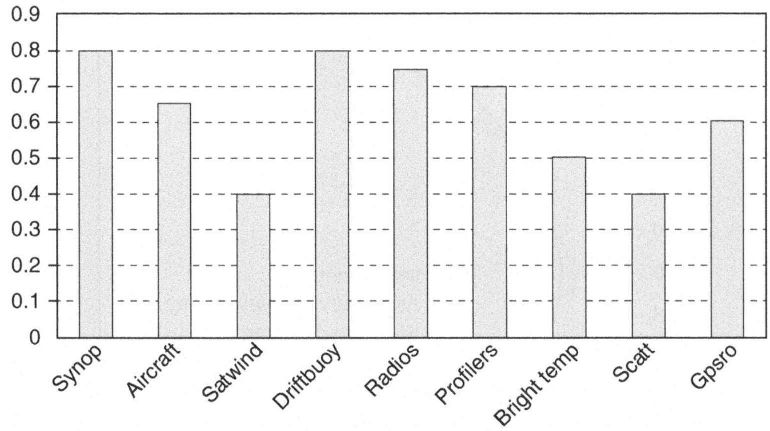

Fig. 8.7 Normalization coefficients of the σ^o values in the French Arpege 4D-Var.

As shown in Desroziers et al. (2005), there is also scope for using such diagnostics for the estimation of correlation between observation errors. Nevertheless, it is clear that the application of the diagnostics still has to be better understood from theoretical and practical points of view.

8.3 Diagnosis of observation error variances

Figure 8.7 shows an example of the application of the diagnostics based on the $J_{\min}$ diagnostics for the French 4D-Var assimilation scheme. The figure shows the diagnosed reduction factor that should be applied to the different observation errors. Indeed, one can see that the σ^o values are rather overestimated in the assimilation scheme. This is especially the case for the SATWIND or the radiance observations, where the diagnostics indicate that the σ^{a0} are overestimated by a factor 2. Of course, it is known that the overestimation of the errors in the assimilation is quite often done on purpose in order to compensate for the lack of correlations in the specified matrix $\mathbf{R}$.

Figure 8.8 shows another example, provided by Bormann et al. (2011a), of the application of different kinds of diagnostics for the σ^o values for AMSUA channels. This figure shows that all diagnostics give similar values and that these values are much lower than the specified values in the ECMWF system. So, it also appears that the σ^o values were rather overestimated in the ECMWF 4D-Var.

8.4 Diagnosis of observation error correlations

Bormann and Bauer at ECMWF (2010) have also produced different diagnostics to measure the interchannel correlations for AMSUA errors (Fig. 8.9). The three diagnostics applied actually show that there are no interchannel correlations for this specific instrument. They also showed that there are no spatial correlations in those

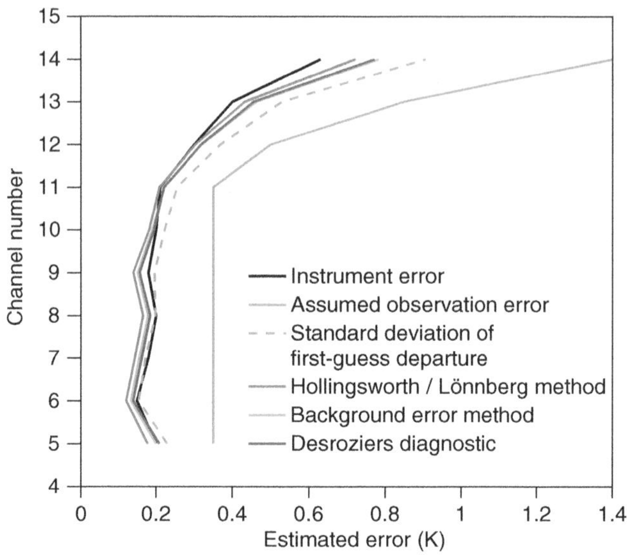

Fig. 8.8 Estimated σ^o values for AMSU-A channels. (From Bormann et al. (2011a). Reproduced with permission of the ECMWF.)

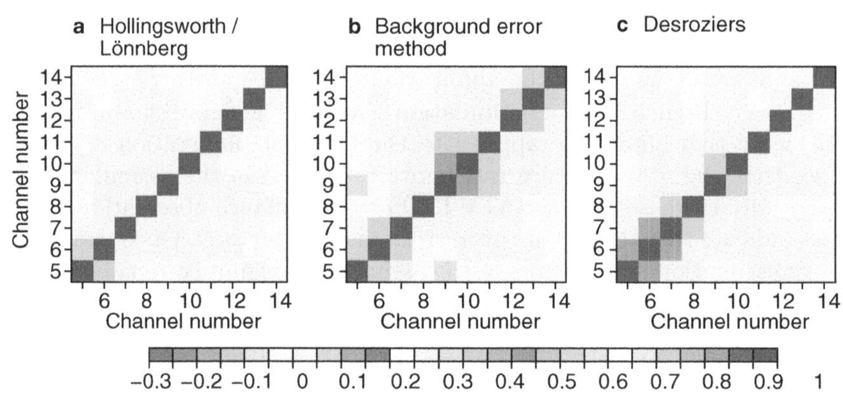

Fig. 8.9 AMSU-A interchannel error correlations. (From Bormann et al. (2010). Copyright © 2010 Royal Meteorological Society.)

observations and then that their variances could be reduced according to the previous diagnostic on the variances. Indeed, ECMWF obtained a very nice positive impact on the forecast skills by just giving more confidence to the AMSUA data. A different situation is found for IASI channels (Bormann et al., 2010), since in this case channels sensitive to water vapor or with strong surface contributions show considerable

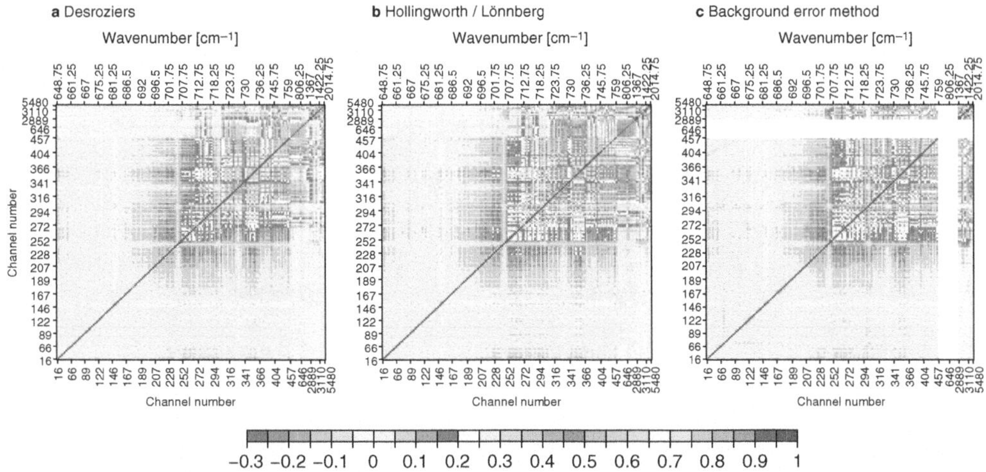

Fig. 8.10 IASI interchannel error correlations. From Bormann et al. (2010). Copyright © 2010 Royal Meteorological Society.)

interchannel correlations (Fig. 8.10). The same kinds of results were obtained by Stewart (2009) at the University of Reading, since she also found strong correlations for specific IASI channels. Similarly, this was also obtained by Garand et al. (2007), at Environment Canada, for AIRS channels, who used a Hollingsworth and Lönnberg-like method to diagnose these interchannel correlations. Then, there is evidence of interchannel correlations for satellite observations, but there is also evidence of spatial correlations for some observations. In particular, Bormann et al. (2003) showed that there are large spatial correlations in the SATWIND observations (Fig. 8.11). As they found no spatial correlations for the AMSUA data, Bormann et al. (2011b) diagnosed, on the contrary, significant spatial correlations in microwave imager radiances, especially for cloudy observations. Similarly, spatial observation error correlations can be found in Doppler radar winds, as shown by Xu et al. (2007) at the NOAA.

8.5 Observation error correlation specification in the assimilation

8.5.1 Time correlations

Other observation errors show temporal correlations as the SYNOP ground stations. Since they especially need to be taken into account in a 4D-Var scheme, Järvinen et al. (1999) at ECMWF proposed to represent them by a simple exponential correlation function:

$$c(t_1, t_2) = a \exp\left[-\left(\frac{t_1 - t_2}{b}\right)^2\right], \tag{8.18}$$

with $b = 6\,\text{h}$.

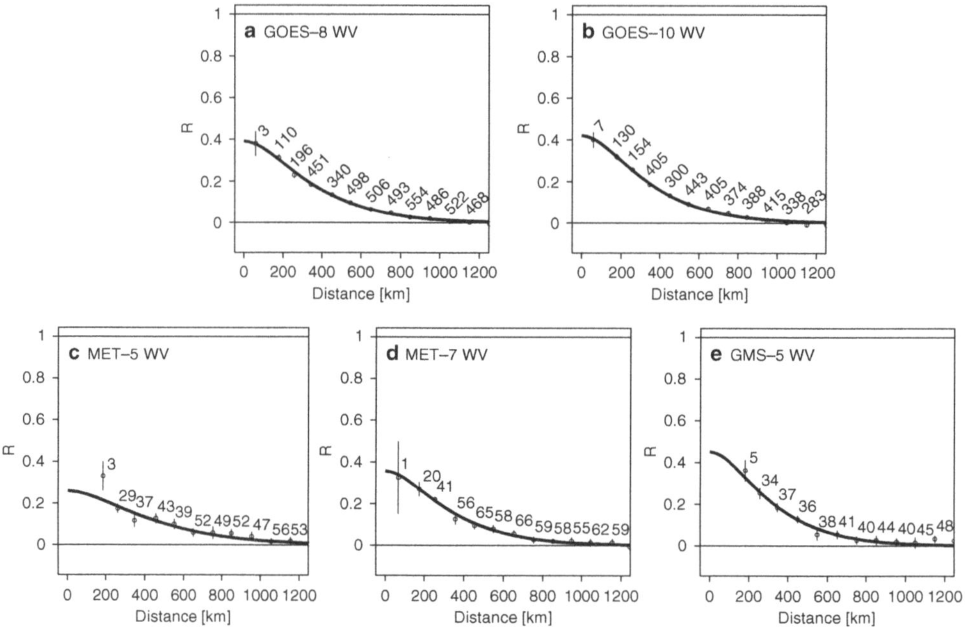

Fig. 8.11 SATWIND spatial error correlations. (From Bormann et al. (2003).)

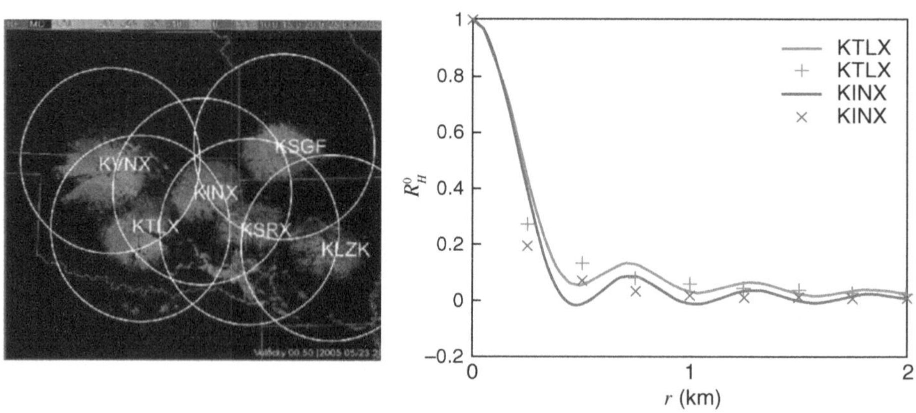

Fig. 8.12 Doppler radar wind spatial error correlations. (From Xu et al. (2007).)

In the 4D-Var minimization used at ECMWF and Météo-France, the current departures between observations and the state are normalized by the observation error standard deviations:

$$\mathbf{z}_i = \sum\nolimits_i^{-1} \left[\mathbf{y}_i^o - H_i(\mathbf{x}^b) - \mathbf{H}_i\delta\mathbf{x}\right], \tag{8.19}$$

where $\sum_i$ is the diagonal matrix containing the error standard deviations for the observation subset i. Then, the J^o term for a subset of observations is written as a simple scalar product of these normalized departures. The way in which time-correlated observations are taken into account is then to compute what are called effective departures $\mathbf{z}_i^{\text{eff}}$ by solving a linear system $\mathbf{z}_i^{\text{eff}}\mathbf{C}^{1/2} = \mathbf{z}_i$ implying the above temporal correlation, and to use these effective departures as for the uncorrelated observations.

8.5.2 Interchannel correlations

Interchannel correlations can also be taken into account in an analysis scheme, at least in a 1D-Var scheme. This has been done, for instance, by Garand et al. (2007) at Environment Canada. Figure 8.13 shows the mean temperature and humidity increments for an ensemble of 1D-Var analyses and proves that taking into account those correlations really makes a difference. Figure 8.13(c) shows those increments in observation space, and it appears logically that increments are smaller when interchannel correlations are represented for the water vapor channels affected by those correlations.

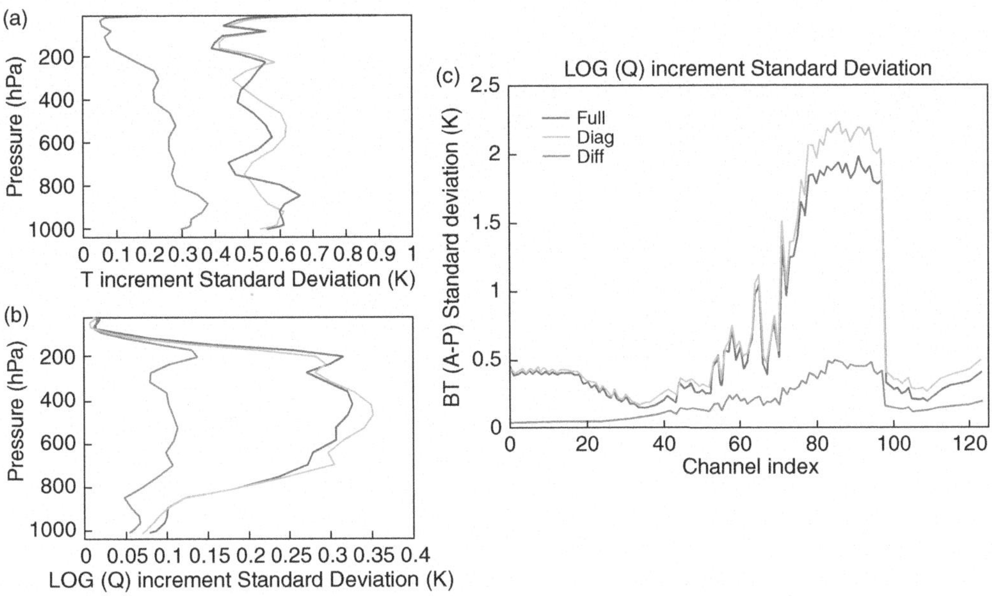

Fig. 8.13 Representation of interchannel error correlations in **R**. (From Garand et al. (2007).)

8.5.3 Spatial correlations

The representation of spatial observation error correlations is more complicated, especially in a variational formulation. Fisher and Radnoti (2006) proposed and implemented an elegant representation of such spatial correlations. It relies on a construction of a square-root correlation model of correlations:

$$\mathbf{R}_i = \mathbf{\Sigma}_i^{-1}\mathbf{C}_i\mathbf{\Sigma}_i^{-\mathrm{T}} \tag{8.20}$$

and

$$\mathbf{C}_i = \mathbf{U}_i\mathbf{U}_i^{\mathrm{T}}. \tag{8.21}$$

The square root $\mathbf{U}_i$ of the correlation matrix $\mathbf{C}_i$ is constructed as a sequence of operators:

$$\mathbf{U}_i = \mathbf{T}_i\mathbf{S}_i^{-1}\mathbf{G}_i^{1/2}, \tag{8.22}$$

where $\mathbf{G}_i$ is the spectral (Hankel) transformation of the correlation function, $\mathbf{S}_i^{-1}$ is the inverse spectral transformation, and $\mathbf{T}_i$ is an interpolation operator at observation locations.

It has to be pointed out that such a formulation is already very useful at this stage to represent realistic observation perturbations in ensemble assimilation, as for instance for SATWIND observations and even if those correlations are not taken into account in the assimilation scheme itself. The code developed at ECMWF is thus used in ensemble assimilation as implemented at Météo-France (Berre et al., 2007) and at ECMWF itself (Isaksen et al., 2010).

The representation of spatial correlations in $\mathbf{R}^{-1}$ is the next step, again proposed by Fisher and Radnoti (2006) at ECMWF. It relies on an eigenpair decomposition of matrix $\mathbf{C}_i$, using a Lanczos algorithm, and on the use of a limited number of eigenpairs in the construction of the inverse of $\mathbf{C}_i$:

$$\mathbf{C}_i = \sum_{k=1}^{K}(1/\lambda_{i,k} - 1)\mathbf{v}_{i,k}\mathbf{v}_{i,k}^{\mathrm{T}}. \tag{8.23}$$

For the time-correlated observations, the practical implementation of these spatially correlated observations relies on the computation of so-called effective departures.

It is also interesting to discuss the use of observations with correlated errors. Figure 8.14 is similar to a result produced by Liu and Rabier (2003). It presents the impact of the observation density on the quality of the retrieved analysis. It shows that if observation errors are uncorrelated, then the use of denser and denser observations always improves the precision of the analysis. On the contrary, if there is observation correlation and if this correlation is not represented in the analysis then the analysis will be degraded if the observations are too dense. This degradation will

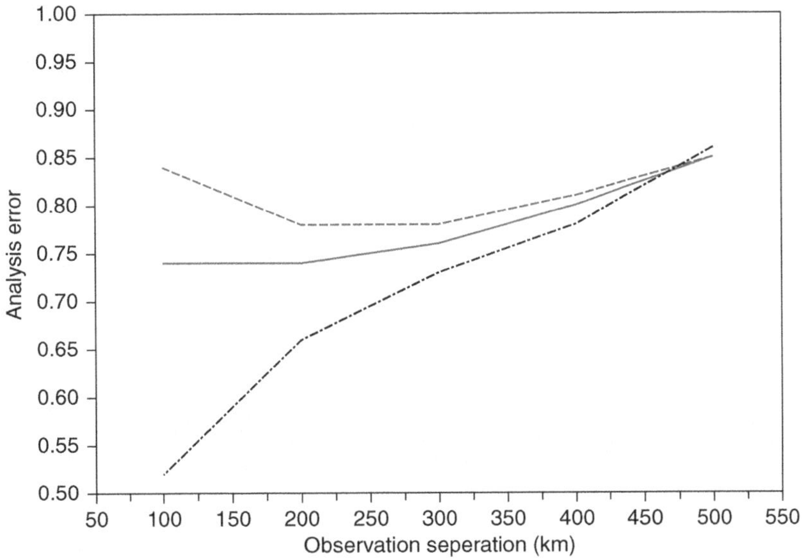

Fig. 8.14 Analysis error as a function of observation density. The dashed–dotted line shows the case where there is no spatial correlation in observation errors. The dashed line shows spatial correlation in observation errors, with a lengthscale $L^o = 200$ km, but not represented in $\mathbf{R}$. The solid line is the same as the dashed line, but with observation error correlation in $\mathbf{R}$. The lengthscale of the background error correlation is $L^b = 200$ km.

start when using observations with a distance between them roughly equal to twice the lengthscale of observation correlation. Another interesting and striking point is that even if observation error correlation is well represented in the matrix $\mathbf{R}$, the precision of the analysis will very quickly saturate when augmenting the density of observations. This is a case where the lengthscale of observation error is close to that of background error.

Figure 8.15 corresponds to a simulated case where there is a correlation in observation error, but where the lengthscale of observation error, L^o, is larger than the that of background error, L^b. This case was suggested by Snyder (2011 personal communication) and could be a likely case in the near future, with increasing resolution of the models and therefore expected decreases in L^b and with the use of denser and denser satellite observations, with likely observation error correlations. In this situation, it appears that the increased density of observations is associated with a decrease in the analysis error, and that this decrease is amplified when the observation error correlation is taken into account. Then, it appears that the most favourable cases are those where background and observation error correlation lengthscales are sufficiently different. In this case, there is a spatial filtering effect, whereas this effect vanishes when the two lengthscales are too close.

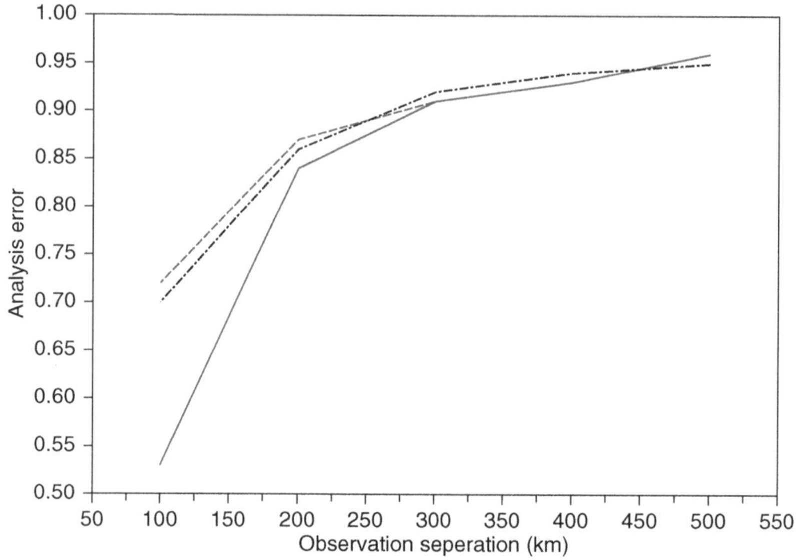

Fig. 8.15 Same as Fig. 8.14, but with a lengthscale of background error correlation $L^b = 20\,\text{km}$.

8.6 Conclusion

Observation errors are not explicitly known. They can be inferred by comparison with other observations or with the background, using innovations. There are diagnostics for observation errors (variances and correlations), but relying on explicit or implicit hypotheses. Correlation of observation errors can be found in many datasets: SYNOP time correlations, AIRS, IASI interchannel correlations, SATWIND, SSM/I, and radar spatial correlations. Those correlations are often neglected, but are accounted for with an empirical thinning and/or an inflation of error variance. Correlations can be more or less easily taken into account. A relevant formulation for spatial error correlation has been proposed and implemented in a real-size system at ECMWF.

In any case, one has to keep in mind that correlated observations are less informative than uncorrelated observations, even if **R** is well specified, when the lengthscale of observation error is close to that of background error. It may thus appear inefficient to add too many correlated observations in this particular case. The situation is different when the lengthscale L^o, of observation error becomes larger than the lengthscale L^b of background error. In this case, the increase in the density of observations amplifies the decrease in the analysis error when the observation error correlation is taken into account. Finally, the tuning of **R** must be consistent with the tuning of **B**, in order to avoid inconsistencies in assimilation schemes.

References

Bennett, A. F., Leslie, L. M., Hagelberg, C. R., and Powers, P. E. (1993). Tropical cyclone prediction using a barotropic model initialized by a generalized inverse method. *Mon. Weather Rev.*, **121**, 1714–1729.

Berre, L., Pannekoucke, O., Desroziers, G., Ştefănescu, S. E., Chapnik, B., and Raynaud, L. (2007). A variational assimilation ensemble and the spatial filtering of its error covariances: increase of sample size by local spatial averaging. In *Proceedings of the ECMWF Workshop on Flow-Dependent Aspects of Data Assimilation, 11–13 June 2007*, pp. 151–168.

Bormann, N. and Bauer, P. (2010). Estimates of spatial and interchannel observation-error characteristics for current sounder radiances for numerical weather prediction. I: Methods and application to ATOVS data. *Q. J. R. Meteorol. Soc.*, **136**, 1036–1050.

Bormann, N., Saarinen, S., Kelly, G., and Thépaut, J.-N. (2003). The spatial structure of observation errors in atmospheric motion vector from geostationnary satellite data. *Mon. Weather Rev.*, **131**, 706–718.

Bormann, N., Collard, A., and Bauer, P. (2010). Estimates of spatial and interchannel observation-error characteristics for current sounder radiances for numerical weather prediction. II: Application to AIRS and IASI data. *Q. J. R. Meteorol. Soc.*, **136**, 1051–1063.

Bormann, N., Collard, A., and Bauer, P. (2011a). Observation errors and their correlations for satellite radiances. *ECMWF Newsletter* 128.

Bormann, N., Geer, A. J., and Bauer, P. (2011b). Estimates of observation-error characteristics in clear and cloudy regions for microwave imager radiances from numerical weather prediction. *Q. J. R. Meteorol. Soc.*, **137**, 2014–2023.

Chapnik, B. (2009). On-line observation covariance matrix tuning based on optimality diagnostic. In *Proceedings of ECMWF Workshop on Diagnostics of Data Assimilation System Performance, Reading, 15-17 June 2009*.

Chapnik, B., Desroziers, G., Rabier, F., and Talagrand, O. (2004). Properties and first application of an error statistics tuning method in variational assimilation. *Q. J. R. Meteor. Soc.*, **130**, 2253–2275.

Courtier, P., Thépaut, J.-N., and Hollingsworth, A. (1994). A strategy for operational implementation of 4D-Var, using an incremental approach. *Q. J. R. Meteor. Soc.*, **120**, 1367–1387.

Dee, D. P. and da Silva, A. (1999). Maximum-likelihood estimation of forecast and observation error covariance parameters. Part I: Methodology. *Mon. Weather Rev.*, **127**, 1822–1834.

Desroziers, G. and Ivanov, S. (2001). Diagnosis and adaptive tuning of observation-error parameters in a variational assimilation. *Q. J. R. Meteor. Soc.*, **127**, 1433–1452.

Desroziers, G., Berre, L., Chapnik, B., and Poli, P. (2005). Diagnosis of observation, background and analysis error statistics in observation space. *Q. J. R. Meteor. Soc.*, **131**, 3385–3396.

Desroziers, G., Berre, L., Chabot, V., and Chapnik, B. (2010). A posteriori diagnostics in an ensemble of perturbed analyses. *Mon. Weather Rev.*, **138**, 3693–3720.

Fisher, M. and Radnoti, G. (2006). Accounting for correlated observation error in a variational analysis system. In *Proceedings of 7th Workshop on Adjoint Applications in Dynamic Meteorology, 8–13 October, Obergurgl, Austria.*

Garand, L., Heilliette, S., and Buehner, M. (2007). Interchannel error correlation associated with AIRS radiance observations: inference and impact in data assimilation. *J. Appl. Meteor.*, **46**, 714–725.

Hollingsworth, A. and Lönnberg, P. (1986). The statistical structure of short-range forecast errors as determined from radiosonde data. Part I: The wind field. *Tellus*, **38A**, 111–136.

Isaksen, L., Haseler, J., Buizza, R., and Leutbecher, M. (2010). The new ensemble of data assimilations. *ECMWF Newsletter* 123, pp. 17–21.

Järvinen, H., Andersson, E., and Bouttier, F. (1999). Variational assimilation of time sequences of surface observations with serially correlated errors. Technical Report ECMWF Reading, 266.

Liu, Z.-Q. and Rabier, F. (2003). The potential of high-density observations for numerical weather prediction: a study with simulated observations. *Q. J. R. Meteor. Soc.*, **129**, 3013–3035.

Ménard, R., Yang, Y., and Rochon, Y. (2009). Convergence and stability of estimated error variances derived from assimilation residuals in observation space. In *Proceedings of the ECMWF Workshop on Diagnostics of Data Assimilation System Performance, Reading, 15–17 June*, pp. 133–144.

Stewart, L. (2009). Correlated observation errors in data assimilation. PhD thesis, University of Reading.

Talagrand, O. (1997). Assimilation of observations, an introduction. *J. Meteor. Soc. Japan*, **75**, 191–209.

Xu, Q., Nai, K., and Wei, L. (2007). An innovation method for estimating radar radial-velocity observation error and background wind error covariances. *Q. J. R. Meteor. Soc.*, **133**, 407–415.

9
Errors. A posteriori diagnostics

O. Talagrand

Laboratoire de Météorologie Dynamique
Ecole Normale Supérieure
Paris, France

Advanced Data Assimilation for Geosciences. First Edition.
Edited by É. Blayo, M. Bocquet, E. Cosme, and L. F. Cugliandolo.

Chapter Contents

9.1 Introduction

The purpose of assimilation is to reconstruct as accurately as possible the state of the observed system from the available data. In geoscience applications, those data essentially consist of the observations proper and of the physical laws that govern the dynamics of the system, available in practice in the form of a discretized numerical model. All these data are affected by some uncertainty, and one may wish to quantify the resulting uncertainty on the final estimate. If one accepts the representation of uncertainty by probability distributions, this leads one to consider the problem of assimilation as a problem in *Bayesian estimation*, namely, determining the probability distribution for the state of the system, conditional to the available data.

A large number of presently existing assimilation algorithms are based on the theory of statistical linear estimation (best linear unbiased estimate, BLUE). Among all linear estimators, the BLUE minimizes the statistical quadratic estimation error. In addition, it achieves Bayesian estimation in the case of additive Gaussian 'errors' in the data. As explained below, the BLUE requires the prior specification of the first- and second-order moments (expectation and covariance matrix) of those errors.

A number of questions naturally arise in this context:

1. How is it possible to objectively evaluate the quality of an assimilation algorithm? In particular, how is it possible to objectively compare the performances of two different assimilation algorithms?
2. How is it possible to objectively determine the expectation and covariance matrix of the errors, whose explicit specification is required for determining the BLUE?
3. Is it possible to objectively verify if an assimilation algorithm is optimal in a given precise sense, for instance in the sense of least error variance?

These questions are discussed in this chapter. Answers, at least partial ones, are given. It is stressed that any procedure for achieving any of the above goals requires hypotheses that cannot be objectively validated on the basis of the data only. Section 9.2 summarizes the main elements of the theory of the BLUE. The three questions above are dealt with in Sections 9.3–9.6, with Section 9.5 being more specifically devoted to objective evaluation of the internal consistency of an assimilation algorithm. Conclusions and comments are given in Section 9.7.

This chapter is an updated version of a previous publication (Talagrand, 2010), to which a few modifications and improvements have been made.

The notation is the same as in Chapter 1 of the present volume, with the exception that the notation E_e will be used, as will be explained below, for denoting a particular type of statistical expectation.

9.2 Reminder on statistical linear estimation

An unknown *state vector* $\mathbf{x}$, belonging to *state space* $\mathcal{S}$, with dimension n, is to be determined from a known *data vector* $\mathbf{z}$, belonging to *data space* $\mathcal{D}$, with dimension m, of the form

$$\mathbf{z} = \mathbf{\Gamma x} + \boldsymbol{\xi}. \tag{9.1}$$

In this expression, $\mathbf{\Gamma}$ is a known operator from $\mathcal{S}$ into $\mathcal{D}$, called the *data operator*, represented by an $m \times n$ matrix, while $\boldsymbol{\xi}$ is an unknown m-dimensional error, assumed to be a realization of a vector random variable in data space. We look for an estimate of $\mathbf{x}$ (the *analysis*) of the form

$$\mathbf{x}^a = \mathbf{a} + \mathbf{A}\mathbf{z},$$

where the n-vector $\mathbf{a}$ and the $n \times m$ matrix $\mathbf{A}$ are to be determined under the following two conditions:

1. The estimate $\mathbf{x}^a$ is independent of the choice of the origin in state space.
2. For any component of $\mathbf{x}$, the statistical expectation of the squared estimation error is minimum.

The solution to this problem is

$$\mathbf{x}^a = \left(\mathbf{\Gamma}^{\mathrm{T}}\mathbf{S}^{-1}\mathbf{\Gamma}\right)^{-1}\mathbf{\Gamma}^{\mathrm{T}}\mathbf{S}^{-1}(\mathbf{z} - \boldsymbol{\mu}), \tag{9.2}$$

i.e.

$$\mathbf{A} = \left(\mathbf{\Gamma}^{\mathrm{T}}\mathbf{S}^{-1}\mathbf{\Gamma}\right)^{-1}\mathbf{\Gamma}^{\mathrm{T}}\mathbf{S}^{-1}, \tag{9.3a}$$

$$\mathbf{a} = -\mathbf{A}\boldsymbol{\mu}, \tag{9.3b}$$

where $\boldsymbol{\mu} = E(\boldsymbol{\xi})$ and $\mathbf{S} = E[(\boldsymbol{\xi} - \boldsymbol{\mu})(\boldsymbol{\xi} - \boldsymbol{\mu})^{\mathrm{T}}]$ are respectively the expectation and covariance matrix of the data error $\boldsymbol{\xi}$.

The corresponding estimation error $\mathbf{x}^a - \mathbf{x}$ has zero expectation,

$$E(\mathbf{x}^a - \mathbf{x}) = 0, \tag{9.4}$$

and has covariance matrix

$$\mathbf{P}^a = E[(\mathbf{x}^a - \mathbf{x})(\mathbf{x}^a - \mathbf{x})^{\mathrm{T}}] = (\mathbf{\Gamma}^{\mathrm{T}}\mathbf{S}^{-1}\mathbf{\Gamma})^{-1}. \tag{9.5}$$

The estimate $\mathbf{x}^a$ is the *best linear unbiased estimate (BLUE)* of $\mathbf{x}$ from $\mathbf{z}$. Its explicit determination requires, at least in view of (9.2), the explicit specification of both the expectation $\boldsymbol{\mu}$ and the covariance matrix $\mathbf{S}$ of the error $\boldsymbol{\xi}$.

The BLUE is unambiguously defined if and only if the matrix $\mathbf{\Gamma}$ is of rank n, i.e. if and only if the condition $\mathbf{\Gamma}\boldsymbol{\xi} = 0$ implies $\boldsymbol{\xi} = 0$. This is called the *determinacy* condition. It implies that $m \geq n$. We set $m = n + p$, $p \geq 0$. The determinacy condition depends only on the data matrix $\mathbf{\Gamma}$, and says nothing about the accuracy of the estimate, which is determined by the covariance matrix $\mathbf{S}$, (9.5).

The BLUE $\mathbf{x}^a$ can also be obtained as the minimizer of the following scalar objective function, defined on state space:

$$\boldsymbol{\xi} \in \mathcal{S} \longrightarrow J(\boldsymbol{\xi}) = \frac{1}{2}\,[\mathbf{\Gamma}\boldsymbol{\xi} - (\mathbf{z} - \boldsymbol{\mu})]^{\mathrm{T}}\,\mathbf{S}^{-1}\,[\mathbf{\Gamma}\boldsymbol{\xi} - (\mathbf{z} - \boldsymbol{\mu})]. \tag{9.6}$$

Variational assimilation is based on minimization of objective functions of the form (9.6). The Kalman filter, although of a totally different algorithmic form, also amounts to minimizing an objective function of the form (9.6).

Equations (9.2), (9.5), and (9.6) are invariant under a change of origin, as well as any invertible linear change of coordinates, in either state or data space. In particular, the product $\boldsymbol{\xi}_1^{\mathrm{T}}\mathbf{S}^{-1}\boldsymbol{\xi}_2$, being invariant under a linear change of coordinates, defines a proper scalar product for any two vectors $\boldsymbol{\xi}_1$ and $\boldsymbol{\xi}_2$ in data space. That scalar product is called the *Mahalanobis scalar product* associated with the covariance matrix $\mathbf{S}$. Expression (9.6) shows that the image $\boldsymbol{\Gamma}\mathbf{x}^a$ of the BLUE $\mathbf{x}^a$ through $\boldsymbol{\Gamma}$ is the point in the image space $\boldsymbol{\Gamma}(\mathcal{S})$ that lies closest, in the sense of the $\mathbf{S}$-Mahalanobis norm, to the unbiased data vector $\mathbf{z} - \boldsymbol{\mu}$. The BLUE is thus seen to be the output of the following three operations:

1. Remove the bias in the data vector $\mathbf{z}$ by subtracting the mean error $\boldsymbol{\mu}$.
2. Project the unbiased data vector onto the subspace $\boldsymbol{\Gamma}(\mathcal{S})$ orthogonally with respect to the $\mathbf{S}$-Mahalanobis scalar product.
3. Take the inverse of the projection through $\boldsymbol{\Gamma}$. The determinacy condition $\text{rank}(\boldsymbol{\Gamma}) = n$ ensures that the inverse is uniquely defined.

It can be seen that the component of $\mathbf{z}$ that is $\mathbf{S}$-Mahalanobis orthogonal to $\boldsymbol{\Gamma}(\mathcal{S})$ has no impact on the result of the estimation process. More precisely, project the data space $\mathcal{D}$ onto the subspace $\boldsymbol{\Gamma}(\mathcal{S})$, and the subspace $\boldsymbol{\Gamma}(\mathcal{S})^{\perp}$ that is $\mathbf{S}$-Mahalanobis orthogonal to $\boldsymbol{\Gamma}(\mathcal{S})$, and denote by $\mathbf{w}_1$ and $\mathbf{w}_2$ the corresponding respective components of a vector $\mathbf{w}$ in $\mathcal{D}$. The data operator $\boldsymbol{\Gamma}$ now becomes

$$\boldsymbol{\Gamma} = \begin{pmatrix} \boldsymbol{\Gamma}_1 \\ 0 \end{pmatrix},$$

where $\boldsymbol{\Gamma}_1$ is an $n \times n$ invertible operator from $\mathcal{S}$ onto $\boldsymbol{\Gamma}(\mathcal{S})$. The data vector $\mathbf{z}$ decomposes into

$$\mathbf{z}_1 = \boldsymbol{\Gamma}_1\mathbf{x} + \boldsymbol{\xi}_1, \tag{9.7a}$$

$$\mathbf{z}_2 = \boldsymbol{\xi}_2. \tag{9.7b}$$

It can now be seen that the analysed estimate is equal to

$$\mathbf{x}^a = \boldsymbol{\Gamma}_1^{-1}(\mathbf{z}_1 - \boldsymbol{\mu}_1) = \mathbf{x} + \boldsymbol{\Gamma}_1^{-1}(\boldsymbol{\xi}_1 - \boldsymbol{\mu}_1).$$

The determination of $\mathbf{x}^a$ therefore requires knowledge of only the orthogonal subspace $\boldsymbol{\Gamma}(\mathcal{S})^{\perp}$ and of the component $\boldsymbol{\mu}_1$ of the mean error $\boldsymbol{\mu}$.

The covariance matrix $\mathbf{S}$ decomposes into the block-diagonal matrix

$$\mathbf{S} = \text{diag}(\mathbf{S}_1, \mathbf{S}_2),$$

where $\mathbf{S}_1$ and $\mathbf{S}_2$ are symmetric definite-positive matrices with respective dimensions $n \times n$ and $p \times p$. It can be seen from (9.5) that the the analysis error covariance matrix is

$$\mathbf{P}^a = \boldsymbol{\Gamma}_1^{-1}\mathbf{S}_1\boldsymbol{\Gamma}_1^{-\mathrm{T}},$$

where $\boldsymbol{\Gamma}_1^{-\mathrm{T}}$ denotes the transpose of $\boldsymbol{\Gamma}_1^{-1}$.

It follows from the above that, contrary to what (9.2), (9.5), or (9.6) apparently suggest, the BLUE $\mathbf{x}^a$ and the associated estimation error covariance matrix $\mathbf{P}^a$ do not depend on the full expectation $\boldsymbol{\mu}$ and covariance matrix $\mathbf{S}$ of the error $\boldsymbol{\xi}$. They depend only on the orthogonal subspace $\boldsymbol{\Gamma}(\mathcal{S})^\perp$ and on the components $\boldsymbol{\mu}_1$ and $\mathbf{S}_1$ of $\boldsymbol{\mu}$ and $\mathbf{S}$ along $\boldsymbol{\Gamma}(\mathcal{S})$. Both $\mathbf{x}^a$ and $\mathbf{P}^a$ are independent of the components $\boldsymbol{\mu}_2$ and $\mathbf{S}_2$ along $\boldsymbol{\Gamma}(\mathcal{S})^\perp$.

We will consider assimilation systems of the general form (9.2), for given, not necessarily exact, bias $\boldsymbol{\mu}$ and covariance $\mathbf{S}$. Such systems also provide an estimate of the corresponding error covariance matrix, in the form (9.5). Since the assumed $\boldsymbol{\mu}$ and $\mathbf{S}$ are not necessarily exact, the corresponding estimate is not necessarily the BLUE. One major point of this chapter is precisely to discuss the possibility of identifying eventual mis-specifications in either the expectation or the covariance of the data error, and of determining exactly that expectation and that covariance.

For convenience, as well as for consistency with the usual notation, we will assume that the mean error $\boldsymbol{\mu}$ (or more precisely, what is thought to be the mean error) has been subtracted from the data vector $\mathbf{z}$. That mean error will therefore no longer appear explicitly in the equations. But the possibility exists that it was not correctly specified in the first place and that a bias may remain in $\mathbf{z}$. The possibility of identifying that bias is one of the questions discussed below.

The matrix $\mathbf{A} = (\boldsymbol{\Gamma}^{\mathrm{T}}\mathbf{S}^{-1}\boldsymbol{\Gamma})^{-1}\boldsymbol{\Gamma}^{\mathrm{T}}\mathbf{S}^{-1}$ is a left-inverse of $\boldsymbol{\Gamma}$. Conversely, any left-inverse $\boldsymbol{\Delta}$ of $\boldsymbol{\Gamma}$ is of the form $(\boldsymbol{\Gamma}^{\mathrm{T}}\boldsymbol{\Sigma}^{-1}\boldsymbol{\Gamma})^{-1}\boldsymbol{\Gamma}^{\mathrm{T}}\boldsymbol{\Sigma}^{-1}$, with an appropriately chosen $m\times m$ positive-definite symmetric matrix $\boldsymbol{\Sigma}$. To see that, let us first note that if the state and data spaces have the same dimension ($m = n$), then $\boldsymbol{\Gamma}$, which has rank n, is exactly invertible, with inverse $\boldsymbol{\Gamma}^{-1}$. $(\boldsymbol{\Gamma}^{\mathrm{T}}\boldsymbol{\Sigma}^{-1}\boldsymbol{\Gamma})^{-1}\boldsymbol{\Gamma}^{\mathrm{T}}\boldsymbol{\Sigma}^{-1}$ is then equal to $\boldsymbol{\Gamma}^{-1}$ for any $\boldsymbol{\Sigma}$. If $m > n$, $\boldsymbol{\Delta}$ has a null-space $\mathrm{Ker}\,\boldsymbol{\Delta}$ with dimension $p = m - n$. In a way similar to what has been done above, we can project the data space onto the image space $\boldsymbol{\Gamma}(\mathcal{S})$ and the kernel $\mathrm{Ker}\,\boldsymbol{\Delta}$. Any positive-definite matrix $\boldsymbol{\Sigma}$ that decomposes in that projection into

$$\boldsymbol{\Sigma} = \mathrm{diag}(\boldsymbol{\Sigma}_1, \boldsymbol{\Sigma}_2)$$

defines an operator $(\boldsymbol{\Gamma}^{\mathrm{T}}\boldsymbol{\Sigma}^{-1}\boldsymbol{\Gamma})^{-1}\boldsymbol{\Gamma}^{\mathrm{T}}\boldsymbol{\Sigma}^{-1}$ that, in addition to being a left-inverse of $\boldsymbol{\Gamma}$, has null-space $\mathrm{Ker}\,\boldsymbol{\Delta}$. That operator is, therefore, identical with $\boldsymbol{\Delta}$. It is seen that $\boldsymbol{\Sigma}_1$ and $\boldsymbol{\Sigma}_2$ can be arbitrary. In particular, if multiplication by $\boldsymbol{\Delta}$ is used for obtaining the BLUE of $\mathbf{x}$, the corresponding estimation error covariance matrix will be equal to

$$\mathbf{P}^a = \boldsymbol{\Gamma}_1^{-1}\boldsymbol{\Sigma}_1\boldsymbol{\Gamma}_1^{-\mathrm{T}}.$$

Since $\boldsymbol{\Sigma}_1$ can be arbitrary, so can $\mathbf{P}^a$. The knowledge of the left-inverse operator that defines the BLUE does not bring any information on the associated estimation error. Contrary to what one might be tempted to think, knowledge of the matrix $(\boldsymbol{\Gamma}^{\mathrm{T}}\boldsymbol{\Sigma}^{-1}\boldsymbol{\Gamma})^{-1}\boldsymbol{\Gamma}^{\mathrm{T}}\boldsymbol{\Sigma}^{-1}$ does not bring, even for a known $\boldsymbol{\Gamma}$, any information on the matrix $(\boldsymbol{\Gamma}^{\mathrm{T}}\boldsymbol{\Sigma}^{-1}\boldsymbol{\Gamma})^{-1}$. Any left-inverse of $\boldsymbol{\Gamma}$ can coexist with any estimation error covariance matrix $\mathbf{P}^a$.

We will therefore consider estimation schemes of the form

$$\mathbf{x}^e = \mathbf{A}^e\mathbf{z}, \tag{9.8}$$

where $\mathbf{A}^e$ is a left-inverse of $\boldsymbol{\Gamma}$ (the superscript e highlights the fact that the estimate $\mathbf{x}^e$ may not be optimal). The scheme will be associated with an estimated error covariance matrix $\mathbf{P}^e$. As mentioned above, one particular purpose of the present study is to determine whether the possible optimality of the scheme (9.8) can be established on objective grounds. In agreement with the fact that there exists no link in the optimal case between the quantities $\mathbf{A}$ and $\mathbf{P}^a$, no link will be assumed here between $\mathbf{A}^e$ and $\mathbf{P}^e$.

Owing to the determinacy condition, it is possible to transform the data vector $\mathbf{z}$, through linear invertible operations, into two components of the form

$$\mathbf{x}^b = \mathbf{x} + \boldsymbol{\xi}^b, \tag{9.9a}$$

$$\mathbf{y} = \mathbf{H}\mathbf{x} + \boldsymbol{\varepsilon}. \tag{9.9b}$$

The vector $\mathbf{x}^b$, which has dimension n, is an explicit estimate of the unknown state vector $\mathbf{x}$, called the background estimate of $\mathbf{x}$. The vector $\mathbf{y}$, which has dimension p, is an additional set of data, linked to the real state vector $\mathbf{x}$ through the (linear) observation operator $\mathbf{H}$, represented by a $p \times n$ matrix.

In (9.9), the data operator is $\boldsymbol{\Gamma} = (\mathbf{I}_n, \mathbf{H}^{\mathrm{T}})^{\mathrm{T}}$. It is in (9.9) that data are usually available in meteorological and oceanographical applications. The expressions (9.2), (9.3), and (9.5) for the BLUE $\mathbf{x}^a$ and the estimation error covariance matrix $\mathbf{P}^a$ then assume the forms

$$\mathbf{x}^a = \mathbf{x}^b - E\left(\boldsymbol{\xi}^b\mathbf{d}^{\mathrm{T}}\right)\left[E\left(\mathbf{d}\mathbf{d}^{\mathrm{T}}\right)\right]^{-1}\left(\mathbf{y} - \mathbf{H}\mathbf{x}^b\right), \tag{9.10a}$$

$$\mathbf{P}^a = E\left(\boldsymbol{\xi}^b\boldsymbol{\xi}^{b\mathrm{T}}\right) - E\left(\boldsymbol{\xi}^b\mathbf{d}^{\mathrm{T}}\right)\left[E\left(\mathbf{d}\mathbf{d}^{\mathrm{T}}\right)\right]^{-1} E\left(\mathbf{d}\boldsymbol{\xi}^{b\mathrm{T}}\right), \tag{9.10b}$$

where $\mathbf{d} = \mathbf{y} - \mathbf{H}\mathbf{x}^b$ is called the *innovation vector* (incidentally, and in order to avoid any confusion with what may be said in other chapter in this volume, it is not assumed here that the errors $\boldsymbol{\xi}^b$ and $\boldsymbol{\varepsilon}$ are uncorrelated).

In (9.7), the left-inverses of $\boldsymbol{\Gamma} = (\mathbf{I}_n, \mathbf{H}^{\mathrm{T}})^{\mathrm{T}}$ are of the form

$$(\mathbf{x}^b, \mathbf{y}) \rightarrow \mathbf{x}^e = \mathbf{x}^b + \mathbf{K}\left(\mathbf{y} - \mathbf{H}\mathbf{x}^b\right), \tag{9.11}$$

where the *gain matrix* $\mathbf{K}$ can be any $n \times p$ matrix. A given gain matrix $\mathbf{K}$ can coexist with any estimated error covariance matrix $\mathbf{P}^a$.

The forms (9.2) and (9.11) are exactly equivalent. At any point in the following, we will use the form that is most convenient for our purpose.

9.3 Objective evaluation of assimilation algorithms

The purpose of assimilation is to estimate as accurately as possible the state of the observed system. The ultimate validation criterion is therefore the accuracy with which the state of the system is estimated, and is naturally quantified by the statistical difference between the estimated values and the corresponding real values of the various parameters that define the state of the system. It is precisely the expectation of the square of that difference that the BLUE is intended to minimize. In most situations, the real values of the quantities to be evaluated will not, however, be available, even a posteriori. The validation can therefore be performed, at best, against observations or estimates that are themselves affected by errors. Consider the simple case when a scalar quantity x ($n = 1$) is to evaluated from two scalar data ($m = 2$) of the form

$$z_1 = x + \xi_1, \tag{9.12a}$$

$$z_2 = x + \xi_2. \tag{9.12b}$$

This is of the form (9.1), with $\mathbf{z} = (z_1, z_2)^{\mathrm{T}}$, $\mathbf{\Gamma} = (1,1)^{\mathrm{T}}$, and $\boldsymbol{\xi} = (\xi_1, \xi_2)^{\mathrm{T}}$. We assume the errors to be unbiased ($E(\xi_1) = E(\xi_2) = 0$) and mutually uncorrelated ($E(\xi_1\xi_2) = 0$) and to have the same variance s ($E(\xi_1^2) = E(\xi_2^2) = s$), so that $\mathbf{S} = s\mathbf{I}_2$. The BLUE is then $x^a = (1/2)(z_1 + z_2)$, and the estimation error covariance matrix $\mathbf{P}^a$ reduces to the scalar $s/2$. Consider a linear estimate x^e of the form

$$x^e = a_1 z_1 + a_2 z_2,$$

with $a_1 + a_2 = 1$ (that is, the condition that $\mathbf{A}^e$ is, in (9.8), a left-inverse of $\mathbf{\Gamma}$). We want to check the possible optimality of x^e by comparison with an observation (or a different estimate) x^o of the form

$$x^o = x + \eta,$$

where η is a random error. In the logic of least-squares statistical minimization taken here, the quantity x^o can legitimately be used for validation if the mean quadratic error $E[(x^e - x^o)^2]$, considered as a function of the estimate x^e, is minimum when x^e is equal to the BLUE x^a. This requires that x^o be unbiased (if it had a non-zero bias $E(\eta)$, the estimate $x^a + E(\eta)$ would achieve a better quadratic fit to x^o than the BLUE x^a). It also requires the error η to be uncorrelated with the data error $\boldsymbol{\xi}$. It is obvious that if η has variance s but is, for instance, strongly and positively correlated with ξ_1, but not with ξ_2, a better fit to x^o will be obtained if $a_1 > a_2$ than if $a_1 = a_2$. More precisely, it can be shown that the linear function of $\mathbf{z}$ that optimally estimates the quantity x^o (in the sense of minimum statistical quadratic error) is equal to

$$x^{oa} = x^a + E(\eta\boldsymbol{\xi}^{\mathrm{T}})\mathbf{S}^{-1}\left(\mathbf{S} - \mathbf{\Gamma}\mathbf{P}^a\mathbf{\Gamma}^{\mathrm{T}}\right)\mathbf{S}^{-1}\mathbf{z} = E\left[\eta(\xi_1 - \xi_2)\right](z_1 - z_2)/(2s).$$

This estimate is different from x^a when the errors η and $\boldsymbol{\xi}$ are mutually correlated, with $E\left[\eta(\xi_1 - \xi_2)\right] \neq 0$.

If the conditions of unbiasedness and decorrelation from data error are verified for the validating observation x^o, then the mean quadratic difference between x^e and x^o is

$$E\left[(x^e - x^o)^2\right] = E\left[(x^e - x)^2\right] + E(\eta^2).$$

It is minimum for $x^e = x^a$.

This shows that an estimate $\mathbf{x}^e$ can be usefully validated only against observations (or other estimates) that are unbiased and affected by errors that are themselves uncorrelated with the errors affecting the data used for producing $\mathbf{x}^e$. In particular, the fit of the analysed fields to the data that has been used in the analysis cannot be a proper diagnostic of the quality of the analysis. It can, in fact, be shown that the fit of the analysed fields to any particular piece of data can be made arbitrarily small by simply decreasing the assumed variance for the error affecting that piece of data.

As a consequence, objective comparison between the results of two different assimilation systems can be performed only against observations or estimates that are uncorrelated with data used in either one of the two systems. A practical difficulty is, of course, that the decorrelation between the data used in the assimilation and in the validation can never be objectively verified, and has to be hypothesized on the basis of, at best, physical knowledge, experience, and good judgement.

9.4 Estimation of the statistics of data errors

How is it possible to objectively determine the expectation $\boldsymbol{\mu} = E(\boldsymbol{\xi})$ and the covariance matrix $\mathbf{S} = E\left[(\boldsymbol{\xi} - \boldsymbol{\mu})(\boldsymbol{\xi} - \boldsymbol{\mu})^{\mathrm{T}}\right]$ whose specification is necessary for determining the BLUE $\mathbf{x}^a$? One way could be to proceed by trial and error, namely to vary $\boldsymbol{\mu}$ and $\mathbf{S}$, and to determine, through comparison against unbiased and independent data, which combination leads to the best statistical fit to those data. One could even envisage that an explicit statistical optimization process could be implemented for determining the optimal values of $\boldsymbol{\mu}$ and $\mathbf{S}$. The sheer size of the numerical dimension of the meteorological or oceanographical problems (or of problems encountered in the study of other systems) clearly shows that there can be no hope to entirely and accurately determine $\boldsymbol{\mu}$ and $\mathbf{S}$ through such procedures. However, empirical tuning of parameters has always been an important part of the development of assimilation systems. Empirical tuning can be systematized in the form of *cross validation*. A typical example is as follows. For a particular class of instrument, assumed to produce unbiased observations with the same constant error variance, the assumed variance is varied in a series of assimilation experiments in order to determine the value for which the fit to independent data is optimized. In spite of a number of instructive studies of cross validation, and of its extension called *generalized cross validation* (see e.g. Wahba et al. (1995) Chapter 20 in this volume), this type of method has not been so far extensively used in meteorological or oceanographical applications.

New meteorological observations are available every day, and a natural question is whether it is possible to determine the quantities $\boldsymbol{\mu}$ and $\mathbf{S}$ through appropriate statistical processing of the observations. It is seen from the background-observation decomposition (9.9) that the only combination of the data that is independent of the unknown state vector $\mathbf{x}$ is the innovation

$$\mathbf{d} = \mathbf{y} - \mathbf{H}\mathbf{x}^b = -\mathbf{H}\boldsymbol{\mu}^b + \boldsymbol{\varepsilon}. \tag{9.13}$$

Within the world of data and assimilation, the innovation is the only objective source of information on the errors affecting the data. The question we consider here is therefore that of which knowledge on $\boldsymbol{\mu}$ and $\mathbf{S}$ can be obtained from statistical processing of the innovation vector.

Consider the *data-minus-analysis* (*DmA*) difference vector

$$\boldsymbol{\delta} = \mathbf{z} - \mathbf{\Gamma x}^e. \tag{9.14}$$

This is the a posteriori misfit between the raw data and the estimated state vector $\mathbf{x}^e$. By the definition (9.8) of $\mathbf{x}^e$, $\boldsymbol{\delta}$ is $\mathbf{\Sigma}$-Mahalanobis orthogonal to the image subspace $\mathbf{\Gamma}(\mathcal{S})$, where $\mathbf{\Sigma}$ is any one of the covariance matrices associated with the left-inverse in (9.8). In the background-observation form (9.9), it decomposes into $\boldsymbol{\delta} = \left[(\mathbf{x}^b - \mathbf{x}^e)^{\mathrm{T}}, (\mathbf{y} - \mathbf{Hx}^e)^{\mathrm{T}}\right]^{\mathrm{T}}$. Now, it can be seen from (9.13) that

$$\begin{aligned} \mathbf{x}^b - \mathbf{x}^e &= -\mathbf{Kd}, \\ \mathbf{y} - \mathbf{Hx}^e &= (\mathbf{I}_p - \mathbf{HK})\mathbf{d}. \end{aligned}$$

Given $\boldsymbol{\delta}$, these equations are invertible for $\mathbf{d}$, and show that, for any analysis scheme (9.11), the innovation and DmA vectors $\mathbf{d}$ and $\boldsymbol{\delta}$ are in one-to-one correspondence. As far as accumulating statistics is concerned, that can be done in two equivalent ways: either a priori on the innovation vector, or (after the analysis has been performed) on the DmA difference. Now, the DmA difference $\mathbf{z} - \mathbf{\Gamma x}^e$ is the component of the data vector that has been seen to be 'rejected' in the analysis, and to have no impact on the analysis or on the estimated analysis error. The conclusion is that no information on the data error that could be useful for the estimation process can be obtained by only statistical processing of the innovation. Prior information obtained from external knowledge of the process that produces the data and from experience, good judgement, or an educated guess will always be necessary.

Now, appropriate external information always exists to some extent. To take a simple example, let us assume that (as has actually happened) the innovation corresponding to one type of observation in a particular meteorological station shows a specific statistical feature (e.g. a systematic bias) that is not present in the innovations corresponding to similar observations performed with similar instruments from stations in the same region. It is obvious that the origin of the bias is to be looked for in the observation, and not in the numerical model that produces the innovation. But that conclusion, as obvious as it is, uses external knowledge relative to the observation and prediction system, and could not be obtained from only a blind statistical processing of the innovation.

We will discuss at some length in Section 9.7 the implications of the conclusion that has been obtained above. We only mention at this stage that the existence of a one-to-one correspondence between the innovation and the DmA difference could have been inferred without computation by simply noting that both those quantities are obtained by eliminating the unknown $\mathbf{x}$ from the data $\mathbf{z}$. The result of the elimination must be independent of how the elimination is performed.

9.5 Diagnostics of internal consistency

The question arises, in view of the conclusion of Section 9.4, of what, if anything, can be done in terms of objective evaluation of the statistics of the data error. It is clear that, if some parameters of those statistics are known, other parameters can be obtained from the accumulated statistics of the innovation. As an example, Daley (1993) considered the horizontal covariance function of the innovation for radiosonde geopotential observations, which he made homogeneous and isotropic through averaging over geographical location and direction. If the observational error is spatially uncorrelated, and uncorrelated with the background error, it will appear in the covariance function as an additional Dirac delta-function term, with correlation distance 0. Extrapolating the observed covariance to a zero distance, Daley thus obtained an estimation of the variance of the observational error. Many similar diagnostic studies can be, and have been, performed. They are necessarily based on a priori assumptions on the parameters of the probability distribution function of the data errors. They will normally lead to new estimates of other pararmeters. This estimation process can be iterated.

A systematic approach is as follows. Any assimilation system of the form (9.8) relies on a priori specification of the expectation $\boldsymbol{\mu}$ and the covariance matrix $\mathbf{S}$. These define in turn an expectation and a covariance matrix for the innovation $\mathbf{d}$. If $\boldsymbol{\mu}$ is, for instance, assumed to be zero, then necessarily $E(\mathbf{d}) = 0$. In addition, if, as is usually done, the background and observation errors are assumed to be uncorrelated, with respective covariance matrices

$$\mathbf{P}^b = E(\boldsymbol{\xi}^b \boldsymbol{\xi}^{b\mathrm{T}}), \quad \mathbf{R} = E(\boldsymbol{\varepsilon}\boldsymbol{\varepsilon}^{\mathrm{T}}), \tag{9.15}$$

then

$$E(\mathbf{d}\mathbf{d}^{\mathrm{T}}) = \mathbf{H}\mathbf{P}^b\mathbf{H}^{\mathrm{T}} + \mathbf{R}.$$

Comparison of the a posteriori observed statistics of the innovation with the a priori assumed statistics may reveal inconsistencies, which one may resolve by appropriately redefining the data error expectation and covariance matrix. In view of the one-to-one correspondence between the innovation and the DmA difference $\boldsymbol{\delta}$, the same diagnostics can be done, alternatively, on the latter. The information will be the same, and the choice is only a matter of convenience. But it must be stressed that, in view of the result proved in Section 9.4, consistency between the a priori assumed statistics and the a posteriori observed statistics is neither a necessary nor a sufficient condition for optimality of the assimilation process. It is not a sufficient condition, because the knowledge of the expectation and covariance of the innovation does not define the covariance matrices $E(\boldsymbol{\xi}^b\mathbf{d}^{\mathrm{T}})$ and $E(\boldsymbol{\xi}^b\boldsymbol{\xi}^{b\mathrm{T}})$ that are necessary for determining $\mathbf{x}^e$ and $\mathbf{P}^a$, (9.10). And it is not a necessary condition, because a possible inconsistency can always be 'explained out' by assuming that it entirely originates in the DmA difference, without modification of the orthogonal space $\boldsymbol{\Gamma}(\mathcal{S})^{\perp}$. As said in Section 9.4, that will modify neither the estimate $\mathbf{x}^e$ nor the associated estimated estimation error covariance matrix $\mathbf{P}^e$. For a fully explicit example, consider again the case of data of the form (9.12):

$$z_1 = x + \xi_1,$$
$$z_2 = x + \xi_2.$$

The estimation is performed under the hypothesis that the errors ξ_1 and ξ_2 are unbiased and mutually uncorrelated, and have the same variance s. The corresponding estimate is

$$x^e = \frac{z_1 + z_2}{2}, \tag{9.16a}$$

with quadratic error

$$p^e = s/2. \tag{9.16b}$$

The innovation, which is here the difference $d = z_1 - z_2$, is expected to have expectation 0 and variance $2s$. Assume the statistics performed on the data show the innovation to have the following expectation and variance:

$$E_e(d) = b, \tag{9.17a}$$
$$E_e(d^2) = b^2 + 2\sigma, \tag{9.17b}$$

where the subscript e means that E_e denotes a posteriori observed statistical means. Equations (9.17) are in contradiction with the hypotheses that have been made on ξ_1 and ξ_2 if $m \neq 0$ and/or $\sigma \neq s$. The image space $\mathbf{\Gamma}(\mathcal{S})$ is in the present case the direction $z_1 = z_2$, while the space that is **S**-Mahalanobis orthogonal to $\mathbf{\Gamma}(\mathcal{S})$ is the direction $z_1 + z_2 = 0$. Projecting the error vector $\boldsymbol{\xi}$ onto those two directions, and assigning (as mentioned above) the inconsistency on the latter, leads to the expectations

$$E(\xi_1) = -E(\xi_2) = -b/2, \tag{9.18a}$$

with the covariance matrix

$$\mathbf{S} = \frac{1}{2}\begin{pmatrix} s+\sigma & s-\sigma \\ s-\sigma & s+\sigma \end{pmatrix}. \tag{9.18b}$$

It is easily verified that these expressions, while being compatible with (9.17a, b), lead to the estimate (9.16a) and the corresponding estimation error (9.16b). This result is absolutely general, and thus it is always possible to specify the error expectation $\boldsymbol{\mu}$ and covariance matrix **S** so as to make them compatible with any expectation and covariance matrix for the innovation, as well as with any expressions for the BLUE $\mathbf{x}^e$ and associated estimation error covariance matrix $\mathbf{P}^e$. That may, on the other hand, require conditions that, in view on the available external knowledge on the data, may be very unlikely, if not impossible. In the above example, accommodation of a bias b requires the biases in ξ_1 and ξ_2 to be exactly the opposites of each other, (9.18a), and accommodation of an a posteriori observed variance σ that is different from the a priori assumed variance s requires correlation between ξ_1 and ξ_2, (9.18b). It may be known from other information sources that this is very implausible, or simply impossible.

The reader may wonder at this stage what would need to be done to change the analysis. To do this, one would need to modify the error covariance matrix **S** in such a way that the orthogonal space $\mathbf{\Gamma}(\mathcal{S})^{\perp}$ is changed. Keeping $\mathbf{\Gamma}(\mathcal{S})^{\perp}$ unchanged, but modifying the component $\mathbf{S}_1$ of $\mathbf{S}$ along $\boldsymbol{\Gamma}(\mathcal{S})$, would not modify the analysis, but would modify the estimation error covariance matrix.

Keeping in mind that reliable interpretation of possible inconsistencies can only come from external knowledge, we describe below a number of diagnostics that can be, and have been, implemented for identifying possible inconsistencies between a priori assumed and a posteriori observed probability distributions of the innovation. Some of these diagnostics are implemented on the innovation itself, others on the DmA difference, and still others on combinations of both.

A first obvious diagnostic is to test for the possible presence of a bias in either the innovation or the DmA difference. The presence of a statistically significant bias in either one of these two quantities is the signature of bias that has been improperly taken into account in either the background or the observations (or both). One can argue that a systematic check of a presence of a residual bias in either the innovation or the DmA difference is likely to be the first consistency diagnostic to be performed on an assimilation system. Identification and removal of biases in assimilation algorithms is a subject of active research (see e.g. Dee and Uppala, 2009). Any solution will always require external hypotheses.

A second simple diagnostic bears on the covariance of the DmA difference $\boldsymbol{\delta}$. It can be seen from (9.13) and (9.14) that $\boldsymbol{\delta}$ is equal to

$$\boldsymbol{\delta} = \left(\mathbf{S} - \boldsymbol{\Gamma}\mathbf{P}^a\boldsymbol{\Gamma}^{\mathrm{T}}\right)\mathbf{S}^{-1}\boldsymbol{\xi}$$

and has covariance matrix

$$E(\boldsymbol{\delta}\boldsymbol{\delta}^{\mathrm{T}}) = \mathbf{S} - \boldsymbol{\Gamma}\mathbf{P}^a\boldsymbol{\Gamma}^{\mathrm{T}}. \tag{9.19}$$

Noting that the second term on the right-hand side of (9.19) is the covariance matrix of the vector $\boldsymbol{\Gamma}(\mathbf{x}^a - \mathbf{x})$, this equation can be written as

$$E\left[(\mathbf{z} - \boldsymbol{\Gamma}\mathbf{x}^a)(\mathbf{z} - \boldsymbol{\Gamma}\mathbf{x}^a)^{\mathrm{T}}\right] = E\left[(\mathbf{z} - \boldsymbol{\Gamma}\mathbf{x})(\mathbf{z} - \boldsymbol{\Gamma}\mathbf{x})^{\mathrm{T}}\right] - E\left[(\boldsymbol{\Gamma}\mathbf{x}^a - \boldsymbol{\Gamma}\mathbf{x})(\boldsymbol{\Gamma}\mathbf{x}^a - \boldsymbol{\Gamma}\mathbf{x})^{\mathrm{T}}\right]. \tag{9.20}$$

The Pythagorean form of this expression shows that the triangle with vertices $\{\mathbf{z}, \boldsymbol{\Gamma}\mathbf{x}^a, \boldsymbol{\Gamma}\mathbf{x}\}$ has a right angle (in the sense of orthogonality defined by the statistical covariance) at point $\boldsymbol{\Gamma}\mathbf{x}^a$, or, equivalently, that the difference $\boldsymbol{\Gamma}(\mathbf{x}^a - \mathbf{x})$ is statistically uncorrelated with the DmA difference, $\mathbf{z} - \boldsymbol{\Gamma}\mathbf{x}^a$.

Equations (9.19) and (9.20) also show that the analysed fields must fit the data to within the accuracy assumed on the latter; Hollingsworth and Lönnberg (1989) have called *efficient* an assimilation system that possesses this particular property. This, with the check of unbiasedness of the innovation or DmA difference, is one basic consistency check that can be performed on an assimilation system. Experience shows that systems that have been used in operations for a long time, and have been progressively improved through, mostly, comparison with independent observations, are consistent concerning the particular diagnostic considered here. Newly developed systems, on the other hand, may be off by a factor as large as one order of magnitude. Such an inconsistency is the signature of a gross mis-specification in the error covariance matrix $\mathbf{S}$, although the check does not say where the mis-specification lies, nor even, in all mathematical rigour, that the system is not optimal because of the mis-specification.

Consider a submatrix of $\mathbf{S}$ that is diagonal, corresponding, for instance, to radiosonde observations that are at mutually sufficiently large distance for the corresponding representativeness errors to be uncorrelated. The analysis error will be spatially correlated, mainly because of the assumed correlation in the background error. The off-diagonal terms in $\mathbf{S}$ will be 0, and (9.19) then shows that the DmA difference will be negatively correlated at short distances. Hollingsworth and Lönnberg (1989) have described an example of a positive short-distance correlation of the DmA difference in the ECMWF (European Centre for Medium-range Weather Forecasts) assimilation system. That was the signature of a mis-specification somewhere in the matrix $\mathbf{S}$. Later checks showed a negative correlation in the DmA difference. To our knowledge, the sign of the DmA difference spatial correlation has not been checked recently in operational assimilation systems.

The check defined by (9.19) does not, of course, provide a measure of the quality of the assimilation system. On the contrary, assume that, as a result of, for instance, an increase in the number of observations, the accuracy of the analysis increases, while observation error variances remain constant. The term that is subtracted in the right-hand side of (9.19), which is a measure of the quality of the analysis, will decrease. As a consequence, the variance of the DmA difference will increase and will tend asymptotically, as it must obviously do in the limit of a perfectly accurate analysis, to the variance of the data error. This constitutes a definitive proof, if one is needed, that the fit of an analysis to the data used in the analysis cannot be a measure of the quality of the analysed fields.

The objective function (9.6) assumes at its minimum the value

$$J_{\min} = J(\mathbf{x}^a) = \frac{1}{2}\,(\mathbf{\Gamma x}^a - \mathbf{z})^{\mathrm{T}}\,\mathbf{S}^{-1}\,(\mathbf{\Gamma x}^a - \mathbf{z}). \tag{9.21}$$

It is (half) the squared $\mathbf{S}$-Mahalanobis norm of the DmA difference $\boldsymbol{\delta}$. In the $\{\mathbf{\Gamma}(\mathcal{S}) - \mathbf{\Gamma}(\mathcal{S})^{\perp}\}$ decomposition of the data space $\mathcal{D}$, (9.21) reads (see (9.7))

$$J_{\min} = \frac{1}{2}\,\boldsymbol{\xi}_2^{\mathrm{T}}\,\mathbf{S}_2^{-1}\,\boldsymbol{\xi}_2.$$

Since $\boldsymbol{\xi}_2$ is in one-to-one linear correspondence with the innovation $\mathbf{d}$, and $\mathbf{S}_2 = E(\boldsymbol{\xi}_2\boldsymbol{\xi}_2^{\mathrm{T}})$, invariance of the Mahalanobis scalar product in a linear transformation implies that

$$J_{\min} = \frac{1}{2}\,\mathbf{d}^{\mathrm{T}}\,\left[E(\mathbf{d}\mathbf{d}^{\mathrm{T}})\right]^{-1}\,\mathbf{d}. \tag{9.22}$$

The value of the objective function at its minimum is (half) the squared Mahalanobis norm of the innovation, with respect to its own covariance matrix. This is a deterministic result, valid for any realization of the minimization process. It is readily seen that $J_{\min}$ is also the value of the dual objective function (see (1.25) of Chapter 1 of this volume on four-dimensional variational assimilation) at its minimum.

Exercise Show by algebraic computation the equality of expressions (9.21) and (9.22).

Using the result that $\mathrm{Tr}(\mathbf{CD}) = \mathrm{Tr}(\mathbf{DC})$ for any two matrices $\mathbf{C}$ and $\mathbf{D}$ such that both products are defined, the expression (9.22) can be rewritten as

$$J_{\min} = \frac{1}{2}\,\mathrm{Tr}\left(\left[E(\mathbf{dd}^{\mathrm{T}})\right]^{-1}\,\mathbf{dd}^{\mathrm{T}}\right)$$

which, on taking an a posteriori observed expectation E_e, becomes

$$E_e(J_{\min}) = \frac{1}{2}\,\mathrm{Tr}\left(\left[E(\mathbf{dd}^{\mathrm{T}})\right]^{-1}\,E_e\left(\mathbf{dd}^{\mathrm{T}}\right)\right) \tag{9.23}$$

For a consistent system, $E_e(\mathbf{dd}^{\mathrm{T}}) = E(\mathbf{dd}^{\mathrm{T}})$, and $J_{\min}$ has expectation

$$E_e(J_{\min}) = \frac{p}{2}. \tag{9.24}$$

The expectation of the objective function at its minimum is half the number of observations. This provides a very simple overall check of consistency of an assimilation system. If the observed expectation of $J_{\min}$ is smaller (respectively larger) than $p/2$, this means that the assimilation system is inconsistent, and that the covariance matrix $E(\mathbf{dd}^{\mathrm{T}})$, as a priori specified, is too large (respectively too small). Note that the presence of a residual bias in the innovation (which can of course be directly checked) would lead to an increase of $J_{\min}$.

$J_{\min}$ is a direct output of variational algorithms, both in their primal and in their dual forms. It can also be computed, although at some numerical cost, in other assimilation algorithms, such as Kalman filtering (see e.g. Ménard and Chang, 2000). The criterion (9.24) seems to have been first described and used, in the context of oceanography and meteorology, by Bennett (1992). Since then, it has been used to test a fairly large number of assimilation systems. One can mention, among others, the works of Ménard and Chang (2000), Talagrand and Bouttier (2000), Cañizares et al. (2001), Muccino et al. (2004), Sadiki and Fischer (2005), Chapnik et al. (2006), and Elbern et al. (2007). A remark similar to that made about (9.19) can also be made here. Systems that have gone through extended operational validation and tuning, even if they have never been subject to the particular check by the criterion (9.24), usually show a value of $E(J_{\min})$ that differs from its theoretical value $p/2$ by a factor of, at most, a few units. Systems that have not gone through extended validation may show a much larger discrepancy in (9.24).

The test associated with the criterion (9.24) is often called the χ^2 test. The χ^2 probability distribution of order p is the distribution of the sum of the squares of p independent Gaussian variables, each with expectation 0 and variance 1. It has expectation p and variance $2p$. It can be seen (for instance by transforming (9.21) to the basis of the principal components of d, in which $E(\mathbf{dd}^{\mathrm{T}})$ is the identity matrix of order p) that, if the data error (and therefore the innovation) is Gaussian, then the quantity $2J_{\min}$ follows a χ^2 distribution of order p. Both the expectation and variance of $J_{\min}$ are then equal to $p/2$. But it is also seen from the above that the expectation of $2J_{\min}$ is equal to $p/2$ independently of whether the data error is Gaussian or not. However, for large p, and even if the innovation is not Gaussian, the central limit theorem (which states that the sum of a large number of independent random variables is Gaussian) ensures that $2J_{\min}$ must approximately follow a χ^2 distribution of order p, The distribution of $J_{\min}$, which has expectation $p/2$ and standard deviation $\sqrt{p/2}$, is then very strongly peaked. Experience shows that a few realizations of an assimilation system are sufficient for reliable estimation of $E(J_{\min})$.

The objective function (9.6) will most often be the sum of a number of independent terms, namely

$$J(\boldsymbol{\xi}) = \sum_{k=1}^{K} J_k(\boldsymbol{\xi}),$$

where

$$J_k(\boldsymbol{\xi}) = \frac{1}{2}\left(\boldsymbol{\Gamma}_k\boldsymbol{\xi} - \mathbf{z}_k\right)^{\mathrm{T}} \mathbf{S}_k^{-1} \left(\boldsymbol{\Gamma}_k\boldsymbol{\xi} - \mathbf{z}_k\right). \tag{9.25}$$

In this equation, $\mathbf{z}_k$ is an m_k-dimensional component of the data vector $\mathbf{z}$ ($\sum_k m_k = m$), and the rest of the notation is obvious. The inverse estimation error covariance matrix is easily obtained from (9.5) as

$$[\mathbf{P}^a]^{-1} = \sum_{k=1}^{K} \boldsymbol{\Gamma}_k^{\mathrm{T}} \mathbf{S}_k^{-1} \boldsymbol{\Gamma}_k. \tag{9.26}$$

Left-multiplying by $\mathbf{P}^a$ and then taking the trace of the result yields

$$\begin{aligned} 1 &= \frac{1}{n}\sum_{k=1}^{K} \mathrm{Tr}\left(\mathbf{P}^a\,\boldsymbol{\Gamma}_k^{\mathrm{T}}\,\mathbf{S}_k^{-1}\,\boldsymbol{\Gamma}_k\right) \\ &= \frac{1}{n}\sum_{k=1}^{K} \mathrm{Tr}\left(\mathbf{S}_k^{-1/2}\boldsymbol{\Gamma}_k\mathbf{P}^a\,\boldsymbol{\Gamma}_k^{\mathrm{T}}\,\mathbf{S}_k^{-1/2}\right), \end{aligned}$$

where to obtain the last equality, use has been made again of the fact that the trace of the product of two matrices is not modified when the order of the factors is reversed. This expression shows that the quantity

$$I(\mathbf{z}_k) = \mathrm{Tr}\left(\mathbf{S}_k^{-1/2}\boldsymbol{\Gamma}_k\mathbf{P}^a\,\boldsymbol{\Gamma}_k^{\mathrm{T}}\,\mathbf{S}_k^{-1/2}\right) \tag{9.27}$$

(which, being the trace of a symmetric positive-definite matrix, is necessarily positive) is a measure of the relative contribution of the subset of data $\mathbf{z}_k$ to the overall accuracy of the analysis, or of the (relative) information content of subset $\mathbf{z}_k$. In particular, in the case of a background-observation decomposition of the form (9.10) (for the background, $\boldsymbol{\Gamma}_k = \mathbf{I}_n$ and $\mathbf{S}_k = \mathbf{P}^b$),

$$I(\mathbf{x}^b) = \frac{1}{n}\,\mathrm{Tr}\left(\mathbf{P}^a\,(\mathbf{P}^b)^{-1}\right) = 1 - \frac{1}{n}\,\mathrm{Tr}\,(\mathbf{K}\,\mathbf{H})$$

$$I(\mathbf{y}) = \frac{1}{n}\,\mathrm{Tr}\,(\mathbf{K}\,\mathbf{H})$$

Rodgers (2000) calls the quantity $I(\mathbf{z}_k)$ the *degrees of freedom for signal (DFS)* or *degrees of freedom for forecast noise (DFN)*, depending on whether the subset $\mathbf{z}_k$ belongs to observations or the background. Equation (9.27) is absolutely general, and is valid for any subset of data, including subsets that may consist of data coming from both the background and the observations. That is clearly seen from the fact that, given any subset $\mathbf{v}$ of the data, the data vector $\mathbf{z}$ can always be transformed, through linear and invertible operations, into $\mathbf{z} \to (\mathbf{v}^{\mathrm{T}}, \mathbf{w}^{\mathrm{T}})^{\mathrm{T}}$, where the errors affecting $\mathbf{w}$ are uncorrelated with the errors affecting $\mathbf{v}$. In that transformation, the objective function $J(\boldsymbol{\xi})$ becomes the sum of two terms corresponding to $\mathbf{v}$ and $\mathbf{w}$, respectively, from which the information contents $I(\mathbf{v})$ and $I(\mathbf{w})$, are clearly defined.

The relative information content $I(\mathbf{z}_k)$ is, essentially, the sum of the weights assigned in the assimilation to the components of $\mathbf{z}_k$, normalized in such a way as to allow consistent comparison between data that have been produced by different operators $\boldsymbol{\Gamma}_k$. Everything else being equal, $I(\mathbf{z}_k)$ increases with decreasing error $\mathbf{z}_k$. Figure 9.1, shows the information content of eight subsets of observations, as determined for the variational assimilation algorithm of the ARPEGE numerical weather prediction system of Météo-France. Each vertical bar is divided into three parts, corresponding respectively, from top to bottom, to observations performed northward of latitude 20°N, between latitudes 20°N and 20°S, and southward of latitude 20°S (for each type of observations, the two bars correspond, as will be explained below, to two numerical algorithms for the computation of the information content).

It is seen that the largest information contents correspond to observation subsets that contain the largest number of observations: radiosonde wind measurements TEMP U (which contain measurements of both horizontal components of the wind) and satellite observations (SATOB and AMSU). The impact of the geographical distribution of the observations is also clearly visible. The information content of the Northern Hemisphere dominates in the radiosonde (TEMP), pilot (PILOT) and aircraft (AIREP) observations, which are much more numerous in the Northern Hemisphere. For satellite observations, the impact of both hemispheres is the same, with larger relative impact of the tropical belt for SATOB (wind estimates for geostationary satellites) than for AMSU observations (infrared radiation measurements performed from satellites on polar orbits).

Figure 9.2 shows the same information contents as in Fig. 9.1, divided now by the number of individual observations in each of the eight subsets. It provides the intrinsic information content of individual observations, independent of the number of observations in a given subset. It can be seen that the radiosonde measurements of humidity have the highest individual information content. It can be noted at this point that recent diagnostics performed at ECMWF, presented in Chapters 5 and 6 of

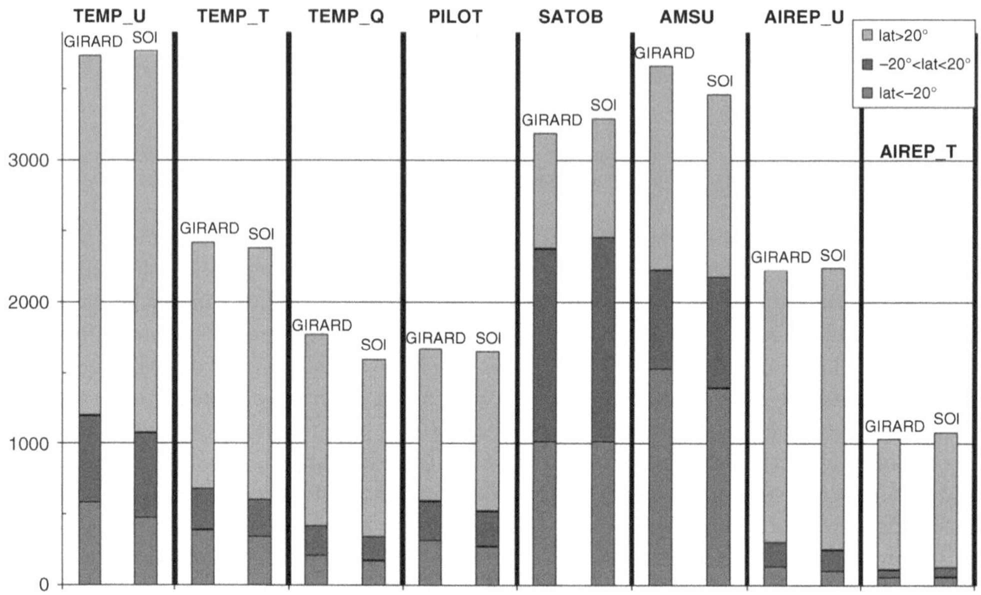

Fig. 9.1 Relative information content (expression (9.27) above) for eight different subsets of observations, as estimated for the variational assimilation system of Météo-France. For each type of observation, the two bars correspond to two different algorithms for computing the relative information content (see text). Each bar is divided into three parts, corresponding respectively, from top to bottom, to the Northern Hemisphere, the tropical belt and the Southern Hemisphere. For TEMP U and AIREP U (wind observations from radiosondes and aircraft, respectively), observations of both horizontal components of the wind vector are included. From Chapnik et al. (2006).

this volume, show drifting buoys (not considered in Fig. 9.1) as the most informative observations as the most informative individually.

Different, but fundamentally similar, diagnostics have been defined and studied by other authors (see e.g. Fisher 2003). These kinds of diagnostics are very useful. Not only do they provide instructive a posteriori information, but they can be used for a priori estimation of information content. They have been used, for instance, by Rabier et al. (2002) to select the most informative channels in satellite radiance measurements. More generally, they can be used as part of *observing system simulation experiments (OSSEs)* for a priori estimation of the gain that can be expected from new instruments.

On the other hand, these diagnostics are based on the assumption that the expectations and variances of data errors have been correctly specified. That is, of course, not necessarily the case, and these diagnostics may, in consequence, be misleading. In particular, they cannot be used in isolation for detecting a possible mis-specification of the required expectations and variances.

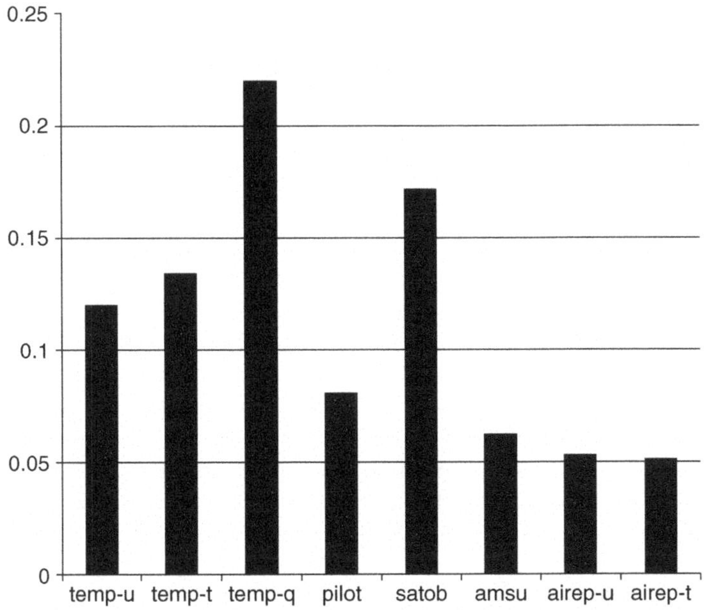

Fig. 9.2 As Fig. 9.1, but averaged for individual observations in each class.

It can be shown (Talagrand, 1999; Desroziers and Ivanov 2001) that the expectation of the term $J_k(\boldsymbol{\xi})$, (9.25), at the minimum of the objective function is equal to

$$E[J_k(\mathbf{x}^a)] = \frac{1}{2}\left[m_k - \mathrm{Tr}\left(\mathbf{S}_k^{-1/2}\boldsymbol{\Gamma}_k\mathbf{P}^a\,\boldsymbol{\Gamma}_k^{\mathrm{T}}\,\mathbf{S}_k^{-1/2}\right)\right], \tag{9.28}$$

where the same trace is present on the right-hand side as in (9.2). Equation (9.28) includes (9.24) as a particular case. It shows in particular that, everything else being equal, $E[J_k(\mathbf{x}^a)]$ will be smaller for more accurate data (smaller $\mathbf{S}_k$). It is obvious that the raw fit of the analysis must be closer with more accurate data. But (9.28) shows that this remains true even when the fit to the data is divided by the covariance matrix of the data error.

Equation (9.28) also provides the basis for further evaluation of the consistency of an assimilation scheme. It suffices to compare the trace of $\mathbf{S}_k^{-1/2}\boldsymbol{\Gamma}_k\mathbf{P}^a\,\boldsymbol{\Gamma}_k^{\mathrm{T}}\,\mathbf{S}_k^{-1/2}$, as computed directly and as determined statistically, (9.28), from results of assimilation experiments. Desroziers and Ivanov (2001) have shown that if the observation error is assumed to be uncorrelated in space and uncorrelated with the background error, then (9.28) can be used to estimate the observation and background error variances. This is, essentially, a systematic extension of the already mentioned work by Hollingsworth and Lönnberg (1986) and Daley (1993). Along the same lines, Chapnik et al. (2006) have used (9.28) to tune the variances of the observational errors in the various channels of the TOVS instrument, carried by satellites of the NOAA series. This has led to a significant change in the estimate of several of the variances (reduction by a factor of

9 in one case). It has also led to a modest, but distinct, improvement in the quality of the ensuing forecasts.

As a side remark, (9.28) also provides what is likely the simplest way of computing the trace $\mathrm{Tr}(\mathbf{S}_k^{-1/2}\mathbf{\Gamma}_k\mathbf{P}^a\mathbf{\Gamma}_k^{\mathrm{T}}\mathbf{S}_k^{-1/2})$. The matrix $\mathbf{S}_k^{-1/2}\mathbf{\Gamma}_k\mathbf{P}^a\mathbf{\Gamma}_k^{\mathrm{T}}\mathbf{S}_k^{-1/2}$ has dimension $m_k \times m_k$, where m_k can reach values of order 10^6. Computing the trace of large matrices that are not explicitly available (which is the case in the assimilation of meteorological or oceanographical observations) raises specific difficulties. A simple way to compute the trace $\mathrm{Tr}(\mathbf{S}_k^{-1/2}\mathbf{\Gamma}_k\mathbf{P}^a\mathbf{\Gamma}_k^{\mathrm{T}}\mathbf{S}_k^{-1/2})$ is to run the assimilation code from unbiased synthetic data affected with errors that have covariance matrix $\mathbf{S}_k$, and to determine the trace from the sample average of $J_k(\mathbf{x}^a)$. Experience shows that for large values of m_k (and similarly to what has been said above concerning the test from the criterion (9.24)), a sample of a few elements is sufficient for determination of $E[J_k(\mathbf{x}^a)]$. It is this particular method that has been used for determining the values marked *SOI* (for *simulated optimal innovation*) in Fig. 9.1 (the method identified as *Girard* is also a Monte Carlo type method, described in Girard, 1987).

Two particular diagnostics have been defined by Desroziers et al. (2005). It is easily deduced from (9.10a) that, for a consistent system,

$$E[(\mathbf{H}(\mathbf{x}^a - \mathbf{x}^b)\mathbf{d}^{\mathrm{T}}] = -\mathbf{H}E(\boldsymbol{\xi}^b\mathbf{d}^{\mathrm{T}}). \tag{9.29}$$

A consistency check can be performed on this equation. If, in addition, the $p \times p$ matrix $\mathbf{H}E(\boldsymbol{\xi}^b\mathbf{d}^{\mathrm{T}})$ is invertible, that check is equivalent to checking the general consistency condition $E(\mathbf{d}\mathbf{d}^{\mathrm{T}}) = E_e(\mathbf{d}\mathbf{d}^{\mathrm{T}})$, i.e. it achieves the most general objective check that can be performed concerning innovation covariances. The rank of the matrix $\mathbf{H}E(\boldsymbol{\xi}^b\mathbf{d}^{\mathrm{T}})$ is equal at most to $\min(p, n)$. In present meteorological applications, p is much smaller than n, and one can assume that $\mathbf{H}E(\boldsymbol{\xi}^b\mathbf{d}^{\mathrm{T}})$ will in general be invertible.

In the case when the background and observation errors $\boldsymbol{\xi}^b$ and $\boldsymbol{\varepsilon}$ are uncorrelated, and with the notation (9.15), the matrix $-\mathbf{H}E(\boldsymbol{\xi}^b\mathbf{d}^{\mathrm{T}})$ is equal to $\mathbf{H}\mathbf{P}^b\mathbf{H}^{\mathrm{T}}$. It is in that form that (9.29) was first given by Desroziers et al. (2005). As we have just seen, the condition for uncorrelation of $\boldsymbol{\xi}^b$ and $\boldsymbol{\varepsilon}$ is actually not necessary.

It can be shown similarly from (9.10a) that, for a consistent system,

$$E\left[(\mathbf{y} - \mathbf{H}\mathbf{x}^a)\,\mathbf{d}^{\mathrm{T}}\right] = E\left(\mathbf{d}\mathbf{d}^{\mathrm{T}}\right) + \mathbf{H}E\left(\boldsymbol{\xi}^b\mathbf{d}^{\mathrm{T}}\right). \tag{9.30}$$

Similarly to what has been said for (9.29), if the matrix $E(\mathbf{d}\mathbf{d}^{\mathrm{T}}) + \mathbf{H}E(\boldsymbol{\xi}^b\mathbf{d}^{\mathrm{T}})$ is invertible, consistency in (9.30) is equivalent to the general consistency condition $E(\mathbf{d}\mathbf{d}^{\mathrm{T}}) = E_e(\mathbf{d}\mathbf{d}^{\mathrm{T}})$.

In the case of uncorrelated background and observation errors, (9.30) reduces to

$$E[(\mathbf{y} - \mathbf{H}\mathbf{x}^a)\mathbf{d}^{\mathrm{T}}] = \mathbf{R}, \tag{9.31}$$

where $\mathbf{R}$ is the covariance matrix of the observation error, (9.15). The matrix $\mathbf{R}$ is usually (at most) block-diagonal with small blocks, and the check (9.31) is easy to implement. Of course, an inconsistency in (9.31) means that there is an inconsistency somewhere in the a priori specification of error statistics, but does not mean that the mis-specification lies only in $\mathbf{R}$. Actually, it does not even mean that $\mathbf{R}$ is mis-specified at all.

Now, the check (9.31) is valid only under the hypothesis that the background and observation errors are uncorrelated. That hypothesis is certainly disputable. If it is not verified, then (9.31) does not even define a consistency check, and is to be used, if at all, with care.

It is worth making a few additional remarks concerning the information content given by the expression (9.27). As a simple example, consider the case of a scalar x that evolves in time t according to the equation

$$x_{t+1} = \alpha\, x_t,$$

with $\alpha > 0$. Assume that two equally accurate observations of x have been performed at times t and $t+1$. The corresponding information contents are easily seen to be in the proportion $(1/\alpha, \alpha)$. For stable systems ($\alpha > 1$), the later observation is more informative, and the earlier observation is less informative. As the two quantities x_t and x_{t+1} are in one-to-one correspondence, this is true independently of the time at which x_t is estimated.

Given two data subsets $\mathbf{v}_1$ and $\mathbf{v}_2$, with respective information contents $I(\mathbf{v}_1)$ and $I(\mathbf{v}_2)$, the information content $I(\mathbf{v})$ of the union set $\mathbf{v} = (\mathbf{v}_1^{\mathrm{T}}, \mathbf{v}_2^{\mathrm{T}})^{\mathrm{T}}$ is equal to $I(\mathbf{v}_1) + I(\mathbf{v}_2)$ if the errors affecting $\mathbf{v}_1$ and $\mathbf{v}_2$ are uncorrelated. If that is not the case, then $\mathbf{v}_1$ and $\mathbf{v}_2$ can be said to be positively, or negatively correlated depending on whether $I(\mathbf{v}) < I(\mathbf{v}_1) + I(\mathbf{v}_2)$ or $I(\mathbf{v}) > I(\mathbf{v}_1) + I(\mathbf{v}_2)$. This defines a sign (and actually a magnitude) for the correlation between two subsets of data. As the information content is invariant under a linear transformation in data space (and, in particular, under a change of sign in any of the individual data), the correlation between the two subsets of data is not systematically related to the sign of the numerical correlations between the components of the errors affecting $\mathbf{v}_1$ and $\mathbf{v}_2$ (in general, these correlations make up a whole matrix, with no unambiguous sign).

The information content $I(\mathbf{z}_k)$ quantifies the relative contribution of subset $\mathbf{z}_k$ to the overall accuracy of the estimate of the state vector $\mathbf{x}$. That notion can be extended to the measure of the contribution of $\mathbf{z}_k$ to the accuracy of the estimate of any subset, say $\mathbf{u}_1$, of $\mathbf{x}$. To see this, denote n_1 the dimension of $\mathbf{u}_1$, and decompose the state vector $\mathbf{x}$ into $(\mathbf{u}_1^{\mathrm{T}}, \mathbf{u}_2^{\mathrm{T}})^{\mathrm{T}}$, where $\mathbf{u}_2$ is the projection of $\mathbf{x}$ onto the subspace that is $\mathbf{P}^a$-Mahalanobis orthogonal to $\mathbf{u}_1$. In this decomposition, the estimate of the error covariance matrix $\mathbf{P}^a$ reads

$$\mathbf{P}^a = \mathrm{diag}(\mathbf{P}_1^a, \mathbf{P}_2^a),$$

with

$$\mathbf{P}_1^a = E\left[(\mathbf{u}_1^a - \mathbf{u}_1)\,(\mathbf{u}_1^a - \mathbf{u}_1)^{\mathrm{T}}\right],$$

$$\mathbf{P}_2^a = E\left[(\mathbf{u}_2^a - \mathbf{u}_2)\,(\mathbf{u}_2^a - \mathbf{u}_2)^{\mathrm{T}}\right]$$

(where the superscript a denotes, as before, analysis). The data operator $\boldsymbol{\Gamma}_k$ decomposes into

$$\mathbf{\Gamma}_k = (\mathbf{\Gamma}_{k,1}, \mathbf{\Gamma}_{k,2}),$$

where $\mathbf{\Gamma}_{k,1}$ (respectively $\mathbf{\Gamma}_{k,2}$) defines the contribution of $\mathbf{u}_1$ (respectively $\mathbf{u}_2$) to the data subset $\mathbf{z}_k$. Equation (9.5) decomposes in turn into

$$[\mathbf{P}_1^a]^{-1} = \sum_k \mathbf{\Gamma}_{k,1}^{\mathrm{T}} \mathbf{S}_k^{-1} \mathbf{\Gamma}_{k,1}, \tag{9.32a}$$

$$[\mathbf{P}_2^a]^{-1} = \sum_k \mathbf{\Gamma}_{k,2}^{\mathrm{T}} \mathbf{S}_k^{-1} \mathbf{\Gamma}_{k,2}. \tag{9.32b}$$

The same derivation that has led from (9.26) to (9.27), but now starting from (9.32), leads to the definition of

$$I_1(\mathbf{z}_k) = \frac{1}{n_1} \mathrm{Tr}\left(\mathbf{S}_k^{-1/2} \mathbf{\Gamma}_{k,1} \mathbf{P}_1^a \mathbf{\Gamma}_{k,1}^{\mathrm{T}} \mathbf{S}_k^{-1/2}\right) \tag{9.33}$$

as the relative contribution of the data subset $\mathbf{z}_k$ to the accuracy of the estimation of $\mathbf{u}_1$. One can thus define the relative contribution of any subset of the data (for instance, the infrared radiances in a given channel over a given geographical area) to the accuracy of the estimate of any subset of the analysed fields (for instance, the estimate of humidity over that same area).

Numerical determination of $I_1(\mathbf{z}_k)$ seems, however, to raise serious problems, since it requires the identification, in one form or another, of the subspace in $\mathcal{S}$ that is $\mathbf{P}^a$-Mahalanobis orthogonal to $\mathbf{u}_1$. It is not clear how this could be achieved in practice in large dimension.

9.6 Diagnostics of optimality of assimilation algorithms

The various diagnostics presented in the preceding sections allow objective comparison of the quality of different assimilation schemes, or evaluation of the internal consistency of a given scheme. They say nothing as to the optimality, or otherwise, of a given scheme. The BLUE is defined for conditions of statistical unbiasedness and minimum estimation error variance. As a consequence, the estimation error $\mathbf{x}^a - \mathbf{x}$, in addition to being unbiased, must be statistically uncorrelated with the DmA difference or, equivalently, with the innovation vector. This is expressed by (9.10a), where the second term on the right-hand side is the orthogonal projection, in the sense of covariance, of (minus) the background error $\mathbf{x} - \mathbf{x}^b$ onto the space spanned by the innovation $\mathbf{y} - \mathbf{H}\mathbf{x}^b$ (see also (9.20)). The optimality condition is often expressed, in an exactly equivalent way, by saying that a sequential algorithm for assimilation is optimal if, and only if, the temporal sequence of innovation vectors is unbiased and uncorrelated (Kailath, 1968).

This optimality condition can be objectively checked against independent observations. Let us consider an observation of the form

$$\mathbf{q} = \mathbf{D}\mathbf{x} + \boldsymbol{\gamma},$$

where $\mathbf{D}$ is a known linear operator and the error γ is assumed to be unbiased and uncorrelated with the data error $\boldsymbol{\xi}$ and therefore with the innovation $\mathbf{d}$. Optimality of the estimate $\mathbf{q}^a = \mathbf{D}\mathbf{x}^a$ of $\mathbf{w}$ is equivalent to the conditions that it be statistically unbiased,

$$E(\mathbf{q} - \mathbf{D}\mathbf{x}^a) = 0, \tag{9.34}$$

and uncorrelated with the innovation,

$$E\left[(\mathbf{q} - \mathbf{D}\mathbf{x}^a)\,\mathbf{d}^{\mathrm{T}}\right] = 0. \tag{9.35}$$

If the unbiasedness condition (9.34) is usually checked in assimilation systems, the uncorrelatenedness condition (9.35), in spite of its simplicity, has so far been rarely used. One of the few examples is by Daley (1992), who computed the correlation of the innovation sequence for the sequential assimilation system that was then in use at the Canadian Meteorological Centre (that system is described by Mitchell et al., 1990). He found significantly non-zero correlations, reaching values of more than 0.4, for the 500 hPa geopotential innovation, at a time lag of 12 hours. Similar tests, performed more recently on a system for assimilation of oceanographical observations, give correlation values around 0.3 (Miller, personal communication).

The diagnostic (9.34), (9.35), if used alone, is actually a 'one-way' diagnostic. If the observed correlation is found to be significantly different from 0, as in the two examples above, then that is proof that the assimilation system is suboptimal, and can be improved. But if the correlation is found to be statistically undistinguishable from 0, then that does not mean that the system cannot be improved. To see this, consider a system that uses as background a short-range forecast produced by a high-quality numerical weather prediction mode. Further suppose that the system uses as background error covariance matrix $\mathbf{P}^b$ the matrix of climatological covariances. This is not erroneous, since the long-term statistical distribution of the background must be close to the climatological distribution. And, provided the covariance matrix of observation error is correctly specified, one can expect that the covariance (9.35) will be 0. However, in view of the quality of present short-range numerical weather forecasts, it is clear that such a system could be significantly improved. Actually, a system that is suboptimal by the criterion (9.35) can very well produce much more accurate estimates than an optimal 'climatological' system.

This first shows that the diagnostic (9.35) does not have much meaning if it is not associated with diagnostics of the magnitude of the difference $\mathbf{q} - \mathbf{D}\mathbf{x}^a$. This is not a problem inasmuch as such diagnostics are performed routinely. But this short discussion also shows that it is impossible to objectively determine, at least on the basis of diagnostics of the form (9.34), (9.35), whether an assimilation system makes the best possible use of the available data.

On the other hand, this certainly does not mean that diagnostics of the form (9.35) should not be used at all. As mentioned above, they have rarely been used so far, but they can objectively detect suboptimality, and would certainly be a useful complement to other commonly used diagnostics.

9.7 Conclusions

We have studied in some detail, in the context of the BLUE, the three questions stated in Section 9.1. The answer to the first question, concerning the possibility of objectively evaluating the quality of an assimilation algorithm, is fairly obvious. Such an evaluation can be made only against unbiased observations that not only have not been used in the assimilation, but also are affected by errors that are uncorrelated with the errors affecting the data that have been used in the assimilation (in the general case of a nonlinear estimation scheme, the condition would be that the errors affecting the verifying observations must be statistically independent of the errors affecting the data that have been used in the assimilation).

The second question concerned the possibility of objectively determining the probability distribution function of the errors affecting the data (the expectation $\boldsymbol{\mu}$ and the covariance matrix $\mathbf{S}$ in the case of the BLUE). It has led to the conclusion that (except for trial-and-error tuning, which cannot be exhaustive in meteorological or oceanographical applications) this will always require external hypotheses, i.e. hypotheses that cannot be objectively validated on the basis of the data only (incidentally, the author does not know if this result, which has been shown here on the basis of a fundamentally linear argument, extends to nonlinear estimation).

Appropriate external information is always available in meteorological and oceanographical applications, but is largely insufficient to entirely define the required quantities $\boldsymbol{\mu}$ and $\mathbf{S}$. Now, there is no other choice in practice than making hypotheses about the statistics of the errors affecting the data. It is important to distinguish as clearly as possible hypotheses that are very unlikely to be ever modified (for instance that errors in radiosonde observations performed a long distance apart are uncorrelated) from hypotheses that are reasonable but probably disputable (for instance that observation errors are statistically uncorrelated with background errors) and from hypotheses that are made for convenience, but are very likely erroneous (for instance that model errors are absent, or even only uncorrelated in time). Ideally, one might wish to define a minimum set of reliable hypotheses such that all remaining necessary error statistics can be objectively determined from statistics of the innovation. That goal seems, however, to be somewhat elusive in the present state of assimilation of meteorological and oceanographical observations. On the other hand, methods such as generalized cross validation (Wahba et al., 1995; Chapter 20 in this volume), which are ultimately trial-and-error experiments but are based on a solid methodological approach, have certainly not received enough attention in meteorological and oceanographical applications.

Systematic comparison between a priori assumed and a posteriori observed statistics of the innovation (or, equivalently, of the DmA difference) can reveal inconsistencies for which there cannot be unambiguous interpretation, but which can, if used with good judgment, help improve the a priori specification of $\boldsymbol{\mu}$ and $\mathbf{S}$.

Concerning objective estimation of the optimality of an assimilation algorithm (the third question), the decorrelation criterion (9.35) is valid only for least-squares estimation (but can be extended to nonlinear least-squares estimation). Although it can prove nothing as to the accuracy of the assimilation, it can nevertheless be useful, and has likely not received enough attention.

Acknowledgements

The author thanks numerous colleagues, in particular F. Bouttier, B. Chapnik, and G. Desroziers, for stimulating discussions. B. Chapnik provided Fig. 9.2.

References

Bennett, A. F. (1992). *Inverse Methods in Physical Oceanography*. Cambridge University Press, Cambridge.

Cañizares, R., Kaplan, A., Cane, M. A., Chen, D., and Zebiak, S. E. (2001). Use of data assimilation via linear low-order models for the initialization of El Niño–Southern Oscillation predictions. *J. Geophys. Res.*, **106**, 30947–30959.

Chapnik, B., Desroziers, G., Rabier, F., and Talagrand, O. (2006). Diagnosis and tuning of observational error statistics in a quasi-operational data assimilation setting. *Q. J. R. Meteorol. Soc.*, **132**, 543–565.

Daley, R. (1992). The lagged innovation covariance: a performance diagnostic for atmospheric data assimilation. *Mon. Weather Rev.*, **120**, 178–196.

Daley, R. (1993). Estimating observation error statistics for atmospheric data assimilation. *Ann. Geophys.*, **11**, 634–647.

Dee, D. and Uppala, S. (2009). Variational bias correction of satellite radiance data in the ERA-Interim reanalysis. *Q. J. R. Meteorol. Soc.*, **135**, 1830–1841.

Desroziers, G., Brousseau, P., and Chapnik, B. (2005). Use of randomization to diagnose the impact of observations on analyses and forecasts. *Q. J. R. Meteorol. Soc.*, **131**, 2821–2837.

Desroziers, G. and Ivanov, S. (2001). Diagnosis and adaptive tuning of observation-error parameters in a variational assimilation. *Q. J. R. Meteorol. Soc.*, **127**, 1433–1452.

Elbern, H., Strunk, A., Schmidt, H. and Talagrand, O. (2007). Emission rate and chemical state estimation by 4-dimensional variational inversion. *Atmos. Chem. Phys.*, **7**, 3749–3769.

Fisher, M. (2003). Estimation of entropy reduction and degrees of freedom for signal for large variational analysis systems. ECMWF Tech. Memo. 397.

Girard, D. (1987). A fast Monte-Carlo cross-validation procedure for large least square problems with noisy data. RR 687-M, IMAG, Université de Grenoble.

Hollingsworth, A. and Lönnberg, P. (1989). The verification of objective analyses: diagnostic of analysis system performance. *Meteorol. Atmos. Phys.*, **40**, 3–27.

Kailath, T. (1968). An innovations approach to least-squares estimation. Part I: Linear filtering in additive white noise. *IEEE Trans. Autom. Control.*, **AC-13**, 646–655.

Ménard, R. and Chang, L. P. (2000). Assimilation of stratospheric chemical tracer observations using a Kalman filter. Part II: χ^2-validated results and analysis of variance and correlation dynamics. *Mon. Weather Rev.*, **128**, 2672–2686.

Mitchell, H., Charette, C., Chouinard, C., and Brasnett, B. (1990). Revised interpolation statistics for the Canadian data assimilation procedure: their derivation and application. *Mon. Weather Rev.*, **118**, 1591–1614.

Muccino, J. C., Hubele, N. F., and Bennett, A. F. (2004). Significance testing for variational assimilation. *Q. J. R. Meteorol. Soc.*, **130**, 1815–1838.

Rabier, F., Fourrié, N., Chafaï D., and Prunet, P. (2002). Channel selection methods for Infrared Atmospheric Sounding Interferometer radiances. *Q. J. R. Meteorol. Soc.*, **128**, 1011–1027.

Rodgers, C. D. (2000). *Inverse Methods for Atmospheric Sounding: Theory and Practice.* World Scientific, London.

Sadiki, W. and Fischer, C. (2005). A posteriori validation applied to the 3D-VAR Arpège and Aladin data assimilation systems. *Tellus*, **57A**, 21–34.

Talagrand, O. (1999). A posteriori evaluation and verification of analysis and assimilation algorithms. In *Proceedings of Workshop on Diagnosis of Data Assimilation Systems*, November 1998, ECMWF, Reading pp. 17–28.

Talagrand, O. (2010). Evaluation of assimilation algorithms. In *Data Assimilation: Making Sense of Observations* (ed. W. A. Lahoz, B. Khattatov, and R. Ménard), pp. 217–240. Springer-Verlag, Berlin.

Talagrand, O. and Bouttier, F. (2000). Internal diagnostics of data assimilation systems. In *Proceedings of Seminar on Diagnosis of Models and Data Assimilation Systems*, September 1999, ECMWF, Reading, pp. 407–409.

Wahba, G., Johnson, D., Gao, F., and Gong, J. (1995). Adaptive tuning of numerical weather prediction models: randomized GCV in three and four dimensional data assimilation. *Mon. Weather Rev.*, **123**, 3358–3369.

10
Error dynamics in ensemble Kalman-filter systems: localization

P. Houtekamer

Meteorological Research Division
Environment Canada
Dorval, Québec, Canada

Advanced Data Assimilation for Geosciences. First Edition.
Edited by É. Blayo, M. Bocquet, E. Cosme, and L. F. Cugliandolo.

Chapter Contents

10.1 Motivation

Localization needs to be used—almost always—in an ensemble Kalman filter (EnKF) owing to restrictions on the size of the ensembles (Hamill et al., 2001). In fact, localization is the key technique that makes the ensemble approximation to the Kalman filter computationally feasible.

How localization is best applied depends on aspects of the model dynamics and the observational network. A reasonable choice often leads to a substantial improvement in performance. Fortunately, as shown in this chapter, the statistics from the ensemble itself can guide us in the selection of a reasonable localization method.

In this chapter, to demonstrate these points, an experimental environment built around the Lorenz III toy model (Lorenz, 2005) is used.

10.2 Estimation of scalars and matrices

In the EnKF, a basic operation is the estimation of the forecast error covariance matrix P^f from an ensemble of N_{ens} random vectors x:

$$P^f_{i,j} = \frac{1}{N_{\text{ens}} - 1} \sum_{k=1}^{N_{\text{ens}}} (x_{i,k} - \overline{x_i})(x_{j,k} - \overline{x_j})^{\text{T}}. \tag{10.1}$$

In atmospheric applications, the dimension of P^f is the dimension of the model state vector and can easily become $N_{\text{model}} = 10\,000\,000$ (number of longitudes $\times$ number of latitudes $\times$ number of vertical levels $\times$ number of variables per level). For practical reasons, however, it is hard to exceed an ensemble size of $N_{\text{ens}} = 100$. Because, with (10.1), we estimate the $0.5N_{\text{model}}(N_{\text{model}} + 1)$ different elements of P^f from only $N_{\text{ens}}N_{\text{model}}$ input numbers, we can a priori expect some estimation problems. In what follows, we will use P^f_e to denote an ensemble-based estimate (using (10.1)) of the true covariance matrix P^f.

For the estimation of scalar quantities such as $P^f_{i,j}$, ensemble sizes $O(100)$ are nevertheless found to be sufficient (Richardson, 2001). Most verification scores for ensemble prediction systems (EPSs), such as Brier scores, converge at an ensemble size of about 50 members. The ensemble of the European Centre for Medium-Range Weather Forecasts (ECMWF) has, for example, 52 members. The North American Ensemble Forecasts System (NAEFS) has 40 members. In a global EPS, once a sufficient number of members has been obtained, further improvement is sought by changing some other critical aspect (such as improving the horizontal resolution or using a more advanced radiation scheme).

In data assimilation, however, we need to estimate covariance structures because we need to spread out the observational information into space. In the Canadian global EnKF system, a substantial gain in quality was noted when the ensemble size was doubled from 96 to 192. The observed gain was bigger than that of either doubling the number of observations or doubling the number of grid points. Now having 192 members for the operational EnKF, we need to re-evaluate what aspect of the system to improve next.

We will first look at a simple example, not yet involving a forecast model, to better appreciate the need for large ensembles in an EnKF system.

10.3 Assimilation of one observation

In the case where one has the optimal Kalman gain matrix K, one can compute the analysis error covariances P^a from

$$P^a = (I - KH)P^f. \tag{10.2}$$

This equation cannot be used when K is suboptimal, as is the case when we use $K = K_e$ as estimated from an ensemble. In this case, we can use the Joseph form for the analysis error covariance:

$$K_e = P_e^f H^{\mathrm{T}} \left(HP_e^f H^{\mathrm{T}} + R\right)^{-1}, \tag{10.3}$$

$$P^a = (I - K_e H)\, P^f \,(I - K_e H)^{\mathrm{T}} + K_e R K_e^{\mathrm{T}}. \tag{10.4}$$

Note that the ensemble-based covariance matrix P_e^f is used in (10.3) to compute the suboptimal K_e whereas the true covariance matrix P^f is used in (10.4). We will use this equation to look at the impact of assimilating one observation in a hypothetical case where P^f is known exactly but a suboptimal ensemble-based K_e is nevertheless used for the data assimilation.

For a simple example, we will take a case with 40 model coordinates. The identity matrix I_{40} is used for the true covariances P^f of the background error. There is one observation, with standard normal distribution, at the first model coordinate.

We use a random number generator to obtain an ensemble of N_{ens} model state vectors $x_i, i = 1, \ldots, N_{\mathrm{ens}}$. Each of the $N_{\mathrm{model}} = 40$ elements of the vector x_i is obtained as an independent random number with standard normal distribution. For a given ensemble size N_{ens}, we can subsequently obtain an ensemble-based estimate P_e^f using (10.1). Next, we use (10.3) and (10.4) to compute the analysis error covariance matrix P^a from the Joseph equation. Since we use the information of an observation, it is legitimate to require that the analysis be more accurate than the trial field. A natural condition for success of the analysis procedure is therefore

$$\mathrm{Tr}(P^a) < \mathrm{Tr}(P^f). \tag{10.5}$$

In the given example with $P^f = I_{40}$, $\mathrm{Tr}(P^f) = 40$. The optimal analysis procedure would use the observation only to correct the first model coordinate, and for this coordinate the error variance would drop from 1 to 0.5. Consequently, we know that the optimal value for $\mathrm{Tr}(P^a)$ is 39.5. As we can see from Fig. 10.1, just to assimilate one observation, and improve with respect to the quality of the trial fields, we need to use at least 40 members. To converge in quality, we need about 10 times more members.

It is remarkable that many papers use an EnKF with ensemble size $N_{\mathrm{ens}} < 100$ and show results of a fair quality. This is essentially possible because of the use of

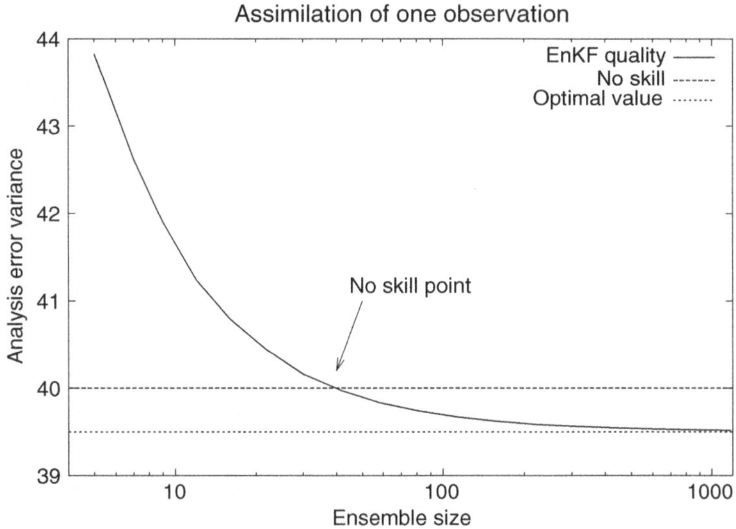

Fig. 10.1 The solid curve shows the trace of the analysis error covariances as a function of ensemble size. For reference, the optimal analytical result is shown short-dashed. The long-dashed curve shows the reference value for the background.

covariance localization. In the current case, with $P^f = I_{40}$, we know that there are no correlations between different variables, and we can thus use covariance localization to force the estimate of the off-diagonal values to zero:

$$P^f_{\mathrm{loc},e} = I \circ P^f_e . \tag{10.6}$$

Here, the Schur or Hadamard product $\circ$ of two matrices A and B is obtained by multiplying the corresponding matrix elements element by element (Gaspari and Cohn, 1999):

$$C_{ij} = A_{ij}B_{ij}, \qquad i = 1, \ldots, N_{\mathrm{model}}, \quad j = 1, \ldots, N_{\mathrm{model}}. \tag{10.7}$$

As shown in Fig. 10.2, this extreme form of localization causes a dramatic improvement in analysis quality. Even with only 2 members, the analysis is now more accurate than the trial field. With 50 members, we converged to a high analysis quality. Thus, since here only scalars need to be estimated in the EnKF, we can get away with using many times fewer ensemble members than previously for the matrix problem.

In practice, there may be substantial correlations between values at nearby grid points. These correlations change with the dynamics of the flow, and with the EnKF we aim at having corresponding flow-dependent analysis increments that are appropriate for the dynamics of the day. Suppressing all covariances in P^f_e would deprive us of the information we are after. Therefore, in practical applications, smooth functions are generally used for the covariance localization (Houtekamer and Mitchell, 2001).

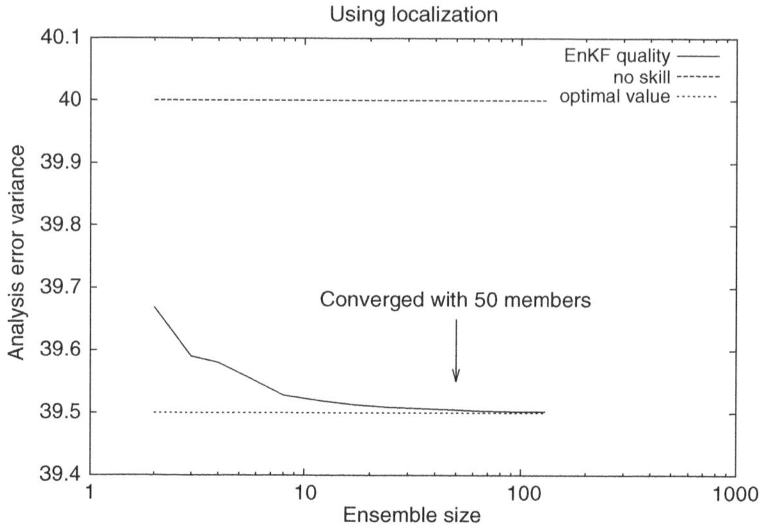

Fig. 10.2 As Fig. 10.1, but using extreme localization

Exercise Quantify the damage done by extreme localization in the case, as before, with a single observation at a point but with $P^f = 1$ (a matrix with all ones).

10.4 Experiments with the Lorenz III model

While, as we have seen, localization can be of critical importance for the quality of an EnKF system, how exactly it should be implemented is often the subject of a fair amount of experimentation. To illustrate some of the issues, in this section, we will perform data assimilation experiments with the Lorenz III model.

This model (Lorenz, 2005) is defined on one latitude band, where it is usually taken to have 960 grid points:

$$X_n = \sum_{i=-I}^{I}{}' (\alpha - \beta|i|) Z_{n+i}, \tag{10.8}$$

$$Y_n = Z_n - X_n \tag{10.9}$$

The large scales X are obtained by applying a smoothing function $(\alpha - \beta|i|)$ and special sum $\sum'$ to the total state Z. Subsequently, the small scales Y are obtained as the remainder.

The large scales X and the small scales Y evolve with a coupled system of equations:

$$\frac{\mathrm{d}X_n}{\mathrm{d}t} = [X, X]_{K,n} - X_n - cY_n + F, \tag{10.10}$$

$$\frac{dY_n}{dt} = b^2[Y, Y]_{1,n} - bY_n + cX_n \tag{10.11}$$

The Y variable evolves $b = 10$ times as fast as the large scale, with an amplitude that is b times smaller. The equations are coupled with c and forced by F. The operator $[\cdot, \cdot]_{\cdot,\cdot}$ is a somewhat complex summation (Lorenz, 2005).

In all experiments, we use an ensemble with 96 members configured in 4 subensembles of 24 members each (Mitchell and Houtekamer, 2009). Different ensemble members use different random perturbations to the observations. Covariance localization is with a finite-support fifth-order piecewise rational function (equation (4.10) in Gaspari and Cohn, 1999).

10.4.1 Sparse and randomly located inaccurate observations

The purpose here is to assimilate only the large scales. No attempt is made to obtain an accurate estimate of the small scales. Parameters of the problem are made to correspond with a large-scale problem:

- Thus, the observational network is sparse, with only 40 observations distributed randomly over the domain of 960 grid points.
- The standard deviation of the observational error is 1.0, which is large compared with the size of the small-scale activity, which has a magnitude of only about 0.3 (Yoon et al., 2010).
- The half-distance for localization is 160 grid points.
- Finally, the observations, and analyses, are every 50 model time steps (0.05 units of time). Comparing with the large-scale dynamics of the atmosphere, this can be thought to correspond to 6 hours.

With these parameters, the EnKF behaves well. For instance, running an assimilation cycle of 400 analysis steps, discarding results from the first 40 analyses, we obtain a value of 0.490 for the root-mean-square (rms) of the ensemble mean. The corresponding value of the mean spread is, with a value of 0.485, essentially no different. We note in passing that the error level remains above the typical amplitude for the small scales, which suggests we had no success at all in assimilating the small scales.

To determine the optimal half-distance for the localization, one can simply perform some experiments with different values for the parameter for the localization, and indeed we found an optimal value near 160. This, however, is not very satisfying. It would be preferable if the optimal localization length could be found from the ensemble itself (Anderson, 2007).

Here, to estimate the noise level in ensemble-based correlation estimates, we follow an earlier approach that assumes that the precision of correlation estimates decreases with distance (Fig. 7 of Houtekamer and Mitchell, 1998) and define the following:

- $\rho_{l,k}(x, y)$ = correlation between values at points x and y, based on ensemble members l to k;
- $\rho_{1,N_{\text{ens}}}(x, y)$ = isotropic correlation(r) = mean correlation with averaging over all points (x, y) separated by a distance r;

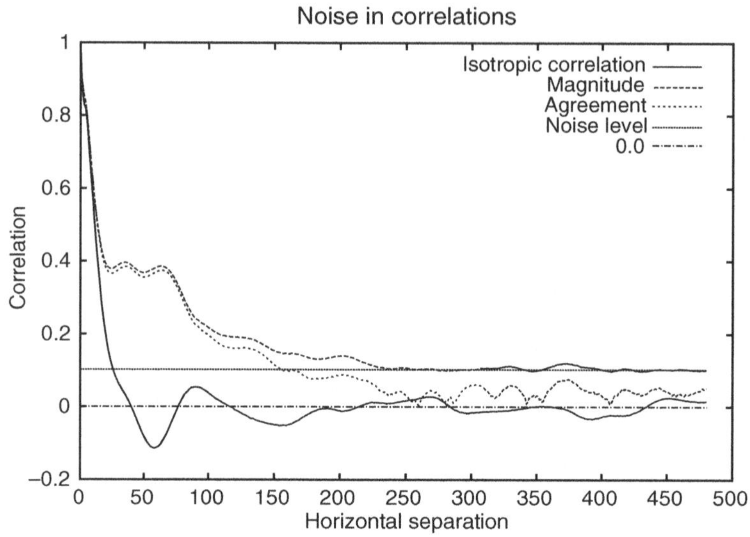

Fig. 10.3 Various measures, defined in the text, for the correlation as a function of horizontal distance as obtained with a sparse observational network.

- $\sqrt{\overline{\rho^2_{1,N_{\rm ens}}(x,y)}}$ = magnitude(r) = rms average correlation amplitude—for large r, the magnitude will converge to $1/\sqrt{N_{\rm ens}}$;
- $\sqrt{\overline{\mid \rho_{1,N_{\rm ens}/2}(x,y)\rho_{N_{\rm ens}/2+1,N_{\rm ens}}(x,y) \mid}}$ = agreement(r) = rms amount of correlation upon which two subensembles agree—for large r, the agreement will converge to zero.

These measures are shown in Fig. 10.3. We note that isotropic correlations drop rapidly and become negative, as expected from the wave-like solutions in the Lorenz III model. The ensemble estimates appear to remain significant until a distance of about 150, when the agreement starts to become smaller than the noise level. With more members in the ensemble, we would have a smaller noise level and we could relax the localization to benefit from small correlations estimated with confidence from the ensemble. Although the criterion to estimate the limit to useful ensemble information could be refined, we have good agreement with the result from the numerical optimization.

10.4.2 Dense and randomly located accurate observations

The same modelling environment can be used to look at the assimilation of small scales. Parameters are now changed to correspond with a small-scale problem:

- Thus, the observational network is dense with 400 observations distributed randomly over the domain with 960 grid points.

- The standard deviation of the observational error is 0.3, which is the typical size of the small scales.
- The half-distance for localization is 20 grid points.
- Finally, the observations, and analyses, are every 10 model time steps (0.01 units of time). Comparing with the large-scale dynamics of the atmosphere, this can be thought to correspond to 1.2 hours.

With these parameters, the EnKF still behaves quite well, but the match between rms and mean spread has degraded a little. For instance, running an assimilation cycle of 800 analysis steps, discarding results from the first 20 analyses, we obtain a value of 0.122 for the rms of the ensemble mean. The corresponding value of the mean spread is, with a value of 0.141, somewhat higher. We note that the error level is now below the typical amplitude for the small scales, which suggests that we had some success with the assimilation of information concerning the small scales.

From Fig. 10.4, we note that, thanks to the frequent, dense, and accurate observations, there is virtually no signal beyond a separation of 20 grid points. An independent optimization experiment did find the optimal value for the half-distance of the localization to be near 20. Interestingly, using a value of 90 or more did lead to divergence of the EnKF. With such values, the negative impact at a distance outweighs the positive impact of nearby observations. The result is an uncontrolled growth of the signal in the small scales. Thus, we note that the localization parameter of 160, which was near-optimal in the case of sparse infrequent and imprecise observations, leads to divergence when used with dense, accurate, and frequent observations.

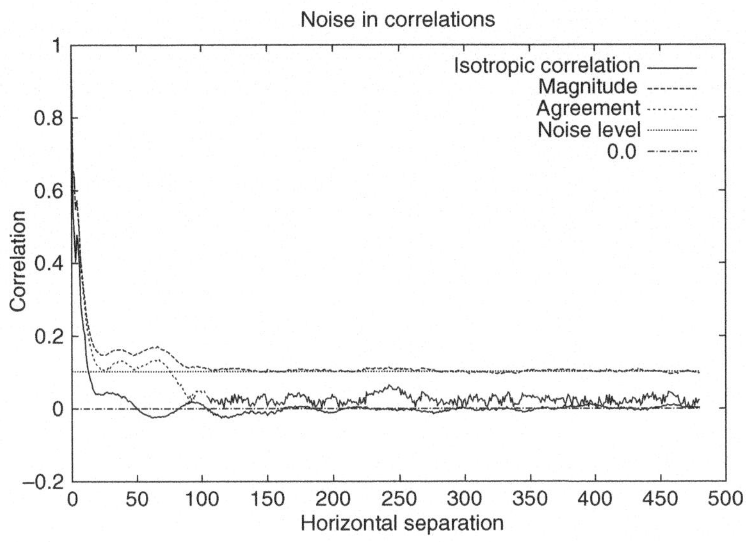

Fig. 10.4 As Fig. 10.3, but for a problem with a dense observational network

10.5 Discussion

As we have seen, for an EnKF to work well—that is, to decrease estimation error with respect to the accuracy of the background—it is necessary to either use an extremely large ensemble or implement covariance localization.

In atmospheric data assimilation, it is certainly possible, and even a standard practice, to tune certain parameters in an ad hoc manner by running a number of experiments. Here, the relevant parameter could be the lengthscale in the finite-support fifth-order piecewise-rational function (equation (4.10) of Gaspari and Cohn, 1999). When using a high model top, one may need to localize both in the horizontal and in the vertical (Houtekamer et al., 2005).

Unfortunately, in practical atmospheric data assimilation, the situation may be some complex hybrid between the sparse and the dense observational networks shown above. For instance, over continents, we often have accurate in situ observations, whereas over the oceans, we have dense but non-local radiance observations.

It is thus necessary to have tools to evaluate whether a proposed localization is actually appropriate for a given problem. One option, demonstrated here, is to split an ensemble into two subensembles to estimate the significance level of the ensemble correlations. It would, however, be desirable to modify the agreement measure that has been used here to make it coherent with the four-subensemble approach actually used in the EnKF (Mitchell and Houtekamer, 2009).

In the atmospheric application, one has to deal with additional problems, related to localization, which cannot be investigated using the Lorenz III model:

- In an atmospheric model, there is a certain balance between different variables. Severe localization will disrupt the balance and deform standard meteorological systems (Lorenc, 2003).
- High volumes of observations may push us towards ever more severe localization to permit extracting all the available information. There is, however, a limit to everything. The practical limit could, for instance, come from the error correlations of the observations.
- The atmosphere has different dynamical regimes. The weather is quite different between the polar and tropical areas or between the troposphere and the stratosphere. The optimal localization procedure could become complex as a consequence.

References

Anderson, J. L. (2007). Exploring the need for localization in ensemble data assimilation using a hierarchical ensemble filter. *Physica D*, **230**, 99–111.

Gaspari, G. and Cohn, S. E. (1999). Construction of correlation functions in two and three dimensions. *Q. J. R. Meteorol. Soc.*, **125**, 723–757.

Hamill, T. M., Whitaker, J. S., and Snyder, C. (2001). Distance-dependent filtering of background error covariance estimates in an ensemble Kalman filter. *Mon. Weather Rev.*, **129**, 2776–2790.

Houtekamer, P. L. and Mitchell, H. L. (1998). Data assimilation using an ensemble Kalman filter technique. *Mon. Weather Rev.*, **126**, 796–811.

Houtekamer, P. L. and Mitchell, H. L. (2001). A sequential ensemble Kalman filter for atmospheric data assimilation. *Mon. Weather Rev.*, **129**, 123–137.

Houtekamer, P. L., Mitchell, H. L., Pellerin, G., Buehner, M., Charron, M., Spacek, L., and Hansen, B. (2005). Atmospheric data assimilation with an ensemble Kalman filter: results with real observations. *Mon. Weather Rev.*, **133**, 604–620.

Lorenc, A. C., (2003). The potential of the ensemble Kalman filter for NWP—a comparison with 4D-Var. *Q. J. R. Meteorol. Soc.*, **129**, 3183–3203.

Lorenz. E. N. (2005). Designing chaotic models. *J. Atmos. Sci.*, **62**, 1574–1587.

Mitchell, H. L. and Houtekamer, P. L. (2009). Ensemble Kalman filter configurations and their performance with the logistic map. *Mon. Weather Rev.*, **137**, 4325–4343.

Richardson, D. S. (2001). Measures of skill and value of ensemble prediction systems, their interrelationship and the effect of ensemble size, *Q. J. R. Meteorol. Soc.*, **127A**, 2473–2489.

Yoon Y.-N., Ott, E., and Szunyogh, I. (2010). On the propagation of information and the use of localization in ensemble Kalman filtering. *J. Atmos. Sci.*, **67**, 3823–3834.

11
Short-range error statistics in an ensemble Kalman filter

P. HOUTEKAMER

Meteorological Research Division
Environment Canada
Dorval, Québec, Canada

Advanced Data Assimilation for Geosciences. First Edition.
Edited by É. Blayo, M. Bocquet, E. Cosme, and L. F. Cugliandolo.

Chapter Contents

11.1 Introduction

To arrive at an optimal data assimilation system, we need to have a good description of the uncertainty in the background field.

Historically, different approaches, with a solid comparison against either a ground truth or observations, have been used to obtain limited descriptions. In the first category, we find observation system simulation experiments (OSSEs) and in the second category we have notably methods based on the statistical analysis of innovation statistics.

The ensemble Kalman filter (EnKF) is a relatively new method that simulates the effect of known sources of error to arrive at a Monte Carlo estimate of the flow-dependent background error statistics. We need to describe and validate the ensemble statistics—in part by comparison with results from the established methods—in order to identify areas of improvement for the EnKF. In this chapter, we first summarize existing methods and subsequently we will study the statistical properties of a research version of the Canadian global EnKF.

11.1.1 Observation system simulation experiments

In an OSSE, one simulates the entire data assimilation process in a controlled manner. As explained by Daley and Mayer (1986), there are four steps in the experiment:

1. Create a 'truth' or nature run from a long integration of a sophisticated general circulation model.
2. Sample the nature run at observation locations that are identical to those in the actual observing system. Add random errors that are 'realistic'.
3. Inject these imperfect pseudo-observations into a state-of-the-art data assimilation scheme and create a set of objective analyses.
4. Subtract the nature run from the set of observed analyses to produce error fields.

Some results of the early OSSE by Daley and Mayer (1986) are as follows:

1. Time and zonally mean temperature errors were found to be relatively small except below 850 hPa and in the tropical stratosphere.
2. Time and zonally mean zonal wind errors were small, except south of 60°S.
3. Divergent wind could only be reliably analysed in waves 1–6.

Although data assimilation methods and forecast models, as well as the observational network, are gradually improving in quality, qualitatively similar results could probably be obtained today.

In view of the technical difficulty of obtaining both a high-quality nature run and realistic error statistics for the observations, a number of centers now collaborate in the joint OSSE project (see <http://gmao.gsfc.nasa.gov/projects/osse/>; (Errico et al., 2007). The nature run, for the joint OSSE project, is provided by the European Centre for Medium-range Weather Forecasts (ECMWF). The main focus is the evaluation of potential future observational networks. In this context, in view of the substantial cost or even the total impossibility of modifying characteristics of the observing system in an on-the-fly manner, it is worthwhile to perform a number of OSSEs to optimize the design before installation of the new observing system.

Oddly, to the knowledge of the author, OSSEs are not used to arrive at a statistical description of the background error in operational data assimilation systems.

11.1.2 Approaches based on radiosonde innovations

A correlation model can be fitted to innovation statistics (O-P) from radiosonde observations (Hollingsworth and Lönnberg, 1986; Lönnberg and Hollingsworth, 1986). The basic assumptions are that radiosonde error and forecast error are independent and that horizontal correlations are due to the forecast error. The above observational studies were at the origin of several correlation models. The method can be applied to other observation types to cover areas with few radiosondes. It is more difficult, however, to justify the basic assumptions in this case. Some results from the classical studies are as follows:

1. Horizontal lengthscales increase with height in the atmosphere.
2. The rotational wind errors are larger than the divergent wind errors (in absolute value).
3. For divergent wind, there are strong negative vertical correlations within the troposphere (perhaps associated with baroclinic processes).
4. For divergent wind in the stratosphere, vertical correlations are positive (perhaps pointing at a tidal origin).

These four properties are also observed in background ensembles used in the Canadian EnKF! In this chapter, only the results corresponding to the first two items will be shown.

11.2 Experimental environment

A global EnKF has been operational since 2005 at Environment Canada to provide initial conditions for a global medium-range ensemble prediction system (EPS) as well as for a regional EPS. Over time, many parameters have been tuned to give best performance.

In this section, we first give a brief description of the pre-operational configuration used to provide results. The results, from 192 members, will be global and for the single analysis time of 10 February 2011 at 12 UT. As we proceed, we compare properties of horizontal and vertical correlations with previous work.

We use the global environmental multiscale (GEM) atmospheric model of our centre (McTaggart-Cowan et al., 2011), with different configurations of the model physics, a 600×300 global horizontal grid, and a 20-minute time step. Model levels are numbered from 1 (for the model top near 2 hPa) to 74 (for wind) or 75 (for temperature and humidity) near the surface. The corresponding hybrid coordinate is 1 at the surface and about 0.002 at the model top.

The 192 ensemble members are grouped into 4 subensembles of 48 members each. After the analysis, isotropic random perturbations are added to the otherwise 'pure'

Table 11.1 Observation count for an experimental EnKF. Observations are for the period 10 February 2011, 9:00–15:00 UTC

LAYER (hPa)	2–14	14–100	100–400	400–1050	2–1050
Surface	0	0	0	16 297	16 297
Upper air	539	5 544	13 757	24 645	44 485
Scatterometer	0	0	0	19 832	19 832
Aircraft reports	0	0	34 670	27 697	62 367
Wind profiler	0	40	1 832	1 670	3 542
GPS-ro	5 736	6 108	4 077	1 103	17 024
tovs radiance	51 395	155 870	115 329	100 219	422 813
Satellite-derived wind	0	0	26 138	22 634	48 772
total	57 670	167 562	195 803	214 097	635 132

EnKF statistics. Note that the NMC method has been used to describe the covariances of the isotropic perturbations (Parrish and Derber, 1992; Pereira and Berre, 2006). The amplitude of these perturbations is, however, reduced before to the addition. A digital filter finalization is used to balance the model integrations (Fillion et al., 1995).

Observations, for the EnKF, have been preprocessed by our centre's deterministic high-resolution (4D-Var) data assimilation system. We benefit from the following:

1. data thinning of, in particular, satellite data to about 150 km;
2. bias correction of the observations;
3. quality control with respect to a high-resolution deterministic background field;
4. a further selection of observations to include only those to be assimilated by the EnKF.

The EnKF system does not use AIRS and IASI observations.

Table 11.1 shows observation counts for a typical 6 h assimilation window. In fact, this particular window is used for the results shown in this chapter. It may be noted that the diversity of the observational network decreases as we go to higher layers in the atmosphere. Above 10 hPa, we are essentially limited to GPS-ro (radio occultation) measurements and tovs radiance measurements. The GPS-ro measurements are processed to provide bias-free temperature observations at point locations. The tovs radiances correspond, via a response function, with some integral of temperature and humidity over a layer in the atmosphere.

11.3 Horizontal correlations

For the horizontal correlation structure, we first want to verify that lengthscales increase with the height in the atmosphere (Hollingsworth and Lönnberg, 1986).

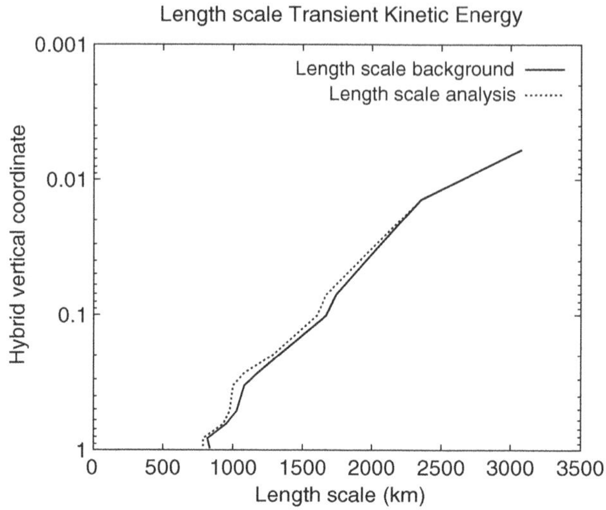

Fig. 11.1 Power spectra are computed for differences within (1) an ensemble of background fields and (2) an ensemble of analyses. The wavenumber below which we have half of the energy is converted into a lengthscale. Results for the trial fields and for the analyses are shown with respectively solid and short-dashed curves.

In Fig. 11.1, we observe, indeed, an increase in lengthscale by about a factor of 3 between the surface and the top of the model.

In the EnKF, both the horizontal and the vertical localization are implemented using a Schur product (Houtekamer et al., 2005):

$$K = \left[\rho_V \circ \rho_H(z) \circ (P^f H^T)\right] \left[\rho_V \circ \rho_H(z) \circ (HP^f H^{\mathrm{T}}) + R\right]^{-1}.$$

To reflect the increase of lengthscales with height, the horizontal localizing function $\rho_H(z)$ in the EnKF changes with height z. As shown in Table 11.2, the parameter for the horizontal localization changes in a stepwise manner. It is noted here that, in the Canadian EnKF, observations are grouped into batches, which are assimilated in a sequential manner (Houtekamer and Mitchell, 2001). In a given batch, all observations are from the same layer in the atmosphere. It is thus straightforward to use different localization parameters for the observations in different atmospheric layers. Comparison of Fig. 11.1 and Table 11.2 suggests that we might obtain improved results from using either a more severe horizontal localization near the surface or a less severe localization for the highest layer.

As has been noted before (Hollingsworth and Lönnberg, 1986), and is seen in Fig. 11.2, the rotational modes carry more error energy than the divergent modes. The current version of our EnKF uses the zonal and meridional wind components in the state vector for the analysis. With this choice of analysis variables, it would be difficult to make use of the predominance of the rotational modes in the algorithm itself (Kepert, 2009). We return to the choice of analysis variables in Section 11.4.

Table 11.2 Maximum distance at which an observation can have an impact as a function of the layer in the atmosphere

LOCALIZATION IN EnKF	
LAYER (hPa)	LENGTH (km)
2–14	3000
14–100	2800
100–400	2500
400–1050	2100

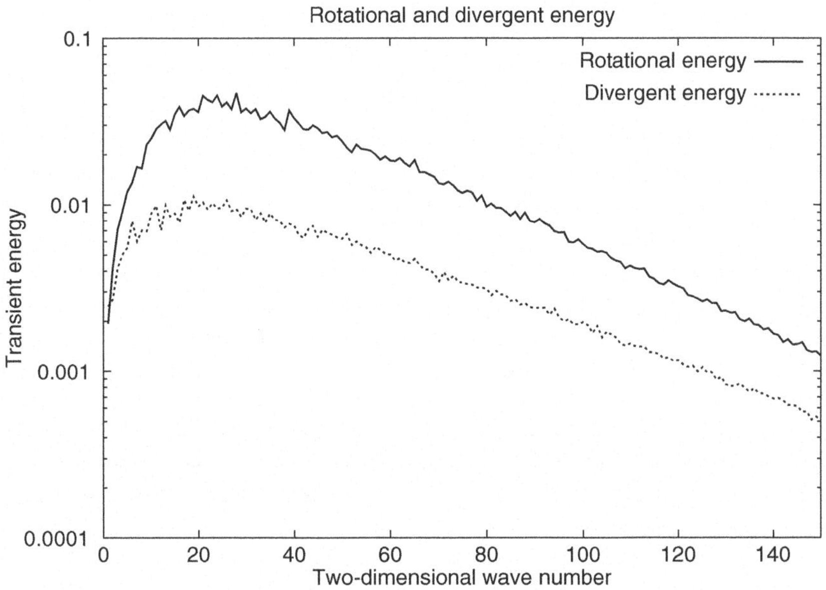

Fig. 11.2 For an ensemble of trial fields, the energy in rotational modes (solid curve) and divergent modes (short-dashed curve) is shown as a function of two-dimensional wavenumber.

11.4 Vertical correlations

In Fig. 11.3, we look at the global mean average for correlation estimates for temperature. A difference with a previous study (Fig. 2 of Houtekamer et al., 2005) is that the number of vertical levels has almost tripled. We also moved to a more recent version of our dynamical model, which now uses Charney–Phillips vertical staggering. In our current results, it would appear that vertical correlation structures are now smooth and relatively well resolved. We still have qualitatively similar structures with negative lobes above and below the central level, located near 420 hPa for the example

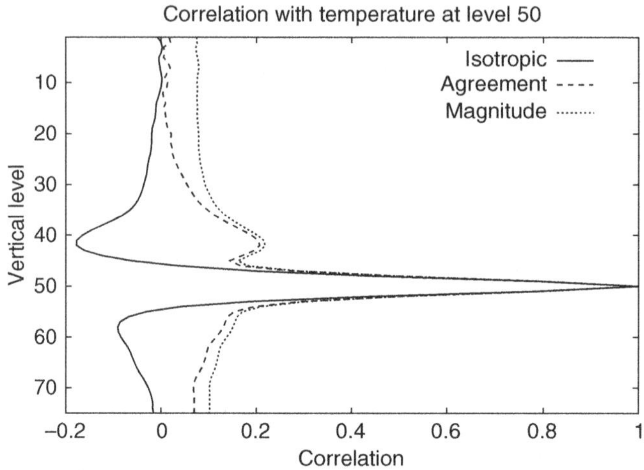

Fig. 11.3 Vertical correlations for temperature with the temperature at level 50 (about 420 hPa). The magnitude measures the root-mean-square (rms) amplitude of estimated correlations and the agreement measures the level of agreement between two halves of the ensemble.

in Fig. 11.3, with respect to which we computed correlations. Assuming, as in Chapter 10, that we can use correlation estimates as long as the level of agreement is higher than the noise level for the magnitude, we note that the ensemble is confident about the significance of the negative lobes.

The negative lobes for temperature are particularly prominent in the higher levels of the atmosphere and are probably related to the 'stratospheric ringing problem' (Healy, 2011). The argument is that radiance observations have a fairly broad response function and cannot resolve narrow vertical structures. Thus, after assimilating large volumes of radiance observations, what remains in an assimilation cycle is a superposition of error structures with a small vertical scale for temperature. Ultimately, this may lead to a large-amplitude vertical oscillation in the null-space of the radiances. To reduce the problem, it is recommended to use radiosonde and GPS-ro observations at high vertical resolution. At our centre, a corresponding change will be implemented in the near future.

In the EnKF, the vertical covariance localization imposes a zero correlation at a distance of two-units of log(pressure) (Houtekamer et al., 2005). This is broadly consistent with ensemble covariances for wind and temperature (see Fig. 11.4), generally becoming insignificant in less than 1.5 units of log(pressure).

We note, again, that the EnKF uses zonal and meridional wind as variables in the state vector. An alternative would be to use stream function and velocity potential (Kepert, 2009). We note from Fig. 11.4, however, that wind is a more local variable than stream function or velocity potential. In agreement with unpublished experiments at our centre, one may thus, when moving to stream function and velocity potential as analysis variables, expect to need more ensemble members in combination with less

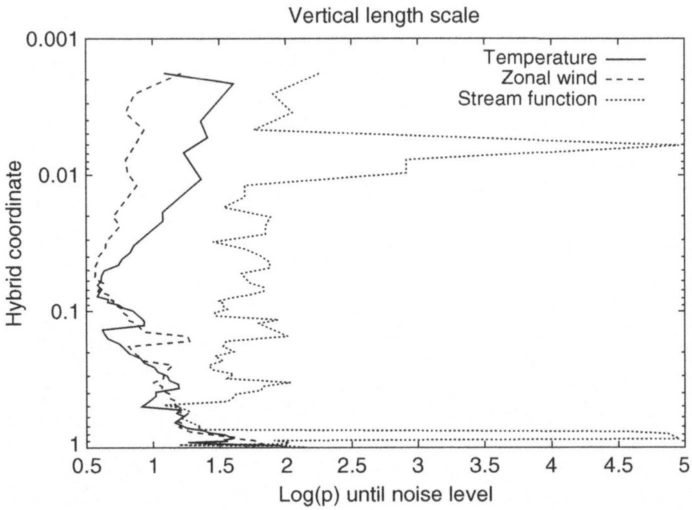

Fig. 11.4 The distance to the level where the agreement drops below the noise level is expressed in units of log(pressure). Results are shown for temperature (solid curve), zonal wind (long-dashed curve), and stream function (short-dashed curve).

covariance localization to obtain an analysis of the same quality. Note, however, that a change in analysis variable could lead to inherently more balanced analyses that no longer need to be followed by an explicit balancing procedure.

11.5 Temporal correlations

The EnKF assimilates data available over a 6 h window to arrive at analyses valid at the central time of the 6 h window. Thus the information can, at most, be advected for 3 h between the observation time and the analysis time. In the EnKF, localization is in space only. Temporal localization might be desirable when the error movement over the time window is comparable to the half-width of the localization function (Bishop and Hodyss, 2009). Temporal localization, however, corresponds to not using observations. If there is an issue, it would be better to reduce the length of the assimilation window or increase spatial localization lengths.

From Fig. 11.5 it appears that, for the highest wind speeds at the top few levels, the time window is perhaps a bit long. This suggests using a less severe horizontal localization near the model top. In general, however, it does not appear that temporal localization would be of critical importance in the global EnKF.

Note that different conclusions might be obtained, for instance, in an assimilation system dealing with deep convection while using severe vertical localization or in multiscale systems using severe horizontal localization for the smallest scales. In general, we can state that, moving to systems with ever increasing horizontal, vertical, and temporal resolution, a reduction of the length of the assimilation window is desirable.

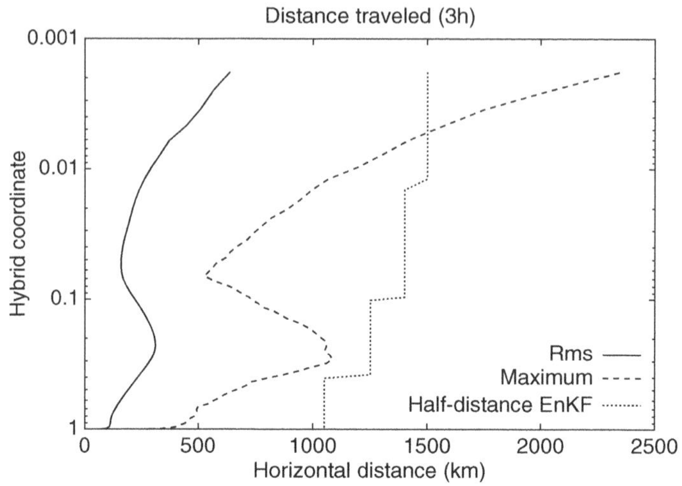

Fig. 11.5 Comparison of the rms advective displacement (solid curve) and the maximum advective displacement (long-dashed curve) with half of the distance for imposed zero correlations (short-dashed curve).

11.6 Stratospheric wind analysis

In the case of a suboptimal ensemble estimate K_e for the gain K, the true analysis error covariance P^a can be obtained using the Joseph form for the analysis error covariance as in (10.4) and

$$P^a = (I - K_e H)P^f(I - K_e H)^{\mathrm{T}} + K_e R K_e^{\mathrm{T}}. \tag{11.1}$$

Here we would need the true covariance matrix of the background error P^f. Since this matrix is generally not available, we can use (10.1) to obtain an ensemble based approximation P_e^f of P^f. Doing the required matrix algebra would give us the estimate $\tilde{P}^a$ of P^a:

$$\tilde{P}^a = (I - K_e H)P_e^f(I - K_e H)^{\mathrm{T}} + K_e R K_e^{\mathrm{T}}. \tag{11.2}$$

In the EnKF, to avoid matrix algebra and given sets of perturbed observations o_i, sets of analyses are computed as follows:

$$x_{a,i} = (I - K_e H)x_{f,i} + K_e o_i, \qquad i = 1, \ldots, N_{\mathrm{ens}} \tag{11.3}$$

The analysis error covariances can subsequently be estimated using the equivalent for P^a of (10.1). This, not writing out the cross terms, gives us

$$P_e^a = \tilde{\tilde{P}}^a = (I - K_e H)P_e^f(I - K_e H)^{\mathrm{T}} + K_e R_e K_e^{\mathrm{T}} + \text{cross-terms} \tag{11.4}$$

Comparing (11.1) and (11.4), we note that P^f and R have been replaced with ensemble-based approximations P_e^f and R_e and cross terms have been added. For

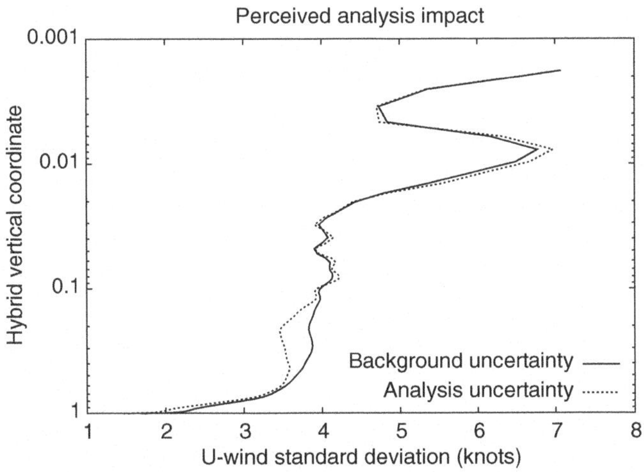

Fig. 11.6 Comparison of the ensemble-based estimates of uncertainty in the background (solid curve) and the analysis (dashed curve) for the zonal component of the wind.

large ensembles and, as assumed, independent forecast and observation errors, the cross terms will be small.

It is the experience of the author that the ensemble-based estimate $P_e^f - P_e^a$ is a useful indicator for the performance of the analysis. In particular for certain problems, such as having an ensemble that is too small for the assimilation problem at hand, one may obtain variances in P_e^a that exceed corresponding values in P_e^f. The comparison of P_e^a and P_e^f is considered an essential step in the development process for the Canadian global EnKF.

From Fig. 11.6, we can, for instance, see that we have a problem with the wind analysis in the stratosphere. Normally, the analysis should have been more accurate than the trial. To explain the lack of improvement, we may note, from Table 11.1, that we have very few direct observations of wind in the stratosphere. Improvements for the wind field must essentially come from GPS-ro and radiance observations via geostrophic coupling. In the tropics and in the stratosphere, however, the geostrophic coupling is not strong and the wind field is thus essentially unconstrained by observations.

11.7 Discussion

An ensemble Kalman filter with $O(100)$ members provides instantaneous informative statistics that can be used to analyse its behavior and optimize its performance. In part, statistical properties of the background ensemble are as in classical studies in the field (Hollingsworth and Lönnberg, 1986). Thus, we see, for instance, that the lengthscale for the errors increases with the height in the atmosphere.

The difference between ensemble spread in the trial fields and analyses can be used to determine if the EnKF responds properly to various types of observations. In the stratosphere, it would seem that the null-space of the observational network is significant. Here, the winds and humidity are largely unobserved. These fields thus evolve with the model dynamics in response to the observed temperature field.

To have a robust system, it is very important to reduce the null-space. Ideally, we would have more direct wind observations in the stratosphere. For temperature, since we largely depend on radiance observations with fairly broad response functions, we lack vertical resolution in the currently used observational network. To alleviate the situation, in the future, we will use GPS-ro and radiosonde observations at higher vertical resolution.

References

Bishop, C. H. and Hodyss, D. (2009). Ensemble covariances adaptively localized with ECO-RAP. Part 1: Tests on simple error models. *Tellus*, **61A**, 84–96.

Daley, R. and Mayer, T. (1986). Estimates of global analysis error from the global weather experiment observational network. *Mon. Weather Rev.*, **114**, 1642–1653.

Errico, R. M., Yang, R., Masutani, M., and Woollen, J. S. (2007). The estimation of analysis error characteristics using an observation system simulation experiment. *Meteorol. Z.*, **16**, 695–708.

Fillion, L., Mitchell, H. L., Ritchie, H., and Staniforth, A. (1995). The impact of a digital filter finalization technique in a global data assimilation system. *Tellus*, **47A**, 304–323.

Healy, S. (2011). Potential of the Radio Occultation Mission. Talk given at 3rd Post-EPS User Consultation Workship, 29–30 September 2011, Darmstadt, Germany.

Hollingsworth, A. and Lönnberg, P. (1986). The statistical structure of short-range forecast errors as determined from radiosonde data. Part I: The wind field. *Tellus*, **38A**, 111–136.

Houtekamer, P. L. and Mitchell, H. L. (2001). A sequential ensemble Kalman filter for atmospheric data assimilation. *Mon. Weather Rev.*, **129**, 123–137.

Houtekamer, P. L., Mitchell, H. L., Pellerin, G., Buehner, M., Charron, M., Spacek, L., and Hansen, B. (2005). Atmospheric data assimilation with an ensemble Kalman filter: results with real observations. *Mon. Weather Rev.*, **133**, 604–620.

Kepert, J. D. (2009). Covariance localisation and balance in an ensemble Kalman filter. *Q. J. R. Meteorol. Soc.*, **135**, 1157–1176.

Lönnberg, P. and Hollingsworth, A. (1986). The statistical structure of short-range forecast errors as determined from radiosonde data. Part II: The covariance of height and wind errors. *Tellus*, **38A**, 137–161.

McTaggart-Cowan, R., Girard, C., Plante, A., and Desgagné, M. (2011). The utility of upper-boundary nesting in NWP. *Mon. Weather Rev.*, **139**, 2117–2144.

Parrish, D. F. and Derber, J. C. (1992). The National Meteorological Center's spectral statistical interpolation analysis system. *Mon. Weather Rev.*, **120**, 1747–1763.

Pereira, M. B. and Berre, L. (2006). The use of an ensemble approach to study the background error covariances in a global NWP model. *Mon. Weather Rev.*, **134**, 2466–2489.

12
Error dynamics in ensemble Kalman filter systems: system error

P. Houtekamer

Meteorological Research Division
Environment Canada
Dorval, Québec, Canada

Advanced Data Assimilation for Geosciences. First Edition.
Edited by É. Blayo, M. Bocquet, E. Cosme, and L. F. Cugliandolo.

Chapter Contents

12.1 Introduction

In this chapter, we will compare results from the Canadian global ensemble Kalman filter (EnKF) with observations. This inevitably leads to discrepancies between the observed real atmosphere and its modelled equivalent. These discrepancies have their origin in *system error.*

In what can be called a system simulation experiment (Houtekamer et al., 1996), one attempts to obtain a coherent picture of the error evolution of a system. Errors can, for instance, be due to things as different as an inappropriate closure assumption in a forecast model and inaccurate observations of surface pressure. In this chapter, we will first describe how Monte Carlo methods can be used to arrive at a definition of 'system error'. Subsequently, we will try to proceed by elimination. First, we use medium-range ensemble forecasts to quantify the understanding of weaknesses of the forecast model. Subsequently, we move to the data assimilation context to see what additional error sources must be present in this context. We end with some speculation on the types of errors we should start to include in our system simulation experiments.

12.2 Monte Carlo methods

In, for instance, Wikipedia (as of May 2012), one can find a definition of what constitutes a Monte Carlo method:

Monte Carlo methods vary, but tend to follow a particular pattern:

1. *Define a domain of possible inputs.*
2. *Generate inputs randomly from a probability distribution over the domain.*
3. *Perform a deterministic computation on the inputs.*
4. *Aggregate the results.*

The most difficult step, in our application, is the first one, where one needs to describe the sources of error in the system. A standard hypothesis in data assimilation is that errors come from observations only. Thus, in step 1, we need to define the uncertainty in the observational network. In a subsequent standard hypothesis, it is assumed that the observations have independent errors with Gaussian distributions. In step 2, we obtain, for each member of the ensemble, a corresponding set of random perturbations. Step 3 is, in our application, a data assimilation cycle where the input is a long sequence of (perturbed) observation values. Note, however, that in some parts of this chapter, the deterministic computation includes performing a medium-range forecast. In the fourth step, we compute statistics over the ensemble of outputs.

It has long been known (Jastrow and Harlem, 1973), that the 'perfect model environment' described above does not lead to a sufficient amount of simulated error. Thus, we have added two items to the above list to detect underdispersion and identify corresponding error sources. In addition, we have modified the second item to include what we call 'system error'. The new list is as follows:

1. Use a domain of possible inputs.
2. • Generate inputs corresponding to unbiased independent Gaussian observations.
   • Obtain inputs corresponding to system error.

3. Perform a deterministic computation on the inputs.
4. Aggregate the results.
5. Compare simulated uncertainty with true error levels.
6. Modify/extend the domain of possible inputs.

As defined here as a remainder, system error can correspond to many types of error. In Kalman filter theory, as well as in the literature on atmospheric data assimilation, the term 'model error' is often used to describe the required additional error sources. As we often use the term 'model' to refer to the dynamical forecast model, this suggests that the model error corresponds to some imperfection of the forecast model. As I will argue in the subsequent sections of this chapter, this is not fair to the model. There are sources of error that are specific to data assimilation.

When one is interested in true uncertainties in the model, i.e. in the errors that can rightly be called 'model errors', it is perhaps most appropriate to validate the spread in medium-range forecasts. It should be noted, however, that in the medium range all errors, whatever their origin or initial structure, tend to project on the same unstable modes (Buizza et al., 2005). It is thus difficult to determine if enough ensemble spread was obtained due to the appropriate mechanisms.

In the short range, for instance when considering statistics in an EnKF, one has to deal with system error, which can be due to either problems with the forecast model or problems with the data assimilation methodology or assumptions. It may be difficult to separate the two aspects. An advantage of the EnKF environment is that, via the data assimilation procedure, the entire covariance description gets evaluated. Thus, the EnKF can potentially be used to validate specific hypotheses for the 'system error'.

12.3 Review of model error

In this section, we will consider errors due to imperfections of the forecast model. To evaluate the magnitude of such errors, we will use the Canadian medium-range global ensemble prediction system (EPS). In the EPS, we account for error in the forecast model using a variety of algorithms (Houtekamer et al., 2009; Charron et al., 2010), which, having being the subject of previous publications, will only briefly be described here:

1. We add isotropic perturbations to the initial conditions for the medium-range forecasts,
2. We use different physical parameterizations for different members of the ensemble,
3. we use stochastic kinetic energy backscatter (SKEB) (Shutts, 2005),
4. we apply ECMWF-like (Buizza et al., 1999) physical tendency perturbations (PTPs).

For more details on the above methods and recent developments, the reader is referred to the proceedings of the 2011 ECMWF workshop on 'Representing Model Uncertainty and Error in Numerical Weather and Climate Prediction Models'. For our current purposes, it is sufficient to know that we use a certain number of methods to account for imperfection of the forecast model. We do want to know, however, if these permit us to describe a substantial fraction of the weaknesses of the dynamical model.

To validate the amount of dispersion, we use the reduced centred random variable (RCRV) (Talagrand et al., 1999; Candille et al., 2007):

$$y = \frac{o - \overline{Hx_i}}{\sqrt{\text{var}(o) + \text{var}(Hx)}}. \tag{12.1}$$

For a perfectly representative ensemble, y will have a standard normal distribution. The mean and standard deviation of y are computed operationally at Environment Canada. The measured standard deviation of y is a measure of systematic over- or underdispersion of the ensemble. Since the ensemble spread, via the var(Hx) term, appears in the denominator, values bigger than one correspond to an underdispersive ensemble.

In Fig. 12.1, we show RCRV verifications for the Canadian operational global ensemble. We see that beyond day 3, for all variables, the ensemble is about 5–10% underdispersive. This fact that the underdispersion remains at a fixed percentage beyond day 3 suggests that the model error sources in the EPS are sufficiently large to account for the deficiencies of the forecast model. Thus, using the SKEB and PTP algorithms and using multiple parameterizations is sufficient to somehow account for all errors in the forecast model.

It is interesting to note that between day 1 and day 3, the ensemble evolves from being slightly overdispersive to slightly underdispersive. In the operational system, the

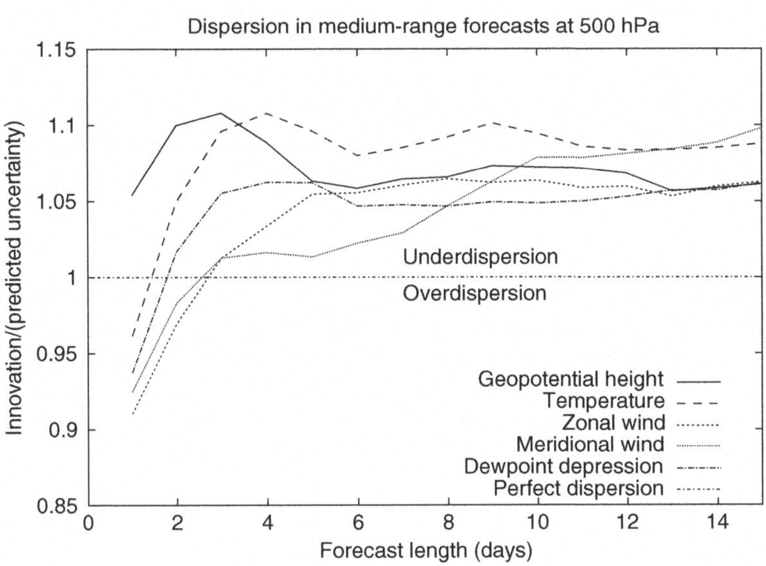

Fig. 12.1 The RCRV is shown for a validation of the ensemble spread at 500 hPa with respect to radiosonde observations. Data are from the operational medium-range ensemble forecast in March 2012.

isotropic perturbations that are added at the beginning of a medium-range forecast are bigger in amplitude than the equivalent perturbations used in the continuous EnKF data assimilation cycle. This odd difference is due to a lack of error growth of the perturbations during the first two forecast days. Since we desire to minimize the underdispersion in the medium range, we compensate for the lack of initial growth by having a larger initial amplitude.

The lack of initial error growth has been noted before (Fig. 9 of Houtekamer et al., 2005). We will speculate on its causes in the subsequent sections.

12.4 Review of data-assimilation error

The strategies used so successfully in the EPS are not efficient for providing spread in context of the EnKF (Fig. 10 of Houtekamer et al., 2009). A similar experience has been obtained at ECMWF (Fig. 9 of Bonavita, 2011). To better understand the dynamics of errors in data-assimilation, we start by considering some classical ideas.

The data assimilation cycle is a breeding ground for the most-unstable modes in the atmosphere (Toth and Kalnay, 1997). The idea here is that unstable modes of the atmosphere are present with some amplitude in the error of the background field for the analysis. The amplitude of these error modes will be reduced by a certain factor by the assimilation of observations. However, after the analysis, they will again grow rapidly and the bred modes will again be of sizable amplitude at the time of the next analysis. In longer-range forecasts, the bred perturbations will continue to grow more rapidly than other modes and they will dominate medium-range forecast errors.

Following a similar line of thought, Pires et al. (1996) show that unstable modes dominate at the end of the time window used for 4-dimensional data assimilation.

Experiments with an EnKF in a perfect-model context (Fig. 3 of Houtekamer et al., 2009) do not clearly support the above picture. While there is some growth of error between the analysis and the subsequent background, it is very modest which suggests that stable modes are just as present as unstable modes with almost no net growth as a consequence.

When we add system error, as in Fig. 12.2, the picture becomes even less attractive. In the troposphere, as discussed before, the analysis reduces the uncertainty for zonal wind (see Fig. 11.6 in Chapter 11 of this volume). The addition of isotropic perturbations raises the uncertainty in both the troposphere and the stratosphere. The subsequent integration with the model will reduce the estimated uncertainty. This must be because the analysis projects on stable modes, which are dissipated by both the dynamical model itself and by the digital filter finalization used in the model (Fillion et al., 1995).

Combining with the evidence from the medium-range verifications (e.g. Fig. 12.1) that we initially lack error growth in the ensemble, we are faced with a puzzle. On the one hand, we have evidence that true forecast errors grow substantially during the first few forecast days, and on the other hand, the EnKF error simulations show that rapidly growing modes do not dominate error growth in the short range. It would thus seem that there are sources of rapidly growing error that are not yet part of our error

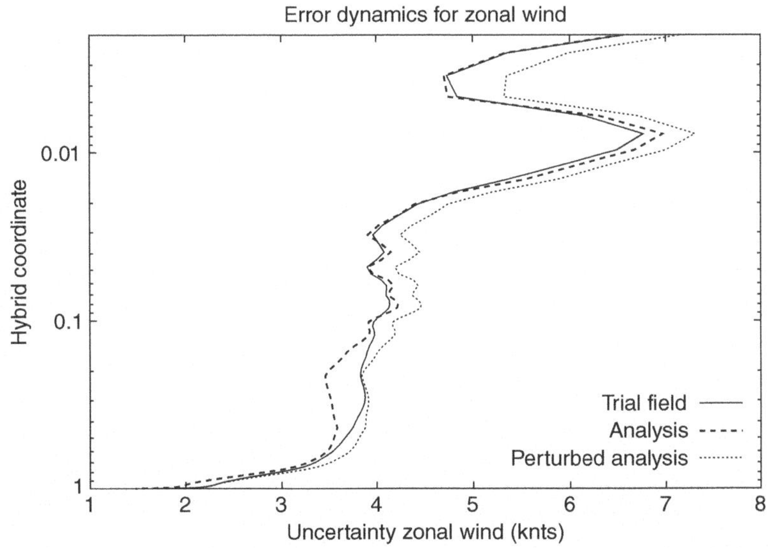

Fig. 12.2 As Fig. 11.6 in Chapter 11 of this volume, but adding a curve (short dashed) for the analysis with isotropic perturbations added. The long-dashed curve shows the uncertainty in the pure analysis. The solid curve is for the uncertainty in the background of the same analysis.

simulation. It is of obvious importance that we better understand the corresponding error dynamics and, in Sections 12.5 and 12.6, we will try to formulate a hypothesis.

12.5 Evidence of bias

Radiance observations, assimilated by the Canadian EnKF, have been previously bias-corrected by the 4D-Var system. In the EnKF, it is subsequently assumed that the bias-corrected observations and the forecast model have no bias. This assumption is idealistic and in reality the bias has not been completely removed.

While the coupling between the EnKF and 4D-Var is ad hoc, it will be used here to demonstrate that the bias-correction procedure does not perfectly remove bias from the observations. In the EnKF, version 4 of the GEM dynamical model is used (McTaggart-Cowan et al., 2011). Version 4 uses a staggered vertical coordinate and has less bias in the stratosphere than version 3. Until recently, the deterministic system (which includes the component to correct observation bias) used version 3 of the model.

The impact of migrating the deterministic system from GEM 3 to GEM 4 on EnKF results is shown in Fig. 12.3. We note an increase (the plotted value is for the observation minus the forecast) of the predicted temperature by about 0.1°C between 100 and 1000 hPa and by even more between 10 and 100 hPa. Note that the same model was used for the EnKF in both experiments. The change in average temperature is essentially due to a different bias correction for the AMSU-A input data. This strongly

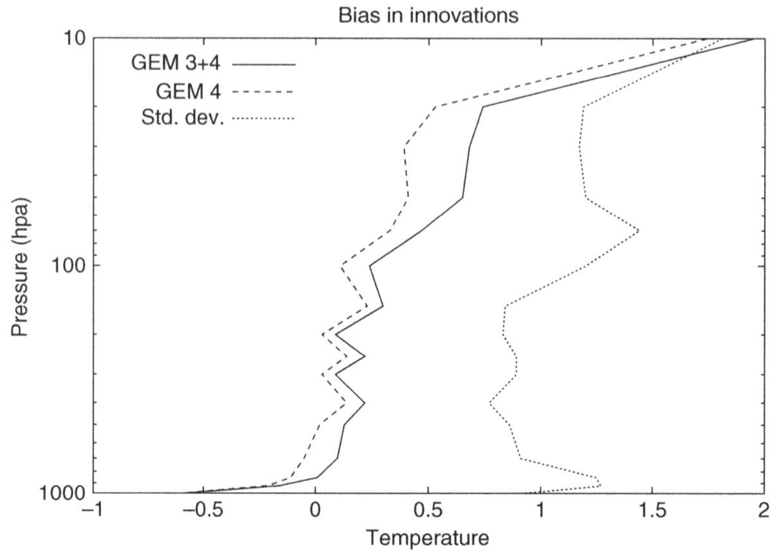

Fig. 12.3 The bias in the verification of the ensemble mean 6 h forecast for temperature against radiosonde observations south of 20°S. In 'GEM 3+4', a different model is used for the bias correction and for the EnKF. In 'GEM4', the same dynamical model is used for both applications. The standard deviation of experiment 'GEM4' is shown for reference.

suggest that, after the bias correction, the input observations have remaining bias with a large horizontal and vertical scale.

In the future, the bias-correction procedure will likely be modified (Mark Buehner, private communication 2012) to give more weight to unbiased observations (Dee and da Silva, 1998).

12.6 Evidence of horizontal error correlations

In most, if not all, operational data assimilation systems it is assumed that different observations have independent errors.

For conventional observing systems (such as the radiosonde network), this assumption is both natural and fairly accurate (with possible exceptions for humidity observations above 200 hPa and radiation effects on some types of sondes).

For space-based observing systems, the assumption is not natural, because the processing of observations from one particular sensor is likely to always suffer from the same problems. If spatial satellite error correlations do exist, their neglect can have a negative impact on analysis quality (Liu and Rabier, 2003). At operational centres, observation thinning and variance inflation procedures are applied to reduce the problem.

A recent study (Bormann and Bauer, 2010) arrives at positive conclusions regarding the error statistics of radiance observations:

1. AMSU-A shows little spatial or interchannel observation-error correlations for all channels used in the assimilation, except for surface-sensitive channels over land.
2. The findings that observation errors for AMSU-A are comparable to the instrument noise suggests that the radiative-transfer error is small after bias correction for these channels.
3. Statistics suggest that AMSU-A may be used more densely or with smaller observation errors.

An even more recent study (Gorin and Tsyrulnikov, 2011) arrives at different conclusions:

1. Satellite-versus-forecast cross correlations cannot be neglected.
2. There are significant temporal AMSU-A error correlations on the time scale of about 1 day.
3. These kinds of error correlations can be expected to exist for any voluminous observations with spatial/temporal error correlations: not only for various satellite observation types, but also for radar data.

Obviously, as can be seen from the above, there is no agreement on the error characteristics of radiance observations. From the experience at our centre (e.g. Fig. 12.3), however, it would appear that after our bias correction procedure, radiance observations have errors with very long temporal and spatial scales. What impact can the discrepancy between assumed and actual error statistics have on the simulation of errors?

Currently, since it is assumed that observation errors have no horizontal correlations, perturbing the observations independently at nearby locations can be expected to introduce simulated errors with short horizontal scales. Such structures likely also have short vertical scales. Structures with short horizontal and vertical scales are not like typical bred modes (Toth and Kalnay, 1997) or modes typically associated with baroclinic instability. It is thus possible that narrow vertical correlations, such as observed in Fig. 11.3, would disappear when taking account of horizontal and inter-channel error correlations. Possibly, the larger error structures would be better supported by the model dynamics.

12.7 Discussion

In medium-range forecasts, a sufficient amount of uncertainty can be simulated with methods based on physical arguments. Currently, in the Canadian medium-range EPS, we use the stochastic kinetic energy backscatter (SKEB) algorithm and we use physical tendency perturbations (PTP) as well as multiple physical parameterizations. Refinement of these methods is obviously possible, but, in the medium-range ensemble forecasts, the underdispersion after day 3 is at a modest 5–10% level, with no clear degradation in time. We note, however, that during the first few days, the perturbations do not grow very rapidly.

In data assimilation, the situation is different in that we have no credible simulations, using first principles only, that arrive at observed error levels. The most successful schemes, such as covariance inflation or the addition of isotropic errors, are ad hoc. It would appear that we need some source of rapidly growing error in the assimilation cycle.

Our current hypothesis is that correlated observation errors can be such a source of rapid error growth. It would appear entirely possible to account for such errors in the Kalman-filter framework (Gorin and Tsyrulnikov, 2011). Thus, the development of methods to sample observation uncertainty is currently the main focus of research for the Canadian global EnKF.

References

Bonavita, M. (2011). Impact and diagnosis of model error in the ECMWF ensemble of data assimilations. In *Proceedings of ECMWF Workshop on Representing Model Uncertainty and Error in Numerical Weather and Climate Prediction Models, Reading, United Kingdom*, pp. 303–318.

Bormann, N. and Bauer, P. (2010). Estimates of spatial and interchannel observation-error characteristics for current sounder radiances for numerical weather prediction. I: methods and application to ATOVS data. *Q. J. R. Meteorol. Soc.*, **136B**, 1036–1050.

Buizza, R., Miller, M., and Palmer, T. N. (1999). Stochastic representation of model uncertainties in the ECMWF ensemble prediction system. *Q. J. R. Meteorol. Soc.*, **125**, 2887–2908.

Buizza, R., Houtekamer, P. L., Toth, Z., Pellerin, G., Wei, M., and Zhu, Y. (2005). A comparison of the ECMWF, MSC, and NCEP global ensemble prediction systems. *Mon. Weather Rev.*, **133**, 1076–1097.

Candille, G., Côté, C., Houtekamer, P. L., and Pellerin, G. (2007). Verification of an ensemble prediction system against observations. *Mon. Weather Rev.*, **135**, 2688–2699.

Charron, M., Pellerin, G., Spacek, L., Houtekamer, P. L., Gagnon, N., Mitchell, H. L., and Michelin, L. (2010). Towards random sampling of model error in the Canadian ensemble prediction system. *Mon. Weather Rev.*, **138**, 1877–1901.

Dee, D. P. and da Silva, A. M. (1998). Data assimilation in the presence of forecast bias, *Q. J. R. Meteorol. Soc.*, **124**, 269–295.

Fillion, L., Mitchell, H. L., Ritchie, H., and Staniforth, A. (1995). The impact of a digital filter finalization technique in a global data assimilation system. *Tellus*, **47A**, 304–323.

Gorin, V. E. and Tsyrulnikov, M. D. (2011). Estimation of multivariate observation-error statistics for AMSU-A data. *Mon. Weather Rev.*, **139**, 3765–3780.

Houtekamer, P. L., Lefaivre, L., Derome, J., Ritchie, H., and Mitchell, H. L. (1996), A system simulation approach to ensemble prediction. *Mon. Weather Rev.*, **124**, 1225–1242.

Houtekamer, P. L., Mitchell, H. L., Pellerin, G., Buehner, M., Charron, M., Spacek, L., and Hansen, B. (2005). Atmospheric data assimilation with an ensemble Kalman filter: results with real observations. *Mon. Weather Rev.*, **133**, 604–620.

Houtekamer, P. L., Mitchell, H. L. and Deng, X. (2009). Model error representation in an operational ensemble Kalman filter. *Mon. Weather Rev.*, **137**, 2126–2143.

Jastrow, R. and Harlem, M. (1973). Simulation studies and the design of the first GARP global experiments. *Bull. Am. Meteorol. Soc.*, **54**, 13–21.

Liu, Z.-Q. and Rabier, F. (2003). The potential of high-density observations for numerical weather prediction: a study with simulated observations. *Q. J. R. Meteorol. Soc.*, **129**, 3013–3035.

McTaggart-Cowan, R., Girard, C., Plante, A., and Desgagné, M. (2011). The utility of upper-boundary nesting in NWP. *Mon. Weather Rev.*, **139**, 2117–2144.

Pires, C., Vautard, R., and Talagrand, O. (1996). On extending the limits of variational assimilation in nonlinear chaotic systems. *Tellus*, **48A**, 96–121.

Shutts, G. J. (2005). A kinetic energy backscatter algorithm for use in ensemble prediction systems. *Q. J. R. Meteorol. Soc.*, **131**, 3079–3102.

Talagrand, O., Vautard, R., and Strauss, B. (1999). Evaluation of probabilistic prediction systems. In *Proceedings of ECMWF Workshop on Predictability, Reading, United Kingdom*, pp. 1–25.

Toth, Z. and Kalnay, E. (1997). Ensemble forecasting at NCEP and the breeding method. *Mon. Weather Rev.*, **125**, 3297–3319.

13
Particle filters for the geosciences

P. J. van Leeuwen

Data Assimilation Research Centre
Department of Meteorology
University of Reading
Earley Gate
P.O. Box 243
Reading RG6 6BB
United Kingdom

Advanced Data Assimilation for Geosciences. First Edition.
Edited by É. Blayo, M. Bocquet, E. Cosme, and L. F. Cugliandolo.

Chapter Contents

13.1 Introduction

In this chapter, we will discuss particle filters and their use in the geosciences. A general review on the application and usefulness of particle filters in geosciences is given in van Leeuwen (2009), and a general overview of particle filtering is given in the excellent book by Doucet et al. (2001). There, it was shown that although interesting progress had been made until 2009, no solution for the degeneracy problem or the curse of dimensionality had been found. However, a lot of progress has been made during the last couple of years.

First, the basic idea behind particle filters is presented, followed by an explanation of why this basic formulation can never work for large-dimensional systems. We discuss resampling as a way to increase the efficiency of particle filters. Then we discuss proposal densities, which form the major part of this chapter. We show that they give us an enormous amount of freedom to build particle filters for very high-dimensional systems, and present an example of a successful approach that works in systems of any dimension by construction. This is the first example of what is undoubtedly an enormous growth in useful methods for extremely high-dimensional systems encountered in the geosciences. To keep the text fluent, I have kept the literature references to a minimum; a more comprehensive, but slightly outdated, literature list for this rapidly emerging field can be found in van Leeuwen (2009).

We discuss the basic particle filter as an importance sampler, and show why straightforward implementation will lead to so-called degeneracy, in which the effective ensemble size reduces to a very small number of particles and the method fails. At the same time, the strength of particle filters will be investigated, namely that particle filters are completely nonlinear they have no problems with model balances after updates (at least not the simple versions), and their performance is not reliant on a good representation of the error covariance of the model state. The latter has been overlooked in the past, but is actually a major selling point.

As real-world examples show us again and again, the data assimilation problem is typically a nonlinear one, especially in the geosciences. The models we use for simulation are almost never linear, and the observation operator that relates model states to observations is quite often nonlinear too. While linearizations have been shown to be very useful to tackle real-world problems, there are several problems that are so nonlinear that these linearizations are just not enough.

As has been discussed in Chapter 3 of this volume, Kalman filters either assume that the update is a linear combination between observations and prior estimate, the best linear unbiased estimate (BLUE), or assume that both the prior and the likelihood are Gaussian-distributed in model state. Of course, when the system is weakly nonlinear, the Kalman filter can be used quite efficiently, and even iterations of the Kalman-filter update can be performed. But when the system if highly nonlinear, these iterations are unlikely to converge, and, if they do, it is unclear to what. Also, the interpretation of the ensemble as a measure for the posterior covariance becomes questionable. It is important to realize that the (ensemble) Kalman filter is not variance-minimizing for a non-Gaussian posterior pdf!

Variational methods such as 4D-Var and the representer method look for the maximum of the posterior probability density function (pdf), or to the minimum of minus the logarithm of this pdf, which amounts to the same state. When the system is linear or Gaussian, it is easy to prove that there is indeed one maximum. Also, for a weakly nonlinear system, variational methods are very useful, and the variational problem can be solved by iteration, sometimes called 'incremental 4D-Var'. However, when the problem is highly nonlinear, it can be expected that the posterior pdf has several local maxima, and the variational methods will converge to one of them. This is not necessarily the global maximum. Another issue is of course the lack of covariance information. Even if the inverse of the Hessian, the local curvature of the pdf at the maximum, can be calculated, it does not represent the covariance of the full posterior pdf.

Nonlinear data assimilation is a whole new ball game, especially when the posterior pdf is multimodal. What does the 'best estimate' mean? Is it the mean of the posterior pdf? Well, definitely not when the posterior is bimodal and the two modes have equal probability mass and are of equal shape. In that case, the mean will fall between the two peaks. Is the global maximum the best estimate? If the posterior pdf has multiple maxima of equal size, the answer is no. Also, when the maximum is related to a relatively small probability mass, it is also not that useful. It becomes clear that the notion of 'best estimate' depends very strongly on the application, and it is perhaps not a very useful concept in nonlinear data assimilation.

The solution to the data assimilation problem is not a best estimate, but the posterior pdf itself. That is exactly what Bayes' theorem tells us—given the prior pdf and the likelihood, we can calculate the posterior pdf, and that is the answer. And the calculation is extremely simple, just a multiplication. So, this little excursion into nonlinear data assimilation learns us that data assimilation is *not an inverse problem*, but a multiplication problem. That is the starting point for this chapter on particle filters.

13.2 A simple particle filter based on importance sampling

The particle filters we will discuss here are based on importance sampling. The most straightforward implementation is what is called basic importance sampling here. (In the statistical literature, one usually finds importance sampling described with a proposal density different from the prior model pdf. However, for pedagogical reasons, we present importance sampling in the following way.) Basic importance sampling is straightforward implementation of Bayes' theorem, as we will show below.

13.2.1 Basic importance sampling

The idea is to represent the prior pdf by a set of particles x_i, which are delta functions centred around state vectors x_i, and from which all statistical measures of interest can be calculated, such as mean and covariance. If one represents the prior pdf by a number of particles, or ensemble members, as in the ensemble Kalman Filter, then

$$p(x) = \sum_{i=1}^{N} \frac{1}{N}\delta(x - x_i), \tag{13.1}$$

and we use this in Bayes' theorem:

$$p(x|y) = \frac{p(y|x)p(x)}{\int p(y|x)p(x)\,\mathrm{d}x}. \tag{13.2}$$

We find

$$p(x|y) = \sum_{i=1}^{N} w_i \delta(x - x_i), \tag{13.3}$$

in which the weights w_i are given by

$$w_i = \frac{p(y|x_i)}{\sum_{j=1}^{N} p(y|x_j)}. \tag{13.4}$$

The density $p(y|x_i)$ is the probability density of the observations given the model state x_i, which is often taken as a Gaussian:

$$p(y|x_i) = A \exp\left\{ -\frac{[y - H(x_i)]^2}{2\sigma^2} \right\}, \tag{13.5}$$

in which $H(x_i)$ is the measurement operator, which is the model equivalent of the observation y, and σ is the standard deviation of the observation error. When more measurements are available, which might have correlated errors, the above should be the joint pdf of all these measurements.

Weighting the particles just means that their relative importance in the probability density changes. For instance, if we want to know the mean of the function $f(x)$, we now have

$$\overline{f(x)} = \int f(x)p(x)\,\mathrm{d}x \approx \sum_{i=1}^{N} w_i f(x_i). \tag{13.6}$$

Common examples for $f(x)$ are x itself, giving the mean of the pdf, and the squared deviation from the mean, giving the covariance.

Up to now, we have not specified what x is. It can be a state vector x^n at a certain time n, or x can be a model trajectory over some time window $(0, n\Delta t)$, so $x = x^{0:n} = (x^0, x^1, \ldots, x^n)$ over n time steps. Here the superscript is the time index, and the subscript is the sample, or particle.

A practical way to implement the particle filter is to calculate the one time or the trajectory sequentially over time, which is where the name 'filter' comes from. The idea is to write the prior density as

$$p(x^{0:n}) = p(x^n|x^{0:n-1})p(x^{0:n-1}). \tag{13.7}$$

Using the fact that the state vector evolution is Markov—i.e. that to predict the future, we only need the present, not the past—we can write

$$p(x^{0:n}) = p(x^n|x^{n-1})p(x^{n-1}|p(x^{n-2})\ldots p(x^1|x^0)p(x^0). \tag{13.8}$$

Before we continue, it is good to realize what the so-called transition densities $p(x^n|x^{n-1})$ actually mean. Consider a model evolution equation given by

$$x^n = f(x^{n-1}) + \beta^n, \tag{13.9}$$

in which β^n is a random term or factor in the model equation that describes the error in the model equation. The idea is that the model is not perfect, i.e. any numerical model used in the geosciences that is used to simulate the real world has errors (and these tend to be significant!). These errors are unknown (otherwise we would include them as deterministic terms in the equations), but we assume we are able to say something about their statistics, (their mean, covariance, etc.). Typically, one assumes the errors in the model equations are Gaussian-distributed with zero mean and known covariance, but that is not always the case. To draw from such a transition density $p(x^n|x^{n-1})$ means to draw β^n from its density and evaluate the model equation given above. In fact, for normally, or Gaussian, distributed model errors β^n with mean zero and covariance Q, we can write

$$p(x^n|x^{n-1}) = N(f(x^{n-1}), Q). \tag{13.10}$$

Note that we assume that the model errors are additive in this chapter. Multiplicative model errors in which the size of the random forcing is dependent on the state x can be accounted for too, but we use additive model errors here for simplicity.

Let us now continue with importance sampling. If we also assume that the observations at different times, conditional on the states at those times, are independent, which is not necessary for the formulation of the theory, but keeps the notation so much simpler, we have for the likelihood

$$p(y^{1:n}|x^{0:n}) = p(y^n|x^n)\dots p(y^1|x^1), \tag{13.11}$$

where we have used that y^j is not dependent on x^k with $j \neq k$ when x^j is known. The posterior density can now be written as

$$\begin{aligned} p(x^{0:n}|y^{1:n}) &= \frac{p(y^{1:n}|x^{0:n})p(x^{0:n})}{p(y^{1:n})} \\ &= \frac{p(y^n|x^n)\dots p(y^1|x^1)p(x^n|x^{n-1})\dots p(x^1|x^0)p(x^0)}{p(y^n)\dots p(y^1)} \\ &= \frac{p(y^n|x^n)p(x^n|x^{n-1})}{p(y^n)}\dots\frac{p(y^1|x^1)p(x^1|x^0)p(x^0)}{p(y^1)}. \end{aligned} \tag{13.12}$$

Realizing that the last ratio in this equation is actually equal to $p(x^{0:1}|y^1)$, we find the following sequential relation:

$$p(x^{0:n}|y^{0:n}) = \frac{p(y^n|x^n)p(x^n|x^{n-1})}{p(y^n)}p(x^{0:n-1}|y^{1:n-1}). \tag{13.13}$$

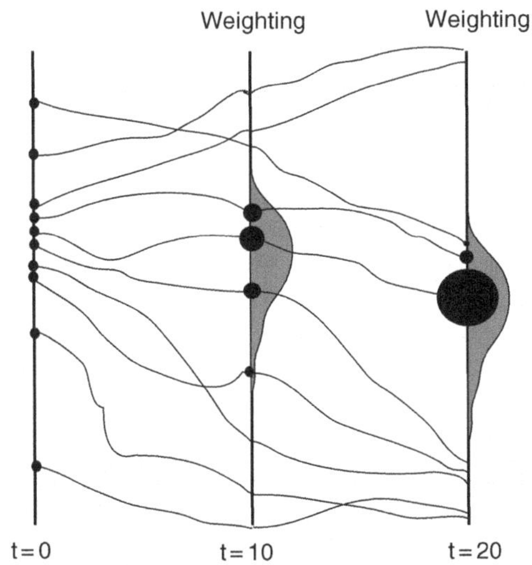

Fig. 13.1 The standard particle filter with importance sampling. The model variable runs along the vertical axis; the weight of each particle corresponds to the size of the bullets on this axis. The horizontal axis denotes time, with observations at a time interval of 10 time units. All particles have equal weight at time 0. At time 10, the likelihood is displayed together with the new weights of each particle. At time 20, only two members have weights different from zero: the filter has become degenerate. (From van Leeuwen (2010).)

This expression allows us to find the full posterior with the following sequential scheme (see Fig. 13.1):

1. Sample N particles x_i from the initial model probability density $p(x^0)$, in which the superscript 0 denotes the time index.
2. Integrate all particles forward in time up to the measurement time. In probabilistic language, we denote this as: sample from $p(x^n|x_i^{n-1})$ for each i, i.e. for each particle x_i run the model forward from time $n-1$ to time n using the nonlinear model equations. The stochastic nature of the forward evolution is implemented by sampling from the density that describes the random forcing of the model.
3. Calculate the weights according to (13.4), normalize them so their sum is equal to 1, and attach these weights to each corresponding particle. Note that the particles are not modified, only their relative weight is changed!
4. Increase n by one and repeat steps 2 and 3 until all observations have been processed.

13.2.2 Why particle filters are so attractive

Despite the problems just discussed, the advantages of particle fitter compared with traditional methods should not be underestimated. First of all, they do solve the complete nonlinear data assimilation problem (see the discussion at the beginning of this chapter).

Furthermore, the good thing about importance sampling is that the particles are not modified, so that dynamical balances are not destroyed by the analysis. The bad thing about importance sampling is that the particles are not modified, so that when all particles move away from the observations, they are not pulled back to the observations. Only their relative weights are changed.

And finally it should be stressed how simple this scheme is compared with traditional methods such as 3D- or 4D-Var and (ensemble) Kalman filters. The success of these scheme depends heavily on the accuracy of the error covariances of the model state vector. In 3D- and 4D-Var, this leads to complicated covariance structures to ensure balances etc. In ensemble Kalman filters, artificial tricks such as covariance inflation and localization are needed to get good results in high-dimensional systems. Particle filters do not have these difficulties.

However, there is (of course) a drawback. Even if the particles manage to follow the observations in time, the weights will differ more and more. Application to even very low-dimensional systems shows that, after a few analysis steps, one particle gets all the weight, while all other particles have very low weights (see Fig. 13.1 at $t = 20$). That means that the statistical information in the ensemble becomes too low to be meaningful. This is called *filter degeneracy.* It gave importance sampling a low profile until resampling was invented, see Section 13.3.

13.3 Reducing the variance in the weights

Several methods exist to reduce the variance in the weights, and we discuss sequential importance resampling here. See van Leeuwen (2009) for other methods. In resampling methods, the posterior ensemble is resampled so that the weights become more equal (Gordon et al., 1993). In Section 13.4, methods are discussed that do change the positions of the prior particles in state space to improve the likelihood of the particles, i.e. to bring them closer to the observations before the weighting with the likelihood is applied.

13.3.1 Resampling

The idea of resampling is simply that particles with very low weights are abandoned, while multiple copies of particles with high weights are kept for the posterior pdf in the sequential implementation. In order to restore the total number of particles N, identical copies of high-weight particles are formed. The higher the weight of a particle, the more copies are generated, such that the total number of particles, becomes N again. Sequential importance resampling does the above, and makes sure that the weights of all posterior particles are equal again, to $1/N$.

Sequential importance resampling is identical to basic importance sampling but for a resampling step after the calculation of the weights. The 'flow chart' reads as follows (see Fig. 13.2):

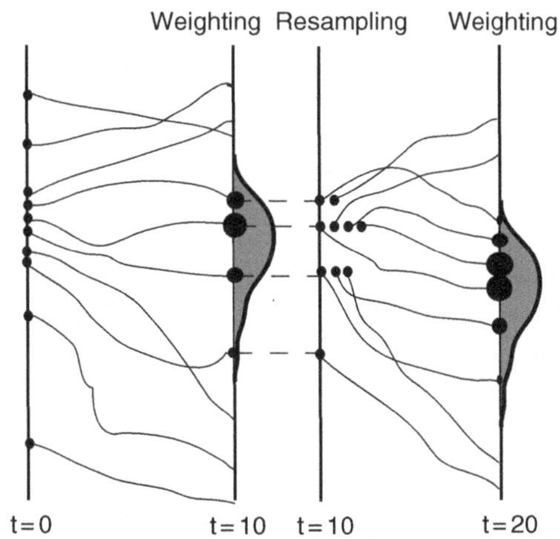

Fig. 13.2 The particle filter with resampling, also called sequential importance resampling. The model variable runs along the vertical axis; the weight of each particle corresponds to the size of the bullets on this axis. The horizontal axis denotes time, with observations at a time interval of 10 time units. All particles have equal weight at time zero. At time 10, the particles are weighted according to the likelihood, and resampled to obtain an equal-weight ensemble. (From van Leeuwen (2010).)

1. Sample N particles x_i from the initial model probability density $p(x^0)$.
2. Integrate all particles forward in time up to the measurement time (so, sample from $p(x^n|x_i^{n-1})$ for each i).
3. Calculate the weights according to (13.4) and attach these weights to each corresponding particle. Note that the particles are not modified—only their relative weight is changed!
4. Resample the particles such that the weights are equal to $1/N$.
5. Repeat steps 2, 3, and 4 sequentially until all observations have been processed.

It is good to realize that the resampling step destroys the smoother character of the method. All particles that are not chosen in the resampling scheme are lost and their evolution is broken. So the smoother estimate is build of fewer and fewer particles over time, until it consists of only one particle, loosing again all statistical meaning.

The resampling can be performed in many ways, and we discuss the most commonly used:

1. *Probabilistic resampling.* Most straightforward is to directly sample randomly from the density given by the weights. Since this density is discrete and one-dimensional, this is an easy task. However, because of the random character of the sampling, so-called sampling noise is introduced. Note that this method is actually generalized Bernoulli, for those versed in sampling techniques.
2. *Residual sampling.* To reduce the sampling noise, residual sampling can be applied. In this re-sampling method, all weights are multiplied with the ensemble

size N. Then n copies are taken of each particle i, in which n is the integer part of Nw_i. After obtaining these copies of all members with $Nw_i \geq 1$, the integer parts of Nw_i are subtracted from Nw_i. The rest of the particles needed to obtain ensemble size N are then drawn randomly from this resulting distribution.

3. *Stochastic universal sampling.* While residual sampling reduces the sampling noise, it can bee shown that stochastic universal sampling has lowest sampling noise. In this method, all weights are put after each other on the unit interval $[0, 1]$. Then a random number is drawn from a uniform density on $[0, 1/N]$, and N line pieces starting from the random number and with interval length $1/N$ are laid on the line $[0, 1]$. A particle is chosen when one of the endpoints of these line pieces falls in the weight bin of that particle. Clearly, particles with high weights span an interval larger than $1/N$ and will be chosen a number of times, while small-weight particles have a negligible change of being chosen.

13.3.2 Is resampling enough?

Snyder et al. (2008) prove that resampling will not be enough to avoid filter collapse. The problem is related to the large number of observations, which make the likelihood peak in only a very small portion of the observation space. The conclusion is that more than simple resampling is needed to solve the degeneracy problem.

13.4 The proposal density

In this section, we will concentrate on recent developments in using the so-called proposal transition density in solving the degeneracy problem. Related to decreasing the variance of the weights is to make sure that all model integrations end up close to the new observations, or, more precisely, ensuring that all posterior particles have similar weights.

First, we discuss what a proposal density is in particle filtering, and how it can be useful. This is then illustrated with using an ensemble Kalman filter as proposal density. This is followed by a discussion of more traditional methods such as the auxiliary particle filter, the backtracking particle filter, and guided sequential importance sampling.

Next, we discuss methods that change the model equations by bringing information on where the future observations are directly into the model equations. We start with the so-called optimal proposal density and show that that idea does not work in high-dimensional spaces with large numbers of independent observations. The optimal proposal density is a one-time-step scheme, assuming observations every time step. The so-called implicit particle filter extends this to multiple time steps between observations. It is shown that the implicit particle filter can be interpreted as a weak-constraint 4D-Var on each particle, with fixed initial condition. We will show that when the number of independent observations is large, this filter will also be problematic. This is followed by a method that will not be degenerate by construction—the equivalent-weights particle filter. Its working is illustrated on a 65 000-dimensional barotropic vorticity model of atmospheric or oceanic flow, hinting that particle filters are now mature enough to explore in, for example, operational numerical weather prediction settings.

We are now to discuss a very interesting property of particle filters that has received little attention in the geophysical community. We start from Bayes' theorem:

$$p(x^{0:n}|y^{0:n}) = \frac{p(y^n|x^n)p(x^n|x^{n-1})}{p(y^n)} p(x^{0:n-1}|y^{1:n-1}). \tag{13.14}$$

To simplify the analysis, and since we concentrate on a filter here, let us first integrate out the past, to get

$$p(x^n|y^{0:n}) = \frac{p(y^n|x^n)}{p(y^n)} \int p(x^n|x^{n-1})p(x^{n-1}|y^{1:n-1})\, \mathrm{d}x^{n-1}. \tag{13.15}$$

This expression does not change when we multiply and divide by a so-called proposal transition density $q(x^n|x^{n-1}, y^n)$, so

$$p(x^n|y^{0:n}) = \frac{p(y^n|x^n)}{p(y^n)} \int \frac{p(x^n|x^{n-1})}{q(x^n|x^{n-1}, y^n)} q(x^n|x^{n-1}, y^n)p(x^{n-1}|y^{1:n-1})\, \mathrm{d}x^{n-1}. \tag{13.16}$$

As long as the support of $q(x^n|x^{n-1}, y^n)$ is equal to or larger than that of $p(x^n|x^{n-1})$ we can always do this. This last condition makes sure we do not divide by zero. Let us now assume that we have an equal-weight ensemble of particles from the previous analysis at time $n-1$, so

$$p(x^{n-1}|y^{1:n-1}) = \sum_{i=1}^{N} \frac{1}{N} \delta(x^{n-1} - x_i^{n-1}). \tag{13.17}$$

Using this in (13.16) gives

$$p(x^n|y^{0:n}) = \sum_{i=1}^{N} \frac{1}{N} \frac{p(y^n|x^n)}{p(y^n)} \frac{p(x^n|x_i^{n-1})}{q(x^n|x_i^{n-1}, y^n)} q(x^n|x_i^{n-1}, y^n). \tag{13.18}$$

As a last step, we run the particles from time $n-1$ to n, i.e. we sample from the transition density. However, instead of drawing from $p(x^n|x_i^{n-1})$, so running the original model, we sample from $q(x^n|x_i^{n-1}, y^n)$, so from a modified model. Let us write this modified model as

$$x^n = g(x^{n-1}, y^n) + \hat{\beta}^n, \tag{13.19}$$

so that we can write for the transition density, assuming $\hat{\beta}^n$ is Gaussian-distributed with covariance $\hat{Q}$:

$$q(x^n|x^{n-1}, y^n) = N(g(x^{n-1}, y^n), \hat{Q}). \tag{13.20}$$

Drawing from this density leads to

$$p(x^n|y^{0:n}) = \sum_{i=1}^{N} \frac{1}{N} \frac{p(y^n|x_i^n)}{p(y^n)} \frac{p(x_i^n|x_i^{n-1})}{q(x_i^n|x_i^{n-1}, y^n)} \delta(x^n - x_i^n), \tag{13.21}$$

so the posterior pdf at time n can be written as

$$p(x^n|y^{1:n}) = \sum_{i=1}^{N} w_i \delta(x^n - x_i^n), \tag{13.22}$$

with weights w_i given by

$$w_i = \frac{1}{N} \frac{p(y^n|x_i^n)}{p(y^n)} \frac{p(x_i^n|x_i^{n-1})}{q(x_i^n|x_i^{n-1}, y^n)}. \tag{13.23}$$

We recognize the first factor in this expression as the likelihood, and the second as a factor related to using the proposal transition density instead of the original transition density to propagate from time $n-1$ to n, so it is related to the use of the proposed model instead of the original model. Note that because the factor $1/N$ and $p(y^n)$ are the same for each particle and we are only interested in relative weights, we will drop them from now on, so

$$w_i = p(y^n|x_i^n) \frac{p(x_i^n|x_i^{n-1})}{q(x_i^n|x_i^{n-1}, y^n)}. \tag{13.24}$$

Finally, let us formulate an expression for the weights when multiple model time steps are present between observation times. Assume the model needs m time steps between observations. This means that the ensemble at time $n-m$ is an equal-weight ensemble, so

$$p(x^{n-m}|y^{1:n-m}) = \sum_{i=1}^{N} \frac{1}{N} \delta(x^{n-m} - x_i^{n-m}). \tag{13.25}$$

We will explore the possibility of a proposal density at each model time step, so for the original model we write

$$p(x^n|y^{0:n}) = \frac{p(y^n|x^n)}{p(y^n)} \int \prod_{j=n-m+1}^{n} p(x^j|x^{j-1}) p(x^{n-m}|y^{1:n-m}) \, \mathrm{d}x^{n-m:n-1}, \tag{13.26}$$

and, introducing a proposal transition density at each time step, we find:

$$p(x^n|y^{0:n}) = \frac{p(y^n|x^n)}{p(y^n)} \int \prod_{j=n-m+1}^{n} \frac{p(x^j|x^{j-1})}{q(x^j|x^{j-1}, y^n)} q(x^j|x^{j-1}, y^n) p(x^{n-m}|y^{1:n-m}) \, \mathrm{d}x^{n-m:n-1}. \tag{13.27}$$

Using the expression for $p(x^{n-m}|y^{1:n-m})$ from (13.25) and choosing randomly from the transition proposal density $q(x^j|x^{j-1}, y^n)$ at each time step leads to

$$w_i = p(y^n|x_i^n) \prod_{j=n-m+1}^{n} \frac{p(x_i^j|x_i^{j-1})}{q(x_i^j|x_i^{j-1}, y^n)}. \tag{13.28}$$

13.4.1 Example: the EnKF as proposal

As an example, we will explore this technique with the Gaussian of the EnKF as the proposal density. First we have to evaluate the prior transition density. Since we know the starting point of the simulation, x_i^{n-1}, and its endpoint, the posterior EnKF sample x_i^n, and we know the model equation, written formally as:

$$x_i^n = f(x_i^{n-1}) + \beta_i^n, \tag{13.29}$$

we can determine β_i^n from this equation directly. We also know the distribution from which this β_i^n is supposed to be drawn, let us say a Gaussian with zero mean and covariance Q. We then find for the transition density:

$$p(x_i^n|x_i^{n-1}) \propto \exp\left\{-\frac{1}{2}\left[x_i^n - f(x_i^{n-1})\right] Q^{-1}\left[x_i^n - f(x_i^{n-1})\right]\right\}. \tag{13.30}$$

This will give us a number for each $[x_i^{n-1},\ x_i^n]$ combination.

Let us now calculate the proposal density $q(x_i^n|x_i^{n-1}, y^n)$. This depends on the ensemble Kalman filter used. For the ensemble Kalman filter with perturbed observations, the situation is as follows. Each particle in the updated ensemble is connected to those before analysis as

$$x_i^n = x_i^{n,old} + K^e\Big[y + \epsilon_i - H\Big(x_i^{n,\mathrm{old}}\Big)\Big], \tag{13.31}$$

in which ϵ_i is the random error drawn from $N(0, R)$ that has to be added to the observations in this variant of the ensemble Kalman filter. K^e is the ensemble Kalman gain, i.e. the Kalman gain using the prior error covariance calculated from the prior ensemble. The particle prior to the analysis comes from that of the previous time step through the stochastic model:

$$x_i^{n,old} = f(x^{n-1}) + \beta_i^n. \tag{13.32}$$

Combining these two gives

$$x_i^n = f(x_i^{n-1}) + \beta_i^n + K^e\big[y + \epsilon_i - H\big(f(x_i^{n-1})\big) - H(\beta_i^n))\big], \tag{13.33}$$

or

$$x_i^n = f(x_i^{n-1}) + K^e\big[y - H\big(f(x_i^{n-1})\big)\big] + (1 - K^e H)\,\beta_i^n + K^e\epsilon_i, \tag{13.34}$$

assuming that H is a linear operator. The right-hand side of this equation has a deterministic part and a stochastic part. The stochastic part provides the transition density going from x_i^{n-1} to x_i^n. Assuming both model and observation errors to be Gaussian-distributed and independent, we find for this transition density

$$q(x_i^n|x_i^{n-1}y^n) \propto \exp\left[-\frac{1}{2}\left(x_i^n - \mu_i^n\right)^{\mathrm{T}} \Sigma_i^{-1} \left(x_i^n - \mu_i^n\right)\right], \tag{13.35}$$

in which μ_i^n is the deterministic 'evolution' of x, given by

$$\mu_i^n = f(x_i^{n-1}) + K^e\big[y - H(x_i^{n-1})\big] \tag{13.36}$$

and the covariance Σ_i is given by

$$\Sigma_i = (1 - K^e H)Q(1 - K^e H)^{\mathrm{T}} + K^e R K^{e\mathrm{T}}, \tag{13.37}$$

where we have assumed that the model and observation errors are uncorrelated. It should be realized that x_i^n does depend on all $x_j^{n,\mathrm{old}}$ via the Kalman gain, which involves the error covariance P^e. Hence we have calculated $q(x_i^n|P^e, x_i^{n-1}, y^n)$ instead of $q(x_i^n|x_i^{n-1}, y^n)$, in which P^e depends on all other particles. The reason why we ignore the dependence on P^e is that in the case of an infinitely large ensemble, P^e would be a variable that depends only on the system, not on specific realizations of that system. This is different from the terms related to x_i^n, which will depend on the specific realization for β_i^n even when the ensemble size is 'infinite'. (Hence another approximation related to the finite size of the ensemble comes into play here, and at this moment it is unclear how large this approximation error is.)

The calculations of $p(x^n|x^{n-1})$ and $q(x_i^n|x_i^{n-1}y^n)$ look like very expensive operations. By realizing that Q and R can be obtained from the ensemble of particles, computationally efficient schemes can easily be derived.

We can now determine the full new weights. Since the normalization factors for the transition and the posterior densities are the same for all particles, the weights are easily calculated. The procedure now is as follows (see Fig. 13.3):

1. Run the ensemble up to the observation time.
2. Perform a (local) EnKF analysis of the particles.
3. Calculate the proposal weights $w_i^* = p(x_i^n|x_i^{n-1})/q(x_i^n|x_i^{n-1}y^n)$.
4. Calculate the likelihood weights $w_i = p(y^n|x_i^n)$.
5. Calculate the full relative weights as $w_i = w_i * w_i^*$ and normalize them.
6. Resample.

It is good to realize that the EnKF step is only used to draw the particles close to the observations. This means that when the weights are still varying too much, one can do the EnKF step with much smaller observational errors. This might look like overfitting but it is not, since the only thing we do in probabilistic sense is to generate particles in those positions in state space where the likelihood is large.

Finally, other variants of the EnKF, such as the adjusted and the transform variants can be used too, as detailed in van Leeuwen (2009). The efficiency of using the EnKF as proposal is under debate at the moment. The conclusions so far seem to be that using the EnKF as proposal in high-dimensional systems does not work. What has

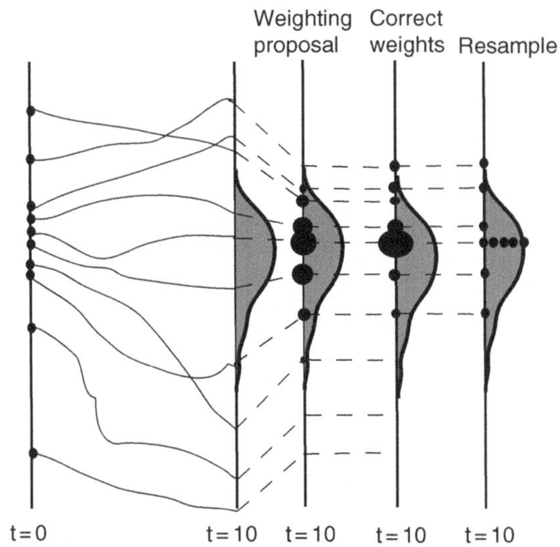

Fig. 13.3 The particle filter with proposal density. The model variable runs along the vertical axis; the weight of each particle corresponds to the size of the bullets on this axis. The horizontal axis denotes time, with observations at a time interval of 10 time units. All particles have equal weight at time zero. At time 10, the particles are brought closer to the observations by using, for example, the EnKF. Then they are weighted with the likelihood and these weights are corrected for the artificial EnKF step. (From van Leeuwen (2010).)

not been tested, however, is to use EnKF proposals with smaller observation matrix R, and more possibilities are still open, like using localisation (see later on in this chapter).

13.4.2 The auxiliary particle filter

In the auxiliary particle filter, the ensemble at time $n-1$ is weighted with information of the likelihood at time n (see Pitt and Shephard, 1999). In this method, one generates a representation of each particle at the time of the new observation, for example by integrating each particle from time $n-1$ to time n using zero model noise. (Depending on the complexity of the stochastic model integrator, this can save considerable time.) Then the particles are weighted with the observations, and a new resampled ensemble is integrated from $n-1$ to arrive closer to the observations. A flow chart reads (see Fig. 13.4) as follows:

1. Integrate each particle from $n-1$ to n with simplified dynamics (e.g. without model noise), producing a representation of the proposal density $q(x^n|x_i^{n-1}, y^n)$.
2. Weight each particle with the new observations as

$$\beta_i \propto p(y^n|x_i^n)w_i^{n-1}. \tag{13.38}$$

These weights are called the 'first-stage weights' or the 'simulation weights'.

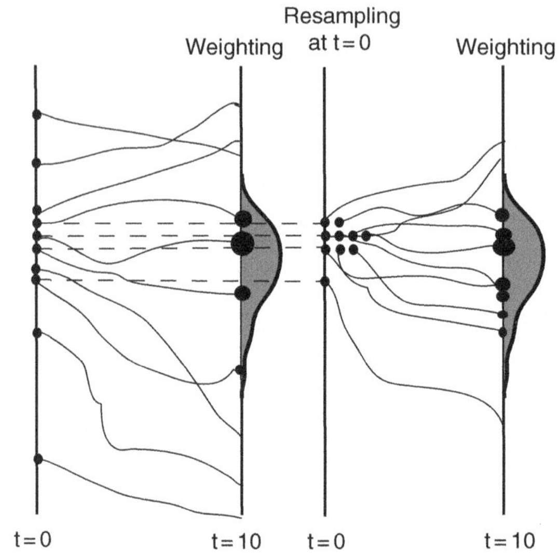

Fig. 13.4 The auxiliary particle filter. The model variable runs along the vertical axis; the weight of each particle corresponds to the size of the bullets on this axis. The horizontal axis denotes time, with observations at a time interval of 10 time units. All particles have equal weight at time zero. At time 10, the particles are weighted according to the likelihood. These weights are used at time 0 to rerun the ensemble up to time 10. (From van Leeuwen (2010).)

3. Resample the particles i at time $n-1$ with these weights, and use these resampled particles j_i as a representation of the proposal density by integrating each forward to n with the full stochastic model, so choosing from $q(x^n|x_{j_i}^{n-1}, y^n)$. Note that j_i connects the original particle i with its new position in state space, that of particle j.
4. Re-weight the members with weights

$$w_i^n = \frac{1}{A} p(y^n|x_i^n) \frac{p(x_i^n|x_{j_i}^{n-1})}{q(x_i^n|x_{j_i}^{n-1}, y^n)\beta_{j_i}}, \tag{13.39}$$

in which A is the normalization factor. A resampling step can be done, but is not really necessary because the actual resampling is done at step 3.

The name 'auxiliary' comes from the introduction of the member index j_i in the formulation. This member index keeps track of the relation between the first-stage weights and the particle sample at $n-1$.

It should be noted that $2N$ integrations have to be performed with this method: one ensemble integration to find the proposal and one for the actual pdf. If adding the stochastic noise is not expensive, step 1 can be done with the stochastic model, which comes down to doing sequential importance resampling twice. However, one could also use a simplified model for the first set of integrations. A geophysical example would be to use a quasi-geostrophic model for the first set of integrations and the full model for the second. One can imagine doing it even more times, zooming in into the likelihood, but at a cost of performing more and more integrations of the model. Figure 13.4 displays how the method works.

13.4.3 Including future observations in the model equations

So far, we have discussed proposal density applications in which the model equations were not changed directly. Of course, in, for example, the auxiliary particle filter, one could use a different model for the first set of integrations to obtain the first-stage weights, but the future observations were not used directly in the model equations. However, much more efficient schemes can be derived that change the model equations such that each particle is pulled towards the future observations at each time step. By keeping track of the weights associated with this, it can be assured that the correct problem is solved and the particles are random samples from the posterior pdf.

As mentioned before, the idea of the proposal transition density is that we draw samples from that density instead of from the original model. Furthermore, these samples can be dependent on the future observations. To see how this works, let us write the stochastic model equation as

$$x_i^n = f(x_i^{n-1}) + \beta_i^n. \tag{13.40}$$

First, we have to understand how this equation is related to the transition density $p(x_i^n|x_i^{n-1})$. The probability to end up in x_i^n starting from x_i^{n-1} is related to β_i^n. For instance, if $\beta_i^n=0$, so there is no model error and, a perfect model, then this probability is 1 if the x_i^n, x_i^{n-1} pair fulfils the perfect model equations and zero otherwise. So, in this case, $p(x_i^n|x_i^{n-1})$ would be a delta function centred on $f(x_i^{n-1})$. However, the more realistic case is that the model error is non-zero. The transition density will now depend on the distribution of the stochastic random forcing. Assuming Gaussian random forcing with zero mean and covariance Q, so $\beta_i^n \sim N(0,Q)$, we find

$$p(x_i^n|x_i^{n-1}) \propto N(f(x_i^{n-1}), Q). \tag{13.41}$$

As mentioned above, we will not use the normal model equation for each particle, but a modified model equation, one that 'knows' about future observations and actually draws the model to those observations. Perhaps the simplest example is to add a term that relaxes the model to the future observation, such as

$$x_i^n = f(x_i^{n-1}) + \beta_i^n + K^n\big[y^{n+m} - H(x_i^{n-1})\big], \tag{13.42}$$

in which $n+m$ is the next observation time. Note that the observation operator H does not contain any model integrations, it is just the evaluation of x_i^{n-1} in observation space. The reason is simple, we do not have x_i^{n+m} yet. Clearly, each particle i will now be pulled towards the future observations, with relaxation strength related to the matrix K^n. In principle, we are free to choose K^n, but it is reasonable to assume that it is related to the error covariance of the future observation R and that of the model equations Q. We will show possible forms in the examples discussed later.

With the simple relaxation, or other techniques, we have ensured that all particles end up closer to the observations. But we cannot just alter the model equations, we have to compensate for this trick. This is why the proposal density turns up in the weights. Each time step, the weight of each particle changes with

$$w_i^n = \frac{p(x_i^n|x_i^{n-1})}{q(x_i^n|x_i^{n-1}, y^n)} \tag{13.43}$$

between observation times. This can be calculated in the following way. Using the modified model equations, we know x_i^{n-1} for each particle, that was our starting point, and also x_i^n. So, assuming the model errors are Gaussian-distributed, this would become

$$p(x_i^n|x_i^{n-1}) \propto \exp\left\{-\frac{1}{2}\left[x_i^n - f(x_i^{n-1})\right]^{\mathrm{T}} Q^{-1}\left[x_i^n - f(x_i^{n-1})\right]\right\}. \tag{13.44}$$

The proportionality constant is not of interest since it is the same for each particle and drops out when the relative weights of the particles are calculated. Note that we have all the ingredients to calculate this and that $p(x_i^n|x_i^{n-1})$ is just a number.

For the proposal transition density we use the same argument, to find:

$$\begin{aligned} q(x_i^n|x_i^{n-1}y^n) &\propto \exp\left\{-\frac{1}{2}\left(x_i^n - f(x_i^{n-1}) - K^n\left[y^n - H\left(x^{n-1}\right)\right]^{\mathrm{T}}\right)\right. \\ &\quad \times Q^{-1}\left(x_i^n - f(x_i^{n-1}) - K^n\left[y^n - H(x^{n-1})\right]\right) \\ &= \exp\left(-\frac{1}{2}\beta_i^{n\mathrm{T}} Q^{-1}\beta_i^n\right). \end{aligned} \tag{13.45}$$

Again, since we did choose β to propagate the model state forward in time, we can calculate this, and it is just a number. In this way, any modified equation can be used, and we know, at least in principle, how to calculate the appropriate weights.

13.4.4 The optimal proposal density

The so-called optimal proposal density is described in the literature, (see e.g. Doucet et al. 2001). It is argued that taking $q(x^n|x^{n-1}, y^n) = p(x^n|x^{n-1}, y^n)$ results in optimal weights. However, it is easy to show that this is not the case. Assume observations every time step and a resampling scheme at every time step, so that a equal-weighted ensemble of particles is present at time $n-1$. Furthermore, assume that model errors are Gaussian-distributed according to $N(0, Q)$ and observation errors are Gaussian-distributed according to $N(0, R)$. First, using the definition of conditional densities, we can write:

$$p(x^n|x^{n-1}, y^n) = \frac{p(y^n|x^n)p(x^n|x^{n-1})}{p(y^n|x^{n-1})}, \tag{13.46}$$

where we have used $p(y^n|x^n, x^{n-1}) = p(y^n|x^n)$. Using this proposal density gives posterior weights

$$\begin{aligned} w_i &= p(y^n|x_i^n)\frac{p(x_i^n|x_i^{n-1})}{q(x_i^n|x_i^{n-1}, y^n)} \\ &= p(y^n|x_i^n)\frac{p(x_i^n|x_i^{n-1})}{p(x_i^n|x_i^{n-1}, y^n)} \\ &= p(y^n|x_i^{n-1}). \end{aligned} \tag{13.47}$$

The latter can be expanded as

$$w_i = \int p(y^n, x^n|x^{n-1})\,\mathrm{d}x^n = \int p(y^n|x^n)p(x^n|x^{n-1})\,\mathrm{d}x^n, \tag{13.48}$$

in which we have again used $p(y^n|x^n, x^{n-1}) = p(y^n|x^n)$. Using the Gaussian assumptions mentioned above (note that the state is never assumed to be Gaussian), we can perform the integration to obtain

$$w_i \propto \exp\left\{-\frac{1}{2}\left[y^n - Hf(x_i^{n-1})\right]^{\mathrm{T}}\left(HQH^{\mathrm{T}} + R\right)^{-1}\left[y^n - Hf(x_i^{n-1})\right]\right\}. \tag{13.49}$$

Note that we have just calculated the probability density of $p(y^n|x_i^{n-1})$.

To estimate the order of magnitude of the first two moments of the distribution of $y^n - Hf(x_i^{n-1})$, it is expanded to $y^n - Hx_t^n + H\left[x_t^n - f(x_i^{n-1})\right]$ in which x_t^n is the true state at time n. If we now use $x_t^n = f(x_t^{n-1}) + \beta_t^n$, this can be expanded further as $y^n - Hx_t^n + H\left[f(x_t^{n-1}) - f(x_i^{n-1})\right] + H\beta_t^n$. To proceed, we make the following restrictive assumptions that will nevertheless allow us to obtain useful order-of-magnitude estimates. Let us assume that both the observation errors R and the observed model errors HQH^{T} are uncorrelated, with variances V_y and V_β, respectively, to find:

$$-\log(w_i) = \frac{1}{2(V_\beta + V_y)}\sum_{j=1}^{M}\left\{y_j^n - H_jx_t^n + H_j\beta_t^n + H_j\left[f(x_t^{n-1}) - f(x_i^{n-1})\right]\right\}^2. \tag{13.50}$$

The variance of w_i arises from varying the ensemble index i. Clearly, the first three terms are given, and we introduce the constant $\gamma_j = y_j^n - H_jx_t^n + H_j\beta_t^n$. To proceed with our order-of-magnitude estimate, we assume that the model can be linearized as $F(x_i^{n-1}) \approx Ax_i^{n-1}$, leading to

$$-\log w_i = \frac{1}{2(V_\beta + V_y)}\sum_{j=1}^{M}\left[\gamma_j + H_jA\left(x_t^{n-1} - x_i^{n-1}\right)\right]^2. \tag{13.51}$$

The next step in our order-of-magnitude estimate is to assume $x_t^{n-1} - x_i^{n-1}$ to be Gaussian-distributed. In that case, the expression above is non-centrally χ_M^2

distributed apart from a constant. This constant comes from the variance of $\gamma_j + H_j A(x_t^{n-1} - x_i^{n-1})$, which is equal to $H_j A P^{n-1} A^{\mathrm{T}} H_j^{\mathrm{T}} = V_x$, in which P^{n-1} is the covariance of the model state at time $n-1$. Hence, we find

$$-\log w_i = \frac{V_x}{2(V_\beta + V_y)} \sum_{j=1}^{M} \frac{\left[\gamma_j + H_j A(x_t^{n-1} - x_i^{n-1})\right]^2}{V_x}. \tag{13.52}$$

Apart from the constant in front, this expression is non-centrally χ_M^2 distributed with variance $a^2 2(M + 2\lambda)$, where $a = V_x/(2(V_\beta + V_y)$ and $\lambda = (\sum_j \gamma_j^2)/V_x$.

We can estimate λ by realizing that for a large enough number of observations we expect $\sum_j (y_j^n - H_j x_t^n)^2 \approx MV_y$, and $\sum_j (y_j^n - H_j x_t^n) \approx 0$. Furthermore, when the dimension of the system under study is large, we expect $\sum_j (H_j \beta_t^n)^2 \approx MV_\beta$. Combining all these estimates, we find that the variance of $-\log w_i$ can be estimated as

$$\frac{M}{2} \left(\frac{V_x}{V_\beta + V_y} \right)^2 \left[1 + 2 \left(\frac{V_\beta + V_y}{V_x} \right) \right]. \tag{13.53}$$

This expression shows that the only way to keep the variance of $-\log w_i$ low when the number of independent observations M is large is to have a very small variance in the ensemble: $V_x \approx (V_\beta + V_y)/M$. Clearly, when the number of observations is large (10 million in typical meteorological applications), this is not very realistic. This expression has been tested in several applications and holds within a factor 0.1 in all tests (van Leeuwen, 2012, unpublished manuscript).

It should be mentioned that a large variance of $-\log w_i$ does not necessarily mean that the weights will be degenerate, because the large variance could be due to a few outliers. However, we have shown that $-\log w_i$ is approximately non-centrally χ_M^2 distributed for a linear model, so the large variance is not due to outliers but intrinsic in the sampling from such a distribution. Furthermore, there is no reason to assume that this variance will behave better for nonlinear models, especially because we did not make any assumptions on the divergent or contracting characteristics of the linear model.

From this analysis, we learn two things: it is the number of independent observations that determines the degeneracy of the filter, and the optimal proposal density cannot be used in systems with a very large number of independent accurate observations.

13.4.5 The implicit particle filter

In 2009, Chorin and Tu (2009) introduced this implicit particle filter. Although their paper is not very clear, with the theory and the application intertwined, discussions with them and later papers (Chorin et al., 2010; Morzfeld and Chorin, 2012) explain the method in more detail. Although they do not formulate their method in terms of a proposal density, to clarify the relation with the other particle filters this is the way it is presented here. In fact, as we shall see, it is closely related to the optimal proposal density discussed before when the observations are available at every time step.

The proposed samples are produced as follows. Assume we have m model time steps between observations. Draw a random vector ξ_i of length the size of the state vector times m. Each element of ξ_i is drawn from $N(0,1)$. The actual samples are now constructed by solving

$$-\log\left[p(y^n|x^n)p\left(x^{n-m+1:n}|x_i^{n-m}\right)\right] = \frac{\xi_i^{\mathrm{T}}\xi_i}{2} + \phi_i \tag{13.54}$$

for each particle x_i. The term ϕ_i is included to ensure that the equation has a solution, so $\phi_i \geq \min\left\{-\log\left[p(y^n|x^n)p(x^{n-m+1:n}|x_i^{n-m})\right]\right\}$. Note that this can be written as

$$p(y^n|x^n)p(x^{n-m+1:n}|x_i^{n-m}) = A\exp\left(-\frac{\xi^{\mathrm{T}}\xi}{2} - \phi_i\right) \tag{13.55}$$

for later reference. One can view this step as drawing from the proposal density $q_x(x^{n-m+1:n}|x_i^{n-m}, y^n)$ via the proposal density $q_\xi(\xi)$, where we have introduced the subscript to clarify the shape of the pdf. These two are related by a transformation of the probability densities as

$$q_x(x^{n-m+1:n}|x_i^{n-m}, y^n)\, dx^{n-m:n} = q_\xi(\xi)\, d\xi, \tag{13.56}$$

so that

$$q_x(x^{n-m:n}|x_i^{n-m}, y^n) = q_\xi(\xi)J, \tag{13.57}$$

in which J is the Jacobian of the transformation $\xi \to x$. We can now write the weights of this scheme as

$$\begin{aligned} w_i &= p(y^n|x_i^n)\frac{p(x_i^{n-m+1:n}|x_i^{n-m})}{q_x(x_i^{n-m+1:n}|x_i^{n-m}, y^n)} \\ &= p(y^n|x_i^n)\frac{p(x^{n-m+1:n}|x_i^{n-m})}{Jq_\xi(\xi_i)}. \end{aligned} \tag{13.58}$$

Using (13.55), we find that the weights are given by

$$w_i = A\frac{\exp(-\phi)}{J}. \tag{13.59}$$

To understand better how this works, let us consider the case of observations every model time step, and Gaussian observation errors, Gaussian model equation errors, and linear observation operator H. In that case, we have

$$\begin{aligned} -\log\left[p(y^n|x^n)p(x^{n-m+1}|x_i^{n-m})\right] &= \frac{1}{2}\left(y^n - Hx_i^n\right)^{\mathrm{T}} R^{-1}\left(y^n - Hx_i^n\right) \\ &\quad + \frac{1}{2}\left[x^n - f(x_i^{n-1})\right]^{\mathrm{T}} Q^{-1}\left[y^n - f(x_i^{n-1})\right] \\ &= \frac{1}{2}\left(x^n - \hat{x}_i^n)\right)^{\mathrm{T}} P^{-1}\left(x^n - \hat{x}_i^n\right) + \phi_i, \end{aligned} \tag{13.60}$$

in which $\hat{x}_i^n = f(x_i^{n-1}) + K[y^n - Hf(x_i^{n-1})]$, the maximum of the posterior pdf, and $P = (1 - KH)Q$, with $K = QH^{\mathrm{T}}(HQH^{\mathrm{T}} + R)^{-1}$. Comparing this with (13.55), we find $x^n = P^{1/2}\xi$, so J is a constant, and

$$\begin{aligned}\phi_i &= \min\left\{-\log\left[p(y^n|x^n)p(x^{n-m+1:n}|x_i^{n-m})\right]\right\}\\ &= \frac{1}{2}\left[y^n - Hf(x_i^{n-1})\right]^{\mathrm{T}}(HQH^{\mathrm{T}} + R)^{-1}\left[y^n - Hf(x_i^{n-1})\right], \end{aligned} \tag{13.61}$$

and finally $w_i \propto \exp(-\phi_i)$. Comparing with the optimal proposal density, we see that when observations are present at every time step, the implicit particle filter is equal to the optimal proposal density, with the same degeneracy problem.

13.4.6 The equivalent-weights particle filter

At the beginning of this chapter, we discussed a very simple nudging scheme to pull the particles towards future observations, written as

$$x_i^n = f(x_i^{n-1}) + \beta_i^n + K^n\left[(y^{n+m} - H(x_i^{n-1})\right], \tag{13.62}$$

in which $n + m$ is the next observation time. Unfortunately, exploring the proposal density by simply nudging will not avoid degeneracy in high-dimensional systems with a large number of observations. Also, more complicated schemes, such as running a 4D-Var on each particle, which is essentially what Chorin and Tu (2009) propose, are likely to lead to strongly varying weights for the particles because its close relation to the optimal proposal density. (However, it must be said that no rigourous proof exists of this statement!) We can expect to have to do something more optimal.

To start, let us recall the expression we found for the proposal density weights in Section 13.4, equation (13.28):

$$w_i = p(y^n|x_i^n)\prod_{j=n-m+1}^{n}\frac{p(x_i^j|x_i^{j-1})}{q(x_i^j|x_i^{j-1}, y^n)}. \tag{13.63}$$

We will use a modified model as explained in Section 13.4.3 for all but the last model time step, which will be different, as explained below. The last time step consists of two stages: first perform a deterministic time step with each particle that ensures that most of the particles have equal weight, and then add a very small random step to ensure that Bayes' theorem is satisfied (for details, see van Leeuwen, 2010, 2011). There are again infinitely many ways to do this. For the first stage we write down the weight for each particle using only a deterministic move, so ignoring the proposal density q for the moment:

$$\begin{aligned}-\log w_i &= -\log w_i^{\mathrm{rest}} + \frac{1}{2}(y^n - Hx_i^n)^{\mathrm{T}}R^{-1}(y^n - Hx_i^n)\\ &\quad + \frac{1}{2}\left[x_i^n - f(x_i^{n-1})\right]^{\mathrm{T}}Q^{-1}\left[x_i^n - f(x_i^{n-1})\right], \end{aligned} \tag{13.64}$$

in which w_i^{rest} is the weight accumulated over the previous time steps between observations, so the p/q factors from each time step. If H is linear, which is not essential but we will assume for simplicity here, this is a quadratic equation in the unknown x_i^n. All other quantities are given. We calculate the minimum of this function for each particle i, which is simply given by

$$-\log w_i = -\log w_i^{\text{rest}} + \frac{1}{2}\left[y^n - Hf(x_i^{n-1})\right]^{\mathrm{T}} (HQH^{\mathrm{T}} + R)^{-1} \left[y^n - Hf(x_i^{n-1})\right]. \tag{13.65}$$

For N particles, this gives rise to N minima. Next, we determine a target weight as the weight that 80% of the particles can reach; i.e. 80% of the minimum $-\log w_i$ is smaller than the target value. (Note that we can choose another percentage; see e.g. Ades and van Leeuwen (2013), who investigate the sensitivity of the filter for values between 70% and 100%.) Define a quantity $C = -\log w_{\text{target}}$, and solve for each particle with a minimum weight larger than the target weight:

$$C = -\log w_i^{\text{rest}} + \frac{1}{2}\left[y^n - Hf(x_i^{n-1})\right]^{\mathrm{T}} (HQH^{\mathrm{T}} + R)^{-1} \left[y^n - Hf(x_i^{n-1})\right]. \tag{13.66}$$

So now we have found the positions of the new particles x_i^n such that all have equal weight. The particles that have a minimum larger than C will come back into the ensemble via a resampling step, to be discussed later.

The equation above has an infinite number of solutions for dimensions larger than 1. To make a choice we assume

$$x_i^n = f(x_i^{n-1}) + \alpha_i K[y^n - Hf(x_i^{n-1})], \tag{13.67}$$

in which $K = QH^{\mathrm{T}}(HQH^{\mathrm{T}} + R)^{-1}$, Q is the error covariance of the model errors, and R is the error covariance of the observations. Clearly, if $\alpha_i = 1$, we find the minimum back. We choose the scalar α_i such that the weights are equal, leading to

$$\alpha_i = 1 - \sqrt{1 - b_i/a_i}, \tag{13.68}$$

in which $a_i = 0.5x_i^{\mathrm{T}}R^{-1}HKx$ and $b_i = 0.5x_i^{\mathrm{T}}R^{-1}x_i - C - \log w_i^{\text{rest}}$. Here $x = y^n - Hf(x_i^{n-1})$, C is the chosen target weight level, and w_i^{rest} denotes the relative weights of each particle i up to this time step, related to the proposal density explained above.

Of course, this last step towards the observations cannot be fully deterministic. A deterministic proposal would mean that the proposal transition density q can be zero while the target transition density p is non-zero, leading to division by zero, because for a deterministic move the transition density is a delta function. The proposal transition density could be chosen as a Gaussian, but since the weights have q in the denominator, a draw from the tail of a Gaussian would lead to a very high weight for a particle that is perturbed by a relatively large amount. To avoid this, q is chosen in the last step before the observations as a mixture density

$$q(x_i^n|x_i') = (1-\gamma)U(-a, a) + \gamma N(0, a^2), \tag{13.69}$$

in which x_i', the particle before the last random step, and γ and a are small. By choosing γ small, the change of having to choose from $N(0, a^2)$ can be made as small as desired. For instance, it can be made dependent on the number of particles N.

To conclude, the equivalent-weight scheme consists of the following steps:

1. Use the modified model equations for each particle for all time steps between observations.
2. Calculate, for each particle i for each of these time steps,

$$w_i^n = w_i^{n-1} \frac{p(x_i^n | x_i^{n-1})}{q(x_i^n | x_i^{n-1}, y^m)}. \tag{13.70}$$

3. At the last time step before the observations, calculate the maximum possible weight for each particle and determine $C = -\log w_{\text{target}}$.
4. Determine the deterministic moves by solving for α_i for each particle as outlined above.
5. Choose a random move for each particle from the proposal density (13.69).
6. Add these random move to each deterministic move, and calculate the full posterior weight.
7. Resample, and include the particles that have been neglected from step 4 on.

Finally, it is stressed again that we do solve the fully nonlinear data assimilation problem with this efficient particle filter, and the only approximation is in the ensemble size. All other steps are completely compensated for in Bayes' theorem via the proposal density freedom.

13.4.7 Application to the barotropic vorticity equations

A few results using the new particle filter with almost equal weights are shown here, (see van Leeuwen and Ades, 2013). Figure 13.5 shows the application of the method

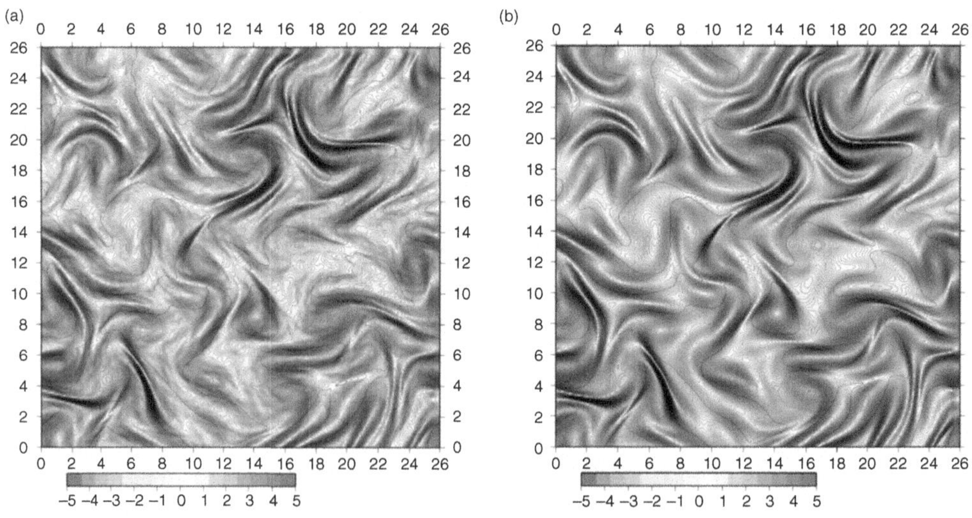

Fig. 13.5 Snapshots of (a) the particle filter mean and (b) the vorticity field of the truth at time 25. Note the highly chaotic state of the fields and the close to perfect tracking.

to the highly chaotic barotropic vorticity equation, governed by:

$$\frac{\partial q}{\partial t} - \frac{\partial \psi}{\partial y}\frac{\partial q}{\partial x} + \frac{\partial \psi}{\partial x}\frac{\partial q}{\partial y} = \beta,$$
$$q = \frac{\partial^2 \psi}{\partial x^2} + \frac{\partial^2 \psi}{\partial y^2}, \tag{13.71}$$

in which q is the vorticity field, ψ is the stream function, and β is a random noise term representing errors in the model equations. It was chosen from a multivariate Gaussian with mean zero, variance 0.01, and decorrelation lengthscale 4 gridpoints. The equations are implemented on a 256×256 grid, using a semi-Lagrangian scheme with time step $\Delta t = 0.04$ and grid spacing $\Delta x = \Delta y = 1/256$, leading to a state dimension close to 65,000. The vorticity field was observed every 50 time steps on every gridpoint. The decorrelation timescale of this system is about 25 time steps, so, even though the full state is observed, this is a very hard highly nonlinear data assimilation problem. The observations were obtained from a truth run, and independent random measurement noise with standard deviation 0.05 was added to each observation.

Only 24(!) particles were used to track the posterior pdf. In the application of the new particle filter, we chose $K = 0.1$ in the nudging term (except for the last time step before the new observations, where the equivalent-weights scheme was used, as explained above), multiplied by a linear function that is zero halfway between the two updates and growing to one at the new observation time. The random forcing was the same as in the original model. This allows the ensemble to spread out owing to the

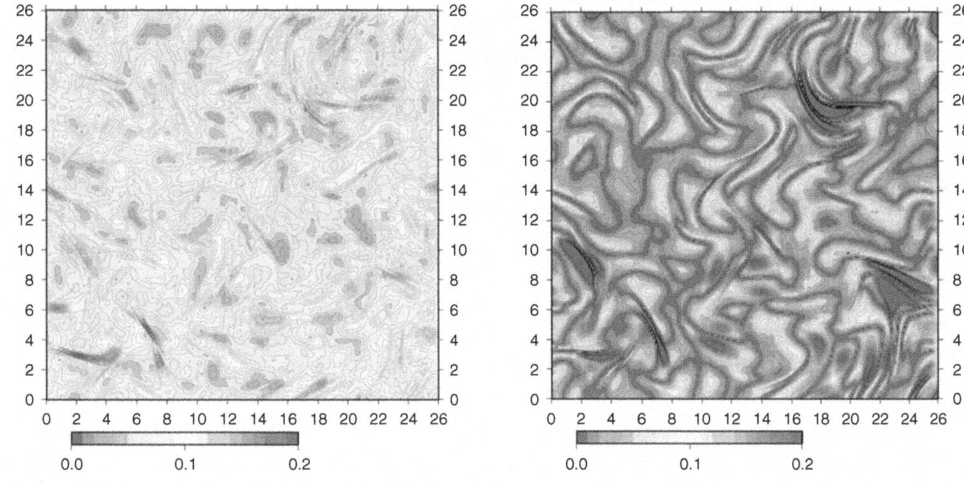

Fig. 13.6 Snapshots of the absolute value of (a) the mean truth misfit and (b) the standard deviation in the ensemble. The ensemble underestimates the spread at several locations, but, averaged over the field, it is slightly higher: 0.074 versus 0.056.

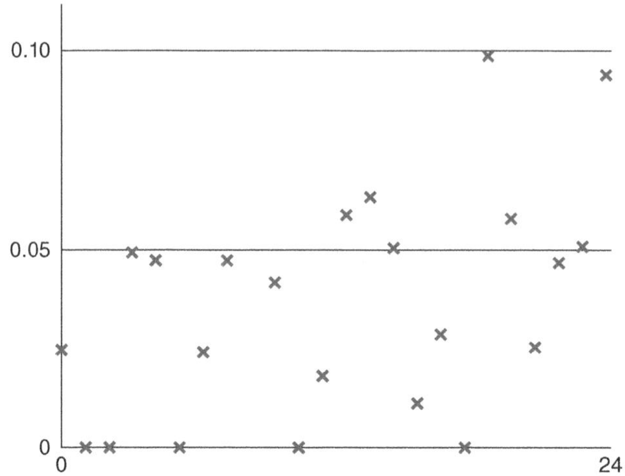

Fig. 13.7 Weights distribution of the particles before resampling. All weights cluster around 0.05, which is close to 1/24 for uniform weights (using 24 particles). The 5 particles with zero weights will be resampled. Note that the other particles form a smoother estimate.

random forcing, and pulling harder and harder towards the new observation the closer it was to the new update time.

Figure 13.6 shows the difference between the mean and the truth after 50 time steps, together with the ensemble standard deviation compared to the absolute value of the mean-truth misfit. Clearly, the truth is well represented by the mean of the ensemble. Figure 13.3 shows that although the spread around the truth is underestimated at several locations, it is overestimated elsewhere.

Finally, Fig. 13.7 shows that the weights are distributed as they should be: they display small variance around the equal weight value 1/24 for the 24 particles. Note that the particles with zero weight had too small a weight to be included in the equivalent weight scheme and will be resampled from the rest.

Because the weights vary so little, they weights can be used back in time, generating a smoother solution for this high-dimensional problem with only 24 particles.

13.5 Conclusions

To try to solve strongly nonlinear data assimilation problems, we have discussed particle filters in this chapter. They have a few strong assets, namely their full nonlinearity, the simplicity of their implementation (although this tends to be lost in more advanced variants), the fact that balances are automatically fulfilled (although, again, more advanced methods might break this), and, quite importantly, that their behaviour does not depend on a correct specification of the model state covariance matrix.

We have also seen their weaknesses in terms of efficiency—the filter degeneracy problem that plagues the simpler implementations. However, recent progress seems to suggest that we are quite close to solving this problem with developments such as the implicit particle filter and the equivalent-weights particle filter. Also, the approximations are becoming more advanced too, and perhaps we do not need a fully nonlinear data assimilation method for real applications.

There is a wealth of new approximate particle filters that typically shift between a full particle filter and an ensemble Kalman filter, depending on the degeneracy encountered. Gaussian mixture models for the prior are especially popular. I have refrained from trying to give an overview here, since there is just too much going on in this area. A brief discussion is given in van Leeuwen (2009)—again not completely up to date. In a few years time, we will have learned what is useful and what is not.

Specifically, I should like to mention the rank histogram filter of Anderson (2010). It approximates the prior ensemble in observation space with a histogram, assuming Gaussian tails at both end members. It then performs Bayes' theorem and multiplies this prior with the likelihood to form the posterior. Samples from this posterior are generated as follows. First, the cumulative probability of the posterior at each prior particle is calculated by integrating the posterior over the regions between the prior particles. We want the posterior particles to have equal probability $1/(N+1)$, and so cumulative probability $n/(N+1)$ for ordered particle n. Therefore, the position of each new particle is found by integrating the posterior pdf $1/(N+1)$ further from the previous new member. As Anderson shows, this entails solving a simple quadratic equation for each particle, with special treatment of the tails.

A few comments are in order. First, the prior is not assumed to be Gaussian, and the likelihood also can be highly non-Gaussian, which is good. However, a potential problem is that the above procedure is performed on each dimension separately, and it is unclear how to combine these dimensions into sensible particles. Localization has to be applied to keep the numerical calculations manageable, and inflation is also needed to avoid ensemble collapse. Also, as far as I can see, when the observations are correlated, the operations explained above have to be done in a higher-dimensional space, making the method more complicated. Finally, the method interpolates in state space, which potentially leads to unbalanced states. Anderson applied the method to a 28 000-dimensional atmospheric model with very promising results.

A word of caution is needed. The contents of this chapter express my present knowledge of the field, and no doubt miss important contributions. Also, the field is developing so rapidly—we have aroused the interest of applied mathematicians and statisticians to our geoscience problems—that it is becoming extremely hard to keep track of all interesting work. (That all these communities publish in their own journals does not make life easy, I am now reviewing particle filter articles in over 20 journals.)

Finally, it must be said that the methods discussed here have a strong bias to state estimation. One could argue that this is fine for prediction purposes, but for model improvement (and thus indirectly forecasting), parameter estimation is of more interest (and then there is the question of parameterisation estimation). Unfortunately, no efficient particle filter schemes exist for that problem. This is a growing field that needs much more input from bright scientists like you, reader!

References

Ades, M. and van Leeuwen, P. J. (2013). An exploration of the equivalent weights particle filter. *Q. J. R. Meterol. Soc.*, **139**, 820–840.

Anderson, J. (2010). A non-Gaussian ensemble filter update for data assimilation. *Mon. Weather. Rev.*, **138**, 4186–4198.

Chorin, A., Morzfeld, M., and Tu, X. (2010). Implicit particle filters for data assimilation. *Commun. Appl. Math. Comput. Sci.*, **5**, 221–240.

Chorin, A. J. and Tu, X. (2009). Implicit sampling for particle filters. *Proc. Nat. Acad. Sci. USA*, **106**, 17249–17254.

Doucet, A., De Freitas, N., and Gordon, N. (2001). *Sequential Monte-Carlo methods in Practice.* Springer-Verlag, Berlin.

Gordon, N. J., Salmond, D. J., and Smith, A. F. M. (1993). Novel-approach to nonlinear non-Gaussian Bayesian state estimation. *IEE Proce. F: Radar Sig. Process.*, **140**, 107–113.

Morzfeld, M. and Chorin, A. J. (2012). Implicit particle filtering for models with partial noise, and an application to geomagnetic data assimilation. *Nonlin. Process. Geophys.*, **19**, 365–382.

Pitt, M. K. and Shephard, N. (1999). Filtering via simulation: auxiliary particle filters. *J. Am. Statist. Assoc.*, **94**, 590–599.

Snyder, C., Bengtsson, T., Bickel, P., and Anderson, J. L. (2008). Obstacles to high-dimensional particle filtering. *Mon. Weather Rev.*, **136**, 4629–4640.

van Leeuwen, P. J. (2009). Particle filtering in geophysical systems. *Mon. Weather. Rev.*, **132**, 4089–4114.

van Leeuwen, P. J. (2010). Nonlinear data assimilation in geosciences: an extremely efficient particle filter. *Q. J. R. Meteoral. Soc.*, **136**, 1991–1999.

van Leeuwen, P. J. (2011). Efficient nonlinear data-assimilation in geophysical fluid dynamics. *Comput. Fluids*, **46**, 52–58.

van Leeuwen, P. J. and Ades, M. (2013). Efficient Fully nonlinear data assimilation for geophysical fluid dynamics. *Comput. Geosciences*, **55**, 16–27.

14
Second-order methods for error propagation in variational data assimilation

F.-X. Le Dimet[1], I. Gejadze[2], and V. Shutyaev[3]

[1]Université Grenoble Alpes
Grenoble, France

[2]Civil and Environmental Engineering
University of Strathclyde
Glasgow, United Kingdom

[3]Institute of Numerical Mathematics
Russian Academy of Sciences
Moscow, Russia

Advanced Data Assimilation for Geosciences. First Edition.
Edited by É. Blayo, M. Bocquet, E. Cosme, and L. F. Cugliandolo.

Chapter Contents

14.1 Introduction

Data assimilation is a first and mandatory step before predicting the evolution of a geophysical flow. A prediction makes sense only if it is associated with an error bar giving an evaluation of the uncertainties on the analysis then on the prediction. For the Kalman-type methods, error propagation is included in the formulation, but this is not the case for variational methods, where the analysis involves the solution of a Euler–Lagrange optimality system that is derived from the model. Therefore error propagation has to be deduced from a derivation of the optimality system and depends on the second-order properties of the model. All the ingredients of a variational methods are sources of errors. These are as follows:

- A model describing the evolution of the flow—realistic models are nonlinear. In the model, there are several type of errors:
  - mathematical errors due to the fact that the model is necessarily an approximation—some of the physical processes involved, such as the full water cycle in atmospheric models, cannot be fully described and are parameterized;
  - numerical errors coming from the discretization of the model;
  - Algorithmic errors—models are nonlinear and a direct solution cannot be computed; instead, an iterative method is used, involving some stopping criterion inducing an error.
- Observations used in the assimilation also contain errors. For in situ observations, there is a physical error; for remote sensing data in many cases, the state variables of the model are not directly observed and pseudo-observations of the state variables can be obtained by solving an inverse problems and therefore involving an approximation. Most of the time, model errors and observation errors are supposed to be independent, but this is not true: the same observation can be used both with a general circulation model and with a very local model, and, according to the context, (i.e. the model), a different confidence will be given to the observation.
- The background term is a first guess of the analysis, and in operational practice it issues from the last prediction and as such has an error that can be considered as the prediction error.

In this chapter, we will show how the variational approach can be used for estimating the propagation of uncertainties in the analysis. We will consider two basic cases:

- In the *deterministic framework*, the uncertainty is a virtual and deterministic perturbation on the parameters of the model and we want to know its impact on some criterion. We will define this case as sensitivity and impact studies.
- In the *stochastic framework*, the uncertainty is a random variable that will be transported by the model as such. The output is a stochastic perturbation on the outputs of the analysis, for which we will have to determine its probabilistic characteristics: mean and variance.

14.2 Variational methods

Variational methods were introduced into meteorology by Sasaki (1958). These methods consider the equations governing the flow as constraints, and the problem is closed by using a variational principle such as minimization of the discrepancy between the model and the observations. In the following, we will consider that the model is discrete with respect to space variables. The use of optimal control techniques (Lions, 1968) was proposed by Le Dimet (1982), Le Dimet and Talagrand (1986), Talagrand and Courtier (1987), Penenko and Obraztsov (1976), and Marchuk and Penenko (1978).

14.2.1 Ingredients

The various ingredients of a variational method are as follows:

- A *state variable*

$$X \in \mathbf{X}$$

that describes the evolution of the medium at the grid points (or some equivalent coefficient). X depends on time and for operational models is of large dimension with around one billion variables.
- A *model* describing the evolution of the fluid. Basically, this is a system of nonlinear differential equations written as

$$\begin{cases} \dfrac{\mathrm{d}X}{\mathrm{d}t} = F\left(X, U\right), \\ X\left(0\right) = V. \end{cases}$$

- A *control variable* $(U, V) \in \mathbf{P}$ (control space). Most of the time the control is the initial condition or/and some internal variables of the model: parameters or boundary conditions. We will assume that when a value has been fixed for the parameter, the model has a unique solution. For the sake of simplicity, we will not consider constraint on the control variable. Nevertheless, humidity and salinity cannot be negative, and therefore the set of controls does not necessarily have the structure of a vector space.
- *Observations* $X_{\text{obs}} \in \mathbf{O}_{\text{obs}}$. These are discrete and depend on time and space, and they are not, from the geographical or physical points of view, in the same space as the state variable. Therefore, we will introduce some operator C mapping the space of states into the space of observations. In practice, this operator can be complex.
- A *cost function* J measuring the discrepancy of the solution of the model associated with (U, V) and the observations:

$$J\left(U, V\right) = \frac{1}{2} \int_0^{\mathrm{T}} \|CX(U, V) - X_{\text{obs}}\|_O^2 \,\mathrm{d}t + \frac{1}{2} \|V - V_b\|_b^2 . \qquad (14.1)$$

The choice of the norm is important because it allows the introduction of some a priori information such as the statistics of the fields through the covariance matrix, which is positive-definite. In the definition of the cost function, $\|.\|_O$ is the norm in the space of observations and is defined by $\|Z\|_O^2 = (B_O Z, Z)$, where B_O is the inverse of the covariance of the error of observation and $(.,.)$ the regular inner product in L^2. By the same token, U_b is the background term and its space is equipped with a norm taking into account B_p the inverse of the covariance of the error of prediction. This term acts like a regularization term in the sense of Tykhonov.

The problem of variational data assimilation can then be set as

$$\text{Find} \quad (U^*, V^*) \in \mathbf{P} \text{ such that } J(U^*, V^*) = \inf_{(U,V)\in\mathbf{P}} J(U, V). \tag{14.2}$$

14.2.2 Optimality system

With respect to (U, V), we have a problem of unconstrained optimization. The problem (14.2) will have a unique solution if J is strictly convex and lower semicontinuous and if

$$\lim_{||(U,V)||\to+\infty} J(U, V) \to +\infty.$$

When J is differentiable, a necessary condition for (U^*, V^*) to be a solution of (14.2) is given by the Euler–Lagrange equation:

$$\nabla J(U^*, V^*) = 0,$$

where ∇J is the gradient of J with respect to (U, V). Furthermore, the determination of ∇J permits one to implement optimization methods of gradient type.

We introduce $P \in \mathbf{X}$, the so-called *adjoint variable*:

$$\begin{cases} \dfrac{\mathrm{d}P}{\mathrm{d}t} + \left[\dfrac{\partial F}{\partial X}\right]^* P = C^* \left(CX - X_{\text{obs}}\right), \\ P\left(T\right) = 0. \end{cases}$$

Here $*$ stands for the transpose of either a vector or an operator. This system is defined as being the adjoint model. Its backward integration leads to the gradient of the cost function, given by

$$\nabla J = \begin{pmatrix} \nabla_U J \\ \nabla_V J \end{pmatrix} = \begin{pmatrix} -\left[\dfrac{\partial F}{\partial U}\right]^* P \\ -P\left(0\right) \end{pmatrix}.$$

The model plus the adjoint model and the condition of nullity of the gradient comprise the optimality system. It worth pointing out that there is no approximation in the derivation of the optimality system and the only assumption is the differentiability of the Jacobian of the model. Some authors consider, at this level, a so-called linear tangent approximation, which is completely unnecessary.

14.2.3 Example: Saint-Venant's equations

Saint-Venant's equations, also known as the shallow-water equations, are used for an incompressible fluid whose depth is small compared with the horizontal dimensions. The general equations of geophysical fluid dynamics are vertically integrated using the hydrostatic hypothesis and therefore vertical acceleration is neglected. In Cartesian coordinates, they are

$$\frac{\partial u}{\partial t} + u\frac{\partial u}{\partial x} + v\frac{\partial u}{\partial y} - fv + \frac{\partial \phi}{\partial x} = 0, \tag{14.3}$$

$$\frac{\partial v}{\partial t} + u\frac{\partial v}{\partial x} + v\frac{\partial v}{\partial y} + fu + \frac{\partial \phi}{\partial y} = 0, \tag{14.4}$$

$$\frac{\partial \phi}{\partial t} + \frac{\partial u\phi}{\partial x} + \frac{\partial v\phi}{\partial y} = 0. \tag{14.5}$$

In this system, $X = (u, v, \phi)^{\mathrm{T}}$ is the state variable, u and v are the components of the horizontal velocity, ϕ is the geopotential (proportional to the height of the free surface), and f is the Coriolis parameter. For the sake of simplicity, the following hypotheses are used:

(a) The error of the model is neglected. Only the initial condition will be considered as a control variable.
(b) Lateral boundary conditions are periodic. This is verified in global models.
(c) Observations are supposed to be continuous with respect to time. Of course, this is not the case in practice.

$\mathbf{C} \equiv \mathbf{I}$, where $\mathbf{I}$ is the identity operator. If $U_0 = (u_0, v_0, \phi_0)^{\mathrm{T}}$ is the initial condition and the cost function is given by

$$J(U_0) = \frac{1}{2}\int_0^{\mathrm{T}} \left(\|u - u_{\mathrm{obs}}\|^2 + \|v - v_{\mathrm{obs}}\|^2 + \gamma\|\phi - \phi_{\mathrm{obs}}\|^2\right) dt, \tag{14.6}$$

where γ is a weight function, then the directional derivatives $\bar{X} = (\bar{u}, \bar{v}, \bar{\phi})^{\mathrm{T}}$ in the direction $h = (h_u, h_v, h_\phi)^{\mathrm{T}}$ (in the control space) will be solutions of the linear tangent model

$$\frac{\partial \bar{u}}{\partial t} + u\frac{\partial \bar{u}}{\partial x} + \bar{u}\frac{\partial u}{\partial x} + v\frac{\partial \bar{u}}{\partial y} + \bar{v}\frac{\partial u}{\partial y} - f\bar{v} + \frac{\partial \bar{\phi}}{\partial x} = 0, \tag{14.7}$$

$$\frac{\partial \bar{v}}{\partial t} + u\frac{\partial \bar{v}}{\partial x} + \bar{u}\frac{\partial v}{\partial x} + v\frac{\partial \bar{v}}{\partial y} + \bar{v}\frac{\partial v}{\partial y} + f\bar{u} + \frac{\partial \bar{\phi}}{\partial y} = 0, \tag{14.8}$$

$$\frac{\partial \bar{\phi}}{\partial t} + \frac{\partial \bar{u}\phi}{\partial x} + \frac{\partial u\bar{\phi}}{\partial x} + \frac{\partial \bar{v}\phi}{\partial y} + \frac{\partial v\bar{\phi}}{\partial y} = 0. \tag{14.9}$$

Introducing three adjoint variables p, q, φ, we can compute the adjoint system, which reads

$$\frac{\partial p}{\partial t} + u\frac{\partial p}{\partial x} + \frac{\partial pv}{\partial y} - q\frac{\partial v}{\partial x} - fq + \frac{\partial \phi\varphi}{\partial x} = u_{\mathrm{obs}} - u, \tag{14.10}$$

$$\frac{\partial q}{\partial t} - p\frac{\partial u}{\partial y} + \frac{\partial qu}{\partial x} + v\frac{\partial q}{\partial y} + fp + \frac{\partial \phi\varphi}{\partial y} = v_{\text{obs}} - v, \tag{14.11}$$

$$\frac{\partial \varphi}{\partial t} + u\frac{\partial \varphi}{\partial x} + v\frac{\partial \varphi}{\partial y} + \frac{\partial p}{\partial x} + \frac{\partial q}{\partial y} = \gamma(\phi_{\text{obs}} - \phi). \tag{14.12}$$

14.3 Second-order methods

The optimality system, the Euler–Lagrange equation, provides only a necessary condition for optimality. In the linear case, the solution is unique if the Hessian is positive-definite. From a general point of view, the information given by the Hessian is important for theoretical, numerical, and practical issues. For operational models, it is impossible to compute the Hessian itself, since it is a square matrix with around 10^{18} terms. Nevertheless, the most important information can be extracted from the spectrum of the Hessian, which can be estimated without explicit determination of this matrix. This information is of importance for estimating the condition number of the Hessian for preparing efficient preconditioning.

A general method to obtain this information is to apply the techniques described above to the couple made by the direct and adjoint models (Le Dimet et al., 2002), leading to a so-called second-order adjoint. The following steps are carried out:

1. Linearization of the direct and adjoint models with respect to the state variable. Since the system is linear with respect to the adjoint variable, no linearization is necessary.
2. Introdution of two second-order adjoint variables.
3. Transposition to reveal the linear dependence with respect to the directions.

The system thus obtained, the second-order adjoint, is used to compute the product of the Hessian by any vector. Of course, if we consider all the vectors of the canonical base, then it will be possible to obtain the complete Hessian.

The determination of this product permits one to access some information:

- By using Lanczos-type methods and deflation, it is possible to compute the eigenvectors and eigenvalues of the Hessian.
- To carry out second-order calculations, optimization methods of Newton-type are used for equations of the form $\nabla G(X) = 0$. The iterations are $X_{n+1} = X_n - H^{-1}(X_n) \cdot \nabla G(X_n)$, where H is the Hessian of G. At each iteration, a linear system should be solved. This is done by carrying out some iterations of a conjugate gradient method, which requires computing the Hessian-vector product.

For the Saint-Venant equations, if the control variables are $Q = (\hat{u}, \hat{v}, \hat{\phi})^{\mathrm{T}}$ and, $R = (\bar{u}, \bar{v}, \bar{\phi})^{\mathrm{T}}$, the second-order system we obtain is

$$\begin{aligned} &\frac{\partial \hat{u}}{\partial t} + u\frac{\partial \hat{u}}{\partial x} + v\frac{\partial \hat{v}}{\partial y} + \hat{u}\frac{\partial v}{\partial y} - \hat{v}\frac{\partial v}{\partial y} - f\hat{v} + \phi\frac{\partial \hat{\phi}}{\partial x} \\ &= \tilde{v}\frac{\partial \bar{v}}{\partial x} - \bar{u}\frac{\partial \tilde{u}}{\partial x} - \bar{v}\frac{\partial \bar{u}}{\partial y} + \tilde{u}\frac{\partial \bar{v}}{\partial y} - \bar{\phi}\frac{\partial \tilde{\phi}}{\partial x} - \bar{u}, \end{aligned} \tag{14.13}$$

$$\frac{\partial \hat{v}}{\partial t} + \hat{u}\frac{\partial u}{\partial y} - u\frac{\partial \hat{v}}{\partial x} + \hat{v}\frac{\partial u}{\partial x} + v\frac{\partial \hat{v}}{\partial y} + f\hat{u} + \phi\frac{\partial \hat{\phi}}{\partial y}$$
$$= \tilde{u}\frac{\partial \bar{u}}{\partial x} - \bar{u}\frac{\partial \tilde{v}}{\partial x} - \tilde{v}\frac{\partial \bar{u}}{\partial y} + \bar{u}\frac{\partial \tilde{v}}{\partial y} - \bar{\phi}\frac{\partial \tilde{\phi}}{\partial y} - \bar{v}, \tag{14.14}$$

$$\frac{\partial \hat{\phi}}{\partial t} + \frac{\partial \hat{u}}{\partial x} + \frac{\partial \hat{v}}{\partial y} + u\frac{\partial \hat{\phi}}{\partial x} + v\frac{\partial \hat{\phi}}{\partial y} = -\bar{u}\frac{\partial \tilde{\phi}}{\partial x} - \bar{v}\frac{\partial \tilde{\phi}}{\partial x} - \gamma\bar{\phi}. \tag{14.15}$$

From the formal point of view, we see that first- and second-order systems differ by second-order terms that do not take into account the adjoint variable. The computation of second derivatives require storing both the trajectories of the direct and adjoint models. For very large models, it could be more economical to recompute these trajectories.

14.3.1 Sensitivity analysis

In the environmental sciences, the mathematical models contain parameters that cannot be estimated very precisely—either because they are difficult to measure or because they represent some subgrid phenomena. Therefore, it is important to be able to estimate the impact of uncertainties on the outputs of the model. Sensitivity analysis is defined by the following:

- X is the state vector of the model and K a vectorial parameter of the model $F(X, K) = 0$.
- $G(X, K)$ is the response function, a real-valued function
- By definition, the sensitivity of the model is the gradient of G with respect to K. The difficulty encountered comes from the implicit dependence of G on K through X, the solution of the model.

Several methods can be used to estimate the sensitivity:

- By finite differences. We obtain

$$\frac{\partial G}{\partial e_i} \simeq \frac{G\left(X(K + \alpha e_i), K + \alpha e_i\right) - G\left(X(K), K\right)}{\alpha}.$$

  The main inconvenience of this method is its computational cost: it requires solving the model as many times as the dimension of the model. Furthermore, the determination of the parameter α may be tricky. If it is too large, the variation of G could be nonlinear; for small values, round-off error may dominate the variation of G. The main advantage of this method is that it is very easy to implement.
- Sensitivity via an adjoint model. Let $F(X, K) = 0$ be the direct model. We introduce its adjoint:

$$\left[\frac{\partial F}{\partial X}\right]^* P = \frac{\partial G}{\partial X}.$$

Then the gradient is given by

$$\nabla G = \frac{\partial G}{\partial K} - \left[\frac{\partial F}{\partial K}\right]^* P.$$

The advantage of this method is that the sensitivity is obtained by only one run of the adjoint model. The price to be paid is the derivation of the adjoint code.

14.3.2 Sensitivity in the presence of data

In geophysics, a usual request is the estimation of the sensitivity with respect to observations. What will be the impact of an uncertainty on the prediction? It is clear that observations are not directly used in the direct model, they take place only as a forcing term in the adjoint model. Therefore, to apply the general formalism of sensitivity analysis, we should apply it not to the model itself but to the optimality system, i.e. the model plus the adjoint model. A very simple example with a scalar ordinary differential equation is given in Le Dimet et al. (2002), showing that the direct model is not sufficient to carry out sensitivity analysis in the presence of data. Deriving the optimality system will introduce second-order derivatives.

Consider a model of the form $F(X, I, K) = 0$, where I stands for the control variables and K for the input parameters of the model. Here F is a steady-state or time-dependent operator, $J(X, I)$ is the cost function in the variational data assimilation, and P is the adjoint variable. Then the optimality system is:

$$\begin{cases} F(X, I, K) = 0, \\ \left[\dfrac{\partial F}{\partial X}\right]^* \cdot P - \dfrac{\partial J}{\partial X} = 0, \\ \left[\dfrac{\partial F}{\partial I}\right]^* \cdot P - \dfrac{\partial J}{\partial I} = 0. \end{cases} \tag{14.16}$$

The optimality system can be considered as a generalized model $\mathcal{F}(Z) = 0$ with the state variable

$$Z = \begin{pmatrix} X \\ P \\ I \end{pmatrix}.$$

If $\hat{K}$ is a perturbation on K, we obtain

$$\frac{\partial F}{\partial X} \cdot \hat{X} + \frac{\partial F}{\partial I} \cdot \hat{I} + \frac{\partial F}{\partial K} \cdot \hat{K} = 0, \tag{14.17}$$

$$\left[\frac{\partial^2 F}{\partial X^2} \cdot \hat{X} + \frac{\partial^2 F}{\partial X \partial I} \cdot \hat{I} + \frac{\partial^2 F}{\partial X \partial K} \cdot \hat{K}\right]^* \cdot P + \left[\frac{\partial F}{\partial X}\right]^* \cdot \hat{P} - \frac{\partial^2 J}{\partial X^2} \cdot \hat{X} - \frac{\partial^2 J}{\partial X \partial I} \cdot \hat{I} = 0, \tag{14.18}$$

$$\left[\frac{\partial^2 F}{\partial I^2}\cdot\hat{I}+\frac{\partial^2 F}{\partial X\partial I}\cdot\hat{X}+\frac{\partial^2 F}{\partial K\partial I}\cdot\hat{K}\right]^*\cdot P+\left[\frac{\partial F}{\partial I}\right]^*\cdot\hat{P}-\frac{\partial^2 J}{\partial I^2}\cdot\hat{I}-\frac{\partial^2 J}{\partial I\partial X}\cdot\hat{X}=0, \tag{14.19}$$

$$\frac{dG}{dK}\cdot\hat{K}=\frac{\partial G}{\partial X}\cdot\hat{X}+\frac{\partial G}{\partial K}\cdot\hat{K}, \tag{14.20}$$

where $\hat{X}$, $\hat{P}$, and $\hat{I}$ are the Gâteaux derivatives of X, P, and I in the direction $\hat{K}$. Let us introduce the second-order adjoint variables Q, R, and L. By taking the inner product of (14.17) with Q, (14.18) with R, and (14.19) with L and adding, we obtain

$$\begin{aligned}
&\left\langle\hat{X},\left[\frac{\partial F}{\partial X}\right]^*\cdot Q+\left[\frac{\partial^2 F}{\partial X^2}\cdot R\right]^*\cdot P-\left[\frac{\partial^2 J}{\partial X^2}\right]^*\cdot R-\left[\frac{\partial^2 J}{\partial I\partial X}\right]^*\cdot L+\left[\frac{\partial^2 F}{\partial X\partial I}\cdot L\right]^*\cdot P\right\rangle\\
&+\left\langle\hat{P},\left[\frac{\partial F}{\partial X}\right]\cdot R+\left[\frac{\partial F}{\partial I}\right]\cdot L\right\rangle\\
&+\left\langle\hat{I},\left[\frac{\partial F}{\partial I}\right]^*\cdot Q+\left[\frac{\partial^2 F}{\partial X\partial I}\cdot R\right]^*\cdot P-\left[\frac{\partial^2 J}{\partial X\partial I}\right]^*\cdot R-\left[\frac{\partial^2 J}{\partial I^2}\right]^*\cdot L+\left[\frac{\partial^2 F}{\partial I^2}\cdot L\right]^*\cdot P\right\rangle\\
&+\left\langle\hat{K},\left[\frac{\partial F}{\partial K}\right]^*\cdot Q+\left[\frac{\partial^2 F}{\partial X\partial K}\cdot R\right]^*\cdot P+\left[\frac{\partial^2 F}{\partial I\partial K}\cdot L\right]^*\cdot P\right\rangle=0.
\end{aligned}$$

By identifying terms in (14.20), it can be seen that if Q, R, and L are defined as solutions of

$$\begin{cases}
\left[\frac{\partial F}{\partial X}\right]^*\cdot Q+\left[\frac{\partial^2 F}{\partial X^2}\cdot R\right]^*\cdot P-\left[\frac{\partial^2 J}{\partial X^2}\right]^*\cdot R-\left[\frac{\partial^2 J}{\partial I\partial X}\right]^*\cdot L+\left[\frac{\partial^2 F}{\partial X\partial I}\cdot L\right]^*\cdot P=\frac{\partial G}{\partial X},\\
\left[\frac{\partial F}{\partial X}\right]\cdot R+\left[\frac{\partial F}{\partial I}\right]\cdot L=0,\\
\left[\frac{\partial F}{\partial I}\right]^*\cdot Q+\left[\frac{\partial^2 F}{\partial X\partial I}\cdot R\right]^*\cdot P-\left[\frac{\partial^2 J}{\partial X\partial I}\right]^*\cdot R-\left[\frac{\partial^2 J}{\partial I^2}\right]^*\cdot L+\left[\frac{\partial^2 F}{\partial I^2}\cdot L\right]^*\cdot P,
\end{cases} \tag{14.21}$$

then we obtain the gradient of G with respect to K (the sensitivity) as:

$$\mathcal{S}=\frac{dG}{dK}=-\left[\frac{\partial F}{\partial K}\right]^*\cdot Q-\left[\frac{\partial^2 F}{\partial X\partial K}\cdot R\right]^*\cdot P-\left[\frac{\partial^2 F}{\partial I\partial K}\cdot L\right]^*\cdot P+\frac{\partial G}{\partial K}. \tag{14.22}$$

To summarize, the algorithm is as follows:

1. Solve the optimality system to find X and P.
2. Solve the system (14.21) to compute Q, R, and L.
3. Then the sensitivity is given by (14.22).

14.4 Sensitivity with respect to sources

14.4.1 Identification of the fields

Let us assume that the flow, described by the state variable X, satisfies between time 0 and time T the differential system

$$\begin{cases} \dfrac{\mathrm{d}X}{\mathrm{d}t} = F(X), \\ X(0) = U. \end{cases} \tag{14.23}$$

A pollutant, considered as a passive tracer, is described by its concentration, whose evolution is given by the equations

$$\begin{cases} \dfrac{\mathrm{d}C}{\mathrm{d}t} = G(X, C, S), \\ C(0) = V, \end{cases} \tag{14.24}$$

where C is the pollutant concentration and S, which is a function of space and time, is the production of pollutant.

The first task is to retrieve the fields from observations $X_{\text{obs}} \in \mathcal{X}_{\text{obs}}$ of the state variable X and $C_{\text{obs}} \in \mathcal{C}_{\text{obs}}$ of the concentration C of the pollutant. We introduce a cost function J defined by

$$J(U, V) = \frac{1}{2}\int_0^T \|EX - X_{\text{obs}}\|^2\,\mathrm{d}t + \frac{1}{2}\int_0^T \|DC - C_{\text{obs}}\|^2\,\mathrm{d}t + \frac{\alpha}{2}\|U - U_b\|^2 + \frac{\beta}{2}\|V - V_b\|^2, \tag{14.25}$$

where E is an operator from the space of the state variable to the space of observations, D is an operator from the space of concentration to the space of observations of concentration, U_b and V_b are the background values of U and V, and α and β are positive constants. The background terms play the role of regularization terms in the cost function. In practice, they are of crucial importance. For retrieving the state variable and the concentration, we have to determine the U^* and V^* that minimize J. They are solutions of the optimality system: if P and Q are defined as the solutions of

$$\begin{cases} \dfrac{dP}{dt} + \left[\dfrac{\partial F}{\partial X}\right]^* P + \left[\dfrac{\partial G}{\partial X}\right]^* Q = E^*(EX - X_{\text{obs}}), \\ P(T) = 0, \end{cases} \tag{14.26}$$

$$\begin{cases} \dfrac{dQ}{dt} + \left[\dfrac{\partial G}{\partial C}\right]^* Q = D^*(DC - C_{\text{obs}}), \\ Q(T) = 0, \end{cases} \tag{14.27}$$

then, from backward integration of the system, we obtain the gradient:

$$\nabla J_U = \alpha(U - U_b) - P(0) = 0, \tag{14.28}$$

$$\nabla J_V = \beta(V - V_b) - Q(0) = 0. \tag{14.29}$$

14.4.2 Evaluation of sensitivities with respect to sources

Let Ω_A, a subdomain of the physical space, be the region of interest (response region) and φ the function giving the measure of the effect of interest. By effect of interest, we mean the effect of the pollutant for which we want to evaluate the sensitivity with respect to the source. φ is a function of the concentration C and of the source S. We define the response function as

$$\Phi_A(C,S) = \int_0^{\mathrm{T}} \int_{\Omega_A} \varphi(C,S)\, dx. \tag{14.30}$$

In the simplest case, φ can be defined as $\varphi = C$, in which case Φ_A is the mean over space and time of the pollutant concentration. By definition, the sensitivity with respect to the source S is the gradient for the response function Φ_A with respect to S. Following the guidelines for the derivation of the gradient as presented in Section 14.3, let $\hat{U}$, $\hat{V}$, $\hat{X}$, $\hat{C}$, $\hat{P}$, and $\hat{Q}$ be the Gâteaux derivatives with respect to S (in the direction s) of the variables of the optimality system given by (14.23), (14.24), (14.26), and (14.27). The functions $\hat{U}$, $\hat{V}$, $\hat{X}$, $\hat{C}$, $\hat{P}$, and $\hat{Q}$ are given by the following systems of equations:

$$\begin{cases} \dfrac{d\hat{X}}{dt} = \dfrac{\partial F}{\partial X} \cdot \hat{X}, \\ \hat{X}(0) = \hat{U}, \end{cases} \tag{14.31}$$

$$\begin{cases} \dfrac{d\hat{C}}{dt} = \dfrac{\partial G}{\partial X} \cdot \hat{X} + \dfrac{\partial G}{\partial C} \cdot \hat{C} + \dfrac{\partial G}{\partial S} \cdot s, \\ \hat{C}(0) = \hat{V}, \end{cases} \tag{14.32}$$

$$\begin{cases} \frac{d\hat{P}}{dt} + \left[\frac{\partial F}{\partial X}\right]^* \cdot \hat{P} + \left[\frac{\partial^2 F}{\partial X^2}\hat{X}\right]^* \cdot P + \left[\frac{\partial G}{\partial X}\right]^* \cdot \hat{Q}, \\ \quad + \left[\frac{\partial^2 G}{\partial X^2} \cdot \hat{X} + \frac{\partial^2 G}{\partial C \partial X} \cdot \hat{C} + \frac{\partial^2 G}{\partial S \partial X} \cdot s\right]^* \cdot Q = E^* E \hat{X}, \\ \hat{P}(T) = 0, \end{cases} \tag{14.33}$$

$$\begin{cases} \dfrac{d\hat{Q}}{dt} + \left[\dfrac{\partial G}{\partial C}\right]^* \cdot \hat{Q} + \left[\dfrac{\partial^2 G}{\partial C^2} \cdot \hat{C} + \dfrac{\partial^2 G}{\partial X \partial C}\hat{X} + \dfrac{\partial^2 G}{\partial S \partial C} s\right]^* \cdot Q = D^* D \cdot \hat{C}, \\ \hat{Q}(T) = 0, \end{cases} \tag{14.34}$$

$$\alpha\hat{U} - \hat{P}(0) = 0, \tag{14.35}$$

$$\beta\hat{V} - \hat{Q}(0) = 0, \tag{14.36}$$

where $\hat{U}$ and $\hat{V}$ are the Gâteaux derivatives of the optimal initial conditions U and V, respectively. These terms are important because the information is transported thanks to the initial condition. For the function $\varphi(C, S)$, the Gâteaux derivative with respect to S is given by

$$\frac{\mathrm{d}\varphi}{\mathrm{d}S} \cdot s = \frac{\partial \varphi}{\partial S} \cdot s + \frac{\partial \varphi}{\partial C} \cdot \hat{C}. \tag{14.37}$$

To compute the gradient $\nabla_S \Phi_A(C, S)$, we introduce six second-order adjoint variables Γ, Λ, Φ, Ψ, R_1, and R_2; the system of equations (14.31) is multiplied by Γ, (14.32) by Λ, (14.33) by Φ, (14.34) by Ψ, (14.35) by R_1, and (14.36) by R_2, all the terms are added together and integrated by parts, and we obtain

$$\mathcal{Z} + \int_0^T \langle \hat{X}, \mathcal{A} \rangle \, dt + \int_0^T \langle \hat{C}, \mathcal{B} \rangle \, dt + \int_0^T \langle \hat{P}, \mathcal{L} \rangle \, dt + \int_0^T \langle \hat{Q}, \mathcal{W} \rangle \, dt + \int_0^T \langle s, \mathcal{K} \rangle \, dt = 0, \tag{14.38}$$

with

$$\begin{aligned} \mathcal{Z} = {} & \langle \Gamma(T), \hat{X}(T) \rangle - \langle \Gamma(0), \hat{X}(0) \rangle + \langle \Lambda(T), \hat{C}(T) \rangle - \langle \Lambda(0), \hat{C}(0) \rangle \\ & + \langle \Phi(T), \hat{P}(T) \rangle - \langle \Phi(0), \hat{P}(0) \rangle + \langle \Psi(T), \hat{Q}(T) \rangle - \langle \Psi(0), \hat{Q}(0) \rangle \\ & + \langle \alpha \hat{U} - \hat{P}(0), R_1 \rangle + \langle \beta \hat{V} - \hat{Q}(0), R_2 \rangle, \end{aligned} \tag{14.39}$$

$$\begin{aligned} \mathcal{A} = {} & -\frac{d\Gamma}{dt} - \left[\frac{\partial F}{\partial X}\right]^* \cdot \Gamma - \left[\frac{\partial G}{\partial X}\right]^* \cdot \Lambda + \left[\frac{\partial^2 F}{\partial X^2} \Phi\right]^* \cdot P \\ & + \left[\frac{\partial^2 G}{\partial X^2} \Phi\right]^* \cdot Q + \left[\frac{\partial^2 G}{\partial C \partial X} \Psi\right]^* \cdot Q - E^* E \Phi, \end{aligned} \tag{14.40}$$

$$\mathcal{B} = -\frac{d\Lambda}{dt} - \left[\frac{\partial G}{\partial C}\right]^* \cdot \Lambda + \left[\frac{\partial^2 G}{\partial C \partial X} \Phi\right]^* \cdot Q + \left[\frac{\partial^2 G}{\partial X^2} \Psi\right]^* \cdot Q - D^* D \Psi, \tag{14.41}$$

$$\mathcal{L} = -\frac{d\Phi}{dt} + \left[\frac{\partial F}{\partial X}\right] \cdot \Phi, \tag{14.42}$$

$$\mathcal{W} = -\frac{d\Psi}{dt} + \left[\frac{\partial G}{\partial C}\right] \cdot \Psi, \tag{14.43}$$

$$\mathcal{K} = -\left[\frac{\partial G}{\partial S}\right]^* \cdot \Lambda + \left[\frac{\partial^2 G}{\partial X^2} \Phi\right]^* \cdot Q + \left[\frac{\partial^2 G}{\partial C \partial S} \Psi\right]^* \cdot Q. \tag{14.44}$$

If the known values are taken into account in the expression of $\mathcal{Z}$, it becomes

$$\begin{aligned}\mathcal{Z} &= \langle \Gamma(T), \hat{X}(T)\rangle - \langle \Gamma(0), \hat{U}\rangle + \langle \Lambda(T), \hat{C}(T)\rangle - \langle \Lambda(0), \hat{V}\rangle - \langle \Phi(0), \hat{P}(0)\rangle \\ &\quad - \langle \Psi(0), \hat{Q}(0)\rangle + \langle \alpha\hat{U} - \hat{P}(0), R_1\rangle + \langle \beta\hat{V} - \hat{Q}(0), R_2\rangle \\ &= \langle \Gamma(T), \hat{X}(T)\rangle + \langle \alpha R_1 - \Gamma(0), \hat{U}\rangle + \langle \Lambda(T), \hat{C}(T)\rangle \\ &\quad + \langle \beta R_2 - \Lambda(0), \hat{V}\rangle - \langle \Phi(0) + R_1, \hat{P}(0)\rangle - \langle \Psi(0) + R_2, \hat{Q}(0)\rangle.\end{aligned} \tag{14.45}$$

To make $\mathcal{Z}$ equal zero, we put

$$\Gamma(T) = 0,\ \alpha R_1 - \Gamma(0) = 0,\ \Lambda(T) = 0,\ \beta R_2 - \Lambda(0) = 0,\ \Phi(0) + R_1 = 0,\ \Psi(0) + R_2 = 0.$$

Excluding $\alpha R_1 = \Gamma(0)$ and $\beta R_2 = \Lambda(0)$, we obtain

$$\alpha\Phi(0) + \Gamma(0) = 0, \quad \beta\Psi(0) + \Lambda(0) = 0.$$

If Γ, Λ, Φ, and Ψ are the solutions of the systems of equations

$$\begin{cases} -\dfrac{d\Gamma}{dt} - \left[\dfrac{\partial F}{\partial X}\right]^* \cdot \Gamma - \left[\dfrac{\partial G}{\partial X}\right]^* \cdot \Lambda + \left[\dfrac{\partial^2 F}{\partial X^2}\Phi\right]^* \cdot P \\ \qquad + \left[\dfrac{\partial^2 G}{\partial X^2}\Phi\right]^* \cdot Q + \left[\dfrac{\partial^2 G}{\partial C \partial X}\Psi\right]^* \cdot Q - E^* E\Phi = 0, \\ \qquad\qquad \Gamma(T) = 0, \end{cases} \tag{14.46}$$

$$\begin{cases} -\dfrac{d\Lambda}{dt} - \left[\dfrac{\partial G}{\partial C}\right]^* \cdot \Lambda + \left[\dfrac{\partial^2 G}{\partial C \partial X}\Phi\right]^* \cdot Q + \left[\dfrac{\partial^2 G}{\partial X^2}\Psi\right]^* \cdot Q - D^* D\Psi = \mathbf{1}_{\Omega_A}\dfrac{\partial \varphi}{\partial C}, \\ \qquad \Lambda(T) = 0, \end{cases} \tag{14.47}$$

$$\begin{cases} \dfrac{d\Phi}{dt} - \left[\dfrac{\partial F}{\partial X}\right] \cdot \Phi = 0, \\ \alpha\Phi(0) + \Gamma(0) = 0, \end{cases} \tag{14.48}$$

$$\begin{cases} \dfrac{d\Psi}{dt} - \left[\dfrac{\partial G}{\partial C}\right] \cdot \Psi = 0, \\ \beta\Psi(0) + \Lambda(0) = 0, \end{cases} \tag{14.49}$$

where $\mathbf{1}_{\Omega_A}$ is the characteristic function of the subdomain Ω_A, then it is found that

$$\nabla\Phi_A = \left[\frac{\partial G}{\partial S}\right]^* \cdot \Lambda - \left[\frac{\partial^2 G}{\partial X^2}\Phi\right]^* \cdot Q - \left[\frac{\partial^2 G}{\partial C \partial S}\Psi\right]^* \cdot Q + \mathbf{1}_{\Omega_A}\frac{\partial \varphi}{\partial S}. \tag{14.50}$$

We have obtained a coupled system of four differential equations (14.46)–(14.49) of first order with respect to time. Two equations have a final condition (backward problems) while the other two have an initial condition (forward problems) depending on the initial values for the first two equations: that is, we have a non-standard problem.

Solving the non-standard problem

For simplicity, we consider a system of two differential equations, the extension to four being straightforward. The method proposed is based on the theory of optimal control (Lions, 1968). We consider the generic system on the time interval $[0, T]$

$$\begin{cases} \dfrac{\mathrm{d}X}{\mathrm{d}t} = K(X, Y), \ t \in [0, T], \\ \\ \dfrac{\mathrm{d}Y}{\mathrm{d}t} = L(X, Y), \ t \in [0, T], \end{cases} \tag{14.51}$$

with

$$\begin{cases} X(T) = 0, \\ \gamma Y(0) + X(0) = 0, \end{cases} \tag{14.52}$$

and $\gamma = \text{const} \geq 0$. Let us transform this into a problem of optimal control. We consider the problem (14.51) with the conditions

$$\begin{cases} X(T) = 0, \\ Y(0) = U. \end{cases} \tag{14.53}$$

We assume that under these conditions, (14.51) has a unique solution for $t \in [0, T]$. Letting $X(0, U)$ be the value of X at time $t = 0$ for the value U of $Y(0)$, we define the cost function

$$J_P(U) = \frac{1}{2} \| \gamma U + X(0, U) \|^2. \tag{14.54}$$

The problem becomes the determination of U^* by minimizing J_P. We can expect that at the optimum, $\gamma Y(0) + X(0, U^*) = 0$ and the problem will be solved. The procedure is similar to that used in Section 14.3. The Gâteaux derivatives with respect to U in the direction u are given by

$$\frac{\mathrm{d}\hat{X}}{\mathrm{d}t} = \frac{\partial K}{\partial X} \cdot \hat{X} + \frac{\partial K}{\partial Y} \cdot \hat{Y}, \tag{14.55}$$

$$\frac{\mathrm{d}\hat{Y}}{\mathrm{d}t} = \frac{\partial L}{\partial X} \cdot \hat{X} + \frac{\partial L}{\partial Y} \cdot \hat{Y}, \tag{14.56}$$

$$\hat{X}(T) = 0, \tag{14.57}$$

$$\hat{Y}(0) = u, \tag{14.58}$$

$$\hat{J}_p(U) = \langle \gamma U + X(0), \gamma u + \hat{X}(0) \rangle. \tag{14.59}$$

Introducing the adjoint variables W and Z and proceeding to the integration by parts, we obtain

$$- \langle \hat{X}(0), W(0) \rangle + \langle \hat{Y}(T), Z(T) \rangle - \langle \hat{Y}(0), Z(0) \rangle$$
$$- \int_0^T \left\langle \hat{X}, \frac{dW}{dt} + \left[\frac{\partial K}{\partial X}\right]^* \cdot W + \left[\frac{\partial L}{\partial X}\right]^* \cdot Z \right\rangle dt$$
$$- \int_0^T \left\langle \hat{Y}, \frac{dZ}{dt} + \left[\frac{\partial K}{\partial Y}\right]^* \cdot W + \left[\frac{\partial L}{\partial Y}\right]^* \cdot Z \right\rangle dt = 0. \tag{14.60}$$

If Z and W are defined as the solution of

$$\frac{dW}{dt} + \left[\frac{\partial K}{\partial X}\right]^* \cdot W + \left[\frac{\partial L}{\partial X}\right]^* \cdot Z = 0, \tag{14.61}$$
$$\frac{dZ}{dt} + \left[\frac{\partial K}{\partial Y}\right]^* \cdot W + \left[\frac{\partial L}{\partial Y}\right]^* \cdot Z = 0, \tag{14.62}$$
$$Z(T) = 0, \; W(0) = \gamma U + X(0), \tag{14.63}$$

then we obtain

$$\nabla J_P(U) = \gamma^2 U + \gamma X(0) - Z(0). \tag{14.64}$$

A theoretical question remains regarding the existence and uniqueness of a solution. Some work in this direction is underway.

Extension and potential development

Without any theoretical difficulty, the development carried out above can be extended as follows:

- In the case of several sources of pollution, the source S becomes a vector $S = (S_1, S_2, \cdots, S_m)$, where S_i is the ith source and m is the total number of sources.
- In the case of several pollutants with kinetic chemistry, the same methodology can be applied, with a vector of concentration place of a single concentration: $C = (C_1, C_2, \cdots, C_m)$. Some numerical difficulties may arise if the characteristic times of the kinetics are very heterogeneous.

14.5 Stochastic methods

We can assume that the errors are independent and that their superposition will give a global Gaussian error. Therefore, observations and predictions are considered as Gaussian and we will try to estimate both the expectation and variances. The basics tools that are used are based on optimal estimation and the Kalman filter. A complete study can be found in Talagrand (2003).

The field $X^v \in \mathbb{R}^n$ and the observation $Z \in \mathbb{R}^m$ are linked by a relation of the form $Z = \Gamma X^v + \zeta$, where ζ is a Gaussian centred random variable with a given variance Σ. The observation Z being known, we seek an estimate X^c of X^v in the form $X^c = AZ$ (A being a linear operator mapping the state space into the observation space).

This operator is unbiased, i.e. $E\left(X^c - X^t\right) = 0$, and minimizes the variance of the observational error, which is the trace of the error covariance matrix:

$$P^c = E\left[\left(X^v - X^c\right)\left(X^v - X^c\right)^*\right].$$

It can be shown that the optimal A is given by

$$A = \left(\Gamma^{\mathrm{T}}\Sigma^{-1}\Gamma\right)^{-1}\Gamma^{\mathrm{T}}\Sigma^{-1},$$

and the associated covariance is given by

$$P^c = \left(\Gamma^{\mathrm{T}}\Sigma^{-1}\Gamma\right)^{-1}.$$

An equivalent variational formulation for determining X^c is minimization of the cost function J defined by

$$J\left(X\right) = \left[\Gamma X - Z\right]^{\mathrm{T}}\Sigma^{-1}\left[\Gamma X - Z\right].$$

14.6 Covariances of the optimal solution error

Consider the mathematical model of a physical process described by the evolution problem

$$\begin{cases} \dfrac{\partial\varphi}{\partial t} = F(\varphi), \quad t \in (0, T), \\ \varphi\big|_{t=0} = u, \end{cases} \tag{14.65}$$

where $\varphi = \varphi(t)$ is an unknown function belonging for any t to a Hilbert space X, $u \in X$, and F is a nonlinear operator mapping X into X. Let $Y = L_2(0, T; X)$ be a space of abstract functions $\varphi(t)$ with values in X, with the norm $\|\varphi\| = \left(\int_0^{\mathrm{T}} \|\varphi\|_X^2 \, \mathrm{d}t\right)^{1/2}$. Suppose that for a given $u \in X$ there exists a unique solution $\varphi \in Y$ to (14.65).

Let $\bar{u}$ be the 'exact' initial state and $\bar{\varphi}$ the solution to the problem (14.65) with $u = \bar{u}$, i.e. the 'exact' state evolution. We define the input data as follows: the background function $u_b \in X$, $u_b = \bar{u} + \xi_b$ and the observations $y \in Y_o$, $y = C\bar{\varphi} + \xi_o$, where $C : Y \to Y_o$ is a linear bounded operator (observation operator) and Y_o is a Hilbert space (observation space), $\xi_b \in X$, $\xi_o \in Y_o$. In particular, Y_o may be finite-dimensional (both in space and in time). The random variables ξ_b and ξ_o may be regarded as the background and the observation error, respectively. Assuming that these errors are normally distributed, unbiased, and mutually uncorrelated, we define the covariance operators $V_b\cdot = E[(\cdot, \xi_b)_X \, \xi_b]$ and $V_o\cdot = E[(\cdot, \xi_o)_{Y_o} \, \xi_o]$, where '$\cdot$' denotes an argument of the respective operator and E is the expectation. We suppose that V_b and V_o are positive-definite and, hence invertible.

Let us introduce a cost function

$$J(u) = \frac{1}{2}(V_b^{-1}(u - u_b), u - u_b)_X + \frac{1}{2}(V_o^{-1}(C\varphi - y), C\varphi - y)_{Y_o} \tag{14.66}$$

and formulate the following data assimilation problem (optimal control problem) with the aim to identify the initial condition: for given $f \in Y$ find $u \in X$ and $\varphi \in Y$ such that they satisfy (14.65) and the cost function $J(u)$ takes its minimum value. We further assume that the optimal solution error $\delta u = u - \bar{u}$ is unbiased, i.e. $E[\delta u] = 0$, with the covariance operator $V_{\delta u} \cdot = E[(\cdot, \delta u)_X \, \delta u]$.

Let us introduce the operator $R : X \to Y_o$ as follows:

$$Rv = C\psi, \quad v \in X, \tag{14.67}$$

where $\psi \in Y$ is the solution of the tangent-linear problem

$$\begin{cases} \dfrac{\partial \psi}{\partial t} - F'(\varphi)\psi = 0, \; t \in (0, T), \\ \psi|_{t=0} = v. \end{cases} \tag{14.68}$$

For a given v, we solve the problem (14.68) and then find Rv by (14.67). The definition of R involves $\varphi = \bar{\varphi} + \delta\varphi$ dependent on $u = \bar{u} + \delta u$ via (14.65); thus, we can write $R = R(\bar{u}, \delta u)$. It has been shown by Gejadze et al. (2008) that the optimal solution error $\delta u = u - \bar{u}$ and the data errors ξ_b and ξ_o are related via the exact operator equation

$$\left[V_b^{-1} + R^*(\bar{u}, \delta u)V_o^{-1}R(\bar{u}, \tau\delta u)\right] \delta u = V_b^{-1}\xi_b + R^*(\bar{u}, \delta u)V_o^{-1}\xi_o, \tag{14.69}$$

where R^* is the adjoint of R and $\tau \in [0, 1]$ is a parameter chosen to make the truncated Taylor series exact.

Let $H(\bar{u}) = V_b^{-1} + R^*(\bar{u}, 0)V_o^{-1}R(\bar{u}, 0)$ be the Hessian of the linearized (auxiliary) control problem (Gejadze et al., 2008). Under the hypothesis that F is twice continuously Fréchet-differentiable, the error equation (14.69) is approximated by

$$H(\bar{u})\delta u = V_b^{-1}\xi_b + R^*(\bar{u}, 0)V_o^{-1}\xi_o. \tag{14.70}$$

It is easy to see from (14.70) that

$$V_{\delta u} = [H(\bar{u})]^{-1}. \tag{14.71}$$

This is a well-established result (Courtier et al., 1994; Rabier and Courtier, 1992; Thacker, 1989), which is usually deduced (without considering (14.69)) by straightforwardly linearizing the original nonlinear data assimilation problem (14.65), (14.66) under the assumption that

$$F(\varphi) - F(\bar{\varphi}) \approx F'(\bar{\varphi})\delta\varphi, \tag{14.72}$$

which is called the tangent-linear hypothesis (TLH). It is said that $V_{\delta u}$ can be approximated by $[H(\bar{u})]^{-1}$ if the TLH (14.72) is valid. That usually happens if the nonlinearity is mild and/or the error δu and, consequently, $\delta\varphi$ are small. We derive (14.71) via (14.69). From this derivation, one can see that the accuracy of (14.71) depends on the accuracy of the approximations $R(\bar{u}, \tau\delta u) \approx R(\bar{u}, 0)$ and $R^*(\bar{u}, \delta u) \approx R^*(\bar{u}, 0)$ in

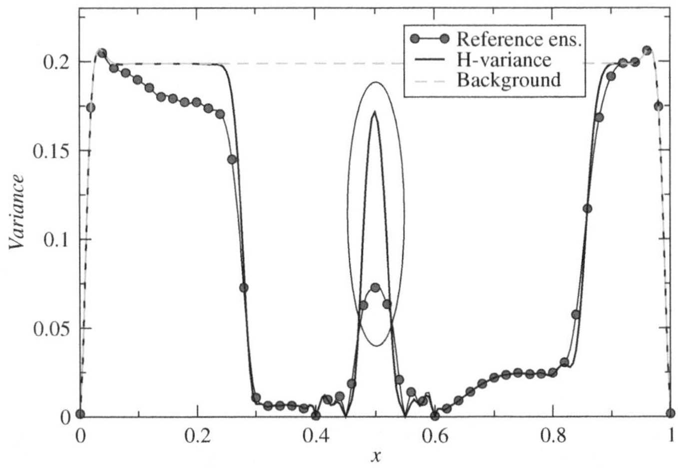

Fig. 14.1 Reference variance, variance by the inverse Hessian, and background variance.

(14.69). Clearly the transition from (14.69) to (14.70) could still be valid even though (14.72) is not satisfied.

As already mentioned, we can use (14.71) if the TLH is valid and, in some cases, beyond the range of its validity. In the general case, however, one *may not* expect that $H^{-1}(\bar{u})$ is always a satisfactory approximation to $V_{\delta u}$. In Fig. 14.1, we present a specially designed example for the evolution model governed by the one-dimensional Burgers equation (for details, see Section 14.8). The difference between the reference value of the variance (circles) and the inverse Hessian-based value (bold solid line) can be clearly seen within the ellipse. The reference variance is obtained by direct Monte Carlo simulation.

Since $R^*(\bar{u}, 0)$ and $H(\bar{u})$ in (14.70) are linear operators and we assume that the errors ξ_b and ξ_o are unbiased and normally distributed, $\delta u \sim \mathbf{N}(0, V_{\delta u})$. Clearly, this result is valid insofar as the TLH and consequently (14.70) itself are satisfied. However, for highly nonlinear dynamical models, the TLH often breaks down (see e.g. Pires et al., 1996). Thus, we have to answer the following question: Can the pdf of δu still be approximated by the normal distribution? If the answer is positive, one should look for a better approximation of the covariance than that given by (14.71).

Let us consider the cost function (14.66), but without the background term. The corresponding error equation is then

$$R^*(\bar{u}, \delta u)V_o^{-1}R(\bar{u}, \tau\delta u)\delta u = R^*(\bar{u}, \delta u)V_o^{-1}\xi_o. \tag{14.73}$$

For a univariate case, the classical result (see Jennrich, 1969) is that δu is asymptotically normal if ξ_o is an independent identically distributed (i.i.d.) random variable with $E[\xi_o] = 0$ and $E[\xi_o^2] = \sigma^2 < \infty$ ('asymptotically' have means that $T \to \infty$ given the finite observation time step dt, or $dt \to 0$ given the finite observation window $[0, T]$). Let us stress that for the asymptotic normality of δu, the error ξ_o is not required to be normal. This original result has been generalized to the multivariate case and to the case of dependent yet identically distributed observations (White and Domowitz, 1984), whereas even more general case is considered in Yuan and Jennrich (1998). Here we consider the complete cost function (14.66) and, correspondingly, the error equation (14.69), which contains terms related to the background term. To analyse a possible impact of these terms, let us follow the reasoning in Amemiya (1983, pp. 337–345), where the error equation equivalent to (14.73) is derived in a slightly different form. It is concluded that the error δu is asymptotically normal when (a) the right-hand side of the error equation is normal and (b) the left-hand side matrix converges in probability to a non-random value. These conditions are met under certain general regularity requirements to the operator R, which are incomparably weaker than the TLH and do not depend on the magnitude of the input errors. Clearly, as applied to (14.69), the first condition holds if ξ_b is normally distributed. Since V_b^{-1} is a constant matrix, the second condition always holds as long as it holds for $R^*(\bar{u}, \delta u)V_o^{-1}R(\bar{u}, \tau\delta u)$. Therefore, one may conclude that δu from (14.69) is bound to remain asymptotically normal. In practice, the observation window $[0, T]$ and time step dt are always finite, implying a finite number of i.i.d. observations. Moreover, it is not easy to assess how large the number of observations must be for the desired asymptotic properties to be reasonably approximated. Some nonlinear least-squares problems in which the normality of the estimation error holds for 'practically relevant' sample sizes are said to exhibit a 'close-to-linear' statistical behavior (Ratkowsky, 1983). The method suggested in Ratkowsky (1983) to verify this behavior is, essentially, a normality test applied to a generated sample of optimal solutions, which is hardly feasible for large-scale applications. Nevertheless, for certain highly nonlinear evolution models, it is reasonable to expect that the distribution of δu might be reasonably close to normal if the number of i.i.d. observations is significant in time and the observation network is sufficiently dense in space. This may happen in assimilation of long time series of satellite observations of ocean surface elevation and temperature, for example.

14.7 Effective inverse Hessian (EIH) method

14.7.1 General consideration

Here we present a new method for estimating the covariance $V_{\delta u}$ to be used in the case of highly nonlinear dynamics, when $[H(\bar{u})]^{-1}$ is not expected to be a good approximation of $V_{\delta u}$. Let us consider the discretized nonlinear error equation (14.69) and denote by $\mathbf{H}$ the left-hand-side operator in (14.69). Then we can write down the expression for δu:

$$\delta u = \mathbf{H}^{-1}\left[V_b^{-1}\xi_b + R^*(\bar{u}, \delta u)V_o^{-1}\xi_o\right],$$

whereas for the covariance $V_{\delta u}$ we obtain

$$V_{\delta u} := E\left[\delta u \delta u^{\mathrm{T}}\right] = E\left[\mathbf{H}^{-1}V_b^{-1}\,\xi_b\xi_b^{\mathrm{T}}\,V_b^{-1}\mathbf{H}^{-1*}\right]$$

$$+\,E\left[\mathbf{H}^{-1}R^*(\bar{u}, \delta u)V_o^{-1}\,\xi_o\xi_o^{\mathrm{T}}\,V_o^{-1}R(\bar{u}, \delta u)\mathbf{H}^{-1*}\right]. \qquad (14.74)$$

As a result of a series of simplifications described in Gejadze et al. (2011), the above equation can be reduced to the form

$$V_{\delta u} \approx V = E\left[[H(\bar{u} + \delta u)]^{-1}\right], \tag{14.75}$$

where $H(\bar{u} + \delta u) = V_b^{-1} + R^*(\bar{u}, \delta u)V_o^{-1}R(\bar{u}, \delta u)$ is the Hessian of the linearized (auxiliary) control problem. The right-hand side of (14.75) may be called the effective inverse Hessian (EIH)—hence the name of the suggested method. In order to compute V directly using this equation, the expectation is substituted by the sample mean:

$$V = \frac{1}{L}\sum_{l=1}^{L}[H(\bar{u} + \delta u_l)]^{-1}. \tag{14.76}$$

The main difficulty with the implementation is a need of computing a sample of optimal solutions $u_l = \bar{u} + \delta u_l$. However, (14.75) does not necessarily require u_l to be an optimal solution. If we denote by $q_{\delta u}$ the pdf of δu, then (14.75) can be rewritten in the form

$$V = \int_{-\infty}^{+\infty} [H(\bar{u} + v)]^{-1} q_{\delta u}(v)\, dv. \tag{14.77}$$

If we assume that in our nonlinear case the covariance matrix V describes meaningfully the pdf of the optimal solution error, then, with the same level of validity, we should also accept the pdf $q_{\delta u}$ as being approximately normal with zero expectation and covariance V, in which case we obtain

$$V = c\int_{-\infty}^{+\infty} [H(\bar{u} + v)]^{-1} \exp\left(-\frac{1}{2}v^{\mathrm{T}}V^{-1}v\right) dv, \tag{14.78}$$

where $c^{-1} = (2\pi)^{M/2}|V|^{1/2}$. The formula (14.76) gives V explicitly, but requires a sample of optimal solutions u_l, $l = 1, \ldots, L$, to be computed. In contrast, the latest expression is a nonlinear matrix integral equation with respect to V, while v is a dummy variable. This equation is actually solved using the iterative process (14.83), as explained in Section 14.7.2. It is also interesting to note that (14.78) is a deterministic equation.

14.7.2 Implementation remarks

Remark 1 Preconditioning is used in variational data assimilation to accelerate the convergence of the conjugate gradient algorithm at the stage of inner iterations of the Gauss–Newton (GN) method, but it also can be used to accelerate formation of the inverse Hessian by the Lanczos algorithm (Fisher et al., 2009) or the BFGS (Gejadze et al., 2010). Since H is self-adjoint, we must consider a projected Hessian in a symmetric form:

$$\tilde{H} = (B^{-1})^* H B^{-1},$$

with some operator $B : X \to X$, defined in such a way that the eigenspectrum of the projected Hessian $\tilde{H}$ is clustered around 1, i.e. the majority of the eigenvalues of $\tilde{H}$ are equal or close to 1. Since the condition number of $\tilde{H}$ is supposed to be much smaller than the condition number of H, a sensible approximation of $\tilde{H}^{-1}$ can usually be obtained (either by Lanczos or BFGS) with a relatively small number of iterations. After that, having $\tilde{H}^{-1}$, one can easily recover H^{-1} using the formula

$$H^{-1} = B^{-1}\tilde{H}^{-1}(B^{-1})^*. \tag{14.79}$$

Assuming that B^{-1} does not depend on δu_l we substitute (14.79) into (14.76) and obtain the version of (14.76) with preconditioning:

$$V = B^{-1}\left(\frac{1}{L}\sum_{l=1}^{L}[\tilde{H}(\bar{u} + \delta u_l)]^{-1}\right)(B^{-1})^*. \tag{14.80}$$

Similarly, assuming that B^{-1} does not depend on the variable of integration, we substitute (14.79) into (14.78) and obtain the version of (14.78) with preconditioning:

$$\begin{aligned} V &= B^{-1}\tilde{V}(B^{-1})^*, \\ \tilde{V} &= c\int_{-\infty}^{+\infty}[\tilde{H}(\bar{u} + v)]^{-1}\exp\left(-\frac{1}{2}v^{\mathrm{T}}V^{-1}v\right)dv. \end{aligned} \tag{14.81}$$

Instead of H^{-1} (14.80) and (14.81) involve $\tilde{H}^{-1}$, which is much less expensive to compute and store in memory. Let us mention here that the suggested method would hardly be feasible for large-scale problems without appropriate preconditioning.

Remark 2 The nonlinear equation (14.81) can be solved, for example, by the fixed-point iterative process as follows:

$$\begin{aligned} V^{p+1} &= B^{-1}\tilde{V}(B^{-1})^*, \\ \tilde{V} &= c^p\int_{-\infty}^{+\infty}[\tilde{H}(\bar{u} + v)]^{-1}\exp\left(-\frac{1}{2}v^{\mathrm{T}}(V^p)^{-1}v\right)dv, \end{aligned} \tag{14.82}$$

for $p = 0, 1, \ldots,$ starting with $V^0 = [H(\bar{u})]^{-1}$. Iterative processes of this type are expected to converge if V^0 is a good initial approximation of V, which is the case in the considered examples. The convergence of (14.82) and other methods for solving (14.81) are subjects for future research.

Remark 3 Different methods can be used for evaluation of the multidimensional integral in (14.82), such as quasi-Monte Carlo (Neiderreiter, 1992). Here, for simplicity, we use the standard Monte Carlo method. This actually implies a return to the formula (14.80). Taking (14.79) into account, the iterative process takes the form

$$V^{p+1} = B^{-1}\left(\frac{1}{L}\sum_{l=1}^{L}[\tilde{H}(\bar{u} + \delta u_l^p)]^{-1}\right)(B^{-1})^*, \tag{14.83}$$

where $\delta u_l^p \sim \mathbf{N}(0, V^p)$. For each l, we compute δu_l^p as

$$\delta u_l^p = (V^p)^{1/2} \xi_l,$$

where $\xi \sim \mathbf{N}(0, I)$ is an independent random series, I is the identity matrix, and $(V^p)^{1/2}$ is a square root of V^p. One can see that for each p, the last formula looks similar to (14.80), with one key difference: δu_l^p in (14.83) is not an optimal solution, but a vector having the statistical properties of the optimal solution.

Remark 4 Let us notice that a few tens of outer iterations by the GN method may be required to obtain one optimal solution, while an approximate evaluation of $\tilde{H}^{-1}$ is equivalent (in terms of computational costs) to just one outer iteration of the GN method. One has to repeat these computations p times; however, only a few iterations on the index p are required in practice. Therefore, one should expect an order-of-magnitude reduction in computational costs when using the method (14.83) compared with (14.80) for the same sample size. Clearly, for realistic large-scale models, the sample size L is going to be limited. Probably, the minimum ensemble size for this method to work is $2L^* + 1$, where L^* is the accepted number of leading eigenvectors of V^p in (14.83).

Remark 5 In order to implement the process (14.83), a sample of vectors $\varphi_l(x, 0) = \delta u_l^p$ must be propagated from $t = 0$ to $t = T$ using the nonlinear model (14.65). Therefore, for each p, one obtains a sample of final states $\varphi_l(x, T)$ consistent with the current approximation of V^p, which can be used to evaluate the forecast and forecast covariance. Since V^p is a better approximation of the analysis error covariance than simply $[H(\bar{u})]^{-1}$, one should expect a better quality of the forecast and covariance (as being consistent with V^p, rather than with $[H(\bar{u})]^{-1}$).

14.8 Numerical examples

14.8.1 Numerical model

As a model, we use the one-dimensional Burgers equation with a nonlinear viscous term,

$$\frac{\partial \varphi}{\partial t} + \frac{1}{2}\frac{\partial(\varphi^2)}{\partial x} = \frac{\partial}{\partial x}\left[\mu(\varphi)\frac{\partial \varphi}{\partial x}\right], \tag{14.84}$$

with $\varphi = \varphi(x, t)$, $t \in (0, T)$, $x \in (0, 1)$, and the Neumann boundary conditions

$$\left.\frac{\partial \varphi}{\partial x}\right|_{x=0} = \left.\frac{\partial \varphi}{\partial x}\right|_{x=1} = 0 \tag{14.85}$$

and viscosity coefficient

$$\mu(\varphi) = \mu_0 + \mu_1 \left(\frac{\partial \varphi}{\partial x}\right)^2, \quad \mu_0, \mu_1 = \text{const} > 0. \tag{14.86}$$

The nonlinear diffusion term with $\mu(\varphi)$ dependent on $\partial\varphi/\partial x$ is introduced to mimic the eddy viscosity (turbulence), which depends on the field gradients (of pressure and temperature), rather than on the field value itself. This type of $\mu(\varphi)$ also allows us to formally qualify the problem (14.84)–(14.86) as strongly nonlinear (Fučik and Kufner, 1980). Let us mention that the Burgers equations are sometimes considered in the data assimilation context as a simple model describing atmospheric flow motion.

We use the implicit time discretization as follows:

$$\frac{\varphi^i - \varphi^{i-1}}{h_t} + \frac{\partial}{\partial x}\left[\frac{1}{2}w(\varphi^i)\varphi^i - \mu(\varphi^i)\frac{\partial \varphi^i}{\partial x}\right] = 0, \quad i = 1, \ldots, N, \ x \in (0,1), \tag{14.87}$$

where i is the time integration index, and $h_t = T/N$ is the time step. The spatial operator is discretized on a uniform grid (h_x is the spatial discretization step, $j = 1, \ldots, M$ is the node number, and M is the total number of grid nodes) using the 'power-law' first-order scheme as described in Patankar (1980), which yields quite a stable discretization scheme (this scheme allows $\mu(\varphi)$ down to 0.5×10^{-4} for $M = 200$ without noticeable oscillations). For each time step, we perform nonlinear iterations on the coefficients $w(\varphi) = \varphi$ and $\mu(\varphi)$, assuming initially that $\mu(\varphi^i) = \mu(\varphi^{i-1})$ and $w(\varphi^i) = \varphi^{i-1}$, and keep iterating until (14.87) is satisfied (i.e. the norm of the left-hand side in (14.87) becomes smaller than the threshold $\epsilon_1 = 10^{-12}\sqrt{M}$). In all the computations presented in this chapter, we use the following parameters: the observation period $T = 0.312$, the discretization steps are $h_t = 0.004$ and $h_x = 0.005$, the state vector dimension $M = 200$, and the parameters in (14.86) are $\mu_0 = 10^{-4}$ and $\mu_1 = 10^{-6}$.

A general property of the Burgers solutions is that a smooth initial state evolves into a state characterized by the areas of severe gradients (or even shocks in the inviscid case). These are precisely the areas of a strong nonlinearity where one might expect violations of the TLH and, consequently, the invalidity of (14.71). For numerical experiments, we choose a certain initial condition that stimulates as highly a nonlinear dynamics as possible. It is given by the formula that

$$\bar{u}(x) = \varphi(x,0) = \begin{cases} 0.5 - 0.5\cos(8\pi x), & 0 \le x \le 0.4, \\ 0, & 0.4 < x \le 0.6, \\ 0.5\cos(4\pi x) - 0.5, & 0.6 < x \le 1. \end{cases}$$

The resulting field evolution $\varphi(x,t)$ is presented in Fig. 14.2.

14.8.2 BFGS for computing the inverse Hessian and other details

The projected inverse Hessian $\tilde{H}(\bar{u} + \delta u)$ is computed as a collateral result of the BFGS iterations while solving the following auxiliary data assimilation problem:

$$\begin{cases} \dfrac{\partial \delta\varphi}{\partial t} - F'(\varphi)\delta\varphi = 0, \quad t \in (0,T), \\ \delta\varphi\big|_{t=0} = B^{-1}\delta u, \\ J_1(\delta u) = \inf\limits_{v} J_1(v), \end{cases} \tag{14.88}$$

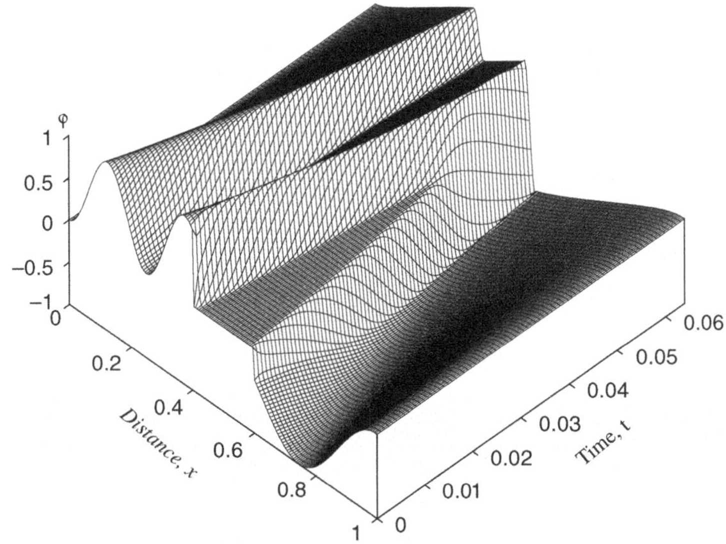

Fig. 14.2 Field evolution.

where

$$J_1(\delta u) = \frac{1}{2}(V_b^{-1}B^{-1}(\delta u - \xi_b), B^{-1}(\delta u - \xi_b))_X + \frac{1}{2}(V_o^{-1}(C\delta\varphi - \xi_o), C\delta\varphi - \xi_o)_{Y_o}. \tag{14.89}$$

The preconditioner used in our method is

$$B^{-1} = V_b^{1/2}[\tilde{H}(\bar{u})]^{-1/2}. \tag{14.90}$$

In order to compute $[\tilde{H}(\bar{u})]^{-1/2}$, we apply the Cholesky factorization of the explicitly formed matrix $\tilde{H}^{-1}$. However, it is important to note that the square-root-vector product $\tilde{H}^{-1/2}w$ can be computed using a recursive procedure based on the accumulated secant pairs (BFGS) or eigenvalues/eigenvectors (Lanczos) as described in Tshimanga et al. (2008), without the need to form $\tilde{H}^{-1}$ or to factorize it. The consistent tangent-linear and adjoint models have been generated from the original forward model by the automatic differentiation tool TAPENADE (Hascoët and Pascual, 2004) and checked using the standard gradient test. The background error covariance V_b is computed assuming that the background error belongs to the Sobolev space $W_2^2[0,1]$ (for details, see Gejadze et al., 2010). The correlation function used in the numerical examples is presented in Fig. 14.3, the background error variance is $\sigma_b^2 = 0.2$, and the observation error variance is $\sigma_o^2 = 10^{-3}$. The observation scheme consists of four sensors located at the points $\hat{x}_k = 0.4,\ 0.45,\ 0.55,\ 0.6$.

14.8.3 Numerical results

First, we compute a large sample ($L = 2500$) of optimal solutions u_l by solving L times the data assimilation problem (14.65), (14.66) with perturbed data $u_b = \bar{u} + \xi_b$

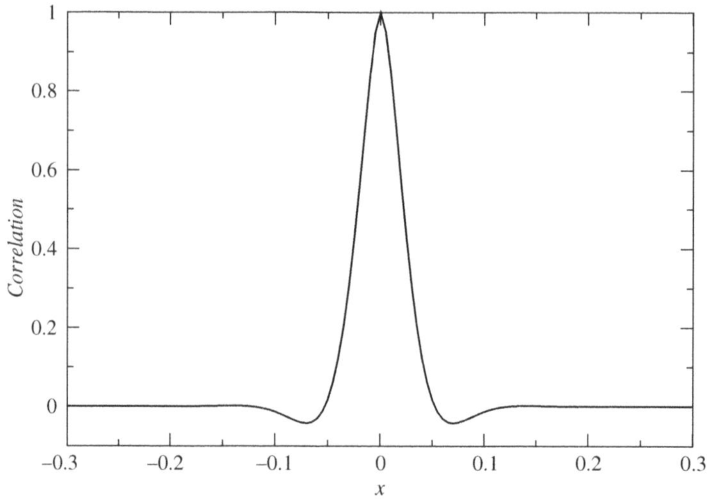

Fig. 14.3 Correlation function.

and $y = C\bar{\varphi} + \xi_o$, where $\xi_b \sim \mathbf{N}(0, V_b)$ and $\xi_o \sim \mathbf{N}\left(0, \sigma_o^2 I\right)$. This large sample is used to evaluate the sample covariance matrix, which is further processed to filter the sampling error (as described in Gejadze et al., 2011); the outcome is considered as a reference value $\hat{V}^\circ$. The original large sample is then partitioned into 100 subsets including $L = 25$ members and into 25 subsets including $L = 100$ members. Let us denote by $\hat{V}_L$ the sample covariance matrix obtained for a subset including L members. Then the relative error in the sample variance (which is the relative sampling error) can be defined as the vector $\hat{\varepsilon}_L$ with components

$$(\hat{\varepsilon}_L)_i = (\hat{V}_L)_{i,i}/\hat{V}^\circ_{i,i} - 1, \qquad i = 1, \ldots, M.$$

The relative error in a given approximation of V is defined as a vector ε with components

$$\varepsilon_i = V_{i,i}/\hat{V}^\circ_{i,i} - 1, \qquad i = 1, \ldots, M. \tag{14.91}$$

We compute this error with V in (14.91) being estimated by one of the following methods:

(1) the inverse Hessian method, i.e. simply using $V_{\delta u} = [H(\bar{u})]^{-1}$;
(2a) the EIH method implemented in the form (14.80), which requires a sample of optimal solutions δu_l to be computed;
(2b) the EIH method implemented as the iterative process (14.83), which requires a sample of δu_l, but does not require that δu_l are optimal solutions.

For the computation of V by methods (2a) or (2b), a sample of δu_l is required, and hence the result depends on the sample size L. The results obtained by methods (2a) and (2b) presented in this chapter are computed with $L = 100$. In method (2b), we

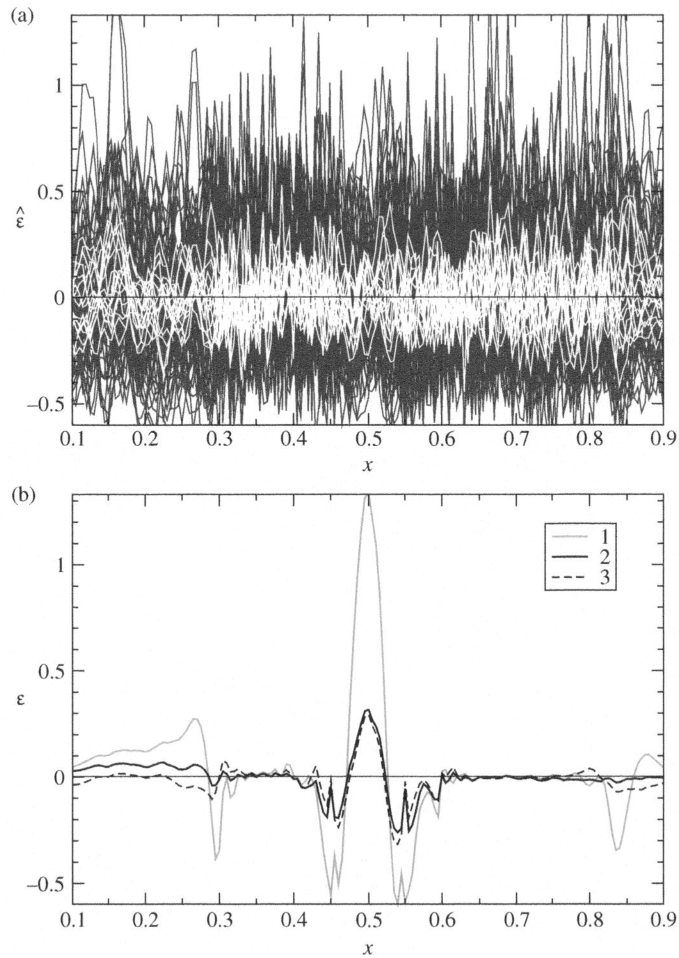

Fig. 14.4 (a) Sample relative error $\hat{\varepsilon}$. Set of $\hat{\varepsilon}$ for $L = 25$, (dark envelope) and set of $\hat{\varepsilon}$ for $L = 100$ (white envelope). (b) Relative error ε by the inverse Hessian (line 1), and by the EIH methods with $L = 100$: method 2a (line 2) and method 2b (line 3).

currently allow enough iterations on the index p for the iterative process (14.83) to converge in terms of the distance between the successive iterates. In practice, this requires just a few iterations, typically 2 or 3.

In Fig. 14.4(a), a set of 100 vectors $\hat{\varepsilon}_{25}$ is presented in dark lines and a set of 25 vectors $\hat{\varepsilon}_{100}$ in the overlaying white lines. These plots reveal the envelopes for the relative error in the sample variance obtained with $L = 25$ and $L = 100$, respectively. The graphs of ε are presented in the Fig. 14.1(b): line 1 corresponds to the method (1) (the inverse Hessian method, see also Fig. 14.1) and lines 2 and 3 to methods (2a) and (2b) (variants of the EIH method).

Looking at Fig. 14.4, we observe that the relative error in the sample variance $\hat{\varepsilon}_{25}$ (dark envelope) exceeds 50% almost everywhere, which is certainly beyond reasonable margins, and $\hat{\varepsilon}_{100}$ (white envelope) is around 25% (which is still fairly large). In order to reduce the white envelope two times, one would need to use the sample size $L = 400$, etc. One should also keep in mind that the envelope for the variance (diagonal of the covariance matrix) is the narrowest compared with its subdiagonals. Thus, developing methods for the estimation of the covariance other than the sampling methods is clearly of value.

Whereas method (1) (the inverse Hessian method) gives an estimate of $V_{\delta u}$ with a small relative error (compared with the sample covariance) in the areas with mild nonlinearity, this error can be much larger in the areas with high nonlinearity. For example, if we imagine that Fig. 14.4(b) is superposed on Fig. 14.4(a), then one could observe line 1 jumping outside the dark envelope in the area surrounding $x = 0.5$; i.e. the relative error by the inverse Hessian is significantly larger here than the sampling error for $L = 25$. At the same time, the relative error obtained by methods (2a) and (2b) is much smaller than the error in line 1, and it would largely remain within the white envelope. The difference between the estimates by methods (2a) and (2b) does not look significant. The best improvement can be achieved for the diagonal elements of $V_{\delta u}$ (the variance). Thus, the covariance estimate obtained by the EIH method is noticeably better than the sample covariance obtained with the equivalent sample size. The suggested algorithm is computationally efficient (in terms of CPU time) if the cost of computing the inverse Hessian is much less than the cost of computing one optimal solution. In the example presented here, one limited-memory inverse Hessian is about 20–30 times less expensive than one optimal solution. Thus, on average, the algorithm (2b) works about 10 times faster than the algorithm (2a), whereas the results of both algorithms are similar in terms of accuracy.

14.9 Conclusions

Error propagation is a key aspect of the modelling large-scale geophysical flows, with the main difficulty being linked to the nonlinearity of the governing equations. In this chapter, we have considered the hind-cast (initialization) data assimilation problem. From the mathematical point of view, this is the initial-value control problem for a nonlinear evolution model governed by partial differential equations. Assuming that the TLH holds, the covariance is often approximated by the inverse Hessian of the objective function. In practice, the same approximation could be valid even though the TLH is clearly violated. However, here we deal with such a highly nonlinear dynamics that the inverse Hessian approach is no longer valid. In this case, a new method for computing the covariance matrix the EIH method, can be used. This method yields a significant improvement of the covariance estimate compared with the inverse Hessian method. The method is potentially feasible for large-scale applications because it can be used in a multiprocessor environment and operates in terms of Hessian-vector products. The software blocks needed for its implementation are the standard blocks of any existing 4D-Var system. All the results deserted in this chapter are consistent with the assumption on a close-to-normal nature of the optimal solution error. This should be expected, taking into account the consistency and asymptotic normality of the estimator and the fact that the observation window in variational data assimilation is usually quite large. In this case, the covariance matrix is a meaningful representative of the pdf. The suggested method may become a valuable option for uncertainty analysis in the framework of the classical 4D-VAR approach when applied to highly nonlinear DA problems.

Acknowledgements

The authors acknowledge support from Région Rhône-Alpes, the INRIA MOISE Project, the Russian Foundation for Basic Research (Project 12-01-00322), and the Russian Federal Research Program Kadry. I. G. acknowledges funding through the Natural Environment Research Council (NERC Grant NE/J018201/1).

References

Amemiya, T. (1983). *Handbook of Econometrics*, Vol. 1. North-Holland, Amsterdam.

Courtier, P., Thépaut, J. N., and Hollingsworth, A. (1994). A strategy for operational implementation of 4D-Var, using an incremental approach. *Q. J. R. Meteorol. Soc.*, **120**, 1367–1388.

Draper, N. R. and Smith, H. (1981). *Applied Regression Analysis*, 2nd edn. Wiley, New York.

Fisher, M., Nocedal, J., Trémolet, Y., and Wright, S. J. (2009). Data assimilation in weather forecasting: a case study in PDE-constrained optimization. *Optim. Eng.*, **10**, 409–426.

Fučik, S. and Kufner, A. (1980). *Nonlinear Differential Equations.* Elsevier, Amsterdam.

Gejadze, I., Le Dimet, F.-X., and Shutyaev, V. (2008). On analysis error covariances in variational data assimilation. *SIAM J. Sci. Comput.*, **30**, 1847–1874.

Gejadze, I., Le Dimet, F.-X. and Shutyaev, V. (2010). On optimal solution error covariances in variational data assimilation problems. *J. Comput. Phys.*, **229**, 2159–2178.

Gejadze, I. Y., Copeland, G. J. M., Le Dimet, F.-X., and Shutyaev, V. (2011). Computation of the analysis error covariance in variational data assimilation problems with nonlinear dynamics. *J. Comput. Phys.*, **230**, 7923–7943.

Hascoët, L. and Pascual, V. (2004). TAPENADE 2.1 User's Guide. INRIA Technical Report 0300.

Jennrich, R. I. (1969). Asymptotic properties of nonlinear least square estimation. *Ann. Math. Statist.*, **40**, 633–643.

Le Dimet, F. X. (1982). A general formalism of variational analysis. Report OK-73091-22-1, CIMMS, Norman, OK.

Le Dimet, F. X. and Talagrand, O. (1986). Variational algorithms for analysis and assimilation of meteorological observations: theoretical aspects. *Tellus*, **38A**, 97–110.

Le Dimet, F.-X., Navon, I. M. and Daescu, D. N. (2002). Second-order information in data assimilation. *Mon. Weather. Rev.*, **130**, 629–648.

Lions, J. L. (1968). *Contrôle optimal des systèmes gouvernés par des équations aux dérivées partielles.* Dunod, Paris.

Liu, D. C. and Nocedal, J. (1989). On the limited memory BFGS method for large scale minimization. *Math. Program.*, **45**, 503–528.

Marchuk, G. I. and Penenko, V. V. (1978). Application of optimization methods to the problem of mathematical simulation of atmospheric processes and environment. In *Modelling and Optimization of Complex Systems. Proceedings of the IFIP-TC7 Conference*, pp. 240–252. Springer, New York.

Marchuk, G. I., Agoshkov, V. I. and Shutyaev, V. P. (1996). *Adjoint Equations and Perturbation Algorithms in Nonlinear Problems.* CRC Press Boca Raton, FL.

Neiderreiter, H. (1992). *Random Number Generation and Quasi-Monte Carlo Methods.* SIAM, Philadelphia.

Patankar, S. V. (1980). *Numerical Heat Transfer and Fluid Flow.* Hemisphere, New York.

Penenko, V. and Obraztsov, N. N. (1976). A variational initialization method for the fields of the meteorological elements. *Sov. Meteorol. Hydrol.*, **11**, 1–11.

Pires, C., Vautard, R., and Talagrand, O. (1996). On extending the limits of variational assimilation in nonlinear chaotic systems. *Tellus*, **A48**, 96–121.

Rabier, F. and Courtier, P. (1992). Four-dimensional assimilation in the presence of baroclinic instability. *Q. J. R. Meteorol. Soc.*, **118**, 649–672.

Ratkowsky, D. A. (1983). *Nonlinear Regression Modelling: A Unified Practical Approach.* Marcel Dekker, New York.

Sasaki, Y. (1958). An objective analysis based on the variational method. *J. Meteorol. Soc. Japan*, **36**, 77–88.

Talagrand, O. (2003). Bayesian estimation, optimal interpolation, statistical linear estimation. In *Data Assimilation for the Earth System*, (ed. R. Swinbank, V. Shutyaev, and W. Lahoz), pp. 21–35. Kluwer, Dordrecht.

Talagrand, O. and Courtier, P. (1987). Variational assimilation of meteorological observations with the adjoint vorticity equation. I: Theory. *Q. J. R. Meteorol. Soc.*, **113**, 1311–1328.

Thacker, W. C. (1989). The role of the Hessian matrix in fitting models to measurements. *J. Geophys. Res.*, **94**, 6177–6196.

Tshimanga, J., Gratton, S., Weaver, A. T. and Sartenaer, A. (2008). Limited-memory preconditioners, with application to incremental four-dimensional variational assimilation. *Q. J. R. Meteorol. Soc.*, **134**, 751–769.

White, H. and Domowitz, I. (1984). Nonlinear regression and dependent observations. *Econometrica*, **52**, 143–162.

Yuan, K.-H. and Jennrich, R. I. (1998). Asymptotics of estimating equations under natural conditions. *J. Multivariate Anal.*, **65**, 245–260.

15
Adjoints by automatic differentiation

L. Hascoët

INRIA, TROPICS team
INRIA Sophia-Antipolis
2004 Route des lucioles, BP 93
06902 Valbonne, France

Advanced Data Assimilation for Geosciences. First Edition.
Edited by É. Blayo, M. Bocquet, E. Cosme, and L. F. Cugliandolo.

Chapter Contents

15.1 Introduction

Computing accurate derivatives of a numerical model is a crucial task in many domains of scientific computing, and in particular for gradient-based minimization. We present automatic differentiation (AD),[1] a software technique to obtain derivatives of functions provided as programs (Griewank and Walther, 2008; Corliss et al., 2001; Bücker et al., 2005; Bischof et al., 2008). Given a numerical model $F : \mathbb{R}^n \to \mathbb{R}^m$ implemented as a program P, AD adapts or transforms P into a new program that computes derivatives of F.

In the context of data assimilation, interest in automatic differentiation lies mainly in its so-called adjoint mode, which computes gradients efficiently. These notes focus on the aspects of AD related to the adjoint mode. In particular, an efficient adjoint mode will probably require AD by source program transformation. This chapter intend to remain at the level of the principles of adjoint AD and of the software techniques that make it efficient. It is not intended to advocate one particular AD tool, although the tool that we are developing (tapenade) is used in places for illustration.

Section 15.2 presents the principles of AD leading to its adjoint mode. Section 15.3 briefly presents the mathematical adjoint approach to computing gradients, underlining similarities and differences with the AD adjoint approach. It presents two applications in oceanography. Section 15.4 focuses on the software techniques from compiler theory (Aho et al., 1986) that modern AD tools use to produce better adjoint code. Section 15.5 presents and contrasts the existing AD tools that are most likely to be useful for data assimilation.

15.2 Elements of AD

Given a computer algorithm P (identified with a piece of program) that implements a function $F : X \in \mathbb{R}^n \mapsto Y \in \mathbb{R}^m$, AD builds a new algorithm (a program piece) P′ that computes derivatives of F by computing the analytical derivative of each elementary mathematical operation in P. The fundamental observation is that any run-time trace of the algorithm P,

$$\{I_1; I_2; \dots I_p;\},$$

computes the composition of elementary mathematical functions, one per instruction I_k,

$$f_p \circ f_{p-1} \circ \cdots \circ f_1,$$

which we can identify with F. This is, of course, assuming that P is a correct implementation of F, i.e. that the discretization and approximation employed in P are sufficiently accurate and do not introduce non-differentiability.

[1] A warning for French-speaking readers. In this chapter,

AD = **A**utomatic **D**ifferentiation ≠ **A**ssimilation de **D**onnées

DA = **D**ata **A**ssimilation ≠ **D**ifférentiation **A**utomatique

Let us clarify the correspondence between the mathematical variables $(X, Y \ldots)$ and the program variables found in `P`. As imperative programs classically overwrite their variables to save memory space, let us call `V` the collection of all the program variables of `P` and consider that each instruction I_k (partly) overwrites `V`. With these conventions, (this run-time trace of) `P` is indeed the program

Original program `P`

	Initialize `V` *with* X
(I_1)	`V` $:= f_1($`V`$)$
	...
(I_k)	`V` $:= f_k($`V`$)$
	...
(I_p)	`V` $:= f_p($`V`$)$
	Retrieve Y *from* `V`

At any given location in `P`, the program variables `V` correspond to one particular set, or vector, of mathematical variables. We will call this vector X_k for the location between instructions I_k and I_{k+1}. The set `V` is actually large enough to accommodate X, Y, or each successive X_k. At each location, `V` may thus 'contain' more than the X_k, but only the X_k play a role in the semantics of the program. The program instruction `V :=` f_k`(V)` actually means taking from `V` the mathematical variables X_{k-1} before the instruction and applying $f_k(X_{k-1})$ to obtain X_k. After I_k, `V` corresponds to X_k. The *Initialize with* and *Retrieve from* instructions in the program sketch define $X_0 = X$ and identify Y to X_p.

Since we identify F with a composition of functions, the chain rule of calculus gives the first-order full derivative, i.e. the Jacobian:

$$F'(X) = f'_p(X_{p-1}) \times f'_{p-1}(X_{p-2}) \times \ldots \times f'_1(X_0).$$

It is thus possible in theory to adapt the algorithm `P` so that it computes $F'(X)$ in addition to $F(X)$. This can be done simply by extending the instruction I_1 that computes $X_1 = f_1(X_0)$ with a piece of code that computes $J_1 = f'_1(X_0) \times Id$, and by extending likewise every instruction I_k by a piece of code that computes $J_k = f'_k(X_{k-1}) \times J_{k-1}$. This transformation is local to each instruction I_k. It is not limited to straight-line code and can be applied to any program `P` with control. The extended algorithm `P'` just reproduces the control decisions taken by `P`. Of course, derivatives are valid only if the control does not change in an open neighborhood around X. Otherwise, the risk is that AD may return a derivative in cases where F is actually non-differentiable. Keeping this caveat in mind, the adapted algorithm can return J_p, the complete Jacobian $F'(X)$. However, the J_k are matrices whose height and width are both of the order of the number of variables in the original `P`, and may require too much memory space.

To work around this difficulty, we observe that the derivative object that is needed for the target application is seldom the full Jacobian matrix, but rather one of the two projections

$$F'(X) \times \dot{X} \quad \text{or} \quad \overline{Y} \times F'(X),$$

where $\dot{X}$ is some vector in $\mathbb{R}^n$, whereas $\overline{Y}$ is some row vector in $\mathbb{R}^m$. Moreover, when $F'(X)$ is needed explicitly, it is very often sparse and can therefore be retrieved from a relatively small number of the above projections. This motivates the so-called tangent and adjoint modes of AD:

- *Tangent mode:* evaluate $\dot{Y} = F'(X) \times \dot{X}$, the directional derivative of F along direction $\dot{X}$. It expands as

$$\dot{Y} = f'_p(X_{p-1}) \times f'_{p-1}(X_{p-2}) \times \ldots \times f'_1(X_0) \times \dot{X}. \tag{15.1}$$

Since $\dot{X}$ is a vector, this formula is most efficiently evaluated from right to left, i.e., using mathematical variables,

$$\begin{aligned}
X_0 &= X, \\
\dot{X}_0 &= \dot{X}, \\
X_1 &= f_1(X_0), \\
\dot{X}_1 &= f'_1(X_0) \times \dot{X}_0, \\
&\vdots \\
X_k &= f_k(X_{k-1}), \\
\dot{X}_k &= f'_k(X_{k-1}) \times \dot{X}_{k-1}, \\
&\vdots \\
X_p &= f_p(X_{p-1}), \\
\dot{X}_p &= f'_p(X_{p-1}) \times \dot{X}_{p-1}, \\
Y &= X_p, \\
\dot{Y} &= \dot{X}_p.
\end{aligned}$$

An algorithm $\dot{\mathtt{P}}$ for this evaluation is relatively easy to construct, as the derivative instructions follow the order of the original instructions. Keeping the original program variables $\mathtt{V}$ to hold the successive X_k, and introducing a set of new program variables $\dot{\mathtt{V}}$ of the same size as $\mathtt{V}$ to hold the successive $\dot{X}_k$, $\dot{\mathtt{P}}$ is written as

Tangent differentiated program $\dot{\mathtt{P}}$

	Initialize $\mathtt{V}$ *with* X *and* $\dot{\mathtt{V}}$ *with* $\dot{X}$
$(\dot{I}_1)$	$\dot{\mathtt{V}} := f'_1(\mathtt{V}) \times \dot{\mathtt{V}}$
(I_1)	$\mathtt{V} := f_1(\mathtt{V})$
	...
$(\dot{I}_k)$	$\dot{\mathtt{V}} := f'_k(\mathtt{V}) \times \dot{\mathtt{V}}$
(I_k)	$\mathtt{V} := f_k(\mathtt{V})$
	...
$(\dot{I}_p)$	$\dot{\mathtt{V}} := f'_p(\mathtt{V}) \times \dot{\mathtt{V}}$
(I_p)	$\mathtt{V} := f_p(\mathtt{V})$
	Retrieve Y *from* $\mathtt{V}$ *and* $\dot{Y}$ *from* $\dot{\mathtt{V}}$

Notice that each derivative statement $\dot{I}_k$ now precedes I_k, because I_k overwrites $\mathtt{V}$.

- *Adjoint mode:* evaluate $\overline{X} = \overline{Y} \times F'(X)$, the gradient of the scalar function $\overline{Y} \times F(X)$ derived from F and weights $\overline{Y}$. It expands as

$$\overline{X} = \overline{Y} \times f'_p(X_{p-1}) \times f'_{p-1}(X_{p-2}) \times \ldots \times f'_1(X_0). \tag{15.2}$$

Since $\overline{Y}$ is a (row) vector, this formula is most efficiently evaluated from left to right, i.e., with mathematical variables,

$$\begin{aligned}
X_0 &= X, \\
X_1 &= f_1(X_0), \\
&\vdots \\
X_k &= f_k(X_{k-1}), \\
&\vdots \\
X_p &= f_p(X_{p-1}), \\
Y &= X_p, \\
\overline{X}_p &= \overline{Y}, \\
\overline{X}_{p-1} &= \overline{X}_p \times f'_p(X_{p-1}), \\
&\vdots \\
\overline{X}_{k-1} &= \overline{X}_k \times f'_k(X_{k-1}), \\
&\vdots \\
\overline{X}_0 &= \overline{X}_1 \times f'_1(X_0), \\
\overline{X} &= \overline{X}_0.
\end{aligned}$$

However, an algorithm that evaluates these formulas cannot be to constructed immediately, since the derivative instructions will follow the *inverse* order of the original instructions. Similarly to the tangent mode, we want the adjoint program to use only the original program's variables V plus a corresponding set of new program variables $\overline{\texttt{V}}$, of the same size as V, to hold the successive $\overline{X}_k$. In that case, we see that, for example, X_{k-1} contained in V will be overwritten by X_k and thus lost, before it is needed to evaluate $\overline{X}_k \times f'_k(X_{k-1})$. We will see later how this problem is solved, but let us keep in mind that there is a *fundamental penalty attached to the adjoint mode* that comes from the need of a data-flow (and control-flow) reversal.

Let us compare the run-time costs of the tangent and adjoint modes. Each run of the tangent differentiated algorithm $\dot{\texttt{P}}$ costs only a small multiple of the run time of the original P. The ratio, which we will call R_t, varies slightly depending on the given P. Typically R_t ranges between 1 and 3. Using a simplified cost model that only counts the number of costly arithmetical operations (only *, /, and transcendentals), R_t is always less than 4. Similarly, for the adjoint differentiated algorithm $\overline{\texttt{P}}$, the run time

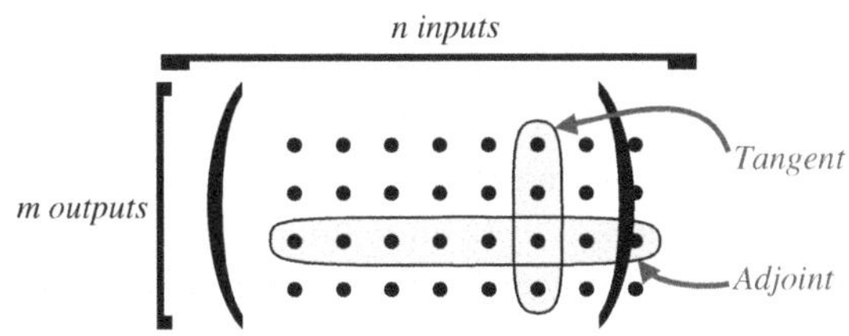

Fig. 15.1 Elements of the Jacobian computable by tangent AD and adjoint AD

is only a small multiple of the run time of P. The ratio, which we will call R_a, varies slightly depending on the given P. In the simplified cost model that only counts costly arithmetical computations, R_t and R_a are identical, but in practice $\overline{\text{P}}$ suffers from the extra penalty coming from the data-flow reversal. Typically, R_a ranges between 5 and 10. Let us compare, with the help of Fig. 15.1, the costs of computing the complete Jacobian $F'(X)$, using no sparsity property, by employing either the tangent mode or the adjoint mode:

- With the tangent mode, we obtain $F'(X)$ column by column by setting $\dot{X}$ successively to each element of the Cartesian basis of the input space $\mathbb{R}^n$. The run time for the full Jacobian is thus $n \times R_t \times$ run time(P).
- With the adjoint mode, we obtain $F'(X)$ row by row by setting $\overline{Y}$ successively equal to each element of the Cartesian basis of the output space $\mathbb{R}^m$. The run time for the full Jacobian is thus $m \times R_a \times$ run time(P).

When n is much larger than m, the adjoint mode is recommended. In particular, this is the case when gradients are needed, such as in optimization or in inverse problems. There are typically very few optimization criteria (often $m = 1$), and on the other hand n is often large, since the optimization parameters may be functions, shapes, or other complex objects. In that case, no matter whether R_a is higher than R_t, the adjoint mode of AD is the only reasonable option. This is the most flagrant situation where adjoint AD can outperform all other strategies, in particular *divided differences* (i.e. evaluating $[F(X + h\dot{X}) - F(X)]/h$) or even tangent AD.

Considering the design and implementation of AD tools, there are two principal ways to code the algorithms $\dot{\text{P}}$ and $\overline{\text{P}}$, namely *operator overloading* and *program transformation*:

- *Operator overloading:* if the language of P permits, we can replace the types of floating-point variables with a new type that contains additional derivative information, and overload the arithmetic operations for this new type so as to propagate this derivative information along. Schematically, the AD tool boils down to a library that defines the overloaded type and arithmetic operations. This approach is both elegant and powerful. The overloaded library can be quickly redefined to compute higher-order derivatives, Taylor expansions, intervals, etc. By its nature, evaluation of the overloaded operations will follow the original order of P. This is fine for the tangent mode, but requires some acrobatics for the adjoint

mode, bearing severe consequences on performance and/or losing a part of the elegance of the approach.
- *Program transformation:* We can instead decide to explicitly build a new source code that computes the derivatives. This implies parsing the original P and building an internal representation, and from it building the differentiated $\dot{\text{P}}$ or $\overline{\text{P}}$. This approach allows the tool to apply some global analysis on P, for instance data flow, to produce more efficient differentiated code. This is very similar to a compiler, except that it produces source code. This approach is more development-intensive than operator overloading, which is one reason why operator overloading AD tools appeared earlier and are more numerous. It also explains why program transformation AD tools are perhaps slightly more fragile and need more effort to follow the continuous evolution of programming constructs and styles. On the other hand, the possibility of global analysis makes program transformation the choice approach for the adjoint mode of AD, which requires control-flow and data-flow reversal and where global analysis is essential to produce efficient code.

These elements of AD are about all the background we need to describe a tangent-mode AD tool. For the adjoint mode, however, we need to address the question of *data-flow reversal.* We saw that the equations of the adjoint mode do not transpose immediately into a program, because the values X_k are overwritten before they are needed by the derivatives. In the context of operator overloading, the natural stat-egy for the adjoint mode is to have the overloaded operations write on a unique 'tape' the long log of program variables being read, combined with differentiable operations, and overwritten. In compiler terminology, this is a kind of *three-address code.* Later, this tape is read from end to beginning by a special code that computes the derivatives. This is indeed a recording not only of the data flow, but also of every arithmetic operation done. There are variations from this scheme, but the tape remains very large. In the context of program transformation, only the data flow need be reversed. There are two ways to solve this problem, and a variety of combinations of them:

- *Recompute-All:* For each derivative instruction $\overleftarrow{I_k}$, we recompute the X_{k-1} that it requires by a repeated execution of the original code, for the stored initial state X_0 to instruction I_{k-1} that computes X_{k-1}. This is sketched in Fig. 15.2. The extra cost in memory is only the storage of X_0. On the other hand, the extra cost in run time is quadratic in p.
- *Store-All:* Each time an instruction I_k overwrites a part of V, we record this part of V into a stack just before overwriting. Later, we restore these values just before executing $\overleftarrow{I_k}$ This is sketched in Fig. 15.3. The extra cost in memory is proportional to p, whereas the extra cost in run time comes from stack manipulation, usually minimal and proportional to p. We sketch a program $\overline{\text{P}}$ that uses the Store-All approach, using `push` and `pop` primitives for stack manipulations, and defining $\mathbf{out}(I_k)$ to be the subset of the variables V that are effectively overwritten by I_k. We see two successive sweeps in $\overline{\text{P}}$. The *forward sweep* $\overrightarrow{\text{P}}$ is essentially a copy of P augmented with storage of overwritten values. The *backward sweep* $\overleftarrow{\text{P}}$

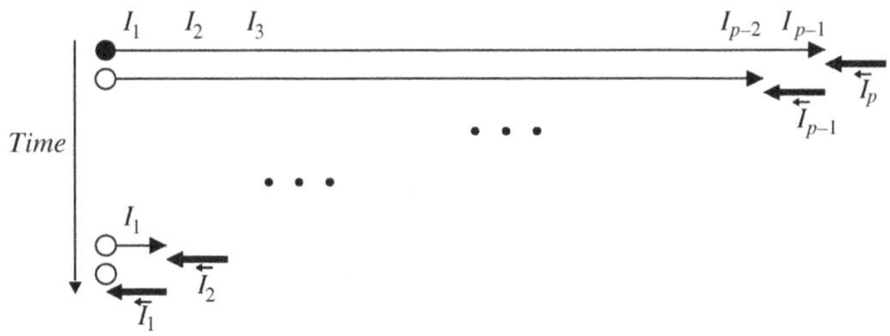

Fig. 15.2 Data-flow reversal with the Recompute-All approach. The large black dot represents storage of X_0, and the large white dots are retrievals

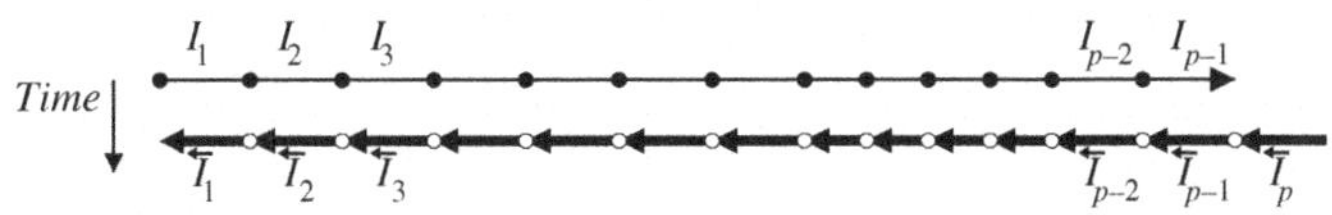

Fig. 15.3 Data-flow reversal with the Store-All approach. The small black dots represent values being recorded before overwriting and the small white dots the corresponding restorations.

is the computation of the derivatives, in reverse order, augmented with retrieval of recorded values. Owing to retrievals, the exit V does not contain the original result Y.

Adjoint differentiated program $\overline{\mathtt{P}}$ (Store-All)

	Initialize V *with* X *and* $\overline{\mathtt{V}}$ *with* $\overline{Y}$
	push(**out**(I_1))
(I_1)	$\mathtt{V} := f_1(\mathtt{V})$
	...
	push(**out**(I_k))
(I_k)	$\mathtt{V} := f_k(\mathtt{V})$
	...
	push(**out**(I_{p-1}))
(I_{p-1})	$\mathtt{V} := f_{p-1}(\mathtt{V})$
($\overleftarrow{I_p}$)	$\overline{\mathtt{V}} := \overline{\mathtt{V}} \times f'_p(\mathtt{V})$
	pop(**out**(I_{p-1}))
	...
	pop(**out**(I_k))
($\overleftarrow{I_k}$)	$\overline{\mathtt{V}} := \overline{\mathtt{V}} \times f'_k(\mathtt{V})$
	...
	pop(**out**(I_1))
($\overleftarrow{I_1}$)	$\overline{\mathtt{V}} := \overline{\mathtt{V}} \times f'_1(\mathtt{V})$
	Retrieve $\overline{X}$ *from* $\dot{\mathtt{V}}$

These data-flow reversal strategies can be adapted to programs P that are no longer straight-line, but that use control. This implies that the control decisions of the original code (branch taken, number of iterations, etc.), or equivalently of the forward sweep, must be made available in reverse for the backward sweep. In a Store-All context, this can be done just as for data: control-flow decisions must be recorded at the exit of the control structure (conditional, loop, etc.) and then retrieved in the backward sweep to control the backward execution. This can use the same stack as data-flow reversal. The conjunction of the recorded control and data values is called the *trajectory*.

In practice, neither pure Recompute-All nor pure Store-All can be applied to large programs, because of their respective cost in run time or memory space. This problem occurs also with overloading-based AD, which behaves like Store-All in this respect. Trade-offs are needed, and the classical trade-off is called *checkpointing*.

- In the Recompute-All approach, checkpointing means choosing a part C of P and storing the state upon exit from this part. Recomputing can then start from this state instead of the initial state. This is sketched in Fig. 15.4(a). At the cost of storing one extra state, the run-time penalty has been divided roughly by two. Checkpoints can be nested to further reduce the run-time penalty, as shown in Fig. 15.4(c).
- In the Store-All approach, checkpointing means choosing a part C of P and *not* recording the overwritten values during C. Before the backward sweep reaches $\overleftarrow{\mathtt{C}}$, C is run again from a stored state this time with recording. This is sketched in Fig. 15.4(b). At the cost of storing one extra state and of running C twice, the peak memory used to record overwritten data is divided roughly by two. Checkpoints can be nested to further reduce the peak memory consumption, as shown in Fig. 15.4(d).

Notice that the execution schemes in Figs. 15.4(c) and (d), for nested checkpoints, become very similar. Recompute-All and Store-All are the two ends of a spectrum, with optimal checkpointing scheme(s) lying somewhere in between. A good placement

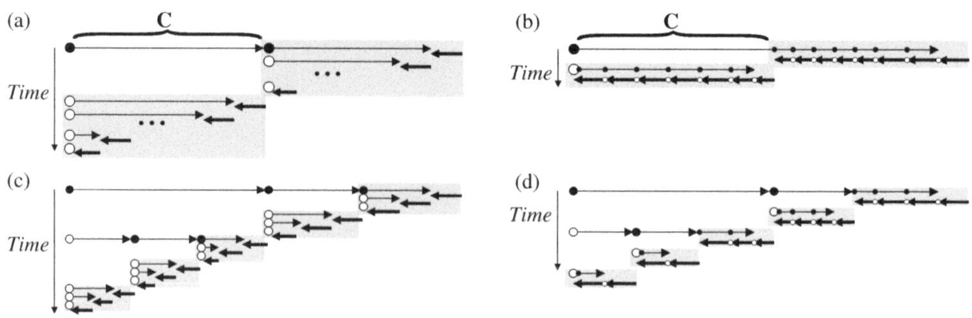

Fig. 15.4 Checkpointing with the Recompute-All (a, c) and Store-All (b, d) approaches. The shaded areas reproduce the basic pattern of the chosen approach. (a, b) Single checkpointing. (c, d) Nested checkpointing.

of (nested) checkpoints is crucial for efficient adjoint differentiation of large codes. Natural candidates to become a checkpointed part are procedure calls and loop bodies, but any piece of code with a single entry point and a single exit point can be chosen. There is no formula nor efficient algorithm to find this optimal placement of checkpoints, except in the case (Griewank, 1992) of a loop with a known number of iterations all of the same cost. A good enough default strategy is to apply checkpointing at the level of each procedure call. In practice, it is important to give the end-user the freedom to place checkpoints by hand. Good placements of checkpoints perform similarly: their memory and run time costs grow with the logarithm of the runtime of P, or, more precisely, the following behaviour is observed:

- The peak memory size during execution, to store states and record data, grows like the logarithm of the run time of P.
- The maximum number of times a checkpointed piece of the program is re-executed, which approximates the slowdown factor of $\overline{\text{P}}$ compared with P, also grows like the logarithm of the run time of P. This explains why the slowdown ratio R_a of $\overline{\text{P}}$ compared with P can be larger by a few units than the ratio R_t of $\dot{\text{P}}$ compared with P.

Actually, we do not need to store the entire memory state to checkpoint a program piece C. What must be stored is called the *snapshot*. In a Store-All context, we can see that a variable need not be in the snapshot if it is not used by C. Likewise, a variable need not be in the snapshot if it is not overwritten between the initial execution of C and the execution of its adjoint $\overline{\text{C}}$.

15.3 Application of adjoint AD to data assimilation

Assume that we have a physical model that represents the way some unknown values determine some measurable values. When this model is complex, its inverse problem is non-trivial. From this physical model, we obtain a mathematical model, which is in general a set of partial differential equations. Let us also assume that by discretization and resolution of the mathematical model, we obtain a program that computes the measurable values from the unknown values.

Using the formalism of *optimal control theory* (le Dimet and Talagrand, 1986), we are studying the state W of some system, W is defined for every point in space, and also—if time is involved—for every instant in an observation period $[0, T]$. The mathematical model relates the state W to a number of external parameters, which are the collection of initial conditions, boundary conditions, model parameters, etc., i.e. all the values that determine the state. Some of these parameters, which we call γ, are the unknown of our inverse problem. This relation between W and γ is implicit in general. It is a set of partial differential equations that we write as

$$\Psi(\gamma, W) = 0. \tag{15.3}$$

Equation (15.3) takes into account all external parameters, but we are only concerned here by the dependence on γ. In optimal control theory, we would call γ our control variable. Any value of γ thus determines a state $W(\gamma)$. We can extract from this state the measurable values, and of course there is very little chance that these values exactly match the values actually measured W_{obs}. Therefore, we start an optimization cycle to modify the unknown values γ until the resulting measurable values match best. We thus define a *cost function* that measures the discrepancy on the measurable values in $W(\gamma)$. In practice, not all values in $W(\gamma)$ can be measured in W_{obs}, but nevertheless we can define this cost function J as the sum at each instant of some squared norm of the discrepancy of each measured value $\|W(\gamma) - W_{obs}\|^2$:

$$j(\gamma) = J(W(\gamma)) = \frac{1}{2}\int_{t=0}^{T} \|W(\gamma)(t) - W_{obs}(t)\|^2 \, dt \tag{15.4}$$

Therefore, the inverse problem is to find the value of γ that minimizes $j(\gamma)$, i.e. such that $j'(\gamma) = 0$. If we use a gradient descent algorithm to find γ, we need to find $j'(\gamma)$ for each γ. To this end, the mathematical approach first applies the chain rule to (15.4), yielding

$$j'(\gamma) = \frac{\partial J(W(\gamma))}{\partial \gamma} = \frac{\partial J}{\partial W}\frac{\partial W}{\partial \gamma}. \tag{15.5}$$

The derivative of W with respect to γ comes from the state implicit equation (15.3), which we differentiate with respect to γ to obtain

$$\frac{\partial \Psi}{\partial \gamma} + \frac{\partial \Psi}{\partial W}\frac{\partial W}{\partial \gamma} = 0. \tag{15.6}$$

Assuming that this can be solved for $\partial W/\partial \gamma$, we can then substitute it into (15.5) to obtain

$$j'(\gamma) = -\frac{\partial J}{\partial W}\frac{\partial \Psi}{\partial W}^{-1}\frac{\partial \Psi}{\partial \gamma}. \tag{15.7}$$

Now is the time to take complexity into account. Equation (15.7) involves one system resolution and then one product. Nowadays, both Ψ and W are discretized with millions of dimensions. $\partial \Psi/\partial W$ is definitely too large to be computed explicitly, and therefore its inverse cannot be computed or indeed stored. We notice that $\partial \Psi/\partial \gamma$ has many columns, whereas $\partial J/\partial W$ has only one row. Therefore, the most efficient way to compute the gradient $j'(\gamma)$ is the *adjoint method*: first compute $\partial J/\partial W$, then run an iterative resolution for

$$\frac{\partial J}{\partial W}\frac{\partial \Psi}{\partial W}^{-1}, \tag{15.8}$$

and then multiply the result (called the *adjoint state* Π) by $\partial \Psi/\partial \gamma$.

Recall now that we already have a resolution program, i.e. a procedure $\mathtt{P}_{\Psi}$ that, given γ, returns W_{γ}, and a procedure $\mathtt{P}_{j}$ that, given a W, evaluates the cost function, i.e. the discrepancy between W and the observed W_{obs}. Adjoint AD of the program that computes

$$j = \mathtt{P}_j(\mathtt{P}_\Psi(\gamma))$$

directly gives the gradient of j, i.e. the desired $j'(\gamma)$. This is indeed very close to the mathematical resolution with the adjoint state. In both cases, $\partial J/\partial W$ is computed first, thus guaranteeing that no large square matrix is stored. Things differ a little for the following stage, because a program is by essence explicit and therefore the resolution for Π and multiplication with $\partial\Psi/\partial\gamma$ are done jointly. Apart from that, adjoint AD can really be considered as the discrete equivalent (on programs) of the above adjoint method.

As a first application, we considered the oceanography code OPA 9.0 (Madec et al., 1998) on a simple configuration known as GYRE. This simulates the behavior of a rectangular basin of water put on the tropics between latitudes 15° and 30°, with the wind blowing to the East. Our control variables γ are the temperature fields in the complete domain and our cost function $j(\gamma)$ is the discrepancy with respect to measurements of the heat flux across some boundary at the northern angle 20 days later. Figure 15.5 shows one gradient $j'(\gamma)$ computed by adjoint AD. This system is discretized with $32 \times 22 \times 31$ nodes and 4320 time steps. The original simulation takes 26 seconds, and the differentiated program computes the gradient above in 205 seconds, which is only 7.9 times as long. Of course, checkpointing is absolutely necessary to reverse this long simulation, yielding several recomputations of the same program

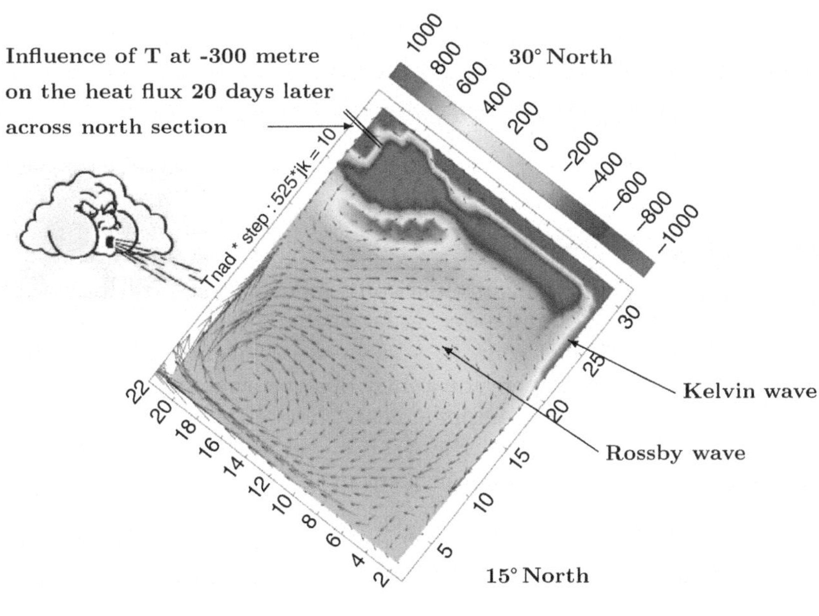

Fig. 15.5 Oceanography gradient by adjoint AD on OPA Gyre.

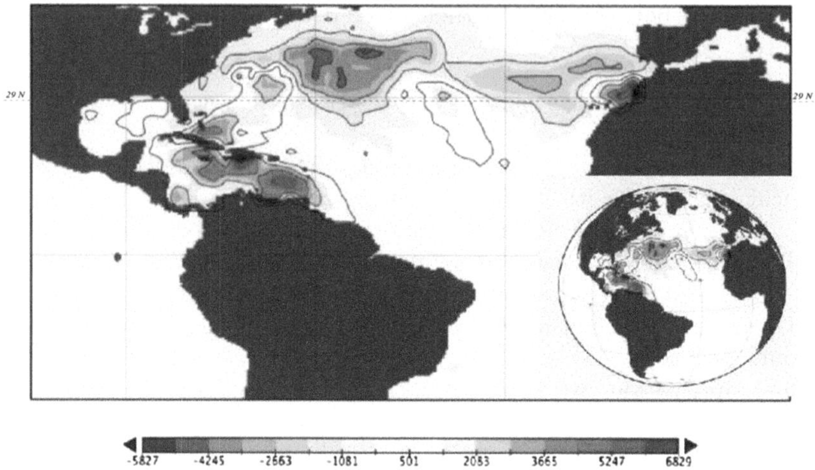

Fig. 15.6 Oceanography gradient by adjoint AD on OPA Nemo.

steps, but nevertheless the factor 7.9 is much better than what tangent AD would require. Checkpointing and storage of the trajectory use a stack that reaches a peak size of 481 Mbytes.

In a second, larger application we considered the NEMO configuration of OPA (2° grid cells, one year simulation) for the north Atlantic. Figure 15.6 shows the gradient of the discrepancy to measurements of the heat flux across the 29th parallel, with respect to the temperature field one year before. With this discretization, there are 9100 control parameters in γ. The adjoint code that computes the gradient with respect to these 9100 parameters takes 6.5 times as long as the original code. The size of the stack that stores the trajectory and the snapshots for checkpointing reaches a peak of 1591 Mbytes.

15.4 Improving the adjoint AD code

AD tools will produce AD adjoints automatically, following the principles described in Section 15.2. The resulting adjoint code can be easily criticized: a skilled programmer can always write a more efficient code, at least in theory, using long and careful work.

Nevertheless, AD tool developers constantly observe what skilled adjoint programmers do, and try to generalize the improvements, model them, and incorporate them into the AD tools. In other situations, the AD tool can be extended with options to let the end-user request a predefined improvement at a given location. We will illustrate here only a few such improvements.

Activity analysis is systematic in many program transformation AD tools. It is useful for both tangent and adjoint AD. Assume that the user has specified the set of *independents*, i.e. the input variables with respect to which differentiation must be done, and the set of *dependents*, i.e. the output variables that must be differentiated.

Activity analysis propagates forward the *varied* variables, i.e. those that depend on some independent in a differentiable way. It also propagates backwards the *useful* variables, i.e. those that influence some dependent in a differentiable way. At each given location in the code, the derivative variable of an original variable needs to be mentioned *only* if this original variable is both varied and useful. In all other situations, the derivative is either certainly null or certainly useless, and therefore need not appear in the differentiated code. Activity analysis is just one of the many data-flow analyses that AD tools use, applying techniques from compiler theory. Figure 15.7 illustrates the benefits of activity analysis. `x` immediately becomes not varied, and `t` is useless. Therefore, the AD tool knows that `xd` and `tb` are null and can be simplified. In some situations, we can even choose not to set them explicitly to zero.

TBR analysis is systematic, and specific to Store-All adjoint AD. The Store-All approach says that the value that is being overwritten by an assignment must be (1) stored just before this assignment in the forward sweep and (2) restored before the adjoint of this assignment in the backward sweep. However, TBR analysis can detect that a particular value is actually not used in the derivative computations. In short, this is the case for every value that is used only in *linear* computations. It is not necessary to store and restore these values, and this saves a significant amount of trajectory in the stack memory. In the example of Fig. 15.8, TBR analysis could prove that neither `x` nor `y` were needed by the differentiated instructions, and therefore these variables need not be `PUSH`'ed on or `POP`'ed from the stack.

original program	**tangent AD**	**adjoint AD**

`x = 1.0`
`z = x*y`
`t = y**2`
`IF (t .GT. 100) ...`
	`x = 1.0`
	`zd = x*yd`
	`z = x*y`
	`t = y**2`
	`IF (t .GT. 100) ...`
		`x = 1.0`
		`z = x*y`
		`t = y**2`
		`IF (t .GT. 100) ...`
		`...`
		`yb = yb + x*zb`

Fig. 15.7 Instructions simplifications due to activity analysis.

original program

original program	**adjoint mode:** **naïve backward sweep**	**adjoint mode:** **backward sweep TBR**

`x = x + EXP(a)`
`y = x + a**2`
`a = 3*z`
	`CALL POPREAL4(a)`
	`zb = zb + 3*ab`
	`ab = 0.0`
	`CALL POPREAL4(y)`
	`ab = ab + 2*a*yb`
	`xb = xb + yb`
	`yb = 0.0`
	`CALL POPREAL4(x)`
	`ab = ab + EXP(a)*xb`
		`CALL POPREAL4(a)`
		`zb = zb + 3*ab`
		`ab = 0.0`
		
		`ab = ab + 2*a*yb`
		`xb = xb + yb`
		`yb = 0.0`
		
		`ab = ab + EXP(a)*xb`

Fig. 15.8 Removing unnecessary storage through TBR analysis.

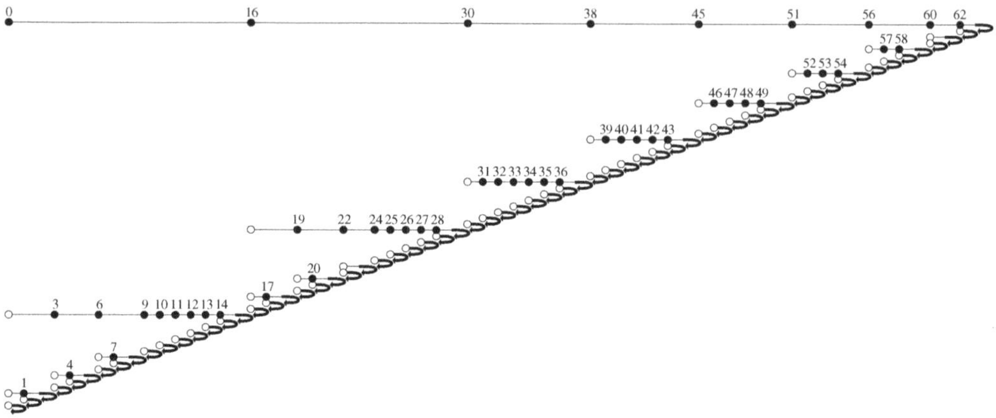

Fig. 15.9 Optimal binomial checkpointing on 64 time steps.

Binomial checkpointing is the optimal way to organize checkpointing for a special class of loops that essentially correspond to time-stepping iterations (Griewank and Walther, 2008). See Fig. 15.9. This is a crucial improvement to adjoint AD of most unsteady simulation codes. It is more powerful than multilevel checkpointing. Basically, binomial checkpointing recursively divides sequences of time steps into two subsequences of unequal length given by a binomial law. The peak memory required by the tape, as well as the slowdown factor coming from recomputations, only grow like the logarithm of the number of time steps. Binomial checkpointing is somewhat intricate to implement in general, so it must be provided by the AD tool. However, the tool cannot discover the time-stepping loops, and the end-user must designate them with a directive to the AD tool.

Linear solvers are better differentiated by hand. In both tangent and adjoint modes, there are simple formulas that propagate derivatives through them very efficiently. In contrast, standard AD will probably do a very poor job differentiating all the minute irrelevant details that are found in an optimized linear solver routine. Here also, there is no hope that an AD tool can detect automatically that a subroutine is a linear solver. This is probably undecidable anyway. Therefore, the AD tool must be guided by user directives. The AD tool will then either apply some predefined differentiation strategy to the linear solver, or just leave a 'hole' in the adjoint code and ask the user to fill the hole with the very short differentiated subroutine for the linear solver. This last approach is often referred to as the *black-box* approach and is not limited to linear solvers: it applies to any procedure for which there is a clever differentiation, or to external routines for which the source code is unavailable.

15.5 AD tools

Here is our (partial) view of available AD tools and of how one might classify them. The best source is the Website <http://www.autodiff.org>, which is the portal managed by our colleagues in Aachen and Argonne in the name of the AD community. It turns out that plenty of experiments have been done that adapt the concepts of AD to particular environments. There was for instance a clever introduction of tangent AD into Microsoft's Excel. However, in the present context, we will focus on the tools that we feel can be applied to inverse problems and data assimilation of industrial size.

As we saw, some AD tools rely on program overloading rather than program transformation. In general, this makes the tool easier to implement. However, some overloading-based AD tools can become very sophisticated and efficient, and represent a fair bit of hard work too. Overloading-based AD tools exist only for target languages that permit some form of overloading (e.g. C++ and Fortran95). Overloading-based AD tools are particularly adapted for differentiations that are mostly local to each statement; i.e. no fancy control flow rescheduling is allowed. On the other hand, these local operations can be very sophisticated—more than what transformation-based AD tools generally provide. For instance, overloading-based AD tools can generally compute not only first, but also second and, third derivatives, and so on, as well as Taylor expansions and interval arithmetic. **Adol-C** (Walther and Griewank, 2012) is an excellent example of an overloading-based AD tool. **FADBAD/TADIFF** are other examples.

The AD tools based on program transformation parse and analyse the original program and generate a new source program. These tools share their general architecture: a front-end very much like a compiler, followed by an analysis component, a differentiation component, and finally a back-end that regenerates the differentiated source. They differ in particular in the language that they recognize and differentiate, and in the AD modes that they provide. They also exhibit some differences in AD strategies—mostly about the adjoint mode. The best known among the other transformation-based AD tools are the following:

- Tapenade (Hascoët and Pascual, 2004) provides tangent and adjoint differentiation of Fortran (77 and 95) and C. Adjoint mode uses the Store-All approach to restore intermediate values. This Store-All approach is selectively replaced by recomputation in simple appropriate situations. Checkpointing is applied by default at the level of procedure calls, but can be triggered or deactivated at other places through used directives.
- Adifor (Carle and Fagan, 2000) differentiates Fortran77 codes in tangent mode. Adifor was once extended towards the adjoint mode (Adjfor), but we believe this know-how has now been re-injected into the OpenAD framework, described below.
- Adic can be seen as the C equivalent of Adifor. However, it has been lately rebased on a completely different architecture, from the OpenAD framework. Adic differentiates ANSI C programs in tangent mode, with the possibility to obtain second derivatives.
- OpenAD/F (Utke et al., 2008) differentiates Fortran codes in tangent and adjoint modes. The general framework of OpenAD claims (like Tapenade) that only front-end and back-end should depend on the particular language, whereas the analysis and differentiation part should work on a language-independent program representation. This is why OpenAD was able to host Adic. OpenAD/F is

made of Adifor and Adjfor components integrated into the OpenAD framework. Its strategy to restore intermediate values in adjoint AD is extremely close to Tapenade's.

- TAMC (Giering, 1997), through its commercial offsprings TAF and TAC++, differentiates Fortran and C files. TAF also differentiates Fortran95 files, under certain restrictions. TAF is commercialized by the FastOpt company in Hamburg, Germany. Differentiation is provided in tangent and adjoint mode, with the Recompute-All approach to restore intermediate values in adjoint AD. Checkpointing and an algorithm to avoid useless recomputations (ERA) are used to avoid explosion of run-time. TAF also provides a mode that efficiently computes the sparsity pattern of Jacobian matrices, using bit-sets.

There are also AD tools that directly interface to an existing compiler. In fact, these are modifications to the compiler such that the compiler performs AD at compile time. This new AD functionality is triggered by new constructs or new directives added into the application language. In a sense, this amounts to modifying the application language so that it has a new 'differentiation' operator. For instance, the NAGWare Fortran95 compiler embeds AD facilities (Naumann and Riehme, 2005) that are triggered by user directives in the Fortran source. To our knowledge, its adjoint mode uses a strategy equivalent to operator overloading.

There are AD tools that target higher-level languages, such as MATLAB. We know of ADiMat, MAD, and INTLAB. Even when they rely on operator overloading, they may embed a fair bit of program analysis to produce efficient differentiated code.

15.6 Conclusion

We have presented automatic differentiation, and more precisely the fundamental notions that are behind the AD tools that use source program transformation. We shall use Fig. 15.10 as a visual support to compare AD with other ways to obtain derivatives. Our strongest claim is that if you need derivatives of functions that are already implemented as programs, then you should seriously consider AD. At first thought, it is simpler to apply the divided differences method (sometimes known also as 'finite differences'), but its inaccuracy is its major drawback.

Notice that divided differences sometimes behave better when the implemented function is not differentiable, because its very inaccuracy has the effect of smoothing discontinuities of the computed function. Therefore, divided differences can be an option when one only has piecewise differentiability. Also, it is true that divided differences may actually cost a little less than the tangent mode, which is their AD equivalent. Nevertheless, when it is possible, it is safer to look for exact analytical derivatives.

Divided differences are definitely a poor choice if you need gradients.

Then again, two options arise: one can consider the problem of finding the derivatives as a new mathematical problem, with mathematical equations (e.g. adjoint equations) that must be discretized and solved numerically. This is a satisfying mathematical approach, but one must be aware of its development cost.

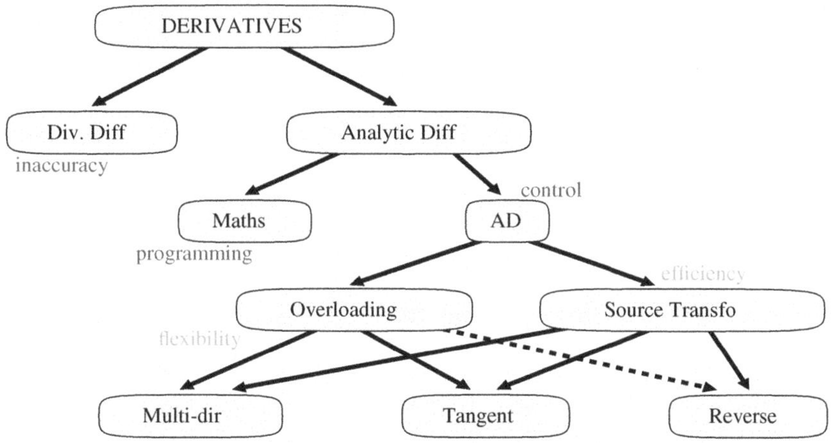

Fig. 15.10 AD and other ways to compute derivatives.

AD is a very promising alternative when the function to be differentiated has already been implemented. In a sense, AD reuses the resolution strategy that has been implemented for the original function into the resolution of its derivatives. When the original model or code changes, AD can be applied again at very little cost.

There are still open questions of non-differentiability introduced by the program's control, or the fact that the iterative resolution of the derivatives is not always guaranteed to converge at the same speed as the original function resolution. But, in practice, AD returns derivatives that are just as good as those returned by the 'mathematical' approach above.

Inside the AD methods, we distinguish overloading-based approaches, which are more flexible and can be adapted to all sorts of derivatives and similar concepts. On the other hand, we advocate source-transformation-based tools for the well-identified goals of tangent and adjoint first-order derivatives. Source transformation gives it full power when it performs global analyses and transformations on the code being differentiated.

Source transformation AD is really the best approach to the adjoint mode of AD, which computes gradients at a remarkably low cost. Adjoint AD is a discrete equivalent of the adjoint methods from control theory. Adjoint AD may appear puzzling and even complex at first sight. But AD tools apply it very reliably, so that a basic understanding of it generally suffices. Adjoint AD is really the method of choice to obtain the gradients required by inverse problems (e.g. data assimilation) and optimization problems.

AD tools can build highly optimized derivative programs in a matter of minutes. They are making steady progress, but the best AD will always require end-user intervention. Moreover, one must keep in mind the limitations of AD in order to make a sensible usage of it. Fundamentally, we have the following to consider:

- Real programs are always only piecewise-differentiable, and only the user can tell if these algorithmic discontinuities will be harmful or not.

- Iterative resolution of the derivatives may not converge as well as the original program, and the knowledge of the numerical scientist is invaluable to study this problem.
- Adjoint AD of large codes will always require a careful profiling to find the best storage/recomputation trade-off.

There are also a number of technical limitations to AD tools, which may be partly lifted in the future, but which are at the current frontier of AD tool development:

- Dynamic memory in the original program is a challenge for the memory restoration mechanism of the adjoint mode.
- Object-oriented languages pose several very practical problems, because they far more intensively use the mechanisms of overloading and dynamic allocation. Data-Flow analysis of object-oriented programs may become harder and return less useful results.
- Parallel communications or other system-related operations may introduce a degree of randomness in the control flow, which is then hard to reproduce, duplicate, or reverse.

References

Aho, A., Sethi, R., and Ullman, J. (1986). *Compilers: Principles, Techniques and Tools*. Addison-Wesley, Reading, MA.

Bischof, C., Bücker, M., Hovland, P., Naumann, U., and Utke, J. (eds.) (2008). *Advances in Automatic Differentiation*. Lecture Notes in Computational Science and Engineering, Vol. 64. Springer, Berlin.

Bücker, M., Corliss, G., Hovland, P., Naumann, U., and Norris, B. (eds.) (2005). *Automatic Differentiation: Applications, Theory, and Implementations*. Lecture Notes in Computational Science and Engineering, Vol. 50. Springer, New York.

Carle, A. and Fagan, M. (2000). ADIFOR 3.0 overview. Technical Report CAAM-TR-00-02, Rice University.

Corliss, G., Faure, C., Griewank, A., Hascoët, L., and Naumann, U. (eds.) (2001). *Automatic Differentiation: from Simulation to Optimization*. Springer, New York, NY.

Giering, R. (1997). TAMC: Tangent Linear and Adjoint Model Compiler, User's Manual. Available at Technical report. <http://www.autodiff.com/tamc>.

Griewank, A. (1992). Achieving logarithmic growth of temporal and spatial complexity in reverse Automatic Differentiation. *Optim. Methods Software*, **1**, 35–54.

Griewank, A. and Walther, A. (2008). *Evaluating Derivatives: Principles and Techniques of Algorithmic Differentiation*, 2nd edn. SIAM, Philadelphia.

Hascoët, L. and Pascual, V. (2004). TAPENADE 2.1 User's Guide. Rapport technique 300, INRIA.

le Dimet, F.-X. and Talagrand, O. (1986). Variational algorithms for analysis and assimilation of meteorological observations: theoretical aspects. *Tellus*, **38A**, 97–110.

Madec, G., Delecluse, P., Imbard, M., and Levy, C. (1998). OPA8.1 Ocean General Circulation Model Reference Manual. Technical Report, Pole de Modelisation, IPSL.

Naumann, U. and Riehme, J. (2005). Computing adjoints with the NAGWare Fortran 95 compiler. In *Automatic Differentiation: Applications, Theory, and implementations* (ed. Bücker, M., Corliss, G., Hovland, P. Naumann, U., and Norriss, B.), pp. 159–169. Lecture Notes in Computational Science and Engineering, Vol. 50. Springer, New York.

Utke, J., Naumann, U., Fagan, M., Tallent, N., Strout, M., Heimbach, P., Hill, C., and Wunsch, C. (2008). OpenAD/F: A modular, open-source tool for automatic differentiation of Fortran codes. *ACM Trans. Math. Software*, **34**, 18:1–18:36.

Walther, A. and Griewank, A. (2012). Getting started with ADOL-C. In *Combinatorial Scientific Computing* (ed. U. Naumann and O. Schenk), pp. 181–202. Chapman & Hall/CRC, Boco Raton, FL.

16
Assimilation of images

A. Vidard[1], O. Titaud[2]

[1]Laboratoire Jean Kuntzmann
INRIA and Université Grenoble Alpes
Grenoble, France
[2]Cerfacs and CLS, Toulouse, France

Advanced Data Assimilation for Geosciences. First Edition.
Edited by É. Blayo, M. Bocquet, E. Cosme, and L. F. Cugliandolo.

Chapter Contents

16.1 Motivations

Since the end of the 1970s, many satellites have been launched to improve our knowledge of the atmosphere and of the oceans by observing the Earth. The geostationary ones provide photographic images of the Earth system, among other data. They provide sequences of such images that show the dynamical evolution of certain meteorological or oceanic 'objects': fronts, clouds, eddies, vortices, etc. Dynamics that the human vision can easily detect in this kind of image sequence clearly has a strong predictive potential (Fig. 16.1). This aspect is clearly favoured by the fact that these data, in contrast to many other measurements, are dense in space, with a resolution close to 1 km and in time with a full image every 15 min for current Meteosat satellites and every 10 min (and even every 2.5 min for Europe only) for the upcoming third generation. This implies a huge amount of data, which can be seen as an asset but can also lead in difficulties for the assimilation system in coping with such an amount of data. In practice, only a tiny percentage (≈ 3–5%) of total satellite data (from polar-orbiting and geostationary satellites) is used in operational numerical weather prediction systems, and images are only used through pseudo-observation (see later) and are given low confidence within the assimilation systems. Considering the cost of the satellite observing systems (the cost of the launch of the Meteosat Third Generation is estimated

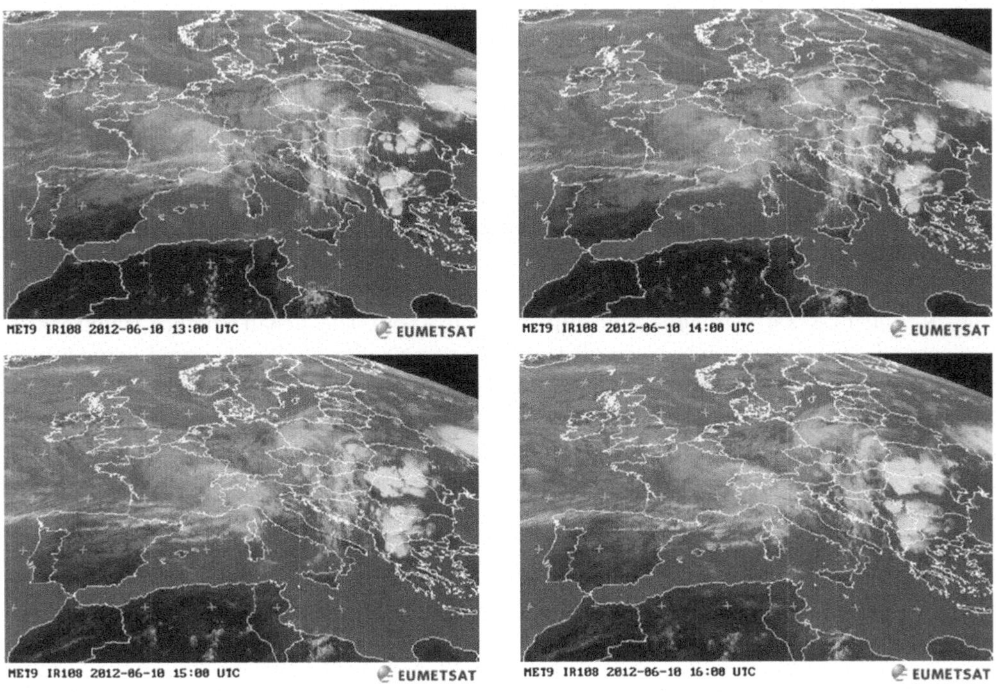

Fig. 16.1 Sequence of Meteosat images over Europe on 12 June 2010 between 13:00 and 16:00 UTC. (Copyright 2010 EUMETSAT.)

at around 2.5 billion Euros) and of the infrastructures required for the collection of the data itself, improving their impact on forecasting systems is an important topic of research.

The purpose of this chapter is therefore to present a brief introduction to image sequence assimilation in a geophysical context. As already mentioned, this is still a topic of research and only basic options are used in operational systems. In relation, two classes of image assimilation techniques can be considered:

- *Pseudo-observation:* an apparent velocity field is estimated from an image sequence using certain image processing techniques. This estimated field can then be used as indirect observations in a classical assimilation scheme. This is what is currently done in a meteorological context (Schmetz et al., 1993).
- *Direct image assimilation (DIA):* Images or image sequences are assimilated through an appropriate observation operator directly into the data assimilation system. This idea follows a general trend to avoid preprocessing of observations before assimilation. This is not yet used operationally, and a selection of possible metrics is presented.

Using images or image sequences together with numerical models may present several difficulties. First, the state variables of the numerical models (e.g. wind, temperature, pressure, humidity in atmosphere modelling and current velocities, temperature, surface elevation, and salinity in ocean modelling) are not directly measured by satellites. What is observed is more or less linked to those variables by diagnostic equations that will have to be included in the assimilation procedure. Moreover, the physical processes that are observed are not always taken into account in the model (e.g. local convection in atmospheric modelling and ocean colour in ocean modelling). These two aspects are not specific to images and may appear for other kinds of observation as well, but they are likely to be systematic in the case of image sequences.

More specifically, images are bidimensional informations, whereas physical processes of geophysical fluids are three-dimensional. From this point of view, a major difference between the ocean and the atmosphere comes from their radiative properties. In the first case, images give information about the ocean surface, whereas in the second case, images integrate all the radiative profile of the observed atmosphere column. Besides, certain meteorological structures (e.g. some kinds of clouds) are located in specific layers. Assimilating information about the dynamical evolution of these structures needs additional information about the altitude of the observed processes (using temperature profile measurements for instance). When, as a result of the radiative properties of the ocean, the images give information about the ocean surface only (typically about a few centimetres—up to a few meters in some cases), the corresponding observed processes may not be depicted well by the model. Additionally, satellite images can be of relatively poor quality. This is mostly true for ocean surface images, which are very often partially occulted by clouds. Moreover, most of the time, they are composite images: It takes Meteosat 15 min to take a whole image chunk after chunk, and the surface SST images come from polar-orbiting satellites and could take a few days for full coverage of a given area (Fig. 16.2). In a sequence, the luminosity of individual images may vary according to the time of the day, and only

Fig. 16.2 Infrared composite image from Metop, 21 January 2008. (From EUMETSAT, O&SI SAF, 2013.)

apparent motion is observed, which could lead to aliasing problems. Another property of images is that, as for human vision, we can presume that the pertinent information coming from an image is mainly provided by its discontinuities or high gradients. Unfortunately, numerical models have a tendency to smooth out these discontinuities. Finally, as already mentioned, handling the massive amount of data included in an image sequence is particularly challenging.

In order to tackle the above-mentioned difficulties, one has to carefully design the assimilation system. It should be possible, despite these difficult conditions, to extract from observations the necessary information about the dynamics of the system. The difficulty in building such systems lies in the adequate definition of the observation space and the related metrics and consequently in the construction of the observation operator.

16.2 Images: level(s) of interpretation

Before considering the assimilation of images, let us give a better definition of what an image is in this context. For the human eye, it is structured information, dense on the domain that it covers; for a computer, it is an array of pixels, and from a mathematical point of view, it can be many things. But, basically, an image is considered as a function of two variables,

$$\begin{aligned} I : \Omega \subset \mathbb{R} \times \mathbb{R} &\to \mathbb{R} \\ (x, y) &\mapsto f(x, y), \end{aligned} \tag{16.1}$$

whose definition can be of a very different nature (see Section 16.2.3).

16.2.1 Numerical raw images

The acquisition device converts the radiation coming from the observed scene into an electrical signal that is digitalized to form the *raw image*. A numerical raw image is an array of pixels whose values represent the quantity of photons that the corresponding photosite received during the acquisition procedure. Raw images pixels can then be considered as radiance measurements. However, from a *mathematical* point of view, an image is commonly considered as a real-valued function of two continuous real variables. This allows the use of powerful mathematical tools developed by the image-processing scientific community. A *numerical image* is the discrete version of a mathematical one that is finally manipulated by computers.

16.2.2 Levels of interpretation

An image can be considered at different levels of interpretation. At the *pixel level*, an image of a fluid in motion is a digital instantaneous expression of the state of the fluid through the radiation that it reflects (visible light) or emits (infrared light). It is then a snapshot of an *apparent state*, which may involve other physical quantities than these that the model simulates, such as tracers (e.g. cloud cover and aerosols in the atmosphere, or chlorophyll in the ocean). The radiances are generally not part of the studied system, and the relationship between this radiation and the state variables is not often trivial. For instance, *cloud cover* is an intricate function of the state variable of meteorological models: Cloud reflectance and intrinsic infrared emission depend both on thermodynamics (temperature and humidity) and on the microphysical processes that occur inside the clouds (involving water, ice, snow, and the size of the particles). At this level, information contained in images is not structured and represents a huge amount of data: typical satellite images are about 5000×5000 in size, i.e. 25 million pixels. This number should be multiplied by the number of frames in the observed sequence.

At the *analysis level(s)*, an image is symbolically described by its contents, typically by the objects and their characterization (geometry, layout, etc.) that appear in the observed scene or by its mathematical description (see further). This kind of interpretation is used for such high-level vision problems as motion estimation, edge detection, and pattern recognition. The advantage of this level is that it may depend-less on the effective radiance emission and it allows emphasis on the structures the image contains (and even their evolution in an image sequence) rather than the actual values of individual pixels.

A good illustration of these different levels of interpretation is provided by the two images from two different MODIS sensors shown in Fig. 16.3. At a pixel level, they represent two very different and somewhat independent quantities of very different nature (temperature and chlorophyll), while at a higher level of interpretation, both images give a similar kind of information about the dynamics of the system. Indeed, one can easily notice the vortices and fronts that are present in the images.

16.2.3 Mathematical definition

There exist several mathematical definitions for images, and it is not the aim here to describe them all, and so only a small subset will be discussed in the following. They essentially depend on the way the image needs to be processed. We can cite the multiscale decomposition approach, which comes from signal processing theory—the

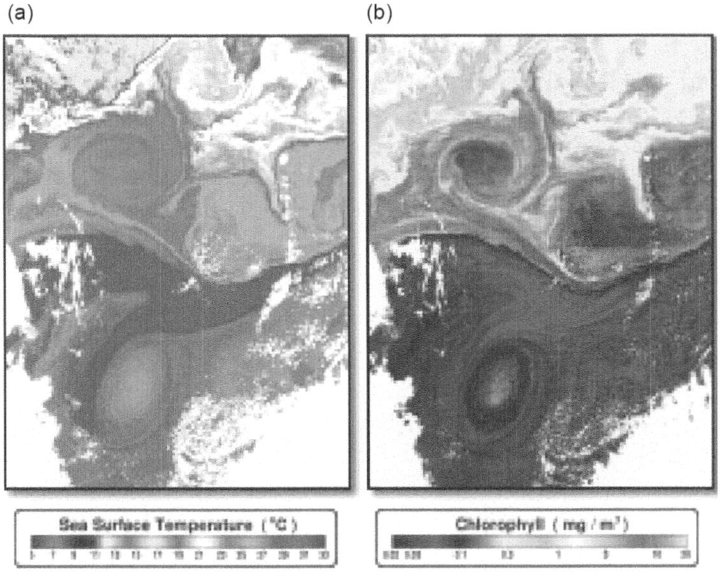

Fig. 16.3 Images over the Gulf Stream on 4 August 2005 from the Aqua satellite (MODIS radiometer). Ocean sea surface temperature (IR channel, a) and chlorophyll concentration (visible channel, b). (Source: NASA, <http://oceancolor.gsfc.nasa.gov/>.)

(wavelet transform (Mallat, 1998), and the more recent curvelet transform (Candès and Donoho, 2003). Another approach uses the partial differential equation framework (Aubert and KornProbst, 2006), where the image is considered as a function of two variables satisfying a partial differential equation that depends on the image-processing problem under consideration. Certain applications use a stochastic approach to define images as a combination of an 'ideal' image (most often deterministic) and a random function (Geman and Geman, 1984). For a good review of image analysis and the underlying mathematical definition, see Chan and Shen (2005).

In the data assimilation framework, the level of interpretation together with the mathematical definition of the image may be represented by an image-type observation space and its corresponding observation operator (see Section 16.4).

16.3 Current use of images in data assimilation: pseudo observation

Sequences of images are widely available in meteorology, oceanography, hydrology, astrophysics, and even medicine. Historically, they have been mainly used for a 'by eye' analysis from experts in these fields. Yet, since the mid 1980s, research has been carried out to derive velocity fields from the sequences, with applications mainly for fluid dynamics (and, very recently, for video compression and medical applications). One way to assimilate images is to use the velocity field derived from image processing techniques such as pseudo-observation of the studied system. This class of motion estimator, directly based on image processing techniques, aims at estimating the velocity field that would 'transport' one frame to another or one part of the frame to the other frame.

There are several ways to extract a velocity field from a sequence of images. These mainly work on two successive frames, even though some extensions allow for the use of more frames at the same time. The following is a short list of the most common methods.

16.3.1 Bogus

The simplest method to assimilate images is to detect specific features in the images and add or displace them in the system. Historically, both steps of this process were done manually by forecasters and modellers, but recently some effort has been made to perform the detection automatically, for instance with the dry intrusions that are important precursors of extreme events (see Fig. 16.4 and Michel and Bouttier, 2006), as well as the assimilation itself, for example with the use of pseudo-observations of potential vorticity structures (Michel, 2011).

16.3.2 Statistical methods

Statistical methods were developed for particle image velocity (PIV) and particle tracking velocity (PTV) experiments, i.e. for laboratory experiments in fluid mechanics

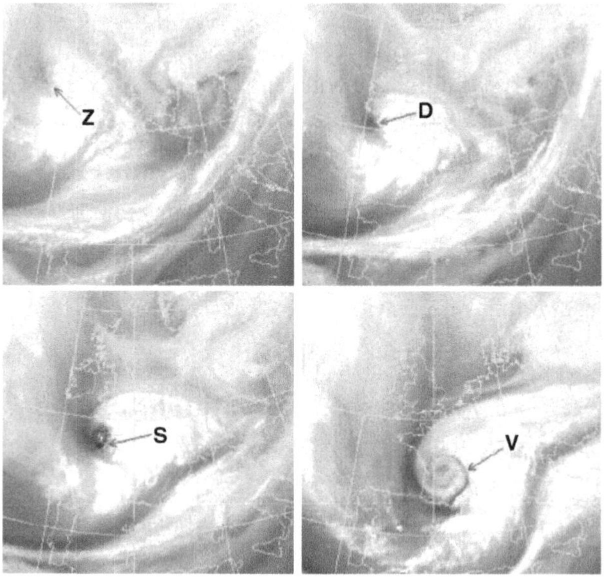

Fig. 16.4 Evolution of a dry intrusion. (From Santurette and Georgiev (2005).)

in which particles are added to the fluid in order to follow the flow Adrian. In this approach, two successive images are divided into small areas, and cross-correlation techniques can be used to match each individual area of the first image with one of the second image:

$$C = \frac{\sum_{i,j}(T_1(i,j) - \bar{T_1})(T_2(i,j) - \bar{T_2})}{\sigma_{T_1}\sigma_{T_2}}, \tag{16.2}$$

with T_1 and T_2 representing the spatial distribution of grey values in the first and second image templates, respectively. The T_2 corresponding to the highest C is selected (if C is larger that a given threshold, typically 0.8).

It is then possible to create a velocity field that would transport all the areas of the first image to the second. These methods are easy to implement, but they can be quite expensive when the number of areas increases. Since the resulting velocity vectors are based on cross-correlating the intensity distributions over small areas of the flow, the resulting velocity field is a spatially averaged representation of the actual velocity field. This obviously has consequences for the accuracy of spatial derivatives of the velocity field, vorticity, and spatial correlation functions that are often derived from PIV velocity fields. PIV methods will obviously not be able to measure vertical velocities. These components might not only be missed—they can also induce errors in the estimated horizontal velocity fields. There also exists stereoscopic PIV, which uses two cameras to estimate all three velocity components (if the observed media allows this), but, for obvious practical reasons they are only available for laboratory experiments at present.

This is the kind of method that is actually currently used in NWP to derive winds from the images from geostationary satellites (Meteosat, GOES and GMS), with the so-called AMV (atmospheric motion vectors). The clouds are used in this process instead of the particles in classical PIV methods, and, owing to the peculiar nature of these tracers, the images require two pre-processing steps before the wind estimations. First, since the tracers (clouds) are not uniformly spread across the image, one has to detect, in each area, features of interest that would be tracked from one image to another with methods generally based on the gradient of the images. A number of quality control procedures are then applied to the selected features to retain only those that will be traceable. For instance, the gradients have to be large enough, cover a significant portion of the area, and avoid areas exhibiting more than two different gradient amplitude.

Second, images are two-dimensional representations of a three-dimensional, system, and therefore one has to allocate a height to a given tracked feature (with the reasonable assumption that clouds will not move up and down between two images). Historically (and still used as a fallback), the window channel brightness temperatures within the target area were analysed, and a mean value for the coldest 20% of the sample was used to represent the temperature at cloud top. This temperature was then compared with a numerical forecast of the vertical temperature profile to arrive at the height of the cloud. More recently AMV producers use so-called CO_2 slicing, where the ratio of the infrared and CO_2 absorption can be linked to the pressure at the location of the observed feature, or the H_2O intercept, which uses the fact that the radiances from a single-level cloud deck for two spectral bands vary linearly with cloud amount. Both CO_2 slicing and H_2O intercept require a model forecast for calibration. This step is an important source of errors in the derivation of AMVs, moreover, these errors are correlated with the model error, since a model forecast is used in the processes. This last aspect is an important breach of the common assumption of uncorrelated errors between model and observation that is made by all the main data assimilation algorithms.

Once this is done, PIV methods can be used to derive wind fields at several altitudes. In order to avoid too much noise in the resulting wind field, three images are used instead of two to derive two wind fields (from image 1 to 2 and from image 2 to 3), and only the sum of the resulting vectors that do not disagree are kept for the final product (for a complete description of these processes, see Schmetz et al. 1993; Nieman et al., 1997).

Because of their indirect nature and the complexity of the pre- and post-processing, describing the errors associated to such wind data is not straightforward—in particular, they are correlated, so complex observation error covariance matrices have to be constracted or the errors have to be significantly inflated—with a consequant reduction in impact. Bormann et al. (2003) found statistically significant spatial error correlations on scales up to about 800 km, and, moreover, they are strongly anisotropic.

16.3.3 Optical flow

Variational methods are probably better adapted to the treatment of the large image sequences available in geophysics, even though they are seldom used in practice. The most common is usually called *optical flow.* This classical approach in computer vision is based on the conservation of the global luminance between two images (Horn and Schunck, 1981), and is aimed at finding the velocity field $\mathbf{w} = (u, v)$ that transports two successive images I_1 and I_2 from one to the other:

$$I_1(x + u(x, y), y + v(x, y)) = I_0(x, y). \tag{16.3}$$

A vector field satisfying this equation is not unique (the aperture problem) and strict conservation is not realistic in general. Therefore, this is transformed into a minimization problem:

$$J(\mathbf{v}) = \frac{1}{2} \int \|I_1(x + u(x, y), y + v(x, y)) - I_0(x, y)\|^2 \, \mathrm{d}x \, \mathrm{d}y + \frac{1}{2}\alpha R(u, v). \tag{16.4}$$

The choice of the regularization term depends on the application field: see (Vigan et al. (2000), for oceanography and Amodei and Benbourhim (1991), for the wind field. A good review of possible choices of R in given by Auroux and Fehrenbach (2010) and some thoughts about preconditioning by Souopgui (2010). The advantage of such a method over PIV is that it can provide a denser velocity field for a fraction of the computing cost (see Fig. 16.5); however the adjustment or the relative weight of the two terms of the cost function (i.e. the value of the parameter α) and the choice of the regularization term are far from trivial (Souopgui, 2010).

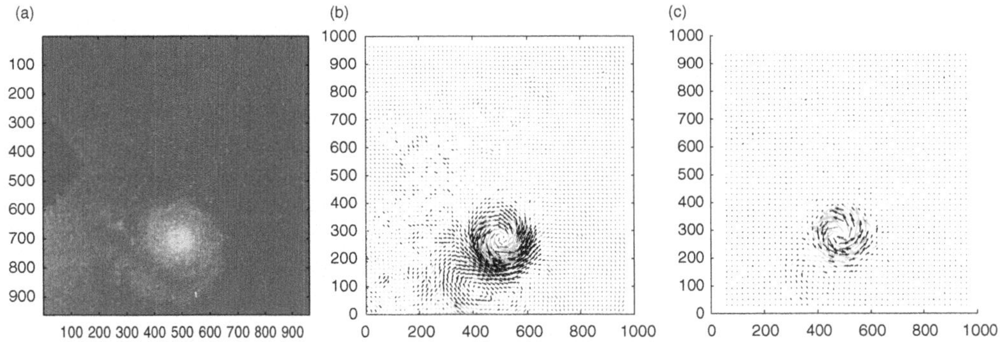

Fig. 16.5 (a) Original image. (b) Velocity field derived from optical flow. (c) Velocity field derived from PIV. Light, medium, and dark represent high, medium, and low confidence, respectively. (From Auroux and Fehrenbach, (2010). Copyright © 2010, Springer-Verlag.)

Once again, this motion estimator is meant to be used for a pair of images; however, as for statistical estimates (but in a more natural manner), it can be extended to use more frames altogether in order to improve consistency, by adding a governing equation for $\mathbf{v}$ to the optimality system generating the pseudo-observations.

As above, it assumes conservation of light and that the image pixels are transported from one image to the other by the velocity field to be estimated. In order to improve the temporal consistency of the estimated velocities, this method requires a simple temporal evolution model $\mathcal{N}$. Along with the transport equation of the pixels, (16.4), it uses

$$\begin{cases} \dfrac{\partial I}{\partial t} + \nabla I \cdot \mathbf{v} = 0, \\[2ex] \dfrac{\partial \mathbf{v}}{\partial t} = \mathcal{N}(\mathbf{v}), \quad I(0) = I_0, \quad \mathbf{v}(0) = \mathbf{v}_0. \end{cases} \tag{16.5}$$

The pixel values of the images I_{obs} are then assimilated in the previous model using variational data assimilation (see Section 2.1 in chapter 2 of this volume), where the control variable is the initial condition $\mathbf{v}_0$ of $\mathcal{N}$:

$$J(\mathbf{v}) = \frac{1}{2}\sum_{i=1}^{m-1} \int \|I_i(x + u(x,y), y + v(x,y)) - I_{i-1}(x,y)\|^2 \, \mathrm{d}x \, \mathrm{d}y + \frac{1}{2}\alpha R(u,v). \tag{16.6}$$

The advantage of this approach is that it takes into account all the images of the sequence at once and therefore can cope with missing data on some of the frames. This has been applied, using a constant velocity ($\mathcal{N} = 0$), in Korotaev et al., (2008) to derive ocean surface velocity fields from sea surface temperature on the Black Sea.

16.3.4 Advantages and limitations of the pseudo-observation technique

The main advantage of the pseudo-observation technique is its simplicity. Indeed, even though the algorithm used to produce pseudo-velocity observations may be sophisticated, the use of the pseudo-observation in an already-existing data assimilation scheme is relatively straightforward. There is no issue of differentiation and no heavy development is generally needed, since this processing is done offline.

However, despite their relative simplicity for implementation and their rapidity, pseudo-observation techniques based on motion estimation may suffer from some limitations. First, a frame-to-frame motion estimator does not take into account a sequence of more than two images. Because of the lack of consistency in time, it cannot capture the dynamical evolution of the image during the entire assimilation window. One of the consequences is that it cannot deal with missing data, which can be quite frequent in some of the applications targeted here. Missing data can be due to a glitch in the observing system or an obstacle (e.g. clouds). Solving (16.6) may mitigate this disadvantage, but the model $\mathcal{N}$ of the temporal evolution of the velocity depends only on $\mathbf{v}$ and therefore is likely to be a somewhat poor representation of the real behaviour of the velocity.

Neither of these two techniques takes into account physical information about the underlying physical processes observed in the images. Furthermore, some assumptions such as the conservation of the luminance are clearly not valid for fluid flow images showing broken-up structures. Finally, velocity fields obtained through the pseudo-observation approach are apparent velocities that can be quite different from the actual velocities. For instance, the absence of a tracer in a subset of an image sequence will lead to zero apparent velocities in that area. Such techniques then need to provide error maps to account for this kind of error in the interpretation of the observation. Therefore, the definition of the observation operators and the specification of the corresponding observation error matrices are likely to be difficult tasks (Bormann et al., 2003).

For all these reasons, an important research effort is being carried out to define a methodology for direct image data assimilation that would combine the information coming from image sequences and that coming from the numerical model, a priori knowledge, and classical observations in a more consistent way (e.g. without the need for pseudo-observations).

16.4 Direct assimilation of images

If one wants to avoid going through the pseudo-observation route, the main difficulty is to define a way of measuring the discrepancy between the model outputs and the images, that is, to define the observation space $\mathcal{I}$ and the associated distance $d_{\mathcal{I}}$: $\mathcal{I} \times \mathcal{I} \to \mathbb{R}$, and the observation operator $H : \mathcal{X} \to \mathcal{I}$. There are many ways to define this triplet, ranges from very basic to highly sophisticated. For instance, to define $\mathcal{I}$, we could consider the following:

1) *Frequency characteristics:* in this case, $\mathcal{I}$ is the image space of a multiscale transformation such as a wavelet on curvelet transform. This approach is interesting because of its well-known efficiency in data compression, denoising, and edge extraction.

2) *Geometric characteristics:* certain image analysis techniques can detect certain geometrical features in the image (e.g. by means of active contours). This feature may take the form of a parameterized curve in $\mathbb{R}^2$, which could be assimilated as Lagrangian observation. This approach will necessitate high-level image analysis.
3) *Qualitative characteristics:* in geophysics, there are several recognizable structures connected to physical phenomena (e.g. cyclones, and fronts,). We could define $\mathcal{I}$ as a dictionary of such pairs of observable structure/physical phenomenon. More specifically in meteorology, clouds may give information about the flow or the physical processes where they are located. For instance, the particular shape of stratus clouds is due to stratified flow, whereas cumulonimbuse clouds are known to involve strong convective processes.

In order to choose the above-defined triplet, one may start from the observation operator. $H_{\mathcal{I}}$ has to extract structures from model state variables (possibly in a differentiable manner), and this can be done, either by creating a synthetic image from the model output and then performing the same extraction on both observed image and synthetic image to define the observation space and its associated distance or by directly extracting these structures from the model output.

In the following, two examples of such definition are presented, one using the synthetic observations route, the second directly extracting dynamical features from the model output and matching them to the one from the images.

16.4.1 Synthetic images and multiscale transform

Features in geophysical images such as eddies or filaments may correspond to tracers in the observed system. A tracer can be defined as any quantity that is transported by the fluid flow. For instance, potential vorticity, temperature, and salinity can be considered as atmospheric and ocean tracers. Modelling and simulating their evolution from the model state variables may help to define the structures observation operator. Indeed, we may use their description maps (e.g. concentration maps) as a *synthetic image sequence* from which comparison with images (in a sense that remain to be defined) can be performed.

In other words, if $\mathbf{q}$ is the tracer concentration, one can model its evolution by transporting it through the model velocity field using a standard transport equation:

$$\frac{\partial \mathbf{q}}{\partial t} + \nabla \mathbf{q} \cdot \mathbf{u} = 0, \tag{16.7}$$

where $\mathbf{u}$ is the model velocity. This allows us to produce a field that will be comparable to observation from the model output. Once again, as for optical flow, this equation can be different, depending on the observed quantity. This way of proceeding requires the availability of the initial tracer field. It can be done thanks to an image at the beginning of the time period, but it cannot cope with missing data. Alternatively, one can augment the state vector with the concentration $\mathbf{q}$ and control its initial condition (or filter its trajectory depending on the assimilation scheme) along with the state vector's.

However, certain features of interest in geophysical fluid images do not correspond to any tracers (such as waves or dry intrusions in atmosphere images). In that case, other techniques should be developed to extract structures from model outputs without synthetic image sequence production.

From this observation operator, there are plenty of possible choices for the observation space. The simplest of these is L^2 with the classical norm: $d^2(\mathbf{x}, I) = \|H(\mathbf{x}) - I\|^2_{L^2}$. In the discrete case, this means comparing the synthetic image $H(\mathbf{x})$ with the observed one, pixel by pixel independently. Papadakis and Memin (2008) showed that assimilating pixel levels directly gives a better result than using an optical flow-based pseudo-observation approach. Also, in Corpetti et al. (2009), such a formalism is used to assimilate images of pressure differences into a three-layer simplified atmospheric model. Nevertheless, we can point out two possible limitations of the use of pixel level for realistic applications in geophysics: first, for large images, the number of observations will become enormous (one per pixel), making any assimilation, difficult; second, it cannot differentiate errors in placement with error in amplitude of the observed/simulated features.

This was the motivation for the approach proposed by Souopgui (2010) and Titaud et al. (2010) and based on a multiscale decomposition of both synthetic and observed images. In the present case, a curvelet transform was used, but similar results can be obtained using other multiscale decompositions, such as wavelets. The transform here actually uses the fact that any image can be decomposed as

$$I = \mathcal{C}(I) \equiv \sum_{j,k,l} \langle \varphi_{j,k,l}, I \rangle \varphi_{j,k,l}, \tag{16.8}$$

where $(\varphi_{j,k,l})_{j,k,l}$ are the elements of the curvelet frame. The inner product $\langle .,. \rangle$ is the usual $L^2(\mathbb{R}^2)$ one. The curvelet transform is a linear isometry, with its adjoint being given by the inverse transformation (reconstruction). This is a convenient feature for practical use in a variational data assimilation framework. A threshold $\mathcal{T}$ is then applied to this decomposition (only a subset of (i, j, k) terms are kept) in order to extract the dominant features of the images. Figure 16.6 shows the effect of two different thresholding techniques on the reconstruction of a cloud image. The observation becomes

$$\mathbf{y} = \mathcal{T} \circ \mathcal{C}(I). \tag{16.9}$$

In order to compare with the model, the same transformation is applied to the synthetic image obtained from the model output,

$$\mathcal{H}(\mathbf{x}) = \mathcal{T} \circ \mathcal{C}(\mathbf{q}), \tag{16.10}$$

with $\mathbf{q}$ linked to $\mathbf{x}$ by (16.7). Then we can define a seminorm on $\mathcal{I}$ from the term-by-term product of the curvelet coefficients that survived the threshold function.

The choice of the threshold $\mathcal{T}$ is of crucial importance, as shown in Souopgui (2010) and illustrated in Fig. 16.7, which compares the velocity field obtained by direct image assimilation in a shallow water model of images coming from a laboratory experiment

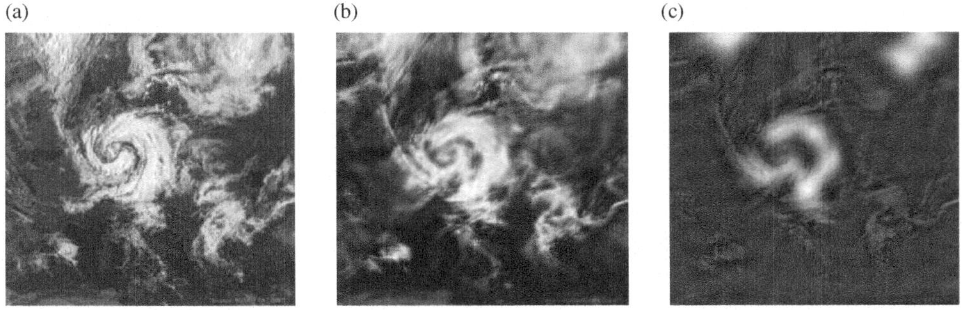

Fig. 16.6 Original image (a) and its curvelet/threshold/reconstruction product for a strong threshold applied to all the coefficients (b) or independently scale by scale (c), both retaining 10% of the coefficients in the curvelet frame.

performed at the CORIOLIS turntable in Grenoble, France (as a watermark behind the vector field). This example emphasizes the robustness of the multiscale approach with respect to the noise on the data. The L^2 norm (Fig. 16.7(a)) gives a noisy result with numerous small displacements, which are actually artefacts of the bad quality of the observation images. On the contrary, a hard threshold (Fig. 16.7(b), where all the curvelet coefficient below a given threshold are discarded) will tend to underestimate the velocity field, while using a finer threshold (Fig. 16.7(c), where a strong threshold is applied scale by scale independently), retaining the same number of coefficients, will lead to a better reconstruction of the isolated vortex, with the right amplitude and without the unwanted small currents. It has to be noted that the convergence of the minimization using the L^2 comparison tends to be much slower than that with the curvelet-based comparison. This is due to the larger number of observations (number of pixels versus number of curvelet coefficients) and to the stronger correlations between the pixels of a same image. This leads to degradation of the conditioning of the minimization problem (Haben et al., 2011). For longer assimilation windows, the L^2-norm-based approach does not even converge, while the truncated curvelet-based one still manages to find a minimum. These results are obtained using images of a relatively modest size (256×256)—the problem is likely to be amplified when switching to Meteosat-type images (5000×5000).

Moreover, the use of the pixel level may then lead to some increase in the representativity error. Note that small-scale structures may be the consequence of mesoscale dynamical processes. This is particularly true for tracer-mixing processes, where spirals and filaments may come from the variability of the meso scale velocity field as it is clearly shown by Lehahn et al. (2007). In a pixel basis, those filaments are not represented as coherent structures while another mathematical modelling could avoid this problem.

Compared with the classical L^2 norm, the use of a higher level of interpretation reduces the size of the observation vector significantly (typically only a few percent of the coefficients are kept). Additionally, it tends to filter out small scales that cannot be represented by the model anyway.

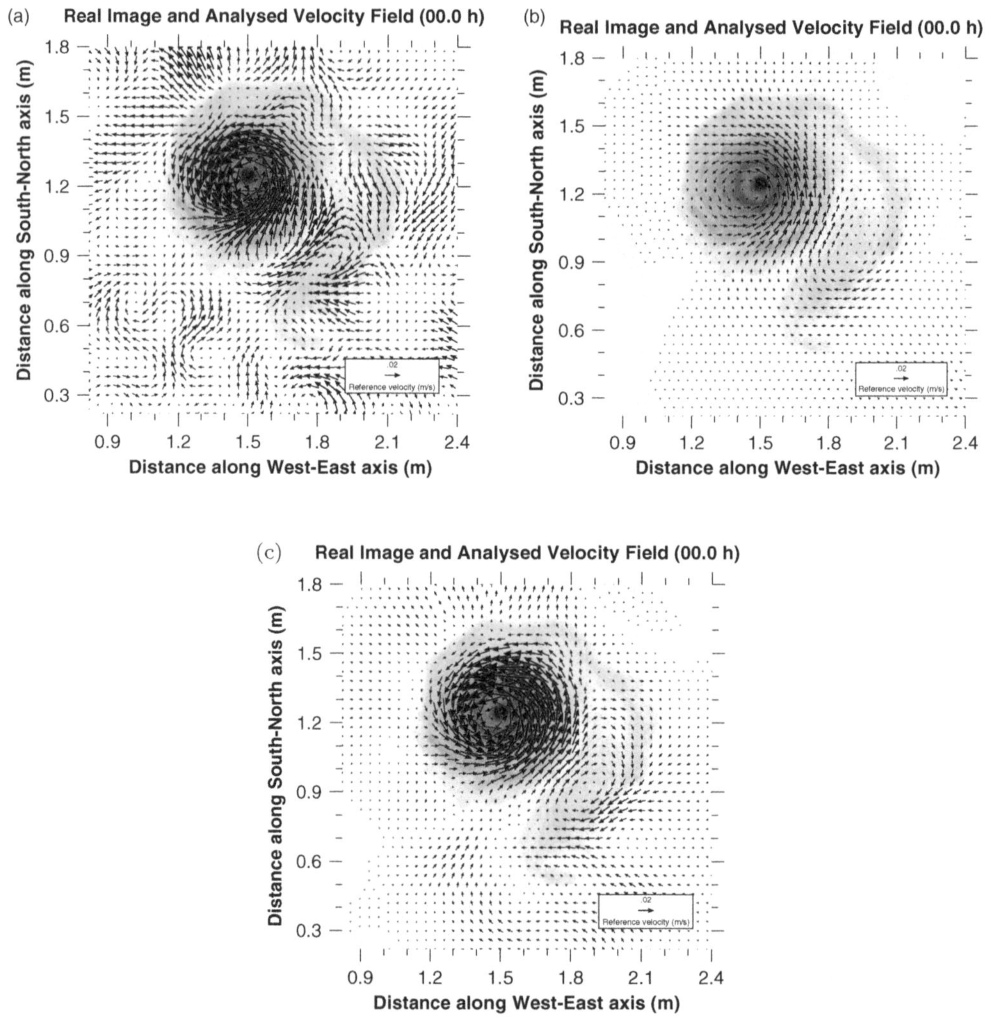

Fig. 16.7 Initial velocity field reconstructed by direct assimilation of a real image sequence using different observation operators and norms. (a) using an L^2 norm; (b) curvelet-based, hard threshold; (c) curvelet-based, fire threshold.

16.4.2 Observation operators based on Lagragian coherent structures

Using a synthetic image sequence to define a structure observation may be difficult because numerical advection schemes are known to smooth discontinuities out. This drawback may lead to representativeness error: observation models may not be able to represent observed image patterns. To overcome this problem, Titaud et al. (2011) suggest to define observation operators based on the computation of Lagrangian coherent structures of the fluid.

Lagrangian coherent structures (LCS) delimit regions of whirls, stretching, or contraction of tracer Ottino. This concept is well defined for time-independent dynamic systems, where these regions correspond to stable and unstable manifolds of hyperbolic trajectories (Wiggins, 1992): contraction is observed along stable manifolds whereas unstable manifolds correspond to divergent directions along which the tracer is stretched. This concept was generalized for flows with general time dependence (particularly geophysical flows) (Haller and Yuan, 2000; Haller, 2011) and they are usually identified in a practical manner as maximizing ridges of a so-called backward finite-time lyapunov exponent (FTLE) field (Haller, 2000, 2001, 2002; Shadden et al., 2005; Mathur et al., 2007. Let $\mathbf{x}(t) = \mathbf{x}(t; \mathbf{x}_0, t_0)$ be the position of a Lagrangian particle at time t, started at $\mathbf{x}_0$ at $t = t_0$ and advected by the time-dependent fluid flow $\mathbf{u}(\mathbf{x}, t)$, $\mathbf{x} \in \Omega \subset \mathbb{R}^2$, $t \in [t_0, t_0 + T]$. The *forward* FTLE $\sigma_{t_0}^{t_0+T}(\mathbf{x}_0)$ at defined as

$$\sigma_{t_0}^{t_0+T}(\mathbf{x}_0) = \frac{1}{|T|} \ln \sqrt{\lambda_{\max}(\Delta)}, \tag{16.11}$$

where $\lambda_{\max}$ is the maximum eigenvalue of the Cauchy–Green strain tensor

$$\Delta = \left[\nabla \phi_{t_0}^{t_0+T}(\mathbf{x}_0)\right]^* \left[\nabla \phi_{t_0}^{t_0+T}(\mathbf{x}_0)\right],$$

where $\phi_{t_0}^{t_0+T}$ is the flow map of the Lagrangian system. The FTLE corresponds to the growth factor of the norm of a perturbation started around $\mathbf{x}_0$ and transported by the flow after the finite advection time T. The associated eigenvector $\varphi_{t_0}^{t_0+T}(\mathbf{x}_0)$ is referred to as the *forward* finite-time Lyapunov vector (FTLV): it corresponds to the direction of maximal stretching of the tracer. FTLE represents the rate of separation of initially neighbouring particles over the finite time window $[t_0, t_0 + T]$. *Backward* FTLE-Vs are similarly defined, with the time direction being inverted in the computation of the Lagrangian trajectory. For more details on the practical computation of FTLE-Vs, see, for example, Shadden et al. (2005, 2009), and Ott (1993) for any types of flows and d'Ovidio et al. (2004) for oceanic flows. FTLE (FTLV) is a scalar (vector) that is computed at a given location $\mathbf{x}_0$. Seeding a domain with particles initially located on a grid leads to the computation of discretized scalar (FTLE) and vector (FTLV) fields. Note that the FTLE-V terminology may be misleading: they are more like practical ways of computing singular vectors of the transport equation than actual Lyapunov vectors.

Backward FTLE fields show contours that correspond reasonably well to the main structures such as filaments, fronts, and spirals that appear in geophysical and biogeochemical tracer fields Beron-Vera and Olascoaga, 2009; Shadden et al., 2009; Olascoaga et al., 2006, 2008). Figure 16.8(a) shows the contours of the FTLE field (16.11) over sea surface temperature (SST) (a) and mixed layer phytoplankton (MLP) (b). The FTLE field is shown in the Fig. 16.8(c).

FTLEs are computed from a sequence of *mesoscale* (1/4°) time-dependent velocity fields filtered from a high-resolution (1/54°) idealized simulation of the North Atlantic Ocean (carried out by Lévy et al., 2009). The high-resolution tracer fields SST and MLP come from the same simulation. FTLE contours of match quite

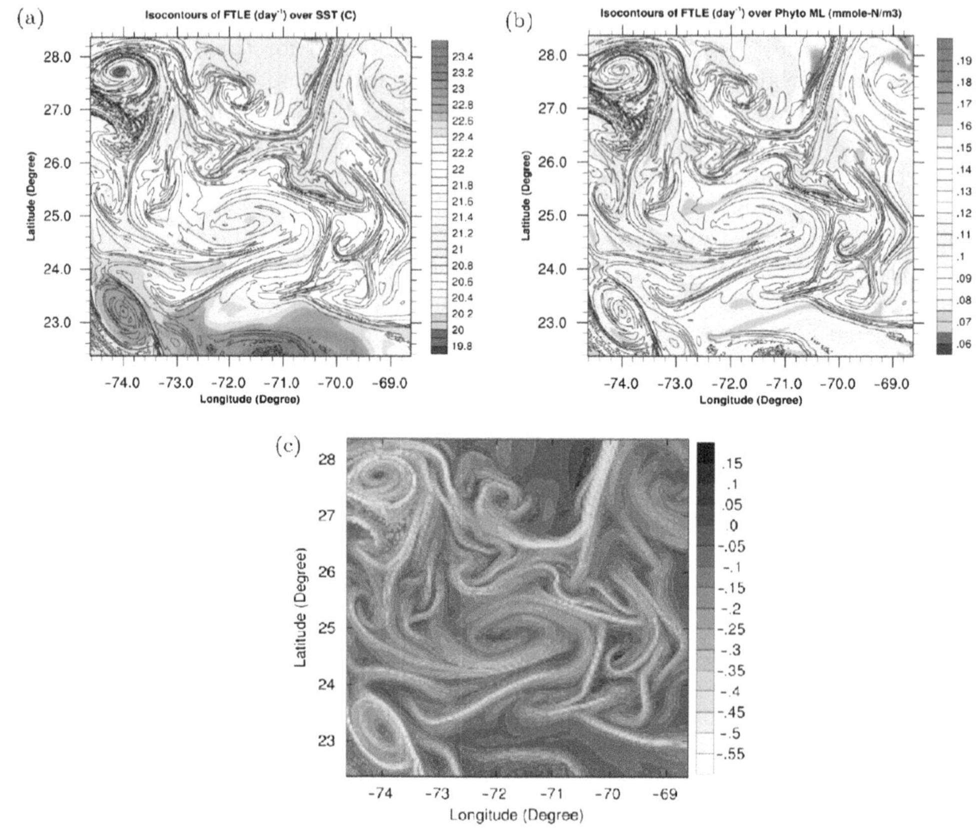

Fig. 16.8 Contours of finite-time lyapunov exponents over (a) sea surface temperature and (b) mixed layer phytoplancton. (c) Finite-Time Lyapunov Exponents field From a 1/54° North Atlantic Ocean simulation.

well the main patterns in the corresponding SST and MLP fields of this simulation. Lapeyre (2002) shows that for a freely decaying two-dimensional turbulence flow, the orientation of the gradient of a *passive* tracer converges to that of backward FTLVs. Such alignment properties have also been observed for realistic oceanic flows and tracers (d'Ovidio et al., 2009b). Figure 16.9(c) shows the orientations of the backward FTLV that correspond to the aforementioned FTLE contour field. Tracer gradients, show similar orientations as can be see in Figs. 16.9(a) for SST and 16.9(b) for MLP. These two properties (pattern matching between tracer and FTLE scalar fields and tracer gradient orientation alignment with FTLV orientation) should be exploited in a direct image assimilation framework. Furthermore, Beron-Vera et al. (2010) showed, using real data, that these properties remain valid with a mesoscale advection, i.e. when the resolution of the velocity field—from which FTLE-V are computed—is much lower than the resolution of the observed tracer field.

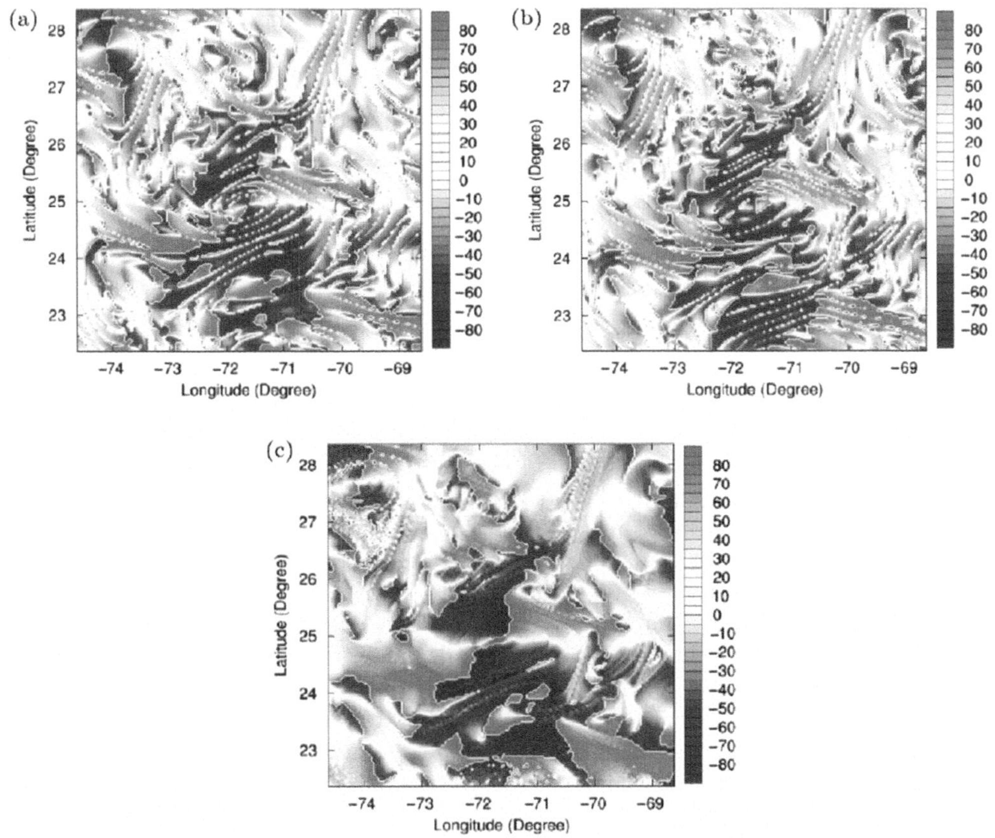

Fig. 16.9 Orientations of (a) sea surface temperature, (b) mixed layer phytoplankton gradients, and (c) finite-time Lyapunov vectors. From a 1/54° North Atlantic Ocean simulation.

This behaviour was also mentioned using another Lagrangian tool (FSLE) by Lehahn et al. (2007) and d'Ovidio et al. (2009b). Our synthetic data exhibit the same behaviour: the FTLE field of Fig. 16.8 is computed from a mesoscale 1/4° velocity field on the same high-resolution (1/54°) grid as that used to compute the SST and MLP fields. This feature is crucial from both a practical and a physical point of view: first, velocity fields obtained from ocean global circulation models do not often provide more than mesoscale information, whereas tracer images contain submesoscale information; second, FTLE-V may be used to quantify and characterize the link between scales.

Titaud et al. (2011) suggest constructing observation operators using information given by the computation of Lagrangian coherent structures. The main idea is to exploit the two aforementioned properties of the FTLE-V to construct a triplet of direct assimilation of images:

- FTLE: the observation space $\mathcal{I}$ is defined as a space of contours. Backward FTLE is viewed as an observation operator acting on the flow field $\mathbf{u}$ (subset of control variable) onto the image space $\mathcal{I}$:

$$\begin{aligned} \mathcal{H}[\mathbf{u}] : \Omega &\to \mathbb{R}, \\ \mathbf{x} &\mapsto \sigma_{t_0}^{t_0+T}(\mathbf{x}). \end{aligned} \tag{16.12}$$

- FTLV: the observation space is the space of function on $\mathbb{R}^2$ with values in the Euclidean unit sphere equipped with an angular distance. The observation operator is simply the FTLV field:

$$\begin{aligned} \mathcal{H}[\mathbf{u}] : \Omega &\to \mathbb{R}, \\ \mathbf{x} &\mapsto \varphi_{t_0}^{t_0+T}(\mathbf{x}). \end{aligned} \tag{16.13}$$

For the moment, few studies exist about the use of this kind of operator in direct image assimilation. In Titaud et al. (2011), a sensibility study proves the potential of this tool by showing that computation of FTLE-V is sensitive (in the direct image assimilation framework) to small perturbations of the velocity field. Figure 16.10 shows the variation of a sensitivity function (cost function) associated with the FTLE-V-based observation operators with respect to the amplitude of nine random perturbations applied to a reference velocity field $\mathbf{u}$. Variations clearly have a minimum at the reference state and exhibit a convex shape that suggests a good situation for the inversion.

Even if not used in this description, we also mention the finite-size Lyapunov exponent (FSLE) (Aurell et al., 1997; Artale et al., 1997), which is another Lagrangian tool that is commonly used in oceanographic contexts for studying mixing processes (d'Ovidio et al., 2004, 2009 a,b; Lehahn et al., 2007 and references therein). In Gaultier et al. (2012), such an FSLE field was used to reconstruct a ocean surface velocity field.

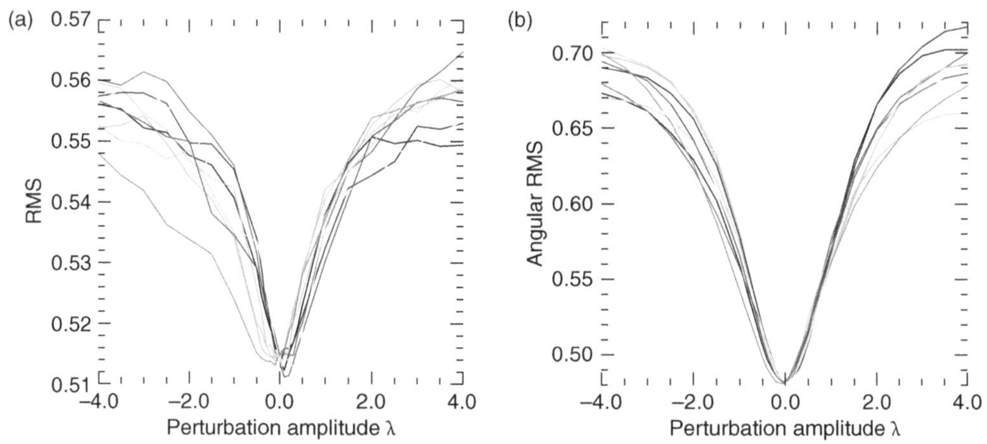

Fig. 16.10 Variation of a sensitivity function (cost function) associated with the FTLE-V-based observation operators with respect to the amplitude of nine random perturbations applied to a reference velocity field $\mathbf{u}$. (a) SST: sensitivity of the FTLE-based observation operator; (b) SST: sensitivity of the FTLV-based observation operator.

References

Adrian, R. J. (1991). Particle-imaging techniques for experimental fluid mechanics. *Ann. Revi. Fluid Mech.*, **23**, 261–304.

Amodei, L. and Benbourhim, M. N. (1991). A vector spline approximation. *J. Approx. Theory*, **67**, 1–29.

Artale, V., Boffetta, G., Celani, A., Cencini, M., and Vulpiani, A. (1997). Dispersion of passive tracers in closed basins: Beyond the diffusion coefficient. *Phys. Fluids*, **9**, 3162–3171.

Aubert, G. and KornProbst, P. (2006). *Mathematical Problems in Image Processing.* Springer, New York.

Aurell, E., Boffetta, G., Crisanti, A., Paladin, G., and Vulpiani, A. (1997). Predictability in the large: an extension of the concept of Lyapunov exponent. *J. Phys. A: Math. Gen.*, **30**, 1–26.

Auroux, D. and Fehrenbach, J. (2010). Identification of velocity fields for geophysical fluids from a sequence of images. *Exp. Fluids*, **50**, 313–328.

Beron-Vera, F. J. (2010). Mixing by low- and high-resolution surface geostrophic currents. *J. Geophys. Res.*, **115**, C10027.

Beron-Vera, F. J. and Olascoaga, M. J. (2009). An assessment of the importance of chaotic stirring and turbulent mixing on the West Florida Shelf. *J. Phys. Oceanogr.*, **39**, 1743–1755.

Beron-Vera, F. J., Olascoaga, M. J., and Goni, G. J. (2010). Surface ocean mixing inferred from different multisatellite altimetry measurements. *J. Phys. Oceanogr.*, **40**, 2466–2480.

Bormann, N., Saarinen, S., and Kelly, G. (2003). The spatial structure of observation errors in atmospheric motion vectors from geostationary satellite data. *Mon. Weather Rev.*, **131**, 706–718.

Candès, E. and Donoho, D. L. (2003). New tight frames of curvelets and optimal representations of objects with piecewise C2 singularities. *Commun. Pure Appl. Math.*, **57**, 219–266.

Chan, T. F. and Shen, J. (2005). *Image Processing and Analysis: Variational, PDE, Wavelet, and Stochastic methods.* SIAM, Philadelphia.

Corpetti, T., Héas, P., Memin, É., and Papadakis, N. (2009). Pressure image assimilation for atmospheric motion estimation. *Tellus A*, **61**, 160–178.

d'Ovidio, F., Fernández, V., Hernández-García, E., and López, C. (2004). Mixing structures in the Mediterranean Sea from finite-size Lyapunov exponents. *Geophys. Res. Lett.*, **31**, L17203.

d'Ovidio, F., Isern-Fontanet, J., López, C., Hernández-García, E., and García-Ladona, E. (2009a). Comparison between Eulerian diagnostics and finite-size Lyapunov exponents computed from altimetry in the Algerian basin. *Deep Sea Res., Part I: Oceanogr. Res. Papers*, **56**, 15–31.

d'Ovidio, F., Taillandier, V., Taupier-Letage, I., and Mortier, L. (2009b). Lagrangian validation of the Mediterranean mean dynamic topography by extraction of tracer frontal structures. *Mercator Ocean Q. Newsletter*, **32**, 24–32.

Gaultier, L., Verron, J., Brankart, J.-M., Titaud, O., and Brasseur, P. (2013). On the inversion of submesoscale tracer fields to estimate the surface ocean circulation. *J. Mar. Syst.*, **126**, 33–42.

Geman, S. and Geman, D. (1984). Stochastic relaxation, Gibbs distributions, and the Bayesian restoration of images. *IEEE Trans. Pattern Anal. Mach. Intell.*, **6**, 721–741.

Haben, S. A., Lawless, A. S., and Nichols, N. K. (2011). Conditioning and preconditioning of the variational data assimilation problem. *Compute. Fluids*, **46**, 252–256.

Haller, G. (2000). Finding finite-time invariant manifolds in two-dimensional velocity fields. *Chaos*, **10**, 99–108.

Haller, G. (2001). Lagrangian structures and the rate of strain in a partition of two-dimensional turbulence. *Phys. Fluids*, **13**, 3365–3385.

Haller, G. (2002). Lagrangian coherent structures from approximate velocity data. *Phys. Fluids*, **14**, 1851–1861.

Haller, G. (2011). A variational theory of hyperbolic Lagrangian coherent structures. *Physica D*, **240**, 574–598.

Haller, G. and Yuan, G. (2000). Lagrangian coherent structures and mixing in two-dimensional turbulence. *Physica D*, **147**, 352–370.

Horn, B. K. P. and Schunck, B. G. (1981). Determining optical flow. *Artif. Intell.*, **17**, 185–203.

Korotaev, G., Huot, E., Le Dimet, F., Herlin, I., Stanichny, S., Solovyev, D. and Wu, L. (2008). Retrieving ocean surface current by 4-D variational assimilation of sea surface temperature images. *Remote Sens. Environ.*, **112**, 1464–1475.

Lapeyre, G. (2002). Characterization of finite-time Lyapunov exponents and vectors in two-dimensional turbulence. *Chaos*, **12**, 688–11.

Lehahn, Y., d'Ovidio, F., Lévy, M., and Heifetz, E. (2007). Stirring of the northeast Atlantic spring bloom: a Lagrangian analysis based on multisatellite data. *J. Geophys. Res.*, **112**, 15.

Lévy, M., Iovino, D., Masson, S., Madec, G., Klein, P., Tréguier, A.-M., and Takahashi, K. (2009). Remote impacts of Sub-Mesoscale Dynamics on new production. *Mercator Ocean Q. Newsletter*, **13**, 13–17.

Mallat, S. G. (1998). *A Wavelet Tour of Signal Processing.* Morgan Kaufmann, Burtengton, MA.

Mathur, M., Haller, G., Peacock, T., Ruppert-Felsot, J. E., and Swinney, H. L. (2007). Uncovering the Lagrangian skeleton of turbulence. *Phys. Rev. Lett.*, **98**, 144502.

Michel, Y. (2011). Displacing potential vorticity structures by the assimilation of pseudo-observations. *Mon. Weather. Rev.*, **139**, 549–565.

Michel, Y. and Bouttier, F. (2006). Automated tracking of dry intrusions on satellite water vapour imagery and model output. *Q. J. R. Meteorol. Soc.*, **132**, 2257–2276.

Nieman, S., Menzel, P., Hayden, C., Gray, D., Wanzong, S., Velden, C., and Daniels, J. (1997). WP Menzel, CM Hayden, D. Gray, ST Wanzong, CS Velden, and J. Daniels, 1997: Fully automated cloud-drift winds in NESDIS operations. *Bull. Am. Meteorol. Soc*, **78**, 1121–1133.

Olascoaga, M. J., Rypina, I. I., Brown, M. G., Beron-Vera, F. J., Koçak, H., Brand, L. E., Halliwell, G. R., and Shay, L. K. (2006). Persistent transport barrier on the West Florida Shelf. *Geophys. Res. Lett.*, **33**, L22603.

Olascoaga, M. J., Beron-Vera, F. J., Brand, L. E., and Koçak, H. (2008). Tracing the early development of harmful algal blooms on the West Florida Shelf with the aid of Lagrangian coherent structures. *J. Geophys. Res.*, **113**, C12014.

O&S.I. SAF Project Team. (2013). Low Earth Orbiter Sea Surface Temperature Product User Manual. *available at <http://www.osi-saf.org>*.

Ott, E. (1993). *Chaos in Dynamical Systems.* Cambridge University Press, Cambridge.

Ottino, J. M. (1989). *The Kinematics of Mixing: Stretching, Chaos, and Transport.* Cambridge University Press, Cambridge.

Papadakis, N. and Memin, É. (2008). Variational assimilation of fluid motion from image sequence. *SIAM J. Imaging Sci.*, **1**, 343–363.

Santurette, P. and Georgiev, C. (2005). *Weather Analysis and Forecasting: Applying Satellite Water Vapor Imagery and Potential Vorticity Analysis.* Elsevier Academic Press. Borlington, MA.

Schmetz, J., Holmlund, K., Hoffman, J., Strauss, B., Mason, B., Gaertner, V., Koch, A., and Van de Berg, L. (1993). Operational cloud-motion winds from Meteosat infrared images. *J. Appli. Meteorolo.*, **32**, 1206–1225.

Shadden, S. C., Lekien, F., and Marsden, J. E. (2005). Definition and properties of Lagrangian coherent structures from finite-time Lyapunov exponents in two-dimensional aperiodic flows. *Physica D*, **212**, 271–304.

Shadden, S. C., Lekien, F., Paduan, J. D., Chavez, F. P., and Marsden, J. E. (2009). The correlation between surface drifters and coherent structures based on high-frequency radar data in Monterey Bay. *Deep Sea Res., Part II: Top. Stud. Oceanogr.*, **56**, 161–172.

Souopgui, I. (2010). Assimilation d'images pour les fluides géophysiques. PhD thesis, Université de Grenoble et Université de Yaoundé I.

Titaud, O., Brankart, J.-M., and Verron, J. (2011). On the use of finite-time Lyapunov exponents and vectors for direct assimilation of tracer images into ocean models. *Tellus A*, **63**, 1038–1051.

Titaud, O., Vidard, A., Souopgui, I., and Le Dimet, F.-X. (2010). Assimilation of image sequences in numerical models. *Tellus A*, **62**, 30–47.

Vigan, X., Provost, C., Bleck, R., and Courtier, P. (2000). Sea surface velocities from sea surface temperature image sequences 1. Method and validation using primitive equation model output. *J. Geophys. Res.*, **105**, 19499–19514.

Wiggins, S. (1992). *Chaotic Transport in Dynamical Systems.* Springer, New York.

17

Multigrid algorithms and local mesh refinement methods in the context of variational data assimilation

L. DEBREU[1], E. NEVEU[2], E. SIMON[3], and F.-X. LE DIMET[4]

[1]INRIA, MOISE project team and Laboratoire Jean Kuntzmann, Grenoble, France

[2]Université de Savoie, LAMA, UMR CNRS 5127, 73376 Le Bourget-du-Lac, France

[3]Nansen Environmental and Remote Sensing Center, Bergen, Norway

[4]MOISE Project Team and Laboratoire Jean Kuntzmann Université Grenoble Alpes Grenoble, France

Advanced Data Assimilation for Geosciences. First Edition.
Edited by É. Blayo, M. Bocquet, E. Cosme, and L. F. Cugliandolo.

Chapter Contents

We investigate the use of multigrid methods and local mesh refinement algorithms in the context of the variational data assimilation method. Section 17.1 recalls basic properties of the traditional variational data assimilation method and focuses on the role of the background error covariance matrix. Section 17.2 shows how multigrid algorithms can efficiently solve the resulting system. Section 17.3 deals with local mesh refinements and Section 17.4 gives some ideas on how to couple the two approaches in the view of local multigrid algorithms.

17.1 Structure of the variational data assimilation problem

17.1.1 Variational data assimilation and background error covariance matrix

We refer to Chapters 1 and 2 of this volume for a general introduction to variational data assimilation methods. We consider the time evolution of a system governed by the following equation:

$$\begin{cases} \dfrac{\mathrm{d}X}{\mathrm{d}t} = F(X), \\ X(t = t_0) = \mathbf{x}, \end{cases} \tag{17.1}$$

where $\mathbf{x}$ is the initial condition at time $t = t_0$ and is our control parameter. The variational data assimilation problem consists in finding the minimum of a cost function $J(\mathbf{x})$ that measures the distance from the numerical model solution to the observations and includes a background or regularization term associated with a first guess $\mathbf{x}_b$,

$$J(\mathbf{x}) = \frac{1}{2}\|\mathbf{x} - \mathbf{x}_b\|^2_{\mathbf{B}^{-1}} + \frac{1}{2}\|H\left(X(\mathbf{x}, t)\right) - \mathbf{y}\|^2_{\mathbf{R}^{-1}}. \tag{17.2}$$

Here $\mathbf{y}$ are the observations. H is the observation operator from the model to the observations space, and $\mathbf{R}$ and $\mathbf{B}$ are respectively the observations and background error covariances matrices. In the following, $\mathbf{R}$ will be assumed to be a diagonal matrix with constant standard deviation equal to σ_{obs}. At a minimum $\mathbf{x}^*$ of J, the gradient is zero and the Euler equation reads

$$\nabla_{\mathbf{x}} J(\mathbf{x}^*) = 0. \tag{17.3}$$

In order to study the main characteristics of the resulting system, we consider the case where the model F and the observations operator H are linear. In that case, the cost function is quadratic and, after discretization, the solution of (17.3) is equivalent to the solution of

$$\mathbf{A}\mathbf{x}^* = \mathbf{b}, \tag{17.4}$$

where $\mathbf{A}$ is the Hessian of the cost function:

$$\mathbf{A} = \mathbf{B}^{-1} + \mathbf{H}^{\mathrm{T}}\mathbf{R}^{-1}\mathbf{H},$$

where $\mathbf{H}$ includes both the model and the observation operators and the right-hand side $\mathbf{b}$ is given by

$$\mathbf{b} = \mathbf{B}^{-1}\mathbf{x}_b + \mathbf{H}^{\mathrm{T}}\mathbf{R}^{-1}\mathbf{y}.$$

It is known that the solution of (17.4) can be put into the form

$$\mathbf{x}^* = \mathbf{x}_b + \mathbf{B}\mathbf{H}^{\mathrm{T}}\left(\mathbf{R} + \mathbf{H}\mathbf{B}\mathbf{H}^{\mathrm{T}}\right)^{-1}(\mathbf{y} - \mathbf{H}\mathbf{x}_b). \tag{17.5}$$

Equation (17.5) shows that a correction is computed in the observation space weighted by the observations and background error covariance matrices. This correction is brought back to the model state and then spatially distributed according to the $\mathbf{B}$ matrix. In practice, (17.4) corresponds to a large system that is solved using a minimization algorithm for which $\mathbf{x}_b$ is the starting point. The scales of the final correction $\mathbf{x}^\star - \mathbf{x}_b$ are thus prescribed by the correlation lengths specified by the $\mathbf{B}$ matrix.

Preconditioning

The performance of an iterative minimization method for the solution of (17.4) is linked to the condition number of this system:

$$\mathcal{K}(\mathbf{A}) = \frac{\lambda_{\max}(\mathbf{A})}{\lambda_{\min}(\mathbf{A})},$$

where $\lambda_{\min}(\mathbf{A})$ and $\lambda_{\max}(\mathbf{A})$ are the smallest and largest eigenvalues of $\mathbf{A}$. The smaller the condition number, the smaller the rate of convergence. When the matrix $\mathbf{B}$ and its square root are available, an alternative, symmetrically preconditioned, form of (17.4) can be deduced (see Courtier (1997) and Chapters 1 and 2 of this volume):

$$\mathbf{A}^P\mathbf{y}^* = \mathbf{B}^{\mathrm{T}/2}\mathbf{A}\mathbf{B}^{1/2}\mathbf{y}^* = \mathbf{B}^{1/2}\mathbf{b}, \tag{17.6}$$

where $\mathbf{x}^* = \mathbf{B}^{1/2}\mathbf{y}^*$. We obtain

$$\mathbf{A}^P = \mathbf{B}^{\mathrm{T}/2}\mathbf{A}\mathbf{B}^{1/2} = \mathbf{I} + (\mathbf{H}\mathbf{B}^{1/2})^{\mathrm{T}}\mathbf{R}^{-1}(\mathbf{H}\mathbf{B}^{1/2}).$$

$\mathbf{A}^P$ has a smallest eigenvalue larger than one and its condition number $\mathcal{K}(\mathbf{A}^P)$ is bounded.

17.1.2 Eigenstructure/ellipticity

From the structure of the matrix $\mathbf{A}$, we can deduce some of the main characteristics of the problem:

1. The background term of the cost function corresponds to the $\mathbf{B}^{-1}$ part of $\mathbf{A}$. While $\mathbf{B}$ is a smoothing operator that distributes the correction over the model space, the $\mathbf{B}^{-1}$ operator is a roughening operator that amplifies small scales. In that sense, it corresponds to an elliptic operator: the small-scale eigenvectors are associated with large eigenvalues while the large-scale eigenvectors are associated with small eigenvalues. This elliptic nature is at the base of the use of multigrid methods as introduced in Section 17.2.

2. If the system is not fully observed ($\mathbf{H}^{\mathrm{T}}\mathbf{H} \neq \mathbf{I}$), but when there is one observation point every $L_{\mathrm{obs}}/\Delta x$ grid points, the eigenvalues may become clustered if the background error correlation length is relatively small. Indeed, in this case, the eigenvalue problem can be approximated in $p = L/L_{\mathrm{obs}}$ similar eigenvalues problems between two observation points. It is known that this clustering of eigenvalues accelerates the convergence of the minimization procedure.

While improving the condition number of the resulting system, the preconditioning (the use of $\mathbf{A}^P$ instead of $\mathbf{A}$) removes the ellipticity. Obviously, the preconditioning is efficient when the relative weight of the background term is not negligible in comparison with the weight on the observation error. For time-dependent problems, the weight of the observations term is also increased accordingly to the size of the assimilation windows. At some point, the preconditioned version behaves worse than the non-preconditioned version.

The notion of ellipticity introduced above is at the starting point of the use of multigrid algorithms. Several results on the eigenstructure of the data assimilation system have been given in Haben et al. (2011). In the case of a fully observed system $\mathbf{H} = \mathbf{I}$, and with a linear advection equation, a detailed study can be done (Neveu et al., 2011).

An example on a linear advection model

We use an elementary advection equation on the one-dimensional periodic domain Ω:

$$\begin{cases} \partial_t \mathbf{X} + c\,\partial_x \mathbf{X} = 0, \quad \text{with } c > 0, \\ \mathbf{X}(x, t = 0) = \mathbf{x}(x), \end{cases} \tag{17.7}$$

with periodic boundary conditions. We suppose that the observations data set $\mathbf{y}^o$ is available at each grid point and we use the following cost function:

$$J(\mathbf{x}) = \frac{1}{2\sigma_o^2 T} \int_0^T \|\mathbf{X}(x,t) - \mathbf{y}^o\|^2 \, dt + \frac{\beta^2}{2\sigma_b^2} \left\| \left(\mathbf{I} - \frac{L_{\mathrm{corr}}^2}{4} \triangle \right) (\mathbf{x} - \mathbf{x}^b) \right\|^2, \tag{17.8}$$

where L_{corr} is the background error correlation length and β is a non-dimensional positive constant that adapts the weight of the regularization term. Using the continuous solution of (17.7), $\mathbf{X}(x,t) = \mathbf{x}(x - ct)$, the expression for the Hessian of J (the discrete form of the Hessian corresponds to the matrix $\mathbf{A}$ introduced above) can be derived:

$$\mathcal{H} = \frac{1}{\sigma_o^2} \left[1 + \gamma^2 \left(\mathbf{I} - \frac{L_{\mathrm{corr}}^2}{4} \frac{\mathrm{d}^2}{\mathrm{d}x^2} \right)^2 \right], \quad \text{where } \gamma = \frac{\beta \sigma_o}{\sigma_b}. \tag{17.9}$$

$\widetilde{\mathcal{H}}(k)$, the Fourier symbol of the Hessian, defined by $\mathcal{H}(e^{ikx}) = \widetilde{\mathcal{H}}(k) e^{ikx}$, k being a wavenumber, is given by the expression

$$\widetilde{\mathcal{H}}(k) = \frac{1}{\sigma_o^2} \left[1 + \gamma^2 \left(1 + \frac{L_{\mathrm{corr}}^2}{2} k^2 + \frac{L_{\mathrm{corr}}^4}{16} k^4 \right) \right]. \tag{17.10}$$

This shows that as soon as $\beta > 0$, the system is elliptic ($|\widetilde{\mathcal{H}}(k)| \geq 1 + \alpha k^2, \alpha > 0$). When (17.7) is discretized using a finite-difference method based on a uniform grid with time step Δt and space step Δx, using an Euler upwind scheme, resulting in a scheme that is first order in space and time, it is well known that the numerical solution produced by the numerical scheme is close at large scales to the exact solution of the following advection–diffusion equation (see Durran, 2010, p. 109):

$$\partial_t \mathbf{X} + c\,\partial_x \mathbf{X} = \epsilon \partial_{xx} \mathbf{X}, \text{ with } \epsilon = \frac{c}{2}\,(\Delta x - c\,\Delta t), \tag{17.11}$$

where ϵ is positive because of the stability constraint $\mu = c\Delta t/\Delta x \leq 1$. The additional term on the right-hand side corresponds to a diffusive term.

Using the same cost function with the advection–diffusion equation (17.11) changes the expression for the Hessian, which in Fourier space is

$$\widetilde{\mathcal{H}}(k) = \frac{1 - e^{-2\epsilon k^2 T}}{\sigma_o^2 2\epsilon k^2} + \frac{\beta^2}{\sigma_b^2}\left(1 + \frac{L_{\text{corr}}^2}{2} k^2 + \frac{L_{\text{corr}}^4}{16} k^4\right). \tag{17.12}$$

Its Taylor expansion at order 2 is

$$\widetilde{\mathcal{H}}(k) = \frac{1}{\sigma_o^2}\left(1 - \epsilon T k^2 + \gamma^2 \left[1 + \frac{L_{\text{corr}}^2}{2} k^2\right)\right] + O(k^4). \tag{17.13}$$

Thus shows that at large scales there is a balance between the artificial numerical diffusion and the regularization term. We note that γ_{lim}, the value of γ that cancels the second-order term, is given by

$$\gamma_{\text{lim}} = \frac{\sqrt{2\epsilon T}}{L_{\text{corr}}}. \tag{17.14}$$

If $\gamma > \gamma_{\text{lim}}$, the Hessian is elliptic (i.e. $|\widetilde{\mathcal{H}}(k)| \geq 1 + \alpha k^2, \alpha > 0$). $\gamma > \gamma_{\text{lim}}$ occurs for

$$\beta > \frac{\sigma_b}{\sigma_o} \frac{\sqrt{2\epsilon T}}{L_{\text{corr}}}.$$

This clearly relates the ellipticity to (i) the ratio between the weights of the observation and background terms (σ_b/σ_o) and (ii) the ratio between the diffusion length and the correlation length.

17.2 Multigrid methods and application to variational data assimilation

Readers can refer to Briggs et al. (2000) for an excellent introduction to multigrid methods and to Hackbusch (2003) for a more detailed review. In Section 17.2.1, basic ideas are reviewed. Section 17.2.2 discusses the adaptation of the multigrid method to the data assimilation problem.

17.2.1 Multigrid methods: algorithm and convergence properties

The general idea of multigrid methods for solving a linear system $\mathbf{Ax} = \mathbf{b}$ is to begin by reducing the small-scale components of the error on the current (high-resolution) grid Ω_f. This is called the pre-smoothing step and should be done in a few iterations according to the ellipticity of the system (large eigenvalues at small scales). The error is then smooth and can be appropriately computed on a coarse-resolution grid Ω_c during the coarse-grid correction step. The correction is then interpolated back to the fine grid. Since the interpolation operator can in turn produce small-scale error components, a post-smoothing step is finally applied.

The basic algorithm with two grid levels, MULTIGRID(ν_1, ν_2, Ω_f, $\mathbf{x}_f$, $\mathbf{b}_f$), reads as follows:

1. *Pre-smoothing*: Apply ν_1 steps of an iterative method S_1 (called a *smoother*) on the fine grid:
$$\mathbf{A}_f \mathbf{x}_f = \mathbf{b}_f, \quad \mathbf{x}_f = S_1^{\nu_1}(\mathbf{x}_f, \mathbf{b}_f).$$
2. *Coarse-grid correction*:
   (i) Transfer the residual to a coarser grid:
$$\mathbf{r}_f = \mathbf{b}_f - \mathbf{A}_f \mathbf{x}_f, \quad \mathbf{r}_c = \mathbf{I}_f^c \mathbf{r}_f, \quad \mathbf{I}_f^c \text{ is the restriction operator.}$$
   (ii) Solve the problem on the coarse grid:
$$\mathbf{A}_c \delta \mathbf{x}_c = \mathbf{r}_c. \tag{17.15}$$
   (iii) Transfer the correction to the fine grid:
$$\mathbf{x}_f = \mathbf{x}_f + \mathbf{I}_c^f \delta \mathbf{x}_c, \quad \mathbf{I}_c^f \text{ is the interpolation operator.}$$
3. *Post-smoothing*: Apply ν_2 steps of an iterative method S_2 on the fine grid:
$$\mathbf{A}_f \mathbf{x}_f = \mathbf{b}_f, \quad \mathbf{x}_f = S_2^{\nu_2}(\mathbf{x}_f, \mathbf{b}_f).$$

The extension of this two-grid algorithm to a multigrid algorithm is done recursively by solving (17.15) by a multigrid algorithm.

Convergence properties

We refer to Hackbusch (2003) for a detailed explanation of the different ingredients of the convergence proof.

If $\mathbf{e}_f = \mathbf{x}_f - \mathbf{x}^*$ is the error, it can be shown that
$$\mathbf{e}_f^{\text{after coarse grid correction}} = \left(\mathbf{I} - \mathbf{I}_c^f \mathbf{A}_c^{-1} \mathbf{I}_f^c \mathbf{A}_f\right) \mathbf{e}_f^{\text{before coarse grid correction}}.$$
Including the smoothing steps (and assuming that only pre-smoothing is applied: $\nu_1 = \nu$, $\nu_2 = 0$),
$$\mathbf{e}_f^{k+1} = (\mathbf{I} - \mathbf{I}_c^f \mathbf{A}_c^{-1} \mathbf{I}_f^c \mathbf{A}_f) S^\nu \mathbf{e}_f^k.$$
The smoothing steps S^ν should remove most of the error at small scales, while the coarse grid correction step should remove the error at large scales. These two properties enable one to show that the spectral radius of the multigrid iteration matrix $\mathbf{M} = (\mathbf{I} - \mathbf{I}_c^f \mathbf{A}_c^{-1} \mathbf{I}_f^c \mathbf{A}_f) S^\nu$ can be made less than one if a sufficient number of smoothing steps ν are applied.

Order of transfer operators and the approximation property

The coarse-grid correction matrix is defined by $(\mathbf{I} - \mathbf{I}_c^f \mathbf{A}_c^{-1} \mathbf{I}_f^c \mathbf{A}_f)$. Let order($\mathbf{A}$) be the order of the differential operator that leads to the matrix $\mathbf{A}$. For the norm of the coarse-grid correction matrix to be bounded, the following conditions on the order of the restriction ($\mathbf{I}_f^c$) and prolongation ($\mathbf{I}_c^f$) operators must hold (see Hemker, 1990):

$$\text{order}(\mathbf{I}_f^c) \geq \text{order}(\mathbf{A}), \quad \text{order}(\mathbf{I}_c^f) \geq \text{order}(\mathbf{A})$$

In addition, the approximation property that states that the coarse-grid matrix $\mathbf{A}_c$ is closed to the fine-grid matrix $\mathbf{A}_f$ and induces a reduction of large-scale error components is maintained by the underlying discretization order of the differential operator. When possible, another approach is to use the so-called Galerkin approximation to deduce the coarse-grid matrix $\mathbf{A}_c$. In this approximation, the restriction and interpolation operators are adjoints,

$$\mathbf{I}_f^c = (\mathbf{I}_c^f)^*, \tag{17.16}$$

and the coarse-grid matrix is defined by

$$\mathbf{A}_c = \mathbf{I}_f^c \mathbf{A}_f \mathbf{I}_c^f. \tag{17.17}$$

It can be shown that if these conditions are fulfilled, then the components of the error that lie in the range of the interpolation operator before the coarse-grid correction step are completely removed by that step.

Full Approximation Scheme variant and nonlinear cases

In the multigrid algorithm introduced above, the coarse-grid correction seeks a solution for the error, (17.15). It is possible to rewrite the algorithm in an equivalent form where a coarse solution is searched instead. This is the full approximation scheme (FAS) (Brandt, 1982), which will have a natural extension to nonlinear operators. The coarse-grid correction step is obtained by expressing the coarse-grid correction $\delta\mathbf{x}_c$ as $\mathbf{x}_c - \mathbf{I}_f^c \mathbf{x}_f$ and the problem is solved for the full approximation $\mathbf{x}_c$ instead of the correction $\delta\mathbf{x}_c$:

1. Solve the problem on the coarse grid:

   $$\mathbf{A}_c \mathbf{x}_c = \mathbf{b}_c, \tag{17.18}$$

   where

   $$\mathbf{b}_c = \mathbf{r}_c + \mathbf{A}_c \mathbf{I}_f^c \mathbf{x}_f = \mathbf{I}_f^c \mathbf{b}_f + \left(\mathbf{A}_c \mathbf{I}_f^c - \mathbf{I}_f^c \mathbf{A}_f\right) \mathbf{x}_f.$$

2. Transfer the correction to the fine grid

   $$\mathbf{x}_f = \mathbf{x}_f + \mathbf{I}_c^f \left(\mathbf{x}_c - \mathbf{I}_f^c \mathbf{x}_f\right), \quad \mathbf{I}_c^f \text{ is the interpolation operator.}$$

The extension to nonlinear operators A is naturally obtained by replacing the coarse-grid correction step by

$$A_c(\mathbf{x}_c) = \mathbf{b}_c, \tag{17.19}$$

where

$$\mathbf{b}_c = \mathbf{I}_f^c \mathbf{b}_f + A_c(\mathbf{I}_f^c \mathbf{x}_f) - \mathbf{I}_f^c A_f(\mathbf{x}_f).$$

Note that when written in this form, it appears that the operator used to restrict the state $\mathbf{I}_f^c \mathbf{x}_f$ might different from the one used to restrict the residual $\mathbf{I}_f^c\,[\mathbf{b}_f - A_f(\mathbf{x}_f)]$. In the following, for he sake of simplicity, we will consider these two operators is be identical.

The convergence study by Yavneh and Dardyk (2006) emphasizes the importance of the coarse-grid operator $A^c(I_f^c \mathbf{x}_f)$ being close enough to the fine-grid operator $A^f(\mathbf{x}_f)$, but this is not a sufficient condition for convergence (see Nash, 2000). An alternative strategy for the treatment of nonlinear problems is to apply a linear multigrid method as a solver inside the successive linearized equations of a Newton method (Hackbusch, 2003, Chapter 9.2).

17.2.2 Multigrid methods for variational data assimilation

In order to apply multigrid methods to variational data assimilation problems, we write the Euler equation (17.3) in the form $A(\mathbf{x}) = \mathbf{b}$, where $A(\mathbf{x}) = \nabla J(\mathbf{x}) - \nabla J(0)$ and $\mathbf{b} = -\nabla J(0)$.

In an FAS algorithm, the coarse-grid correction step (17.19), $A_c(\mathbf{x}_c) = \mathbf{I}_f^c \mathbf{b}_f + A_c(\mathbf{I}_f^c \mathbf{x}_f) - \mathbf{I}_f^c A_f(\mathbf{x}_f)$, leads to

$$\nabla J^c_{\mathrm{FAS}}(\mathbf{x}_c) = \nabla J^c(\mathbf{x}_c) + \mathbf{I}_f^c \nabla J^f(\mathbf{x}_f) - \nabla J^c(\mathbf{I}_f^c \mathbf{x}_f) = 0. \tag{17.20}$$

By defining

$$L = \mathbf{I}_f^c \nabla J^f(\mathbf{x}_f) - \nabla J^c(\mathbf{I}_f^c \mathbf{x}_f), \tag{17.21}$$

we have to minimize on the coarse grid the new cost function J^c_{FAS}:

$$J^c_{\mathrm{FAS}}(\mathbf{x}_c) = J^c(\mathbf{x}_c) + \langle L, \mathbf{x}_c \rangle. \tag{17.22}$$

Definition of coarse-grid background error covariance matrix

The coarse-grid background error covariance matrix $\mathbf{B}_c^{-1}$ can be deduced from the Galerkin condition (17.17):

$$\mathbf{B}_c^{-1} = \mathbf{I}_f^c \mathbf{B}_f^{-1} \mathbf{I}_c^f.$$

Application of the Galerkin condition for the observation term is not affordable, since it would require that the model be run at high resolution on each grid on the hierarchy.

Coarse-grid correction matrix for a linear advection equation

The coarse-grid correction matrix is given by $\mathbf{K} = \left(\mathbf{I} - \mathbf{I}_c^f \mathbf{A}_c^{-1} \mathbf{I}_f^c \mathbf{A}_f\right)$ where for our data assimilation problem the $\mathbf{A}$ matrices correspond to the coarse and fine-grid Hessians. For the model problem introduced in Section 17.1.2 on a linear advection equation, Fig. 17.1 represents the typical behaviour of this operator at the different scales.

We observe a local maximum at large scales ($\theta = k\Delta x \ll 1$) that has to be linked to the diffusion inherent to the numerical scheme (the Courant number $\mu = c\Delta t/\Delta x$ is here equal to 0.9) (see (17.13)). Details can be found in Neveu et al. (2011).

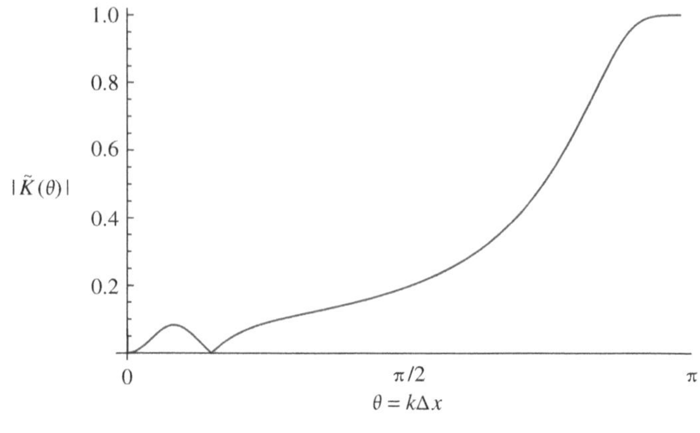

Fig. 17.1 The coarse-grid correction in Fourier space according to $\theta = k\Delta x$ (k is the wavenumber and Δx the space step), for different values of γ.

17.2.3 Additional remarks

Smoothers

Smoothers have been introduced in the multigrid algorithm in section 17.2.1. Their role is to smooth the small-scale components of the error before the coarse-grid correction step. Typical optimization methods that rely on matrix vector products only do not have good smoothing properties. Design of such smoothers is thus an important concern. However, in the context of successive assimilations, the construction of approximate smoothers (Tang and Wan, 2000) and/or the use of deflated conjugate gradient algorithms (Tang et al., 2009) can be investigated.

Multigrid preconditioning

We have seen that the approximation of the fine grid by the coarse grid can be altered by errors in the numerical model. To render the use of multigrid methods more robust, multigrid iterations can be used as a preconditioner inside an optimization method (Oostelee and Washio, 1998; Debreu et al., 2013). In Debreu et al. (2013), it is shown how to efficiently couple the multigrid preconditioner and the traditional $\mathbf{B}^{1/2}$ preconditioner introduced in Section 17.1.

Design of coarse-grid background and observation error covariance matrices

In Section 17.2.2, the coarse-grid background error covariance matrix is deduced from the fine grid one following the Galerkin condition. This choice can be called a deterministic choice and is in agreement with the multigrid convergence theory when applied to the background term only. Nothing has been said about the coarse-grid observation error covariance matrix $\mathbf{R}_c$, which has actually been taken to be identical to the fine-grid one in our experiments. From a statistical point of view, this matrix should probably be different on the coarse grid(s) owing the difference in representativeness errors. The design of coarse-grid background- and observation-error covariance matrices is one of the main subjects of Chapter 18 and is also important to consider in the context of multigrid methods.

17.3 Data assimilation and local mesh refinement

Nested models are commonly used in meteorology and oceanography. Such systems, developed within the framework of structured grids, allow a local increase in mesh resolution in areas where it seems to be necessary, by running the same model on a hierarchy of grids. These methods have been applied successfully in atmosphere and ocean modelling (for a review see Debreu and Blayo, 2008). The location of the high-resolution grids can be dictated by physical considerations or, in the context of data assimilation, by the presence of high-resolution observations.

17.3.1 Formulation of the nested models

We consider the general case of a high-resolution model, covering the local domain ω, embedded in a coarser-resolution model covering the larger domain Ω. With obvious notation, the local high-resolution grid and the global coarse-resolution grid are denoted respectively by ω_f and Ω_c. The corresponding state vectors are denoted respectively by $\mathbf{X}_f$ and $\mathbf{X}_c$. We also denote by ω_c the part of the grid Ω_c corresponding to the local domain ω (see Fig. 17.2).

In the case of *one-way interaction*, the coarse-grid model provides boundary conditions to the high-resolution model using an interpolation operator $\mathbf{I}_c^f$. In the case of *two-way interaction*, a feedback term from the fine grid onto the coarse grid is added. The coarse solution is updated locally (in $\mathring{\omega}_c$, the interior of ω_c) by the

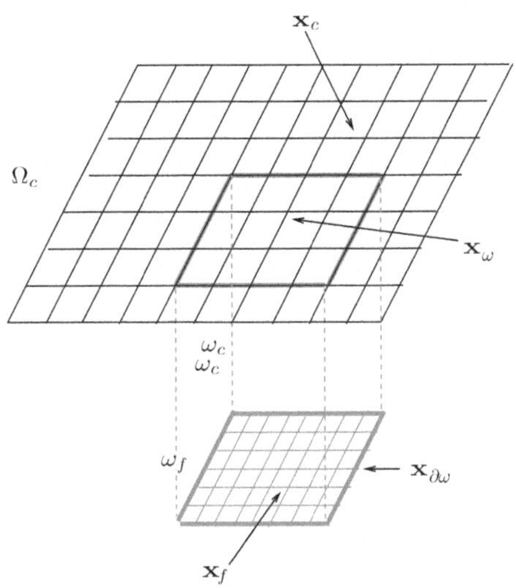

Fig. 17.2 Notation used in the definition of the nested models.

high-resolution solution using a restriction operator $\mathbf{I}_f^c$. Semi-discretized equations of the nested system can be written as follows:

$$\textbf{domain } \Omega_c\textbf{:}$$
$$\begin{cases} \dfrac{\partial \mathbf{X}_c}{\partial t} = F(\mathbf{X}_c, \mathbf{X}_\omega) \quad \text{on} \quad \Omega_c \times [0,T], \\ \mathbf{X}_c(t=0) = \mathbf{x}_c, \\ \mathbf{X}_\omega = \mathbf{I}_f^c(\mathbf{X}_f) \quad \text{on} \quad \mathring{\omega}_c \times [0,T]; \end{cases}$$
$$\textbf{domain } \omega_f\textbf{:} \tag{17.23}$$
$$\begin{cases} \dfrac{\partial \mathbf{X}_h}{\partial t} = F(\mathbf{X}_h, \mathbf{X}_{\partial\omega}) \quad \text{on} \quad \omega_f \times [0,T], \\ \mathbf{X}_f(t=0) = \mathbf{x}_f, \\ \mathbf{X}_{\partial\omega} = \mathbf{I}_c^f(\mathbf{X}_c) \quad \text{on} \quad \partial\omega_f \times [0,T]. \end{cases}$$

After discretization, the problems have to be integrated in time in a specific order. The model is first integrated on the coarse grid Ω_c and then on the high-resolution ω_f grid with boundary conditions given by a spatial and temporal interpolation of the coarse values. Finally, feedback can be applied in the two-way case. An example of this integration order with a time refinement factor of 2 is shown in Fig. 17.3.

17.3.2 The two-grid optimality systems

In order to obtain a formulation of variational data assimilation in the specific context of a two-grid system, we have first to redefine several fundamental notions.

State vector

The state vector $\mathbf{X}$ of the two-grid model is composed of the state vectors of both grids:

$$\mathbf{X} = \begin{bmatrix} \mathbf{X}_c \\ \mathbf{X}_f \end{bmatrix}.$$

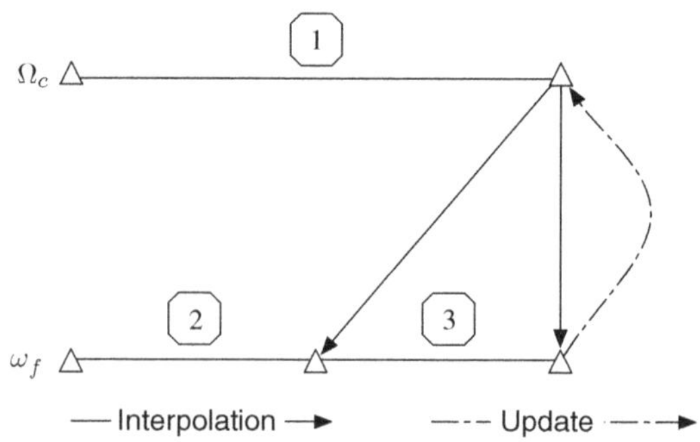

Fig. 17.3 Integration of the two grids over one coarse-grid time step in the case of a time refinement ratio equal to 2.

It corresponds to the 'multigrid state vector' proposed in Barth et al., (2007) for the application of a SEEK filter in a two-way nested model of the Ligurian Sea (see also Melet et al., 2012). Therefore, the initial condition $\mathbf{x}$, which will be the control variable for the minimization problem, and its background value $\mathbf{x}^b$ read, with obvious notation,

$$\mathbf{x} = \begin{bmatrix} \mathbf{x}_c \\ \mathbf{x}_f \end{bmatrix}, \qquad \mathbf{x}^b = \begin{bmatrix} \mathbf{x}_c^b \\ \mathbf{x}_f^b \end{bmatrix}.$$

Cost function

The cost function is defined as the sum of the misfit to the first guess and the misfits to the observations on both grids:

$$\begin{aligned} J(\mathbf{x}) &= J^b(\mathbf{x}) + J_c^{\text{obs}}(\mathbf{x}) + J_f^{\text{obs}}(\mathbf{x}) \\ &= \frac{1}{2}(\mathbf{x} - \mathbf{x}^b)^T B^{-1}(\mathbf{x} - \mathbf{x}^b) \\ &+ \frac{1}{2}\sum_{i=0}^{N}[H_c^i(\mathbf{X}_c^i(\mathbf{x})) - \mathbf{y}_c^i]^{\mathrm{T}} \mathbf{R}_c^{i\,-1}[H_c^i(\mathbf{X}_c^i(\mathbf{x})) - \mathbf{y}_c^i] \\ &+ \frac{1}{2}\sum_{i=0}^{N}[H_f^i(\mathbf{X}_f^i(\mathbf{x})) - \mathbf{y}_f^i]^{\mathrm{T}} \mathbf{R}_f^{i\,-1}[H_f^i(\mathbf{X}_f^i(\mathbf{x})) - \mathbf{y}_f^i], \end{aligned}$$

where B is a now a two-grid background error covariance matrix. The gradient of the background term J^b reads $\nabla J^b(\mathbf{x}) = B^{-1}(\mathbf{x} - \mathbf{x}^b)$. The gradients of the observation terms J_c^{obs} and J_f^{obs} can be obtained using the adjoint method (see e.g. Le Dimet and Talagrand, 1986) applied to the nested system. We note that $\mathbf{P}$ is the coarse-resolution component of the adjoint vector ($\mathbf{P}$ is the adjoint variable of $\mathbf{X}_c$) and $\mathbf{Q}$ is the high-resolution companant ($\mathbf{Q}$ is the adjoint variable of $\mathbf{X}_f$). In the case of two-way interaction, the adjoint system reads as follows:

domain Ω_c:

$$\begin{cases} \dfrac{\partial \mathbf{X}_c}{\partial t} = F(\mathbf{X}_c, \mathbf{X}_\omega) \quad \text{on} \quad \Omega_c \times [0, T], \\ \mathbf{X}_c(x, 0) = \mathbf{x}_c, \\ \mathbf{X}_\omega = \mathbf{I}_f^c(\mathbf{X}_f) \quad \text{on} \quad \mathring{\omega}_f \times [0, T], \end{cases}$$

$$\begin{cases} \dfrac{\partial \mathbf{P}}{\partial t} + \left[\dfrac{\partial F}{\partial \mathbf{X}_c}\right]^* \cdot \mathbf{P} + \mathbf{I}_f^c \left[\dfrac{\partial F}{\partial \mathbf{X}_{\partial\omega}}\right]^* \cdot \mathbf{Q} = \mathbf{H}_c^* \mathbf{R}_c^{-1}[\mathbf{H}_c \mathbf{X}_c(t) - \mathbf{y}_c(t)], \\ \mathbf{P}(T) = 0, \end{cases}$$

$$\nabla_{\mathbf{x}_c} J^{obs} = -\mathbf{P}(0);$$

domain ω_f:

$$\begin{cases} \dfrac{\partial \mathbf{X}_f}{\partial t} = F(\mathbf{X}_f, \mathbf{X}_{\partial\omega}) \quad \text{on} \quad \omega_f \times [0, T], \\ \mathbf{X}_f(x, 0) = \mathbf{x}_f, \\ \mathbf{X}_{\partial\omega} = \mathbf{I}_c^f(\mathbf{X}_c) \quad \text{on} \quad \partial\omega_f \times [0, T], \end{cases}$$

$$\begin{cases} \dfrac{\partial \mathbf{Q}}{\partial t} + \left[\dfrac{\partial F}{\partial \mathbf{X}_f}\right]^* \cdot \mathbf{Q} + \mathbf{I}_c^f \left[\dfrac{\partial F}{\partial \mathbf{X}_\omega}\right]^* \cdot \mathbf{P} = \mathbf{H}_f^* \mathbf{R}_f^{-1} [\mathbf{H}_f \mathbf{X}_f(t) - \mathbf{y}_f(t)], \\ \mathbf{Q}(T) = 0, \end{cases}$$

$$\nabla_{\mathbf{x}_f} J^{obs} = -\mathbf{Q}(0).$$

here the operator $\mathbf{I}_c^f = \mathbf{I}_c^f(\mathbf{X}_c)$ is the linearization of $\mathbf{I}_c^f$ in $\mathbf{X}_c$, the operator $\mathbf{I}_f^c$ is the adjoint of $\mathbf{I}_c^f$, the operator $\mathbf{I}_f^c = \mathbf{I}_f^c(\mathbf{X}_f)$ is the linearization of $\mathbf{I}_f^c$ in $\mathbf{X}_f$, and the operator $\mathbf{I}_c^f$ is the adjoint of $\mathbf{I}_f^c$. We observe a feedback from the high resolution adjoint model onto the coarse-resolution one, in the opposite direction of the interactions existing in the direct formulation. In the two-way case, the addition of feedback from the high-resolution solution to the coarse-resolution solution leads to an interaction in the adjoint model from the coarse-resolution adjoint solution to the high-resolution adjoint solution. As for the direct system, the two grids must be integrated simultaneously in this adjoint system.

In Simon et al. (2011), the importance of the design of the background error covariance matrices have been addressed. The $\mathbf{B}$ matrix is modelled by

$$\mathbf{B} = \left(\begin{array}{c|c} \mathbf{B}_{cc} & \mathbf{B}_{cf} \\ \hline \mathbf{B}_{fc} & \mathbf{B}_{ff} \end{array}\right)$$

where the matrices $\mathbf{B}_{cc}$, and $\mathbf{B}_{ff}$ traditional background error covariance matrices approximated using a Laplacian diffusion operator and computed independently on each grid. The cross-terms $\mathbf{B}_{cf}$ and $\mathbf{B}_{fc}$ are there to include constrains on the coarse and fine analysis increments using the grid interactions operators (see Simon et al., 2011). Figure 17.4 shows the increments obtained for single-observation experiments using a shallow-water model and with different formulations of the two-grid background error covariance matrix. The $\mathbf{B}_{\text{multi}}$ formulation includes the cross-terms, whereas the $\mathbf{B}_{\text{indep}}$ formulation does not.

The structures of the coarse grid increments differ fundamentally between the two formulation in both magnitude and geometry. In particular, as expected, the increments obtained on the coarse grid using the $\mathbf{B}_{\text{multi}}$ formulation match very well those computed on the fine grid.

When using a correlation operator based on a diffusion operator, a fully two-grid $\mathbf{B}$ covariance matrix can also be obtained by applying the Laplacian diffusion in a two-grid context (i.e. with coarse–fine grid interactions at the level of each step of the diffusion operator). Preliminary experiments in the 3D-Var version of the NEMO ocean model have proved the superiority of this approach.

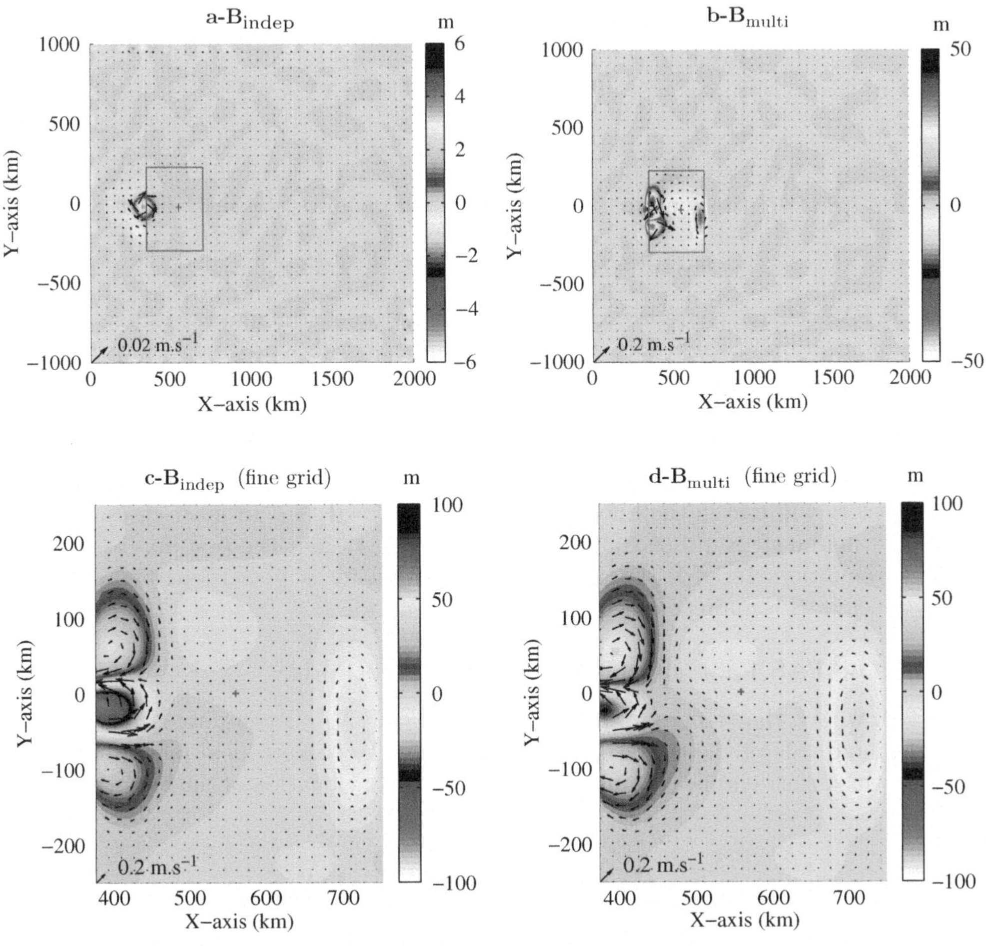

Fig. 17.4 Single-observation experiments. (a,b) Maps of coarse-resolution analysis increment for (a) the matrix $\mathbf{B}_{\text{indep}}$ and (b) the matrix $\mathbf{B}_{\text{multi}}$. (c, d) Maps of fine-resolution analysis increments for (c) $\mathbf{B}_{\text{indep}}$ and (d) $\mathbf{B}_{\text{multi}}$. The high-resolution grid is localized by the box. The observation is localized by a cross in the centre of fine grid.

17.4 Coupling the two approaches

In this final section, we introduce some ideas on how to couple the two preceding approaches into a local multigrid algorithm for variational data assimilation.

17.4.1 The FAS applied to the optimality system

Instead of applying the FAS to the Euler equation, it can be applied to the optimality system:

$$\begin{cases} \dfrac{\partial X}{\partial t} = F(X), \\ \dfrac{\partial P}{\partial t} = -\dfrac{\partial F}{\partial X}(X)P + \mathbf{H}^T\mathbf{R}^{-1}\left[H(X) - \mathbf{y}\right], \\ \nabla J = -P(0) + \mathbf{B}^{-1}(\mathbf{x} - \mathbf{x}_b) = 0. \end{cases}$$

The coarse-grid equation derived from the FAS algorithm is given by (17.19):

$$A_c(\mathbf{x}_c) = \mathbf{I}_f^c\mathbf{b}_f + A_c(\mathbf{I}_f^c\mathbf{x}_f) - \mathbf{I}^c A_f(\mathbf{x}_f).$$

When the FAS scheme is applied to the optimality system, the main modification goes into the integration of the direct model on the coarse grid, which now reads

$$\frac{\partial X_c}{\partial t} = F_c(X_c) + \left[\mathbf{I}_f^c F_f(X_f) - F_c(\mathbf{I}_f^c X_f)\right]. \tag{17.24}$$

As a result of this correction, at convergence of the minimization process, the coarse-grid trajectory will match the fine-grid trajectory: $X_c = \mathbf{I}_f^c X_f$. In most cases, this will render the nonlinear multigrid algorithm more robust than the one applied only to the Euler equation. Indeed, the approximation property is better satisfied since the coarse-grid model remains close to the fine-grid solution.

17.4.2 The FAS for local mesh refinement methods

In (17.24), we see that a defect $\tau_f^c = \mathbf{I}_f^c F_f(X_f) - F_c(\mathbf{I}_f^c X_f)$ is computed and added to the right-hand side of the coarse-grid equations. In the case of local mesh refinement methods, this defect can be computed only locally at the location of the high-resolution grid:

$$\frac{\partial X_c}{\partial t} = F_c(X_c) + \left[\mathbf{I}_f^c F_f(X_f) - F_c(\mathbf{I}_f^c X_f)\right]_\omega$$

At convergence, the solution now corresponds to that of a two-way nesting algorithm: the coarse-grid solution is the restriction of the fine-grid solution. Similarly, the coarse-grid gradient is taken to be the one given by the FAS expression

$$\nabla J_{\text{FAS}}^c(\mathbf{x}_c) = \nabla J^c(\mathbf{x}_c) + [\mathbf{I}_f^c \nabla J^f(\mathbf{x}_f) - \nabla J^c(\mathbf{I}_f^c\mathbf{x}_f)]_\omega = 0, \tag{17.25}$$

where again the difference with (17.20) is that the restriction occurs only inside the fine-grid domain ω.

17.5 Conclusions and perspectives

Multigrid algorithms provide ways of accelerating the minimization process associated with variational data assimilation methods. The full approximation scheme allows one to apply the multigrid algorithms in nested models with locally high resolution, for example in regions with a high density of observations. There are many other potential applications where multigrid methods can be effective. Among them, the problem of a full state control (not just the initial) condition may become feasible in a multigrid context where the control would be decomposed in a hierarchical basis (e.g. wavelets) and where most of its large-scale components could be efficiently solved on coarse-resolution grids. The use of nonlinear multigrid methods as continuation algorithms (i.e. starting the minimization on coarse grids where there are fewer nonlinearities) is also very attractive and may enable the solution of highly nonlinear problems.

References

Barth, A., Alvera-Azcárate, A., Beckers, J.-M., Rixen, M., and Vandenbulcke, L. (2007). Multigrid state vector for data assimilation in a two-way nested model of the ligurian sea. *J. Mar. Syst.*, **65**, 41–59.

Brandt, A. (1982). Guide to multigrid development. In *Multigrid Methods* (ed. W. Hackbusch and U. Trottenberg). Lecture Notes in Mathematics, Vol. 960, pp. 220–312. Springer, Berlin.

Briggs, W. L., Henson, V. E., and McCormick, S. F. (2000). *A Multigrid Tutorial*, 2nd edn. SIAM, Philadelphia.

Courtier, P. (1997). Dual formulation of four-dimensional variational assimilation. *Q. J. R. Meteorol. Soc.*, **123**, 2449–2461.

Debreu, L. and Blayo, E. (2008). Two-way embedding algorithms: a review. *Ocean Dyn.*, **58**, 415–428.

Debreu, L., Neveu, E., Simon, E., Le Dimet, F.-X., and Vidard, A. (2013). Multigrid solvers and multigrid preconditioners for the solution of variational data assimilation problems. Submitted to *Q. J. R. Meteorol. Soc.*

Durran, D. R. (2010). *Numerical Methods for Fluid Dynamics with Applications to Geophysics*. Springer, New York.

Haben, S. A., Lawless, A. S., and Nichols, N. K. (2011). Conditioning and preconditioning of the variational data assimilation problem. *Comput. Fluids*, **46**, 252–256.

Hackbusch, W. (2003). *Multigrid Methods and Applications*. Springer-Berlin.

Hemker, P. W. (1990). On the order of prolongations and restrictions in multigrid procedures. *J. Comput. Appl. Math.*, **32**, 423–429.

Le Dimet, F.-X. and Talagrand, O. (1986). Variational algorithms for analysis and assimilation of meteorological observations: theoretical aspects. *Tellus A*, **38**, 97–110.

Melet, A., Verron, J., and Brankart, J.-M. (2012). Potential outcomes of glider data assimilation in the Solomon sea: control of the water mass properties and parameter estimation. *J. Mar. Syst.*, **94**, 232–246.

Nash, S. G. (2000). A multigrid approach to discretized optimization problems. *Optim. Methods Software*, **14**, 99–116.

Neveu, E., Debreu, L., and Le Dimet, F.-X. (2011). Multigrid methods and data assimilation, convergence study and first experiments on non-linear equations. *ARIMA*, **14**, 63–80.

Oostelee, C and Washio, T (1998). On the use of multigrid as a preconditioner. In *Proceedings of Ninth International Conference on Domain Decomposition Methods, Bergen.*

Simon, E., Debreu, L., and Blayo, E. (2011). 4D variational data assimilation for locally nested models: complementary theoretical aspects and application to a 2D shallow water model. *Int. J. Numer. Methods. Fluids*, **66**(2), 135–161.

Tang, J. M., Nabben, R., Vuik, C., and Erlangga, Y.A. (2009). Comparison of two-level preconditioners derived from deflation, domain decomposition and multigrid methods. *J. Sci. Comput.*, **39**, 340–370.

Tang, W. and Wan, W. (2000). Sparse approximate inverse smoother for multigrid. *SIAM J. Matrix Anal. Appl.*, **21**, 1236–1252.

Yavneh, I. and Dardyk, G. (2006). A multilevel nonlinear method. *SIAM J. Sci. Comput.*, **28**, 24–46.

18
Selected topics in multiscale data assimilation

M. Bocquet[1], L. Wu[1], F. Chevallier[2], and M. R. Kookhan[1]

[1]Université Paris-Est, CEREA joint laboratory École des Ponts ParisTech
and EdF R&D, INRIA, Paris Rocquencourt Research Center
École des Ponts ParisTech
Champs-sur-Marne, France

[2]Laboratoire des Sciences du Climat et de l'Environnement
CEA-CNRS-UVSQ
Gif-sur-Yvette, France

Advanced Data Assimilation for Geosciences. First Edition.
Edited by É. Blayo, M. Bocquet, E. Cosme, and L. F. Cugliandolo.

Chapter Contents

18.1 Introduction

In geophysical sciences, physics unfolds and should be modelled at many spatial and temporal scales. This stems from the very large scope of the systems often considered in ambitious applications, from the global synoptic motions of the atmosphere accounting for heat transfer and dominant winds, down to the microphysics accounting for clouds, hydrometeors, and aerosols. This is also intrinsic to the physics of turbulence that shapes geophysical flows and that correlates scales so that it becomes difficult to model them separately. In addition to the complexity of multiscale models that need to be dealt with in geophysical data assimilation systems, one needs to incorporate observations pertaining to many scales, such as in situ observations, satellite observations, radar, and lidar. Even though all of this is to a large extent ignored in current data assimilation systems, it is easy to envision future data assimilation systems accounting for these multiscale aspects, not only at a technical level by interfacing models and observations at different scales, but also at a more fundamental mathematical level in that respect. Potentially useful analysis tools exist, such as multigrid methods and wavelets, but developing a fully consistent data assimilation system based on them is still a challenge.

This chapter provides an exposition of a selection of problems considered in this field. Most of the results were obtained in the context of the ANR (French National Science Foundation) MSDAG project.

There are a few insightful papers in the geophysical literature addressing mathematical approaches to multiscale data assimilation, for instance Liu and Rabier (2002), Janjić and Cohn (2006), Oke and Sakov (2008), and Koohkan and Bocquet (2012). They offer a different, often complementary, view on the topic to what is presented here, although several concepts are common.

This chapter deals with a multiscale extension of the best linear unbiased estimate (BLUE) inference, and its application to several problems. It also deals with a more pragmatic and perhaps more rewarding approach of the same problem, focusing on the multiple-scales treatment of innovation statistics. These methods are applied to atmospheric transport and chemistry problems.

18.2 Bayesian multiscale analysis

In the following, we will discuss how to account for multiple scales in the BLUE estimation principle at the heart of most current data assimilation methods. Our starting point is the observation equation

$$\boldsymbol{\mu} = \mathbf{H}\boldsymbol{\sigma} + \boldsymbol{\epsilon}, \tag{18.1}$$

with the simplest assumptions, where $\boldsymbol{\mu}$ and $\boldsymbol{\epsilon}$ are the observation and error vectors defined in $\mathbb{R}^p$ and $\boldsymbol{\sigma}$ is the control space state vector in $\mathbb{R}^N$. In the following discussion, $\boldsymbol{\sigma}$ also coincides with the state vector. The observation errors are assumed to follow a normal distribution, $\boldsymbol{\epsilon} \sim \mathcal{N}(\mathbf{0}, \mathbf{R})$, whereas the background errors on the state vector $\boldsymbol{\sigma}$ are assumed to be distributed according to a Gaussian, $\boldsymbol{\sigma} \sim \mathcal{N}(\boldsymbol{\sigma}_b, \mathbf{B})$.

We present some partial results drawn from Bocquet (2009), Bocquet et al. (2011), Wu et al. (2011), and Koohkan et al. (2012).

18.2.1 Scaling operators

To climb up or down the spatial or temporal scales, one needs to define a *restriction operator* that tells how a state vector $\boldsymbol{\sigma}$ is coarse-grained and a *prolongation operator* that tells how the state vector is refined through the scales. See Rodgers (2000) for an in-depth discussion of the topic.

First, let us consider the restriction operator. Assume $\boldsymbol{\sigma}$ is a state vector that is known in the finest regular grid Ω isomorphic to $\mathbb{R}^{N_{\mathrm{fg}}}$. Let ω be a discretization of the control space, which we also call a *representation.* It is isomorphic to $\mathbb{R}^N$, meaning that it has N grid cells. These N grid cells are possibly of different sizes and shapes, but they form a partition of control space. It could be a regular grid, or an adaptive grid that partitions the space and possibly the time domain. The coarse-graining of $\boldsymbol{\sigma}$ in ω is defined by $\boldsymbol{\sigma}_\omega = \boldsymbol{\Gamma}_\omega \boldsymbol{\sigma}$, where $\boldsymbol{\Gamma}_\omega : \mathbb{R}^{N_{\mathrm{fg}}} \rightarrow \mathbb{R}^N$ stands for the linear coarse-graining operator. This operator is supposed to be unambiguously defined. It could stand for a simple averaging. But it could also be a more complex coarse-graining, with an associated prolongation operator given by a spline interpolation, or model-specific coarser Jacobian matrices, such as $\mathbf{H}$.

A state vector can also be refined thanks to a prolongation operator $\boldsymbol{\Gamma}_\omega^\star : \mathbb{R}^N \rightarrow \mathbb{R}^{N_{\mathrm{fg}}}$ that refines $\boldsymbol{\sigma}_\omega$ into $\boldsymbol{\sigma} = \boldsymbol{\Gamma}_\omega^\star \boldsymbol{\sigma}_\omega$. This linear operator is ambiguous, since additional information is needed to reconstruct a source at higher resolution. A schematic of the use of the restriction and prolongation operators is displayed in Fig. 18.1

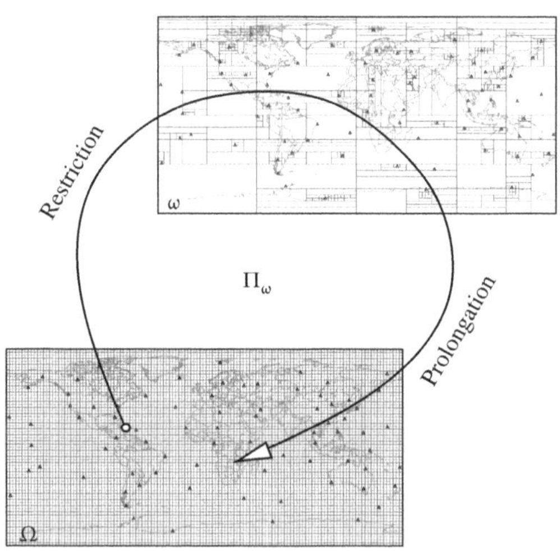

Fig. 18.1 Schematic of the restriction and prolongation operators from the finest regular grid to a representation (adaptive grid) ω, and vice versa.

One simple choice, which we shall call the deterministic one, is to set $\boldsymbol{\Gamma}^\star_\omega = \boldsymbol{\Gamma}^{\mathrm{T}}_\omega$. However, in this data assimilation framework, one has prior information on the control state vectors that may be exploited, in the form of a probability density function $q(\boldsymbol{\sigma})$. Following our initial assumption, q is Gaussian: $q(\boldsymbol{\sigma}) \sim \mathcal{N}(\boldsymbol{\sigma}_b, \mathbf{B})$. From this prior defined in the finest regular grid, one can infer, thanks to $\boldsymbol{\Gamma}_\omega$, the prior pdf of $\boldsymbol{\sigma}$ in the representation ω:

$$q_\omega(\boldsymbol{\sigma}_\omega) \sim \mathcal{N}(\boldsymbol{\sigma}^b_\omega, \mathbf{B}_\omega)\,, \tag{18.2}$$

with

$$\boldsymbol{\sigma}^b_\omega = \boldsymbol{\Gamma}_\omega \boldsymbol{\sigma}_b\,, \qquad \mathbf{B}_\omega = \boldsymbol{\Gamma}_\omega \mathbf{B} \boldsymbol{\Gamma}^{\mathrm{T}}_\omega\,. \tag{18.3}$$

Conversely, assume that one knows $\boldsymbol{\sigma}_\omega$ in the representation ω. Since the problem is underdetermined, one could opt for the most likely refinement. This is given by the mode of $q(\boldsymbol{\sigma}|\boldsymbol{\sigma}_\omega)$. From Bayes' rule, it is clear that

$$q(\boldsymbol{\sigma}|\boldsymbol{\sigma}_\omega) = \frac{q(\boldsymbol{\sigma})}{q_\omega(\boldsymbol{\sigma}_\omega)} \delta\left(\boldsymbol{\sigma}_\omega - \boldsymbol{\Gamma}_\omega \boldsymbol{\sigma}\right)\,, \tag{18.4}$$

where δ is the Dirac distribution in $\mathbb{R}^N$. Then the mode of this posterior Gaussian distribution is given by

$$\boldsymbol{\sigma}^\star = \boldsymbol{\sigma}_b + \mathbf{B}\boldsymbol{\Gamma}^{\mathrm{T}}_\omega (\boldsymbol{\Gamma}_\omega \mathbf{B} \boldsymbol{\Gamma}^{\mathrm{T}}_\omega)^{-1} \left(\boldsymbol{\sigma}_\omega - \boldsymbol{\Gamma}_\omega \boldsymbol{\sigma}_b\right)\,. \tag{18.5}$$

Thus, $\boldsymbol{\Gamma}^\star_\omega$ would be an affine operator. We denote by $\boldsymbol{\Lambda}^\star_\omega$ its tangent linear component:

$$\boldsymbol{\Lambda}^\star_\omega \equiv \mathbf{B}\boldsymbol{\Gamma}^{\mathrm{T}}_\omega (\boldsymbol{\Gamma}_\omega \mathbf{B} \boldsymbol{\Gamma}^{\mathrm{T}}_\omega)^{-1}\,. \tag{18.6}$$

Moreover, we define

$$\boldsymbol{\Pi}_\omega \equiv \boldsymbol{\Lambda}^\star_\omega \boldsymbol{\Gamma}_\omega = \mathbf{B}\boldsymbol{\Gamma}^{\mathrm{T}}_\omega (\boldsymbol{\Gamma}_\omega \mathbf{B} \boldsymbol{\Gamma}^{\mathrm{T}}_\omega)^{-1} \boldsymbol{\Gamma}_\omega\,, \tag{18.7}$$

so that we can choose as a prolongation operator

$$\boldsymbol{\Gamma}^\star_\omega \equiv (\mathbf{I}_{N_{\mathrm{fg}}} - \boldsymbol{\Pi}_\omega)\boldsymbol{\sigma}_b + \boldsymbol{\Lambda}^\star_\omega\,, \tag{18.8}$$

where $\mathbf{I}_{N_{\mathrm{fg}}}$ is the identity operator in the finest grid control space. Since the refinement is now a probabilistic process, errors are attached to it and can be estimated. The corresponding error covariance matrix is

$$\mathbf{P}^\star_\omega = \mathbf{B} - \mathbf{B}\boldsymbol{\Gamma}^{\mathrm{T}}_\omega (\boldsymbol{\Gamma}_\omega \mathbf{B} \boldsymbol{\Gamma}^{\mathrm{T}}_\omega)^{-1} \boldsymbol{\Gamma}_\omega \mathbf{B} = \left(\mathbf{I}_{N_{\mathrm{fg}}} - \boldsymbol{\Pi}_\omega\right)\mathbf{B}\,.$$

As expected, if the representation ω is close to the finest grid

$$\frac{N_{\mathrm{fg}} - \mathrm{Rank}(\boldsymbol{\Pi}_\omega)}{N_{\mathrm{fg}} \ll 1}\,, \tag{18.9}$$

then the refinement error is negligible. If the representation is coarse, $\text{Rank}(\boldsymbol{\Pi}_\omega)/N_{\text{fg}} \ll 1$, the refinement error is limited by that of the background.

These scaling operators first satisfy

$$\boldsymbol{\Gamma}_\omega \boldsymbol{\Gamma}_\omega^\star = \mathbf{I}_N\,, \tag{18.10}$$

which is a consistency identity. Any reasonable prolongation operator should satisfy it. Then, one verifies that

$$\boldsymbol{\Gamma}_\omega^\star \boldsymbol{\Gamma}_\omega = \left(\mathbf{I}_{N_{\text{fg}}} - \boldsymbol{\Pi}_\omega\right)\boldsymbol{\sigma}_b + \boldsymbol{\Pi}_\omega\,. \tag{18.11}$$

The linear operator $\boldsymbol{\Pi}_\omega$ is a projector, since it can be checked that $\boldsymbol{\Pi}_\omega^2 = \boldsymbol{\Pi}_\omega$. Besides, it is $\mathbf{B}^{-1}$-symmetric, since

$$\begin{aligned} \langle \mathbf{x}, \boldsymbol{\Pi}_\omega \mathbf{y}\rangle_{\mathbf{B}^{-1}} &= \mathbf{x}^{\mathrm{T}}\mathbf{B}^{-1}\mathbf{B}\boldsymbol{\Gamma}_\omega^{\mathrm{T}}(\boldsymbol{\Gamma}_\omega \mathbf{B}\boldsymbol{\Gamma}_\omega^{\mathrm{T}})^{-1}\boldsymbol{\Gamma}_\omega \mathbf{y} \\ &= \mathbf{x}^{\mathrm{T}}\boldsymbol{\Pi}_\omega^{\mathrm{T}}\mathbf{B}^{-1}\mathbf{y} = \langle \boldsymbol{\Pi}_\omega^{\mathrm{T}}\mathbf{x}, \mathbf{y}\rangle_{\mathbf{B}^{-1}}\,, \end{aligned} \tag{18.12}$$

where $\langle\,,\,\rangle_{\mathbf{B}^{-1}}$ is the scalar product built on $\mathbf{B}^{-1}$. In matrix form, this is equivalent to

$$\boldsymbol{\Pi}_\omega \mathbf{B} = \mathbf{B}\boldsymbol{\Pi}_\omega^{\mathrm{T}}\,. \tag{18.13}$$

$\boldsymbol{\Pi}_\omega$ cannot be the identity, because the coarse-graining implies a loss of information that, in general, cannot be fully recovered.

Schematically, we have the following:

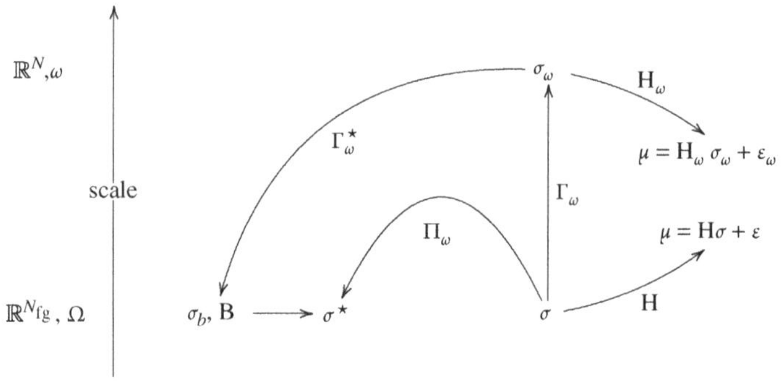

18.2.2 Observation equation in any representation

The mathematical formalism having been laid, the observation equation (18.1) can be written in any representation ω. The Jacobian $\mathbf{H}$ becomes $\boldsymbol{H}_\omega = \mathbf{H}\boldsymbol{\Gamma}_\omega^\star$. By inheritance from $\boldsymbol{\Gamma}_\omega^\star$, $\boldsymbol{H}_\omega$ is an affine operator. The observation equation reads

$$\boldsymbol{\mu} = \boldsymbol{H}_\omega \boldsymbol{\sigma}_\omega + \boldsymbol{\epsilon}_\omega = \mathbf{H}\boldsymbol{\Gamma}_\omega^\star \boldsymbol{\Gamma}_\omega \boldsymbol{\sigma} + \boldsymbol{\epsilon}_\omega = \mathbf{H}\boldsymbol{\sigma}_b + \mathbf{H}\boldsymbol{\Pi}_\omega(\boldsymbol{\sigma} - \boldsymbol{\sigma}_b) + \boldsymbol{\epsilon}_\omega\,. \tag{18.14}$$

The error $\boldsymbol{\epsilon}_\omega$ has been made scale-dependent, because several sources of errors depend on the scale, such as the aggregation errors (to be defined soon), or the errors in model subgrid physical parameterizations.

18.2.3 General scale-dependent errors

Indeed, a general decomposition of the error that account for scales could involve (i) the scale-independent observation error $\boldsymbol{\epsilon}$, which would also include model error that could be scale-independent, (ii) an error due to discretization $\boldsymbol{\epsilon}_\omega^c$ (coarse-graining), and (iii) a model error that would be scale-dependent $\boldsymbol{\epsilon}_\omega^m$:

$$\boldsymbol{\epsilon}_\omega = \boldsymbol{\epsilon} + \boldsymbol{\epsilon}_\omega^c + \boldsymbol{\epsilon}_\omega^m . \tag{18.15}$$

18.2.4 Representativeness/aggregation errors

Let us assume that errors are specified in the finest grid level, $\boldsymbol{\epsilon} = \boldsymbol{\mu} - \mathbf{H}\boldsymbol{\sigma}$, and that they may originate from many sources, including scale-independent model error. One assumes that there is no scale-dependent model error ($\boldsymbol{\epsilon}_\omega^m = 0$), so that errors at larger scales $\boldsymbol{\epsilon}_\omega = \boldsymbol{\mu} - \mathbf{H}_\omega\boldsymbol{\sigma}_\omega$ are supposed to be due solely to this original error, plus an error due entirely to coarsening, or aggregation error, which is a form of representativeness error, albeit from the control-space viewpoint. In that case, the model scaling is entirely described by the coarsening $\boldsymbol{H}_\omega = \mathbf{H}\boldsymbol{\Gamma}_\omega^\star$. Since $\boldsymbol{\mu} = \mathbf{H}\boldsymbol{\sigma} + \boldsymbol{\epsilon} = \mathbf{H}\boldsymbol{\sigma}_b + \mathbf{H}\boldsymbol{\Pi}_\omega(\boldsymbol{\sigma} - \boldsymbol{\sigma}_b) + \boldsymbol{\epsilon}_\omega$, the scale-dependent error reads

$$\boldsymbol{\epsilon}_\omega = \boldsymbol{\epsilon} + \boldsymbol{\epsilon}_\omega^c = \boldsymbol{\epsilon} + \mathbf{H}\left(\mathbf{I}_{N_{\text{fg}}} - \boldsymbol{\Pi}_\omega\right)(\boldsymbol{\sigma} - \boldsymbol{\sigma}_b), \tag{18.16}$$

with $\boldsymbol{\epsilon}_\omega^c = \mathbf{H}\left(\mathbf{I}_{N_{\text{fg}}} - \boldsymbol{\Pi}_\omega\right)(\boldsymbol{\sigma} - \boldsymbol{\sigma}_b)$ identified as the *scale-covariant* representativeness/aggregation error.

Assuming independence of the error and source error priors, the computation of the covariance matrix of these errors yields

$$\mathbf{R}_\omega = \mathbf{R} + \mathbf{R}^c = \mathbf{R} + \mathbf{H}\left(\mathbf{I}_{N_{\text{fg}}} - \boldsymbol{\Pi}_\omega\right)\mathbf{B}\left(\mathbf{I}_{N_{\text{fg}}} - \boldsymbol{\Pi}_\omega\right)\mathbf{H}^{\mathrm{T}} = \mathbf{R} + \mathbf{H}\left(\mathbf{I}_{N_{\text{fg}}} - \boldsymbol{\Pi}_\omega\right)\mathbf{B}\mathbf{H}^{\mathrm{T}}. \tag{18.17}$$

Since $\mathbf{H}\left(\mathbf{I}_{N_{\text{fg}}} - \boldsymbol{\Pi}_\omega\right)\mathbf{B}\mathbf{H}^{\mathrm{T}}$ is a positive matrix, the mean variance of the errors always increases because of the aggregation.

Intuitively, the statistics of the innovation vector $\boldsymbol{\mu} - \mathbf{H}\boldsymbol{\sigma}_b$ should not depend on the scale. However, when written in terms of errors, the innovation depends formally on the representation ω:

$$\begin{aligned}\boldsymbol{\mu} - \mathbf{H}\boldsymbol{\sigma}_b &= \boldsymbol{\mu} - \boldsymbol{H}_\omega\boldsymbol{\sigma}_\omega + \boldsymbol{H}_\omega\boldsymbol{\sigma}_\omega - \mathbf{H}\boldsymbol{\sigma}_b \\ &= \boldsymbol{\epsilon}_\omega + \boldsymbol{H}_\omega\boldsymbol{\sigma}_\omega - \boldsymbol{H}_\omega\boldsymbol{\sigma}_\omega^b = \boldsymbol{\epsilon}_\omega + \boldsymbol{H}_\omega\left(\boldsymbol{\sigma}_\omega - \boldsymbol{\sigma}_\omega^b\right).\end{aligned} \tag{18.18}$$

We have used the fact that

$$\boldsymbol{\mu}_b = \boldsymbol{H}_\omega\boldsymbol{\sigma}_\omega^b = \mathbf{H}\boldsymbol{\Gamma}_\omega^*\boldsymbol{\Gamma}_\omega\boldsymbol{\sigma}_b = \mathbf{H}\boldsymbol{\sigma}_b + \mathbf{H}\boldsymbol{\Pi}_\omega(\boldsymbol{\sigma}_b - \boldsymbol{\sigma}_b) = \mathbf{H}\boldsymbol{\sigma}_b. \tag{18.19}$$

Fortunately, this paradox is only superficial, since one can check that the statistics of the innovation are truly scale-independent:

$$\mathbf{R}_\omega + \mathbf{H}_\omega\mathbf{B}_\omega\mathbf{H}_\omega^{\mathrm{T}} = \mathbf{R} + \mathbf{H}(\mathbf{I}_{N_{\text{fg}}} - \boldsymbol{\Pi}_\omega)\mathbf{B}\mathbf{H}^{\mathrm{T}} + \mathbf{H}\boldsymbol{\Pi}_\omega\mathbf{B}\boldsymbol{\Pi}_\omega^{\mathrm{T}}\mathbf{H}^{\mathrm{T}} = \mathbf{R} + \mathbf{H}\mathbf{B}\mathbf{H}^{\mathrm{T}}. \tag{18.20}$$

18.2.5 Estimation of the performance in the finest grid Ω

More generally, an analysis performed in the representation ω is obtained by coarsening the analysis at the finest scale. Hence, in this case, the multiscale formalism has no theoretical benefit compared with performing data assimilation in the finest grid. (There are major practical advantages, though, such as dividing the computational load.) This can be understood by directly applying Bayes' rule, using Gaussian statistics:

$$q(\boldsymbol{\sigma}_\omega|\boldsymbol{\mu}) = \frac{q(\boldsymbol{\mu}|\boldsymbol{\sigma}_\omega)q(\boldsymbol{\sigma}_\omega)}{q(\boldsymbol{\mu})} \propto \exp\left[-\frac{1}{2}(\boldsymbol{\sigma}_\omega - \boldsymbol{\sigma}^b_\omega)^{\mathrm{T}}\mathbf{B}_\omega^{-1}(\boldsymbol{\sigma}_\omega - \boldsymbol{\sigma}^b_\omega) \right.$$
$$\left. -\frac{1}{2}(\boldsymbol{\mu} - \boldsymbol{H}_\omega\boldsymbol{\sigma}_\omega)^{\mathrm{T}}\mathbf{R}_\omega^{-1}(\boldsymbol{\mu} - \boldsymbol{H}_\omega\boldsymbol{\sigma}_\omega)\right]. \tag{18.21}$$

This leads to the estimate

$$\begin{aligned}\boldsymbol{\sigma}^a_\omega &= \boldsymbol{\sigma}^b_\omega + \mathbf{B}_\omega\mathbf{H}_\omega^{\mathrm{T}}\left(\mathbf{R}_\omega + \mathbf{H}_\omega\mathbf{B}_\omega\mathbf{H}_\omega^{\mathrm{T}}\right)^{-1}(\boldsymbol{\mu} - \mathbf{H}\boldsymbol{\sigma}_b) \\ &= \boldsymbol{\Gamma}_\omega\left[\boldsymbol{\sigma}_b + \mathbf{B}\mathbf{H}^{\mathrm{T}}\left(\mathbf{R} + \mathbf{H}\mathbf{B}\mathbf{H}^{\mathrm{T}}\right)^{-1}(\boldsymbol{\mu} - \mathbf{H}\boldsymbol{\sigma}_b)\right] = \boldsymbol{\Gamma}_\omega\boldsymbol{\sigma}_a\,, \end{aligned} \tag{18.22}$$

with $\boldsymbol{\sigma}^a$ the state estimation in the finest grid. The analysis error covariance matrix transforms similarly according to

$$\mathbf{P}^a_\omega = \boldsymbol{\Gamma}_\omega\left[\mathbf{B} - \mathbf{B}\mathbf{H}^{\mathrm{T}}\left(\mathbf{R} + \mathbf{H}\mathbf{B}\mathbf{H}^{\mathrm{T}}\right)^{-1}\mathbf{H}\mathbf{B}\right]\boldsymbol{\Gamma}_\omega^{\mathrm{T}} = \boldsymbol{\Gamma}_\omega\mathbf{P}^a\boldsymbol{\Gamma}_\omega^{\mathrm{T}}\,, \tag{18.23}$$

where $\mathbf{P}^a$ is the analysis error covariance matrix in the finest grid. This can also be consistently obtained, through the finest scale:

$$q(\boldsymbol{\sigma}_\omega|\boldsymbol{\mu}) = \int \mathrm{d}\boldsymbol{\sigma}\, q(\boldsymbol{\sigma}_\omega|\boldsymbol{\sigma},\boldsymbol{\mu})q(\boldsymbol{\sigma}|\boldsymbol{\mu}) = \int \mathrm{d}\boldsymbol{\sigma}\,\delta\left(\boldsymbol{\sigma}_\omega - \boldsymbol{\Gamma}_\omega\boldsymbol{\sigma}\right)q(\boldsymbol{\sigma}|\boldsymbol{\mu})\,, \tag{18.24}$$

which yields (18.22) and (18.23) by a simple convolution of Gaussian probability density functions.

18.2.6 Estimation of the performance in the coarser grid ω

We have just shown that, theoretically, it is better to estimate the truth $\boldsymbol{\sigma}_t$ (living in the finest grid) within the finest grid. First, that was assuming that (i) it is numerically affordable, and (ii) there is no scale-dependent model error. Therefore, in practice and with real data, the finest grid will not necessary be the best space in which to perform the analysis, and we believe that there should be an optimal degree of refinement of control space. The concept is illustrated Fig. 18.2.

Secondly, we could be interested in the more modest but safer objective to estimate the truth projected in a coarser grid. Therefore, we are interested in the analysis of the error $\|\boldsymbol{\sigma}^a_\omega - \boldsymbol{\sigma}^t_\omega\|$, within the targeted representation ω coarser than Ω.

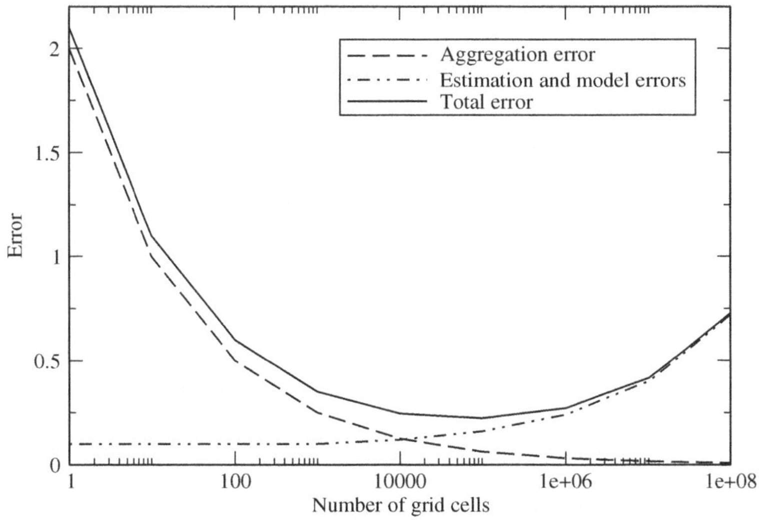

Fig. 18.2 Schematic of the posterior error (arbitrary units) or of a reasonable criterion as a function of the resolution resulting from the aggregation, model, and estimation errors.

We evaluate the inversion errors at different scales by decomposing the errors at the finest scale in $\mathbf{B}^{-1}$-norm into two parts: one for the variability at the finest scale and the other for coarser variations. Substituting the equation

$$\boldsymbol{\sigma}^a = \boldsymbol{\Gamma}^\star_\omega \boldsymbol{\sigma}^a_\omega = \boldsymbol{\sigma}^b + \boldsymbol{\Lambda}^\star_\omega \left(\boldsymbol{\sigma}^a_\omega - \boldsymbol{\sigma}^b_\omega\right), \tag{18.25}$$

into $\|\boldsymbol{\sigma}^a - \boldsymbol{\sigma}^t\|^2_{\mathbf{B}^{-1}}$, we obtain

$$\left\|\boldsymbol{\sigma}^a - \boldsymbol{\sigma}^t\right\|^2_{\mathbf{B}^{-1}} = \|\left(\mathbf{I}_{N_{\text{fg}}} - \boldsymbol{\Pi}_\omega\right)\left(\boldsymbol{\sigma}^b - \boldsymbol{\sigma}^t\right) + \boldsymbol{\Lambda}^*_\omega \boldsymbol{\sigma}^a_\omega - \boldsymbol{\Pi}_\omega \boldsymbol{\sigma}^t\|^2_{\mathbf{B}^{-1}} \tag{18.26}$$

$$= \|\left(\mathbf{I}_{N_{\text{fg}}} - \boldsymbol{\Pi}_\omega\right)\left(\boldsymbol{\sigma}^b - \boldsymbol{\sigma}^t\right) + \boldsymbol{\Lambda}^*_\omega \left(\boldsymbol{\sigma}^a_\omega - \boldsymbol{\sigma}^t_\omega\right)\|^2_{\mathbf{B}^{-1}}. \tag{18.27}$$

Letting $\boldsymbol{\gamma}^f = \left(\mathbf{I}_{N_{\text{fg}}} - \boldsymbol{\Pi}_\omega\right)\left(\boldsymbol{\sigma}^b - \boldsymbol{\sigma}^t\right)$ and $\boldsymbol{\gamma}^c = \boldsymbol{\Lambda}^*_\omega\left(\boldsymbol{\sigma}^a_\omega - \boldsymbol{\sigma}^t_\omega\right)$, one can verify that $\boldsymbol{\gamma}^f$ and $\boldsymbol{\gamma}^c$ are $\mathbf{B}^{-1}$-orthogonal, that is, $\boldsymbol{\gamma}^{f,\mathrm{T}}\mathbf{B}^{-1}\boldsymbol{\gamma}^c = 0$. Therefore, we have

$$\left\|\boldsymbol{\sigma}^a - \boldsymbol{\sigma}^t\right\|^2_{\mathbf{B}^{-1}} = \|\boldsymbol{\gamma}^f\|^2_{\mathbf{B}^{-1}} + \|\boldsymbol{\gamma}^c\|^2_{\mathbf{B}^{-1}}. \tag{18.28}$$

Here $\boldsymbol{\gamma}^f$ characterizes the variability at the finest scale and $\boldsymbol{\gamma}^c$ is related to the departure $\boldsymbol{\sigma}^a_\omega - \boldsymbol{\sigma}^t_\omega$ at coarse scales. As the representation ω is refined, the error $\|\boldsymbol{\sigma}^a - \boldsymbol{\sigma}^t\|^2_{\mathbf{B}^{-1}}$ is bound to decrease as fine details of the background state are better and better exploited. Meanwhile, the error $\|\boldsymbol{\gamma}^c\|^2_{\mathbf{B}^{-1}}$ is bound to increase, since retrieval at finer and finer scale is becoming erroneous.

18.3 Application to Bayesian control space design

One application of the above theoretical results is the Bayesian design of adaptive representations of control space. A typical problem is to know whether there is an optimal representation ω of control space using N grid cells. The above mathematical formalism can help us solve this problem. However, we first need to define 'optimal', by choosing an optimality criterion.

18.3.1 Design criteria

The Fisher criterion

Given our original incentive, to construct an adaptive grid of control space that is optimal for data assimilation, the optimality criterion must be a measure of the quality of the analysis. In Bocquet (2009), the following criterion was chosen:

$$\mathcal{J} = \mathrm{Tr}\left(\mathbf{B}\mathbf{H}^{\mathrm{T}}\mathbf{R}^{-1}\mathbf{H}\right). \tag{18.29}$$

It is inspired by the Fisher information matrix,[1] normalized by the background error covariance matrix, so that the criterion is invariant under a change of coordinate in control space (for a given grid). Specifically, it measures the reduction of uncertainty granted by the observations.

In a representation ω, the criterion reads

$$\mathcal{J}_\omega = \mathrm{Tr}\left(\mathbf{B}_\omega\mathbf{H}_\omega^{\mathrm{T}}\mathbf{R}_\omega^{-1}\mathbf{H}_\omega\right). \tag{18.30}$$

The operator $\mathbf{H}_\omega = \mathbf{H}\boldsymbol{\Lambda}_\omega^\star$ is the tangent linear operator of the affine operator $\boldsymbol{H}_\omega$ (which explains the difference of notation). Because only the linear part of $\boldsymbol{H}_\omega$ survives when averaging over the errors to obtain second-order moments, $\mathbf{H}_\omega$ appears in the criterion rather than $\boldsymbol{H}_\omega$.

If one assumes that the errors are essentially scale-independent, then $\mathbf{R}_\omega \simeq \mathbf{R}$. In that case, $\mathcal{J}_\omega$ can be written in terms of $\boldsymbol{\Pi}_\omega$ using the machinery developed earlier:

$$\mathcal{J}_\omega = \mathrm{Tr}\left[\boldsymbol{\Gamma}_\omega\mathbf{B}\boldsymbol{\Gamma}_\omega^{\mathrm{T}}\left(\boldsymbol{\Lambda}_\omega^\star\right)^{\mathrm{T}}\mathbf{H}^{\mathrm{T}}\mathbf{R}^{-1}\mathbf{H}\boldsymbol{\Lambda}_\omega^\star\right] = \mathrm{Tr}\left(\boldsymbol{\Pi}_\omega\mathbf{B}\boldsymbol{\Pi}_\omega^{\mathrm{T}}\mathbf{H}^{\mathrm{T}}\mathbf{R}^{-1}\mathbf{H}\right). \tag{18.31}$$

Using the Bayesian prolongation operator $\boldsymbol{\Gamma}_\omega^\star$, which makes use of the prior, one further obtains

$$\mathcal{J}_\omega = \mathrm{Tr}\left(\boldsymbol{\Pi}_\omega\mathbf{B}\mathbf{H}^{\mathrm{T}}\mathbf{R}^{-1}\mathbf{H}\right), \tag{18.32}$$

owing to the $\mathbf{B}^{-1}$-symmetry of $\boldsymbol{\Pi}_\omega$.

But, if the errors are scale-covariant following (18.16), the Fisher criterion (18.29) reads

$$\mathcal{J}_\omega = \mathrm{Tr}\left(\mathbf{B}_\omega\mathbf{H}_\omega^{\mathrm{T}}\mathbf{R}_\omega^{-1}\mathbf{H}_\omega\right) = \mathrm{Tr}\left\{\boldsymbol{\Pi}_\omega\mathbf{B}\mathbf{H}^{\mathrm{T}}\left[\mathbf{R} + \mathbf{H}(\mathbf{I}_N - \boldsymbol{\Pi}_\omega)\mathbf{B}\mathbf{H}^{\mathrm{T}}\right]^{-1}\mathbf{H}\right\}, \tag{18.33}$$

which is more difficult to optimize because of the nonlinear dependence of $\mathcal{J}_\omega$ on $\boldsymbol{\Pi}_\omega$.

[1] Ronald Aylmer Fisher was a famous English statistician (1890–1962) who did pioneering work in fundamental statistics, mathematical genetics, and evolutionary biology.

Fisher criterion and representativeness error

In the absence of scale-dependent model error, one can identify the normalized scale-covariant aggregation error, which is a measure of the representativeness error induced by the aggregation of grid cells:

$$\underbrace{\mathrm{Tr}\left[\mathbf{R}^{-1}\left(\mathbf{R}_\omega - \mathbf{R}\right)\right]}_{\text{normalized aggregation error}} = \underbrace{\mathrm{Tr}\left(\mathbf{BH}^{\mathrm{T}}\mathbf{R}^{-1}\mathbf{H}\right)}_{\text{Fisher criterion in }\Omega} - \underbrace{\mathrm{Tr}\left(\boldsymbol{\Pi}_\omega\mathbf{BH}^{\mathrm{T}}\mathbf{R}^{-1}\mathbf{H}\right)}_{\text{Fisher criterion in }\omega}. \tag{18.34}$$

It is clear that the larger the Fisher criterion, the smaller the aggregation error. Hence, optimal representations for the Fisher criterion are representations that minimize the representativeness errors.

Degrees of freedom for the signal

This criterion $\mathcal{J}_\omega = \mathrm{Tr}\left(\mathbf{I}_N - \mathbf{B}_\omega^{-1}\mathbf{P}_\omega^a\right)$ is known to measure the number of degrees of freedom for the signal (DFS), i.e. the information load that helps resolve the parameter space. It is actually more common in data assimilation literature than the cost function (18.32). In the absence of any source of errors, the DFS are equal to the number of scalar observations that are assimilated (p here). In the presence of errors, the DFS range between 0 and the number of observations p, because the information of the observations is also used to resolve the noise (Rodgers, 2000). So the maximization of $\mathcal{J}_\omega$ entails maximizing these degrees of freedom, which seems very natural.

Assuming scale-covariant errors of the aggregation form described earlier, the dependence on $\boldsymbol{\Pi}_\omega$ of the DFS criterion is actually simple:

$$\begin{aligned}\mathcal{J}_\omega &= -\mathrm{Tr}\left(\mathbf{B}_\omega^{-1}\mathbf{P}_\omega^a - \mathbf{I}_N\right) = \mathrm{Tr}\left[\mathbf{H}_\omega\mathbf{B}_\omega\mathbf{H}_\omega^{\mathrm{T}}\left(\mathbf{R}_\omega + \mathbf{H}_\omega\mathbf{B}_\omega\mathbf{H}_\omega^{\mathrm{T}}\right)^{-1}\right] \\ &= \mathrm{Tr}\left[\boldsymbol{\Pi}_\omega\mathbf{BH}^{\mathrm{T}}\left(\mathbf{R} + \mathbf{HBH}^{\mathrm{T}}\right)^{-1}\mathbf{H}\right],\end{aligned} \tag{18.35}$$

using the innovation statistics scaling (18.20).

Note that the criterion (18.32) is the limiting case of this DFS criterion when $\mathbf{R}$ is inflated or when $\mathbf{B}$ vanishes.

18.3.2 Numerical solution

The explicit dependence of the criteria on the representation ω, through $\boldsymbol{\Pi}_\omega$ is the key to an optimization of the criteria on the ω in a *library* of potential discretizations of control space. This permits an algebraic formulation of any adaptive discretization that follows a hierarchy of discretization. This formulation enables the definition of a conventional objective function to be minimized. All the mathematical details can be found in Bocquet (2009), Bocquet et al. (2011), and Bocquet and Wu (2011).

18.3.3 Example: the United Nations International Monitoring Network

Optimal representations of the control space can be constructed by the Bayesian multiscale framework and the design methodology introduced above. These optimal representations, under the information criteria such as the Fisher criterion or DFS, are in fact information maps indicating how information from observation sites could be propagated towards control space. One direct application of such information maps is to evaluate the performance of a monitoring network in the context of inverse modelling. In this section, we summarize one such example applied to the International Monitoring System (IMS) radionuclide network (Koohkan et al., 2012).

The IMS radionuclide network is operated by the United Nations CTBT Organisation (CTBTO) to enforce the Comprehensive Nuclear-Test-Ban Treaty (CTBT), which bans nuclear explosions. There is a total of 80 stations, of which 79 have assigned locations. We shall consider these 79 stations. As of June 2011, 60 stations were certified and operational to collect seismic, infrasound, and hydroacoustic data as well as radionuclide (particulate matter and noble gases) activity concentrations (triangles in Fig. 18.4 mark their locations). This IMS network is usually assessed by its detection capability using global atmospheric transport models (ATM), whereas, in our approach, we evaluate the potential of the IMS network for inverse modelling of the source.

The year 2009 is the focus of the study. Activity concentrations measurements are integrated over 24 hours. Therefore, $79 \times 365 = 28\,835$ observations are considered. The relationship between the emission source parameters and the observations is derived from the influence functions (also known as adjoint solutions or footprints or retroplumes) obtained using the FLEXPART Lagrangian transport model. The temporal extent of each influence function is 14 days, with a time step of $\Delta t = 3$ hours. In the absence of significant correlations between observation errors, and between background errors, we assume diagonal observation and background error covariance matrices.

We compute optimal adaptive grids of the source parameter space by maximizing the Fisher criterion or, alternatively, the DFS. This optimization takes into account the monitoring network, the uncertainty prescription, and the relationship between the source parameters and the observations derived from atmospheric transport over 2009.

The comparison of the performance of the regular and optimal representations shown in Fig. 18.3 demonstrates the efficiency of the optimal grids in terms of uncertainty reduction. We present one resulting optimal grid under the Fisher criterion in Fig. 18.4. Areas of the domain where the grid cells of the optimal adaptive grid are large emphasize zones where the retrieval is more uncertain, whereas areas where the grid cells are smaller and denser stress regions where more source variables can be resolved.

The observability of the globe through inverse modelling has been studied in cases with strong, realistic, and small model error. The strong-error and realistic-error cases yield heterogeneous adaptive grids, indicating that information does not propagate far from the monitoring stations, whereas in the small-error case, the grid is much more

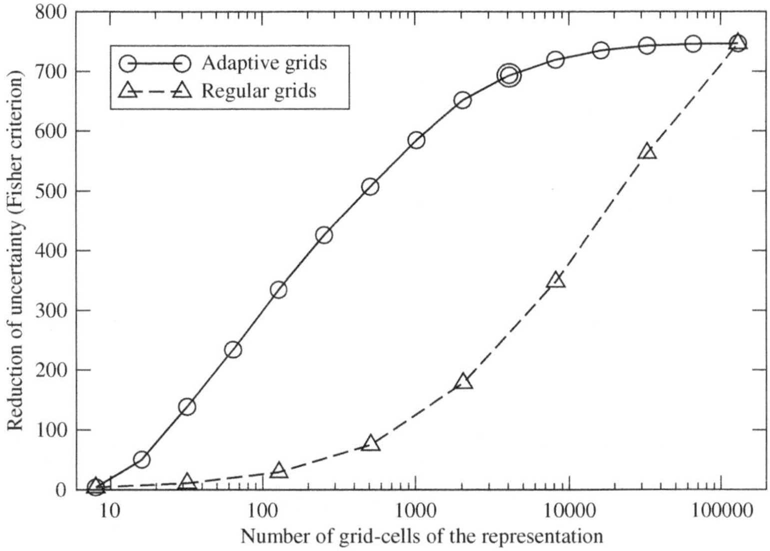

Fig. 18.3 Fisher criterion of optimal adaptive grids and regular grids against the number of grid cells in the representation. The representation in Fig. 18.4 corresponds to the point indicated by a double circle.

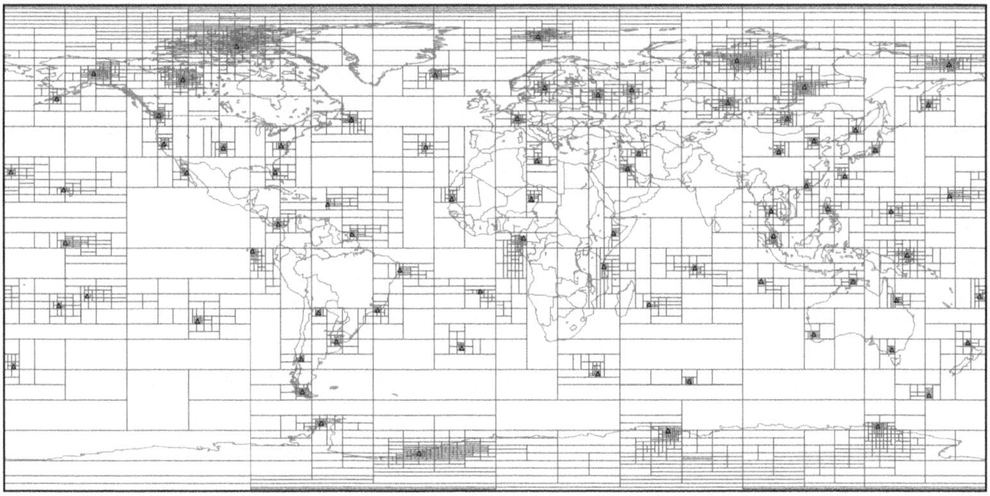

Fig. 18.4 An optimal adaptive grid for $N = 4096$ grid cells derived from the optimization of the Fisher criterion.

homogeneous. In all cases, several specific continental regions remain poorly observed, such as Africa and the tropics, because of the trade winds. The Northern Hemisphere is better observed through inverse modelling, mostly because it contains more IMS stations. This unbalance leads to a better performance of inverse modelling in the Northern Hemisphere winter.

18.3.4 Example: Ring 2

The inversion of CO_2 surface fluxes from atmospheric concentration measurements involves discretizing the flux domain in time and space. Because of the sparsity of the available concentration observations, the spatial extent of fluxes, and the dispersive nature of the atmospheric transport, the inversion of carbon fluxes is an ill-posed inverse problem. One popular remedy addressing this observation–flux imbalance is to reduce the effective degrees of freedom of fluxes by aggregation of flux variables within large regions (so-called ecoregions). However, the failure to take into account the flux variations at fine scales leads to the aggregation error that, in some cases, can be of the same order as the flux magnitude.

Thanks to the Bayesian multiscale framework and the control space design methodology, optimal multiscale representations can be obtained by maximizing the DFS. Moreover scale-dependent aggregation errors can be identified and explicitly formulated for more reliable inversions (Wu et al., 2011). We demonstrate this by performing inversions using synthetic continuous hourly CO_2 concentration data in the context of the Ring 2 experiment in support of the North American Carbon Program Mid Continent Intensive (MCI, <http://www.ring2.psu.edu>).

A ring of eight towers around the state of Iowa collects hourly averaged CO_2 concentration observations (in ppm) in and out of the corn belt area. The locations of these towers are shown in Fig. 18.5. The time period of the experiment is from 1 June 2007 at 0000 UTC to 16 June 2007 at 0000 UTC. The time length is 15 days. The total number of hourly observations p is thus 2880 ($8 \times 24 \times 15$). The two-dimensional spatial domain (an area of 980 km $\times$ 980 km) is discretized into a finest regular grid of 128×128 points. Simulations of a vegetation model SiBcrop within this spatio-temporal domain are used as the reference true fluxes. Backward particle trajectories over 15 days are generated using the Lagrangian transport model LPDM driven by the atmospheric transport simulated by the meteorological model WRF. These particles within the surface grid cells are recorded to compute the influences of the fluxes on concentration observations to form the Jacobian matrix $\mathbf{H}$. The background error covariance matrix $\mathbf{B}$ either is diagonal or follows a Balgovind model that parameterizes the isotropic exponential decaying correlation. We assume $\mathbf{R}$ to be diagonal; that is, the observation errors are spatio-temporally independent.

We plot optimal representations obtained with the DFS criterion in Fig. 18.5, which demonstrate the heterogeneity of optimal propagation of information from observation sites to the whole domain. The Balgovind correlation in background errors implicitly imposes aggregation on fluxes. This results in a more uniformly distributed optimal

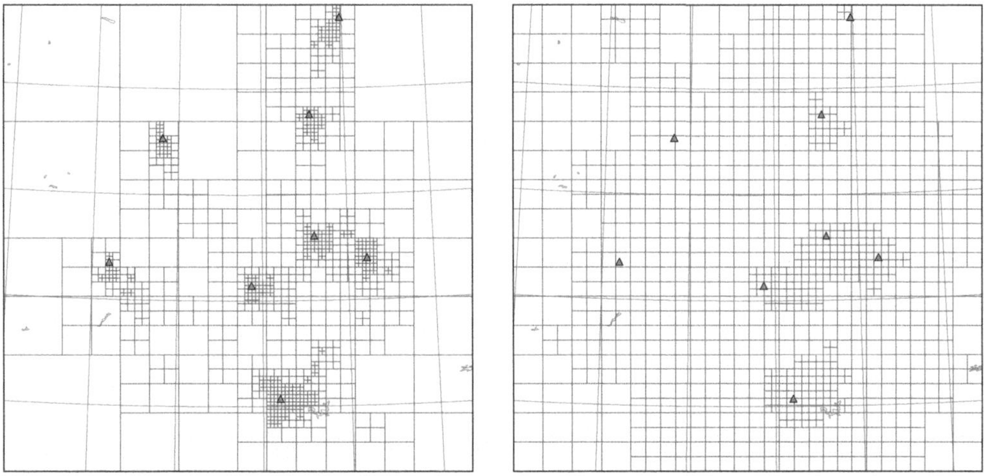

(a) BD-R3-N1024 (b) B50-R3-N1024

Fig. 18.5 Optimal multiscale representations under the DFS criterion. There is a total of 1024 grid cells. The observation error standard deviation is 3 ppm. (a) Diagonal background error covariance matrix. (b) Balgovind background error covariance matrix with correlation length equal to 50 km.

grid than that in the case of diagonal **B**. We compare the performance of the regular and optimal representations with or without taking into account the aggregation error in Fig. 18.6. For diagonal **B**, the optimal grids are more efficient than regular grids in terms of information gain (similar to Fig. 18.3). For Balgovind **B**, this gain of optimal grids is less obvious (Fig. 18.6(a)), since those optimal grids are more uniformly distributed. When the correlations in the errors of a priori fluxes are physically unrealistic (Fig. 18.6(b)), it is recommended that the aggregation errors should be taken into account explicitly. In general, only a small part of observations from the sparse Ring 2 network are effectively assimilated (DFS/$p < 20\%$).

18.4 Empirical multiscale statistics

In previous sections, we have introduced restriction and prolongation operators to transfer fluxes between scales. The aggregation error related to these operators has been identified and formulated for viable inversions. In CO_2 inversions, we have shown the important role of the background error covariance matrix.

In this section, we sketch a quite different approach (Chevallier et al., 2012) addressing the multiscale issue, in which objective statistics are first explored to estimate the background error correlations at the finest site-level scale. Then the background error covariance matrices at coarser scales are constructed by averaging using established statistical models at the finest scale.

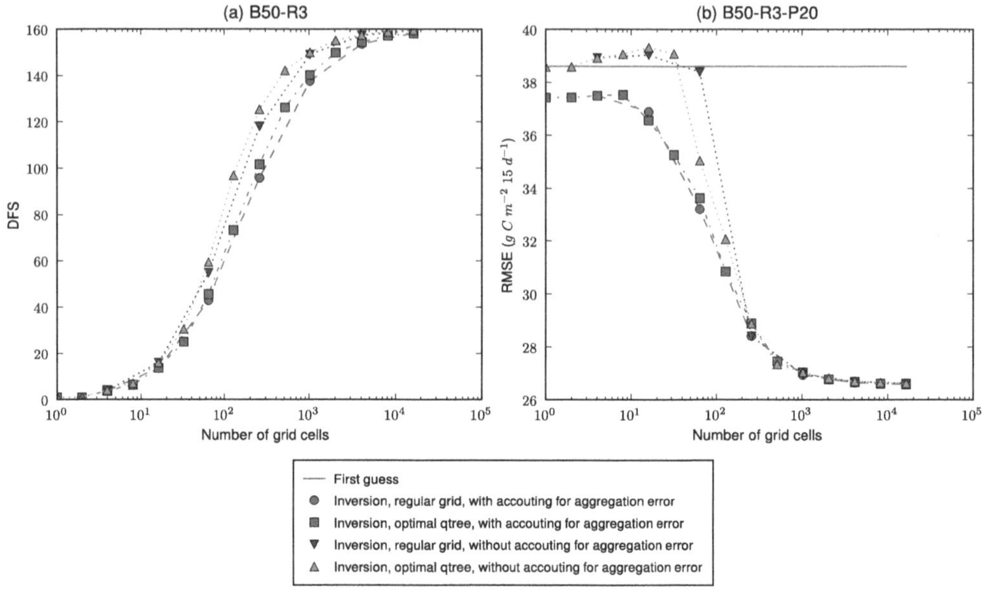

Fig. 18.6 Performance of the regular and optimal multiscale grids. The background error covariance matrix **B** takes the form of a Balgovind correlation with correlation length equal to 50 km. The observation error standard derivation is 3 ppm. (a) DFS value. (b) Root-mean-square Error (RMSE) with a first guess generated by perturbing the true flux with a smaller correlation length of 20 km.

18.4.1 Objective statistics at the finest scale

The objective statistics are based on the comparison between the surface fluxes calculated by a process-based terrestrial ecosystem model ORCHIDEE and daily averages of CO_2 flux measurements at 156 sites across the world in the global FLUXNET network. We use flux observations collected between 1991 and 2007. ORCHIDEE simulations are performed for these sites and are driven by meteorological data (Interim Reanalysis of the European Centre for Medium-Range Weather Forecasts). The background flux errors have been estimated by the statistics of the model–observation differences (the innovation).

At the daily scale, the standard deviation of the model–data fit is 2.5 gC m^{-2} d^{-1}. Temporal autocorrelations (Fig. 18.7(b)) are significant at the weekly scale (>0.3 for lags less than four weeks). The spatial structure of the error appears to be a function of the lag distance between pairs of sites based on the Pearson correlation coefficient of the model–observation differences (Fig. 18.7(a)). The median reveals spatial structure at short distances (<100 km). The correlation median is 0.33 for distances less than 100 km, 0.26 for distances between 100 and 200 km, and negligible beyond 400 km.

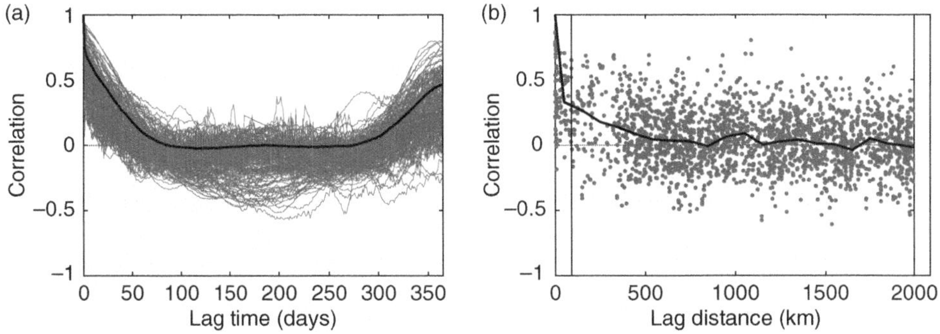

Fig. 18.7 (a) Time correlogram, i.e. autocorrelation between the errors of the ORCHIDEE model at distant times for the same site. Each thin line corresponds to a different site. The thick line represents the all-site autocorrelation; (b) Distance correlogram, i.e. correlation between the daily net ecosystem exchange (NEE) errors of the ORCHIDEE model at pairs of distant sites for the same time. Each point includes all the common years of one of the site pairs. The thick line represents the median of the points per 100 km bin.

The short spatial correlation length estimated from FLUXNET data does not seem to justify the aggregations by imposing biome-dependent correlations.

A statistical model of all-site autocorrelation can be established by fitting the thick lines in Fig. 18.7,

$$\begin{aligned} r_\tau(\tau) &= 0.767 - 0.018\tau + (0.01222\tau)^2 - (0.00809\tau)^3 \\ &\quad + (0.00513\tau)^4 - (0.0012\tau)^5 + 0.216/(1+10\tau), \qquad (18.36) \\ r_\delta(\delta) &= 1/(0.032\delta + 1)\,, \qquad (18.37) \end{aligned}$$

with r_τ, the dependence of the correlation as a function of lag time τ in days, and r_δ, the correlation as a function of the lag distance δ in km.

18.4.2 Background error at coarser scales

If each flux variable at a coarse scale is assumed to be the arithmetic mean of fluxes at finest pixels of 1 day and 1 km^2, then the variance of the flux error on a coarser grid cell can be computed by multiplying the error variance of the finest pixel with the arithmetic mean of correlations of flux errors between two arbitrary finest pixels in that coarser grid cell. The correlation of errors of fluxes between two coarser-scale grid cells can be computed in a similar way.

In Fig. 18.8, we show the effect of temporal and spatial aggregation on the error statistics at various aggregation scales. In general, aggregation reduces error variances, while increasing correlations. As an example, for a typical inversion at a global scale of grid-point (300 km × 300 km) monthly fluxes, the prior flux error follows an approximate e-folding correlation length of 500 km only, with correlations from one month to the next as large as 0.6.

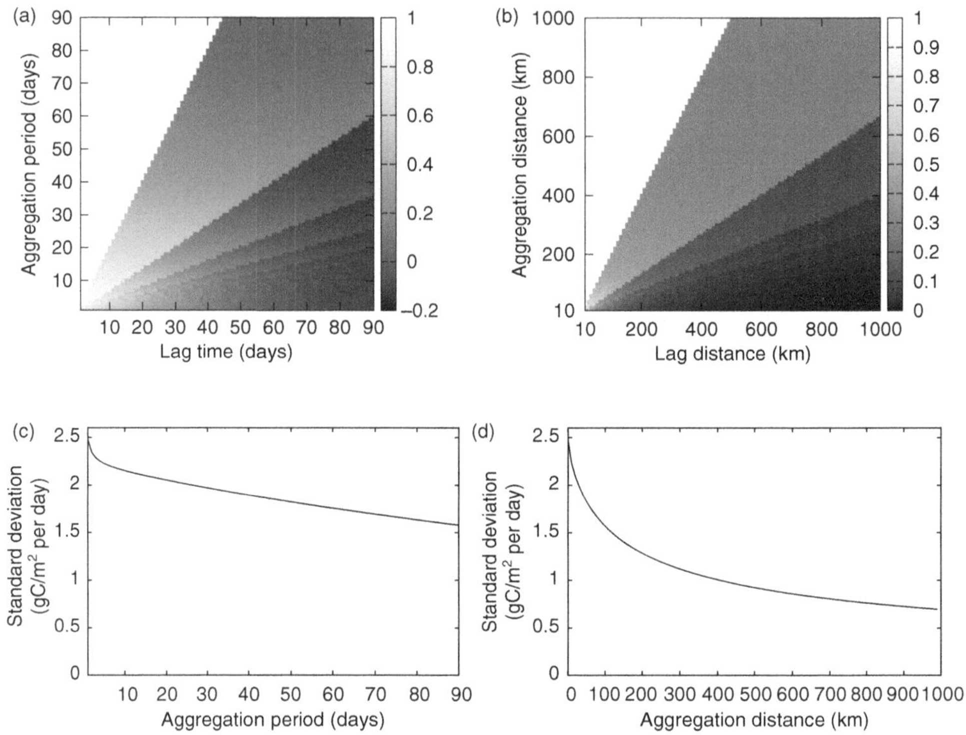

Fig. 18.8 Effect of (a, c) temporal and (b, d) spatial aggregation of the fluxes on error correlation (a, b) and error standard deviation (c, d). The aggregation distance (b, d) is defined as the length of the side of a square on which the aggregation is performed.

18.5 Conclusion

These selected topics have offered only a restricted view on a few facets of the multiscale aspects of geophysical data assimilation. In spite of the apparent heterogeneity of the formalism and concepts, a few pivotal ideas and requirements have emerged. The first is the need for properly formalized operators for changing scales and converting/transferring information through the scales. These operators need to be able to consistently transfer not only the fields but also their statistics. Another key idea is the importance of accounting and formalizing representativeness errors or alternatively aggregation errors. Finally, the paradigm of information budget that underlies modern data assimilation is even more pregnant when dealing with the multiscale aspects of data assimilation, since changing scales may involve losing or gaining information, such as the forecast and analysis steps of standard data assimilation.

References

Bocquet, M. (2009). Towards optimal choices of control space representation for geophysical data assimilation. *Mon. Weather. Rev.*, **137**, 2331–2348.

Bocquet, M. and Wu, L. (2011). Bayesian design of control space for optimal assimilation of observations. II: Asymptotics solution. *Q. J. R. Meteorol. Soc.*, **137**, 1357–1368.

Bocquet, M., Wu, L., and Chevallier, F. (2011). Bayesian design of control space for optimal assimilation of observations. I: Consistent multiscale formalism. *Q. J. R. Meteorol. Soc.*, **137**, 1340–1356.

Chevallier, F., Wang, T., Ciais, P., Maignan, F., Bocquet, M., Arain, A, Cescatti, A., Chen, J.-Q., Dolman, H., Law, B. E., Margolis, H. A., Montagni, L., and Moors, E. J. (2012). What eddy-covariance flux measurements tell us about prior errors in co_2-flux inversion schemes. *Global Biogeochem. Cycles*, **26**, GB1021.

Janjić, T. and Cohn, S. E. (2006). Treatment of observation error due to unresolved scales in atmospheric data assimilation. *Mon. Weather. Rev.*, **134**, 2900–2915.

Koohkan, M. R. and Bocquet, M. (2012). Accounting for representativeness errors in the inversion of atmospheric constituent emissions: application to the retrieval of regional carbon monoxide fluxes. *Tellus B*, **64**, 19047.

Koohkan, M. R., Bocquet, M., Wu, L., and Krysta, M. (2012). Potential of the international monitoring system radionuclide network for inverse modelling. *Atmos. Environ.*, **54**, 557–567.

Liu, Z.-Q. and Rabier, F. (2002). The interaction between model resolution, observation resolution and observation density in data assimilation: a one-dimensional study. *Q. J. R. Meteorol. Soc.*, **128**, 1367–1386.

Oke, P. R. and Sakov, P. (2008). Representation error of oceanic observations for data assimilation. *J. Atmos. Ocean. Tech.*, **25**, 1004–1017.

Rodgers, C. D. (2000). *Inverse Methods for Atmospheric Sounding*, Vol. 2. World Scientific, Singapore.

Wu, L., Bocquet, M., Lauvaux, T., Chevallier, F., Rayner, P., and Davis, K. (2011). Optimal representation of source–sink fluxes for mesoscale carbon dioxide inversion with synthetic data. *J. Geophys. Res.*, **116**, D21304.

19
Data assimilation in meteorology

F. Rabier[1] and M. Fisher[2]

[1]CNRM-GAME, Météo-France and CNRS, Toulouse, France
[2]ECMWF, Reading, UK

Advanced Data Assimilation for Geosciences. First Edition.
Edited by É. Blayo, M. Bocquet, E. Cosme, and L. F. Cugliandolo.

Chapter Contents

Data assimilation is the art of combining model and observations. It relies on a set of equations grounded on sound statistics. Many theoretical studies have investigated optimal ways to define various quantities (observation operators, and observation and background error covariances) and how to combine all the flow-dependent information (time evolution of the model, using linearized models in 4D-Var or ensembles in the ensemble Kalman filter (EnKF)). But more is needed to get a reliable and optimal analysis. In practice, a lot of attention has to be paid to details in the handling of observations, the specification of (possibly non-Gaussian) error statistics, and the filtering of the resulting analysis.

This chapter discusses some of the implementation details (including the pre- and post-processing of observations) that are necessary to apply data assimilation in the context of numerical weather prediction (NWP). These details are often mentioned only in passing in data assimilation talks or articles.

The chapter is divided into three parts. In the first part, we discuss the processing of observations. We start in Section 19.1 by describing the transformation of raw data into a form that can be processed by a data assimilation system. In Section 19.2, we discuss how observations and models are compared. An important part of this comparison is to identify and eliminate from the analysis observations that have been corrupted by gross errors that are not described by the assumed statistics of observation error.

Thinning of data is an important aspect of data assimilation for NWP, since it is not possible to make use of all the observations that are available. We discuss this in Section 19.3.

In the second part of the paper, we discuss two aspects of data assimilation for NWP that are not related to observations, but are nonetheless important. Filtering of the analysis to remove spurious inertia–gravity waves is presented in Section 19.4, while methods to handle nonlinearities and non-Gaussian error statistics are discussed in Section 19.5.

One of the identifying characteristics of data assimilation for NWP is that it is a real-time activity. It is necessary to perform an analysis within the short period of time between the arrival of the last observation and the start of the forecast. As data assimilation becomes ever more computationally demanding (because of increased model resolution and larger numbers of observations), it will be neccessary to devise new algorithms that better exploit the parallel nature of the computers on which they run. The final part of the paper (Section 19.6) discusses the degree to which 4D-Var can be made more parallel. Some preliminary, but encouraging, results are presented that suggest that 4D-Var can be made significantly more parallel than current implementations suggest, and that 4D-Var will remain an important and viable NWP data assimilation algorithm into the forseeable future.

19.1 Transforming data

First of all, it is necessary to process raw data, i.e. to transform them so as to get observations that can be easily integrated into a data assimilation system. Among the different kinds of observations, radiosondes take simple measurements of such atmospheric parameters as wind, temperature, pressure, and humidity. Other observations are very indirect measurements. For example, satellite operators process several series of images from satellites to derive atmospheric motion vectors, tracking a target in an image at consecutive times. Radiances, widely used in NWP, measure the electromagnetic spectrum and provide very indirect information on temperature, humidity, surface, clouds, ozone, and other constituents. These radiances can be used directly in data assimilation schemes, including a radiative transfer model in the observation operator, or via retrievals, where the transformation to atmospheric parameters is performed individually for each vertical profile before the analysis. Some NWP centres combine these two approaches, starting with a one-dimensional retrieval of some parameters at the observation points, also useful for quality control, followed by the assimilation of radiances in a full 3D- or 4D-Var scheme using some of the retrieved parameters such as surface temperature, emissivity, or cloud parameters as input. The second step of data pre-processing consists of spatial averaging. Some of the horizontal averaging can be performed by data producers as is the case for the clear sky radiances, averaged in boxes typically of 16×16 pixels for SEVIRI on board MSG at EUMETSAT. Some averaging can also be done at the user level. In the pre-processing of all-sky radiances at the European Centre for Medium-Range Weather Forecasts (ECMWF), it was found that averaging observations to create AMSR-E super-observations at a 80 km scale was beneficial (Geer and Bauer, 2010). Another example is the processing of satellite winds at the Naval Research Laboratory (NRL). NRL produces super-observations with a complex algorithm, averaging in boxes (prisms of about 2° side), then performing prism-quartering when there is a high degree of variability within the box, and checking that observations agree within a certain range (Pauley, 2003). Their processing of satellite winds is more elaborate than at other NWP centres, and satellite winds are found to be more influential on the analysis/forecast quality at NRL than at other centres (Gelaro et al., 2010). At meso-scale, a similar averaging is performed for radar data. Radar winds are pre-processed in a similar fashion in the HIRLAM 3D-Var (Lindskog et al., 2004) and at NCAR for WRF (Zhang et al., 2009). Super-observations are created from radial winds by averaging in polar space. These super-observation winds are smoother than the raw data and are representative on spatial scales characteristic for the averaging procedure (Fig. 19.1). To avoid averaging radial winds with significantly different directions, fewer polar bins are used near the radar than at longer ranges. At Météo-France, a median concept is used rather than a mean value (Montmerle and Faccani, 2009). A median filter is applied on boxes of 5×5 pixels, replacing observed values by the median of neighbouring points.

Averaging data can also be performed in the time dimension. For example, the GPS ground-based system provides integrated water vapour at high temporal frequencies. In Europe, depending on the operators, GPS observation time periods can range from 5 to 60 minutes. For these data, a time averaging of observations is performed, with a 30–60 minute period compatible with the length of the 4D-Var slots (Poli et al., 2007; Macpherson et al., 2008). Similarly, time averaging can be performed for NCEP Stage IV radar and gauge precipitation data, used at ECMWF for example. Averaging the data over different accumulation periods can produce different results

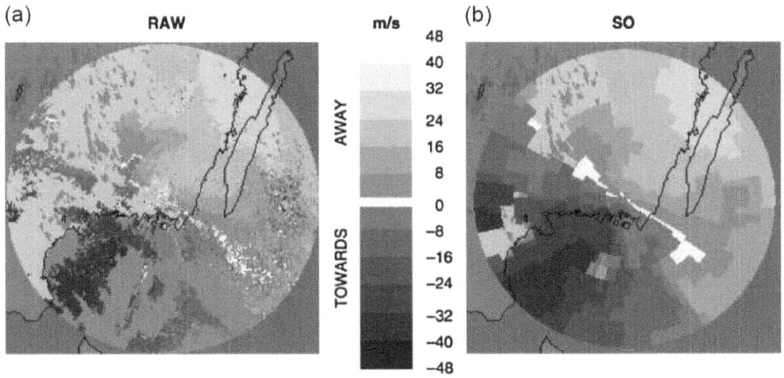

Fig. 19.1 Averaging of radial winds from raw data (a) into super-observations (b). (From Lindskog et al. (2004).)

in the assimilation. In particular, although hourly data can be used, 6-hourly accumulations are found to perform better. The correlation between departures computed in the full trajectory (high horizontal resolution with full physics) and in first minimization (low horizontal resolution, with simplified physics) increases from 0.2 to 0.7. The use of 6-hourly accumulations can be seen as a compromise between linearity and observation usage over the 4D-Var 12-hour window (Lopez, 2011).

Finally, the treatment of the observations in the vertical will be discussed. A vertical choice of data is performed for some observations: typically, one might select only one observation per model level (this is usually the case for aircraft or radio-occultation observations). For radiosonde observations, the transmission of data to NWP centres follows some rules: observations are transmitted either if they are on mandatory levels (fixed pressure levels) or if they are at a significant level in the vertical profile corresponding to a change in slope. As these observations are the backbone of the global observing system, all levels are usually kept in the assimilation. At meso-scale, one might even interpolate between significant and mandatory levels to yield additional data points (Benjamin et al., 2004). This forces the analysis to fit the near-linear structures implied by the absence of intermediate significant levels. In the future, more data points will be transmitted and the processing might change, possibly forcing centres to average data in the vertical or to select only a subset of points. Processing of hyper-spectral sounder data provides new challenges. At each pixel, thousands of radiances describe the infrared part of the spectrum. Each channel is sensitive to various quantities (mainly temperature or humidity) and to a certain vertical slice of the atmosphere. One needs to compress the information contained in the hyperspectral sounder data to make their assimilation possible. The idea then is to select the most relevant channels to describe adequately all parameters of interest. For instance, such a channel selection was studied for the Atmospheric Infra Red Sounder (AIRS) and for the Infrared Atmospheric Sounding Interferometer (IASI) (Fourrié and Thépaut, 2003; Rabier et al., 2002; Collard, 2007) and an approach based on the information content was found to be particularly effective. Let us introduce the 'degrees of freedom for signal' (DFS). Writing the analysis $\mathbf{x}^a$ as a function of the background $\mathbf{x}^b$ and of the observations $\mathbf{y}$,

$$\mathbf{x}^a = \mathbf{x}^b + \mathbf{K}(\mathbf{y} - \mathbf{H}\mathbf{x}^b), \tag{19.1}$$

where $\mathbf{K}$ is the Kalman gain matrix and $\mathbf{H}$ the observation operator (assumed linear for the sake of simplicity). We have

$$\mathbf{K} = \mathbf{B}\mathbf{H}^{\mathrm{T}}(\mathbf{H}\mathbf{B}\mathbf{H}^{\mathrm{T}} + \mathbf{R})^{-1}, \tag{19.2}$$

or, equivalently,

$$\mathbf{K} = (\mathbf{B}^{-1} + \mathbf{H}^{\mathrm{T}}\mathbf{R}^{-1}\mathbf{H})^{-1}\mathbf{H}^{\mathrm{T}}\,\mathbf{R}^{-1}, \tag{19.3}$$

where $\mathbf{B}$ is the background error covariance matrix and $\mathbf{R}$ the observation error covariance matrix. One can write the analysis in observation space:

$$\mathbf{H}\mathbf{x}^a = \mathbf{H}\mathbf{x}^b + \mathbf{H}\mathbf{K}(\mathbf{y} - \mathbf{H}\mathbf{x}^b). \tag{19.4}$$

One can compute the derivative of $\mathbf{H}\mathbf{x}^a$ with respect to the observations $\mathbf{y}$,

$$\frac{\partial(\mathbf{H}\mathbf{x}^a)}{\partial\mathbf{y}} = (\mathbf{H}\mathbf{K})^{\mathrm{T}}, \tag{19.5}$$

define the DFS as the trace of this matrix:

$$\mathrm{DFS} = \mathrm{Tr}(\mathbf{H}\mathbf{K})^{\mathrm{T}} = \mathrm{Tr}(\mathbf{H}\mathbf{K}) = \mathrm{Tr}(\mathbf{K}\mathbf{H}), \tag{19.6}$$

and use the fact that

$$\mathbf{K}\mathbf{H} = \mathbf{A}\mathbf{H}^{\mathrm{T}}\mathbf{R}^{-1}\mathbf{H} = \mathbf{I} - \mathbf{A}\mathbf{B}^{-1}, \tag{19.7}$$

where $\mathbf{A}$ is the analysis error covariance matrix. The DFS is then the trace of the matrix $\mathbf{I} - \mathbf{A}\mathbf{B}^{-1}$; it quantifies the gain brought by the observations. In the channel selection procedure, the channel that most improves the DFS is selected. Once this most informative channel has been chosen, the $\mathbf{A}$ matrix is updated and the next best channel chosen, and so forth. This process is applied to a set of representative atmospheric profiles in order to produce an optimal set of channels, valid in most conditions. These are the channels that are going to be assimilated in NWP (Fig. 19.2). Principal component compression is also envisaged (Collard et al., 2010). Principal components are computed from a large set of spectra, and principal component amplitudes can then be related to observed radiances. It is found that around 200 principal components are required to represent the signal, the rest being considered as noise. Using the leading principal components is then efficient for data transfer and noise filtering. It is a method well suited to represent the relatively small amount of information contained in spectra. But, as pointed out by Collard et al. (2010), this new way of handling data means that we will have to adapt the way we deal with these observations. In particular, there are still pending issues, such as the introduction of error correlations, extending the current practice of selecting channels not affected by surface/cloud, and monitoring (identifying faulty channels that are sensitive to given levels in the atmosphere).

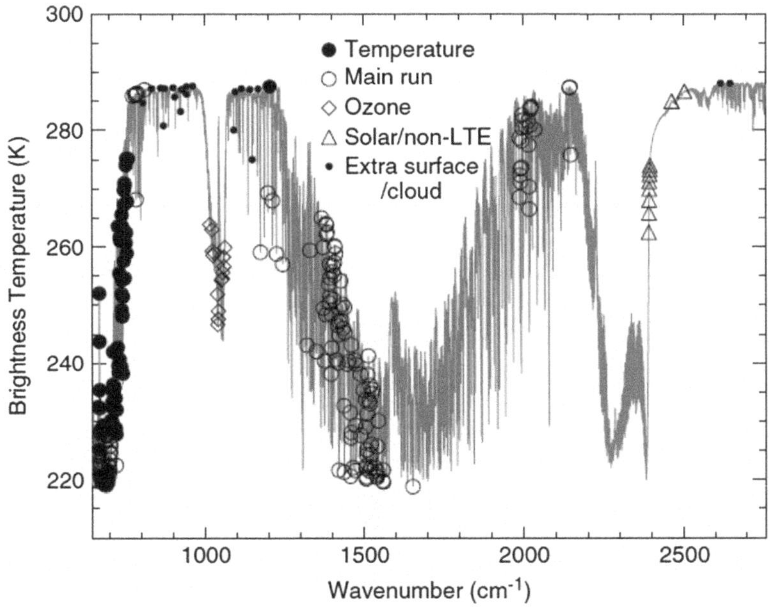

Fig. 19.2 Channel selection (300 channels) for the IASI hyperspectral sounder, illustrated on an observed spectrum. (From Collard (2007). Copyright © 2007 Royal Meteorological Society.)

19.2 Comparing data and models

The comparison of the observations with the model and the choice of the relevant data to be used usually make up the second part of the post-processing step. This process is called the monitoring of observations. Examples at NWP centres show that changes in the vertical or horizontal resolution of the model can, for instance, allow the model to fit more closely high-peaking channels or ground-based synoptic stations. These might then be considered for assimilation, as they can now be represented by the model. The presence of bias is another issue to be dealt with. The monitoring procedure usually shows the presence of systematic differences between model and observations, and part of this bias seen in the monitoring can be attributed to observations. It is then adequate to remove the observation bias before performing the analysis, and simple to elaborate bias correction schemes have been devised to this purpose. On the simple side, ground-based GPS data are usually treated with a bias correction simply based on averaged deviations from the model (Poli et al., 2007) or on a 10-day running mean (Macpherson et al., 2008). For other types of observations, such as radiosondes, the origin of the bias can be attributed to physical processes. The bias then depends on a few factors, such as sonde type, sonde age, solar elevation, and pressure level. Using this information on the type of bias can help to parameterize the bias correction. If one can identify a more precise, almost bias-free sonde type, one can then deduce the bias from the other sonde types by comparison. This is what was done by Nuret et al. (2008) during the AMMA field campaign over Africa. For a limited time period in 2006, a scattered sampling was achieved at Niamey (Niger) for Vaisala RS-80A and RS-92 sonde types, using four sondes of each type, every day, for 25 days. As RS-92 is the newest, more precise type of radiosonde, it served as the basis for the RS-80A one. Augusti-Panareda et al. (2009) went a step further, still considering that the RS-92 was almost bias-free (at least at night) and using the model as an intermediate by double-differencing as explained below, thus assuming that the model biases have the same characteristics everywhere. A cumulative distribution function (CDF) matching is performed, followed by a fitting of four sine-wave components of a Fourier series depending on the observed humidity, for each pressure level, solar elevation, and sonde type (Fig. 19.3). The monitoring biases are computed both for the reference sonde $\mathbf{y}^{\text{ref}}$ (Vaisala RS-92) and for any sonde $\mathbf{y}$. Then, the bias of the sonde $\mathbf{y}$ is written as $\text{Bias}(\mathbf{y} - \mathbf{y}^{\text{ref}}) = \text{Bias}(\mathbf{y} - \mathbf{y}^b) - \text{Bias}(\mathbf{y}^{\text{ref}} - \mathbf{y}^b)$, where $\mathbf{y}^b$ is the model background (short-range forecast) in observation space. As both biases were computed beforehand, the bias for sonde $\mathbf{y}$ can now be computed. For satellite radiances, an a priori knowledge about the parameters affecting the observation bias is known. Harris and Kelly (2001) used scan-dependence and air-mass predictors to describe the bias, with regression coefficients computed over a long time series. The selected air-mass predictors are the model thicknesses (1000–300 hPa, 200–50 hPa, and so on), the surface temperature, and the integrated water content. The regression coefficients can be adapted offline before any analysis, or inside the assimilation (the so-called variational bias correction; see Auligné et al., 2007; Dee and Uppala, 2009). Comparing model and observations is also useful in removing erroneous data. Each observation is subject to a variety of errors: biases from calibration, random errors, and representativeness errors, but also gross errors coming from instrument malfunction or transmission error for instance. Data with gross errors are useless and need to be eliminated by a quality control step.

19.2.1 Quality control

One of the main differences between real-world applications of data assimilation and theoretical studies using simulated data is the unruly nature of real observations.

Real observations often have poorly known error characteristics, and these characteristics may vary from one type of observation to the next. For example, aircraft-based measurements have different error characteristics on take-off and landing than they do in level flight, and satellite-based instruments have errors that are sensitive to the presence of cloud, land, ice, etc. in the field of view.

Observations are frequently biased, and this bias must be identified and corrected before the observations are assimilated. Furthermore, in any set of observations, we can expect to find a few outliers that are contaminated by gross errors. These observations have errors that are not described by the assumed error statistics.

There are many possible causes of gross error, ranging from malfunctioning mechanical parts to data transmission error to simple human error (e.g. in reading a thermometer).

In current meteorological data assimilation systems, analysis error is significantly smaller than typical observation errors. Consequently, a single bad observation can

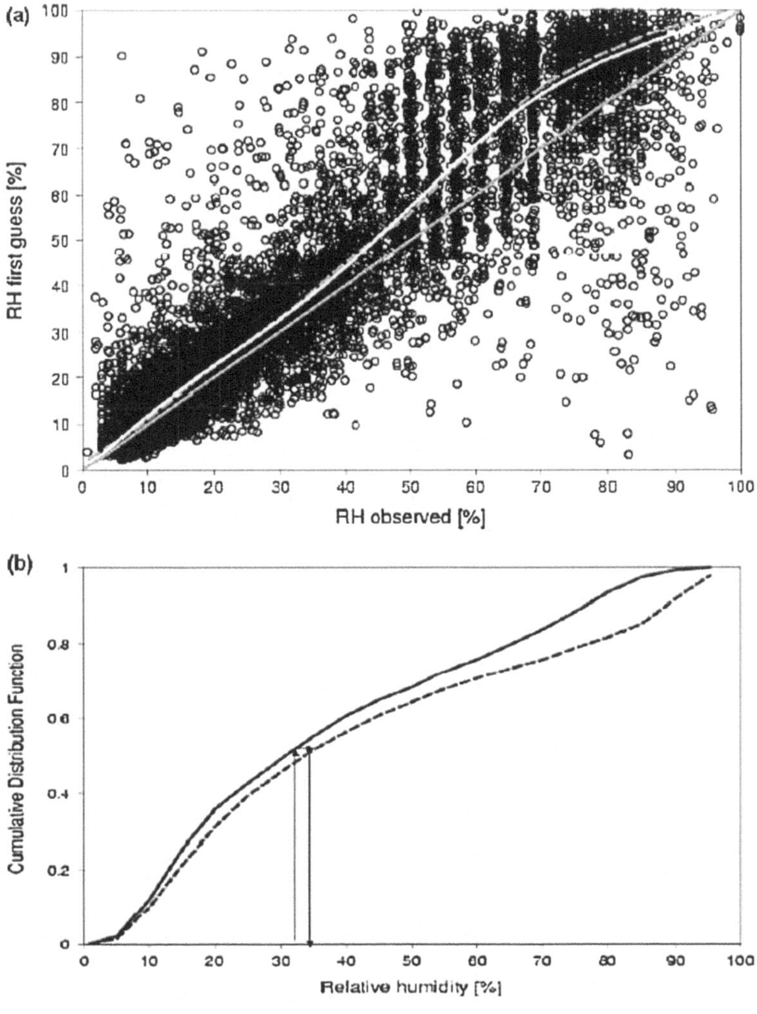

Fig. 19.3 (a) Scatter plot of first guess versus observed relative humidity for Vaisala Digicora I, II radiosondes at 925 hPa and positive solar elevation larger than 27.5°. The grey solid line is the identity line. The grey dashed curve shows the bias obtained from CDF matching and the white solid curve is the best fit using an expansion of the bias formulation on four sine waves. (b) Cumulative distribution function for observation (solid line) and first guess (dashed line). Arrows illustrate the CDF matching technique. (From Augusti-Panareda et al. (2009). Copyright © 2009 Royal Meteorological Society.)

have a drastic impact on the quality of the analysis. This makes quality control of observations a vital component of any real-world, meteorological data assimilation system.

An example of the importance of quality control

An example of how a single bad observation can damage an analysis was identified in the ECMWF data assimilation system in July 2011, when a poor analysis was made of the developing typhoon Ma-on. This typhoon went on to strike southern Japan. It caused the deaths of five people and damage that has been estimated at US$50 million.

The ECMWF analysis correctly identified the development of the typhoon, producing (with one glaring exception) a sequence of analyses with successively deeper central pressure. The exception in this sequence was the analysis for 12UTC on 13 July, in which the cyclone had almost disappeared.

The cause of the bad analysis was traced to a single drifting buoy whose pressure sensor had become stuck. The buoy reported a surface pressure of 1011 hPa, regardless of the actual surface pressure.

Stages of quality control

Quality control in a typical NWP system can be divided into three stages.

The first stage consists of a set of non-statistical 'sanity' checks, aimed at removing obviously nonsensical data. Examples of such checks include tests to identify unphysical values (such as impossibly warm or cold temperatures), observations that are missing some information (e.g. their location), multiple copies of the same observation, or reports from ships or aircraft that have moved impossibly large distances since they last reported. In addition to the above, a list of known poor observation types or stations is maintained and regularly updated. This is used to remove observations that have a history of problems.

The second stage of quality control applies statistical tests to each observation, independently of other observations. The tests identify observations whose errors are unlikely to fit the assumed statistics. A standard test of this type is the 'first-guess check'. This compares each observation against its model equivalent, computed from the prior. The difference between the observed value and its model equivalent has a variance equal to the sum of the assumed variances of the prior and of the observation errors. The first-guess check rejects observations for which the difference between the observation and its model equivalent exceeds a few standard deviations (according to the assumed statistics). The idea is that a difference of such magnitude is highly unlikely to result from random error, suggesting that the observation has been contaminated by a gross error that is not described by the assumed error statistics.

The first-guess check is widely used. However, it suffers from two important limitations: First, it skews the pdf of observation error towards the background. This introduces correlation between the prior and the observations that is difficult to take into account in the analysis. Second, it does not take into account supporting evidence from nearby observations. It is capable of rejecting large numbers of mutually consistent observations. (In this situation, it is likely that the assumed statistics of the prior are in error, not those of the observations.)

For these reasons, rejection limits are typically set wide, so that the first-guess check is implemented as a rather weak test that identifies a small number of very bad observations.

The final stage of quality control applies what is referred to by Gandin (1988) as 'consistency checks'. These checks apply statistical tests that take into account nearby observations, together with the spatial structure of the error in the prior. Two common approaches to consistency checking are 'buddy checking' and variational quality control.

Consistency checking takes into account the spatial structure of the errors in the prior. This can be done (similarly to the sanity and first-guess checks) as a filtering step that is performed before data assimilation, and which consists of a statistical test of the multivariate distribution of sets of neighbouring observation departures, d_i. In this approach, the aim is to filter out bad observations, leaving a well-behaved, Gaussian-distributed set of observations that is passed to the assimilation algorithm. This makes the approach well suited to linear, Gaussian analysis techniques. The filtering approach to consistency testing is often called 'buddy checking'.

A typical implementation (e.g. Dee et al., 2001) performs an initial classification of observations as either 'suspect' or 'buddies'. The buddies are then used, together with an estimate of the spatial statistics of the observation departures from the prior, to predict the corresponding departures for the suspect observations. The differences between the actual and predicted departures for the suspect observations are used to determine whether to accept or reject the suspect observations.

Buddy checking is a pre-processing steps that can be performed before the analysis. It is an important technique for use with linear analysis methods, since, after quality control, it is reasonable to assume that the errors of the retained observations are Gaussian.

However, the tests depend on the prior, and therefore introduce correlations between observation error and prior error that are difficult to take into account (and are usually neglected).

An alternative approach is to take into account the non-Gaussian nature of observation errors within the assimilation algorithm itself. This approach is well suited to variational methods, where non-Gaussian error distributions can be handled directly via non-quadratic terms in the cost function (Lorenc, 1988; Andersson et al., 1999).

Variational quality control is based on the interpretation of the observation term of the cost function as (minus the log of) a conditional probability:

$$J_o(x) \propto -\log\left[p(y|x)\right]. \tag{19.8}$$

For Gaussian observation errors, J_o is quadratic, but more complex error distributions can also be taken into account.

Ingleby and Lorenc (1993) (see also Andersson and Järvinen, 1999) assumed $p(y|x)$ to be the sum of a Gaussian (representing 'normal' observation error) and a uniform distribution (representing gross error). However, this distribution produces a cost function that is very flat away from the minimum (see Fig. 19.4). There is no incentive for

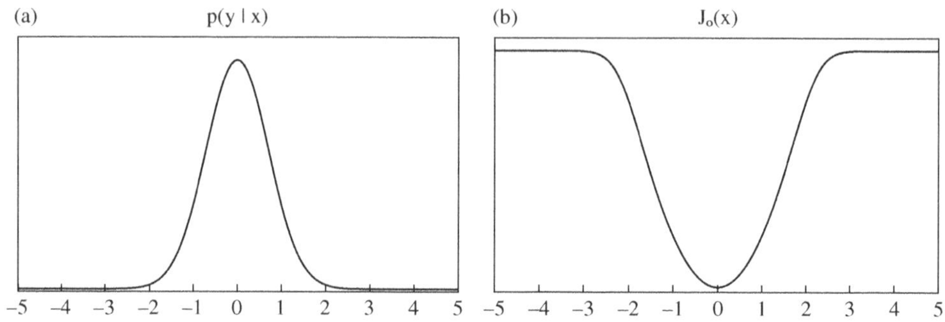

Fig. 19.4 (a) A probability density function constructed as a linear combination of a Gaussian and a uniform distribution. (b) The corresponding cost function.

the analysis to move closer to observations in the flat part of the cost function (even if there are large numbers of supporting observations there), since this does not result in a reduction in cost. Hence, this distribution shares with the first-guess check the possibility of rejecting large numbers of mutually supporting observations.

This problem can be overcome by replacing the Gaussian-plus-uniform distribution by a 'fat-tailed' distribution that is Gaussian for small values of d_i, but exponential for larger values. This type of distribution was introduced by Huber (1964). The corresponding cost function switches between an L_2 (quadratic) metric for small d_i and an L_1 (linear) metric for large d_i. As well as providing a robust estimate that allows outlying observations to support each other, there is some evidence that the fat-tailed pdf is a good representation of the actual statistics of observation errors.

19.3 Thinning the dataset

Different analysis schemes use different temporal thinning of data. In 4D-Var, one groups observations in 30 or 60 minute time-slots and thins observations within each time-slot. In a 3D-Var, one usually selects data closer to the central analysis time (e.g. ± 1.5 h for aircraft data). In non-cycled schemes, data are chosen to be really representative of analysis time. For example, for the hourly real-time mesoscale analysis (De Pondeca et al., 2011), the time window is selected so as to be within ± 12 minutes of the analysis time. For dense datasets such as those provided by satellite sounding data, a horizontal thinning is performed, both for practical cost reasons and to avoid observation error correlations between adjacent observations, which are not accounted for in assimilation schemes. Generally, the thinning is performed in a simple manner, by latitude–longitude boxes. Within each box, the observation is selected according to a given quality criterion, for example distance to guess, quality indicator provided by the data producer, small value of radial wind variance in the super-observation, or maximum number of elevations that pass the quality control in radar profiles. More elaborate adaptive thinning techniques have been devised by Ochotta et al. (2005), where either observations representative of clusters are inserted iteratively, or observations from the full set are removed if they are considered redundant.

In a simulated observation context, Liu and Rabier (2002, 2003) showed that, for uncorrelated errors, using more observations always improves the analysis, but this does not hold for correlated errors. To neglect the observation error correlation can indeed deteriorate both the analysis and the forecast if it becomes too large. Depending on the strength of the observation correlation, an optimal thinning distance was found to preserve a reasonable quality in the analysis. An example is shown in Fig. 19.5. As can be seen on both the analysis errors (Fig. 19.5(a)) and the forecast errors (Fig. 19.5(b)), for weak to moderate correlation intensities, there is a thinning interval that leads to the best results. This optimum thinning is obtained around 0.5° for the weak correlations and around 0.6–0.7° for the moderate correlations. These values correspond to the interval for which the correlation between adjacent observations

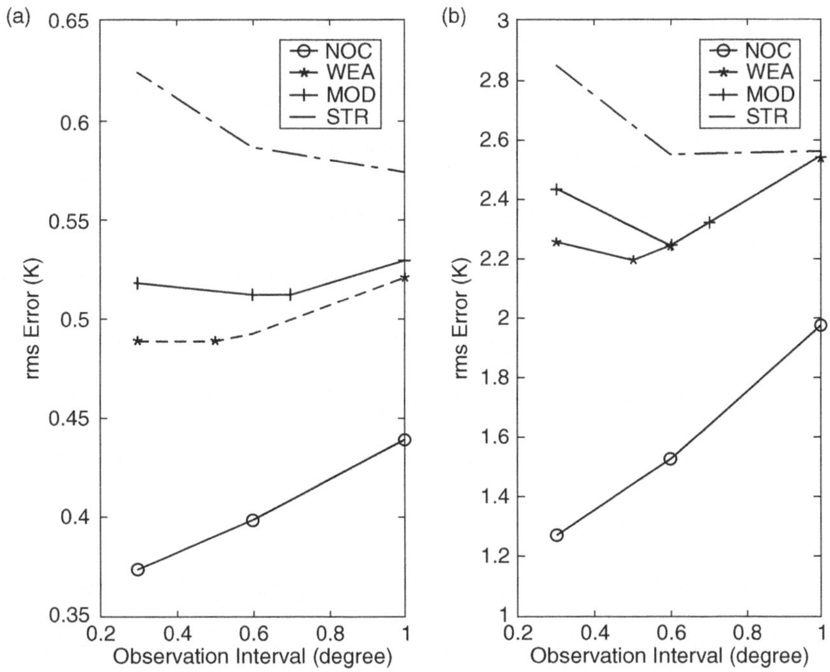

Fig. 19.5 Root-mean-square analysis (a) and 48-hour forecast (b) errors as functions of observation interval in the context of simulated AMSU-A radiances, for which the observation error correlation is zero (NOC = no correlations, solid lines with circles), weak (WEA, dashed lines with stars), moderate (MOD, solid lines with crosses), and strong (STR, dash–dotted lines). For a distance of 0.6° between observations, the respective values of correlations are 0.6 for the strongly correlated observations, 0.3 for the moderately correlated observations, and 0.15 for the weakly correlated observations. (From Liu and Rabier (2003). Copyright © 2003 Royal Meteorological Society.)

becomes less than 0.2. It seems that the suboptimal system neglecting correlations in the observation errors can give reasonable results, as long as the correlation between adjacent observations is not larger than values of the order of 0.2.

In fact, such correlations exist in meteorological satellite observations, as demonstrated for atmospheric motion vectors by Bormann et al. (2003) and for satellite radiances by Bormann and Bauer (2010) and Bormann et al. (2010), which justify the practice of thinning in NWP centres. In a real-observations context, the optimal thinning distance was investigated in the Met Office NWP system by Dando et al. (2007) over a three-week period.

The control experiment used a thinning distance of 308 km for ATOVS data. Then, this distance was altered for other experiments, where it was ranging from 40 to 600 km, and the quality of the resulting analysis was judged by the quality of the associated forecasts. An optimal distance was found to be between 100 and 150 km, as seen in Fig. 19.6. In particular, it was found detrimental to use thinning at the finest 40 km distance, especially in the Tropics, where weak gradients are present in the meteorological fields. The reason put forward is that at high observation densities, the outcome may be determined by the meteorological features associated with the most influential analysis errors. If the analysis errors responsible for the majority of the forecast error are in a region of strong gradient, additional ATOVS observations may be beneficial. With active weather systems in the extra-tropical regions, the highest observation densities may have a positive impact on the analysis. This reasoning was used in another context at ECMWF, in which optimal AMSU-A thinning using singular-vector information was performed in the Southern Hemisphere (Bauer et al., 2001). Several configurations were tested over two seasons: an experiment with a global density of 1.25°, an experiment with a global high density of 0.625°, an experiment with high-density only in areas determined by the singular vectors of the day

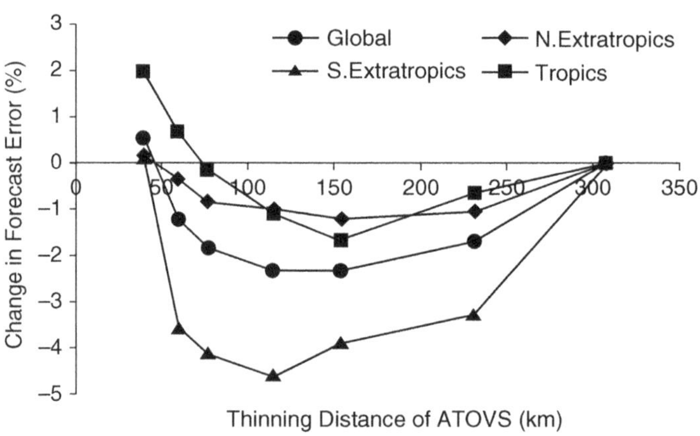

Fig. 19.6 Change in forecast error based on different ATOVS thinning distances in the analysis, for different regions, compared with a reference (308 km thinning). (From Dando et al. (2007). Copyright © 2007 Royal Meteorological Society.)

(where small perturbations of the flow might grow the most rapidly during a finite time period; Buizza et al., 2000), an experiment with high density only in areas determined by the climatological singular vectors, and an experiment with high density in random areas. First, it was shown that the high-density experiment always performed better than any others. It was further demonstrated that in the warm season, singular-vector-based thinning performs significantly better than random or climatological thinning. Finally, the global density enhancement is found to be very positive at high latitudes and locally negative in the Tropics. Singular-vector-based thinning initially produced a smaller impact that gradually amplified to a consistent positive impact. In winter, results were more mixed, probably because more data are rejected in cloudy areas and over sea-ice, lessening differences between scenarios. In the EnKF context, the radius of influence of observations (ROI) also plays a role in the quality of the analysis. An example is given by Zhang et al. (2009). In WRF, radar wind data assimilation is performed in three domains: D1 (40.5 km), D2 (13.5 km), and D3 (4.5 km) over the USA in a case of tropical cyclone with a 30-member EnKF. Several experiments are performed with either fixed ROIs in each domain, from 1215 to 135 km. Other experiments are performed with ROI varying from 1215 to 135 km from the largest domain D1 to the smallest domain D3. The best experiment uses a successive covariance localization, with a different localization radius of influence used for different groups of observations by random sampling. First, one tries to dynamically remove important aspects of the large-scale errors by assimilating a relatively small subset of observations with a large radius of influence. Next, this radius is decreased and higher-density observations are used to constrain both smaller-scale errors and remaining large-scale errors. This approach is motivated by the fact that with the EnKF performing serial observation processing, the error correlation lengthscale decreases, since large scales are better defined by the previously assimilated observations.

19.4 Filtering the analysis

During the forecast, the model can take time to adjust initial fields with respect to the model equations. This spin-up contains processes of dynamical adjustment with excess energy radiated away by inertia–gravity waves, and processes of diabatic adjustment where model physics tendencies converge with respect to other model forcings. Ideally, the analysis should create balanced increments through the background error covariance matrix. There is also a possibility to include constraint terms in the analysis (Gauthier and Thépaut, 2001). But, in practice, it is still often necessary to perform a posterior filtering of the analysis. With this procedure, the initial state cannot generate model tendencies that project onto high-frequency model solutions. The filtering step removes spurious waves that can affect the first hours in the forecast and thus the assimilation cycle. Among the different methods that can be used, the two most popular are the digital filter initialization (DFI; Lynch and Huang, 1992; Huang and Lynch, 1993) and the incremental analysis update (IAU; Lorenc et al., 1991; Bloom et al., 1996). DFI implies running the model backward and forward and then filtering the high frequencies and choosing a cut-off frequency. In the forward step, the model can be run in diabatic mode, with the parameterizations of the physical processes activated, as in normal conditions, whereas in the backward step, the model has to be run in adiabatic mode in order to be revertible. IAU consists in spreading the insertion of the increment in a time window around the nominal analysis time, with direct insertion over a few time-steps. Initial imbalance can be measured by the time tendency of the surface pressure over the domain of interest for example. This imbalance depends on the quality of the analysis. Chen and Huang (2006) showed that the Cressman analysis points to more imbalance in the initial fields than the 3D-Var analysis (Fig. 19.7). Similarly, in the rapid update cycle, a high-frequency data assimilation system, Benjamin et al. (2004) found less balance in fields initialized with optimum interpolation than with a 3D-Var. Initialization can quite drastically change the resulting initial condition fields. One way to counteract this major impact is to perform incremental initialization (Fischer and Auger, 2011). Let us denote by $\mathbf{x}^{af}$ the filtered

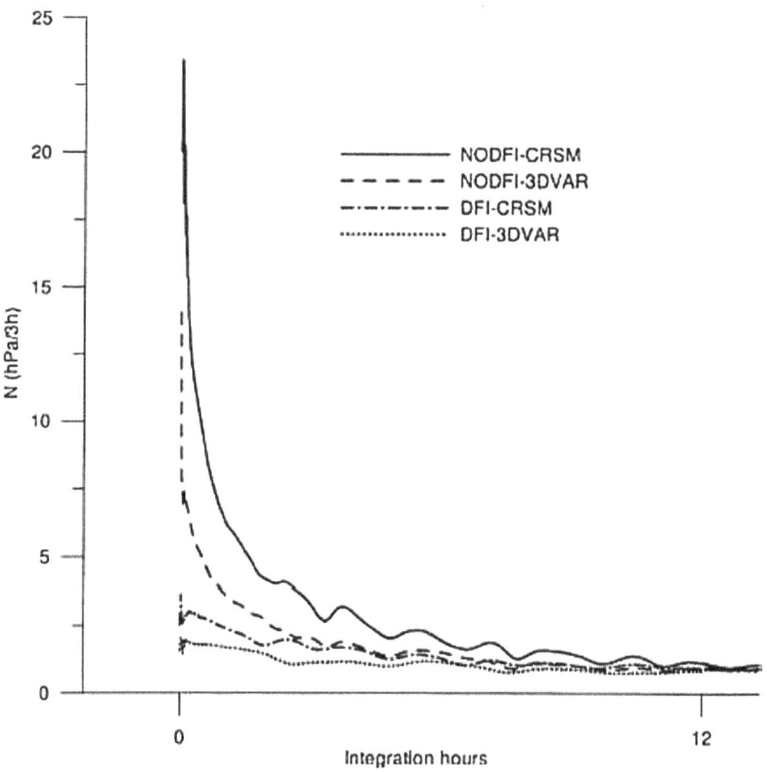

Fig. 19.7 Imbalance measures from different initial conditions, before and after the application of DFI. The analyses are performed with a Cressman or an OI method. This figure shows the evolution of the mean absolute surface pressure tendency N [hPa/(3 h)] in the first 12 h forecast averaged from 14 cycles from 0000 UTC 21 August to 1200 UTC 27 August 2002. (From Chen and Huang (2006).)

analysis state. It is simply equal to $\mathrm{DFI}(\mathbf{x}^a)$ after standard DFI. The increment for standard DFI is then

$$\mathbf{x}^{af} - \mathbf{x}^b = \mathrm{DFI}(\mathbf{x}^a) - \mathbf{x}^b = [\mathrm{DFI}(\mathbf{x}^a) - \mathrm{DFI}(\mathbf{x}^b)] + [\mathrm{DFI}(\mathbf{x}^b) - \mathbf{x}^b]. \tag{19.9}$$

It is the sum of a balanced increment and of the result of an operation removing the high frequencies in $\mathbf{x}^b$. In the incremental DFI, the filtered state is

$$\mathbf{x}^{af} = \mathbf{x}^b + [\mathrm{DFI}(\mathbf{x}^a) - \mathrm{DFI}(\mathbf{x}^b)], \tag{19.10}$$

and the total increment is then

$$\mathbf{x}^{af} - \mathbf{x}^b = \mathrm{DFI}(\mathbf{x}^a) - \mathrm{DFI}(\mathbf{x}^b). \tag{19.11}$$

If high frequencies exist in the background, in relation to rapidly developing meteorological features for example, it is better not to filter them out, and to privilege the incremental DFI approach. Filtering is effective not only for forecasts, but also for the assimilation. For rapid cycles, the assimilation cycle can be adversely affected by spurious waves, which have to be filtered out before the next cycle is performed, so as to compare the model fields with new observations from a noise-free forecast. This is becoming more important as the tendency is to go to hourly cycles for fine-scale data assimilation.

19.5 Nonlinearities and non-Gaussian densities in variational data assimilation

In variational data assimilation, the cost function is minus the log of the conditional pdf. Assuming that observation errors are independent of errors in the prior, Bayes' theorem gives

$$\begin{aligned} J(x) &= -\log[p(x|y)] + \text{const} \\ &= -\log\left[\frac{p(y|x)p(x)}{p(y)}\right] + \text{const} \\ &= -\log[p(y|x)] - \log[p(x)] + \text{const}, \end{aligned} \tag{19.12}$$

where, in the final line, we have absorbed $p(y)$ into the additive constant.

This simple relationship allows considerable flexibility in specifying non-Gaussian pdfs for both observation error and background error. In practice, there are many cases where the ability to specify non-Gaussian pdfs is very useful. Particular examples include non-Gaussian observation error (for example for directionally ambiguous scatterometer-derived winds); variational quality control; bounded variables such as humidity and trace gases; and nonlinear dynamical balances.

Accounting for non-Gaussian pdfs makes the cost function non-quadratic. Although algorithms exist to directly minimize such functions, the problem is typically formulated as a sequence of quadratic approximations. The sequence of quadratic cost functions corresponds to a sequence of local Gaussian approximations to the pdfs.

In meteorology, a further approximation is made by solving the quadratic problems at reduced spatial resolution. This approach is known as the incremental algorithm.

19.5.1 Symmetrizing transformations

Humidity is a difficult variable to handle in data assimilation for NWP. The amount of water vapour in a parcel of air cannot be less than zero, and cannot exceed a threshold determined by the saturated vapour pressure (corresponding to 100% relative humidity).

The presence of these thresholds guarantees that errors in the prior and observations are non-Gaussian. Moreover, the PDF of humidity error is typically highly skewed. For example, if the prior state at a given point has low relative humidity r_b, then for a given increment $\delta r > 0$, the state $r_b + \delta r$ is more likely than the state $r_b - \delta r$, whereas for high prior humidity, $r_b - \delta r$ is more likely than $r_b + \delta r$.

A novel method of removing this asymmetry was introduced by Hólm (2003) and has been implemented in the ECMWF data assimilation system.

Let us consider an arbitrary pair of members of an ensemble whose states are believed to be samples from the prior distribution. We will denote the relative humidities of the two members at a given grid-point by r_1 and r_2, and the difference between these two values by δr.

The conditional pdf $p(r_2|r_1)$ is highly asymmetric for values of r_1 close to zero or 100%. However, since the two ensemble members were chosen arbitrarily, they are interchangeable. That is, the joint pdf satisfies, $p(r_1, r_2) = p(r_2, r_1)$ and is therefore symmetric about the line $r_1 = r_2$. We illustrate this schematically in Fig. 19.8.

To exploit this symmetry, we consider the conditional pdf along the diagonal line $r_2 + r_1 = \text{const}$. The quantity that varies along this line is $\delta r = r_2 - r_1$, so that the conditional pdf $p(\delta r|r_2 + r_1)$ must be symmetric. Writing $r_2 + r_1 = 2r_1 + \delta r$, we see that $p(\delta r|2r_1 + \delta r))$ (or, equivalently, $p(\delta r|r_1 + \frac{1}{2}\delta r))$ is symmetric.

Now, r_1 is a sample from the prior distribution. Let us consider it to be the prior, r_b, to be used in an analysis. Let us also define a nonlinear transformation of δr:

$$\delta\tilde{r} = \frac{\delta r}{\sigma\left(r_b + \frac{1}{2}\delta r\right)}, \tag{19.13}$$

where σ is a standard deviation that can be empirically determined from the ensemble.

The result of this transformation is that $\delta\tilde{r}$ has a symmetric distribution with unit variance. In practice, Hólm (2003) found that this pdf is well modelled by a unit-variance Gaussian, as shown in Fig. 19.9.

19.5.2 Nonlinear dynamics

The atmosphere is a nonlinear system. However, at least on the larger scales, for which there is predictability out to a few days, the dynamics are reasonably close to linear.

Gilmour et al. (2001) estimated that the evolution of the atmosphere becomes dominated by nonlinearity after about two days, whereas Pires et al. (1996) concluded that the linear approximation is valid for two to three days for large spatial scales, and for around 36 hours at the mesoscale. A detailed study of the accuracy of the linear approximation in 4D-Var was carried out by Trémolet (2004).

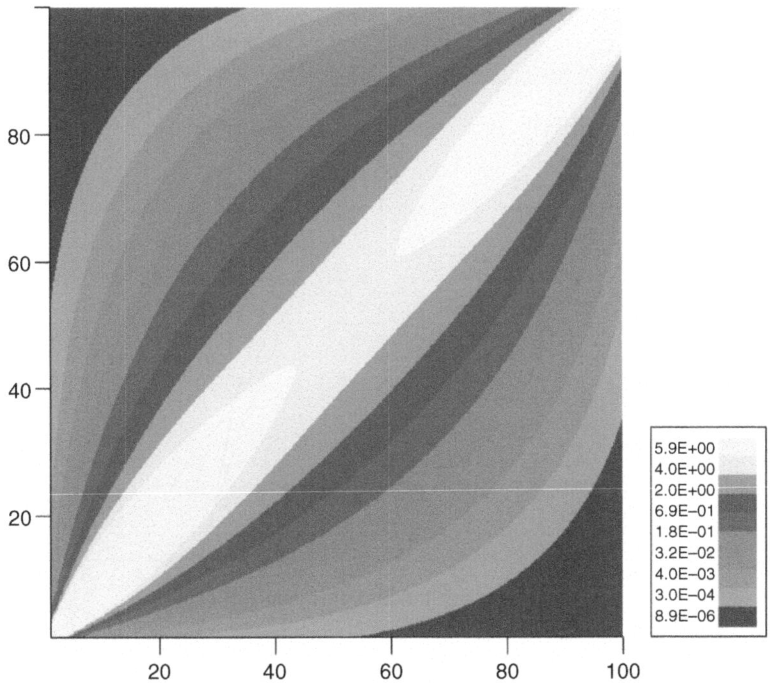

Fig. 19.8 Schematic representation of the joint pdf of relative humidity prior states for two members of an ensemble.

Nonlinearity is one of the main factors that limits the length of the 'assimilation window' in current implementations of 4D-Var for NWP to, typically, 6–12 hours.

There are good reasons for believing that longer assimilation windows would be beneficial. However, it is well known (e.g. Roulston, 1999; Pires et al., 1996) that the cost function in strong-constraint 4D-Var demonstrates increasingly pathological behaviour, with increasingly many local minima, as the analysis window is increased in length. This behaviour is due to the model integration that appears in the observation term of the cost function:

$$J_o(\mathbf{x}_0) = \frac{1}{2}\sum_{k=0}^{N} \left[\mathbf{y}_k - \mathcal{H}_k(\mathcal{M}_{0,k}(\mathbf{x}_0))\right]^{\mathrm{T}} \mathbf{R}_k^{-1} \left[\mathbf{y}_k - \mathcal{H}_k(\mathcal{M}_{0,k}(\mathbf{x}_0))\right]. \tag{19.14}$$

Here, $\mathbf{x}_0$ is the state at the start of the 4D-Var window and $\mathcal{M}_{0,k}$ represents a model integration from the start of the window to the time t_k at which $\mathbf{y}_k$ is valid.

Clearly, if $t_k \gg t_0$, we can expect $\mathcal{M}_k$ to be highly nonlinear and highly sensitive to initial conditions. In this case, the cost function will fluctuate wildly in response to small changes in the initial state, $\mathbf{x}_0$.

Weak-constraint 4D-Var does not suffer from this problem, provided the cost function is expressed as a function of the four-dimensional state. In this formulation, the cost function may be written as

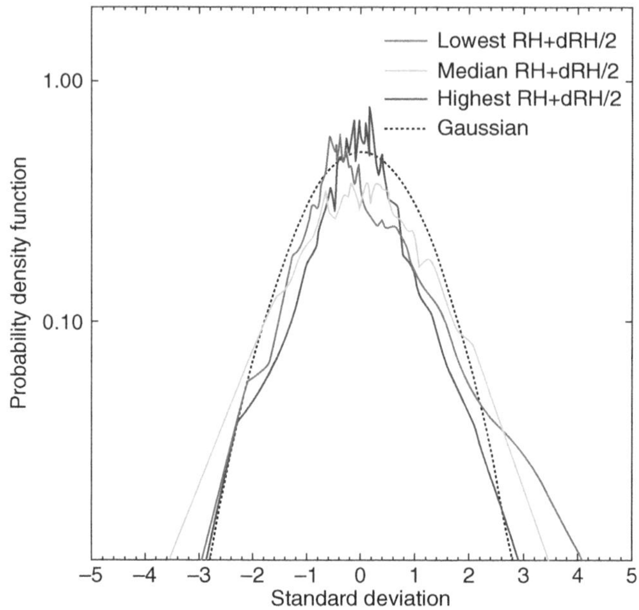

Fig. 19.9 The pdf of the transformed humidity varaible $\delta\tilde{r}$ calculated from samples stratified according to $r_b + \frac{1}{2}\delta r$. The transformed variable is reasonably well approximated by a Gaussian for all values of $r_b + \frac{1}{2}\delta r$.

$$\begin{aligned} J(\mathbf{x}_0, \mathbf{x}_1, \ldots, \mathbf{x}_N) &= \frac{1}{2}\left(\mathbf{x}_0 - \mathbf{x}_b\right)^{\mathrm{T}} \mathbf{B}^{-1}\left(\mathbf{x}_0 - \mathbf{x}_b\right) \\ &+ \frac{1}{2}\sum_{k=0}^{N}\left[\mathbf{y}_k - \mathcal{H}_k(\mathbf{x}_k)\right]^{\mathrm{T}} \mathbf{R}_k^{-1}\left[\mathbf{y}_k - \mathcal{H}_k(\mathbf{x}_k)\right] \\ &+ \frac{1}{2}\sum_{k=1}^{N}\left[\mathbf{x}_k - \mathcal{M}_{k-1,k}(\mathbf{x}_{k-1})\right]^{\mathrm{T}} \mathbf{Q}_k^{-1}\left[\mathbf{x}_k - \mathcal{M}_{k-1,k}(\mathbf{x}_{k-1})\right]. \end{aligned} \tag{19.15}$$

The only nonlinear operators in this equation are $\mathcal{H}_k$ and $\mathcal{M}_{k-1,k}$. The former is likely to be reasonably linear. The latter represents an integration over a short time interval, and is also nearly linear.

The difference between strong-constraint 4D-Var and weak-constraint 4D-Var can easily be seen by considering what happens if we change a single $\mathbf{x}_i$. Only three terms of the cost function change:

$$\left[\mathbf{y}_i - \mathcal{H}_i(\mathbf{x}_i)\right]^{\mathrm{T}} \mathbf{R}_i^{-1}\left[\mathbf{y}_i - \mathcal{H}_i(\mathbf{x}_i)\right], \tag{19.16}$$

$$\left[\mathbf{x}_{i+1} - \mathcal{M}_{i,i+1}(\mathbf{x}_i)\right]^{\mathrm{T}} \mathbf{Q}_{i+1}^{-1}\left[\mathbf{x}_{i+1} - \mathcal{M}_{i,i+1}(\mathbf{x}_i)\right], \tag{19.17}$$

$$\left[\mathbf{x}_i - \mathcal{M}_{i-1,i}(\mathbf{x}_{i-1})\right]^{\mathrm{T}} \mathbf{Q}_i^{-1}\left[\mathbf{x}_i - \mathcal{M}_{i-1,i}(\mathbf{x}_{i-1})\right]. \tag{19.18}$$

Since the operators in these terms are nearly linear, the change in the cost function in response to the change in $\mathbf{x}_i$ is well behaved.

In practice, it has been shown in experiments with simple nonlinear models (Fisher and Auvinen, 2012) that weak-constraint 4D-Var can be used with analysis windows many times longer than the timescale over which nonlinear effects begin to dominate (as determined, for example, by the nonlinearity index of Gilmour et al., 2001).

19.6 Parallel algorithms for 4D-Var

In this section, we speculate on a future direction for 4D-Var, as it is applied in NWP. There are two important issues that will drive this development. First, as analyses become increasingly accurate, owing in part to more and better observations, it will no longer be possible to ignore the errors of the assimilating model. Second, the real-time constraints of NWP, coupled with the tendency towards ever more computationally expensive algorithms and the predicted developments in computer architectures, will push a move towards more parallel computational methods.

It is likely that the need to take model error into account will engender a move away from strong-constraint 4D-Var towards weak-constraint formulations. In principle, this should also aid efforts to parallelize 4D-Var, since it opens the possibility of parallelisation in the time dimension.

In incremental, weak-constraint 4D-Var, we solve a sequence of quadratic problems of the form

$$\begin{aligned}\hat{J}(\delta\mathbf{x}_0, \delta\mathbf{x}_1, \ldots, \delta\mathbf{x}_N) &= \frac{1}{2}\left(\delta\mathbf{x}_0 - \mathbf{b}_0\right)^{\mathrm{T}} \mathbf{B}^{-1}\left(\delta\mathbf{x}_0 - \mathbf{b}_0\right) \\ &+ \frac{1}{2}\sum_{k=0}^{N}\left(\mathbf{d}_k - \mathbf{H}_k\delta\mathbf{x}_k\right)^{\mathrm{T}} \mathbf{R}_k^{-1}\left(\mathbf{d}_k - \mathbf{H}_k\delta\mathbf{x}_k\right) \\ &+ \frac{1}{2}\sum_{k=1}^{N}\left(\delta\mathbf{x}_k - \mathbf{M}_{k-1,k}\delta\mathbf{x}_{k-1} - \mathbf{c}_k\right)^{\mathrm{T}} \mathbf{Q}_k^{-1}\left(\delta\mathbf{x}_k - \mathbf{M}_{k-1,k}\delta\mathbf{x}_{k-1} - \mathbf{c}_k\right),\end{aligned} \tag{19.19}$$

where (using superscript t to denote linearization state)

$$\mathbf{b}_0 = \mathbf{x}_b - \mathbf{x}_0^{\mathrm{t}}, \qquad \mathbf{c}_k = \mathcal{M}_{k-1,k}(\mathbf{x}_{k-1}^{\mathrm{t}}) - \mathbf{x}_k^{\mathrm{t}}, \qquad \mathbf{d}_k = \mathbf{y}_k - \mathcal{H}_k(\mathbf{x}_k^{\mathrm{t}}).$$

We can simplify the notation by defining some four-dimensional vectors and matrices:

$$\delta\mathbf{x} = \begin{pmatrix} \delta\mathbf{x}_0 \\ \delta\mathbf{x}_1 \\ \vdots \\ \delta\mathbf{x}_N \end{pmatrix}, \qquad \delta\mathbf{p} = \begin{pmatrix} \delta\mathbf{x}_0 \\ \delta\mathbf{q}_1 \\ \vdots \\ \delta\mathbf{q}_N \end{pmatrix}. \tag{19.20}$$

These vectors are related through $\delta\mathbf{q}_k = \delta\mathbf{x}_k - \mathbf{M}_k\delta\mathbf{x}_{k-1}$. We can write this relationship in matrix form as

$$\delta\mathbf{p} = \mathbf{L}\delta\mathbf{x}, \tag{19.21}$$

where

$$\mathbf{L} = \begin{pmatrix} \mathbf{I} & & & & \\ -\mathbf{M}_1 & \mathbf{I} & & & \\ & -\mathbf{M}_2 & \mathbf{I} & & \\ & & \ddots & \ddots & \\ & & & -\mathbf{M}_N & \mathbf{I} \end{pmatrix}.$$

Now consider applying the matrix $\mathbf{L}$ to a vector $\delta\mathbf{x}$. Since $\delta\mathbf{x}$ contains the initial conditions for each of the model integrations required to apply $\mathbf{L}$, these model integrations can be performed in parallel. In contrast, application of $\mathbf{L}^{-1}$ to a vector $\delta\mathbf{p}$ requires a sequence of model integrations, corresponding to a forced integration of the model:

$$\delta\mathbf{x}_k = \mathbf{M}_k \delta\mathbf{x}_{k-1} + \delta\mathbf{q}_k. \tag{19.22}$$

To complete the notation, we define:

$$\mathbf{R} = \begin{pmatrix} \mathbf{R}_0 & & & \\ & \mathbf{R}_1 & & \\ & & \ddots & \\ & & & \mathbf{R}_N \end{pmatrix}, \quad \mathbf{D} = \begin{pmatrix} \mathbf{B} & & & \\ & \mathbf{Q}_1 & & \\ & & \ddots & \\ & & & \mathbf{Q}_N \end{pmatrix}, \tag{19.23}$$

$$\mathbf{H} = \begin{pmatrix} \mathbf{H}_0 & & & \\ & \mathbf{H}_1 & & \\ & & \ddots & \\ & & & \mathbf{H}_N \end{pmatrix}, \quad \mathbf{b} = \begin{pmatrix} \mathbf{b}_0 \\ \mathbf{c}_1 \\ \vdots \\ \mathbf{c}_N \end{pmatrix}, \quad \mathbf{d} = \begin{pmatrix} \mathbf{d}_0 \\ \mathbf{d}_1 \\ \vdots \\ \mathbf{d}_N \end{pmatrix}.$$

We can then write the quadratic cost function (19.19) either as a function of $\delta\mathbf{x}$,

$$J(\delta\mathbf{x}) = (\mathbf{L}\delta\mathbf{x} - \mathbf{b})^{\mathrm{T}}\mathbf{D}^{-1}(\mathbf{L}\delta\mathbf{x} - \mathbf{b}) + (\mathbf{H}\delta\mathbf{x} - \mathbf{d})^{\mathrm{T}}\mathbf{R}^{-1}(\mathbf{H}\delta\mathbf{x} - \mathbf{d}), \tag{19.24}$$

or as a function of $\delta\mathbf{p}$,

$$J(\delta\mathbf{p}) = (\delta\mathbf{p} - \mathbf{b})^{\mathrm{T}}\mathbf{D}^{-1}(\delta\mathbf{p} - \mathbf{b}) + (\mathbf{H}\mathbf{L}^{-1}\delta\mathbf{p} - \mathbf{d})^{\mathrm{T}}\mathbf{R}^{-1}(\mathbf{H}\mathbf{L}^{-1}\delta\mathbf{p} - \mathbf{d}). \tag{19.25}$$

It is clear that (19.25), which we call the 'forcing formulation', is a sequential algorithm. Evaluation of the cost function (or its gradient) requires application of $\mathbf{L}^{-1}$. On the other hand, (19.24) (which we call the 4D-state formulation) represents a parallel algorithm, since it requires application of $\mathbf{L}$ rather than its inverse. Unfortunately, it has proved difficult to find an effective preconditioner for this formulation. (Typically, preconditioning of 4D-Var is achieved using the inverse of the matrix which appears in the first term of the cost function. In the forcing formulation, the required preconditioner is the matrix $\mathbf{D}$, which is known. The equivalent preconditioner for the 4D-state formulation is $\mathbf{L}^{-1}\mathbf{D}\mathbf{L}^{-\mathrm{T}}$. Since this preconditioner contains the matrix $\mathbf{L}^{-1}$, its application requires a sequential integration of the model over the analysis window, thereby destroying the parallel nature of the 4D-state formulation. Attempts to restore the parallel nature of the algorithm by approximating $\mathbf{L}^{-1}$ in the preconditioner have, so far, been unsuccessful.)

19.6.1 The saddle-point formulation

The saddle-point formulation (Fisher and Auvinen, 2012) provides an interesting alternative to the two formulations discussed above. It is most easily derived by considering the equation for the gradient of the cost function. At the minimum,

$$\nabla J = \mathbf{L}^{\mathrm{T}}\mathbf{D}^{-1}(\mathbf{L}\delta\mathbf{x} - \mathbf{b}) + \mathbf{H}^{\mathrm{T}}\mathbf{R}^{-1}(\mathbf{H}\delta\mathbf{x} - \mathbf{d}) = \mathbf{0}.$$

If we define:

$$\lambda = \mathbf{D}^{-1}(\mathbf{b} - \mathbf{L}\delta\mathbf{x}), \qquad \mu = \mathbf{R}^{-1}(\mathbf{d} - \mathbf{H}\delta\mathbf{x}),$$

then

$$\left.\begin{array}{r} \mathbf{D}\lambda + \mathbf{L}\delta\mathbf{x} = \mathbf{b} \\ \mathbf{R}\mu + \mathbf{H}\delta\mathbf{x} = \mathbf{d} \\ \mathbf{L}^{\mathrm{T}}\lambda + \mathbf{H}^{\mathrm{T}}\mu = \mathbf{0} \end{array}\right\} \Longrightarrow \begin{pmatrix} \mathbf{D} & \mathbf{0} & \mathbf{L} \\ \mathbf{0} & \mathbf{R} & \mathbf{H} \\ \mathbf{L}^{\mathrm{T}} & \mathbf{H}^{\mathrm{T}} & \mathbf{0} \end{pmatrix} \begin{pmatrix} \lambda \\ \mu \\ \delta\mathbf{x} \end{pmatrix} = \begin{pmatrix} \mathbf{b} \\ \mathbf{d} \\ \mathbf{0} \end{pmatrix}.$$

The matrix of the saddle-point system is real, symmetric, but indefinite. Note that the matrix contains no inverse sub-matrices, and each of the matrices and vectors that appear in the more usual formulations of 4D-Var appear as separate blocks in the matrix or right hand side of the equation.

The saddle-point system does not contain $\mathbf{L}^{-1}$. Hence it represents a parallel algorithm. Moreover, it has been possible to find preconditioners that allow the system to be solved in a reasonable number of iterations.

One preconditioner we have tested is (cf. Bergamaschi et al., 2012):

$$\tilde{\mathcal{P}} = \begin{pmatrix} \mathbf{D} & \mathbf{0} & \tilde{\mathbf{L}} \\ \mathbf{0} & \mathbf{R} & \mathbf{0} \\ \tilde{\mathbf{L}}^{\mathrm{T}} & \mathbf{0} & \mathbf{0} \end{pmatrix} \quad \Rightarrow \quad \tilde{\mathcal{P}}^{-1} = \begin{pmatrix} \mathbf{0} & \mathbf{0} & \tilde{\mathbf{L}}^{-\mathrm{T}} \\ \mathbf{0} & \mathbf{R}^{-1} & \mathbf{0} \\ \tilde{\mathbf{L}}^{-1} & \mathbf{0} & -\tilde{\mathbf{L}}^{-1}\mathbf{D}\tilde{\mathbf{L}}^{-\mathrm{T}} \end{pmatrix} \tag{19.26}$$

Here, $\tilde{\mathbf{L}}$ is an easily-inverted approximation of $\mathbf{L}$. Of the approximations we have tried, $\tilde{\mathbf{L}} = \mathbf{I}$ gave the best results in a parallel context.

The saddle-point algorithm has been tested in a two-level quasigeostrophic system with a 24-hour window divided into eight subwindows. The numerical cost of the algorithm, measured in necessarily sequential model integrations, is illustrated in Fig. 19.10. Although the saddle-point formulation requires a larger number of iterations of the numerical solver than the forcing formulation of 4D-Var, this is more than offset by the ability to run model integrations in parallel. As a result, the wall-clock time required for the saddle point algorithm for this problem is less than half that required for the forcing formulation of 4D-Var.

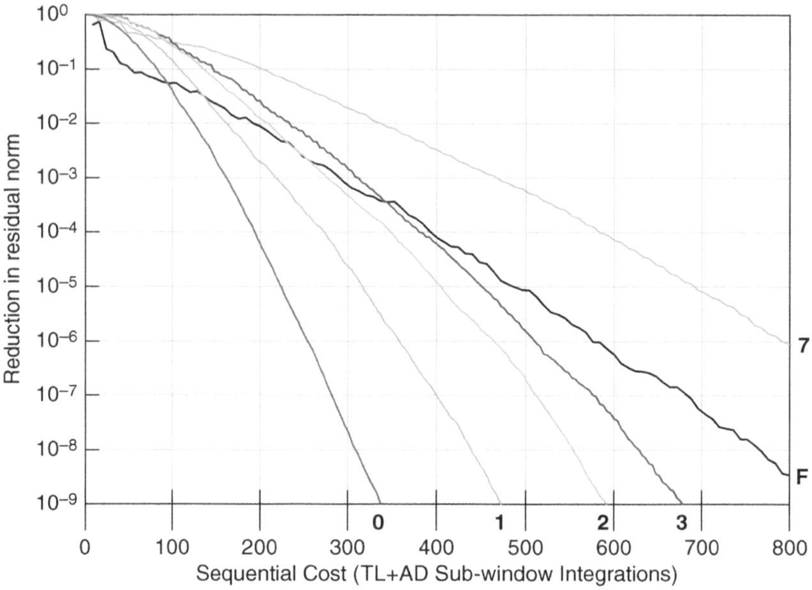

Fig. 19.10 Convergence of the saddle-point algorithm as a function of sequential subwindow integrations for preconditioners corresponding to different approximations $\tilde{\mathbf{L}}$ to $\mathbf{L}$ in (19.26). The line labelled '0' is for $\tilde{\mathbf{L}} = \mathbf{I}$; the line labelled '7' is for $\tilde{\mathbf{L}} = \mathbf{L}$. For comparison, the convergence of the forcing formulation is shown, labelled 'F'.

19.7 Conclusion

As has been illustrated above, in a NWP real-world context, there is much more to data assimilation than solving a mathematical equation with an optimality criterion. A whole area of research is devoted to more practical issues such as the processing of observations or the filtering of the analysis. This is justified by the facts that a single erroneous observation can destroy the structure of an important meteorological feature such as a tropical cyclone and that noise in the forecast can make it impossible for the forecasters to use it in the first forecast ranges. Therefore, the importance of these issues should not be underestimated.

Variational methods allow considerable flexibility in the specification of non-Gaussian PDFs for observation and prior errors. This flexibility can be exploited, for example to take into account non-Gaussian observation error, or to apply variational quality control.

Although strong-constraint 4D-Var is the most commonly used assimilation algorithm for operational weather prediction, it is likely to become increasingly necessary to take into account the errors of the assimilating model. This is expected to lead to the increased adoption of weak-constraint versions of 4D-Var. An attractive feature of some weak-constraint formulations is that they provide an opportunity for more parallel solution algorithms. This will become increasingly important as computers become more parallel.

References

Agusti-Panareda A., Vasiljevic, D., Beljaars, A., Bock, O., Guichard, F., Nuret, M., Garcia Mendez, A., Andersson, E., Bechtold, P., Fink, A., Hersbach, H., Lafore, J.-P., Ngamini, J.-B., Parker, D. J., Redelsperger, J.-L., and Tompkins, A. (2009). Radiosonde humidity bias correction over West African region for the special AMMA reanalysis at ECMWF. *Q. J. R. Meteorol. Soc.*, **135**, 595–617.

Andersson E. and Järvinen, H. (1999). Variational quality control. *Q. J. R. Meteorol. Soc.*, **125**, 697–722.

Auligné, T., McNally, A. P., and Dee, D. P. (2007). Adaptive bias correction for satellite data in a numerical weather prediction system. *Q. J. R. Meteorol. Soc.*, **133**, 631–642.

Bauer, P., Buizza, R., Cardinali, C., and Thépaut, J.-N. (2001). Impact of singular-vector-based satellite data thinning on NWP. *Q. J. R. Meteorol. Soc.*, **137**, 286–302.

Benjamin S. G., Devenyi, D., Weigandt, S. S., Brundage, K. J., Brown, J. M., Grell, G. A., Kim, D., Schwartz, B. E., Smirnova, T. G., and Smith, T. L. (2004). An Hourly Assimilation-Forecast Cycle: The RUC. *Mon. Weather Rev.*, **132**, 495–518.

Bergamaschi, L. (2012). On eigenvalue distribution of constraint-preconditioned symmetric saddle point matrices. *Namer. Linear Alg. Appl.*, **19**, 754–772.

Bloom S. C., Takacs, L. L., Da Silva, A. M., and Ledvina, D. (1996). Data Assimilation Using Incremental Analysis Updates. *Mon. Weather. Rev.*, **124**, 1256–1271.

Bormann, N. and Bauer, P. (2010). Estimates of spatial and interchannel observation-error characteristics for current sounder radiances for numerical weather prediction. I: Methods and application to ATOVS data. *Q. J. R. Meteorol. Soc.*, **136**, 1036–1050.

Bormann, N., Saarinen, S., Kelly, G., and Thépaut, J.-N. (2003). The spatial structure of observation errors in atmospheric motion vectors from geostationary satellite data. *Mon. Weather Rev.*, **131**, 706–718.

Bormann N., Collard, A., and Bauer, P. (2010). Estimates of spatial and interchannel observation-error characteristics for current sounder radiances for numerical weather prediction. II: Application to AIRS and IASI data. *Q. J. R. Meteorol. Soc.*, **136**, 1051–1063.

Buizza, R., Barkmeijer, J., Palmer, T. N., and Richardson, D. S. (2000). Current status and future developments of the ECMWF Ensemble Prediction System. *Meteorol. Applic.*, **7**, 163–175.

Chen, M. and Huang, X. Y. (2006). Digital filter initialization for MM5, *Mon. Weather Rev.*, **134**, 1222–1236.

Collard, A. D. (2007). Selection of IASI channels for use in numerical weather prediction. *Q. J. R. Meteorol. Soc.*, **133**, 1977–1991.

Collard, A. D., McNally, A. P., Hilton, F. I., Healy, S. B., and Atkinson, N. C. (2010). The use of principal component analysis for the assimilation of high-resolution infrared sounder observations for numerical weather prediction. *Q. J. R. Meteorol. Soc.*, **136**, 2038–2050.

Dando M. L., Thorpe, A. J., and Eyre, J. R. (2007). The optimal density of atmospheric sounder observations in the Met Office NWP system. *Q. J. R. Meteorol. Soc.*, **133**, 1933–1943.

Dee, D. P. and Uppala, S. (2009). Variational bias correction of satellite radiance data in the ERA-Interim reanalysis. *Q. J. R. Meteorol. Soc.*, **135**, 1830–1841.

Dee, D. P., Rukhovets, L., Todling, R., Da Silva, A. M., and Larson, J. W. (2001). An adaptive buddy check for observational quality control. *Q. J. R. Meteorol. Soc.*, **127**, 2451–2471.

De Pondeca M. S. F. V., Manikin, G. S., DiMego, G., Benjamin, S. G., Parrish, D. F., Purser, R. J., Wu, W.-S., Horel, J. D., Myrick, D. T., Lin, Y., Aune, R. M., Keyser, D., Colman, B., Mann, G., and Vavra, J. (2011). The real-time mesoscale analysis at NOAAs National Centers for Environmental Prediction: current status and development. *Weather Forecasting*, **26**, 593–612.

Fisher, M. and Auvinen, H. (2012). Long Window 4D-Var. In *Proceedings of ECMWF Seminar on Data Assimilation for Atmosphere and Ocean, 6–9 September 2011*, pp. 189–202.

Fischer, C. and Auger, L. (2011). Some experimental lessons on digital filtering in the ALADIN-France 3DVAR based on near-ground examination, *Mon. Weather Rev.*, **139**, 774–785.

Fourrié, N. and Thépaut, J.-N. (2003). Evaluation of the AIRS near-real-time channel selection for application to numerical weather prediction. *Q. J. R. Meteorol. Soc.*, **129**, 2425–2439.

Gandin, L. (1988). Complex quality control of meteorological observations, *Mon. Weather Rev.*, **116**, 1137–1156.

Gauthier, P. and Thépaut, J.-N. (2001). Impact of the digital filter as a weak constraint in the preoperational 4DVAR assimilation system of Meteo-France. *Mon. Weather Rev.,* **129**, 2089–2102.

Gelaro, R., Langland, R. H., Pellerin, S., and Todling, R. (2010). The THORPEX observation impact intercomparison experiment. *Mon. Weather Rev.*, **138**, 4009–4025.

Gilmour, I., Smith, L. A., and Beizza, R. (2001). On the duration of the linear regime: Is 24 hours a long time in weather forecasting? *J. Atros. Sci.*, **58**, 3525–3539.

Harris, B. A. and Kelly, G. (2001). A satellite radiance-bias correction scheme for data assimilation. *Q. J. R. Meteorol. Soc.*, **127**, 453–1468.

Hólm, E. V. (2003). Revision of the ECMWF humidity analysis: construction of a gaussian control variable. In *Proceedings of ECMWF/GEWEX Workshop on Humidity Analysis*.

Huang, X.-Y. and Lynch, P. (1993). Diabatic digital-filtering initialization: application to the HIRLAM model. *Mon. Weather Rev.*, **121**, 589–603.

Huber, P. J. (1964). Robust estimation of a location parameter. *Ann. Statist.*, **53**, 73–101.

Ingleby, N. B. and Lorenc, A. C. (1993). Bayesian quality control using multivariate normal distributions. *Q. J. R. Meteorol. Soc.*, **119**, 1195–1225.

Lindskog, M., Salonen, K., Jarvinen, H., and Michelson, D. B. (2004). Doppler radar wind data assimilation with HIRLAM 3DVAR. *Mon. Weather Rev.*, 132, 1081–1092.

Liu, Z. Q. and Rabier, F. (2003). The potential of high-density observations for numerical weather prediction: a study with simulated observations. *Q. J. R. Meteorol. Soc.*, **129**, 3013–3035.

Liu, Z. Q. and Rabier, F. (2002). The interaction between model resolution, observation resolution and observation density in data assimilation: a one-dimensional study. *Q. J. R. Meteorol. Soc.*, **128**, 1367–1386.

Lopez, P. (2011). Direct 4D-Var assimilation of NCEP Stage IV radar and gauge precipitation data at ECMWF. *Mon. Weather Rev.*, **139**. 2098–2116.

Lorenc, A. C. (1988). Optimal nonlinear objective analysis. *Q. J. R. Meteorol. Soc.*, **114**, 205–240.

Lorenc, A. C., Bell, R. S., and McPherson, B. (1991). The Meteorological-Office analysis correction data assimilation scheme *Q. J. R. Meteorol. Soc.*, **117**, 59–89.

Lynch, P. and Huang, X.-Y. (1992). Initialization of the HIRLAM model using a digital filter. *Mon. Weather Rev.*, **120**, 1019–1034.

Macpherson, S. R., Deblonde, G., and Aparicio, J. M. (2008). Impact of NOAA ground-based GPS observations on the Canadian regional analysis and forecast system. *Mon. Weather Rev.*, **136**, 2727–2746.

Montmerle, T. and Faccani, C. (2009). Mesoscale assimilation of radial velocities from Doppler radars in a preoperational framework. *Mon. Weather Rev.*, **137**, 1939–1953.

Nuret M., Lafore, J.-P., Bock, O., Guichard, F., Agusti-Panareda, A., N'Gamini, J. B., and Redelsperger, J.-L. (2008). Correction of humidity bias for Vaisala RS80-A sondes during the AMMA 2006 observing period. *J. Atmos. Ocean. Tech.*, **25**, 2152–2158.

Ochotta, T., Gebhardt, C., Saupe, D., and Wergen, W. (2005). Adaptive thinning of atmospheric observations in data assimilation with vector quantization and filtering methods. *Q. J. R. Meteorol. Soc.*, **131**, 3427–3437.

Pauley, P. (2003). Superobbing satellite winds for NAVDAS. Report from the Naval Research Laboratory.

Pires, C., Vautard, R., and Talagrand, O. (1996). On extending the limits of variational assimilation in nonlinear chaotic systems. *Tellus*, **48A**, 96–121.

Poli, P., Moll, P., Rabier, F., Desroziers, G., Chapnik, B., Berre, L., Healy, S. B., Andersson, E., and El Guelai, F. Z. (2007). Forecast impact studies of zenith total delay data from European near real-time GPS stations in Meteo France 4DVAR. *J. Geophys. Res.*, **112**, D06114.

Rabier, F., Fourrié, N., Chafai, D., and Prunet, P. (2002). Channel selection methods for infrared atmospheric sounding interferometer radiances. *Q. J. R. Meteorol. Soc.*, **128**, 1011–1027.

Roulston, M. S. (1999). Estimating the errors on measured entropy and mutual information. *Physica D*, **125**, 285–294.

Trémolet, Y. (2004). Diagnostics of linear and incremental approximations in 4D-Var. *Q. J. R. Meteorol. Soc.*, **130**, 2233–2251.

Zhang, F., Weng, Y., Sippel, J. A., Meng, Z., and Bishop, C. H. (2009). Cloud-resolving hurricane initialization and prediction through assimilation of Doppler radar observations with an ensemble Kalman filter. *Mon. Weather Rev.*, **137**, 2105–2125.

20

An introduction to inverse modelling and parameter estimation for atmosphere and ocean sciences

M. Bocquet

Université Paris-Est, CEREA Joint Laboratory École des Ponts ParisTech
and EdF R&D
INRIA, Paris Rocquencourt Research Centre
École des Ponts ParisTech
Champs-sur-Marne, France

Advanced Data Assimilation for Geosciences. First Edition.
Edited by É. Blayo, M. Bocquet, E. Cosme, and L. F. Cugliandolo.

Chapter Contents

20.1 Introduction

This chapter is a short survey of the inverse modelling and parameter estimation methods currently used in the atmosphere and ocean sciences. The focus is not on the retrieval of the initial system condition or on the estimation of the full state vector, but on the estimation of some of the model parameters. This can go as far as large parameter fields (forcing fields) that may drive or constrain a model. For instance, in atmospheric chemistry, the species concentration fields often depend strongly on the emission (possibly sink) fields. Ocean models significantly depend on friction and diffusion field parameters. Efficient data assimilation should probably incorporate these fields as part of the control space.

Our exposition of the inverse modelling principles will follow the geophysical data assimilation standpoint. In that view, estimation of the initial state or of model parameters can both be considered inverse modelling (or data assimilation) problems and the same generic tools should theoretically apply.

The underlying framework of geophysical data assimilation is Bayesian and that is why we shall introduce it from the start, whereas it is usually introduced at a later stage in inverse modelling textbooks. Besides common grounds, there are differences between data assimilation and inverse modelling problems. Obviously, the former are often focused on systems' forecast, whereas the latter are focused on parameter retrieval. There are also differences in the error magnitudes on the uncertain parameters to be retrieved. In inverse modelling problems, the emphasis is on the lack of data or the lack of observability of these parameters. As a consequence, great care is usually taken in the use of background information (coined *regularization* in the general inverse modelling literature), which is crucial and often missing. The careful use of every bit of information was emphasized early in the seminal work of Tarantola and Valette (1982). Bayesian inference offers a framework to account for this information.

There are highly recommended textbooks on inverse modelling problems. Several are listed in the references (Tarantola, 1987, 2005; Bennett, 1992, 2002; Groestsch, 1993; Rodgers, 2000; Enting, 2002; Vogel, 2002; Kaipio and Somersalo, 2005; Wunsch, 2006; Aster et al., 2006; Hansen, 2010; Doicu et al., 2010). Atmosphere and ocean inverse modelling problems are benefiting from advances in adjacent communities such as solid earth science, where data assimilation problems are essentially inverse problems, and remote sensing, where the community has developed very advanced one-dimensional retrieval schemes.

20.2 Bayesian approach to inverse problems

20.2.1 Posing an inverse problem

Assume one has a set of n model parameters, represented by a vector $\boldsymbol{\sigma} \in \mathbb{R}^n$ that one wishes to infer from a set of p observations, represented by a vector $\boldsymbol{\mu} \in \mathbb{R}^p$. The two vectors are related by a model $\mathbf{H}$, which is assumed linear for now. So, $\mathbf{H}$ is a matrix of $\mathbb{R}^{p\times n}$. This model $\mathbf{H}$ may be the composition of several models, such as a dynamical model and an observation model:

$$\boldsymbol{\mu} = \mathbf{H}\boldsymbol{\sigma} + \boldsymbol{\epsilon}\,. \tag{20.1}$$

As an essential piece of the puzzle, a vector of errors, $\boldsymbol{\epsilon} \in \mathbb{R}^p$, has been introduced to account for the possible mismatch between the observations and the model's predictions.

The goal of inverse modelling is to estimate the parameter vector $\boldsymbol{\sigma}$, knowing the observation vector $\boldsymbol{\mu}$. In the vast majority of geophysical problems, the exploitation of the observational data is far from sufficient to provide a reasonable answer to this problem. Here are several reasons why:

- The problem is often largely underdetermined: $p \ll n$. This may be a matter of design of the inverse modelling problem. A way around is to aggregate the control variables $[\boldsymbol{\sigma}]_k$ into coarser variables so as to reduce the number of effective parameters. An alternative is to project $\boldsymbol{\sigma}$ onto a vector space generated by a set of predefined solutions $\boldsymbol{\sigma}^1, \ldots, \boldsymbol{\sigma}^m$, with $m \ll n$:

$$\boldsymbol{\sigma} = \sum_{l=1}^{m} \omega_l \boldsymbol{\sigma}^l \,. \tag{20.2}$$

  This has been largely used in remote sensing inverse problems (Rodgers, 2000). However, the aggregation comes with a price: the loss of resolution.
- The physics (encoded in $\mathbf{H}$) may be such that a perturbation in some of the variables to retrieve has very little impact on the observations, so that it is difficult to extract significant information from the observations to inform about these variables. It is quite unavoidable whenever diffusion (or effective diffusion, possibly generated by turbulence or mixing) is present. At the extreme, it may go as far as the unobservability of some of the parameters. For instance, in atmospheric physics, the boundary-layer top, or the thermocline in ocean sciences, is an obstacle to the propagation of information that may lead to poor observability beyond these barriers.
- The errors in $\mathbf{H}$ and their incorrect a priori estimation also significantly complicate the search for a reliable solution of (20.1).

In these circumstances, that are very often met in the atmospheric and oceanic sciences, the inverse problem is said to be *ill-posed.*

As a consequence, it is mandatory to take into account a priori information, or even make physical or mathematical assumptions on the solution. A reliable framework to do so is a probabilistic one that mathematically handles the representation of uncertainty. Moreover, within a probabilistic framework, the Bayesian formalism allows us to process all pieces of information that may enter the inverse modelling problem. This approach to inverse modelling problem is often called *statistical inversion* (Kaipio and Somersalo, 2005).

Chernobyl case study: the problem

The theoretical elements of this overview will be illustrated by the inverse modelling of the Chernobyl accident source term. The accident took place on 26 April 1986 (local time), in Reactor 4 of the Chernobyl nuclear power plant, near Pripyat, Ukraine (formerly in the USSR). Among the radionuclides that could be transported away across

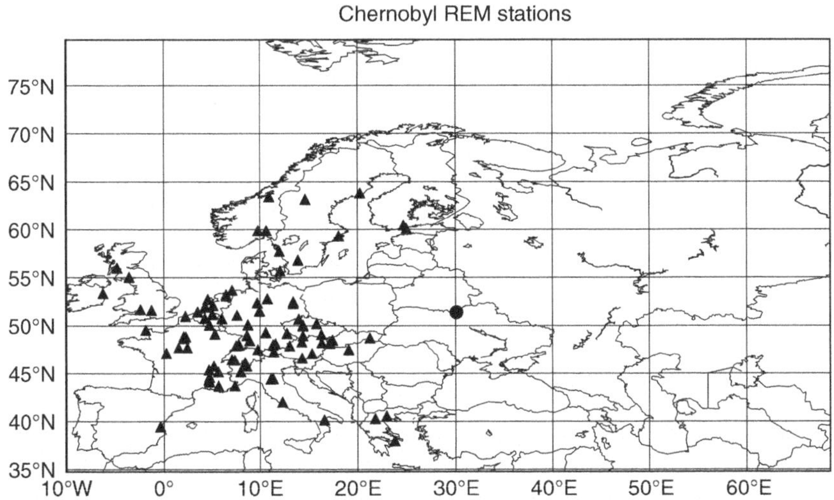

Fig. 20.1 Domain of simulation needed to build the Jacobian matrix **H**. Stations of the REM network that monitored the Chernobyl plume are indicated by black triangles. The location of the power plant is indicated by a black circle.

thousands of kilometres with significant health impact were iodine-131 or caesium-137, and we shall focus on the latter. The caesium-137 radioactive plume was monitored by activity concentrations in the air throughout western Europe at the REM network's stations (see Fig. 20.1). In the following, we shall exploit 1283 measurements over 10 days. The source term is assumed to be located at the Chernobyl power plant. It is discretized as a two-dimensional (2D) field, with a time-step of 6 hours over 10 days, that, is 40 time-steps. It is also vertically discretized with 12 vertical levels, since the heat plume rise has been considerable for Chernobyl (initial cover blast and lasting fire). Therefore, $480 = 12 \times 40$ control variables are considered. The source term and the measurements are related using a 3D chemistry and transport model. For caesium-137, the transport, physics and chemistry can be assumed linear, so that the matrix **H** of (20.1) can easily be computed. Note that the observation sites are far away from Chernobyl, which leads to a difficult ill-conditioned inverse modelling problem. Inverse modelling of the Chernobyl source term has been studied in Davoine and Bocquet (2007) and Bocquet (2012).

20.2.2 Bayes' inference

Bayes' inference implies making statistical assumptions about the prior errors and the parameters to retrieve. Ideally, one would have determined a priori a probability density function (pdf) for $\boldsymbol{\sigma}$, denoted by $p_b(\boldsymbol{\sigma})$ (b stands for background). When background information is available in the form of a previous forecast, this seems natural. It is much less natural for a specific set of model parameters, which cannot be easily *forecast*. One can surely use a previous estimation, but its uncertainty is generally not given or is difficult to establish. Moreover, one needs to model the error pdf, as $p_e(\boldsymbol{\epsilon}) = p_e(\boldsymbol{\mu} - \mathbf{H}\boldsymbol{\sigma})$, using (20.1).

Bayes' rule specifies the a posteriori pdf, that is, the pdf of the retrieved parameters knowing the observation data and prior information:

$$p(\boldsymbol{\sigma}|\boldsymbol{\mu}) = \frac{p_e(\boldsymbol{\mu} - \mathbf{H}\boldsymbol{\sigma})p_b(\boldsymbol{\sigma})}{p(\boldsymbol{\mu})}, \tag{20.3}$$

where

$$p(\boldsymbol{\mu}) = \int \mathrm{d}\boldsymbol{\sigma}\, p_e(\boldsymbol{\mu} - \mathbf{H}\boldsymbol{\sigma})p_b(\boldsymbol{\sigma}) \tag{20.4}$$

enforces a normalization of $p(\boldsymbol{\sigma}|\boldsymbol{\mu})$ to 1, but does not depend on $\boldsymbol{\sigma}$. Since $p(\boldsymbol{\sigma}|\boldsymbol{\mu})$ is a quite complex and rich mathematical object, and does not single out an estimate for $\boldsymbol{\sigma}$, one additionally needs to specify an *estimator*. That means we need to select one particular solution. Examples are the mean, $\overline{\boldsymbol{\sigma}} = \mathrm{E}_{p(\boldsymbol{\sigma}|\boldsymbol{\mu})}\left[\boldsymbol{\sigma}\right]$, or the maximum a posteriori estimator (MAP): $\boldsymbol{\sigma}_a = \operatorname{argmax} p(\boldsymbol{\sigma}|\boldsymbol{\mu})$. In most circumstances, the MAP estimator is the most natural choice since its maximizes the likelihood of $\boldsymbol{\sigma}$. From Bayes' rule (20.3), it is clear that the MAP $\boldsymbol{\sigma}_a$ is the argument of the minimum of the cost function

$$\mathcal{J}(\boldsymbol{\sigma}) = -\ln p_e(\boldsymbol{\mu} - \mathbf{H}\boldsymbol{\sigma}) - \ln p_b(\boldsymbol{\sigma}). \tag{20.5}$$

This underlying Bayesian framework is also explicitly or implicitly used in the data assimilation literature (Lorenc, 1986). For instance, 3D-Var and 4D-Var cost functions can be derived along these lines.

20.2.3 Physical regularization

Assume that the prior pdf, p_b, is Gaussian-distributed and peaked at a *first-guess* $\boldsymbol{\sigma}_b$, with a prior error covariance matrix $\mathbf{B}$:

$$p_b(\boldsymbol{\sigma}) = \frac{e^{-\frac{1}{2}(\boldsymbol{\sigma}-\boldsymbol{\sigma}_b)^{\mathrm{T}}\mathbf{B}^{-1}(\boldsymbol{\sigma}-\boldsymbol{\sigma}_b)}}{\sqrt{(2\pi)^n|\mathbf{B}|}}, \tag{20.6}$$

where $|\mathbf{B}|$ is the determinant of the matrix $\mathbf{B}$. The distribution of the observation errors $\boldsymbol{\epsilon}$ is also assumed Gaussian, with an error covariance matrix denoted by $\mathbf{R}$ and without any bias: $\mathrm{E}\left[\boldsymbol{\epsilon}\right] = \mathbf{0}$. As a result the pdf p_e is

$$p_e(\boldsymbol{\epsilon}) = \frac{e^{-\frac{1}{2}\boldsymbol{\epsilon}^{\mathrm{T}}\mathbf{R}^{-1}\boldsymbol{\epsilon}}}{\sqrt{(2\pi)^p|\mathbf{R}|}}. \tag{20.7}$$

Using these two pdfs, the corresponding cost function for the MAP estimator is

$$\mathcal{J}(\boldsymbol{\sigma}) = \frac{1}{2}\left(\boldsymbol{\mu} - \mathbf{H}\boldsymbol{\sigma}\right)^{\mathrm{T}}\mathbf{R}^{-1}\left(\boldsymbol{\mu} - \mathbf{H}\boldsymbol{\sigma}\right) + \frac{1}{2}\left(\boldsymbol{\sigma} - \boldsymbol{\sigma}_b\right)^{\mathrm{T}}\mathbf{B}^{-1}\left(\boldsymbol{\sigma} - \boldsymbol{\sigma}_b\right). \tag{20.8}$$

Hence, the resulting cost function is made of a misfit term plus a background term. This latter term is a *regularization term* of *Tikhonov* type (Tikhonov and Arsenin,

1977). Even if the system is not well observed, and the inverse problem is ill-defined, this term guarantees the existence of a unique solution to the problem. The minimum of the cost function is

$$\boldsymbol{\sigma}_a = \boldsymbol{\sigma}_b + \mathbf{B}\mathbf{H}^{\mathrm{T}}\left(\mathbf{R} + \mathbf{H}\mathbf{B}\mathbf{H}^{\mathrm{T}}\right)^{-1}(\boldsymbol{\mu} - \mathbf{H}\boldsymbol{\sigma}_b)\,, \tag{20.9}$$

which is known as *the best linear unbiased estimator* (BLUE).

Chernobyl case study: an example of inversion

The BLUE estimator can be applied to our Chernobyl inverse problem. Even though the total released retrieved activity is of the right order of magnitude, the retrieved field and temporal profile is unsatisfying because it entails significant negative fluxes. Theoretically, more observations would help in constraining the problem, but we do not have them. Besides, they would not guarantee the positivity of the retrieval. One must constrain the solution to be positive. This entails mathematical complications that will be addressed in this chapter. A state-of-the-art positive retrieval (using hyper-parameters obtained in a non-Gaussian maximum-likelihood study to be discussed later) is displayed in Fig. 20.2.

It could be useful to derive the posterior pdf, beyond obtaining the argument of its maximum. Using elementary algebra, and in particular Gaussian integral analytical results, these two assumptions yield the normalization constant through Bayes' formula:

$$p(\boldsymbol{\mu}) = \frac{e^{-\frac{1}{2}(\boldsymbol{\mu}-\mathbf{H}\boldsymbol{\sigma}_b)^{\mathrm{T}}\left(\mathbf{R}+\mathbf{H}\mathbf{B}\mathbf{H}^{\mathrm{T}}\right)^{-1}(\boldsymbol{\mu}-\mathbf{H}\boldsymbol{\sigma}_b)}}{\sqrt{(2\pi)^p|\mathbf{R}+\mathbf{H}\mathbf{B}\mathbf{H}^{\mathrm{T}}|}}\,. \tag{20.10}$$

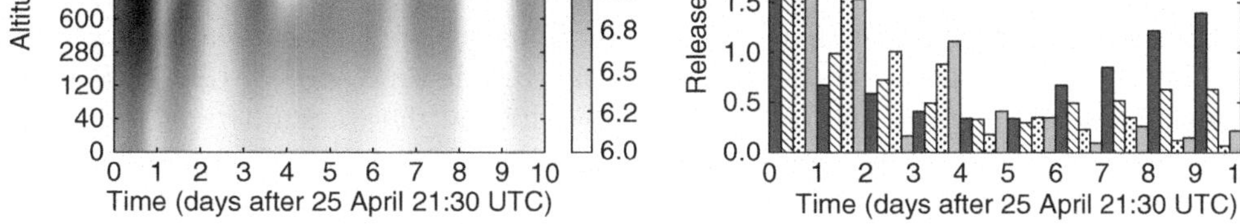

Fig. 20.2 (a) Density plot of the caesium-137 retrieval. The density represents the decimal log of the rates in $\mathrm{Bq\,m^{-1}\,s^{-1}}$. (b) Comparison of the daily temporal profile of our solution to previous direct estimation of the caesium-137 source term: Organisation for Economic Co-operation and Development (OECD), United Nations Scientific Committee on the Effects of Atomic Radiation (UNSCEAR) (Talerko, 2005).

This pdf is quite useful, since it gives the likelihood of the observation set $\boldsymbol{\mu}$, under the statistical prior assumption made on the parameters and the errors.

The a posteriori pdf, which constitutes the probabilistic Bayes' inference result, reads:

$$p(\boldsymbol{\sigma}|\boldsymbol{\mu}) = \frac{e^{-\frac{1}{2}(\boldsymbol{\sigma}-\boldsymbol{\sigma}_a)^{\mathrm{T}}\mathbf{P}_a^{-1}(\boldsymbol{\sigma}-\boldsymbol{\sigma}_a)}}{\sqrt{(2\pi)^n|\mathbf{P}_a|}}, \tag{20.11}$$

where $\mathbf{P}_a$ is the posterior error covariance matrix of the errors of the BLUE:

$$\mathbf{P}_a = \mathbf{B} - \mathbf{B}\mathbf{H}^{\mathrm{T}}\left(\mathbf{R}+\mathbf{H}\mathbf{B}\mathbf{H}^{\mathrm{T}}\right)^{-1}\mathbf{H}\mathbf{B} = \left(\mathbf{B}^{-1}+\mathbf{H}^{\mathrm{T}}\mathbf{R}^{-1}\mathbf{H}\right)^{-1}. \tag{20.12}$$

20.2.4 Averaging kernel

The BLUE can be rewritten using the Sherman–Morrison–Woodbury formula:

$$\boldsymbol{\sigma}_a = \boldsymbol{\sigma}_b + \mathbf{K}\left(\boldsymbol{\mu} - \mathbf{H}\boldsymbol{\sigma}_b\right), \tag{20.13}$$

where the gain $\mathbf{K}$ is given by

$$\mathbf{K} = \mathbf{P}_a\mathbf{H}^{\mathrm{T}}\mathbf{R}^{-1} \quad \text{and} \quad \mathbf{P}_a^{-1} = \mathbf{B}^{-1}+\mathbf{H}^{\mathrm{T}}\mathbf{R}^{-1}\mathbf{H}. \tag{20.14}$$

Adopting this BLUE inference, how far are we from the truth (call it $\boldsymbol{\sigma}_t$ and $\boldsymbol{\epsilon}_t$)? The truth satisfies $\boldsymbol{\mu} = \mathbf{H}\boldsymbol{\sigma}_t + \boldsymbol{\epsilon}_t$, and we have, from (20.13),

$$\boldsymbol{\sigma}_a - \boldsymbol{\sigma}_b = \mathbf{K}\left(\boldsymbol{\mu} - \mathbf{H}\boldsymbol{\sigma}_b\right) = \mathbf{K}\mathbf{H}\left(\boldsymbol{\sigma}_t - \boldsymbol{\sigma}_b\right) + \mathbf{K}\boldsymbol{\epsilon}_t. \tag{20.15}$$

With an ideal inverse modelling system, one would have $\mathbf{KH}$ close to the identity operator $\mathbf{I}_n$ of control space. Indeed, that would allow a much more faithful reconstruction of the truth: any variation in the truth $\boldsymbol{\sigma}_t$ would then result in the same variation in the analysis $\boldsymbol{\sigma}_a$. This matrix $\mathbf{A} = \mathbf{KH}$ is called the *averaging kernel.* It is heavily emphasized and a cornerstone of remote sensing (Rodgers, 2000).

20.2.5 Singular-value decomposition and mathematical regularization

In this section, it is assumed that the prior distributions are Gaussian, and we choose to solve the inverse problem using the Bayesian framework and a MAP estimator. According to the physical interpretation of geophysical data assimilation, the MAP estimator provides a compromise between the observations and a first guess from the background. Mathematically, the Tikhonov regularization term is doing more than just making the solution unique. To see what it does, it is useful to make a singular-value decomposition (SVD) of $\mathbf{H}$, which contains everything one knows about the physics and observation models:

$$\mathbf{H} = \mathbf{U}\boldsymbol{\Lambda}\mathbf{V}^{\mathrm{T}} = \sum_{l=1}^{q}\mathbf{u}_l\lambda_l\mathbf{v}_l^{\mathrm{T}}. \tag{20.16}$$

$\mathbf{U}$ is an orthogonal matrix of $\mathbb{R}^{p\times p}$, that is, $\mathbf{U}^{\mathrm{T}}\mathbf{U} = \mathbf{U}\mathbf{U}^{\mathrm{T}} = \mathbf{I}_p$. Its columns (the left singular vectors $\mathbf{u}_l$) form an orthonormal basis of the observation space. $\mathbf{V}$ is a matrix of $\mathbb{R}^{n\times p}$ that satisfies $\mathbf{V}^{\mathrm{T}}\mathbf{V} = \mathbf{I}_p$. Its columns (the right singular vectors $\mathbf{v}_l$) form a set of orthonormal vectors of control space. $\mathbf{\Lambda}$ is a diagonal matrix in $\mathbb{R}^{p\times p}$, with non-negative entries $\lambda_l \geq 0$. The singular values are ordered such that the first q values are positive $\lambda_l > 0$ for $l = 1, \ldots, q$, with $q \leq \min(p, n)$.

One can now perform a change of basis in both observation and control space, in order to statistically decorrelate (or make independent in this Gaussian context) the error variables. The following transformations are applied:

$$\boldsymbol{\sigma} \longrightarrow \mathbf{B}^{-\frac{1}{2}}\boldsymbol{\sigma} \quad \text{and} \quad \boldsymbol{\epsilon} \longrightarrow \mathbf{R}^{-\frac{1}{2}}\boldsymbol{\epsilon}\,. \tag{20.17}$$

That is why we can assume $\mathbf{B} = m^2\mathbf{I}_n$ and $\mathbf{R} = \chi^2\mathbf{I}_p$, without loss of generality. In order to keep a sense of the physical dimensions, the standard deviations m and χ have been introduced. With these transformations, the BLUE estimator reads

$$\boldsymbol{\sigma}_a = \boldsymbol{\sigma}_b + \mathbf{H}^{\mathrm{T}}\left[(\chi/m)^2\mathbf{I}_p + \mathbf{H}\mathbf{H}^{\mathrm{T}}\right]^{-1}(\boldsymbol{\mu} - \mathbf{H}\boldsymbol{\sigma}_b)\,, \tag{20.18}$$

so that the error covariance matrices can be taken as the identity.

With these choices, and inserting the SVD of $\mathbf{H}$ into (20.18), one finds

$$\boldsymbol{\sigma}_a = \boldsymbol{\sigma}_b + \sum_{l=1}^{q} \frac{\lambda_l^2}{(\chi/m)^2 + \lambda_l^2}\,\frac{\mathbf{u}_l^{\mathrm{T}}(\boldsymbol{\mu} - \mathbf{H}\boldsymbol{\sigma}_b)}{\lambda_l}\mathbf{v}_l\,. \tag{20.19}$$

Comparison with the naive SVD solution

Let us look in detail at this expansion. A straightforward solution of the inverse problem

$$\boldsymbol{\mu} - \mathbf{H}\boldsymbol{\sigma}_b = \mathbf{H}(\boldsymbol{\sigma} - \boldsymbol{\sigma}_b) + \boldsymbol{\epsilon}\,, \tag{20.20}$$

without errors and statistical consideration, consists in using the singular-value decomposition of $\mathbf{H}$ and looking for a solution in the vector space spanned by the right singular vectors of $\mathbf{H}$, which is

$$\boldsymbol{\sigma}_a^{\mathrm{naive}} = \boldsymbol{\sigma}_b + \mathbf{V}\mathbf{\Lambda}^{-1}\mathbf{U}^{\mathrm{T}}(\boldsymbol{\mu} - \mathbf{H}\boldsymbol{\sigma}_b) = \boldsymbol{\sigma}_b + \sum_{l=1}^{q} \frac{\mathbf{u}_l^{\mathrm{T}}(\boldsymbol{\mu} - \mathbf{H}\boldsymbol{\sigma}_b)}{\lambda_l}\mathbf{v}_l\,. \tag{20.21}$$

It is similar to (20.19), but the vectors of the expansion are not multiplied by the non-dimensional coefficients

$$\varphi_l = \frac{\lambda_l^2}{(\chi/m)^2 + \lambda_l^2} \tag{20.22}$$

that filter out the SVD expansion. For singular values of $\mathbf{H}$ such that $\lambda_l \gg \chi/m$, one has $\varphi_l \simeq 1$: the regularized solution is given by the naive SVD solution. These singular values correspond to useful degrees of freedom from which information can be extracted to build up a solution. For singular values such that $\lambda_l \ll \chi/m$, $\varphi_l \simeq 0$: the higher-order expansion terms of the SVD decomposition are filtered out. These correspond to noisy degrees of freedom, from which it is difficult to extract useful information. Therefore, the Tikhonov regularization acts as a low-pass filter on the SVD expansion of $\mathbf{H}$, which represents the physics of our inverse problem.

Why regularization is often necessary

We have justified the Tikhonov regularization from its physical (data assimilation framework) interpretation. To what does this latter mathematical standpoint correspond? Why should we ignore some of the details of the physics, the degrees of freedom such that $\lambda_l \ll \chi/m$, to properly solve the inverse problem? Why not use the naive SVD expansion without filtering?

Assume the singular values have been ordered: $\lambda_1 \geq \lambda_2 \geq \ldots \geq \lambda_m > 0$. The SVD of $\mathbf{H}$ decomposes the physics (concatenated to the observation operator) into as many trends given by the singular vectors that are more and more fluctuating with increasing order. In a large majority of physical systems (that includes most geophysical problems with diffusion/dissipation), the singular values λ_l are swiftly decreasing with their order l. If the coefficients $\mathbf{u}_l^{\mathrm{T}}(\boldsymbol{\mu} - \mathbf{H}\boldsymbol{\sigma}_b)$ are not decreasing fast enough, then the high-order highly oscillating singular vectors will make a strong contribution to the inverse problem solution. Any error in the innovation will be severely amplified with high-order singular values, leading to an erroneous oscillating solution.

If the physics and the innovation are such that the coefficients of the projection of the innovation decrease faster than the singular-values sequence, then the naive SVD solution is probably reliable. This is known as the (discrete) Picard condition. Clearly, in the presence of noise, even if the first coefficients $\mathbf{u}_l^{\mathrm{T}}(\boldsymbol{\mu} - \mathbf{H}\boldsymbol{\sigma}_b)$ indeed decrease faster than the few first singular values, this condition must be breached beyond some l, when the coefficient of the innovation projection is of the magnitude of the noise. Therefore, in the context of real noisy systems, the Picard condition is limited to the first elements of the SVD decomposition down to some order representative of the noise.

Chernobyl case study: Picard condition

The Picard condition is illustrated on the Chernobyl inverse problem. The test is carried out using *synthetic observations.* The curves of the singular values and the scalar product of the observation vector with the left singular vectors are plotted in Fig. 20.3. The two curves are rescaled so that they match at the first singular value $l = 1$. It is clear that for $l \leq 50$, the Picard condition is satisfied (which is not bad for a significantly ill-conditioned problem).

20.2.6 DFS, entropy, and resolution

Degrees of freedom for the signal

In the SVD basis, the averaging kernel reads

$$\mathbf{A} = \sum_{l=1}^{q} \mathbf{v}_l \varphi_l \mathbf{v}_l^{\mathrm{T}} . \qquad (20.23)$$

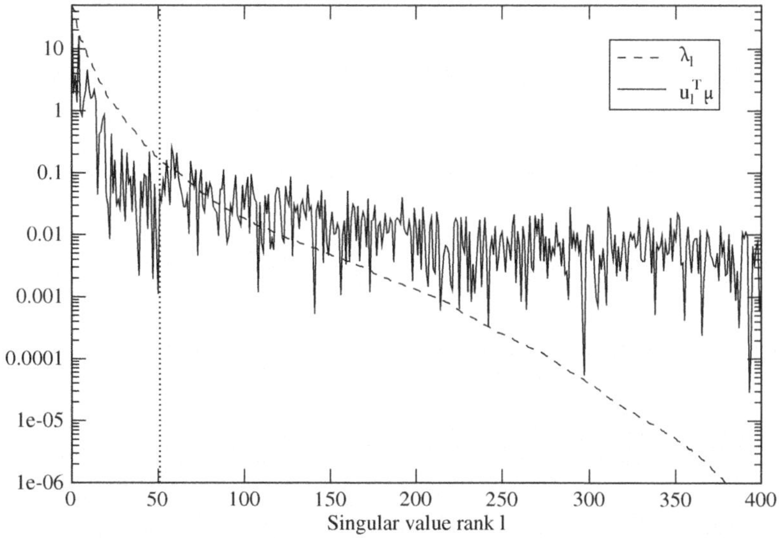

Fig. 20.3 Illustration of the Picard condition for the Chernobyl inverse modelling problem.

The trace of $\mathbf{A}$,

$$\operatorname{Tr}\mathbf{A} = \sum_{l=1}^{q} \varphi_l, \tag{20.24}$$

sums up the *degrees of freedom for the signal* (DFS), that is, the number of individual scalar pieces of information available to retrieve the solution.

The total DFS can be written

$$\text{DFS} = \operatorname{Tr}\mathbf{A} = \operatorname{Tr}(\mathbf{KH}) = \operatorname{Tr}(\mathbf{HK}) \leq p. \tag{20.25}$$

So the total DFS is either the sum of the n diagonal entries $[\mathbf{KH}]_{kk}$ or the sum of the p diagonal entries $[\mathbf{HK}]_{ii}$. It is therefore tempting to interpret $[\mathbf{HK}]_{ii}$ as a proxy for the number of degrees of freedom for the signal that one can extract from observation μ_i. $\mathbf{G} = \mathbf{HK}$ is often called the *resolution* matrix. This has been used in sensitivity studies and allows one to qualify the contribution of individual instruments in a data assimilation system.

Chernobyl case study: DFS

Assuming $m = 3.13 \times 10^9\,\mathrm{Bq\,m^{-1}\,s^{-1}}$ and $\chi = 1.12\,\mathrm{Bq\,m^{-3}}$ (see later), the BLUE estimation of the caesium-137 source yields DFS $\simeq 52$ out of 1283 observations. If one chooses parameters obtained from a non-Gaussian inversion that assume the positivity of the source, the DFS drops to about 10. The approach is inconsistent, since the DFS estimations are based on Gaussian assumptions. However, the tendency seems correct: enforcing the positivity of the source allows a better and broader identification of the errors, diminishing the degrees of freedom for the signal. Yet, this could be balanced by a better extraction of information from an observation, since the prior that enforces positivity of the source is closer to the truth. In a non-Gaussian context, there are better tools to measure an equivalent to the DFS, such as the relative entropy (Bocquet, 2008) (to be introduced in the next paragraph), which boils down to the DFS when the error statistics are Gaussian.

Relative entropy

Another objective measure of the gain of information obtained from the observation is the entropy. In particular, the relative entropy, which we shall later use as an alternative Bayesian inference tool, allows us to quantify the information contained in the posterior distribution that was not contained in the prior distribution. From information theory, information contained in the pdf q, relatively to the pdf p, is quantified by the Kullback–Leibler divergence, or relative entropy:

$$\mathcal{K}(q,p) = \int \mathrm{d}\boldsymbol{\sigma}\, q(\boldsymbol{\sigma}) \ln \frac{q(\boldsymbol{\sigma})}{p(\boldsymbol{\sigma})} . \tag{20.26}$$

The integral on the control variables $\boldsymbol{\sigma}$ is formal. Its genuine nature is case-dependent. It can be a discrete sum if the control space is discrete. This quantity follows from several axioms that make it a perfect candidate to measure information (but other less attractive functions are possible). It has nice mathematical properties, such as positivity, and convexity with respect to both p and q, and invariance under reparameterization of the pdf. More can be learnt about it from Cover and Thomas (1991).

With Gaussian pdf assumptions, taking q as the prior and p as the posterior of the BLUE analysis, one can show that, in our context and with our assumptions, this gain in information is

$$\Delta\mathcal{I} = \frac{1}{2}\ln|\mathbf{BP}_a^{-1}| + \frac{1}{2}(\boldsymbol{\sigma}^a - \boldsymbol{\sigma}^b)^{\mathrm{T}}\mathbf{B}^{-1}(\boldsymbol{\sigma}^a - \boldsymbol{\sigma}^b) + \frac{1}{2}\mathrm{Tr}\left(\mathbf{P}_a\mathbf{B}^{-1} - \mathbf{I}_n\right). \tag{20.27}$$

Averaging over all possible outcomes of $\boldsymbol{\mu}$, one has

$$\mathrm{E}\left[\Delta\mathcal{I}\right] = -\frac{1}{2}\mathrm{Tr}\left(\mathbf{I}_n - \mathbf{A}\right) = \frac{1}{2}\sum_{l=1}^{q}\ln\left(1 + \lambda_l^2\right). \tag{20.28}$$

Resolution

Assume that control space corresponds to a spatial field of parameters. From (20.25), one could interpret $[\mathbf{A}]_{kk}$ as the number of degrees of freedom used to retrieve the local parameter $[\boldsymbol{\sigma}]_k$. Its inverse can then be interpreted as the number of local parameters that one can retrieve with one DFS. It gives an indication of the local *resolution* of the inverse method (*the power to resolve*). However, it is an approximate estimation, because in most complex cases $\mathbf{A}$ is not diagonal.

Bocquet (2009) has defined a rigorous estimator for the resolution whether $\mathbf{A}$ is diagonal or not. It consists in partitioning the control space into adaptive grid cells such that the partition maximizes the DFS. The finer a grid cell, the higher is the resolution locally (see Chapter 18 in this volume).

20.2.7 Are the results physical?

Regularization is necessary not only to obtain a mathematically sound solution, but also to obtain a physical solution of the inverse problem. We saw that, through the choice of χ/m, the regularization filters out oscillating contributions to the SVD expansion of the solution. If χ/m is not chosen well and is much lower than the true or realistic ratio $(\chi/m)_t$, then the solutions will incorporate unphysical oscillating components, which might better explain the observations but contradict the physics. This is called *overshooting*. If χ/m is not chosen well and is much higher than the true or realistic ratio $(\chi/m)_t$, then the solution will lose part of the signal contained in the observations. The solution is generally too smooth. This is called *undershooting*. A physically balanced solution could be obtained in between these two regimes when $\chi/m \simeq (\chi/m)_t$.

When considering the naive SVD expansion, and progressively incorporating higher-order singular vectors, the solution starts in the undershooting regime, passes a physically optimal point, and finally begins overshooting when incorporating unphysically oscillating vectors. This phenomenon is often called *semiconvergence*.

Chernobyl case study: semiconvergence

This is illustrated on the Chernobyl case study in Fig. 20.4.

20.3 Alternative approaches

20.3.1 Backus–Gilbert method

One of the historical inverse modelling methods is related to the concept of resolution we introduced in Section 20.2.6. The Backus–Gilbert method (Backus and Gilbert, 1968) seeks to design the inverse method so as to make some resolution proxy as close to the identity as possible. It assumes that the estimator is linear in the observation vector: $\boldsymbol{\sigma}_a = \boldsymbol{\Omega}\boldsymbol{\mu}$, where $\boldsymbol{\Omega}$ is a matrix to be optimally defined. Since $\boldsymbol{\mu}_t = \mathbf{H}\boldsymbol{\sigma}_t$, we have $\boldsymbol{\sigma}_a = \boldsymbol{\Omega}\mathbf{H}\boldsymbol{\sigma}_t$. So $\mathbf{A} = \boldsymbol{\Omega}\mathbf{H}$ is essentially the averaging kernel of the method. Assuming the inverse of the averaging kernel provides a notion of resolution, a simplified form of the Backus–Gilbert method seeks to minimize

$$\mathrm{Tr}\left[(\mathbf{A}-\mathbf{I}_n)(\mathbf{A}-\mathbf{I}_n)^{\mathrm{T}}\right] \tag{20.29}$$

over all possible $\boldsymbol{\Omega}$. Its variation under a variation $\delta\boldsymbol{\Omega}$ of $\boldsymbol{\Omega}$ is

$$\delta\left[\mathrm{Tr}\,(\mathbf{A}-\mathbf{I}_n)(\mathbf{A}-\mathbf{I}_n)^{\mathrm{T}}\right] = 2\mathrm{Tr}\left\{\delta\boldsymbol{\Omega}\left[(\mathbf{H}\mathbf{H}^{\mathrm{T}})\boldsymbol{\Omega}^{\mathrm{T}}-\mathbf{H}\right]\right\}. \tag{20.30}$$

Assuming $p \ll n$, an $\boldsymbol{\Omega}$ matrix that satisfies the optimality condition is

$$\boldsymbol{\Omega}_a = \mathbf{H}^{\mathrm{T}}\left(\mathbf{H}\mathbf{H}^{\mathrm{T}}\right)^{-1}. \tag{20.31}$$

Using this optimal $\boldsymbol{\Omega}_a$, the solution is the *Moore–Penrose generalized inverse*:

$$\boldsymbol{\sigma}_a = \mathbf{H}^{\mathrm{T}}\left(\mathbf{H}\mathbf{H}^{\mathrm{T}}\right)^{-1}\boldsymbol{\mu}\,. \tag{20.32}$$

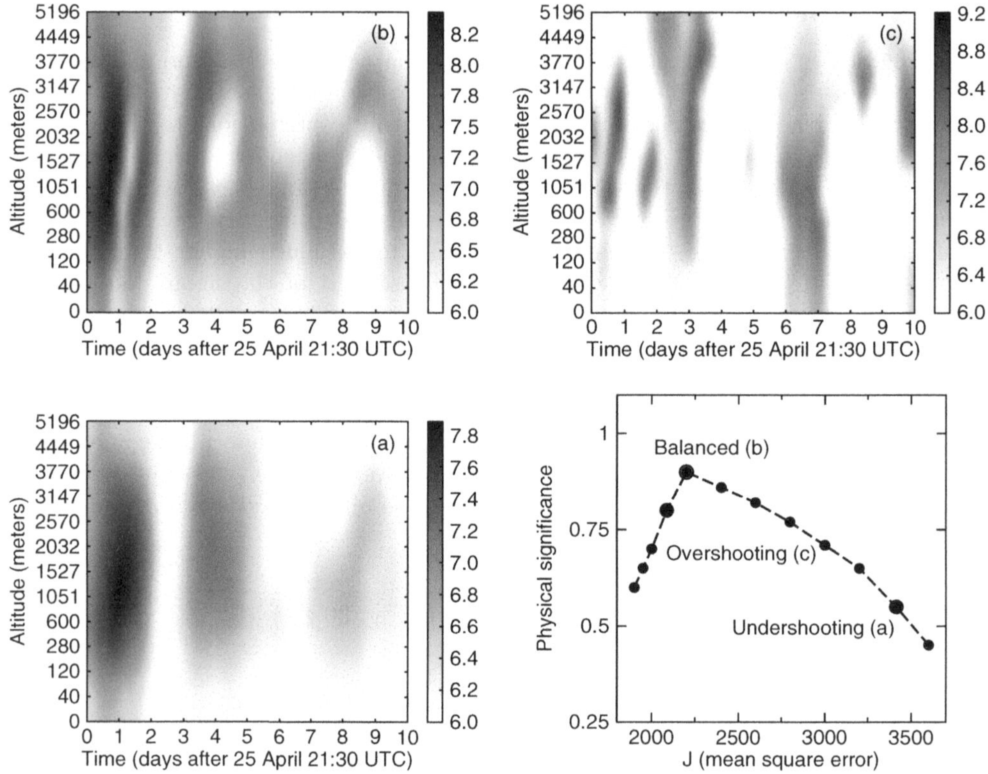

Fig. 20.4 Illustration of the semi-convergence and the need for the appropriate level of regularisation. The density represents the decimal log of the rates in $\mathrm{Bq\,m^{-1}\,s^{-1}}$.

20.3.2 Maximum entropy on the mean

Let us now describe a quite different inference for the inverse problem. It is of Bayesian nature, in the sense that the goal is to make an inference based on all available information. It is different from the Backus–Gilbert method, which is not of Bayesian nature. This approach is called the *maximum entropy on the mean* (MEM). It can be formulated in terms of the pdfs that were used in the Bayesian inference, that is, the prior pdf $p(\boldsymbol{\sigma})$ of the control variables, a prior pdf $p(\boldsymbol{\epsilon})$ of the errors, and an observation set encoded in vector $\boldsymbol{\mu} \in \mathbb{R}^p$. Instead of using Bayes' formula to infer the posterior pdf $p(\boldsymbol{\sigma}|\boldsymbol{\mu})$, we look for the posterior pdf that *minimizes the gain of information, except for what is gained from the observations.*

At first, this may sound paradoxical, since we may have the feeling that we should maximize the information in the inference. In fact, the goal is to wisely exploit the information from the background and the observations but not spurious information that has no relation to them. And this is what is meant by *minimizing the information.* In information theory language, this is equivalent to maximizing the *entropy.* This type of inference has been convincingly advocated by Jaynes (1957a,b). The second problem is to know how to quantify this information or, alternatively, entropy. We could use the relative entropy earlier introduced as an axiomatic measure of information:

$$\mathcal{K}(q,\, p) = \int \mathrm{d}\boldsymbol{\sigma}\, q(\boldsymbol{\sigma}) \ln \frac{q(\boldsymbol{\sigma})}{p(\boldsymbol{\sigma})}\,. \tag{20.33}$$

We would like to minimize the quantity on the condition that, on average on the posterior pdf, one has $\boldsymbol{\mu} = \mathbf{H}\boldsymbol{\sigma} + \boldsymbol{\epsilon}$. In the MEM framework, this relationship is only statistically enforced and one needs to make $\boldsymbol{\epsilon}$ part of the inference since it is not strictly (value-wise) enforced as $\boldsymbol{\epsilon} = \boldsymbol{\mu} - \mathbf{H}\boldsymbol{\sigma}$. The prior is the product $p(\boldsymbol{\sigma})p(\boldsymbol{\epsilon})$, but we can make it even more general and define the general prior pdf $\nu(\boldsymbol{\sigma}, \boldsymbol{\epsilon})$. To do so, we define the Lagrangian

$$\mathcal{L}(p) = \int \mathrm{d}\boldsymbol{\sigma}\, \mathrm{d}\boldsymbol{\epsilon}\, p(\boldsymbol{\sigma}, \boldsymbol{\epsilon}|\boldsymbol{\mu}) \ln \frac{p(\boldsymbol{\sigma}, \boldsymbol{\epsilon}|\boldsymbol{\mu})}{\nu(\boldsymbol{\sigma}, \boldsymbol{\epsilon})} + \boldsymbol{\beta}^{\mathrm{T}} \int \mathrm{d}\boldsymbol{\sigma}\, \mathrm{d}\boldsymbol{\epsilon}\, p(\boldsymbol{\sigma}, \boldsymbol{\epsilon}|\boldsymbol{\mu})\, (\boldsymbol{\mu} - \mathbf{H}\boldsymbol{\sigma} - \boldsymbol{\epsilon}). \tag{20.34}$$

The vector $\boldsymbol{\beta}$ is made of p Lagrange multipliers that enforce the observation equation on the mean. It is then easy to formally optimize this Lagrangian on the posterior pdf. In doing so, one should not forget to enforce the normalization of the posterior pdf to 1. The solution is

$$p(\boldsymbol{\sigma}, \boldsymbol{\epsilon}|\boldsymbol{\mu}) = \frac{\nu(\boldsymbol{\sigma}, \boldsymbol{\epsilon}) \exp\left[\boldsymbol{\beta}^{\mathrm{T}} (\mathbf{H}\boldsymbol{\sigma} + \boldsymbol{\epsilon})\right]}{Z(\boldsymbol{\beta})}\,, \tag{20.35}$$

where $Z(\boldsymbol{\beta})$ is the normalization factor, universally known in (statistical) physics as the *partition function*:

$$Z(\boldsymbol{\beta}) = \int \mathrm{d}\boldsymbol{\sigma}\, \mathrm{d}\boldsymbol{\epsilon}\, \nu(\boldsymbol{\sigma}, \boldsymbol{\epsilon}) \exp\left[\boldsymbol{\beta}^{\mathrm{T}} (\mathbf{H}\boldsymbol{\sigma} + \boldsymbol{\epsilon})\right]. \tag{20.36}$$

We have obtained the functional form of the posterior pdf, but $\boldsymbol{\beta}$ is unknown. This partial solution should be inserted into the Lagrangian, so as to obtain the dual cost function

$$\widehat{\mathcal{L}}(\boldsymbol{\beta}) = \ln Z(\boldsymbol{\beta}) - \boldsymbol{\beta}^{\mathrm{T}} \boldsymbol{\mu}\,. \tag{20.37}$$

It is a dual cost function in the sense that it is to be minimized in the space of Lagrange parameters, which is isomorphic to the space of observations: $\boldsymbol{\beta} \in \mathbb{R}^p$. Beware that this dual cost function is actually the negative of the original Lagrangian, called in this context the *level-1 primal cost function*. And yet, it is to be minimized and not maximized, because the primal and the dual cost functions are geometrically related through a saddle-point problem.

Because the Kullback–Leibler divergence is convex, and because one seeks the solution in a functional vector space of pdfs that satisfy the observation equation on the mean (a convex set), the dual cost function is also convex. This property is very useful for high-dimensional data assimilation problems. In addition, it mathematically guarantees that the primal and dual problems are tightly related and that the minimum of one is exactly the opposite of the minimum of the other (there is no *duality gap*):

$$\min_p \mathcal{L}(p) = -\min_{\boldsymbol{\beta}} \widehat{\mathcal{L}}(\boldsymbol{\beta})\,. \tag{20.38}$$

Just as for Bayes' formula, one needs an estimator to make some sense of the posterior pdf. However, following this MEM inference, only the mean estimator really makes sense:

$$\overline{\boldsymbol{\sigma}} = \int \mathrm{d}\boldsymbol{\sigma}\, \mathrm{d}\boldsymbol{\epsilon}\, \boldsymbol{\sigma} p(\boldsymbol{\sigma}, \boldsymbol{\epsilon}|\boldsymbol{\mu})\,, \qquad \overline{\boldsymbol{\epsilon}} = \int \mathrm{d}\boldsymbol{\sigma}\, \mathrm{d}\boldsymbol{\epsilon}\, \boldsymbol{\epsilon} p(\boldsymbol{\sigma}, \boldsymbol{\epsilon}|\boldsymbol{\mu})\,. \tag{20.39}$$

Let us consider the Gaussian case. The prior pdf and likelihood are given by (20.6) and (20.7). Then one needs to compute the partition function, summing on all possible prior realizations of the control variables and errors:

$$Z(\boldsymbol{\beta}) = \int \mathrm{d}\boldsymbol{\sigma}\, \mathrm{d}\boldsymbol{\epsilon}\, p_b(\boldsymbol{\sigma}) p_e(\boldsymbol{\epsilon}) \exp\left[\boldsymbol{\beta}^{\mathrm{T}}(\mathbf{H}\boldsymbol{\sigma} + \boldsymbol{\epsilon})\right] \tag{20.40}$$

$$= \int \mathrm{d}\boldsymbol{\sigma}\, \mathrm{d}\boldsymbol{\epsilon}\, \frac{e^{-\frac{1}{2}(\boldsymbol{\sigma}-\boldsymbol{\sigma}_b)^{\mathrm{T}}\mathbf{B}^{-1}(\boldsymbol{\sigma}-\boldsymbol{\sigma}_b) - \frac{1}{2}\boldsymbol{\epsilon}^{\mathrm{T}}\mathbf{R}^{-1}\boldsymbol{\epsilon} + \boldsymbol{\beta}^{\mathrm{T}}(\mathbf{H}\boldsymbol{\sigma}+\boldsymbol{\epsilon})}}{\sqrt{(2\pi)^{n+p}|\mathbf{B}||\mathbf{R}|}} \tag{20.41}$$

$$= \exp\left[\frac{1}{2}\boldsymbol{\beta}^{\mathrm{T}}\left(\mathbf{R} + \mathbf{H}\mathbf{B}\mathbf{H}^{\mathrm{T}}\right)\boldsymbol{\beta} + \boldsymbol{\beta}^{\mathrm{T}}\mathbf{H}\boldsymbol{\sigma}_b\right]. \tag{20.42}$$

As a consequence, the dual cost function is

$$\widehat{\mathcal{L}}(\boldsymbol{\beta}) = \frac{1}{2}\boldsymbol{\beta}^{\mathrm{T}}\left(\mathbf{R} + \mathbf{H}\mathbf{B}\mathbf{H}^{\mathrm{T}}\right)\boldsymbol{\beta} - \boldsymbol{\beta}^{\mathrm{T}}(\boldsymbol{\mu} - \mathbf{H}\boldsymbol{\sigma}_b). \tag{20.43}$$

One recovers the physical-space statistical analysis system (PSAS) cost function of data assimilation (Courtier, 1997). The minimization of $\widehat{\mathcal{L}}(\boldsymbol{\beta})$ will yield the optimal $\overline{\boldsymbol{\beta}}$, and then $p(\boldsymbol{\sigma}, \boldsymbol{\epsilon})$ thanks to (20.35). It is then easy to check that the mean estimator yields the BLUE estimator: $\overline{\boldsymbol{\sigma}} = \boldsymbol{\sigma}_a$, and $\overline{\boldsymbol{\epsilon}} = \boldsymbol{\epsilon}_a$ Note that the optimal $\overline{\boldsymbol{\beta}}$ gives the second-order sensitivity of the cost function to each one of the observation.

This type of inference has been pioneered in geophysics by Rietsch (1977), but for a small-dimensional problem. MEM as an inference method for inverse problems has been advocated by Shore (1984). It has drawn the attention of mathematicians considering its connection with large deviation theory (Ellis, 1985), and for its usefulness in inverse problems (Le Besnerais et al., 1999). The connection with geophysical data assimilation, emphasizing the role of the reference pdf of the Kullback discrepancy as a prior, and some possible generalizations of PSAS with non-Gaussian priors have been highlighted in Bocquet (2005b,c) and Krysta and Bocquet (2007).

For linear models, it allows us to obtain a rigorously equivalent dual cost function, even if the prior is non-Gaussian, or even if it is non-convex. For instance, assume that we want to retrieve positive emission factors of an atmospheric species. Typical background pdfs for this source are positive pdfs, such as log-normal, exponential,

Bernoulli, gamma, or Poisson distributions. Let us consider the Poisson distribution. In grid cell k, the prior pdf, parameterize by θ_k, is

$$P(x_k) = e^{-\theta_k} \frac{\theta_k^{x_k}}{x_k!} . \tag{20.44}$$

x_k is an integer and $P(x_k)$ is seen as the probability that a mass of pollutant $m\,x_k$ is emitted in grid cell k, where m is the mass scale. Because we assumed that the variables of the emission field are independent from one another, we can factorize the partition function: $Z(\boldsymbol{\beta}) = \prod_{k=1}^{n} Z_k(\boldsymbol{\beta})$. Locally, we have

$$Z_k(\boldsymbol{\beta}) = \exp\left\{-\theta_k + \theta_k \exp\left(m \left[\boldsymbol{\beta}^{\mathrm{T}}\mathbf{H}\right]_k\right)\right\}, \tag{20.45}$$

so that

$$\widehat{\mathcal{J}}(\boldsymbol{\beta}) = \sum_{k=1}^{n} \theta_k \left\{\exp\left(m \left[\boldsymbol{\beta}^{\mathrm{T}}\mathbf{H}\right]_k\right) - 1\right\} - \boldsymbol{\beta}^{\mathrm{T}}\boldsymbol{\mu} . \tag{20.46}$$

The reconstructed emission field is

$$\overline{\sigma}_k = m\,\theta_k \exp\left(m \left[\boldsymbol{\beta}^{\mathrm{T}}\mathbf{H}\right]_k\right). \tag{20.47}$$

It is not difficult to generalize this reconstruction in the presence of observation errors (Bocquet, 2005a).

The method's main strength is in the ease with which the convex (dual or primal) cost functions can be built. Its main drawback is that the MEM principle is not meant to provide higher-order posterior statistics beyond the mean. By making a connection between the MEM inference and the traditional Bayesian inference, it is nevertheless possible to do so (Bocquet, 2008).

Chernobyl case study: a MEM inversion

We have applied the MEM approach to the Chernobyl inverse modelling problem. We have chosen the Poisson distribution for the source prior, with $m = 1.0 \times 10^8\,\mathrm{Bq\,m^{-1}\,s^{-1}}$, and $\theta = 0.1$, and a normal distribution for the error prior with $\chi = 1.33\,\mathrm{Bq\,m^{-3}}$. The m and χ parameters are those that will later be estimated in a non-Gaussian framework. The θ parameter tells us how close the source is to vanishing a priori (Davoine and Bocquet, 2007). Those three values are fairly realistic, but they have not been optimally adjusted, as opposed to what was done in Davoine and Bocquet (2007). The retrieved activity is 9.51×10^{16} Bq. The source is shown in Fig. 20.5. The DFS from BLUE are approximate and not a completely consistent measure of the degrees of freedom in this non-Gaussian context. Similarly to the definitions of the observation error term J_o and the background error term J_b in the usual Gaussian cost function, one can define the J_o and J_b terms in the non-Gaussian case, still assuming independence of the two error types. As a substitute for the DFS, the values of the J_o and J_b terms at the minimum are objective measures (entropies) of the information extracted either to reconstruct noise or to reconstruct the source respectively (Bocquet, 2005a). Here, we have $J_b = 114$ and $J_o = 1232$.

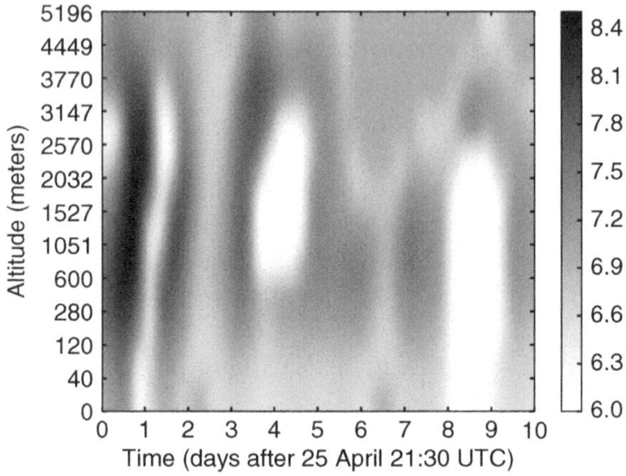

Fig. 20.5 Retrieved caesium-137 source of the Chernobyl accident obtained from the MEM inference using a Poisson prior for the source and a normal prior for the errors.

Quality of the analysis with a MEM inference

Let us assume the prior statistics are Gaussian: $\boldsymbol{\sigma} \sim \mathcal{N}(\boldsymbol{\sigma}^b, \mathbf{B})$ and $\boldsymbol{\epsilon} \sim \mathcal{N}(\mathbf{0}, \mathbf{R})$. One indicator that measures the reduction of uncertainty carried out in an analysis is given by the ratio

$$\rho = 1 - \frac{||\boldsymbol{\sigma}^t - \overline{\boldsymbol{\sigma}}||^2_{\mathbf{B}^{-1}} + ||\boldsymbol{\epsilon}^t - \overline{\boldsymbol{\epsilon}}||^2_{\mathbf{R}^{-1}}}{||\boldsymbol{\sigma}^t - \boldsymbol{\sigma}_b||^2_{\mathbf{B}^{-1}} + ||\boldsymbol{\epsilon}^t||^2_{\mathbf{R}^{-1}}} = \frac{||\overline{\boldsymbol{\sigma}} - \boldsymbol{\sigma}^b||^2_{\mathbf{B}^{-1}} + ||\overline{\boldsymbol{\epsilon}}||^2_{\mathbf{R}^{-1}}}{||\boldsymbol{\sigma}^t - \boldsymbol{\sigma}^b||^2_{\mathbf{B}^{-1}} + ||\boldsymbol{\epsilon}^t||^2_{\mathbf{R}^{-1}}}, \tag{20.48}$$

where $\boldsymbol{\sigma}^t$ and $\boldsymbol{\epsilon}^t$ are the true state and error vectors. The norm $||.||_{\mathbf{G}}$, with $\mathbf{G}$ a positive-definite matrix, is defined by $||\boldsymbol{\sigma}||^2_{\mathbf{G}} = \boldsymbol{\sigma}^{\mathrm{T}}\mathbf{G}\boldsymbol{\sigma}$. The transformation from the first equality to the second is due to the Pythagorean theorem. The analysis $(\overline{\boldsymbol{\sigma}}, \overline{\boldsymbol{\epsilon}})$ in $\mathbb{R}^n \oplus \mathbb{R}^p$ is the orthogonal projection of $(\boldsymbol{\sigma}_b, \mathbf{0})$ (with respect to the scalar product defined by $\mathbf{B}^{-1} \oplus \mathbf{R}^{-1}$) on the hyperplane of couples $(\boldsymbol{\sigma}, \boldsymbol{\epsilon})$ such that $\boldsymbol{\mu} = \mathbf{H}\boldsymbol{\sigma} + \boldsymbol{\epsilon}$. This is equivalent to the geometrical interpretation of the analysis in terms of projection using the Mahalanobis norm as a scalar product (Desroziers et al., 2005; Chapnik et al., 2006). This ensures that $0 \leq \rho \leq 1$, with $\rho = 1$ when the reduction of uncertainty is maximum.

Bocquet (2005a) has generalized without approximation this score to non-Gaussian priors provided that the analysis is performed using the MEM principle. In a condensed form, the generalization of (20.48) is

$$\rho = \frac{\mathcal{K}(p_{\overline{\boldsymbol{\sigma}},\overline{\boldsymbol{\epsilon}}}, \nu)}{\mathcal{K}(p_{\boldsymbol{\sigma}_t,\boldsymbol{\epsilon}_t}, \nu)}. \tag{20.49}$$

$p_{\overline{\boldsymbol{\sigma}},\overline{\boldsymbol{\epsilon}}}$ and $p_{\boldsymbol{\sigma}_t,\boldsymbol{\epsilon}_t}$ are the pdfs, belonging to the exponential family, (20.35), whose state and error averages are $\overline{\boldsymbol{\sigma}}$ and $\overline{\boldsymbol{\epsilon}}$ in the former case, and $\boldsymbol{\sigma}_t$ and $\boldsymbol{\epsilon}_t$ in the latter case.

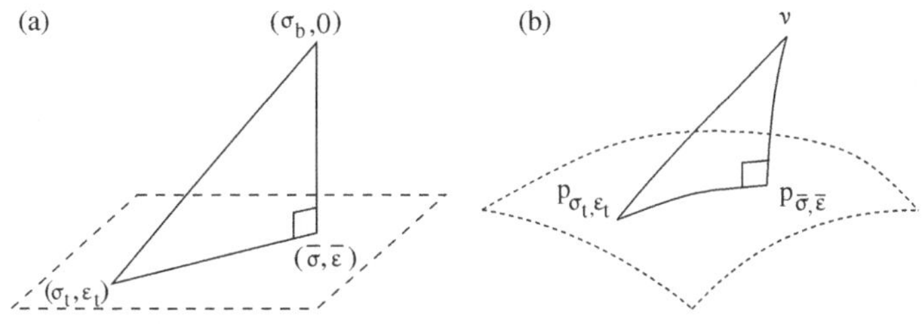

Fig. 20.6 Projection of the (generalized) background $(\boldsymbol{\sigma}_b, \mathbf{0})$ onto the manifold of the $(\boldsymbol{\sigma}, \boldsymbol{\epsilon})$ couples that satisfy $\boldsymbol{\epsilon} = \mathbf{H}\boldsymbol{\sigma} + \boldsymbol{\epsilon}$. (a) Gaussian case. (b) Non-Gaussian case.

Remember that ν is the joint prior pdf. To derive this result, one need the Czizar lemma, which corresponds to a Pythagorean identity in curved space:

$$\mathcal{K}(p_{(\boldsymbol{\sigma}_t,\boldsymbol{\epsilon}_t)}, p_{(\overline{\boldsymbol{\sigma}},\overline{\boldsymbol{\epsilon}})}) = \mathcal{K}(p_{(\boldsymbol{\sigma}_t,\boldsymbol{\epsilon}_t)}, \nu) - \mathcal{K}(p_{(\overline{\boldsymbol{\sigma}},\overline{\boldsymbol{\epsilon}})}, \nu). \tag{20.50}$$

The Kullback–Leibler terms, $\mathcal{K}(p_{\overline{\boldsymbol{\sigma}},\overline{\boldsymbol{\epsilon}}}, \nu)$ and $\mathcal{K}(p_{\boldsymbol{\sigma}_t,\boldsymbol{\epsilon}_t}, \nu)$ are expressed here in their abstract form, but they can be estimated numerically using their equivalent explicit expressions. In particular, $\mathcal{K}(p_{\overline{\boldsymbol{\sigma}},\overline{\boldsymbol{\epsilon}}}, \nu)$ is often identified with the minimum of the objective function, and is just a numerical byproduct of a data assimilation variational scheme. In the Gaussian case and with independent background and error priors, (20.49) simplifies to (20.48). Figure 20.6 gives a schematic view of these projections in a Gaussian and a non-Gaussian framework.

20.4 Estimation of second-order statistics

The BLUE method of data assimilation is based on the often incorrect assumption that the second-order statistics of the errors are known. Hence, an improvement of the data assimilation schemes requires an (at least partial) estimation of the second-oder statistics. Actually, this problem was identified very early on in the inverse problem community. Parameters of the error covariance matrices are often called *hyper-parameters* in this context. There is a vast literature about this topic: Engl et al., (2000), Vogel (2002), Hansen (2010), and (Doicu et al., 2010), to mention only a few textbooks! This does not cover the endeavours to solve the same problem in a forecasting context (see Chapters 7–9 in this volume).

Reading the data assimilation or inverse problem literature, one may have the feeling that this is a refinement, or an addendum to the state/parameter problems. My point of view is that it is more a problem of fitting the errors of our inverse problem to a statistical model. This point of view is that of environmental statistics, where the models are not necessarily as numerically complex as those met in geophysical data assimilation. Following this standpoint, we should solve for the mean state as well as the higher-order moments all together. Of course, this may not be numerically affordable or there may not be enough observations to provide a reliable result.

From basic data assimilation theory, the estimation of error statistics has been done a priori (building the error-covariance matrices). This is quite difficult in inverse

Table 20.1 Summary of the estimated values of the hyper-parameters for several methods.

	Analysis	m/χ	m (Bq m^{-1} s^{-1})	χ (Bq m^{-3})	M(Bq)
L-curve	Gaussian	5.42×10^7	irr.	irr.	7.38×10^{16}
GCV	Gaussian	fail	irr.	irr.	fail
ML	Gaussian	3.13×10^9	3.51×10^9	1.12	15.4×10^{16}
Desroziers	Gaussian	3.13×10^9	3.51×10^9	1.12	15.4×10^{16}
L-curve	Non-Gaussian	7.31×10^7	irr.	irr.	8.04×10^{16}
GCV	Non-Gaussian	3.00×10^8	irr.	irr.	10.9×10^{16}
ML	Non-Gaussian	7.52×10^7	1.00×10^8	1.33	8.29×10^{16}

'irr.' stands for irrelevant, and 'fail' means that the method did not permit an estimation.

modelling problems, where the uncertainty is often difficult to quantify and cannot be inferred from a forecast. Therefore, it is mandatory to make use of the observations to infer the hyper-parameters. These methods are qualified as *a posteriori*. We can consider two classes of a posteriori methods, depending on whether one has a prior idea of the residual error (Engl et al., 2000). We shall only mention the most popular ones. Several of these methods will be applied to the Chernobyl case. (The numerical results are collected in Table 20.1 for comparison.)

20.4.1 Having an estimation of the residual error

This first set of methods assumes that one knows the order of magnitude of the errors in the system. In particular, this is only possible if model errors are quantitatively characterized.

Discrepancy principle

The idea is to truncate the SVD expansion at index l_η when the residue is of the order η of the error present in the system:

$$||\boldsymbol{\mu} - \mathbf{H}\boldsymbol{\sigma}^a_{l_\eta}|| \simeq \eta\,, \tag{20.51}$$

where

$$\boldsymbol{\sigma}^a_{l_\eta} = \boldsymbol{\sigma}^b + \sum_{l=1}^{l_\eta} \frac{\mathbf{u}_l^{\mathrm{T}}(\boldsymbol{\mu} - \mathbf{H}\boldsymbol{\sigma}_b)}{\lambda_l}\mathbf{v}_l\,. \tag{20.52}$$

Instead of the regularization by truncation, the same inverse problem can be regularized using the Tikhonov approach:

$$\mathcal{J}(\boldsymbol{\sigma}) = \frac{1}{2}\left(\boldsymbol{\mu} - \mathbf{H}\boldsymbol{\sigma}\right)^{\mathrm{T}}\mathbf{R}^{-1}\left(\boldsymbol{\mu} - \mathbf{H}\boldsymbol{\sigma}\right) + \frac{1}{2\lambda^2}\left(\boldsymbol{\sigma} - \boldsymbol{\sigma}_b\right)^{\mathrm{T}}\mathbf{B}^{-1}\left(\boldsymbol{\sigma} - \boldsymbol{\sigma}_b\right), \tag{20.53}$$

with the addition of the hyper-parameter λ that parameterizes $\mathbf{B}$. The criterion translates into choosing λ_η such that

$$||\boldsymbol{\mu} - \mathbf{H}\boldsymbol{\sigma}^a_\eta|| \simeq \eta\,, \tag{20.54}$$

where $\boldsymbol{\sigma}^a_\eta$ is the solution of the inverse problems using Tikhonov regularization with $\lambda^2_\eta \mathbf{B}$. This is known as the *discrepancy principle* or Morozov's principle (Morozov, 1966).

20.4.2 In the absence of an estimation of the residual error

Maximum likelihood

Let us enforce scales in the error covariance matrices:

$$\mathbf{R} = \chi^2 \mathbf{R}_0\,, \qquad \mathbf{B} = m^2 \mathbf{B}_0\,, \tag{20.55}$$

where $\mathbf{R}_0$ and $\mathbf{B}_0$ are considered fixed in the following.

Of Bayesian inspiration, the maximum likelihood principle consists in computing the pdf $p(\boldsymbol{\mu}|m,\chi)$ of the observation vector conditional on the choice of hyper-parameters for m and χ. To obtain this pdf, one can use

$$p(\boldsymbol{\mu}|m,\chi) = \int \mathrm{d}\boldsymbol{\sigma}\,\mathrm{d}\boldsymbol{\epsilon}\,\delta(\boldsymbol{\mu} - \mathbf{H}\boldsymbol{\sigma})p(\boldsymbol{\epsilon})p(\boldsymbol{\sigma}) = \int \mathrm{d}\boldsymbol{\sigma}\,p(\boldsymbol{\mu}|\boldsymbol{\sigma})p(\boldsymbol{\sigma})\,. \tag{20.56}$$

This integral can be formidable to compute. However, under Gaussian assumptions for the errors, (20.6) and (20.7), it can be computed analytically and leads to (20.10):

$$p(\boldsymbol{\mu}|m,\chi) = \frac{e^{-\frac{1}{2}(\boldsymbol{\mu}-\mathbf{H}\boldsymbol{\sigma}_b)^{\mathrm{T}}\left(\mathbf{R}+\mathbf{H}\mathbf{B}\mathbf{H}^{\mathrm{T}}\right)^{-1}(\boldsymbol{\mu}-\mathbf{H}\boldsymbol{\sigma}_b)}}{\sqrt{(2\pi)^p|\mathbf{R}+\mathbf{H}\mathbf{B}\mathbf{H}^{\mathrm{T}}|}}\,. \tag{20.57}$$

This is the likelihood of $\boldsymbol{\mu}$. But, assuming a uniform non-informative prior for the hyper-parameters m and χ, this is also the likelihood for m and χ given the observation vector $\boldsymbol{\mu}$. Hence, to obtain the more likely values, one has to minimize

$$\frac{1}{2}(\boldsymbol{\mu} - \mathbf{H}\boldsymbol{\sigma}_b)^{\mathrm{T}}\left(\mathbf{R} + \mathbf{H}\mathbf{B}\mathbf{H}^{\mathrm{T}}\right)^{-1}(\boldsymbol{\mu} - \mathbf{H}\boldsymbol{\sigma}_b) + \frac{1}{2}\ln|\mathbf{R} + \mathbf{H}\mathbf{B}\mathbf{H}^{\mathrm{T}}|\,. \tag{20.58}$$

For the method to work well, the matrices $\mathbf{R}$ and $\mathbf{H}\mathbf{B}\mathbf{H}^{\mathrm{T}}$ should have significantly different structures, in order to avoid degeneracy in the optimization of m and χ.

This method for estimating the hyper-parameters is the widespread method adopted in geophysical data assimilation. The use of this principle was promoted by Dee (1995). Desroziers' iterative method (Desroziers and Ivanov, 2001; Chapnik et al., 2004) for the estimation of the error covariance matrix is a fixed-point method to minimize this function. It has been used by Winiarek et al. (2012) for the estimation of the error in the reconstruction of the Fukushima–Daiichi radionuclides source term.

The semi-Gaussian generalization of (20.57) leads to the likelihood (Winiarek et al., 2012):

$$p(\boldsymbol{\mu}|m,\chi) = \frac{e^{-\frac{1}{2}(\boldsymbol{\mu}-\mathbf{H}\boldsymbol{\sigma}_b)^{\mathrm{T}}\left(\mathbf{R}+\mathbf{HBH}^{\mathrm{T}}\right)^{-1}(\boldsymbol{\mu}-\mathbf{H}\boldsymbol{\sigma}_b)}}{\sqrt{(2\pi)^p|\mathbf{R}+\mathbf{HBH}^{\mathrm{T}}|}} \times \left(\int_{\boldsymbol{\sigma}\geq\mathbf{0}} \mathrm{d}\boldsymbol{\sigma}\, \frac{e^{-\frac{1}{2}(\boldsymbol{\sigma}-\boldsymbol{\sigma}_b)^{\mathrm{T}}\mathbf{B}^{-1}(\boldsymbol{\sigma}-\boldsymbol{\sigma}_b)}}{\sqrt{(2\pi)^n|\mathbf{B}|}}\right)^{-1} \times \int_{\boldsymbol{\sigma}\geq\mathbf{0}} \mathrm{d}\boldsymbol{\sigma}\, \frac{e^{-\frac{1}{2}(\boldsymbol{\sigma}-\boldsymbol{\sigma}_a)^{\mathrm{T}}\mathbf{P}_a^{-1}(\boldsymbol{\sigma}-\boldsymbol{\sigma}_a)}}{\sqrt{(2\pi)^n|\mathbf{P}_a|}}. \tag{20.59}$$

which is more difficult to compute efficiently. However, this can be done for moderately large systems (say a few thousand control variables); see Lauvernet et al. (2009) and Winiarek et al. (2012). This approach to estimating the hyper-parameters is statistically consistent with the inversion under the positivity constraint.

Chernobyl case study: maximum likelihood(s)

We apply the maximum-likelihood principle to the Chernobyl case study. First, the pdf $p(\boldsymbol{\mu}|m,\chi)$ of (20.57), whose density is plotted in Fig. 20.7, is minimized on both m and χ. There is only one minimum in this square of the hyper-parameter space. The result is $m^* = 3.51 \times 10^9\,\mathrm{Bq\,m^{-1}\,s^{-1}}$ and $\chi^* = 1.12\,\mathrm{Bq\,m^{-3}}$, with a total retrieved source activity $M = 5.94 \times 10^{16}$ Bq. It can be compared with Desroziers's iterative scheme, which converges to one minimum of the likelihood. Indeed, we found exactly the same result. However, this approach ignores the positivity of the source. We have derived and computed the non-Gaussian pdf $p(\boldsymbol{\mu}|m,\chi)$ of (20.59), which

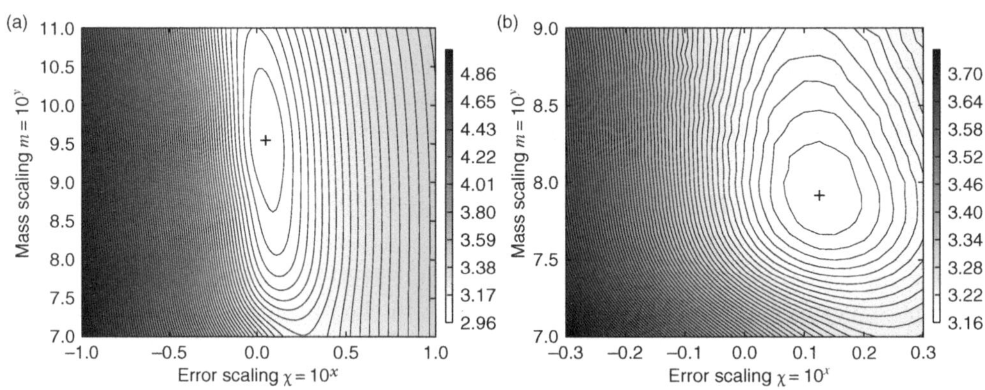

Fig. 20.7 Log-likelihood density plot for the linear case (a) and the non-Gaussian case (b), in the (m,χ) plane. The location of the minimum is indicated by a cross. Its solution is shown in Fig. 20.2.

takes into account the positivity of the source without approximation. The corresponding density plot is shown in Fig. 20.7. The optimal hyper-parameters are $m^\star = 1.0 \times 10^8\,\mathrm{Bq\,m^{-1}\,s^{-1}}$ and $\chi^\star = 1.33\,\mathrm{Bq\,m^{-3}}$, with a total activity $M = 8.29 \times 10^{16}\,\mathrm{Bq}$.

Generalized cross-validation

The generalized cross-validation principle is of statistical inspiration. The goal is to choose a regularization hyper-parameter that makes the data (that is, $\boldsymbol{\mu}$) consistent. This method has been pioneered and extensively studied by Wahba (1990) and colleagues.

The idea is to remove one scalar observation μ_i indexed by i from $\boldsymbol{\mu}$, leaving the observation vector $\boldsymbol{\mu}^{[i]} \in \mathbb{R}^{p-1}$, and to predict the value μ_i using $\boldsymbol{\mu}^{[i]}$. With Tikhonov regularization, we obtain

$$\boldsymbol{\sigma}_a^{[i]} = \boldsymbol{\sigma}_b + \mathbf{B}\mathbf{H}_{[i]}^{\mathrm{T}} \left(\mathbf{R}_{[i]} + \mathbf{H}_{[i]}\mathbf{B}\mathbf{H}_{[i]}^{\mathrm{T}}\right)^{-1} \left(\boldsymbol{\mu}^{[i]} - \mathbf{H}_{[i]}\boldsymbol{\sigma}_b\right), \tag{20.60}$$

The index $[i]$ on control vectors indicate that they are the specific solutions with observation vector $\boldsymbol{\mu}^{[i]}$. $\mathbf{H}_{[i]}$ is the observation operator without row i. $\mathbf{R}_{[i]}$ is the observation error covariance matrix without row i and column i.

Assuming the scale dependence of the error covariance matrices given in (20.55), the following expressions will depend only on the scaling parameter $\lambda = m/\chi$. The prediction error is

$$e_{\mathrm{CV}}(\lambda) = \frac{1}{p}\sum_{i=1}^{p}\left[\mathbf{H}_i\boldsymbol{\sigma}_a^{[i]}(\lambda) - \mu_i\right]^2, \tag{20.61}$$

where $\mathbf{H}_i$ is row i of $\mathbf{H}$. The optimal regularization parameter λ minimises e_{CV}. An algebraic result, known as the *leave-one-out lemma*, allows us to rewrite the prediction error in term of the solution $\boldsymbol{\sigma}_a(\lambda)$, rather than the p solutions $\boldsymbol{\sigma}_a^{[i]}(\lambda)$:

$$e_{\mathrm{CV}}(\lambda) \simeq \frac{1}{p}\sum_{i=1}^{p}\left[\frac{\mathbf{H}_i\boldsymbol{\sigma}_a(\lambda) - \mu_i}{1 - [\mathbf{G}]_{ii}}\right]^2, \tag{20.62}$$

where $\mathbf{G} = \mathbf{HK}$ is the resolution matrix. This expression is much more tractable. However, it is not an invariant expression under permutations of the scalar observations. To avoid this issue, the generalized cross-validation technique assumes that the $[\mathbf{G}]_{ii}$ can be approximated by their average value, $p^{-1}\mathrm{Tr}\,\mathbf{G}$, so that

$$e_{\mathrm{CV}}(\lambda) \simeq \frac{1}{p}\sum_{i=1}^{p}\left[\frac{\mathbf{H}_i\boldsymbol{\sigma}_a(\lambda) - \mu_i}{1 - p^{-1}\mathrm{Tr}\,\mathbf{G}}\right]^2 = \frac{||\mathbf{H}\boldsymbol{\sigma}_a(\lambda) - \boldsymbol{\mu}||^2}{p\left(1 - p^{-1}\mathrm{Tr}\,\mathbf{G}\right)^2}. \tag{20.63}$$

Using the SVD expansion of the Tikhonov-regularized solution, we obtain the GCV criterion to be minimized:

$$\mathcal{G}(\lambda) = \frac{||\mathbf{H}\boldsymbol{\sigma}_a(\lambda) - \boldsymbol{\mu}||^2}{\left(p - \sum_{l=1}^{q}\varphi_l(\lambda)\right)^2}. \tag{20.64}$$

Chernobyl case study: GCV

The GCV principle is first applied using $\boldsymbol{\sigma}_a$ obtained from the BLUE analysis. The GCV curve is displayed in Fig. 20.8. The GCV diagnostic fails because there is no minimum of the GCV curve. This is known to happen when the Gaussianity of errors that is used in the GCV derivation is far from satisfied (Hansen, 2010). However, if we choose the analysis $\boldsymbol{\sigma}_a$ as the minimum of (20.8) under the assumption that the source term is positive, the GCV curve does exhibit a minimum, corresponding to the optimal $\lambda^\star = 3.0 \times 10^8$ and a total activity of $M = 1.09 \times 10^{17}$ Bq.

L-curve

The L-curve method is empirical but quite intuitive, and it is efficient in practice. The inverse problem is solved using Tikhonov regularization, but we do not know a priori the error magnitude η. Let us consider again the regularized cost function:

$$\mathcal{J}(\boldsymbol{\sigma}) = \frac{1}{2}\left(\boldsymbol{\mu} - \mathbf{H}\boldsymbol{\sigma}\right)^{\mathrm{T}} \mathbf{R}^{-1} \left(\boldsymbol{\mu} - \mathbf{H}\boldsymbol{\sigma}\right) + \frac{1}{2}\left(\boldsymbol{\sigma} - \boldsymbol{\sigma}_b\right)^{\mathrm{T}} \mathbf{B}^{-1} \left(\boldsymbol{\sigma} - \boldsymbol{\sigma}_b\right) \tag{20.65}$$

$$\propto \frac{1}{2}\left(\boldsymbol{\mu} - \mathbf{H}\boldsymbol{\sigma}\right)^{\mathrm{T}} \mathbf{R}_0^{-1} \left(\boldsymbol{\mu} - \mathbf{H}\boldsymbol{\sigma}\right) + \frac{1}{2\lambda^2}\left(\boldsymbol{\sigma} - \boldsymbol{\sigma}_b\right)^{\mathrm{T}} \mathbf{B}_0^{-1} \left(\boldsymbol{\sigma} - \boldsymbol{\sigma}_b\right), \tag{20.66}$$

The goal is to estimate the hyper-parameter λ. We consider the two departures

$$\Delta_\epsilon = ||\boldsymbol{\mu} - \mathbf{H}\boldsymbol{\sigma}^a(\lambda)||, \qquad \Delta_\sigma = ||\boldsymbol{\sigma}^a(\lambda) - \boldsymbol{\sigma}_b||, \tag{20.67}$$

when λ varies. For large λ, the constraint of the prior is emphasized in the cost function, and $\boldsymbol{\sigma}^a_\lambda$ will get closer to $\boldsymbol{\sigma}_b$. Therefore, Δ_ϵ will increase, while Δ_σ will get smaller and

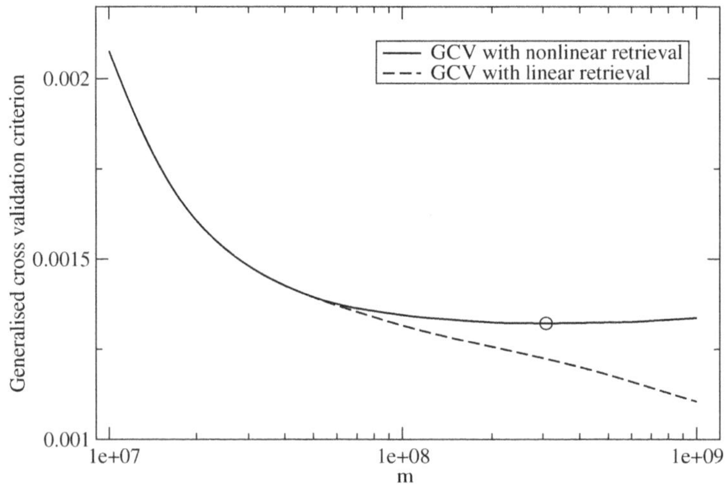

Fig. 20.8 Generalized cross-validation plot for the Chernobyl case study, in the linear and nonlinear retrieval cases.

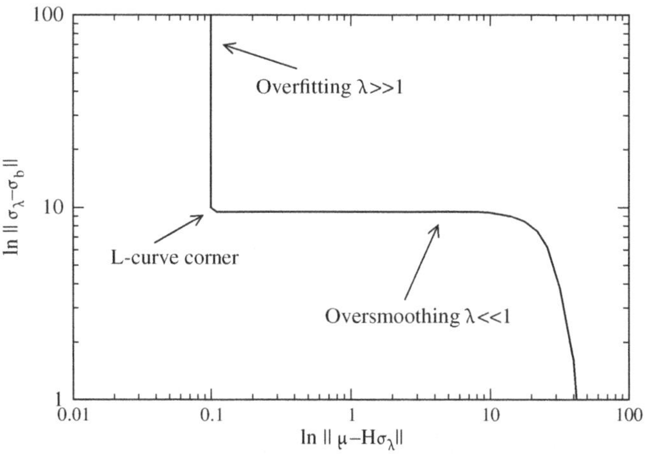

Fig. 20.9 Ideal L-curve schematic.

smaller. For small λ, the background constraint is less and less important, and the inversion will seek to minimize the residue Δ_ϵ. This overshoot is, in most problems, likely to create oscillating or fluctuating solutions that have a large norm, and hence a large Δ_σ.

If one plots the curve parameterized by λ in the $(\Delta_\epsilon, \Delta_\sigma)$-plane, two asymptotes should appear, depending on whether the solution is over-smoothed by a too strong regularisation or over-fits the observation because of too weak a regularization. These two asymptotes are quite generic and form an L-curve. It is often even more striking in a log scale, in the $(\ln \Delta_\epsilon, \ln \Delta_\sigma)$-plane. Intuitively, the right balance between the two regimes is in between the two asymptotes, in the corner of the L-curve. Figure 20.9 is a schematic of the L-curve. In practice, this ideal point is chosen as the point of maximal curvature of the L-curve. This method has been pioneered by Hansen (1992).

Chernobyl case study: L-curve

We apply the L-curve principle to the Chernobyl problem in two cases: when the BLUE solution for $\boldsymbol{\sigma}_a$ is used and when the positive solution for $\boldsymbol{\sigma}_a$ is used. This leads to two distinct L-curves plotted in Fig. 20.10. The corner is diagnosed by computing the curvature. In the Gaussian case, we find $\lambda^\star = 5.42 \times 10^7$, for a total reconstructed released activity of $M = 7.38 \times 10^{16}$ Bq. In the non-Gaussian case, we find $\lambda^\star = 7.31 \times 10^7$, for a total reconstructed released activity of $M = 8.04 \times 10^{16}$ Bq.

20.4.3 Nonlinearity in models, non-Gaussianity in priors

So far, we have only considered linear models $\mathbf{H}$. This simplification does allow us to cover most of inverse modelling issues. However, many models in geophysics are nonlinear, and nonlinearities and non-Gaussianity do considerably increase the difficulty of data assimilation and inverse modelling.

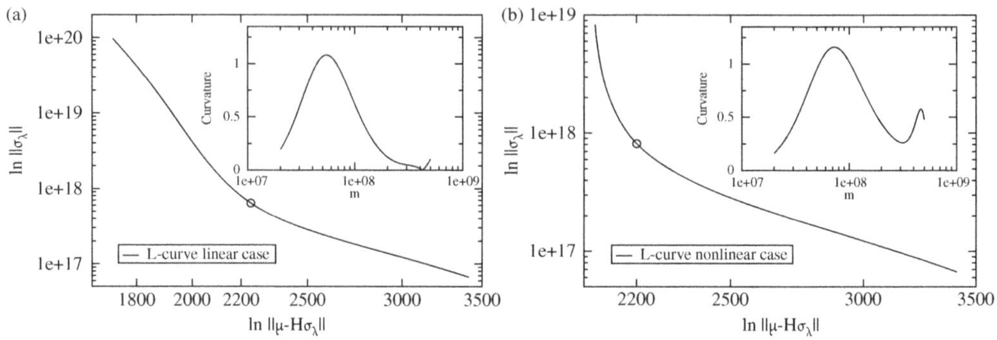

Fig. 20.10 L-curve when the $\boldsymbol{\sigma}_a$ are given by the BLUE solutions (a) and when they are given by the minimization of the cost function under a positivity constraint (b). In both cases, the curvature is plotted in the inset. The diagnosed corner is reported on the L-curve as a circle.

The nonlinear model will be denoted by $\mathcal{H}$, and its tangent linear by $\mathbf{H}$. The goal is to solve the nonlinear inverse problem

$$\boldsymbol{\mu} = \mathcal{H}(\boldsymbol{\sigma}) + \boldsymbol{\epsilon}\,, \tag{20.68}$$

with the same Gaussian assumptions on the prior errors as in (20.6) and (20.7). One approximate solution is to linearize the model around the first guess:

$$\mathcal{H}(\boldsymbol{\sigma}) \simeq \mathcal{H}(\boldsymbol{\sigma}_b) + \mathbf{H}(\boldsymbol{\sigma} - \boldsymbol{\sigma}_b)\,. \tag{20.69}$$

However, this may often be unsatisfying because in many inverse problems the prior is quite far from the truth, and the linearization is not performed at the right location, yielding a bias in the analysis.

One enlightening point of view to solve the nonlinear inverse problem is to consider its variational formulation:

$$\mathcal{J}(\boldsymbol{\sigma}) = \frac{1}{2}\left[\boldsymbol{\mu} - \mathcal{H}(\boldsymbol{\sigma})\right]^{\mathrm{T}} \mathbf{R}^{-1} \left[\boldsymbol{\mu} - \mathcal{H}(\boldsymbol{\sigma})\right] + \frac{1}{2}\left(\boldsymbol{\sigma} - \boldsymbol{\sigma}_b\right)^{\mathrm{T}} \mathbf{B}^{-1} \left(\boldsymbol{\sigma} - \boldsymbol{\sigma}_b\right). \tag{20.70}$$

This point of view connects with the considerable literature on numerical optimisation (for an accessible textbook, see Nocedal and Wright, 2006).

To minimise this cost function with high-dimensional models in mind, deterministic descent iterative methods are usually preferred. They make use of the gradient $\nabla\mathcal{J}$ of the cost function to find a proper direction of descent, or both the gradient $\nabla\mathcal{J}$ and the Hessian Hess($\mathcal{J}$). Here, we have

$$\nabla\mathcal{J} = -\mathbf{H}^{\mathrm{T}}\mathbf{R}^{-1}\left[\boldsymbol{\mu} - \mathcal{H}(\boldsymbol{\sigma})\right] + \mathbf{B}^{-1}\left(\boldsymbol{\sigma} - \boldsymbol{\sigma}_b\right) \tag{20.71}$$

$$\mathrm{Hess}(\mathcal{J}) = \mathbf{H}^{\mathrm{T}}\mathbf{R}^{-1}\mathbf{H} + \mathbf{B}^{-1}\,. \tag{20.72}$$

One of the simplest methods is the steepest-descent method that uses the gradient to define a search direction. Starting with $\boldsymbol{\sigma}^{(0)} = \boldsymbol{\sigma}_b$, it leads to the following iterations:

$$\boldsymbol{\sigma}^{(q+1)} = \boldsymbol{\sigma}^{(q)} - \theta^{(q)}\nabla\mathcal{J}^{(q)}\,. \tag{20.73}$$

The superscript (q) means that the operators are computed, possibly linearized, at $\boldsymbol{\sigma}^{(q)}$. $\theta^{(q)}$ can be determined using a line-search method, which does not usually requires one to compute the gradient but only the cost function.

When the computation of the successive Hessians is affordable, one can use the Gauss–Newton method:

$$\boldsymbol{\sigma}^{(q+1)} = \boldsymbol{\sigma}^{(q)} - \left[\text{Hess}(\mathcal{J}^{(q)})\right]^{-1} \nabla\mathcal{J}^{(q)}. \tag{20.74}$$

This translates into

$$\boldsymbol{\sigma}^{(q+1)} = \boldsymbol{\sigma}^{(q)} - \left(\mathbf{H}_{(q)}^{\mathrm{T}}\mathbf{R}^{-1}\mathbf{H}_{(q)} + \mathbf{B}^{-1}\right)^{-1} \mathbf{B}^{-1}\left(\boldsymbol{\sigma}^{(q)} - \boldsymbol{\sigma}_b\right) \tag{20.75}$$

$$+ \left(\mathbf{H}_{(q)}^{\mathrm{T}}\mathbf{R}^{-1}\mathbf{H}_{(q)} + \mathbf{B}^{-1}\right)^{-1} \mathbf{H}_{(q)}^{\mathrm{T}}\mathbf{R}^{-1}\left(\boldsymbol{\mu} - \mathcal{H}(\boldsymbol{\sigma}^{(q)})\right) \tag{20.76}$$

$$= \boldsymbol{\sigma}_b + \mathbf{K}_{(q)}\left[\boldsymbol{\mu} - \mathcal{H}(\boldsymbol{\sigma}^{(q)}) + \mathbf{H}_{(q)}\left(\boldsymbol{\sigma}^{(q)} - \boldsymbol{\sigma}_b\right)\right], \tag{20.77}$$

where

$$\mathbf{K}_{(q)} = (\mathbf{B}^{-1} + \mathbf{H}_{(q)}^{\mathrm{T}}\mathbf{R}^{-1}\mathbf{H}_{(q)})^{-1}\mathbf{H}_{(q)}^{\mathrm{T}}\mathbf{R}^{-1}. \tag{20.78}$$

This method is very efficient for mildly nonlinear problems of moderate dimensionality. Under loose and often satisfied conditions, its convergence is known to be *superlinear*. With higher-dimensional systems, one can resort to a quasi-Newton method to avoid the use of the Hessian. For very non-Gaussian problems, where multiple minima are present, there is a risk of overshooting, and the cost function is not always well approximated by its local quadratic expansion. Modern methods rather use *trust-region* methods, which have not often been used in geophysical problems.

This principle is used in the iterative ensemble Kalman filter (IEnKF), whose goal is to better account for some nonlinear effects neglected by the traditional EnKF (Wishner et al., 1969; Gu and Oliver, 2007; Sakov et al., 2012; Bocquet and Sakov, 2012).

20.4.4 Iterative regularizing methods

Finding the proper hyper-parameters (properly adjusting the prior error covariance matrices) requires one to perform an inversion for each sample value of the hyperparameters. This may be too costly for very high-dimensional problems.

On the other hand, computing such an estimate can require the use of iterative methods, whether the model is linear or nonlinear. Indeed, for very high-dimensional systems, it is expansive or difficult to compute $\mathbf{HBH}^{\mathrm{T}}$ and its SDV. Therefore, only the application of $\mathbf{H}$, and possibly its adjoint $\mathbf{H}^{\mathrm{T}}$, is affordable, so that iterative methods are used.

The idea is to relate the iteration number to some degree of regularisation. For a practitioner of inverse modelling techniques, the idea seems intuitive. However, it is only correct for a subclass of iterative methods.

Assume that the model $\mathbf{H}$ is linear and that the prior error covariance matrices $\mathbf{R}_0$ and $\mathbf{B}_0$ are unit matrices, so that the variational problem is

$$\mathcal{J}(\boldsymbol{\sigma}) = \frac{1}{2}||\boldsymbol{\mu} - \mathbf{H}\boldsymbol{\sigma}||^2 + \lambda||\boldsymbol{\sigma}||^2 . \tag{20.79}$$

where $\boldsymbol{\sigma}$ represents the deviation from the prior $\boldsymbol{\sigma} \longleftarrow \boldsymbol{\sigma} - \boldsymbol{\sigma}_b$ and $\boldsymbol{\mu}$ the innovation vector, $\boldsymbol{\mu} \longleftarrow \boldsymbol{\mu} - \mathbf{H}\boldsymbol{\sigma}_b$. $||.||$ are the canonical norms of $\mathbb{R}^p$ and $\mathbb{R}^n$.

Instead of solving multiple times for different values of λ, we would like to iteratively minimize $||\boldsymbol{\mu} - \mathbf{H}\boldsymbol{\sigma}||^2$, but stopping at a proper stage of the iterations. Iterative methods that could allow us to perform this iterative regularization are the Landweber method, the ART method, and the conjugate gradient method (Hestenes and Stiefel, 1952). If the first two have slow convergence rates, especially for high-dimensional problems, the conjugate gradient method is currently used in geophysical data assimilation, as a competitive substitute for BFGS for very large quadratic optimisation problem (such as an inner loop).

The conjugate gradient method is a Krylov subspace iterative method. At iteration q, the estimate is found in the Krylov subspace

$$\mathcal{K}^{(q)} = \left\{\mathbf{H}^{\mathrm{T}}\boldsymbol{\mu}, \left(\mathbf{H}^{\mathrm{T}}\mathbf{H}\right)\mathbf{H}^{\mathrm{T}}\boldsymbol{\mu}, \ldots, \left(\mathbf{H}^{\mathrm{T}}\mathbf{H}\right)^{q-1}\mathbf{H}^{\mathrm{T}}\boldsymbol{\mu}\right\}. \tag{20.80}$$

The search directions of the descent algorithm are orthogonal and collinear to the vectors of the orthonormal basis incrementally built within the Krylov space. One main consequence is that the residue $||\boldsymbol{\mu} - \mathbf{H}\boldsymbol{\sigma}^{(q)}||$ is monotonically decreasing while the norm of the estimate $||\boldsymbol{\sigma}^{(q)}||$ is monotonically increasing:

$$||\boldsymbol{\mu} - \mathbf{H}\boldsymbol{\sigma}^{(q+1)}|| \leq ||\boldsymbol{\mu} - \mathbf{H}\boldsymbol{\sigma}^{(q)}||, \qquad ||\boldsymbol{\sigma}^{(q+1)}|| \geq ||\boldsymbol{\sigma}^{(q)}|| . \tag{20.81}$$

Therefore it is possible to incrementally build an L-curve, with the iteration index as the parameter of the curve. When the corner is identified, the minimization can be stopped.

This application of the conjugate gradient algorithm to the least-squares problem is known as the conjugate gradient least-squares method (CGLS).

I am not aware of complex 3D geophysical data assimilation contributions using controlled iterative minimization, but this set of methods seems promising. On the downside, as with variational methods, it is not simple to optimize their parallel implementation, whereas multiple estimations with different hyper-parameter values can be trivially parallelized.

Chernobyl case study: CGLS

The iterative regularization CGLS method is applied to the Chernobyl problem. The successive iterations are plotted in Fig. 20.11. Interpreted as an L-curve, the fitting curve corner yields the optimal conjugate gradient iteration, which is between $q^\star = 3$ and $q^\star = 4$. This corresponds to a total reconstructed activity of $M = (7.69\text{–}11.0) \times 10^{16}$ Bq.

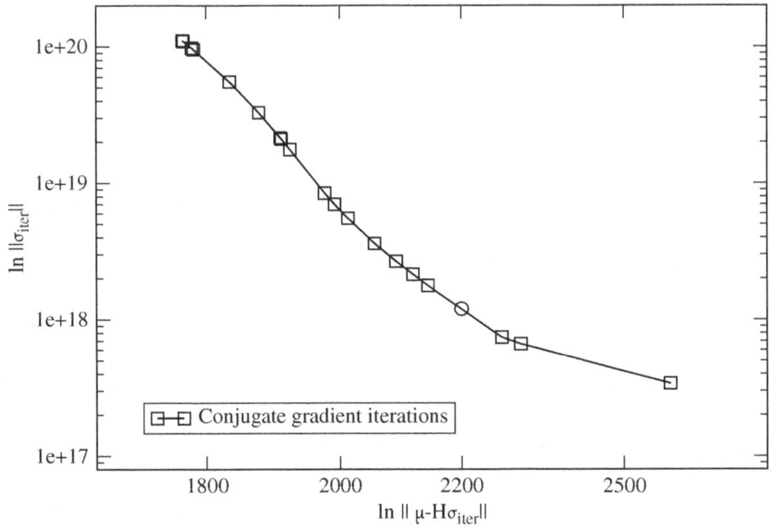

Fig. 20.11 L-curve obtained from the CGLS iterations.

20.5 Inverse modelling in atmospheric and ocean sciences: a selection

20.5.1 Linear models are not always boring

In inverse modelling studies, the linearity of the model makes a significant practical difference. Above all, it allows one to compute the Jacobian matrix $\mathbf{H}$ that relates once and for all the observation space to the control space. In atmospheric tracer studies, this trick is widely used: accidental release of atmospheric pollutant (Bophal, Chernobyl, Algeciras, Fukushima), greenhouse gases (CO_2), volcano ashes, etc.

Once the Jacobian has been computed, which may be costly, inversions are almost cost-free in comparison. A posteriori studies that, for instance, involve the computation of second-order statistics are computationally easy to perform. This cannot be so with a non-linear model, which would require an update of the Jacobian in a proper inversion.

Some studies are mildly nonlinear, so that it may be worth updating the Jacobian a limited number of times (a typical case is the inverse modelling of methane sources or any mildly nonlinear chemical species).

The Jacobian $\mathbf{H} \in \mathbb{R}^{p \times n}$ may be computed either column by column or row by row, depending whether $p \ll n$ or $n \ll p$. Computing it column by column implies using the forward model. Each column corresponds to one simulation driven by one elementary source in control space. Computing it row by row implies using the adjoint model. Each row is related to an observation and is given by a solution of the adjoint model, known as an adjoint solution, an influence function, a footprint, or a retroplume. This is useful in the case $p \ll n$. It requires either generating the exact adjoint model, or a fast and approximate adjoint model that needs to be validated.

With linear models given by their Jacobian matrix, one can safely resort to the BLUE estimator computed by matrix algebra or one can use the MEM approach with possibly non-Gaussian priors.

20.5.2 Inverse modelling of atmospheric constituents

One of the most active subjects related to the inverse modelling of parameters in atmospheric sciences is the inverse modelling of atmospheric constituent sources (Enting, 2002). The source (and possibly sinks) of chemical species is a dominant forcing of most atmospheric chemistry and climate simulations. The following are typical problems:

- estimation of greenhouse gases emission/sink fields for climate simulation (carbon monoxide, methane, etc.);
- estimation of air quality pollutant emission for sanitary and regulatory purposes—NO_x, CO, SO_2, VOC, particulate matter, dust, pollen;
- estimation of sources of accidental pollution (from the local to global scale): radionuclides (Chernobyl, Fukushima–Daiichi, etc), chemical (Bohpal), ashes (volcanoes), wildfires, etc.

20.5.3 Inverse modelling in meteorology, oceanography, and climate

In meteorology and oceanography, where the forcing fields are of less importance, the literature is not as vast. However, interest in the inverse modelling of parameters is increasing alongside the focus on model error. I mention several of those papers in the next section according to the methodology used. Ruiz et al. (2013) have reviewed the subject, with a focus on ensemble-based methods.

20.5.4 Parameter estimation: stochastic and variational methods

Interest is growing in the estimation of other parameters in the physical and chemical parameterizations, along with the technical capabilities of doing so. Several of the more or less common data assimilation techniques are being used. Here is a summary of just a limited selection of articles, excluding source and emission atmospheric chemistry papers. (Hundreds of papers deal with the inverse modelling of sources/sinks in atmospheric chemistry, cf. Chapters 21 and 22 in this volume. One solid reference is Elbern et al. (2007).)

Kalman filter, extended Kalman filter, ensemble Kalman filter, and particle filter

In sequential filtering, the parameters usually incorporate the state estimation scheme by augmenting the state vector to include those parameters. The filter gradually builds up correlations between the state variables and the parameters that are usually unobserved. This also requires an evolution model for the parameters. It often coincides with the persistence model or some simple ad hoc propagation law.

Aksoy et al. (2006) use an augmented state EnKF to optimize vertical diffusivity coefficients in the MM5 model. Vossepoel and van Leeuwen (2007) estimate eddy diffusivity coefficients in an ocean model, using a particle filter. That study also relies on augmentation of the state vector selected by likelihoods during the Bayesian update. Kondrashov et al. (2008) jointly optimize state and ocean-atmosphere coupling parameters using an extended Kalman filter. Wirth and Verron (2008) estimate friction coefficients at the bottom of the ocean for a better assessment of gravity currents, using an ensemble Kalman filter. Barbu et al. (2009) use an ensemble Kalman filter to estimate the chemical rates of conversion between SO_2 and SO_4.

From the inverse modelling standpoint, smoothers should actually be better candidates than filters. Perhaps the additional technical difficulties with smoothers have made them not the first choice of most authors.

Simulated annealing and other stochastic methods

Other computationally demanding methods that can be applied on a low-dimensional space of parameters include simulated annealing, Metropolis–Hastings, neural networks, machine learning, among others. Most of these methods are simple to implement and do not necessitate an adjoint. They are efficient in retrieving a few parameters, but are often inefficient for high-dimensional problems.

For instance, Jackson et al. (2004) optimize climate model parameters using an accelerated simulated annealing algorithm. Liu et al. (2005) constrain land surface and atmospheric flux parameters using the stochastic optimization of an objective function. Bocquet (2012) uses simulated annealing (besides 4D-Var) to rescale the vertical eddy diffusivities of a transport model used in the Chernobyl simulation.

Variational methods

Variational methods allow us to express inverse problems in simple terms. They can also be computationally efficient for large state, parameter, and observation spaces. As a smoothing method, they can handle nonlinear deviation of models. However, they do require an adjoint or some substitute for it. Pulido and Thuburn (2006) estimate atmospheric gravity-wave drag with 4D-Var, using meteorological analyses as observations. Bocquet (2012) estimates source, vertical, and horizontal diffusivities, wet scavenging and dry deposition parameters, or fields of parameters using 4D-Var in a transport model.

20.5.5 Parameter estimation: fundamental issue

When estimating a parameter of a model by inverse modelling, a fundamental issue is to know whether one is really estimating the unobserved parameter or if one is actually tuning this parameter to compensate for some other model error. Note that even in this latter case, this tuning may have some value in forecasting, even if the parameter value is unphysical. This is discussed in Bocquet (2012) and generalized to fields of parameters. In this latter case, the model of the data assimilation system becomes the combination of a deterministic model and a statistical model whose variables correspond to possibly unphysical parameters.

Chernobyl case study: estimating parameters and field of parameters

Let us give two examples of parameter field estimations for the Chernobyl dispersion accident event.

Figure. 20.12(a) shows optimal vertical diffusivity coefficients. In the first case, they are obtained with a reference transport model, so that they are presumably tuned to compensate for other sources of model error. Consistently, the K_z coefficients depart significantly from the vertical diffusion model values. In the second case, they are obtained with the reference transport model but from which several biases (source of model errors) have been removed.

Figure. 20.12(b) shows an optimal field of dry deposition velocities for caesium-137. The values that are obtained and their spatial distribution are unrealistic. We believe this approach mostly corrects representativeness errors, because the surroundings of the stations are significantly affected. However, this dry deposition correction on caesium-137 not only leads to a much better analysis in terms of statistical indicators, which is not surprising, but when used on caesium-134 and validated against non-assimilated observations, the improvement is just as spectacular. That is why it must correct for model errors that are common to the two species, such as representativeness errors.

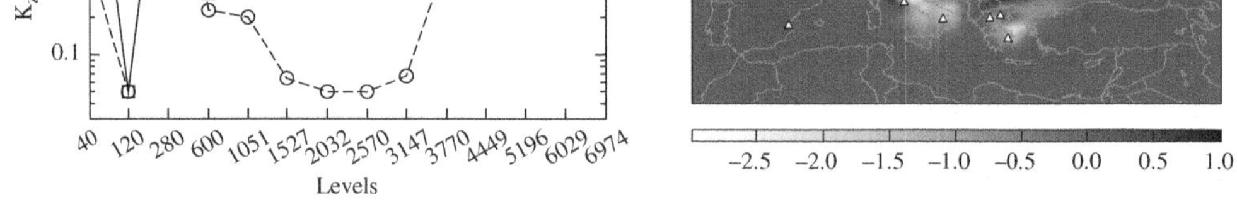

Fig. 20.12 (a) Values of the optimal daily scaling factors of the vertical diffusivity for each layer interface (1–14) of a reference transport model and for the same transport model but with the optimized removal processes and source term (the unbiased case). The terrain-following altitudes of these interfaces are indicated in the horizontal axis (in metres). (b) Map of the dry deposition velocities, optimized for caesium-137. More precisely, $\log_{10}(v^{\mathrm{dep}}/v^{\mathrm{dep}}_{\mathrm{ref}})$ is plot, where $v^{\mathrm{dep}}_{\mathrm{ref}} = 0.145\ \mathrm{cm\,s^{-1}}$ is the reference value of the dry deposition velocity. The white triangles indicate the REM stations that have measured caesium-137. The black cross indicates the location of the Chernobyl power plant.

20.6 Conclusion

In this overview of inverse modelling and parameter estimation methods, I have tried to put across a few messages.

- The inverse modelling standpoint of data assimilation used in geophysical problems is often more focussed on the retrieval of model parameters. Prior statistics on these parameters are often missing or unreliable. But the two have the same statistical Bayesian basis and they share many mathematical methods and interpretations.
- There is a huge literature on inverse problems that is often relevant to geophysical problems. Even though the standpoints of inverse problems and geophysical data assimilation problems are converging, there are still relevant methods and results of inverse modelling that are ignored, or left aside, in geophysical data assimilation.
- Because the error magnitude and the error structure are themselves often erroneous, it is often necessary to characterize the errors and estimate a few (hyper-)parameter values that constrain their statistics.
- As stressed by Albert Tarantola, inverse modelling is a quest for information, and quantitative characterization of this information (such as the relative entropy) is bound to offer clear means to design new inference schemes.
- As for geophysics, there is a fascinating and rich emerging problem/opportunity with the interplay of the characterization of model error and the inverse modelling of fields of model parameters.

Acknowledgements

Like data assimilation, inverse modelling is always an experience that has to be learnt (considering real data) through failures, mistakes, and disappointment. I am glad that I was not alone in the face of these obstacles and I wish to thank my colleagues and students Lin Wu, Mohammad Reza Koohkan, Victor Winiarek, Monika Krysta, Yelva Roustan, Olivier Saunier, Xavier Davoine, Pablo Saide, Anne Mathieu, and Bruno Sportisse.

This chapter has benefited from the useful comments and suggestions of Anna Shlyaeva, Laura Slivinski, and Giovanni Ruggiero, three active members of this Les Houches session.

References

Aksoy, A., Zhang, F., and Nielsen-Gammon, J. (2006). Ensemble-based simultaneous state and parameter estimation in a two-dimensional sea-breeze model. *Mon. Weather Rev.*, **134**, 2951–2969.

Aster, R. C., Borchers, B., and Thuber, C. H. (2006). *Parameter Estimation and Inverse Problems*. Elsevier Academic Press, Walthan, MA.

Backus, G. E. and Gilbert, J. F. (1968). The resolving power of gross earth data. *Geophys. J. R. Astron. Soc.*, **16**, 169–205.

Barbu, A. L., Segers, A. J., Schaap, M., Heemink, A. W., and Builtjes, P. J. H. (2009). A multi-component data assimilation experiment directed to sulphur dioxide and sulphate over Europe. *Atmos. Environ.*, **43**, 1622–1631.

Bennett, A. F. (1992). *Inverse Methods in Physical Oceanography.* Cambridge University Press, Cambridge.

Bennett, A. F. (2002). *Inverse Modeling of the Ocean and Atmosphere.* Cambridge University Press, Cambridge.

Bocquet, M. (2005*a*). Grid resolution dependence in the reconstruction of an atmospheric tracer source. *Nonlin. Process. Geophys.*, **12**, 219–233.

Bocquet, M. (2005*b*). Reconstruction of an atmospheric tracer source using the principle of maximum entropy. I: Theory. *Q. J. R. Meteorol. Soc.*, **131**, 2191–2208.

Bocquet, M. (2005*c*). Reconstruction of an atmospheric tracer source using the principle of maximum entropy. II: Applications. *Q. J. R. Meteorol. Soc.*, **131**, 2209–2223.

Bocquet, M. (2008). Inverse modelling of atmospheric tracers: Non-Gaussian methods and second-order sensitivity analysis. *Nonlin. Process. Geophys.*, **15**, 127–143.

Bocquet, M. (2009). Towards optimal choices of control space representation for geophysical data assimilation. *Mon. Weather Rev.*, **137**, 2331–2348.

Bocquet, M. (2012). Parameter field estimation for atmospheric dispersion: Application to the Chernobyl accident using 4D-Var. *Q. J. R. Meteorol. Soc.*, **138**, 664–681.

Bocquet, M. and Sakov, P. (2012). Combining inflation-free and iterative ensemble Kalman filters for strongly nonlinear systems. *Nonlin. Process. Geophys.*, **19**, 383–399.

Chapnik, B., Desroziers, G., Rabier, F., and Talagrand, O. (2004). Properties and first application of an error-statistics tuning method in variational assimilation. *Q. J. R. Meteorol. Soc.*, **130**, 2253–2275.

Chapnik, B., Desroziers, G., Rabier, F., and Talagrand, O. (2006). Diagnosis and tuning of observational error in a quasi-operational data assimilation setting. *Q. J. R. Meteorol. Soc.*, **132**, 543–565.

Courtier, P. (1997). Dual formulation of four-dimensional variational assimilation. *Q. J. R. Meteorol. Soc.*, **123**, 2449–2461.

Cover, T. M. and Thomas, J. A. (1991). *Elements of Information Theory.* Wiley-Interscience, New York.

Davoine, X. and Bocquet, M. (2007). Inverse modelling-based reconstruction of the Chernobyl source term available for long-range transport. *Atmos. Chem. Phys.*, **7**, 1549–1564.

Dee, D. P. (1995). On-line estimation of error covariance parameters for atmospheric data assimilation. *Mon. Weather Rev.*, **123**, 1128–1145.

Desroziers, G., Berre, L., Chapnik, B., and Poli, P. (2005). Diagnosis of observation, background and analysis-error statistics in observation space. *Q. J. R. Meteorol. Soc.*, **131**, 3385–3396.

Desroziers, G. and Ivanov, S. (2001). Diagnosis and adaptive tuning of observation-error parameters in a variational assimilation. *Q. J. R. Meteorol. Soc.*, **127**, 1433–1452.

Doicu, A., Trautmann, T., and Schreier, F. (2010). *Numerical Regularization for Atmospheric Inverse Problems*. Springer, New York.

Elbern, H., Strunk, A., Schmidt, H., and Talagrand, O. (2007). Emission rate and chemical state estimation by 4-dimensional variational inversion. *Atmos. Chem. Phys.*, **7**, 3749–3769.

Ellis, R. S. (1985). *Entropy, Large Deviations, and Statistical Mechanics*. Springer.

Engl, H. W., Hanke, M., and Neubauer, A. (2000). *Regularization of Inverse Problems*. Kluwer Academic Publishers, Dordrecht.

Enting, I. G. (2002). *Inverse Problems in Atmospheric Constituent Transport*. Cambridge University Press, Cambridge.

Groestsch, C. W. (1993). *Inverse Problems in the Mathematical Sciences*. Vieweg, Wiesbaden.

Gu, Y. and Oliver, D. S. (2007). An iterative ensemble Kalman filter for multiphase fluid flow data assimilation. *SPE J.*, **12**, 438–446.

Hansen, P. C. (1992). Analysis of discrete ill-posed problems by means of the L-curve. *SIAM Rev.*, **34**, 561–580.

Hansen, P. C. (2010). *Discrete Inverse Problems: Insight and Algorithms*. SIAM, Philadelphia.

Hestenes, M. R. and Stiefel, E. (1952). Methods of conjugate graidents for solvinf linear systems. *J. Res. Nat. Bur. Stand.*, **49**, 409–436.

Jackson, C., Sen, M. K., and Stoffa, P. L. (2004). An efficient stochastic Bayesian approach to optimal parameter and uncertainty estimation for climate model predictsions. *J. Climate*, **17**, 2828–2841.

Jaynes, E. T. (1957a). Information statistics and statistical mechanics. *Phys. Rev.*, **106**, 620–630.

Jaynes, E. T. (1957b). Information statistics and statistical mechanics ii. *Phys. Rev.*, **108**, 171–190.

Kaipio, J. and Somersalo, E. (2005). *Statistical and Computational Inverse Problems*. Springer, Berlin.

Kondrashov, D., Sun, C., and Ghil, M. (2008). Data assimilation for a coupled ocean atmosphere model. Part II: Parameter estimation. *Mon. Weather Rev.*, **5**, 5062–5076.

Krysta, M. and Bocquet, M. (2007). Source reconstruction of an accidental radionuclide release at European scale. *Q. J. R. Meteorol. Soc.*, **133**, 529–544.

Lauvernet, C., Brankart, J.-M., Castruccio, F., Broquet, G., Brasseur, P., and Verron, J. (2009). A truncated Gaussian filter for data assimilation with inequality constraints: Application to the hydrostatic stability condition in ocean models. *Ocean Model.*, **27**, 1–17.

Le Besnerais, G., Bercher, J.-F., and Demoment, G. (1999). A new look at entropy for solving linear inverse problems. *IEEE Trans. Inf. Theory*, **45**, 1565–1578.

Liu, Y., Gupta, H. V., Sorooshian, S., Bastidas, L. A., and Shuttlewort, W. J. (2005). Constraining land surface and atmospheric parameters of a locally coupled model using observational data. *J. Hydrometeorol.*, **6**, 156–172.

Lorenc, A. C. (1986). Analysis methods for numerical weather prediction. *Q. J. R. Meteorol. Soc.*, **112**, 1177–1194.

Morozov, V. A. (1966). On the solution of functional equations by the method of regularization. *Sov. Math. Dokl.*, **7**, 414–417.

Nocedal, J. and Wright, S. J. (2006). *Numerical Optimization.* Springer, Berlin.

Pulido, M. and Thuburn, J. (2006). Gravity wave drag estimation from global analyses using variational data assimilation principles. Part II: A case study. *Q. J. R. Meteorol. Soc.*, **132**, 1527–1543.

Rietsch, E. (1977). The maximum entropy approach to inverse problems: spectral analysis of short data records and density structure of the earth. *J. Geophys.*, **42**, 489–506.

Rodgers, C. D. (2000). *Inverse Methods for Atmospheric Sounding*, Vol. 2. World Scientific, Singapore.

Ruiz, J. J., Pulido, M., and Miyoshi, T. (2013). Estimating model parameters with ensemble-based data assimilation: a review. *J. Meteorol. Soc. Japan*, **91**, 79–99.

Sakov, P., Oliver, D., and Bertino, L. (2012). An iterative EnKF for strongly nonlinear systems. *Mon. Weather Rev.*, **140**, 1988–2004.

Shore, J. E. (1984). Inversion as logical inference—theory and applications of maximum entropy and minimum cross-entropy. *SIAM-AMS Proc.*, **14**, 139–149.

Talerko, N. (2005). Mesoscale modelling of radioactive contamination formation in ukraine caused by the chernobyl accident. *J. Environ. Radioactivity*, **78**, 311–329.

Tarantola, A. (1987). *Inverse Problem Theory.* Elsevier, Amsterdam.

Tarantola, A. (2005). *Inverse Problem Theory and Methods for Model Parameter Estimation.* SIAM, Philadelphia.

Tarantola, A. and Valette, B. (1982). Inverse problems = quest for information. *J. Geophys.*, **50**, 159–170.

Tikhonov, A. N. and Arsenin, V., Y. (1977). *Solutions of Ill-Posed Problems.* Winston, Washington DC.

Vogel, C. R. (2002). *Computational Methods for Inverse Problems.* SIAM, Philadelphia.

Vossepoel, F. C. and van Leeuwen, P. J. (2007). Parameter estimation using a particle method: inferring mixing coefficients from sea level observations. *Mon. Weather Rev.*, **135**, 1006–1020.

Wahba, G. (1990). *Spline Models for Observational Data.* SIAM, Philadelphia.

Winiarek, V., Bocquet, M., Saunier, O., and Mathieu, A. (2012). Estimation of errors in the inverse modeling of accidental release of atmospheric pollutant: Application to the reconstruction of the cesium-137 and iodine-131 source terms from the Fukushima Daiichi power plant. *J. Geophys. Res.*, **117**, D05122.

Wirth, A. and Verron, J. (2008). Estimation of friction parameters and laws in 1.5D shallow-water gravity currents on the f-plane, by data assimilation. *Ocean Dyn.*, **58**, 247–257.

Wishner, R. P., Tabaczynski, J. A., and Athans, M. (1969). A comparison of three non-linear filters. *Automatica*, **5**, 487–496.

Wunsch, C. (2006). *Discrete Inverse and State Estimation Problems with Geophysical Fluid Applications.* Cambridge University Press, Cambridge.

21
Greenhouse gas flux inversion

F. Chevallier

Laboratoire des Sciences du Climat et de l'Environnement
CEA-CNRS-UVSQ
Gif-sur-Yvette, France

Advanced Data Assimilation for Geosciences. First Edition.
Edited by É. Blayo, M. Bocquet, E. Cosme, and L. F. Cugliandolo.

Chapter Contents

21.1 Introduction

Measurements of the CO_2 mixing ratio in the atmosphere carry the imprint of CO_2 surface fluxes. However, reversing the sign of time to infer the latter based on the former is an ill-posed mathematical problem because both atmospheric mixing and sparse observation sampling make some of the flux information vanish: a given set of measurements is consistent with an infinite number of CO_2 flux maps, not all of which would be judged realistic by carbon experts. The underdetermination has to be lifted with some regularization constraint, i.e. by the introduction of some prior knowledge of the flux maps, in the form of flux model outputs, flux climatologies, or emission inventories, with associated uncertainties. This is expressed in the most generic form with Bayes' theorem and is implemented in *atmospheric inversion systems*. To link the observation space and the space of CO_2 surface fluxes, atmospheric inversion systems rely on a numerical model of atmospheric transport, which is the main component of the observation operator.

In the following, we successively introduce the assimilated measurements, with emphasis on remote sensing, the error statistics that the inversion systems assume, and the inversion methods.

21.2 Observations

CO_2 is a long-lived gas in the atmosphere, so its background concentrations are large compared with the gradients generated by recent surface fluxes. Monitoring such gradients requires very demanding relative accuracy that is much more challenging to meet than for shorter-lived species, such as carbon monoxide or aerosols.

The reference in situ measurements are calibrated with precision better than 0.1 part per million (ppm) (Zhao and Tans, 2006). Such measurements have been performed by a series of networks over the globe maintained by state-funded institutions[1] for decades. There is now a growing interest from private investors.[2] Not all collected data can be assimilated, because some stations (e.g. in the presence of steep orography or close to local sources) or some data acquisition periods (e.g. when the planetary boundary layer is stable) are poorly modelled by the underlying transport model and have to be excluded. Each month, the number of observations over the globe that can be assimilated is of the order of 10 000.

For remote sensing, the European Space Agency's Climate Change Initiative on Greenhouse Gases defined threshold requirements for systematic errors of 0.5 ppm ($\sim 0.13\%$) for CO_2 (GHG_CCI, 2011), which is very challenging because many significant error sources are involved in the retrieval process. As a direct consequence of this stringent requirement, there is very limited experience in inferring CO_2 sources and sinks from satellite retrievals using real data, because most efforts have stumbled on too-large retrieval biases. Two papers report failed attempts

[1] See e.g., ⟨http://www.esrl.noaa.gov/gmd/ccgg/trends/, https://icos-atc-demo.lsce.ipsl.fr/⟩.
[2] See e.g., ⟨http://earthnetworks.com⟩.

with upper-tropospheric retrievals from the TIROS Operational Vertical Sounder (TOVS) (Chevallier et al., 2005) and Atmospheric InfraRed Sounder (AIRS) (Chevallier et al., 2009). A third paper presents mixed results with mid-tropospheric retrievals from the Tropospheric Emission Spectrometer (TES) (Nassar et al., 2011): a better fit with independent measurements is seen only when TES data are used in conjunction with surface air sample measurements. An ongoing intercomparison of CO_2 inversion products from the Greenhouse Gases Observing Satellite (GOSAT) recently showed that the level of robustness of the results is currently low (Houweling et al., 2012). Despite the difficult progress of this research, the usefulness of space-borne monitoring of the CO_2 column has been experimentally confirmed using the calibrated surface measurements of the CO_2 column from the Total Carbon Column Observing Network (TCCON) (Chevallier et al., 2011).

The inversion of methane fluxes from CH_4 retrievals faces the same challenge as for CO_2. In this field, it is worth noting the approach of Bergamaschi et al. (2009), who explicitly take biases of the Scanning Imaging Absorption Spectrometer for atmospheric Chartography (SCIAMACHY) retrievals into account using a parametric (latitude-dependent) bias-correction model: they use surface air sample measurements to remove them.

Next to these studies with real data, many studies report observation system simulation experiments (OSSEs); i.e. they rely on simulations to define inversion strategies and assess the impact of the retrievals (e.g. Houweling et al., 2004; Hungershöfer et al., 2010).

21.3 Uncertainties

Inversion systems merge two types of information: the observations of CO_2 concentrations and the prior surface fluxes. Multivariate Gaussian-shaped uncertainties are usually associated to each information piece.

The observation errors are defined with respect to the inversion system and combine the measurement errors, the errors of the underlying transport model, and representation errors. The characteristics of measurement errors are known from experimental validation. Transport model errors are evaluated from comparisons of model simulations (e.g. Gurney et al., 2002). They are usually very significant, with large correlations in space and time, which motivated the creation of the Atmospheric Tracer Transport Model Intercomparison Project[3] in the 1990s to tackle this issue: this international effort is still going on. Representation errors are characterized from theoretical work (Kaminski et al., 2001; Chapter 18 of this volume) and experimental measurements of subgrid-scale mixing ratio variability (Lin et al., 2004).

The errors of the prior fluxes have been mostly assigned by expert judgement. Two studies are exceptions in that they have attempted to shed light on the characteristics of the prior errors based on observations. In the first study, Michalak et al. (2005) used CO_2 mixing ratio measurements within a flux inversion system by introducing some poorly known characteristics of the prior errors in the state vector. In the second

[3] ⟨http://transcom.project.asu.edu/index.php⟩.

study, Chevallier et al. (2006) relied on local CO_2 flux measurements to constrain the prior error distribution. They showed (i) a heavy-tail distribution that contradicts the usual assumption of a multivariate Gaussian distribution and that makes it difficult to realistically represent the prior error budget at all space–time scales simultaneously, (ii) a large temporal error structure that legitimates the inversion of surface fluxes at coarse temporal resolution (typically weekly or monthly), and (iii) a small spatial error structure that imposes inversion of surface fluxes at high spatial resolution (ideally a few tens of kilometres). Points (ii) and (iii) also indicate that, everything else being equal, spatially dense measurements, such as those provided by satellite retrievals, are much more informative for the inversion than temporally dense data, such as those provided by a surface station measuring continuously.

Since Bayes' solution to the inference problem is a probability density function, the inverse method directly quantifies the uncertainty of the flux estimate, which is an obvious advantage compared with alternative (bottom-up) methods for flux estimation (e.g. Zaehle et al., 2005; IPCC, 2000). This capability is rarely highlighted, because the Bayesian uncertainty estimates are considered as very uncertain themselves, with a tendency towards overconfidence (e.g. Tolk et al., 2011). Indeed, it is usually felt that there is not enough evidence to reliably fill the large covariance matrices that describe each input error component listed above. Therefore, Bayesian posterior errors are often complemented by the spread of sensitivity tests (e.g. Gurney et al., 2002) or are replaced by this one (e.g. Peters et al., 2007) when uncertainties are described. However, given the structure of Bayes' theorem, the realism of the inverted fluxes is tied to the realism of their Bayesian error bars: the credibility of the posterior errors challenges the credibility of the posterior fluxes.

21.4 Methods

The exceptionally long lifetime of CO_2 links any concentration measurement to global CO_2 surface fluxes in the distant past. This long memory of the measurements imposes very long inversion windows (typically months or years; see e.g. Bruhwiler et al., 2005). Combined with the high variability of atmospheric transport at short space–time scales, it hinders the implementation of Bayes' theorem at reasonable computational cost. In practice, compromises have been chosen by the various scientific teams.

Flux inversion has been traditionally performed on the basis of a closed-form (i.e. analytical) solution to Bayes' theorem (e.g., Gurney et al., 2002), but the computational cost of this analytical solution restricts it to small-size problems (typically less than 10^4 variables to infer or less than 10^4 measurements to exploit), while there is evidence that the global inference problem for CO_2 vegetation fluxes is intrinsically of much larger size because of the above-mentioned small spatial correlation structures of the prior errors and long lifetime of CO_2. In the context of high-performance parallel computing, ensemble forms of Bayes' theorem (i.e. using a statistical ensemble of simulations) have become increasingly popular (e.g. Peters et al., 2007; Feng et al., 2009), but in practice the relatively small sizes that can be currently afforded for the statistical ensembles do not allow one to tackle problems of higher resolution than the analytical approach can handle and the aggregation error remains a critical issue. Imported from the data assimilation systems of numerical weather prediction, the variational form of Bayes' theorem (i.e. adjoint-based) has been introduced over the last decade to lift this restriction (Chevallier et al., 2005; Rödenbeck, 2005). For example, Chevallier et al. (2010) performed a global 21-year inversion with a single minimization, at grid point (3.75° longitude and 2.5° latitude) and weekly scale with this approach, but they still needed six weeks of sustained parallel computing, the parallelization being for the code of the transport model.

Two types of CO_2 inversion studies do not refer to the Bayesian paradigm. The first comes from the geostatistical framework (e.g. Gourdji et al., 2010). It can actually be written in a Bayesian form, and therefore does not differ from the previous studies in this respect, except when the prior fluxes are scaled with unknown drift coefficients that are estimated together with the fluxes. With these coefficients, the inversion equations rise in complexity and one of their inputs rises in abstraction (the prior error covariance matrix becomes a matrix that characterizes how real fluxes deviate from the product of the prior fluxes and of the drift coefficients). The introduction of drift coefficients makes the uncertainty of the prior fluxes infinite (which does not reflect the reality of prior errors). The second type of non-Bayesian studies is the limiting case when the observational constraint is large enough not to necessitate any regularization: the flux can be estimated directly from mass balance considerations. This strategy was used for the Northern American carbon budget by Crevoisier et al. (2010) using in situ measurements, and for the estimation of power plant emissions by Krings et al. (2011) using retrievals from an aircraft-borne spectrometer.

The dimension of the control vector can be artificially reduced by optimizing some internal parameters of dynamical models of the carbon cycle rather than the fluxes directly (Kaminski et al., 2002; Knorr et al., 2010). Such inversion systems are usually called *carbon cycle data assimilation systems* (CCDASs). As for the standard inversion approach, CCDASs allow one to infer the space–time distribution of surface CO_2 fluxes, but their underlying process-based model allows them to use a much larger diversity of measurements, for example local flux measurements, satellite observations of vegetation activity, or biomass inventories. Their model prognostic equations can also spread the observational information well beyond measurement time (Rayner et al., 2011). The downside of this strategy is the interference from the errors of the model equations in the data assimilation process. Indeed, these equations are imposed on the inversion as a strong constraint, even though many processes cannot yet be described from first principles at the resolution of global models, in particular for vegetation, and empirical equations have to model them.

21.5 Conclusion

This short and subjective overview of the inversion field has aimed at introducing the richness of this application of data assimilation techniques. Similar *top-down* methods are being implemented for other greenhouse gases, with the additional complexity brought by the chemical reactions of these gases in the atmosphere (e.g. Bergamaschi et al., 2009; Corazza et al., 2011). These developments have been stimulated by the societal need for better source/sink accounting methods in order to (i) improve process models of the greenhouse gas fluxes (which can then be used to evaluate scenarios of the future climate) and (ii) help to mitigate climate change. Indeed, the top-down method offers promising advantages: (i) it is a *full accounting* approach that accounts for all the gas flows of the biosphere and of society; (ii) it can yield results weeks only after real time, while bottom-up approaches lag by months or years; (iii) it processes all measurements in a standardized way; (iv) it is not intrusive; (v) it directly quantifies the uncertainty of its computations; (vi) it can ingest all available information, including inventories. These advantages are balanced with two main limitations. First, the method precision is related to observation density, while measurements of the CO_2 mixing ratio are still expensive compared with those of other species routinely measured in the atmosphere. Second, the inversion involves physical (chemistry–transport) and statistical (error) models that are uncertain and, to some extent, subjective. Additionally, the sectorization of the results by emission process is needed by some main users (such as policy makers), but is not straightforward. As a consequence, these techniques are the topic of increasingly active research, with several pilot studies at local or regional scale being conducted in vegetation landscapes (Lauvaux et al., 2009, 2011) and in cities,[4] and a series of satellite missions being prepared by space agencies in the USA, in Europe, and in Asia.

References

Bergamaschi, P. et al. (2009). Inverse modeling of global and regional CH_4 emissions using SCIAMACHY satellite retrievals. *J. Geophys. Res.*, **114**, D22301.

Bruhwiler, L. M. P., Michalak, A. M., Peters, W., Baker, D. F., and Tans, P. P. (2005). An improved Kalman smoother for atmospheric inversions. *Atmos. Chem. Phys.*, **5**, 2691–2702.

Chevallier, F. et al. (2005). Inferring CO_2 sources and sinks from satellite observations: method and application to TOVS data. *J. Geophys. Res.*, **110**, D24309.

Chevallier, F., Viovy, N., Reichstein, M., and Ciais, P. P. (2006). On the assignment of prior errors in Bayesian inversions of CO_2 surface fluxes. *Geophys. Res. Lett.*, **33**, L13802.

Chevallier, F., Bréon, F.-M., and Rayner, P. J. (2007). The contribution of the Orbiting Carbon Observatory to the estimation of CO_2 sources and sinks: theoretical study in a variational data assimilation framework. *J. Geophys. Res.*, **112**, D09307.

Chevallier, F. et al. (2009). AIRS-based vs. surface-based estimation of carbon surface fluxes. *J. Geophys. Res.*, **114**, D20303.

[4] See e.g. ⟨http://influx.psu.edu/, http://co2-megaparis.lsce.ipsl.fr/, http://citycarbon.picarro.com/⟩.

Chevallier, F. et al. (2010). CO_2 surface fluxes at grid point scale estimated from a global 21-year reanalysis of atmospheric measurements. *J. Geophys. Res.*, **115**, D21307.

Chevallier, F. et al. (2011). Global CO_2 fluxes inferred from surface air-sample measurements and from TCCON retrievals of the CO_2 total column. *Geophys. Res. Lett.*, **38**, L24810.

Corazza, M. et al. (2011). Inverse modelling of European N_2O emissions: assimilating observations from different networks. *Atmos. Chem. Phys.*, **11**, 2381–2398.

Crevoisier, C., Sweeney, C., Gloor, M., Sarmiento, J. L., and Tans, P. P. (2010). Regional US carbon sinks from three-dimensional atmospheric CO_2 sampling. *Proc. Natl Acad. Sci. USA*, **107**, 18348–18353.

Feng, L., Palmer, P. I., Bösch, H., and Dance, S. (2009). Estimating surface CO_2 fluxes from space-borne CO_2 dry air mole fraction observations using an ensemble Kalman filter. *Atmos. Chem. Phys.*, **9**, 2619–2633.

GHG_CCI (2011). User Requirements Document for the Essential Climate Variable Greenhouse Gases, Version 1. Available at: ⟨http://www.esa-ghg-cci.org/sites/default/files/documents/public/documents/URDv1_GHG-CCI_final.pdf⟩.

Gourdji, S. M., Hirsch, A. I., Mueller, K. L., Andrews, A. E., and Michalak, A. M. (2010). Regional-scale geostatistical inverse modeling of North American CO_2 fluxes: a synthetic data study. *Atmos. Chem. Phys.*, **10**, 6151–6167.

Gurney, K. R., et al. (2002). Towards robust regional estimates of CO_2 sources and sinks using atmospheric transport models. *Nature*, **415**, 626–630.

Houweling, S., Bréon, F.-M., Aben, I., Rödenbeck, C., Gloor, M., Heimann, M. and Ciais, P. (2004). Inverse modeling of CO_2 sources and sinks using satellite data: a synthetic inter-comparison of measurement techniques and their performance as a function of space and time. *Atmos. Chem. Phys.*, **4**, 523–538.

Houweling, S. et al. (2012). Global CO_2 fluxes from GOSAT: first results from an inter-comparison of inverse models. *Geophys. Res. Abst.*, **14**, EGU2012-12978, 2012.

Hungershöfer, K., Bréon, F.-M., Peylin, P., Chevallier, F., Rayner, P., Klonecki, A., and Houweling, S. (2010). Evaluation of various observing systems for the global monitoring of CO_2 surface fluxes. *Atmos. Chem. Phys.*, **10**, 10503–10520.

IPCC (2000). Good Practice Guidance and Uncertainty Management in National Greenhouse Gas Inventories. Available at: ⟨http://www.ipcc-nggip.iges.or.jp/public/gp/english/⟩.

Kaminski, T., Rayner, P. J., Heimann, M., and Enting, I. G. (2001). On aggregation errors in atmospheric transport inversions. *J. Geophys. Res.*, **106**, 4703–4715.

Kaminski, T. et al. (2002). Assimilating atmospheric data into a terrestrial biosphere model: A case study of the seasonal cycle. *Global Biogeochem. Cycles*, **16**, 4.

Knorr, W. et al. (2010). Carbon cycle data assimilation with a generic phenology model. *J. Geophys. Res.*, **115**, G04017.

Krings, T. et al. (2011). MAMAP—a new spectrometer system for column-averaged methane and carbon dioxide observations from aircraft: retrieval algorithm and first inversions for point source emission rates. *Atmos. Meas. Tech.*, **4**, 1735–1758.

Lauvaux, T. et al. (2009). Bridging the gap between atmospheric concentrations and local ecosystem measurements. *Geophys. Res. Lett.*, **36**, L19809.

Lauvaux, T. et al. (2011). Constraining the CO_2 budget of the corn belt: exploring uncertainties from the assumptions in a mesoscale inverse system. *Atmos. Chem. Phys.*, **12**, 337–354.

Lin, J. C. et al. (2004). An empirical analysis of the spatial variability of atmospheric CO_2: implications for inverse analyses and space-borne sensors. *Geophys. Res. Lett.*, **31**, L23104.

Michalak, A. M., Hirsch, A., Bruhwiler, L., Gurney, K. R., Peters, W., and Tans, P. P. (2005). Maximum likelihood estimation of covariance parameters for Bayesian atmospheric trace gas surface flux inversions. *J. Geophys. Res.*, **110**, D24107.

Nassar, R. et al. (2011). Inverse modeling of CO_2 sources and sinks using satellite observations of CO_2 from TES and surface flask measurements. *Atmos. Chem. Phys.*, **11**, 6029–6047.

Peters, W. et al. (2007), An atmospheric perspective on North American carbon dioxide exchange: Carbon Tracker. *Proc. Natl Acad. Sci. USA*, **48**, 18925–18930.

Rayner, P. J., Koffi, E., Scholze, M., Kaminski, T., and Dufresne, J.-L. (2011). Constraining predictions of the carbon cycle using data. *Phil. Trans. R. Soc. Lond. A*, **369**, 1955–1966.

Rödenbeck, C. (2005). Estimating CO_2 sources and sinks from atmospheric mixing ratio measurements using a global inversion of atmospheric transport. Available at: ⟨http://www.bgc-jena.mpg.de/mpg/websiteBiogeochemie/Publikationen/Technical_Reports/tech_report6.pdf⟩.

Tolk, L. F., Dolman, A. J., Meesters, A., and Peters, W. (2011). A comparison of different inverse carbon flux estimation approaches for application on a regional domain. *Atmos. Chem. Phys.*, **11**, 10349–10365.

Zaehle, S., Sitch, S., Smith, B., and Hatterman, F. (2005). Effects of parameter uncertainties on the modeling of terrestrial biosphere dynamics. *Global Biogeochem. Cycles*, **19**, GB3020.

Zhao, C. and Tans, P. P. (2006). Estimating uncertainty of the WMO mole fraction scale for carbon dioxide in air. *J. Geophys. Res.*, **111**, D08S09.

22

Data assimilation in atmospheric chemistry and air quality

H. ELBERN[1,2], E. FRIESE[1], L. NIERADZIK[1], and J. SCHWINGER[1]

[1]Rhenish Institute for Environmental Research at the University of Cologne (RIU) Cologne, Germany

[2]Institute for Energy and Climate Research-8, Research Centre Jülich Jülich, Germany

Advanced Data Assimilation for Geosciences. First Edition.
Edited by É. Blayo, M. Bocquet, E. Cosme, and L. F. Cugliandolo.

Chapter Contents

22.1 Introduction

The assimilation of trace gas observations in atmospheric chemistry models dates back as early as the 1960s, when sulphur concentrations were estimated by a variational technique. Whereas in the early days, kriging methods were applied, simply to estimate surface concentrations from a network of in situ observations, chemical data assimilation later evolved to provide chemical state estimates in three realms: reacting tropospheric chemistry, with a focus on air quality in the boundary layer; reacting stratosphere chemistry, with a focus on ozone hole mechanisms; and source–sink inversion of greenhouse gases. While this lecture places emphasis on the first two–areas, the third combines techniques of data assimilation and inverse modelling (see Chapter 21).

Early attempts to analyse tracer fields were based on monovariate kriging techniques of surface concentrations (e. g. Fedorov, 1998) and chemical state estimates, frequently referred to as analyses. There is no significant distinction between chemistry data assimilation and meteorological data assimilation, when only purely spatial techniques are applied. This is in contrast to tropospheric chemistry spatio-*temporal* data assimilation, where notable differences occur. By attempting to combine observations of different times, intermittently applied spatial data assimilation procedures cannot make use of the known physical and chemical laws as a very useful constraint. The ability to do so would not only enlarge the observational data base by measurements collected over a full time interval, but also enforce some chemical consistency, dependent on the model design.

From a theoretical viewpoint, only advanced spatio-temporal data assimilation or inversion techniques are candidates for a viable solution: they are able to combine model information with data in a chemically and dynamically consistent way, while at the same time, being able to provide a best linear unbiased estimate (BLUE). One underlying assumption of a BLUE is the validity of Gaussian probability density functions. Given that all requested theoretical assumptions are satisfied, the BLUE property is provided by the four-dimensional variational data assimilation technique and the Kalman filter, including appropriate variants thereof.

22.1.1 Advanced chemical data assimilation

With the advent of spaceborne sensors for Earth obserbation, chemical data assimilation has rapidly evolved in recent years. A broad survey of examples is given by Lahoz and Errera (2010). An application of the 4D-Var technique with a heavily reduced stratospheric reaction mechanism in connection with a trajectory model has been presented by Fisher and Lary (1995). This was the first study to consider the assimilation of chemically active stratospheric constituents.

Errera and Fonteyn (2001) presented the first application of the 4D-Var method in the stratosphere with a comprehensive stratospheric chemistry–transport model (CTM), now the Belgian Assimilation System of Chemical Observations from ENVISAT (BASCOE; see also Errera et al. (2008) for further developments).

22.1.2 Tropospheric applications

Tropospheric data assimilation is distinct not only from meteorological data assimilation but also from stratospheric data assimilation, since not only the initial values but also the emission rates are of like importance for the simulation quality. The fact that they are not sufficiently well known makes them an additional candidate for optimization by observations. Hence, we have a coupled data assimilation and inversion problem, which must be addressed.

Within the scope of an identical-twin set-up, a first implementation of the 4D-variational technique for emission optimization including reactive chemistry is described in Elbern et al. (2000). A real-world application with a fully fledged CTM, the EURAD model, is given in Elbern and Schmidt (2001). Including all emitted species at each surface grid point, the typical optimization space of initial values by atmospheric chemical state constituents is replaced by a scaled emission rate space in Elbern et al. (2007). On the microscale, a practical application has been presented by Quélo et al. (2005) for NO_x emissions and their diurnal profile, using the Polair3D model. Another regional tropospheric 4D-Var assimilation system is STEM–2K1 (Chai et al., 2006). Dedicated flight missions also serve for spatio-temporal data assimilation. The measurements obtained during the NASA Transport and Chemical Evolution over the Pacific (TRACE-P) airborne mission (see e.g. Talbot et al., 2003) were the first to be assimilated by chemical 4D-Var. Chai et al. (2006) made use of the limited-area Sulphur Transport Eulerian Model, version 2K1 (STEM-2K1) and its adjoint. The background error covariance matrix is simplified to a diagonal formulation. To contrasts its favourable results with a study without data assimilation, see Carmichael et al. (2003) for a TRACE-P-related model installation and performance. Comparison of model simulations of the TRACE–P mission without and with advanced data assimilation revealed numerous interesting features. It was found that assimilating ozone observations from one of two independent flights improves model prediction of the other flight by ozone measurements, which are withheld for validation. Specifically, the adjusted initial fields after only assimilating NO_y observations lead to better predictions of NO, NO_2, and PAN, based on their agreement with the withheld measurements. In addition, the model predictions of NO_y improve significantly after assimilating the aforementioned multiple observation species, which are independent of the withheld NO_y measurements.

By using of the variational inversion technique on a global scale, Müller and Stavrakou (2005) assimilated tropospheric column retrievals of CO and NO_2 to assess bulk emission rates of continental scales.

Recently, Vira and Sofiev (2012) applied the variational approach to estimate sulphur emissions for Mount Etna.

The reduced-rank square-root Kalman filter of the Long Term Ozone Simulation (LOTOS) model (van Loon et al., 2000) and the EUROS model (Hanea et al., 2004) were furnished with sophisticated implementations of complexity-reduced Kalman filters. This approach was selected to factorize covariance matrices by a few principal components (Verlaan and Heemink, 1995). Combination with an ensemble Kalman filter method resulted in additional capability (Hanea et al., 2004). Here, optimization parameters include emission rates, photolysis rates, and deposition rates, the correction quantities of which are formally introduced as 'noise' parameters in the Kalman filter formulation.

Further recent applications of the ensemble Kalman filter are due to Constantinescu et al. (2006a, b). The set-up of the correlation matrix is not left to the ensemble directly, but is constructed by autoregressive modelling. Still, special measures proved necessary to avoid 'ensemble collapse' effects with 'filter divergence'—an effect due to the insufficient spread of the ensemble members and missing model error covariance formulation.

Research on the solution of inversion problems aiming at source and sink estimates has become well established over the last few decades (see Chapter 21). In most cases, inversion with respect to quasi-passive tracers has been performed. Newsam and Enting (1988) and Enting and Newsam (1990) addressed the global problem of the distribution of sources and sinks of carbon dioxide by the inversion of a diffusion equation, formally solved by associated Legendre functions.

In the sequel to this work, a variety of other studies were made, all based on a very limited number of flask measurements (Bousquet et al., 1999a,b; Enting et al., 1995; Fan et al., 1998; Gloor et al., 1999; Gurney, 2002). The variational approach too has been adopted for source and sink estimates aiming to contribute to better specification of greenhouse gas budgets (Kaminski et al., 1999a,b; Houweling et al., 1999).

Optimizing model parameters, Kaminski et al. (2002) assimilated 41 CO_2 measurement data in a simplified terrestrial biosphere model by the 4D-Var technique, achieving more realistic flux simulations. Attempting to overcome the limitations of CO_2 in situ observations, satellite data from the Atmospheric Infrared Sounder (AIRS) have been assimilated into the ECMWF model using the 4D-Var technique by Engelen et al. (2004). As the results were satisfying only in tropical regions, improved global source and sink estimates cannot be expected with the present data base and assimilation system configuration.

Robertson and Langner (1992) used variational data assimilation for source estimation in the frame of the ETEX experiment on the mesoscale. Issartel (2003) applied the concept of retroplumes for source identification and estimates within the framework of 'illumination'. A sophisticated and trail-blazing approach has been selected by Bocquet (2005a,b), where the maximum-entropy principle has been invoked to estimate position, time, and strengths of emission sources.

All emission source studies cited above remained focused on source or sink estimates of a single passive tracer, which is observed in some way, without modelling reactive chemistry. Only a few attempts have been made to address the general feasibility of solving the source inversion problem for reactive chemistry, ideally estimating precursor sources by observational data from product pollutants. With chemically reactive emittands to be estimated, merging of data assimilation and inverse modlling will proceed in the future.

22.1.3 Tropospheric aerosol data assimilation

The attempt to combine state-of-the-art spatio-temporal data assimilation methods with state-of-the-art aerosol modules, combined with advanced satellite retrieval methods, is challenging. All these lines of development have evolved separately, and aerosol assimilation studies so far have focused on one aspect or the other, but never on both together. Collins et al. (2001) applied the MATCH model, in which sulphate, black carbon, organic carbon, and mineral dust are predicted, while sea salt aerosols are diagnosed by optimal interpolation. Aerosol optical depth (AOD) retrieved from NOAA AVHRR over oceans is the target analysis. Studies on a more fundamental basis on the feasibility of variational aerosol data assimilation have been presented by Henze et al. (2004) and Sander et al. (2005). Whereas in Collins et al. (2001), emphasis is placed on modelling of a more sophisticated aerosol optical depth by state-of-the-art modules, the OI assimilation scheme applied here provides for the BLUE property only for a time instant, without coupling with other species at that time. Data assimilation has also been extended to inverse modelling of biomass burning emissions. Zhang et al. (2005) applied a Bayesian inversion technique to find special sensitivity of the results to a priori emissions and to the altitude of the aerosol layer. In a 4D-Var context, Benedetti and Fisher (2007) introduced the NMC method to assess the background error statistics of global aerosol distribution for Earth system monitoring, using satellite aerosol retrievals and in situ data. Similarly motivated, Zhang et al. (2008) developed an operational variational AOD data assimilation system. Yumimoto et al. (2008) applied a 4D-Var data assimilation system for a regional dust model RAMS/CFORS-4DVAR to an adjoint inversion of a heavy dust event over eastern Asia, where the vertical profiles of the dust extinction coefficients derived from a Lidar network are directly incorporated. The authors could demonstrate significant improvements for dust emission inversion.

An overview of the emerging field of advanced chemical data assimilation is provided by Carmichael et al. (2007).

22.1.4 Background error covariance modelling

Apart from moisture patterns, in meteorological data assimilation, spatial correlations are mainly determined by typically smooth pressure and flow fields. In contrast, tropospheric chemistry data assimilation has to deal with confined urban plumes with different chemical scenarios than these close to rural air masses on a local spatial scale. Given typically sparse observation networks, best use must be made by data assimilation techniques to analyse fields completely. As optimal analyses require a realistic representation of error statistics, the treatment of the background or forecast error covariance matrix (BECM) constitutes a core task in designing an assimilation system. A properly specified BECM not only balances the forecast or background error with respect to observation errors, but also guides the spreading of measurement information given a statistically well-estimated influence or decorrelation length. Using known correlations, the BECM can therefore exploit the information contents of a retrieval or observation as thoroughly as possible. The treatment of the BECM involves two independent general problems. First, an algorithmic formulation must be found to model an extremely high-dimensional matrix that is generally too big to be represented explicitly. Second, statistically useful entries must be inferred in some suitable way.

In the past, mostly either only variances were considered (diagonal BECM, Errera et al., 2008) or the specification of covariances in chemical data assimilation mainly rested on assumptions of homogeneity (constant horizontal correlation lengths all over the globe) and isotropy (constant horizontal correlation lengths in each direction). Despite the necessary simplicity, a skilful parameterization should be capable of representing the relevant structures of the background error covariances. This includes the possibility of modelling inhomogeneous and anisotropic correlation lengthscales.

In meteorological data assimilation, the formulation of the BECM has attracted much research (for a comprehensive survey, see Bannister, 2008a,b). An early attempt to move to anisotropic and inhomogeneous background errors was presented by Thiebaux (1976) using a suitably defined autoregressive scheme. Thépaut et al. (1996) and Otte et al. (2001) further demonstrated the need to relax the constraint of isotropy. A latitudinally dependent correlation function was defined by Wu et al. (2002) through recursive filtering. Purser et al. (2003) generalized the approach to variable anisotropy and inhomogeneity adaptive to geographic location. Weaver and Courtier (2001) applied a diffusion method, providing the same statistical properties as the recursive filtering. While the former authors claimed higher numerical efficiency, the work of the latter showed promise to better account for abrupt correlation changes, such as those at fronts and other air mass boundaries, owing to the local control by diffusion coefficients.

For tropospheric chemical data assimilation, Hölzemann et al. (2001) introduced inhomogeneous BECMs, to account for urban-to-rural chemical regime changes of the boundary layer. On the global scale, Segers et al. (2005) applied a complexity reduced global stratospheric Kalman-filter system with anisotropic covariance formulation by a parameterization of correlations. MIMOSA (Modèle Isentrope de transport Méso échelle de Ozone Stratosphérique par Advection), a horizontally high-resolution transport model with 16 isentropic levels, demonstrated preservation of fine-scale structures in the analysed ozone field (Fierli et al., 2002). Background error correlations were flow-dependent and anisotropic, specified in terms of distance and by PV field.

22.2 Advanced chemistry data assimilation

For a gas-phase chemistry–transport model, the differential equation can be written as

$$\frac{\partial c_i}{\partial t} + \nabla \cdot (\vec{v} c_i) - \nabla \cdot \left(\rho \mathbf{G} \nabla \frac{c_i}{\rho} \right) - \sum_{r=1}^{R} \left\{ k(r) \left[s_i(r_+) - s_i(r_-) \right] \prod_{j=1}^{U} c_j^{s_j(r_-)} \right\} = E_i + D_i \tag{22.1}$$

where c_i is the concentration of species i, $\vec{v}$ is the wind velocity, $s \in \mathbb{N}_0$ is the stoichiometric coefficient, $k(r)$ is the reaction rate of reaction r, being either productive (r_+) or destructive (r_-) for species i, U is the number of species in the mechanism, E_i is the emission rate of species i, D_i is the deposition rate of species i, ρ is the air density, and $\mathbf{G}$ is the symmetric eddy diffusivity tensor.

As an example, in EURAD–IM, a comprehensive tropospheric Eulerian model operating on continental to local scale, the chemistry–transport model calculates the transport, diffusion, and gas-phase transformations of about 60 chemical species with

158 reactions. Today's advanced chemistry–transport models include comprehensive aerosol schemes, which simulate numerous additional physical chemical processes with various types and sizes of aerosols. See, for example, the Modal Aerosol Dynamics Model for Europe (MADE/SORGAM), the Model for Simulating Aerosol Interactions and Chemistry (MOSAIC, with 4 or 8 bins) sectional model aerosol parameterization, or the GOCART aerosol model (see e.g. Ackermann et al., 1998; Schell et al., 2001). The treatment of these components is addressed in Section 22.5. Mineral dust and sea salt simulations are typically added by separate aerosol modules.

Chemical mechanisms as part of (22.1) are described by solvers of stiff ordinary differential equations. Therefore, assimilation of reactive chemical constituents, most prominently those that are in photochemical equilibrium, generally perturbs the chemical balance, and the chemical state after ensuing relaxation is beyond the control of purely spatial data assimilation techniques. Hence, data assimilation techniques that include the full dynamics should be selected and full models introduced as constraints. This is why our further discussion will focus on advanced spatio-temporal chemical data assimilation, represented by complexity-reduced Kalman filtering and the 4D-Var technique.

22.2.1 Kalman-filter equations

Kalman filtering is detailed in Chapter 3, and the standard formulas are repeated here only for later reference.

The forecast equation propagating the model state $\mathbf{x}^f$ from time $i-1$ to i by the model resolvent or integration operator $\mathbf{M}(t_i, t_{i-1})$ reads

$$\mathbf{x}^f(t_i) = \mathbf{M}(t_i, t_{i-1})\mathbf{x}^a(t_{i-1}) + \eta, \tag{22.2}$$

where indices a and f indicate analysis and forecast, respectively. The observation error is η.

The forecast error covariance matrix $\mathbf{P}_i^f$ at time step i reads

$$\mathbf{P}_i^f = \mathbf{M}(t_i, t_{i-1})\mathbf{P}_i^a\mathbf{M}^{\mathrm{T}}(t_i, t_{i-1}) + \mathbf{Q}, \tag{22.3}$$

with analysis error covariance matrix $\mathbf{P}_i^a$ and model error covariance matrix $\mathbf{Q}$. The innovation, which is the difference between the vector of observations $\mathbf{y}_i$ and the forecasted model equivalent obtained by the observation operator $\mathbf{H}$ applied to the forecasted state $\mathbf{x}^f(t_i)$, is weighted by the Kalman gain matrix

$$\mathbf{K}_i := \mathbf{P}_i^b\mathbf{H}_i^{\mathrm{T}}(\mathbf{H}_i\mathbf{P}_i^b\mathbf{H}_i^{\mathrm{T}} + \mathbf{R}_i)^{-1}, \tag{22.4}$$

resulting in the analysis increment

$$\mathbf{x}^a(t_i) - \mathbf{x}^f(t_i) = \mathbf{K}_i[\mathbf{y}_i - \mathbf{H}\ \mathbf{x}^f(t_i)]. \tag{22.5}$$

The analysis error covariance matrix $\mathbf{P}_i^a$ is given by

$$\mathbf{P}_i^a = (\mathbf{I} - \mathbf{K}_i\mathbf{H})\mathbf{P}_i^b. \tag{22.6}$$

The above Kalman-filter equations, particularly (22.2), are computationally too costly, and therefore the ensemble Kalman filter (EnKF) and the reduced-rank square-root Kalman filter (RRSRKF), which allow for a feasible spatio–temporal data assimilation approach while approximating the BLUE property, are used. Hanea et al. (2004) successfully implemented both methods for tropospheric chemistry data assimilation, and Constantinescu et al. (2007a,b) the EaKF.

For the sake of conciseness, only the RRSRKF is considered here, with a discussion of its application in atmospheric chemistry.

22.2.2 Reduced-rank square-root Kalman filter

Here, the approach of Hanea et al. (2004) is chosen. This technique approximates the $n \times n$ covariance matrices $\mathbf{P}^{f,a}$ by a product of suitably selected $n \times q$, $q << n$ low-rank matrices $\mathbf{S}^{f,a}$. The q leading eigenvectors are the basis for the determination of $\mathbf{S}$. A key question is the selection of parameters, which control the computation of eigenvectors. The choice must be made according to a double criterion: paucity of knowledge of the parameter and its impact on the evolution of the model. As mentioned above, the parameters are not necessarily the initial values alone—emission rates and deposition velocities may also have a similar impact. These should be included in the calculation of the model eigenvectors. With the eigenvector decomposition $\mathbf{P}^{f,a} = \mathbf{V}^{f,a}(\mathbf{D}^{f,a})(\mathbf{V}^{f,a})^{\mathrm{T}}$, where $\mathbf{D}$ denotes the diagonal matrix of eigenvalues, the matrix $\mathbf{S}^{f,a} = \mathbf{V}^{f,a}(\mathbf{D}^{f,a})^{1/2}$ complies with the requirement. The system noise matrix $\mathbf{Q}$ is treated by a similar procedure, and factorized by an $n \times r$ $\mathbf{T}$, with $r << n$. Hence, we have

$$\mathbf{P} \approx \mathbf{S}\mathbf{S}^{\mathrm{T}}, \qquad \mathbf{Q} \approx \mathbf{T}\mathbf{T}^{\mathrm{T}}. \tag{22.7}$$

Calculation of the forecast error covariance matrix rests on $2 \times p$ model integrations only:

$$\mathbf{S}^f\mathbf{S}^{f\mathrm{T}} = \mathbf{M}\mathbf{S}^a\mathbf{S}^{a\mathrm{T}}\mathbf{M}^{\mathrm{T}} + \mathbf{T}\mathbf{T}^{\mathrm{T}}. \tag{22.8}$$

Including model errors, we have $\mathbf{S}^f = [\mathbf{M}\mathbf{S}^a, \mathbf{T}]$, and, with the definition $\Psi := \mathbf{H}\ \mathbf{S}^f$, the Kalman gain matrix reads

$$\mathbf{K} \ = \mathbf{S}^f\Psi^{\mathrm{T}}(\Psi\Psi^{\mathrm{T}} + \mathbf{R})^{-1}.$$

The analysis error covariance matrix can be written as

$$\begin{aligned} \mathbf{S}\ ^a\mathbf{S}^{a\mathrm{T}} &= (\mathbf{I} - \mathbf{K}\ \mathbf{H})\mathbf{S}^f\mathbf{S}^{f\mathrm{T}} \\ &= \mathbf{S}^f\left[\mathbf{I} - \Psi^{\mathrm{T}}(\Psi\Psi^{\mathrm{T}} + \mathbf{R})^{-1}\psi\right]\mathbf{S}^{f\mathrm{T}}. \end{aligned} \tag{22.9}$$

All calculations can be performed without actually calculating the full matrices P; only a square-root representation is needed, and is supported by the algorithm. The positive-semidefinite property of covariance matrices is used. This is implemented in the square-root form

$$\mathbf{S}^a = \mathbf{S}^f\left[\mathbf{I} - \Psi^{\mathrm{T}}(\Psi\Psi^{\mathrm{T}} + \mathbf{R})^{-1}\psi\right]^{1/2}. \tag{22.10}$$

Methods exist to realize the measurement updates with special procedures. Hanea et al. (2004) adopted a *scalar update* formalism, where each measurement is processed individually. Defining $\gamma := (\Psi^{\mathrm{T}}\Psi + \mathbf{R})^{-1}$ for each time step, the latter formula can be rearranged to read

$$\mathbf{S}^a = \mathbf{S}^f - \mathbf{K}\Psi[1 + (\gamma\mathbf{R})^{1/2}]^{-1}. \tag{22.11}$$

22.2.3 4D variational data assimilation

An example of 4D-Var in tropospheric chemistry is the work of Elbern et al. (2007), with the EURAD–IM (European Air Pollution Dispersion–Inverse Model).

In this implementation, the inversion for emission rate optimization is included, extending also to unobserved species. Deviations of the background chemical state $\mathbf{x}(t_0) - \mathbf{x}_b = \delta\mathbf{x}(t_0)$ and the emission inventory $\mathbf{e}(t_0) - \mathbf{e}_b = \delta\mathbf{e}(t_0)$ may be combined to define an incremental formulation of an extended cost function J as follows (for a more detailed description, see e.g. Elbern et al., 2000):

$$\begin{aligned} J(\delta\mathbf{x}(t_0), \delta\mathbf{e}) &= \frac{1}{2}(\delta\mathbf{x})^{\mathrm{T}}\mathbf{B}^{-1}\delta\mathbf{x} + \frac{1}{2}\int_{t_0}^{t_N} (\delta\mathbf{e})^{\mathrm{T}}\mathbf{K}^{-1}\delta\mathbf{e}\,\mathrm{d}t \\ &+ \frac{1}{2}\int_{t_0}^{t_N} [\mathbf{d}(t) - \mathbf{H}(t)\delta\mathbf{x}(t)]^{\mathrm{T}}\mathbf{R}^{-1}[\mathbf{d}(t) - \mathbf{H}(t)\delta\mathbf{x}(t)]\,\mathrm{d}t. \end{aligned} \tag{22.12}$$

J is a scalar functional defined on the time interval $t_0 \leq t \leq t_N$ and dependent on the vector-valued state variable $\mathbf{x}(t)$. $\mathbf{d}(t) := \mathbf{y}(t) - \mathbf{H}(t)\delta\ \mathbf{x}_b(t)$ is the observation minus model discrepancy at time t, when first-guess initial values and emission inventory values are taken. The background and emission error covariance matrices are defined as follows: for the first guess or background values, $\mathbf{B}$ is $N \times N$ with N the number of model variables; for the emission factors, $\mathbf{K}$ is $E \times E$, with E the number of emitting grid points times emitted species; and for the observation errors the matrix is $M(t) \times M(t)$, with $M(t)$ the number of available observations at time t. The operator $\mathbf{H}(t)$ calculates the model equivalent to each observation.

The gradient of the cost function J with respect to the joint chemical state and emission rate variable $\mathbf{z} = (\delta\mathbf{x}, \delta\mathbf{e})^{\mathrm{T}}$ is

$$\begin{aligned} \frac{\partial J}{\partial(\delta\mathbf{x}, \delta\mathbf{e})^{\mathrm{T}}} &= -\mathbf{B}^{-1}(\delta\mathbf{x}(t)) \\ &- \sum_{t_0}^{t_N} \mathbf{M}^{\mathrm{T}}(t_0, t)\mathbf{H}^{\mathrm{T}}(t)\mathbf{R}^{-1}[\mathbf{d} - \mathbf{H}(t)\delta\mathbf{x}(t)] \\ &- \sum_{t_0}^{t_N} \mathbf{K}^{-1}[\mathbf{e}_b(t) - \mathbf{e}(t)], \end{aligned} \tag{22.13}$$

where $\mathbf{M}^{\mathrm{T}}(t_0, t)$ denotes the adjoint (= transpose) model operator, formally integration from time t backward in time to the initial time t_0.

The development of the adjoint chemistry–transport model $\mathbf{M}^{\mathrm{T}}(t_0, t)$ is one of the major challenges in 4D-Var. In its differential equation form, starting from (22.1), the adjoint model equations then read, after application of the calculus of variations,

$$-\frac{\partial \delta c_i^*}{\partial t} - \vec{v}\nabla\delta c_i^* - \frac{1}{\rho}\nabla\cdot(\rho\mathbf{K}\nabla\delta c_i^*)$$
$$+\sum_{r=1}^{R}\left\{k(r)\frac{s_i(r_-)}{c_i}\prod_{j=1}^{U}c_j{}^{s_j(r_-)}\sum_{n=1}^{U}[s_n(r_+) - s_n(r_-)]\delta c_n^*\right\} = 0, \tag{22.14}$$

with δc_i^* being the adjoint variable of c_i, while D_i is held fixed.

The associated adjoint operators include the gas-phase mechanism, the transport schemes, and an implicit vertical diffusion scheme. The emission data in this study are taken from EMEP (a cooperative programme for monitoring and evaluation of the long-range transmission of air pollutants in Europe).

With square-root factorizations $\mathbf{B} = \mathbf{B}^{1/2}\mathbf{B}^{\mathrm{T}/2}$ and $\mathbf{K} = \mathbf{K}^{1/2}\mathbf{K}^{\mathrm{T}/2}$, we define new variables $\mathbf{v}$ and $\mathbf{w}$ by

$$\mathbf{v} := \mathbf{B}^{-1/2}\delta\mathbf{x}, \qquad \mathbf{w} := \mathbf{K}^{-1/2}\delta\mathbf{u}, \tag{22.15}$$

leading to a minimization problem equivalent to (22.12).

The cost function then reads

$$J(\mathbf{v},\mathbf{w}) = \frac{1}{2}\mathbf{v}^{\mathrm{T}}\mathbf{v} + \frac{1}{2}\mathbf{w}^{\mathrm{T}}\mathbf{w} + \frac{1}{2}\sum_{i=0}^{T}(\mathbf{d}_i - \mathbf{H}\delta\mathbf{x}_i)^{\mathrm{T}}\ \mathbf{R}^{-1}\ (\mathbf{d}_i - \mathbf{H}\delta\mathbf{x}_i). \tag{22.16}$$

The gradient of J with respect to $(\mathbf{v},\mathbf{w})^{\mathrm{T}}$ can be shown to be

$$\nabla_{(\mathbf{v},\mathbf{w})^{\mathrm{T}}}J = -\begin{pmatrix}\mathbf{v}\\ \mathbf{w}\end{pmatrix} - \begin{pmatrix}\mathbf{B}^{1/2}\mathbf{0}\\ \mathbf{0}\mathbf{K}^{1/2}\end{pmatrix}$$
$$\times\sum_{m=0}^{T}\mathbf{M}^{\mathrm{T}}(t_0, t_m)\mathbf{H}^{\mathrm{T}}\mathbf{R}^{-1}[\mathbf{d}(t_m) - \mathbf{H}\delta\mathbf{x}(t_m)], \tag{22.17}$$

The minimization problem is typically solved by some quasi-Newton minimization procedure, such as L–BFGS.

The variational chemistry data assimilation algorithm is composed of four components:

(1) the forward model;
(2) the adjoint of its tangent–linear form;
(3) the background error covariance matrix;
(4) the minimization routine.

The estimated observation accuracy, to be included in the observation error covariance matrix, is about 10%, or at least 2 ppbv for ozone. An additional portion of the error is assumed to be due to the poor spatial representativeness, as implied by a horizontal resolution of 54 km. This is a special problem arising with the use of urban and suburban measurement sites, although observations close to areas with elevated traffic load were omitted. For more details, see Elbern et al. (2000) and Elbern and Schmidt (2001).

22.3 A posteriori validation in atmospheric chemistry

A posteriori validation of data assimilation results in atmospheric chemistry is possible, as in meteorological applications (see Chapters 7–9). An example is given for stratospheric chemistry data assimilation, applying the Desroziers et al. (2005) observation space validation. With $\mathbf{y}$, $\mathbf{x}^{\mathrm{b}}$, and $\mathbf{x}^{\mathrm{a}}$ being the set of p observations, the forecast (background) state, and the analysed state, respectively, Desroziers et al. (2005) have shown that the differences between the analysis and background fields, projected into observation space by the generally nonlinear operator H, and the observations

$$\begin{aligned}\mathbf{d}_{\mathrm{b}}^{\mathrm{o}} &:= \mathbf{y} - H(\mathbf{x}^{\mathrm{b}})\,,\\ \mathbf{d}_{\mathrm{b}}^{\mathrm{a}} &:= H(\mathbf{x}^{\mathrm{a}}) - H(\mathbf{x}^{\mathrm{b}})\,,\\ \mathbf{d}_{\mathrm{a}}^{\mathrm{o}} &:= \mathbf{y} - H(\mathbf{x}^{\mathrm{a}})\,,\end{aligned}$$

can be used to diagnose the consistency of covariance matrices. Assuming a perfect assimilation system, the following relations hold:

$$E\left[\mathbf{d}_{\mathrm{b}}^{\mathrm{a}}(\mathbf{d}_{\mathrm{b}}^{\mathrm{o}})^{\mathrm{T}}\right] = \mathbf{H}\tilde{\mathbf{B}}\mathbf{H}^{\mathrm{T}}\,, \tag{22.18}$$

$$E\left[\mathbf{d}_{\mathrm{a}}^{\mathrm{o}}(\mathbf{d}_{\mathrm{b}}^{\mathrm{o}})^{\mathrm{T}}\right] = \tilde{\mathbf{R}}\,, \tag{22.19}$$

$$E\left[\mathbf{d}_{\mathrm{b}}^{\mathrm{a}}(\mathbf{d}_{\mathrm{a}}^{\mathrm{o}})^{\mathrm{T}}\right] = \mathbf{H}\tilde{\mathbf{A}}\mathbf{H}^{\mathrm{T}}, \tag{22.20}$$

$$E\left[\mathbf{d}_{\mathrm{b}}^{\mathrm{o}}(\mathbf{d}_{\mathrm{b}}^{\mathrm{o}})^{\mathrm{T}}\right] = \mathbf{H}\tilde{\mathbf{B}}\mathbf{H}^{\mathrm{T}} + \tilde{\mathbf{R}}\,, \tag{22.21}$$

where $E[\,]$ is the expectation operator, and $\tilde{\mathbf{R}}$, $\tilde{\mathbf{B}}$, and $\tilde{\mathbf{A}}$ are the observation, background, and analysis error covariance matrices, respectively. In the nonlinear case, the matrix $\mathbf{H}$ is the Jacobian of H at the analysis point, i. e.

$$\mathbf{H} := \left.\frac{\partial H}{\partial \mathbf{x}}\right|_{\mathbf{x}=\mathbf{x}^{\mathrm{a}}}.$$

In practice, the tuning of the background error covariance matrix $\mathbf{B}$ must be realized by some low parameter variation. The following procedure is performed. For details, see Schwinger and Elbern (2010). $\mathbf{B}$ is decomposed into a correlation matrix $\mathbf{C}$ and a diagonal matrix $\mathbf{\Sigma} := \mathrm{diag}(\sigma_1^{\mathrm{b}}, \ldots, \sigma_n^{\mathrm{b}})$ containing the background error standard deviations:

$$\mathbf{B} = \mathbf{\Sigma}\mathbf{C}\mathbf{\Sigma}\,. \tag{22.22}$$

The problem of finding a suitable BECM formulation is split into (i) specifying correct standard deviation values σ_k^{b} for each model grid point $k = 1, \ldots, n$ and (ii) finding an appropriate spatial correlation length L in the parameterisation of $\mathbf{C}$. For simplicity, it was decided to use a isotropic correlation structure, defined by a global horizontal correlation lengthscale L.

The first step is to use (22.18) and (22.19) to adjust σ_k^{b} iteratively. This leads to consistent, but not necessarily correct, error standard deviations σ^{b} and σ^{o}; that is, these conditions are satisfied for mean diagonal elements of **R** and **B**. In order to avoid degradation of the results due to model errors and/or misspecification of the correlation lengthscale L (which is still untuned at this stage), the tuning is performed on a subset of observations only.

Thereupon, a suitable correlation lengthscale L for the parameterisation of **C** is estimated, based on (22.18). For this, a set of assimilation experiments has been conducted, using the tuned background variances, with varying values of L. A special subset of observations has to be chosen again, in order to obtain meaningful results. The χ^2-criterion and an inspection of forecast skill are used to validate the optimized lengthscale.

Before the results are presented, the following considerations should be taken into account, as a potential explanation for overly optimistic interpretation of the forecast error covariances. The diagnosed values of background and observation error variances can be estimated on the basis of (22.18) and (22.19) by

$$(\tilde{\sigma}_i^{\mathrm{b}})^2 = \frac{1}{p_i}(\mathbf{d}_{\mathrm{b}}^{\mathrm{a}})_i^{\mathrm{T}}(\mathbf{d}_{\mathrm{b}}^{\mathrm{o}})_i = \frac{1}{p_i}\sum_{j=1}^{p_i}(y_j^{\mathrm{a}} - y_j^{\mathrm{b}})(y_j^{\mathrm{o}} - y_j^{\mathrm{b}}), \tag{22.23}$$

$$(\tilde{\sigma}_i^{\mathrm{o}})^2 = \frac{1}{p_i}(\mathbf{d}_{\mathrm{a}}^{\mathrm{o}})_i^{\mathrm{T}}(\mathbf{d}_{\mathrm{b}}^{\mathrm{o}})_i = \frac{1}{p_i}\sum_{j=1}^{p_i}(y_j^{\mathrm{o}} - y_j^{\mathrm{a}})(y_j^{\mathrm{o}} - y_j^{\mathrm{b}}), \tag{22.24}$$

for a subset of observations i with a number of p_i observations, with obvious indexing.

Uncritical application of (22.23) leads to a significant underestimation of background errors. A closer look at (22.23) and (22.24) reveals that, given fixed background and observation values $y_j^{\mathrm{o}} - y_j^{\mathrm{b}}$, $y_j^{\mathrm{a}} - y_j^{\mathrm{b}}$ controls the values of $(\tilde{\sigma}_i^{\mathrm{b}})^2$. If applied in a real-world 4D-Var assimilation system, the term $y_j^{\mathrm{a}} - y_j^{\mathrm{b}}$ can remain erroneously small for the following reasons:

1. *Violation of the perfect model assumption.* Owing to some unresolved bias, imperfect model physics or chemistry, or insufficient model resolution, it may not be possible to reproduce some observations within the assimilation window properly for the analysis run. Hence, the analysis y^{a} often stays closer to y^{b} than it would if all theoretical assumptions for a BLUE were satisfied. In other words, it is closer to y^{b} than it would be were the assimilating model was perfect. An example of a kind of model error that is likely to occur in chemical modelling is depicted in Fig. 22.1(a): relaxation towards a model-described chemical equilibrium different from the true state. This evolves to an analysis degrading to the background state with increasing time.
2. *Misspecified correlation lengths.* If the true correlation length L^{t} is small ($L^{\mathrm{t}} \ll r$), but the correlation lengthscale L, taken in the assimilation procedure, is overestimated, i.e. $L \geq r$, then the assimilation algorithm will smooth out spatial variability and only small analysis increments will emerge. Again, the analysis will stay too close to the background state, as illustrated in Fig. 22.1(b).

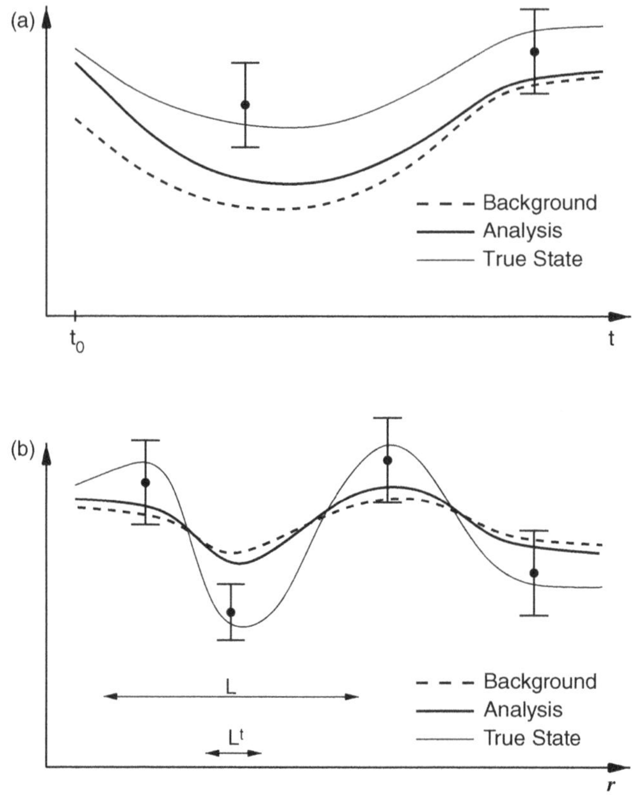

Fig. 22.1 Illustration of two specific cases for which the straightforward application of the method of Desroziers et al. leads to an underestimation of background errors and overestimation of observation errors. (a) Due to model error, and (b) due to miss-specification of the correlation length scale. See text for further explanation.

As an example result, the diagnosed analysis error valid in observation space can be approximated on the basis of (22.20) by

$$(\tilde{\sigma}_i^{\mathrm{a}})^2 = \frac{1}{p_i}(\mathbf{d}_{\mathrm{b}}^{\mathrm{a}})_i^{\mathrm{T}}(\mathbf{d}_{\mathrm{a}}^{\mathrm{o}})_i = \frac{1}{p_i}\sum_{j=1}^{p_i}(y_j^{\mathrm{a}} - y_j^{\mathrm{b}})(y_j^{\mathrm{o}} - y_j^{\mathrm{a}})\,. \tag{22.25}$$

The analysis error calculated by means of (22.25) is shown together with the diagnosed background and observation errors in Fig. 22.2.

22.4 Tropospheric chemical data assimilation

22.4.1 Assimilation of column data

Stratospheric satellite data assimilation benefits from many limb-viewing scanning geometries (MIPAS, MLS, and many others), which provide height-resolved values

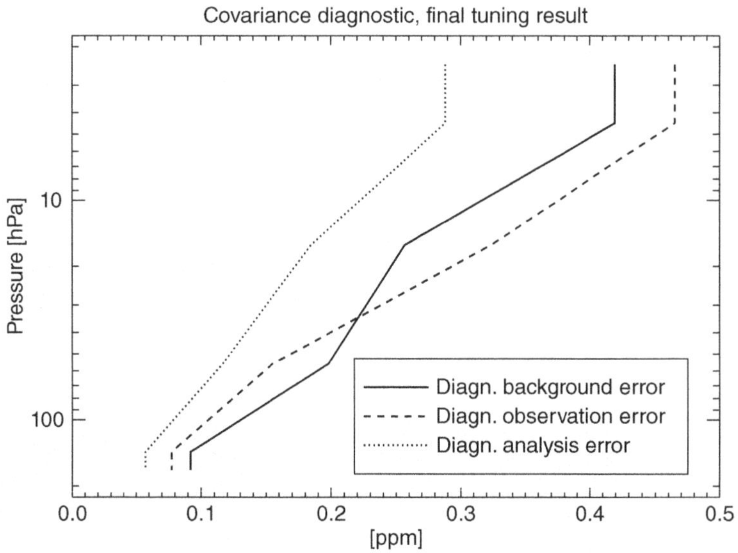

Fig. 22.2 Diagnosed analysis error in observation space $\tilde{\sigma}^{\mathrm{a}}$ (dotted), diagnosed background error $\tilde{\sigma}^{\mathrm{b}}$ (solid), and the diagnosed observation error $\tilde{\sigma}^{\mathrm{o}}$ (dashed) for the final tuning result. Line-styles are also indicated in the box at the bottom right of the figure.

down to 7 km height, if missing cloud cover permits. In contrast, and as an analogue of assimilation liquid water path data or humidity columns, tropospheric chemistry satellite data assimilation is largely dependent on integral values of columns or slant columns. To use this information, the scalar value must be redistributed to model height levels.

Formally, an integral equation has to be solved for vertical NO_2 molecule density function x (σ is the vertical coordinate):

$$y = \int_1^0 h(\sigma)x(\sigma)\,\mathrm{d}\sigma$$

or, in height-level-dependent discretized form,

$$y = \sum_{k=1}^{K} h_k x_k,$$

with h being the averaging kernel, giving the sensitivity of the sensor with respect to height.

With a scalar column retrieval operator h, mapping from model colum subgrid to the column observation space, we find

$$d\mathbf{x}_a^c = \underbrace{\mathbf{Bh}^{\mathrm{T}}}(r+b)^{-1}\left\{y^0 - h[\mathbf{x}_b]\right\},$$

where r and b are the column retrieval error and the forecast error variance of the model of the column data, and the underbraced part $\mathbf{B}\mathbf{h}^{\mathrm{T}}$ is the adjoint representer. The latter expression controls the redistribution of the innovation on the model column grid according to both sensor sensitivity h and model skill by $\mathbf{B}$.

Figure 22.3 shows an example of OMI NO_2 column assimilation on a nested domain with 5 km resolution. Clearly visible is the analysis with column values that are enhanced when compared with the background run, but are markedly below the

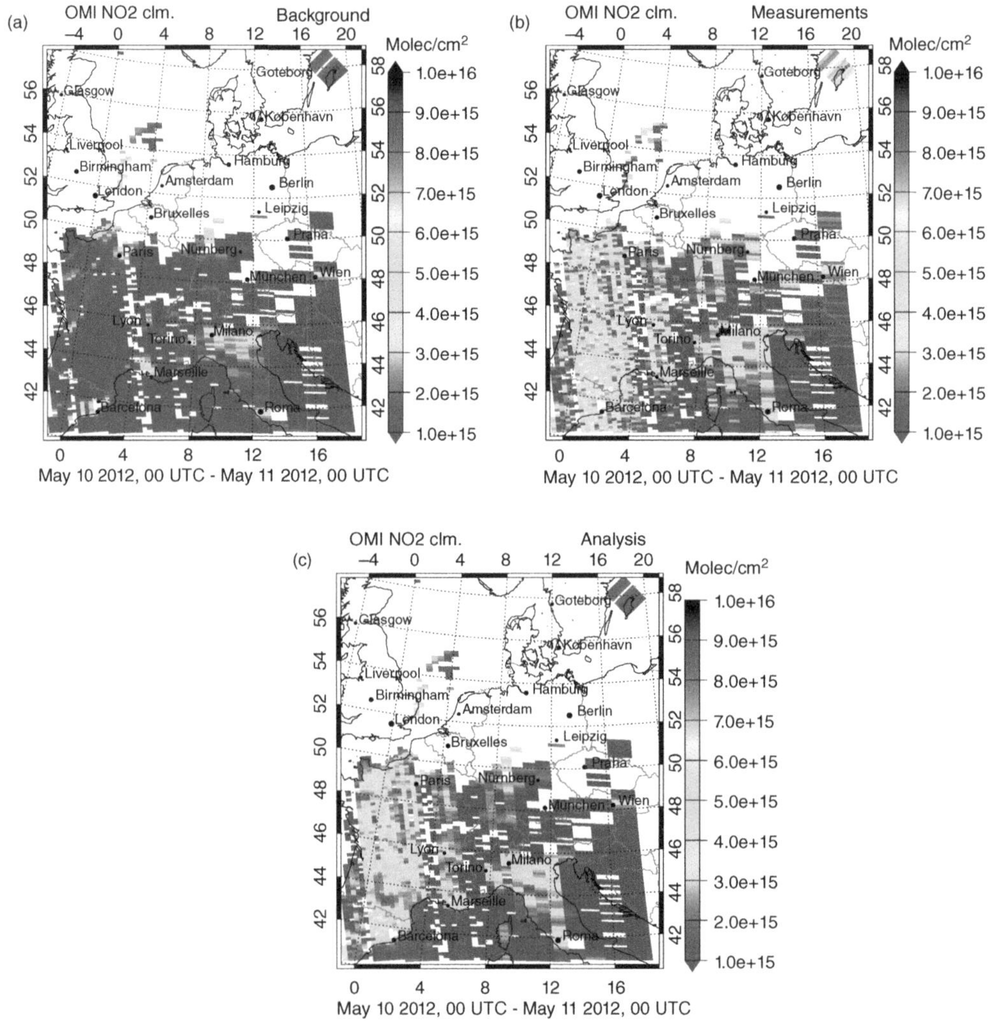

Fig. 22.3 A nested domain data assimilation of OMI NO_2 column retrieval assimilation for central Europe at 5 km resolution. (a) Model equivalent for the background without assimilation. (b) Assimilated OMI NO_2 column retrievals from KNMI. (c) Model equivalent for the analysis.

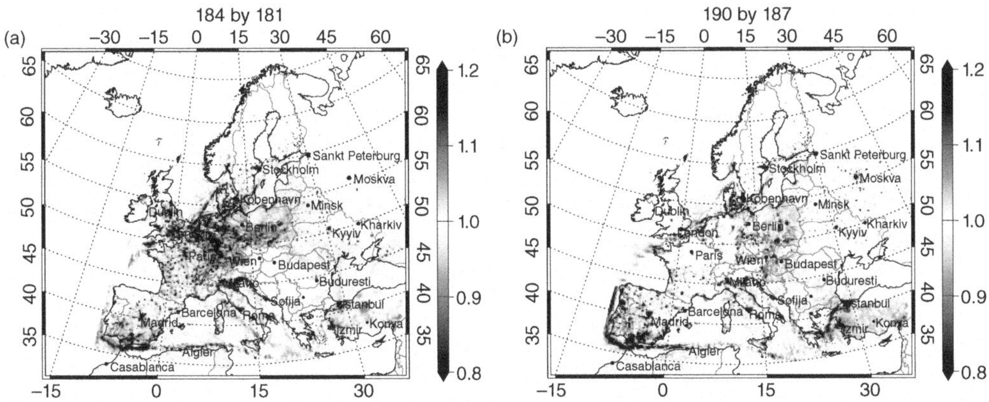

Fig. 22.4 Emission rate optimisation factors for NO_2 after assimilation of OMI retrieved NO_2 tropospheric columns from KNMI for 4 days assimilation sequence. (a) Results after assimilation procedures from 1–4 July 2006. (b) Results for 7–11 July 2006.

observations, except for the Barcelona area. The enhanced value retrievals east of Sweden in the northeast corner of the domain are hardly reflected in the analyses. This is due to the minor observation weight as a consequence of observation error attribution by the data provider KNMI.

A study focused on a numerical grid with horizontal model resolution comparable to the minimal OMI 24×13 km^2 footprint; the horizontal model resolution was refined to 15 km^2. OMI retrievals were assimilated by averaging kernels, using error information from the data provider. The data assimilation system, including the formulation of the background error covariances, is described in Elbern et al. (2007). After assimilation, a 24-hour forecast was made, starting at 08:00 UTC. The analysis produced by the assimilation was the initial field, and an emission rate correction factor was applied. Numerical experiments suggest that 12 iterations are sufficient to ensure convergence to the observations.

As a final result, Fig. 22.4 displays emission correction factors to be applied to the emission inventory data.

22.5 Aerosol data assimilation

Analog to the redistribution problem for NO_2 column assimilation, where a scalar quantity is to be redistributed to a vertical profile, typical aerosol data pose similar problems, with an even more problematic grade. Operational in situ data are typically given in terms of PM_{10} values, integrating all particle up to a diameter of $10\,\mu$m. Given the complex composition of aerosol types, including mineral dust and sea salt, a redistribution to the high-dimensional aerosol module representation, which is of order 100, must be provided for each observed and observation affected grid cell. The importance of the background formulation is again obvious.

Given aerosol optical depths data from MODIS or similar sensors, an even more challenging problem is posed, with the vertical column redistribution problem being combined with the aerosol-type problem.

For aerosols, a case study with high wildfire activity over the Iberian peninsula is presented. The first example combines standard operational PM10 data with SYNAER retrievals (Holzer-Popp et al., 2002) for a case with frequently occurring wild fires. A two-week episode was chosen to evaluate the performance of the aerosol data assimilation system with wildfires.

Three individual assimilation experiments were performed for this episode: (i) assimilation of SYNAER retrievals only; (ii) assimilation of ground-based in situ PM_{10} only; (iii) assimilation of SYNAER retrievals and ground-based in situ PM_{10}.

The following example was confined to the water-soluble species sulphate as a representative for this group. The accumulation mode concentrations are shown in Fig. 22.5. Comparing the SYNAER-only (Fig. 22.5(b)) and EEA-only (Fig. 22.5(c)) analysis increments reveals that these are of similar magnitude. This is also reflected in the SYNAER and in situ data plot (Fig. 22.5(d)), where the signature of assimilated SYNAER retrievals can clearly be seen over Scandinavia and the Iberian Peninsula, and from the Kola Peninsula to Greece. The analysis map (Fig. 22.5(a)) shows that the impact of data assimilation in the areas observed is rather high in terms of a strong increase, whereas over Central Europe, where there is already a higher sulphate load, an increase as well as a decrease can be observed.

The second example presents the assimilation of data from condensation particle counters from the Research Centre Jülich, which are used for special campaigns only. The ZEPTER-2 campaign was conducted from 17 October to 8 November 2008, with a Zeppelin NT ('New Technology') being used as an instrument platform for a measurement campaign investigating photochemistry, chemical composition, and particle properties of the lower troposphere. The flight missions operated over areas with different land use in the region around Lake Constance in southern Germany.

Among a number of instruments to measure photolysis rates and concentrations of gas-phase species such as O_3, CO, NO_x, HCHO, HONO, and HO_2, a condensation particle counter (CPC) was mounted during 12 flights to measure particle number densities *PND*.

Here, the rationale for the assimilation problem was to estimate particle number densities with three different horizontal resolutions. The airship's flight path along with the observed PND_3 (number density of particles with a diameter between 5 nm and 3 μm) are shown in Fig. 22.6. Time series of this flight for each of the three model domains are plotted in Fig. 22.7. Each of the three plots shows the same observations. Peak values with particle number densities up to $2000\,\mathrm{cm}^{-3}$ occur while crossing streets. However, not only do the background values differ owing to distinct resolutions, but also the model layer in which the instrument is located at a certain time can vary.

Owing to the complex meteorological conditions described above, there is no influence from preceeding assimilation instances that day, and thus there is no difference between the First Guess and the Reference Run here; i.e. the First Guess is in compliance with the Reference Run. Assimilation on the *EUR* grid (Fig. 22.7(a))

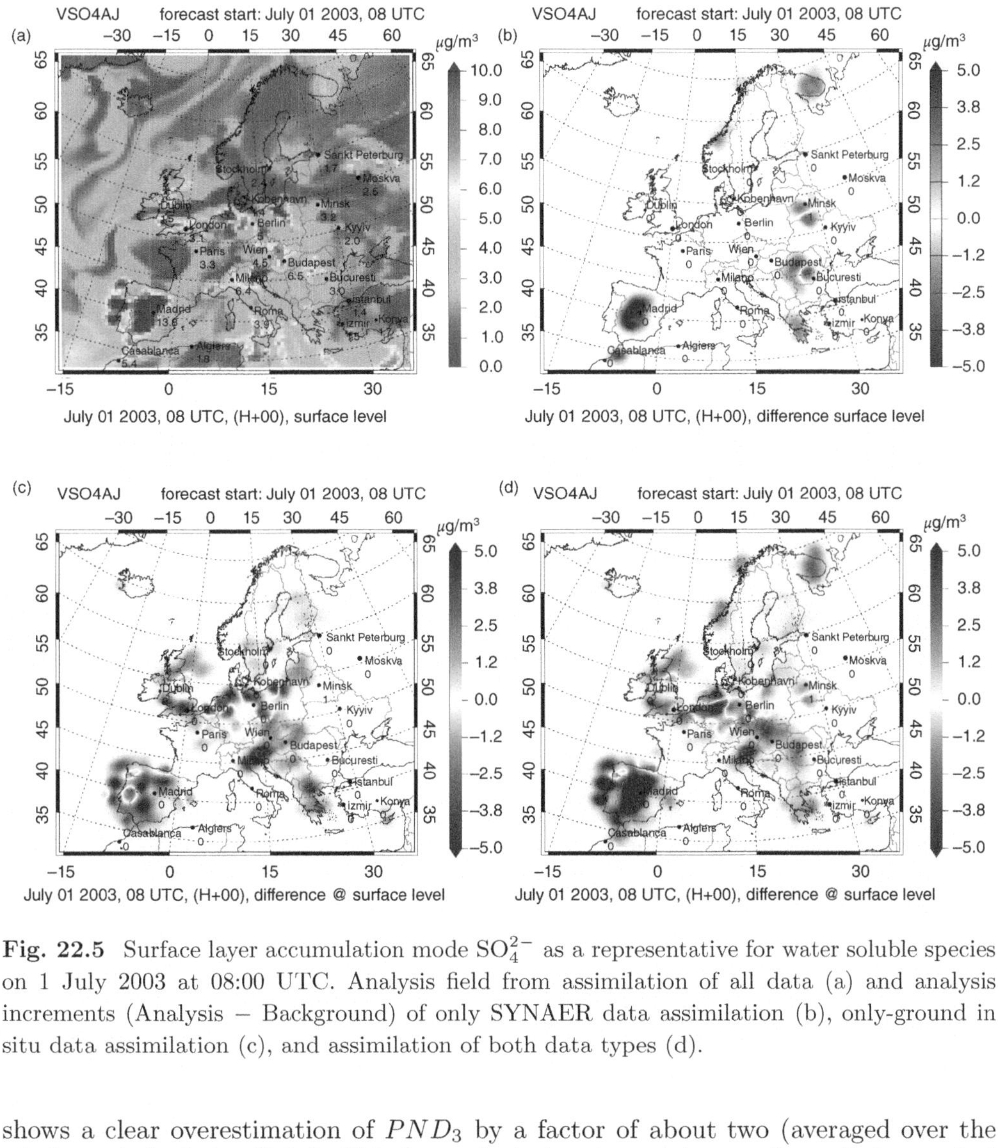

Fig. 22.5 Surface layer accumulation mode SO_4^{2-} as a representative for water soluble species on 1 July 2003 at 08:00 UTC. Analysis field from assimilation of all data (a) and analysis increments (Analysis − Background) of only SYNAER data assimilation (b), only-ground in situ data assimilation (c), and assimilation of both data types (d).

shows a clear overestimation of PND_3 by a factor of about two (averaged over the whole period of the flight). This large deviation is mainly an effect of the vertical mismatch of the model. While the accumulation mode particle number densities do not vary very much within the boundary layer, one can find a strong variation in the Aitken mode, whose particles are mostly formed near the surface by emissions from biogenic or combustion processes. This is principally due to aerosol aging, i.e. particle growth by condensation and coagulation, thus reducing the amount of Aitken-mode particles rather quickly. Since PND_3 has a large contribution from the Aitken-mode particles, it is also subject to aerosol aging. This strong height dependence is also reflected in the obvious strong anticorrelation between the model layer (light/dark

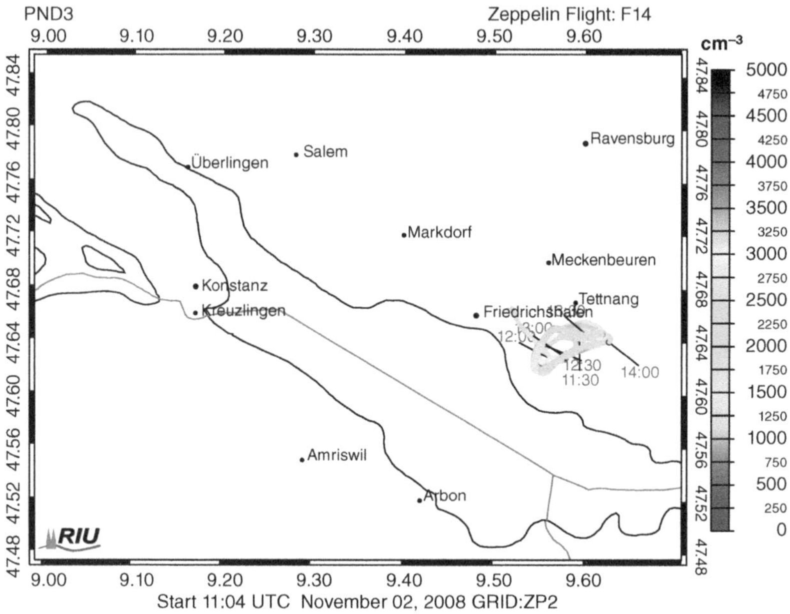

Fig. 22.6 Measured PND_3 along the track of flight F14 on 2 November 2008 over the Forest of Tettnang.

bars) and the Reference Run PND_3 (black solid line) on the *EUR* domain, where the Zeppelin is located mainly in model layers three and four (approximately 110 m above ground) because the terrain level is an average over an area of $45 \times 45\,\text{km}^2$ which here is 700 m above see level. In the *CEN* domain, the model already represents the proper magnitude, although still one layer too low (mostly layer six (approximately 260 m above ground) and with a misfit in the curve (model values decrease, while measurements increase). This misfit is almost removed on the *ZP2* domain with its higher resolution of 5 km. Here, the observations are mainly located in model layer seven (approximately 350 m above ground), which is exactly the height given as the difference between the altitude of the Zeppelin (the upper dark solid line in the lower parts of the plots) and the terrain height (the lower dark solid line). Although the conditions on the different domains were very different, the assimilation system managed to reproduce the observed values very well on each single grid (the light solid line), except for the peaks in the measurements that are related to plumes from traffic emissions.

Figure 22.8 gives insight into the vertical distribution of increments. Here, negative increments can be found in the Lake Constance area and in elevated layers over the Alps, where stronger winds prevail. These negative increments at 11 UTC have a maximum backward travel time of two hours because afterwards there is an underestimation of the simulation (the upper dark solid line in Fig. 22.7(c)).

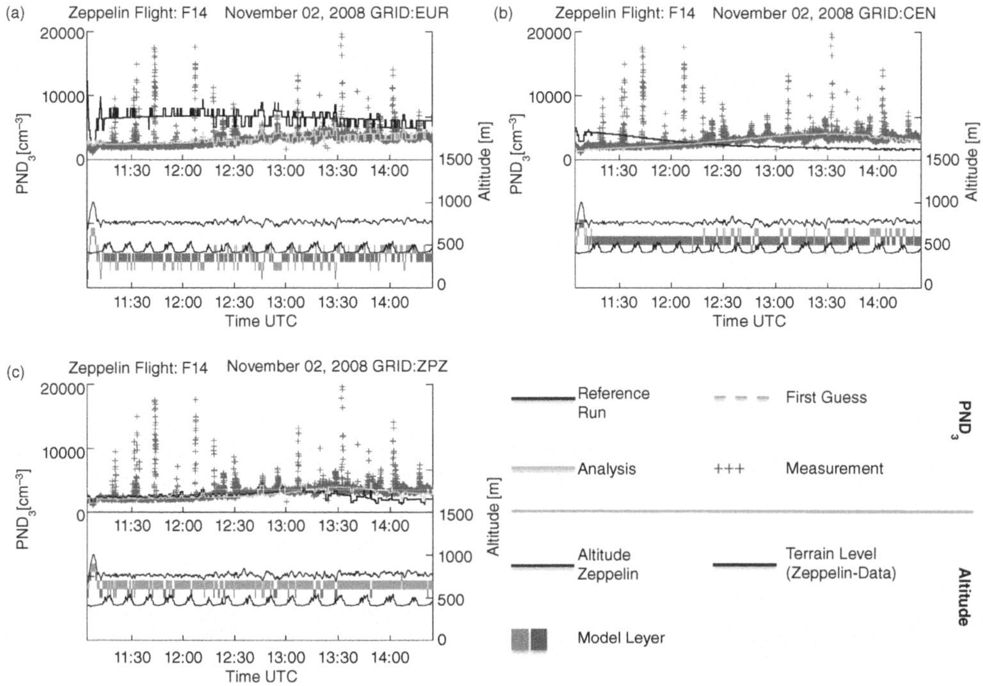

Fig. 22.7 Particle number densities PND_3 of flight F14 assimilated from 11:00 to 15:00 UTC on 2 November 2008, on the different domains *EUR* (a), *CEN* (b), and *ZP2* (c). Light and dark bars in the lower half denote the model layer of the Zeppelin. The layer number is given in the centred column.

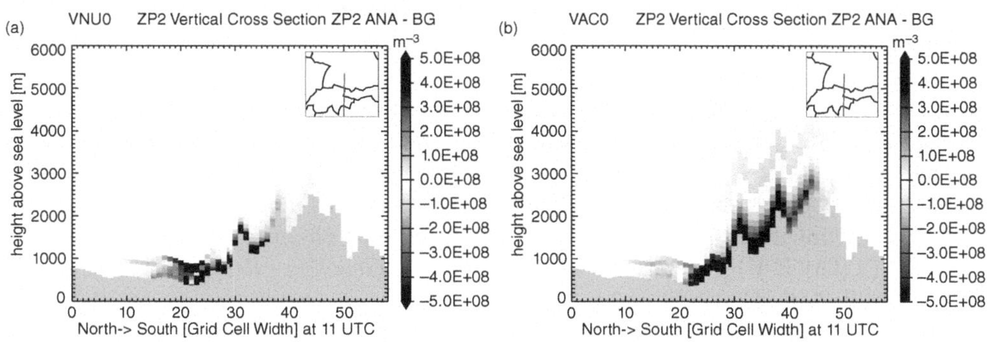

Fig. 22.8 Vertical cross section of analysis increments of particle number densities for Aitken (a) and accumulation mode (b) from assimilation of flight 14 on 2 November 2008, 11:00 UTC. The cross-section is taken along the vertical line in the small map inlay from north to south. Terrain height is plotted in grey. (a) Aitken mode number density. (b) Accumulation mode number density.

References

Ackermann, I. J., Hass, H., Memmesheimer, M., Ebel, A., Binkowski, F. S., and Shankar, U. (1998). Modal aerosol dynamics model for Europe: development and first applications. *Atmos. Environ.*, **32**, 2981–2999.

Bannister, R. N. (2008a). A review of forecast error covariance statistics in atmospheric variational data assimilation. I: Characteristics and measurements of forecast error covariances. *Q. J. R. Meteorol. Soc.*, **134**, 1951–1970.

Bannister, R. N. (2008b). A review of forecast error covariance statistics in atmospheric variational data assimilation. II: Modelling the forecast error covariance statistics. *Q. J. R. Meteorol. Soc.*, **134**, 1971–1996.

Benedetti, A. and Fisher, M. (2007). Background error statistics for aerosols, *Q. J. R. Meteorol. Soc.*, **133**, 391–405.

Bernhofer, C. and Köstner, B. (eds.) (2005). *Vertical Transport of Energy and Trace Gases at Anchor Stations and Their Spatial/Temporal Extrapolation under Complex Natural Conditions (VERTIKO): Project Summary*, Vol. 12. Tharandter Klimaprotokolle, Technische Universität Dresden.

Bocquet, M. (2005a). Reconstruction of an atmospheric tracer source using the principle of maximum entropy. I: Theory. *Q. J. R. Meteorol. Soc.*, **131**, 2191–2208.

Bocquet, M. (2005b). Reconstruction of an atmospheric tracer source using the principle of maximum entropy. II: Applications. *Q. J. R. Meteorol. Soc.*, **131**, 2209–2223.

Bousquet, P., Ciais, P., Peylin, P., Ramonet, M., and Monfray, P. (1999a). Inverse modeling of annual CO_2 sources and sinks, 1. Method and control inversion. *J. Geophys. Res.*, **104**, 26161–26178.

Bousquet, P., Ciais, P., Peylin, P., Ramonet, M., and Monfray, P. (1999b). Inverse modeling of annual CO_2 sources and sinks, 2. Sensitivity study. *J. Geophys. Res.* **104**, 26179–26193.

Carmichael, G. R. et al. (2003). Regional-scale chemical transport modeling in support of the analysis of observations obtained during the TRACE-P experiment. *J. Geophys. Res.*, **108**, 8823.

Carmichael, G. R., Sandu, A., Chai, T., Daescu, D. N., Constantinescu, E. M., and Tang, Y. (2007). Predicting air quality: Improvements through advanced methods to integrate models and measurements. *J. Comput. Phys.*, **227**, 3540–3571.

Chai, T. F., Carmichael, G. R., Sandu, A., Tang, Y. H., and Daescu, D. N. (2006). Chemical data assimilation of transport and chemical evolution over the Pacific (TRACE-P) aircraft measurements. *J. Geophys. Res.*, **111**, D02301, doi:10.1029/2005JD005883.

Collins, W., Rasch, P., Eaton, B., Khattatov, B., Lamarque, J.-F., and Zender, C. (2001). Simulating aerosols using a chemical transport model with assimilation of satellite aerosol retrievals: methodology for INDOEX. *J. Geophys. Res.*, **106**, 7313–7336.

Constantinescu, E. M., Sandu, A., Chai, T., and Carmichael, G. R. (2007a). Ensemble-based chemical data assimilation. I: General approach. *Q. J. R. Meteorol. Soc.*, **133**, 1229–1243.

Constantinescu, E. M., Sandu, A., Chai, T., and Carmichael, G. R. (2007b). Ensemble-based chemical data assimilation. II: Covariance localization, *Q. J. R. Meteorol. Soc.*, **133**, 1245–1256.

Deeter, M. N. et al. (2003). Operational carbon monoxide retrieval algorithm and selected results for the MOPITT instrument, *J. Geophys. Res.*, **108**, 4399.

Desroziers, G., Berre, L., Chapnik, B., and Poli, P. (2005). Diagnosis of observation, background and analysis-error statistics in observation space. *Q. J. R. Meteorol. Soc.*, **131**, 3385–3396.

Elbern, H. and Schmidt, H. (2001). Ozone episode analysis by four-dimensional variational chemistry data assimilation. *J. Geophys. Res.*, **106**, 3569–3590.

Elbern, H., Schmidt, H., Talagrand, O., and Ebel, A. (2000). 4D–variational data assimilation with an adjoint air quality model for emission analysis, *Environ. Model. Software*, **15**, 539–548.

Elbern, H., Strunk, A., Schmidt, H., and Talagrand, O. (2007). Emission rate and chemical state estimation by 4-dimensional variational inversion. *Atmos. Chem. Phys.*, **7**, 1–59.

Engelen, R., Andersson, E., Chevallier, F., Hollingsworth, A., Matricardi, M., McNally, A. P., Thépaut, J. N., and Watts, P. D. (2004). Estimating atmospheric CO_2 from advanced infrared satellite radiances within an operational 4D-Var data assimilation system: methodology and first results. *J. Geophys. Res.*, **109**, D19309, doi:10.1029/2004JD004777, 2004.

Enting, I. G. and Newsam, G. N. (1990). Inverse problems in atmospheric constituent studies: II. Sources in the free atmosphere. *Inverse Problems*, **6**, 349–362.

Enting, I. G., Trudinger, C. M., and Francey, R. J. (1995). A synthesis inversion of the concentration and $\delta^{13}C$ of atmospheric CO_2. *Tellus*, **47B**, 35–52.

Errera, Q. and Fonteyn, D. (2001). Four-dimensional variational chemical assimilation of CRISTA stratospheric measurements. *J. Geophys. Res.*, **106**, 12,253–12,265.

Errera, Q., Daerden, F., Chabrillat, S., Lambert, J. C., Lahoz, W. A., Viscardy, S., Bonjean, S., and Fonteyn, D. (2008). 4D-Var assimilation of MIPAS chemical observations: ozone and nitrogen dioxide analyses. *Atmos. Chem. Phys.*, **8**, 6169–6187.

Eskes, H. J. and Boersma, K. F. (2003). Averaging kernels for DOAS total-column satellite retrievals. *Atmos. Chem. Phys.*, **3**, 1285–1291.

Fan, S. M., Gloor, M., Mahlman, J., Pacala, S., Sarmiento, J., Takahashi, T., and Tans, P. (1998). A large terrestrial carbon sink in North America implied by atmospheric data and oceanic carbon dioxide data and models. *Science*, **282**, 442–446.

Fedorov, V. (1998). Kriging and other estimators of spatial field characteristics (with special reference to environmental studies). *Atmos. Environ.*, **23**, 174–184.

Fierli, F., Hauchecorne, A., Bekki, S., Theodore, B., and d'Andon, O. F. (2002). Data assimilation of stratospheric ozone using a high-resolution transport model. *Geophys. Res. Lett.*, **29**, 19.1–19.4.

Fisher, M. and Lary, D. (1995). Lagrangian four-dimensional variational data assimilation of chemical species. *Q. J. R. Meteorol. Soc.*, **121**, 1681–1704.

Gloor, M., Fan, S. M., Pacala, S., Sarmiento, J., and Ramonet, M. (1999). A model-based evaluation of inversions of atmospheric transport, using annual mean mixing ratios, as a tool to monitor fluxes of nonreactive trace substances like CO_2 on a continental scale. *J. Geophys. Res.*, **104**, 14245–14260.

Gurney, K. R. (2002). Towards robust regional estimates of CO_2 sources and sinks using atmospheric transport models. *Nature*, **415**, 626–630.

Hamill, T. M. (2006). *Predictability of Weather and Climate*, Chapter 6: Ensemble-based atmospheric data assimilation, pp. 124–156. Cambridge University Press, Cambridge.

Hanea, R. G., Velders, G. J. M., and Heemink, A. (2004). Data assimilation of ground-level ozone in Europe with a Kalman filter and chemistry transport. *J. Geophys. Res.*, **109**, D10302, doi:10.1029/2003JD004283.

Hanna, S. R., Chang, J. C., and Fernau, M. E. (1998). Monte Carlo estimates of uncertainties in predictions by a photochemical grid model (UAM-IV) due to uncertainties in input variables. *Atmos. Environ.*, **32**, 3619–3628.

Hanna, S. R., Lu, Z. G., Frey, H. C., Wheeler, N., Vukovich, J. Arunachalam, S., Fernau, M., and Hansen, D. A. (2001). Uncertainties in predicted ozone concentrations due to input uncertainties for the UAM-V photochemical grid model applied to the July 1995 OTAG domain. *Atmos. Environ.*, **35**, 891–903.

Henze, D., Seinfeld, J. H., Liao, W., Sandu, A., and Carmichael, G. R. (2004). Inverse modeling of aerosol dynamics: condensational growth. *J. Geophys. Res.*, **109**, D14201.

Heue, K. P., Richter, A., Bruns, M., Burrows, J. P., von Friedeburg, C., Platt, U., Pundt, I., Wang, P., and Wagner, T. (2005). Validation of SCIAMACHY tropospheric NO_2-columns with AMAXDOAS measurements. *Atmos. Chem. Phys.*, **5**, 1039–1051.

Hölzemann, J., Elbern, H., and Ebel, A. (2001). Psas and 4D-Var data assimilation for chemical state analysis by urban and rural observation sites. *Phys. Chem.* Earth, **26**, 807–812.

Holzer-Popp, T., Schroedter, M., and Gesell, G. (2002). Retrieving aerosol optical depth and type in the boundary layer over land and ocean from simultaneous GOME spectrometer and ATSR-2 radiometer measurements. 1, Method description. *J. Geophys. Res.*, **107**, 4578.

Houweling, S., Kaminski, T., Dentener, F., Lelieveld, J., and Heimann, M. (1999). Inverse modeling of methane sources and sinks using the adjoint of a global transport model. *J. Geophys. Res.*, **104**, 26137–26160.

Issartel, J. P. (2003). Rebuilding sources of linear tracers after atmospheric concentration measurements. *Atmos. Chem. Phys.*, **3**, 2111–2125.

Kalnay, E. (2003). *Atmospheric Modeling, Data Assimilation and Predictability.* Cambridge University Press, Cambridge.

Kaminski, T., Heimann, M., and Giering, R. (1999a). A coarse grid three-dimensional global inverse model of the atmospheric transport: 1. Adjoint model and Jacobian matrix. *J. Geophys. Res.*, **104**, 18535–18553.

Kaminski, T., Heimann, M., and Giering, R. (1996b). A coarse grid three-dimensional global inverse model of the atmospheric transport: 2. Inversion of the transport of CO_2 in the 1980s. *J. Geophys. Res.*, **104**, 18555–18581.

Kaminski, T., Knorr, W., Rayner, P. J., and Heimann, M. (2002). Assimilating atmospheric data into a terrestrial biosphere model: A case study of the seasonal cycle, *Glob. Biogeochem. Cycles*, **16**, 14.1–14.16.

Lahoz, W. A., Geer, A. J., Bekki, S., Bormann, N., Ceccherini, S., Elbern, H., Errera, Q., Eskes, H. J., Fonteyn, D., Massart, D. R. J. B. K. S., Peuch, V.-H., Rharmili, S., Ridolfi, M., Segers, A., Talagrand, O., Vik, H. T. A. F., and von Clarmann, T. (2007). The Assimilation of Envisat data (ASSET) project. *Atmos. Chem. Phys.*, **7**, 1773–1796.

Lahoz, W. A. and Errera, Q. (2010). Constituent assimilation. In *Data Assimilation: Making Sense of Observations* (ed. W. Lahoz, B. Khattatov, and R. Menard,), pp. 449–490. Springer-Verlag, Berlin.

Marenco, A. et al. (1998). Measurement of ozone and water vapor by Airbus in-service aircraft: the MOZAIC airborne program, An overview. *J. Geophys. Res.*, **103**, 25631–25642.

Müller, J. F. and Stavrakou, T. (2005). Inversion of CO and NO_x emissions using the adjoint of the IMAGES model. *Atmos. Chem. Phys.*, **5**, 1157–1186.

Müller, M. D., Kaifel, A. K., Weber, M., Tellmann, S., Burrows, J. P., and Loyola, D. (2003). Ozone profile retrievel from Global Ozone Monitoring Experiment (GOME) data using a neural network approach (Neural Network Ozone Retrieval System (NNORSY)). *J. Geophys. Res.*, **108**, 4497, doi:10.1029/2002JD002784.

Newsam, G. N. and Enting, I. G. (1988). Inverse problems in atmospheric constituent studies: I. Determination of surface sources under a diffusive transport approximation, *Inverse Problems*, **4**, 1037–1054.

Nieradzik, L. P. and Elbern, H. (2006). Variational assimilation of combined satellite retrieved and in situ aerosol data in an advanced chemistry transport model. In *Proceedings of ESA Atmospheric Science Conference*, ESA, ESA-ESRIN, Frascati, 2006.

Otte, T. L., Seaman, N. L., and Stauffer, D. R. (2001). A heuristic study on the importance of anisotropic error distributions in data assimilation. *Mon. Weather. Rev.*, **129**, 766–783.

Purser, R. J., Wu, W.-S., Parrish, D. F., and Roberts, N. M. (2003). Numerical aspects of the application of recursive filters to variational statistical analysis. Part II: Spatially inhomogeneous and anisotropic general covariances. *Mon. Weather Rev.*, **131**, 1536–1548.

Quélo, D., Mallet, V., and Sportisse, B. (2005). Inverse modeling of NO_x emissions at regional scale over northern France. preliminary investigation of the second-order sensitivity. *J. Geophys. Res.*, **110**, D24310, doi:10.1029/2005JD006151.

Riishojgaard, L. P. (1998). A direct way of specifying flow-dependent background error correlations for meteorological analysis systems. *Tellus*, **50A**, 42–57.

Robertson, L. and Langner, J. (1992). Source function estimate by means of variational data assimilation applied to the ETEX-I tracer experiment. *Atmos. Environ.*, **32**, 4219–4225.

Sander, R., Kerkweg, A., Jöckel, P., and Lelieveld, J. (2005). The new comprehensive atmospheric chemistry module MECCA. *Atmos. Chem. Phys.*, **5**, 445–450.

Schell, B., Ackermann, I. J., Hass, H., and Ebel, A. (2001). SORGAM: a secondary organic aerosol module for comprehensive air quality models. In *Proceedings of EUROTRAC Symposium 2000*. Springer-Verlag, Berlin.

Schmidt, H. and Martin, D. (2003). Adjoint sensitivity of episodic ozone in the Paris area to emissions on the continental scale. *J. Geophys. Res.*, **108**, 8561, doi:10.1029/2001JD00158.

Schwinger, J. and Elbern, H. (2010). Chemical state estimation for the middle atmosphere by four-dimensional variational data assimilation: a posteriori validation of error statistics in observation space. *J. Geophys. Res.*, **115**, D18307, doi:10.1029/2009JD013115.

Segers, A. J., Eskes, H. J., van der, R. J., van Oss, A, R. F., and van Velthoven, P. F. J. (2005). Assimilation of GOME ozone profiles and a global chemistry–transport model using a Kalman filter with anisotropic covariance. *Q. J. R. Meteorol. Soc.*, **131**, 477–502.

Talagrand, O. (1998). A posteriori evaluation and verification of analysis and assimilation algorithms. In *Proceedings of Workshop on Diagnosis of Data Assimilation Systems, European Centre for Medium-Range Weather Forecasts Reading, England*, 2–4 November, 1998.

Talbot, R. et al. (2003). Reactive nitrogen in asian continental outflow over the western pacific: results from the NASA Transport and Chemical Evolution over the Pacific (TRACE-P) airborne mission. *J. Geophys. Res.*, **108**, 8803.

Thépaut, J.-N., Courtier, P., Belaud, G., and Lematre, G. (1996). Dynamical structure functions in a four-dimensional variational assimilation: a case study. *Q. J. R. Meteorol. Soc.*, **122**, 535–561.

Thiebaux, H. J. (1976). Anisotropic correlation functions for objective analysis. *Month. Weather Rev.*, **104**, 994.

Thouret, V., Cho, J., Newell, R., Marenco, A., and Smit, H. (2000). General characteristics of tropospheric trace constituent layers observed in the mozaic program. *J. Geophys. Res.*, **105**, 17379–17392.

van Loon, M., Builtjes, P. J. H., and Segers, A. J. (2000). Data assimilation of ozone in the atmospheric transport chemistry model LOTOS, *Environ. Model. Software*, **15**, 603–609.

Verlaan, M. and Heemink, A. W. (1995). Reduced rank square root filters for large scale data assimilation problems. In *Proceedings of Second International Symposium on Assimilation of Observations in Meteorology and Oceanography.*

Vira, J. and Sofiev, M. (2012). On variational data assimilation for estimating the model initial conditions and emission fluxes for short-term forecasting of SO_x concentrations. *Atmos. Environ.* **46**, 318–328.

Volz-Thomas, A., Geiss, H., Hofzumahaus, A., and Becker, K. H. (2003). Introduction to special section: Photochemistry experiment in berlioz, *J. Geophys. Res.*, **108**, 8252.

Weaver, A. and Courtier, P. (2001). Correlation modelling on the sphere using a generalized diffusion equation. *Q. J. R. Meteorol. Soc.*, **127**, 1815–1846.

Wu, W.-S., Purser, R. J., and Parrish, D. F. (2002). Three-dimensional variational analysis with spatially inhomogeneous covariances. *Mon. Weather Rev.*, **130**, 2905–2916.

Yumimoto, K., Uno, I., Sugimoto, N., Shimizu, A., Liu, Z., and Winker, D. M. (2008). Adjoint inversion modeling of Asian dust emission using lidar observations. *Atmos. Chem. Phys.*, **8**, 2869–2884.

Zhang, S., Penner, J. E., and Torres, O. (2005). Inverse modeling of biomass burning emissions using Total Ozone Mapping Spectrometer aerosol index for 1997. *J. Geophys. Res.*, **110**, D21306.

Zhang, J., Reid, J. S., Westphal, D. L., Baker, N. L., and Hyer, E. J. (2008). A system for operational aerosol optical depth data assimilation over global oceans. *J. Geophys. Res.*, **113**, D10208.

23

Combining models and data in large-scale oceanography: examples from the consortium for Estimating the Circulation and Climate of the Ocean (ECCO)

I. FUKUMORI

Jet Propulsion Laboratory, California Institute of Technology
Pasadena, CA 91109, USA

Advanced Data Assimilation for Geosciences. First Edition.
Edited by É. Blayo, M. Bocquet, E. Cosme, and L. F. Cugliandolo.

Chapter Contents

23.1 Introduction

Data assimilation, or, more generally, the systematic combination of observations and models, has become an increasingly common tool for studying ocean circulation. Inspired by examples from numerical weather forecasting, observations are combined with ocean models to correct model errors on the one hand and to extrapolate the data information in space and time on the other. The results provide descriptions of the ocean that are more complete and more accurate than what are available from either observations or model simulations alone.

Here, examples from the Consortium for Estimating the Circulation and Climate of the Ocean (ECCO) are described that illustrate applications of such model–data syntheses to studies of large-scale ocean circulation and their utility in other disciplines. In comparison with applications in weather forecasting, often the objective in many oceanographic studies, and in the examples described here, is to analyse processes and mechanisms of ocean circulation as opposed to forecasting the ocean per se. Disparate purposes entail different requirements that must be taken into consideration in deriving the estimates as well as in utilizing the results. A brief discussion of some of the practical considerations in applying advanced estimation methods to large-scale ocean circulation problems is also provided.

23.2 Physical consistency

ECCO was founded in the early 1990s to advance ocean state estimation into a practical tool for studying ocean circulation (Wunsch et al., 2009). The effort was motivated in part by the unprecedented volume and diversity of ocean observations from the World Ocean Circulation Experiment and its related programmes at that time and by a recognition that a truly global coherent description of the ocean would only be possible by using general circulation models to combine the sparse and irregular data. Today, the effort has been expanded in extent and in scope to investigate various mechanisms controlling ocean circulation and to study the ocean's impact on Earth's climate system. In comparison with many other model–data syntheses, most of the ECCO estimates are characterized by their physically consistent temporal evolution. This distinction is fundamental and makes ECCO's estimates particularly conducive to studying processes of ocean circulation, as it allows quantitative analyses of, for instance, property budgets, circulation pathways, and causal mechanisms. The issue is illustrated in Fig. 23.1.

In typical data assimilation, which has been developed primarily for numerical weather forecasting purposes, the aim is to correct the model's initial condition such that the model, when integrated further in time, will provide the most accurate prediction. For instance, a prior model state estimate at time t, $\mathbf{x}_t^f$ (with superscript f denoting a prior estimate, or, equivalently, a previous 'forecast' and the subscript t denoting time), which was obtained by model integration from an earlier estimate $\mathbf{x}_{t-1}^a$, is corrected to $\mathbf{x}_t^a$ (superscript a for 'analysis') that is closer to observations $\mathbf{y}_t$ as denoted by the crosses and the curves that connect them in Fig. 23.1. The cycle is repeated until new observations become available at some future time $t+1$. Although the corrected

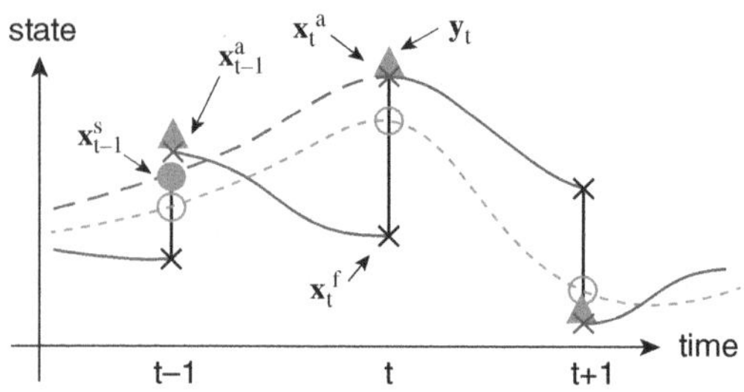

Fig. 23.1 Schematic illustration of a model state's temporal evolution in typical model–data syntheses. The abscissa is time and the ordinate is the state. Data are denoted by grey triangles. Light and dark crosses and the corresponding curve and line segments illustrate typical data assimilation estimates. The circles and the long- and short-dashed curves describe smoothed estimates. See the text for further description.

estimate $\mathbf{x}_t^a$ may be closer to observations and provide more accurate forecasts than otherwise, the state's evolution from one analysis time to another (e.g. from $\mathbf{x}_{t-1}^a$ to $\mathbf{x}_t^a$) cannot be physically accounted for as the data increments (differences between $\mathbf{x}_t^f$ and $\mathbf{x}_t^a$), which reflect errors in the model (initial condition $\mathbf{x}_{t-1}^a$ and trajectory connecting $\mathbf{x}_{t-1}^a$ to $\mathbf{x}_t^f$), are not explicitly attributed to any physical process.

In comparison, an explicit physically consistent temporal evolution from $t-1$ to t can be obtained by inverting (fitting) the model itself:

$$\mathbf{x}_t = \mathbf{A}\ \mathbf{x}_{t-1} + \mathbf{G}\mathbf{c}_{t-1}^0 \tag{23.1}$$

Here, a linear model is assumed for simplicity and the vector $\mathbf{c}^0$ denotes the model's prior control, which includes forcing and boundary conditions. Bold upper- and lower-case variables denote matrices and (column) vectors, respectively. Given the estimate $\mathbf{x}_t^a$, one can invert (23.1) to correct the state and controls at time $t-1$, using $\mathbf{x}_{t-1}^a$ and $\mathbf{c}_{t-1}^0$ as prior estimates such as

$$\begin{pmatrix} \mathbf{x}_{t-1}^s \\ \mathbf{c}_{t-1}^s \end{pmatrix} = \begin{pmatrix} \mathbf{x}_{t-1}^a \\ \mathbf{c}_{t-1}^0 \end{pmatrix} + \begin{pmatrix} \mathbf{P}_{t-1}^a\mathbf{A}^{\mathrm{T}}\left(\mathbf{A}\mathbf{P}_{t-1}^a\mathbf{A}^{\mathrm{T}} + \mathbf{G}\mathbf{Q}_{t-1}\mathbf{G}^{\mathrm{T}}\right)^{-1} \\ \mathbf{Q}_{t-1}\mathbf{G}^{\mathrm{T}}\left(\mathbf{G}\mathbf{Q}_{t-1}\mathbf{G}^{\mathrm{T}} + \mathbf{A}\mathbf{P}_{t-1}^a\mathbf{A}^{\mathrm{T}}\right)^{-1} \end{pmatrix}$$
$$\left(\mathbf{x}_t^a - \mathbf{A}\mathbf{x}_{t-1}^a - \mathbf{G}\mathbf{c}_{t-1}^0\right). \tag{23.2}$$

The new estimates are denoted with superscript s on the left-hand side and are illustrated by the filled circle and the long-dashed curve in Fig. 23.1. The matrices $\mathbf{P}_{t-1}^a$ and $\mathbf{Q}_{t-1}$ denote error covariance matrices of $\mathbf{x}_{t-1}^a$ and $\mathbf{c}_{t-1}^0$, respectively. Given the new estimate $\mathbf{x}_{t-1}^s$ from this inversion, the model can also be inverted further back in time to yield a solution of general form

$$\begin{pmatrix} \mathbf{x}_i^s \\ \mathbf{c}_i^s \end{pmatrix} = \begin{pmatrix} \mathbf{x}_i^a \\ \mathbf{c}_i^0 \end{pmatrix} + \begin{pmatrix} \mathbf{P}_i^a \mathbf{A}^{\mathrm{T}} \left(\mathbf{A} \mathbf{P}_i^a \mathbf{A}^{\mathrm{T}} + \mathbf{G} \mathbf{Q}_i \mathbf{G}^{\mathrm{T}} \right)^{-1} \\ \mathbf{Q}_i \mathbf{G}^{\mathrm{T}} \left(\mathbf{G} \mathbf{Q}_i \mathbf{G}^{\mathrm{T}} + \mathbf{A} \mathbf{P}_i^a \mathbf{A}^{\mathrm{T}} \right)^{-1} \end{pmatrix} \left(\mathbf{x}_{i+1}^s - \mathbf{A} \mathbf{x}_i^a - \mathbf{G} \mathbf{c}_i^0 \right) \quad (23.3)$$

(note the use of $\mathbf{x}_{i+1}^s$ in the last term of (23.3), with superscript s in place of $\mathbf{x}_t^a$ in (23.2) with superscript a). The estimate (23.3) of the state and control, $\mathbf{x}_i^s$ and $\mathbf{c}_i^s$, illustrated by the open circles and short-dashed curve in Fig. 23.1 extending backward in time from some future instant, is the so-called 'smoothed estimate' described by estimation and control theory (Bryson and Ho, 1975). That this smoothed solution (23.3) satisfies the model (23.1) and therefore has an explicit physically consistent temporal evolution can be shown by substitution.

A fundamental element in a smoothed solution, besides the state itself, is the smoothed control estimate, $\mathbf{c}_i^s$. ECCO employs various smoothing algorithms for different purposes, including the adjoint method (also known as 4D-Var), the Rauch–Tung–Strievel (RTS) fixed interval smoother, and Green's functions. It is instructive to note that the smoothing described above is an inversion of the model (**A** and **G** of (23.1)), whereas Kalman filtering can be recognized as an inversion of the observations (**H** of (23.16) described later), as can be seen by comparing (23.3) with (23.15), also discussed later (Wunsch, 1996).

23.3 ECCO products

Most of the ECCO analyses are global or near-global estimates of the ocean and employ various smoothing algorithms (Section 23.2) using the Massachusetts Institute of Technology General Circulation Model (MITgcm; Marshall et al., 1997). Estimated controls include external forcings of the ocean (e.g. winds, air–sea heat and freshwater fluxes), mixing parameters, and the model's initial condition. The ECCO website ⟨http://ecco-group.org⟩ has descriptions and links to its products, which can be grouped into two: the Central Production Solutions and the Development and Experimental Solutions.

The Central Production is the flagship analysis of the ECCO Consortium and consists of multidecadal analyses of (near-) global ocean circulation, employing nearly all extant observations (Table 23.1) using the adjoint method. The various versions of the Central Production differ in model resolution, model physics, spatial extent, and controls. The Version 2 Central Production (also known as ECCO-GODAE) has been employed in a number of applications, some of which are described in Section 3. The Version 2 estimate has a 1° spatial resolution with 23 vertical levels spanning the period from 1992 to 2004. The ongoing Version 3 Central Production extends the Version 2 analysis period by employing further improvements in the model (e.g., prognostic sea-ice model) and optimization schemes (e.g., modelling correlated error), while the Version 4 Central Production that is under development includes the Arctic Ocean in its model domain, thus making the estimates truly global.

The Development and Experimental Solutions provide various targeted analyses with particular foci and include test-bed syntheses designed for future incorporation into the Central Production. For instance, the ECCO Near Real-Time Analysis (also

Table 23.1 Observations employed in ECCO Central Production

VARIABLE	SOURCE (DURATION)/REFERENCE (YEAR)
Sea level	TOPEX/Poseidon, Jason-1 and -2 (1992–2011), Geosat-Follow-On (2001–2005), ERS-1/2 (1992–2001), ENVISAT (2002–2008), tide gauges (1992–2011)
Climatology	Gouretski and Koltermann (2004)
Temperature	CTDs (1992–2011), XBTs (1992–2011), Argo floats (1997–2011), TOGA/TAO+PIRATA moorings (1992–2011), SEaOS (2006–08), RAPID moorings (2004–2010)
Salinity	CTDs (1992–2011), Argo floats (1997–2011), SEaOS (2006–2008)
SST	AVHRR (1992–2011), TMI (1998–2011), AMSR-E (2002–2011)
Transports	Florida Strait transport (1992–2011)
Sea ice	SSM/I (1992–2011)
Mean SSH	GRACE (2002–2011), altimetry (1993–2011)
Bottom Pressure	GRACE (2002–2011)
Wind Stress	NSCAT (1996–1997), QuikSCAT (1999–2008), ERS-1/2 (1992–2001)

known as ECCO-JPL) provides estimates from 1993 to the present that are extended monthly on a near real-time basis. The Near Real-Time Analysis employs satellite sea surface height measurements and in situ temperature profiles using a computationally efficient Kalman filter and RTS smoother (see e.g. Section 23.5.1). The near real-time nature of this analysis is useful in many applications, especially those that require timely estimates, some of which are also described in Section 23.4. The Development and Experimental Solutions also include products of the ECCO Ocean and Ice Interactions (also known as ECCO-2) and the German ECCO (GECCO) projects. The Ocean and Ice Interactions effort aims to estimate the ocean at spatial resolutions that begin to resolve ocean eddies on a global basis and to develop coupled ocean, sea-ice, and ice-sheet data synthesis capabilities. The GECCO project extends an earlier version of the Central Production (Version 1) to cover the full 50-year period (1950–2000) of the NCEP/NCAR atmospheric re-analysis.

23.4 Examples of ECCO applications

23.4.1 Analyses of ocean circulation

Model–data syntheses' complete physical descriptions of the ocean provide effective means to analyse the ocean as a whole. For instance, several studies have employed ocean state estimates to investigate the meridional overturning circulations (MOCs) of the ocean, which play vital roles in regulating climate by redistributing heat and other properties across the globe. Lee and Fukumori (2003) found interannual meridional heat-transport variations of the so-called subtropical cell having an opposite phase between the western-boundary current and the basin interior circulation, demonstrating the significance of monitoring the ocean in its entirety as opposed to one particular component alone. Wunsch and Heimbach (2006) analysed the depth-integrated heat and volume transports of the North Atlantic Ocean and identified significant month-to-month variability that ship-based observations do not resolve, illustrating the importance of estimates with adequate temporal resolution.

Cabanes et al. (2008) attributed interannual changes in North Atlantic meridional overturning circulation to fluctuations in pycnocline depth near the basin's western boundary driven by variations in local wind. In comparison, Köhl and Stammer (2008) found decadal changes in the overturning circulation to be associated with remotely forced density anomalies such as those from the Denmark Strait overflow propagating southward along the western boundary and baroclinic Rossby waves propagating westward in the interior of the basin.

The physically consistent nature of smoothed estimates also allows complete budget studies of various properties that help discern processes and mechanisms underlying the state of the ocean and its variabilities. In addition to the model state itself (temperature, salinity, velocity, sea level), individual terms of the property budgets, such as separate flux components of advection and mixing, are available for many of the ECCO products to facilitate such analyses. Such flux output also allows for accounting high-frequency variability in the circulation (e.g. Reynolds fluxes) that time-averaged state output cannot resolve.

For instance, Kim et al. (2007) analysed the temperature budget of the equatorial Pacific to study interannual variations of the surface mixed layer associated with El Niño and La Niña events. In comparison with most studies, in which one or more components of the budget are estimated as residuals of others, Kim et al. (2007) explicitly evaluated every component of the mixed-layer temperature budget separately from one another, the sum of which equals the estimated temperature change. The study reveals that contributions of vertical mixing and zonal advection are comparable to each other and are the largest among all processes that give rise to heat content variations and the resulting air–sea heat-flux anomalies, but do not explain the entire temperature change by themselves. Entrainment by vertical advection through the mixed-layer base and meridional advection by Ekman transport are smaller, but also contribute significantly to the variation. Additionally, the study demonstrates the utility of explicit flux estimates by quantifying the importance of submonthly variations in the circulation (e.g. tropical instability waves) that cannot be resolved adequately by typical time-averaged state output.

Similar budget analyses can also be found, for example, in Kim et al. (2004), who linked an abrupt warming of upper ocean temperature in the late 1990s in the North Pacific Ocean to coherent variations in atmospheric forcing and ocean circulation. Halkides et al. (2009, 2011) found ocean circulation playing a major role in surface mixed-layer temperature variations in tropical regions of the Indian and Pacific Oceans. Qu et al. (2011) analysed the near-surface salinity budget to investigate the role of ocean circulation in explaining the spatial offset between the maxima of sea surface salinity and net air–sea freshwater fluxes in the subtropical North Atlantic Ocean.

Budgets are of particular interest in understanding sea level variations. Global mean sea level rise provides one of the most compelling pieces of observational evidence for global climate change. Satellite measurements available since the early 1990s show a global mean sea level rise of approximately 3 mm/year, which can be ascribed to the sum of thermal expansion of the ocean due to an increase in ocean heat content and additional ocean mass due to runoff from increased melting of glaciers and ice sheets on land. Identifying the relative contributions of the two is fundamental to understanding climate change and its underlying processes. But because of inaccuracies in observations and incompleteness in available measurements, closing the sea level budget is not trivial from data alone.

Using the ECCO estimate constrained by hydrographic data and altimetric variability, but purposely excluding global mean sea level observations and satellite ocean bottom pressure measurements, Wunsch et al. (2007) estimate a 1.6 mm/year global mean sea level rise that amounts to approximately 60% of the pure altimetric measurement. Of this 1.6 mm/year rise, they attribute 1.1 mm/year to freshwater input and the rest to an increase in ocean heat content. Although the values are somewhat smaller than other estimates, the model–data synthesis estimate represents one of the first complete estimates of the sea level budget constrained by observations.

Regional sea level variations are often much larger than that of the global mean due to water mass redistribution by ocean circulation, which is well resolved by model–data syntheses. Piecuch and Ponte (2011) analysed the budget of sea level variability by decomposing the variations into their different components. In particular, local variations of sea level are shown to be almost entirely steric changes as opposed to mass-induced changes of sea level. The steric changes in turn are found to be dominated by variations in advection as opposed to surface buoyancy changes or mixing.

23.4.2 Applications of the model adjoint

In addition to being useful in combining observations with models, a model's adjoint, which provides a computationally efficient means to evaluate the sensitivity of the model to its various elements, is particularly effective in identifying and quantifying mechanisms of ocean circulation. For instance, Fukumori et al. (2007) employed the ECCO adjoint to identify the causal mechanism of a near-uniform basin-wide sea level fluctuation of the Mediterranean Sea observed by satellite altimetry. The sensitivity of a model property at a time t, say the mean sea level of the Mediterranean Sea, J_t, to the model's control vector at an earlier instant $t-n$, $\mathbf{c}_{t-n}$, can be written, using the chain rule and the model defined by (23.1), as

$$\frac{\partial J_t}{\partial \mathbf{c}_{t-n}} = \frac{\partial \mathbf{x}_{t-n+1}^{\mathrm{T}}}{\partial \mathbf{c}_{t-n}} \frac{\partial \mathbf{x}_{t-n+2}^{\mathrm{T}}}{\partial \mathbf{x}_{t-n+1}} \cdots \frac{\partial \mathbf{x}_t^{\mathrm{T}}}{\partial \mathbf{x}_{t-1}} \frac{\partial J_t}{\partial \mathbf{x}_t} = \mathbf{G}_{t-n}^{\mathrm{T}} \mathbf{A}_{t-n+1}^{\mathrm{T}} \cdots \mathbf{A}_{t-1}^{\mathrm{T}} \frac{\partial J_t}{\partial \mathbf{x}_t}, \tag{23.4}$$

which can be recognized as integrating the model adjoint $\mathbf{A}^{\mathrm{T}}$ (and $\mathbf{G}^{\mathrm{T}}$) backwards in time (subscripts for $\mathbf{A}$ and $\mathbf{G}$ are introduced for clarity in (23.4)). Such sensitivity can be used to quantify specific contributions of different controls to the cost function J_t, which can be written as

$$J_t = \sum_{i=0}^{\infty} \frac{\partial J_t}{\partial \mathbf{c}_{t-i}^{\mathrm{T}}} \mathbf{c}_{t-i}. \tag{23.5}$$

Equation (23.5) can be interpreted as a decomposition of J_t among contributions from different individual controls. For instance, the contribution to J_t from wind at a particular location, $\tau(x, y)$, can be written as

$$\sum_{i=0}^{\infty} \frac{\partial J_t}{\partial \tau_{t-i}(x, y)} \tau_{t-i}(x, y). \tag{23.6}$$

Then the fraction of the mean sea level variance of the Mediterranean Sea accounted for by zonal wind at each model grid point (x, y) can be defined as

$$1 - \frac{\mathrm{Var}\left(J_t - \sum_{i=0}^{\infty} \frac{\partial J_t}{\partial \tau_{t-i}(x, y)} \tau_{t-i}(x, y) \right)}{\mathrm{Var}(J_t)}, \tag{23.7}$$

where Var denotes the variance of its argument. In Fukumori et al. (2007), zonal wind is found to account for 74% of the mean sea level variance of the Mediterranean Sea. Moreover, the analysis demonstrates that the winds responsible for this are almost entirely those in the vicinity of the Strait of Gibraltar.

The model adjoint can also be employed to identify the origin of different water masses and their properties. Specifically, just as the evolution of a passive tracer can be interpreted as describing where a tracer-tagged water mass goes to (its fate), the evolution of its adjoint can be recognized as identifying where an adjoint passive tracer-tagged water comes from (its origin), with their trajectories defining their respective circulation pathways. Fukumori et al. (2007) employed such a passive tracer and its adjoint to study the circulation pathway of water in the equatorial Pacific Ocean associated with El Niño and La Niña events. The analysis reveals differences between the water's origin and fate that are not apparent in the zonally averaged Eulerian circulation known as the Subtropical Cell. The study also shows the significance of eddy transport in short-circuiting the circuitous tropical–subtropical circulation pathway, which helps account for observed tracer distributions in the ocean.

Unlike trajectories of finite number of floats, the evolution of passive tracers and their adjoint account for the entire circulation of water masses, including effects of mixing, providing additional insight into the workings of the ocean. ECCO has available a web-based interface ⟨http://ecco.jpl.nasa.gov/tracer⟩ to conduct such tracer analyses. Tracers are initialized to a unit value in a user-defined region and integrated over a user-specified time interval. Forward tracer and backward adjoint tracer integrations are conducted when the specified start time precedes the end time and vice versa, respectively.

Employing this tool, Gierach et al. (2012) analysed how changes in ocean circulation between different El Niño events affect the biological response in the equatorial Pacific Ocean. A simulated adjoint tracer was also employed by Wang et al. (2004) to trace the origin of temperature and salinity anomalies found in the tropical Pacific Ocean during El Niños to regions off the equator. Qu et al. (2009) analysed the origin of 13°C Water, a mode water found in the eastern equatorial Pacific Ocean that forms a thermostat.

The efficient model sensitivity analysis afforded by model adjoints is also effective in observing system simulation experiments (OSSE), such as those by Köhl and Stammer (2004), who employed the ECCO adjoint to analyse optimal observations in constraining meridional heat transport across the Greenland–Scotland ridge. Losch and Heimbach (2007) employed the adjoint to demonstrate how a model's bottom topography should be treated with uncertainty as in other model parameters by investigating the sensitivity of various key circulation quantities, such as the strength of meridional overturning circulation, to the model's representation of bottom topography.

23.4.3 Applications in other disciplines

Ocean state estimates are also of fundamental utility in understanding processes dependent on or affected by ocean circulation. For instance, McKinley et al. (2003, 2004) employed a biogeochemical model driven by ECCO ocean state estimates to study interannual air–sea flux variabilities of O_2 and CO_2. In both cases, temporal changes in ocean circulation, in particular the El Niño/Southern Oscillation (ENSO) cycle and wintertime convection, are found to be dominant controlling mechanisms of global variations in the air–sea fluxes.

Changes in large-scale ocean circulation also have significant effects on geodetic variabilities. For example, Gross et al. (2003) found changes in ocean circulation affecting polar motion in a wide range of frequencies. Moreover, a much higher coherence is found between observed and modelled polar motion excitation based on data-constrained ocean state estimates than that based on model simulations without data, demonstrating the utility of model–data syntheses. Changes in the Earth's oblateness (Dickey et al., 2002) and variations in the length of the day (Marcus et al., 2012) have also been attributed to variabilities in ocean circulation based on applications of similar ocean state estimates. Wu et al. (2010) employed an ocean state estimate in combination with other models and observations to separate effects in the geodetic measurements of present-day water transport across the globe and glacial isostatic adjustment.

23.5 Practical considerations in employing advanced estimation methods

23.5.1 State approximations

The ECCO Kalman filter and RTS smoother employ a series of approximations to evaluate the filter and smoother gains, in particular the model state error covariance matrices, so as to render the calculation computationally feasible. These approximations include a time-asymptotic approximation (Fukumori et al., 1993) that does away with integrating the error covariance continuously in time by deriving and employing its time-asymptotic solution in the estimation instead. Additionally, this calculation is further simplified by partitioned reduced-state approximations (Fukumori, 2002; Fukumori and Malanotte-Rizzoli, 1995) that approximate the model state with others with fewer degrees of freedom. By combining such approximations with the numerical model, the model's state error covariance matrix is derived using the Riccati equation with the corresponding control error covariance matrices, but with smaller associated computational requirements than otherwise, owing to the approximated state's smaller effective dimension.

Reduced-state approximations and their variants have been employed in various applications of Kalman filtering and related smoothing. However, care must be taken in defining the approximations so as to ensure consistency of the results. In particular, the state reduction must be designed to form a closed dynamic system. For instance, the model state and its controls can be written in terms of their approximations, $\mathbf{x}'$ and $\mathbf{c}'$, as

$$\begin{aligned} \mathbf{x} &= \mathbf{B}\mathbf{x}' + \mathbf{N}\mathbf{n}, \\ \mathbf{c} &= \tilde{\mathbf{B}}\mathbf{c}' + \tilde{\mathbf{N}}\mathbf{m}, \end{aligned} \tag{23.8}$$

where $\mathbf{B}$ and $\tilde{\mathbf{B}}$ define the corresponding approximations. Here, $\mathbf{N}$ and $\tilde{\mathbf{N}}$ are the approximation's null spaces with amplitudes $\mathbf{n}$ and $\mathbf{m}$, respectively. Substituting (23.8) into (23.1) and sorting terms, the approximated state and its corresponding controls satisfy

$$\begin{aligned} \mathbf{B}\mathbf{x}'_{t+1} + \mathbf{N}\mathbf{n}_{t+1} &= \mathbf{A}(\mathbf{B}\mathbf{x}'_{t+1} + \mathbf{N}\mathbf{n}_{t+1}) + \mathbf{G}(\tilde{\mathbf{B}}\mathbf{c}'_{t+1} + \tilde{\mathbf{N}}\mathbf{m}_{t+1}) \\ &= (\mathbf{A}\mathbf{B}\mathbf{x}'_{t+1} + \mathbf{G}\tilde{\mathbf{B}}\mathbf{c}'_{t+1}) + (\mathbf{A}\mathbf{N}\mathbf{n}_{t+1} + \mathbf{G}\tilde{\mathbf{N}}\mathbf{m}_{t+1}). \end{aligned} \tag{23.9}$$

By multiplying both sides of this equation with the pseudo-inverse of $\mathbf{B}$, denoted by $\mathbf{B}^*$, and noting the orthogonality between $\mathbf{B}$ and $\mathbf{N}$, the approximated state's evolution is governed by

$$\mathbf{x}'_{t+1} = \mathbf{B}^*\mathbf{A}\mathbf{B}\mathbf{x}'_{t+1} + \mathbf{B}^*\mathbf{G}\tilde{\mathbf{B}}\mathbf{c}'_{t+1} + (\mathbf{B}^*\mathbf{A}\mathbf{N}\mathbf{n}_{t+1} + \mathbf{B}^*\mathbf{G}\tilde{\mathbf{N}}\mathbf{m}_{t+1}). \tag{23.10}$$

The last term in parentheses, which depends on the null space ($\mathbf{n}$ and $\mathbf{m}$), is a source of inaccuracy in defining a reduced-state model and can undermine the consistency of the resulting estimate (Section 23.2). However, because of their orthogonality, this dependence could be ignored if range and null-space perturbations remain within their respective domain through the model's integration (i.e. a closed system); that is,

$$\begin{aligned} \mathbf{B}^*\mathbf{A}\mathbf{N} &\approx \mathbf{B}^*\mathbf{N} = 0, \\ \mathbf{B}^*\mathbf{G}\tilde{\mathbf{N}} &\approx \mathbf{B}^*\mathbf{N} = 0. \end{aligned} \tag{23.11}$$

In the ECCO filter and smoother, the reduced state consists of the gravest vertical dynamic modes defined on a coarse subset of the model grid. To first approximation, a perturbation's modal structure is time-invariant and large-scale perturbations remain large-scale, thus satisfying this requirement.

23.5.2 Approaches in implementing Kalman filters and related smoothers

Here we review some considerations in implementing Kalman filters and related smoothers. Derivation and implementation of Kalman filters and related smoothers are often described in terms of matrix algebra. System matrices, such as $\mathbf{A}$ and $\mathbf{G}$ in (23.1) or their equivalent in approximated models (e.g. (23.10)), can be explicitly derived numerically as Green's functions. For instance, for a general model $\mathcal{G}$ that implicitly integrates the state as (23.1),

$$\mathbf{x}_t = \mathbf{A}\mathbf{x}_{t-1} + \mathbf{G}\mathbf{c}_{t-1} \equiv \mathcal{G}(\mathbf{x}_{t-1}, \mathbf{c}_{t-1}), \tag{23.12}$$

the ith column of $\mathbf{A}$, $(\mathbf{A})_i$, can be evaluated as

$$(\mathbf{A})_i = \mathbf{A}\mathbf{e}_i + \mathbf{G}\mathbf{0} = \mathcal{G}(\mathbf{x} + \mathbf{e}_i, \mathbf{c}) - \mathcal{G}(\mathbf{x}_i, \mathbf{c}), \tag{23.13}$$

where $\mathbf{e}_i$ is the ith column of the identity matrix. Other equivalent matrix operators of the model can be derived similarly.

An alternative form of the Kalman gain based on the analysis error covariance matrix $\mathbf{P}_t^a$,

$$\mathbf{x}_t^a = \mathbf{x}_t^f + \mathbf{P}_t^a \mathbf{H}^{\mathrm{T}} \mathbf{R}_t^{-1} (\mathbf{y}_t - \mathbf{H}\mathbf{x}_t^f), \tag{23.14}$$

can be computationally more convenient than the more common form based on the forecast error covariance matrix $\mathbf{P}_t^f$,

$$\mathbf{x}_t^a = \mathbf{x}_t^f + \mathbf{P}_t^f \mathbf{H}^{\mathrm{T}} (\mathbf{H}\mathbf{P}_t^f \mathbf{H}^{\mathrm{T}} + \mathbf{R}_t)^{-1} (\mathbf{y}_t - \mathbf{H}\mathbf{x}_t^f), \tag{23.15}$$

especially when $\mathbf{P}_t^a$ is available, such as with the asymptotic approximation, since (23.14) involves fewer operations than (23.15) does. Here, $\mathbf{H}\mathbf{x}_t$ is the model equivalent of the data $\mathbf{y}_t$ and $\mathbf{H}$ its operator:

$$\mathbf{y}_t \approx \mathbf{H}\mathbf{x}_t. \tag{23.16}$$

As in any matrix–vector operation, the computations are carried out from right to left in (23.14) as a sequence of matrix–vector operations, as opposed to evaluating any of the matrix–matrix products, since the former involves fewer computations than the latter.

Transpose operators, such as $\mathbf{H}^{\mathrm{T}}$ in (23.14), can conveniently be implemented using the adjoint of the operator defined algorithmically when the matrices are not available explicitly. Similar approaches can be taken to implement the pseudo-inverse of a state approximation $\mathbf{B}^*$ (Section 23.5.1), which can be defined as

$$\mathbf{B}^* = \left(\mathbf{B}^{\mathrm{T}}\mathbf{B}\right)^{-1} \mathbf{B}^{\mathrm{T}}. \tag{23.17}$$

The operator in parentheses in (23.17) can be evaluated numerically using a Green's function approach (e.g. (23.13)) and inverted explicitly, since it has a small dimension by design (i.e. state reduction).

23.5.3 Approaches in implementing adjoint methods

Besides generating an adjoint code, which is a significant undertaking in itself, some considerations are also called for in carrying out actual adjoint model integrations in large-scale modelling studies. For instance, in typical model calculations, past model states that are needed in the adjoint calculation are overwritten as the model is integrated forwards in time and are no longer available during the adjoint's integration backwards in time. Saving model states at every single model time-step is usually not feasible in either computer memory or on disk. Instead, so-called checkpointing schemes (also known as divided adjoints) can be employed in such calculations to minimize the computational requirements (Fig. 23.2), whereby model restart files are output to disk at certain intervals and are read back in during adjoint integrations to recompute the intervening model states as necessary.

In ECCO calculations, a three-level checkpointing scheme that consists of a nested three-level loop is employed, as illustrated in Fig. 23.2. Restart files are first written to disk at N intervals during the model's forward integration step spanning the entire analysis period, denoted as the outer level (level 1). The adjoint integration then proceeds backwards one interval of the outer loop at a time, which is further split into two levels. In the middle level (level 2), a particular outer level restart file is read in (t_k) and the model is integrated forwards again to compute and temporarily save another sequence of restart files at M intervals spanning the particular outer level interval. Then, in the inner level (level 3), a middle level restart file (s_i) is read back in and the model is integrated forwards once again over L model time-steps to the next middle level restart, with the model state and its ancillary variables computed and stored entirely in memory for every time-step to allow the adjoint integration between the successive middle level restarts (s_{i+1} to s_i).

After the inner level adjoint integration over the L time-steps, the next inner level calculation is performed over the preceding middle level interval (from s_i to s_{i-1}),

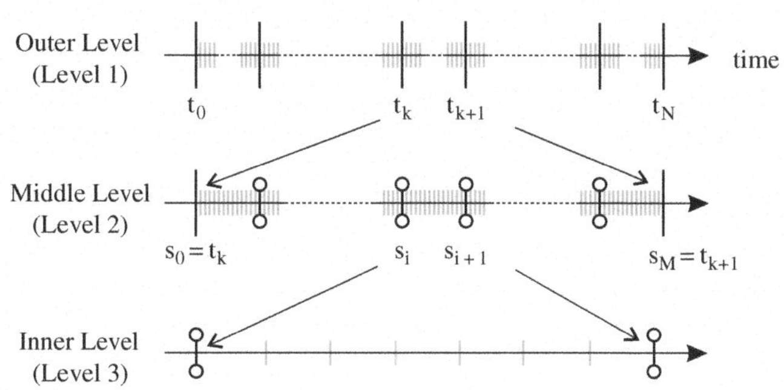

Fig. 23.2 A schematic illustration of a three-level checkpointing scheme. Black tick marks denote instances of restart files saved to disk. Grey tick marks represent other individual model time-steps retained in memory. The middle level restarts (black with circles) and inner level states (grey) are overwritten as adjoint integration proceeds. See the text for further description.

replacing the previous L time-steps in memory, and the procedure is repeated until the beginning of the entire existing middle level is reached ($s_0 = t_k$). Then, a new sequence of middle level restarts are computed and saved for the next preceding outer level interval ($s_0 = t_{k-1}$ to $s_M = t_k$), writing over the previous M restart files, and the whole level 2/level 1 process repeated until the beginning of the entire analysis period (t_0) is reached.

The three-level checkpointing scheme requires $N + M$ restart files at any given time on disk and L time-steps in memory, which is far smaller than storing the entire number of time-steps, which amounts to $N \times M \times L$. In practice, L is chosen to be as large as the available computer memory allows and N and M are chosen to be as close to each other as possible, which results in minimum storage requirements. For instance, the 19-year ECCO Version 3 Central Solution model has a one-hour time-step, which amounts to over 166 000 time-steps in 19 years, but with a three-level checkpointing scheme with $(N, M, L) = (84, 84, 24)$, the disk and memory storage only amount to an equivalent of $N + M + L = 192$ time-steps.

Strong nonlinearities in the model can result in inaccuracies in the model adjoint and require careful treatment. For instance, mixing events, such as convection, are episodic and highly nonlinear, and so a gradient cannot easily be defined. In this particular situation, the exact mixing sequence (e.g. state-dependent mixing parameters) can be stored during the forward calculation and utilized during the adjoint integration step, as opposed to recomputing it to avoid inaccuracies associated with the adjoint's linearization. Similarly, utilizing stored ancillary variables where possible as opposed to recomputing them during checkpointing can also reduce the computational requirements of the adjoint and warrant consideration.

23.5.4 Prescribing errors

The data and control error covariance matrices, $\mathbf{R}$ and $\mathbf{Q}$ respectively in (23.3) and (23.14), and implicitly in $\mathbf{P}$, are fundamental in combining models and data and require proper specification. Data error requires some clarification, since it includes what some might consider to be model error. Besides inaccuracies inherent to observing systems, the 'data error' in this context includes so-called 'representation error', which corresponds to aspects of the natural system (i.e. the ocean in this discussion) that the model cannot fundamentally simulate.

Following Cohn (1997), the nature of data error can best be understood by considering the differences among the true states of the natural system $\mathbf{w}_t$, the model $\bar{\mathbf{x}}_t$, and the observations $\mathbf{y}_t$. For instance, the model's true state $\bar{\mathbf{x}}_t$ (the overbar is to distinguish it from estimates) can be regarded as representing the true ocean in finite dimensions:

$$\bar{\mathbf{x}}_t \equiv \mathbf{\Pi} \mathbf{w}_t. \tag{23.18}$$

The function $\mathbf{\Pi}$ defines the model state given the complete state of the ocean $\mathbf{w}_t$, which is continuous with complete physics that the model does not necessarily resolve. Similarly, observations $\mathbf{y}_t$,, which also have a finite dimension, are samples of this true ocean $\mathbf{w}_t$, which could be written as

$$\mathbf{y}_t = \mathbf{E}\mathbf{w}_t + \varepsilon, \tag{23.19}$$

where $\mathbf{E}$ describes the sampling operation and ε denotes measurement errors (e.g. instrument error).

In terms of the model (23.16), (23.19) can be rewritten using (23.18) as

$$\mathbf{y}_t = \mathbf{H}\bar{\mathbf{x}}_t + (\mathbf{E}\mathbf{w}_t - \mathbf{H}\boldsymbol{\Pi}\mathbf{w}_t) + \varepsilon. \tag{23.20}$$

The last two terms of (23.20),

$$(\mathbf{E}\mathbf{w}_t - \mathbf{H}\boldsymbol{\Pi}\mathbf{w}_t) + \varepsilon, \tag{23.21}$$

can be identified as the data error. The first part of (23.21) is the representation error and is the difference between error-free observations ($\mathbf{E}\mathbf{w}_t$) and the error-free equivalent of the model ($\mathbf{H}\boldsymbol{\Pi}\mathbf{w}_t$). The two are generally different because the model does not simulate the entire spectrum of the ocean but only parts of it as defined by (23.18). For instance, a finite-resolution model does not simulate variability that occurs on scales smaller than its resolution, and therefore such small-scale variability that is part of the observations $\mathbf{y}_t$ should be considered data error.

Data error and control error can be deduced empirically. In ECCO, the so-called 'covariance matching' method described by Fu et al. (1993) is employed. For instance, observations $\mathbf{y}$ and their model simulation's equivalent $\mathbf{m}$ could be written as the sum of the true signal $\mathbf{s}$ (the first term on the right-hand side of (23.19)) and their respective uncertainties $\mathbf{r}$ and $\mathbf{p}$:

$$\begin{aligned} \mathbf{y} &= \mathbf{s} + \mathbf{r}, \\ \mathbf{m} &= \mathbf{s} + \mathbf{p}. \end{aligned} \tag{23.22}$$

For simplicity, we assume $\mathbf{s}$, $\mathbf{r}$, and $\mathbf{p}$ to have zero means and to be uncorrelated with each other. Then, the covariance among these elements can be written as

$$\begin{aligned} \langle \mathbf{y}\mathbf{y}^{\mathrm{T}} \rangle &= \langle \mathbf{s}\mathbf{s}^{\mathrm{T}} \rangle + \langle \mathbf{r}\mathbf{r}^{\mathrm{T}} \rangle, \\ \langle \mathbf{m}\mathbf{m}^{\mathrm{T}} \rangle &= \langle \mathbf{s}\mathbf{s}^{\mathrm{T}} \rangle + \langle \mathbf{p}\mathbf{p}^{\mathrm{T}} \rangle, \\ \langle \mathbf{y}\mathbf{m}^{\mathrm{T}} \rangle &= \langle \mathbf{s}\mathbf{s}^{\mathrm{T}} \rangle, \end{aligned} \tag{23.23}$$

where $\langle\,\rangle$ denotes statistical expectation. Assuming ergodicity and stationarity, the quantities on the left-hand side could be estimated by averaging the data and model estimates in time. Then,

$$\begin{aligned} \langle \mathbf{r}\mathbf{r}^{\mathrm{T}} \rangle &\approx \left[\mathbf{y}\mathbf{y}^{\mathrm{T}}\right] - \left[\mathbf{y}\mathbf{m}^{\mathrm{T}}\right], \\ \langle \mathbf{p}\mathbf{p}^{\mathrm{T}} \rangle &\approx \left[\mathbf{m}\mathbf{m}^{\mathrm{T}}\right] - \left[\mathbf{y}\mathbf{m}^{\mathrm{T}}\right], \end{aligned} \tag{23.24}$$

where $[\,]$ denotes sample average. The former is a direct estimate of the data error covariance matrix $\mathbf{R}$, while the latter provides an indirect estimate of the control error covariance matrix $\mathbf{Q}$. An example of such estimates and their application can be found, for example, in Fukumori et al. (1999).

23.5.5 Correlated errors

For computational convenience, data errors are often assumed to be uncorrelated with each other both temporally and spatially, as well as among different types of observations, by employing diagonal error covariance matrices $\mathbf{R}$ and defining cost functions J with a summation of model–data differences over time, for example,

$$J = \sum_t \left(\mathbf{c}_t - \mathbf{c}_t^0\right)^{\mathrm{T}} \mathbf{Q}^{-1} \left(\mathbf{c}_t - \mathbf{c}_t^0\right) + \sum_t \left(\mathbf{y}_t - \mathbf{H}\mathbf{x}_t\right)^{\mathrm{T}} \mathbf{R}^{-1} \left(\mathbf{y}_t - \mathbf{H}\mathbf{x}_t\right). \qquad (23.25)$$

Consideration is warranted in utilizing observations to ensure consistency between what the actual data error correlations are relative to what is assumed, so as to minimize inaccuracies in the resulting model–data syntheses. For instance, observations at much higher spatial and/or temporal resolution than what a model can simulate can be strongly correlated with each other because of representation error. Decimating the data in advance to resolutions commensurate with the model can minimize correlation among such errors and can be effective when assuming uncorrelated data errors.

Similarly, control errors are also often assumed to be uncorrelated, in particular in time. However, some controls are known to have significant errors in their time-mean and/or their trend that are correlated in time by definition. Defining separate controls for time-mean and trend errors, distinct from their time-variable components, provides one way to account for such correlation but still permits the separate controls to be treated as uncorrelated with each other without inconsistencies.

Spatial and intervariable correlation among control errors can be accounted for in reduced-state filters and smoothers by explicitly specifying covariances among the reduce-state components, which is computationally feasible owing to the relatively small dimension of the reduced-state. Additionally, operators defining the reduced state approximation $\tilde{\mathbf{B}}$, (23.8), such as 'objective mapping' (Bretherton et al., 1976), implicitly account for correlation among the model variables through their mapping operations. Spatial correlation among controls can also be accounted for, without explicitly forming covariance matrices, by a variable transformation that employs a diffusion operator (Weaver and Courtier, 2001).

23.6 Summary

Examples of applying advanced estimation methods to study large-scale ocean circulation problems have been described. Understanding what is being solved in such estimates is fundamental to effectively utilizing their results. In particular, syntheses designed for numerical prediction often do not provide a complete accounting of the physics controlling the estimate's temporal evolution. Explicit estimation of model controls, i.e. the sources of model error, is essential to obtaining a physically consistent description of the ocean.

The complete and consistent estimates of the ocean, together with advanced estimation and modelling tools, provide a powerful means to gain insight into the workings of the ocean and its impact on climate. Model adjoints are especially useful, not only in conducting model–data syntheses, but also in identifying causal mechanisms and circulation pathways, and in quantifying sensitivities of the ocean to various elements of the dynamic system.

Recognizing underlying assumptions employed in advanced estimation methods and ensuring consistencies with what is employed are essential for conducting effective model–data syntheses and in devising additional approaches and approximations to further improve the fidelity of the analyses.

Acknowledgements

This work was carried out in part at the Jet Propulsion Laboratory (JPL), California Institute of Technology, under contract with the National Aeronautics and Space Administration.

References

Bretherton, F. P., Davis, R. E., and Fandry, C. B. (1976). A technique for objective analysis and design of oceanographic experiments applied to MODE-73. *Deep-Sea Res.*, **23**, 559–582.

Bryson, A. E., Jr and Ho, Y.-C. (1975). *Applied Optimal Control.* Hemisphere, New York.

Cabanes, C., Lee, T. and Fu, L.-L. (2008). Mechanisms of interannual variations of the meridional overturning circulation of the North Atlantic Ocean. *J. Phys. Oceanogr.*, **38**, 467–480.

Cohn, S. E. (1997). An introduction to estimation theory. *J. Meteorol. Soc. Japan*, **75**, 257–288.

Dickey, J. O., Marcus, S. L., de Viron, O., and Fukumori, I. (2002). Recent Earth oblateness variations: Unraveling climate and postglacial rebound effects. *Science*, **298**, 1975–1977.

Fu, L.-L., Fukumori, I. and Miller, R. N. (1993). Fitting dynamic models to the Geosat sea level observations in the Tropical Pacific Ocean. Part II: A linear, wind-driven model. *J. Phys. Oceanogr.*, **23**, 2162–2181.

Fukumori, I. (2002). A partitioned Kalman filter and smoother. *Mon. Weather Rev.*, **130**, 1370–1383.

Fukumori, I. and Malanotte-Rizzoli, P. (1995). An approximate Kalman filter for ocean data assimilation; an example with an idealized Gulf Stream model. *J. Geophys. Res.*, **100**, 6777–6793.

Fukumori, I., Benveniste, J., Wunsch, C., and Haidvogel, D. B. (1993). Assimilation of sea surface topography into an ocean circulation model using a steady-state smoother. *J. Phys. Oceanogr.*, **23**, 1831–1855.

Fukumori, I., Raghunath, R., Fu, L., and Chao, Y. (1999). Assimilation of TOPEX/POSEIDON data into a global ocean circulation model: How good are the results? *J. Geophys. Res.*, **104**, 25647–25665.

Fukumori, I., Lee, T., Cheng, B., and Menemenlis, D. (2004). The origin, pathway, and destination of Niño-3 water estimated by a simulated passive tracer and its adjoint. *J. Phys. Oceanogr.*, **34**, 582–604.

Fukumori, I., Menemenlis, D., and Lee, T. (2007). A near-uniform basin-wide sea level fluctuation of the Mediterranean Sea. *J. Phys. Oceanogr.*, **37**, 338–358.

Gierach, M. M., Lee, T., Turk, D., and McPhaden, M. J. (2012). Biological response to the 1997-98 and 2009-10 El Niño events in the equatorial Pacific Ocean. *Geophys. Res. Lett.*, **39**, L10602, doi: 10.1029/2012GL051103.

Gross, R. S., Fukumori, I. and Menemenlis, D. (2003). Atmospheric and oceanic excitation of the Earth's wobbles during 1980–2000. *J. Geophys. Res.*, **108**, 2370, doi:10.1029/ 2002JB002143.

Halkides, D. J. and Lee, T. (2009). Mechanisms controlling seasonal-to-interannual mixed layer temperature variability in the southeastern tropical Indian Ocean. *J. Geophys. Res.*, **114**, C02012, doi:10.1029/2008JC004949.

Halkides, D. J., Lucas, L. E., Waliser, D. E., Lee, T., and Murtugudde, R. (2011). Mechanisms controlling mixed-layer temperature variability in the eastern tropical Pacific on the intraseasonal timescale. *Geophys. Res. Lett.*, **38**, L17602, doi:10.1029/ 2011GL048545

Kim, S.-B., Lee, T., and Fukumori, I. (2004). The 1997–99 abrupt change of the upper ocean temperature in the Northcentral Pacific. *Geophys. Res. Lett.*, **31**, L22304, doi:10.1029/2004GL021142

Kim, S.-B., Lee, T., and Fukumori, I. (2007). Mechanisms controlling the interannual variation of mixed layer temperature averaged over the Niño3 region. *J. Climate*, **20**, 3822–3843.

Köhl, A., and Stammer, D. (2004). Optimal observations for variational data assimilation. *J. Phys. Oceanogr.*, **34**, 529–542.

Köhl, A. and Stammer, D. (2008). Variability of the meridional overturning in the North Atlantic from 50-year GECCO state estimation. *J. Phys. Oceanogr.*, **38**, 1913–1930.

Lee, T. and Fukumori, I. (2003). Interannual to decadal variation of tropical–subtropical exchange in the Pacific Ocean: boundary versus interior pycnocline transports. *J. Climate*, **16**, 4022–4042.

Losch, M. and Heimbach, P. (2007). Adjoint sensitivity of an ocean general circulation model to bottom topography. *J. Phys. Oceanogr.*, **37**, 377–393.

Marcus, S. L., Dickey, J. O., Fukumori, I., and de Viron, O. (2012). Detection of the Earth rotation response to a rapid fluctuation of Southern Ocean circulation in November 2009. *Geophys. Res. Lett.*, **39**, doi:10.1029/2011GL050671.

Marshall, J. C., Adcroft, A., Hill, C., Perelman, L., and Heisey, C. (1997). A finite-volume, incompressible Navier Stokes model for studies of the ocean on parallel computers. *J. Geophys. Res.*, **102**, 5753–5766.

McKinley, G. A., Follows, M. J., Marshall, J. and Fan, S. M. (2003). Interannual variability of air–sea O_2 fluxes and the determination of CO_2 sinks using atmospheric O_2/N_2. *Geophys. Res. Lett.*, **30**, doi: 10.1029/2002GL016044.

McKinley, G. A., Follows, M. J. and Marshall, J. (2004). Mechanisms of air–sea CO_2 flux variability in the equatorial Pacific and the North Atlantic. *Global Biogeochem. Cycl.*, **18**, GB2011, doi: 10.1029/2003GB002179.

Piecuch, C. G. and Ponte, R. M. (2011). Mechanisms of interannual steric sea level variability. *Geophys. Res. Lett.*, **38**, L15605, doi:10.1029/2011GL048440.

Qu, T., Gao, S., Fukumori, I., Fine, R. A., and Lindstrom, E. J. (2009). Origin and pathway of equatorial 13°C Water in the Pacific identified by a simulated passive tracer and its adjoint. *J. Phys. Oceanogr.*, **39**, 1836–1853.

Qu, T., Gao, S., and Fukumori, I. (2011). What governs the North Atlantic salinity maximum in a global GCM? *Geophys. Res. Lett.*, **38**, L07602, doi:10.1029/2011GL046757.

Wang, O., Fukumori, I., Lee, T., and Cheng, B. (2004). On the cause of eastern equatorial Pacific Ocean T–S variations associated with El Niño. *Geophys. Res. Lett.*, **31**, L15309, doi:10.1029/2004GL020188.

Weaver, A. and Courtier, P. (2001). Correlation modelling on the sphere using a generalized diffusion equation, *Q. J. R. Met. Soc.*, **127**, 1815–1846.

Wu, X. P. et al. (2010). Simultaneous estimation of global present-day water transport and glacial isostatic adjustment. *Nature Geosci.*, **3**, 642–646.

Wunsch, C. (1996). *The Ocean Circulation Inverse Problem.* Cambridge University Press, Cambridge.

Wunsch, C. and Heimbach, P. (2006). Estimated decadal changes in the North Atlantic meridional overturning circulation and heat flux 1993–2004. *J. Phys. Oceanogr.*, **36**, 2012–2024.

Wunsch, C., Ponte, R. M., and Heimbach, P. (2007). Decadal trends in sea level patterns: 1993–2004. *J. Climate*, **20**, 5889–5911.

Wunsch, C., Heimbach, P., Ponte, R. M., and Fukumori, I. (2009). The global general circulation of the ocean estimated by the ECCO-Consortium. *Oceanography*, **22**, 88–103.

24
Data assimilation in coastal oceanography: IS4DVAR in the Regional Ocean Modelling System (ROMS)

J. Zavala-Garay, J. Wilkin, and J. Levin

Ocean Modeling Group, Institute of Marine and Coastal Sciences
Rutgers University, USA

Advanced Data Assimilation for Geosciences. First Edition.
Edited by É. Blayo, M. Bocquet, E. Cosme, and L. F. Cugliandolo.

Chapter Contents

One of the many applications of data assimilation is the estimation of an analysis to be used as initial conditions for model hindcasts/forecasts. Realizing that satellite data is the only source of information in real time in most parts of the world ocean, the Ocean Modeling group of Rutgers University in the United States has developed methodologies that correctly exploit the information content in remotely sensed observations. In this chapter, we evaluate the extent to which incremental, strong-constraint, four-dimensional variational (IS4DVAR) data assimilation can improve prediction of mesoscale and submesoscale variability using the Regional Ocean Modelling System (ROMS). We present examples of two applications of IS4DVAR in two very different dynamical regimes: the East Australia Current (EAC) and the Middle Atlantic Bight (MAB). The two main sources of satellite information, namely sea surface temperature (SST) and sea surface height anomaly (SSHA), are found to be highly complementary, and therefore both need to be assimilated in order to approximate the three-dimensional structure of the ocean. We also present practical aspects for the implementation of IS4DVAR such as bias correction due mainly to strong biases present in the global ocean models, adequate modelling of the background error covariances, observation processing, and assimilation of along-track data of SSHA. The EAC application is characteristic of a deep ocean domain where quasi-geostrophy dominates the observed variability with highly isotropic decorrelation scales, justifying the use of gridded SSHA analysis such as AVISO and 1-day composites of SST. The MAB application is characteristic of coastal oceans with wide shelves and abrupt shelf breaks where scales can be more anisotropic owing to flow/bathymetry interactions, and consequently assimilation of along-track data and individual passes of infrared SST are preferred. Using reprocessed coastal altimeter data derived using dealiasing corrections and other required near-coast data filtering and error-correction approaches, we evaluate the extent to which the satellite data can constrain and improve the observed dynamics. The evaluation in both cases is performed by comparing not assimilated subsurface hydrographic observations (from XBTs, CTD, and autonomous gliders) with model hindcast. We find in both applications that the information content in the satellite data can be effectively propagated by the assimilation system helping to constrain the model to resemble the observed variability.

24.1 The Regional Ocean Modelling System and the IS4DVAR data assimilation algorithm

24.1.1 The Forward Model

The numerical ocean model used here is the Regional Ocean Modelling System (ROMS; ⟨http://www.myroms.org⟩), a three-dimensional numerical ocean circulation model developed principally for regional applications (Haidvogel et al., 2000, 2008). It solves the Reynolds-averaged Navier–Stokes equations using the hydrostatic and Boussinesq approximations. ROMS employs split-explicit separation of fast barotropic and slow baroclinic modes, and it is formulated in vertically stretched terrain following 'sigma') coordinates. Shchepetkin and McWilliams (2003, 2005) describe in detail the algorithms that comprise the ROMS computational kernel. It includes high-order advection and time-stepping schemes, weighted temporal averaging of the barotropic mode to reduce aliasing into the slow baroclinic motions, and conservative parabolic splines for vertical discretization. The barotropic pressure-gradient formulation used reduces the pressure gradient truncation error that has previously limited the accuracy of terrain following coordinate models (Shchepetkin and McWilliams, 2005). Momentum, scalar advection, and diffusive processes are represented using transport equations, while the vertical velocity is computed diagnostically using volume conservation.

24.1.2 The IS4DVAR driver

ROMS offers three variants of 4D-Var data assimilation as described by Moore et al. (2011): a primal formulation of incremental strong constraint 4D-Var (I4DVAR), a dual formulation based on a physical-space statistical analysis system (4D-PSAS), and a dual formulation representer-based variant of 4D-Var (R4D-Var). In its more general form, I4D-Var finds adjustments to the initial conditions, boundary conditions, and surface forcing in order to find the best estimate of the circulation in the least-squares sense (model error can be also included in 4D-PSAS and R4D-Var). In this chapter, we use the I4DVAR formalism under the assumption that the surface and lateral boundary forcings are error-free and therefore the only control variable adjusted is the initial condition. Following Sasaki (1970), assuming that the surface and lateral boundary forcings are error-free is equivalent to imposing the model as a 'strong constraint', and we therefore refer to this case as incremental, strong-constraint, four-dimensional variational (IS4DVAR) to avoid confusion with the more general capability of ROMS I4DVar, where boundary conditions and surface forcing are also part of the control vector. The variational data assimilation procedure operates by finding the minimum of a cost function representing the misfit between the model solution and observations subject to the condition that the solution obtained must satisfy exactly the equations defining the numerical model.

Let $\mathbf{x}(t) = \{\mathbf{u},\mathbf{v},\mathbf{T},\mathbf{S},\zeta\}$ be the column state vector of ROMS comprising the concatenation of velocities $(\mathbf{u},\mathbf{v})$, temperature $(\mathbf{T})$, salinity $(\mathbf{S})$, and SSH (ζ) vectors at all model grid points at a given time. We wish to combine the state vector $\mathbf{x}$ with all the observations available in a given period of time $t \in (t_{i-1}, t_i)$ to produce an analysis $\mathbf{x}^a(t)$ given a prior or background estimate $\mathbf{x}^b(t)$ of $\mathbf{x}$. In the experiments here, $\mathbf{x}^b$ is obtained from a forward ROMS run initialized from a previous analysis. Since $\mathbf{x}^b(t)$ will be close in some sense to the optimum state we are trying to estimate, it is convenient to formulate the problem in terms of 'increments' or perturbations $\delta\mathbf{x}$ to the background field: $\mathbf{x}^a = \mathbf{x}^b + \delta\mathbf{x}$.

The time evolution of the state vector is given by $\mathbf{x}(t) = M(t, t_{i-1})(\mathbf{x}(t_{i-1}))$, where M is the nonlinear propagator that advances the initial condition $\mathbf{x}(t_{i-1})$ in time. A first-order expansion of the analysis $\mathbf{x}^a(t) = M(t, t_{i-1})[\mathbf{x}^b(t_{i-1}) + \delta\mathbf{x}(t_{i-1})]$ is given by $\mathbf{x}^a(t) = \mathbf{x}^b(t) + \mathbf{M}(t, t_{i-1})\delta\mathbf{x}(t_{i-1})$, where $\mathbf{M}$ is a linearization of M about the nonlinear trajectory $\mathbf{x}^b(t)$ or the tangent-linear model of ROMS.

In the IS4DVAR implementation used, here the only control variable adjusted is the initial condition, and therefore the increment $\delta\mathbf{x}(t)$ is uniquely determined by

$\delta\mathbf{x}(t_{i-1})$. The incremental formulation of Courtier et al. (1994) minimizes the quadratic cost function $J = J_b + J_o$, with

$$J_b = \frac{1}{2}\delta\mathbf{x}^{\mathrm{T}}(t_{i-1})\mathbf{B}^{-1}\delta\mathbf{x}(t_{i-1}), \tag{24.1}$$

$$J_o = \frac{1}{2}\sum_{i=1}^{n}[\mathbf{HM}_i\delta\mathbf{x}(t_{i-1}) - \mathbf{d}_i]^{\mathrm{T}}\mathbf{O}_i^{-1}[\mathbf{HM}_i\delta\mathbf{x}(t_{i-1}) - \mathbf{d}_i], \tag{24.2}$$

where n is the number of observations, and the information provided by the observations $\mathbf{y}_i^o$ at time t_i is introduced through an innovation vector

$$\mathbf{d}_i = \mathbf{y}_i^o - HM(t_i, t_{i-1})(\mathbf{x}^b(t_{i-1})), \tag{24.3}$$

where H is a (possibly nonlinear) *observation operator* and $\mathbf{H}$ its linear counterpart. In the experiments here, H is simply a spatial interpolation, since the observed variables are elements of the state vector. H could, however, represent a more complicated formulation, such as that extracting the projection of model surface velocities along the radial paths observed by high-frequency coastal radars, or the transfer function between an observable biological variable and the state vector. The term J_b, (24.1), measures the total variation of the initial condition, which is weighted in the dot product by the inverse of the background error covariance matrix $\mathbf{B}$. The second term J_o, (24.2), measures the difference between the observations and model predictions in its incremental form, which is weighted by the inverse of the observation error covariance $\mathbf{O}_i$. This matrix is assumed here to be diagonal (i.e. observation errors are assumed to be spatially uncorrelated) and time-independent.

In practice, the problem is solved in an iterative procedure that exploits the fact that the solution of the adjoint model at time t_{i-1} gives the gradient of J_o with respect to changes in the initial conditions when forced by the weighted misfit $\mathbf{O}_i^{-1}[\mathbf{HM}_i\delta\mathbf{x}(t_{i-1}) - \mathbf{d}_i]$. This gradient information is then combined with $\partial J_b/\partial\mathbf{x}(t_{i-1}) = \mathbf{B}^{-1}\delta\mathbf{x}(t_{i-1})$ to find the total gradient, and used in a conjugate gradient algorithm to define a new increment $\delta\mathbf{x}$ in the direction that minimizes the total cost function J.

The advantage of defining the minimization of a problem in terms of increments is that the resulting J is a quadratic function of the control variable $\delta\mathbf{x}$ and therefore we deal with a simplified problem that is guaranteed to have a unique minimum, as opposed to the nonlinear formulation with potentially many minima. As suggested by Courtier et al. (1994), a practical approach to account for nonlinearities while retaining the practical advantages of the quadratic minimization problem is to allow the basic state trajectory used in the tangent linear and adjoint models to be regularly updated with the most recent state estimate of the minimization of the quadratic J. These nonlinear updates are performed in the so-called *outer loops*, while the iterative quadratic minimizations are found within the so-called *inner loops*. When data assimilation is applied repeatedly in a sequence of consecutive analysis intervals, the first guess is obtained from the previous assimilation cycle and is therefore reasonably close to the optimal estimate.

24.2 ROMS IS4DVAR in a quasi-geostrophic domain: the East Australia Current

24.2.1 The East Australia Current ROMS application

The East Australia Current (EAC; Fig. 24.1) is a southward-flowing western boundary current that separates from the coast of Australia at around 32°S latitude, flowing eastwards into the South Pacific. Extending from the Coral Sea to the Tasman Sea, the EAC system generates strong mesoscale variability as it branches into the Tasman Front (Tiburg et al., 2001). Although most of the EAC flows eastwards at this separation point, as much as one-third of the original EAC flows southwards trapped against the western boundary (Ridgway and Godfrey, 1997). The origin of the strong mesoscale variability seems to be associated with several factors, including nonlinear instabilities and upper-ocean–topographic coupling (Tiburg et al., 2001), oscillation and propagation of the Tasman Front, and formation of cold-core frontal eddies and baroclinic Rossby waves (Marchesiello and Middleton, 2000). The spatial and temporal variability of the EAC was characterized using 6 years of atlimeter-derived geostrophic stream-function and velocities derived from tracking thermal features in satellite imagery by Bowen et al. (2005), revealing cyclones and anticyclones propagating southwards and westwards with periods between 90 and 180 days. They found no evidence of mesoscale signals propagating westwards from the South Pacific basin, consistent with the notion that mesoscale variability is generated by intrinsic instabilities. Therefore, the EAC is an attractive region for evaluating the ROMS IS4DVAR because it is characterized by strong mesoscale variability (Stammer, 1997; Mata et al., 2000) that seems to be generated principally by internal instabilities (Bowen et al., 2005), making the area an ideal scenario for the study of an assimilation system where the control variables are the initial conditions.

The model domain extends from the Australian coast to 162°E, and from 26°S to 46°S (Fig. 24.1). The spatial resolution used in this study was one-quarter of a degree (18 km to 30 km) in the horizontal and 30 levels in terrain-following coordinates in the vertical, with increased resolution towards the surface (an average resolution of 1.7 m). We consider two years of simulation/assimilation (2001 and 2002) where ROMS was forced at the boundaries with output from the data-assimilative US Navy Coastal Ocean Model (NCOM). At the boundaries, temperature and salinity were clamped to the NCOM solution and the sea level and depth-integrated velocities were imposed following the Chapman (1985) and Flather (1976) formulations, respectively, which permit surface gravity waves to radiate out of the model domain. Air–sea heat and momentum fluxes were prescribed from daily averaged US Navy NOGAPS analyses (Rasmond, 1992) and the vertical turbulent mixing closure was the k-profile parameterization scheme (Large et al., 1994). No tidal forcing was imposed.

When we integrated the model forwards for 30 years, the water mass properties remained stationary and the seasonal cycle compared well with the climatology derived from observations. The model, however, failed to reproduce the observed mesoscale. Figure 24.1(b) compares the standard deviation of the SSHA from the AVISO analysis with that from the model for the two years of study (2001 and 2002). There is considerable disagreement in the spatial distribution of the variance (Fig. 24.1), and

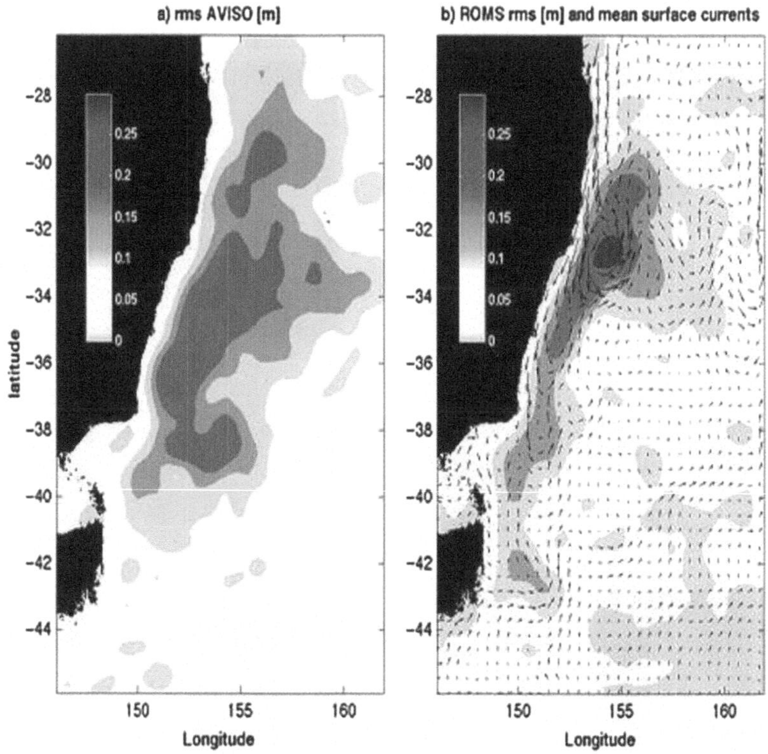

Fig. 24.1 (a) Variance in the AVISO SSHA gridded analysis for the EAC. (b) Forward model variance in SSH and mean surface currents.

its temporal evolution results in low correlation and large root-mean-square (RMS) error when compared with the AVISO analysis (Fig. 24.2(a) and (c)).

24.2.2 IS4DVAR data assimilation in the EAC

Given the disagreement with the observed mesoscale variability as depicted by the SSHA (similar characteristics were found with the mesoscale SST anomalies), we applied IS4DVAR in a sequential way. Details of all the assimilation experiments summarized next can be found in Zavala-Garay et al. (2012).

Experimentation suggested an optimal value of three outer and five inner loops for the present application. The initial increments are scaled by the specification of a standard deviation file of all state variables in the modelling of the background error covariance **B**, reflecting the statistics of physically realizable increments. As noted above, the EAC ROMS application produces reliable seasonal variability but fails to produce observed mesoscale events. To establish bounds on plausible mesoscale increments, therefore, the standard deviations were computed from the anomalies around the mean seasonal cycle from a 30-year forward run. The lengthscales chosen

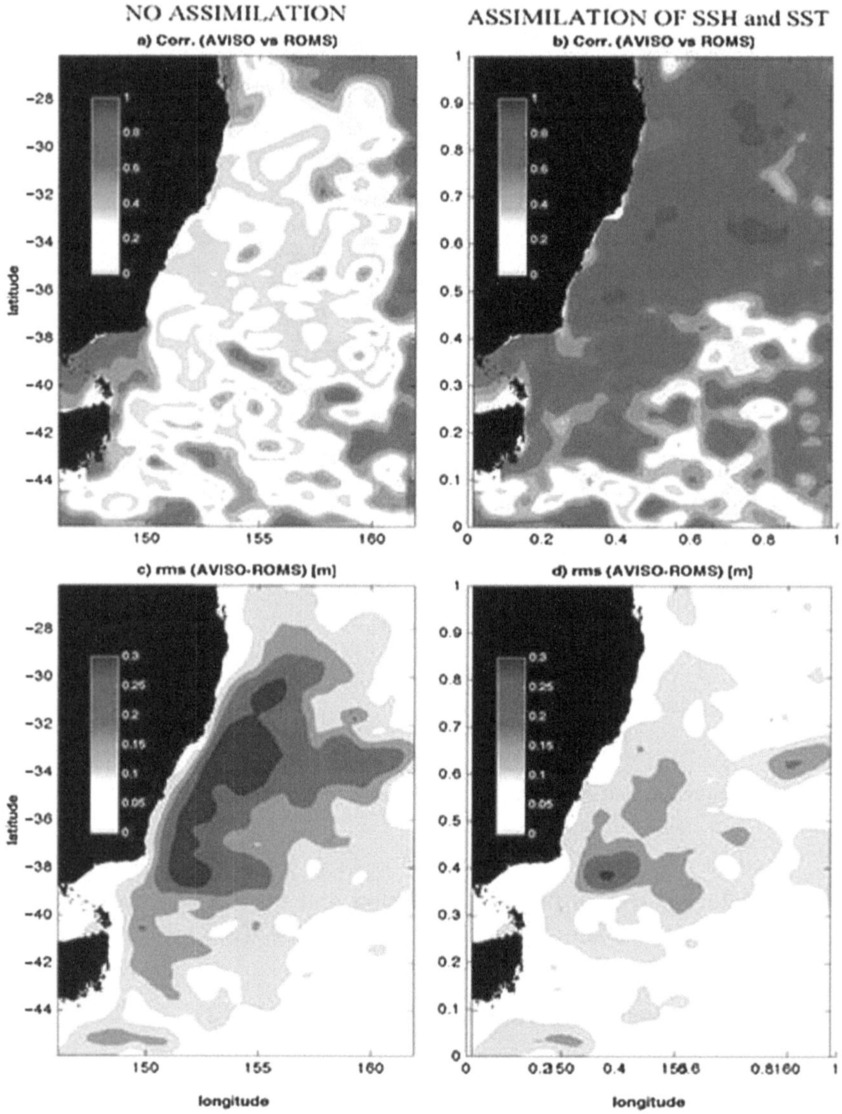

Fig. 24.2 Correlation (a, b) and RMS error (c, d) between AVISO SSHA and ROMS solutions. (a) and (c) without data assimilation; (b) and (d) with assimilation of SSH and SST.

for the background covariance were 100 km in the horizontal and 150 m in the vertical direction, being representative spatial decorrelation scales of the mesoscale anomalies from the same forward run. The horizontal scales used are also in agreement with the decorrelation scales estimated from along-track satellite for the area by Wilkin et al. (2002).

The observations considered in the assimilation experiments are daily composites of satellite sea surface temperature, seven-day average reanalysis of satellite altimeter sea level anomalies from AVISO, and subsurface temperature profiles from high-resolution expendable bathythermograph (XBT). The error assigned to the observations was 3 cm for the gridded SSHA and 1°C for the one-day composites of SST. In order to assimilate the SSHA data, a background field or mean dynamic topography (MDT) needs to be added to the anomaly filed, since ROMS produces total SSH. For the experiments presented here, the MDT was assumed to be the long-term mean of the forward run. This is a reasonable assumption, since a 30-year integration preserved the water mass properties and compared well with the observed climatology.

Considering all available observations for the years 2001 and 2002, ROMS analysis initial conditions were found every week where the model state at the end of each assimilation window was the first guess or background state for the subsequent assimilation experiment (the first guess for the first assimilation window was obtained from the NCOM model). Assimilation of surface information only (SST and SSH) resulted in poor estimates of the true subsurface ocean state (as depicted by the XBTs) and therefore poor forecast skill of subsurface conditions. Including the XBTs in the assimilation experiments improves the ocean state estimation in the vicinity of the XBT transects. By introducing subsurface pseudo-observations—which we term 'synthetic CTD'—based on an empirical relationship between satellite surface observations and subsurface variability resulted in a significant improvement in ocean state estimates that lead to skilful forecasts for up to two weeks in the domain considered.

Figures 24.2(b) and (d) show the correlation and RMS error after sequential IS4DVAR data assimilation for the third case (i.e. assimilation of SSHA and SST and the use of the synthetic-CTD). High correlations and low RMS error are obtained specially in the region of maximum observed variance (cf. with Fig. 24.1(a)). More importantly, the recovered vertical structure is in good agreement with the subsurface temperature measured by the XBTs, resulting also in a high correlation, and low RMS error and bias down to 1000 m (continuous black lines in Fig. 24.3).

24.2.3 Impct in nowcast/forecast skill

In order to evaluate the impact and persistence of the corrections after IS4DVAR, a sequence of ocean forecasts were produced by initializing the model with the state at the end of each assimilation window and comparing with the available XBT observations within 1 week, 2 weeks, etc. Figure 24.3 shows the correlation, RMS error, and bias for the different forecast lags. The results suggest that reliable subsurface forecasts can be obtained up to two weeks in advance (correlations of 0.6 and RMS error smaller than 2°C) before the model skill relaxes to the skill of the model without data assimilation. It should be noted, however, that there are at least two possible limitations if we want to apply this methodology in other parts of the coastal ocean: (1) the need of an empirical relationship, if it exists at all, has to be ideally derived from a large amount of historical observations that are concurrent with the satellite era, and (2) the optimal interpolation used to produce the gridded SSH is based on the assumption of isotropic horizontal correlation scales. One alternative for the first requirement is to derive the

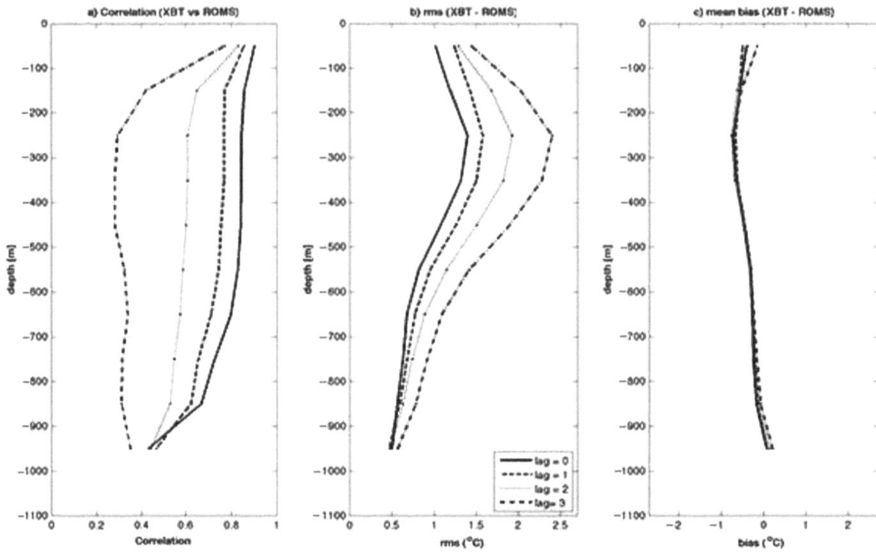

Fig. 24.3 Skill of ROMS in forecasting all the available XBT data. Skill is presented in terms of correlation between the temperature observed by the XBT and that predicted by the model (a), the RMS of the difference between the two (b), and the mean bias (c) within depth bins of 100 m width (thin horizontal dashed lines). The different lags of the forecasts (in weeks) for all panels are indicated in the key at the bottom of panel (b).

empirical relationship from a long integration of the forward model, provided that there is no drift in the state of the model and that its statistics is stationary. In the more general case, however, inferring the correct subsurface structure will depend crucially on the efficiency of the adjoint model in propagating the information content in the satellite observations to the subsurface, and in the correct modelling of the background error covariance. In Section 24.3, we present the application of IS4DVAR in a more complex coastal domain.

24.3 ROMS IS4DVAR in a complex coastal domain: the Middle Atlantic Bight

Section 24.2 demonstrated how IS4DVAR can be used to improve ocean state estimation and forecast in the EAC via assimilation of gridded maps of SSH and SST and the use of an empirical vertical projection of the surface information. The EAC region is characterized by a narrow continental shelf and most of the domain features highly isotropic horizontal correlation scales, and therefore the use of gridded products is justified as it is in many other parts of the deep ocean where quasi-geostrophic dynamics explains a good part of the mesoscale variability. However, this is not the case as we approach the coast, especially in areas with wide shelves and complex bathymetry, and therefore the methodology outlined in Sections 24.1 and 24.2 has to be refined to account for other complications such as tides, anisotropic correlation scales, etc. In this section, we present the practical aspects of applying variational data assimilation techniques (IS4DVAR in this case) in a more complex coastal domain such as the Middle Atlantic Bight (MAB). The dominant circulation in the MAB has short, anisotropic scales due to the presence of a shelf–slope front along the continental shelf break, a significant Slope Sea gyre circulation, and Gulf Stream (GS) warm-core ring interactions with the shelf–slope front that drive episodic exchanges of coastal and deep ocean waters. The advantage of the variational approach is that it allows a combination of data from multiple satellites and in situ observing systems with the model physics, therefore introducing dynamic and kinematic constraints into the analysis.

24.3.1 The Middle Atlantic Bight ROMS application: ESPRESSO

The MAB (Fig. 24.4) is characterized by a permanent shelf–slope front that separates the relatively fresh and cool waters in the broad (100–200 km) and shallow (100–200 m) shelf from saltier, warmer slope water (Chapman and Lentz, 1994). The continental shelf break exhibits large horizontal and vertical gradients in water properties associated with the shelf-break front, a feature susceptible to nonlinear instabilities and strong interactions with Gulf Stream warm-core rings that impinge onto the continental slope (Ryan et al., 2001). Long-term studies of the front suggest both persistence

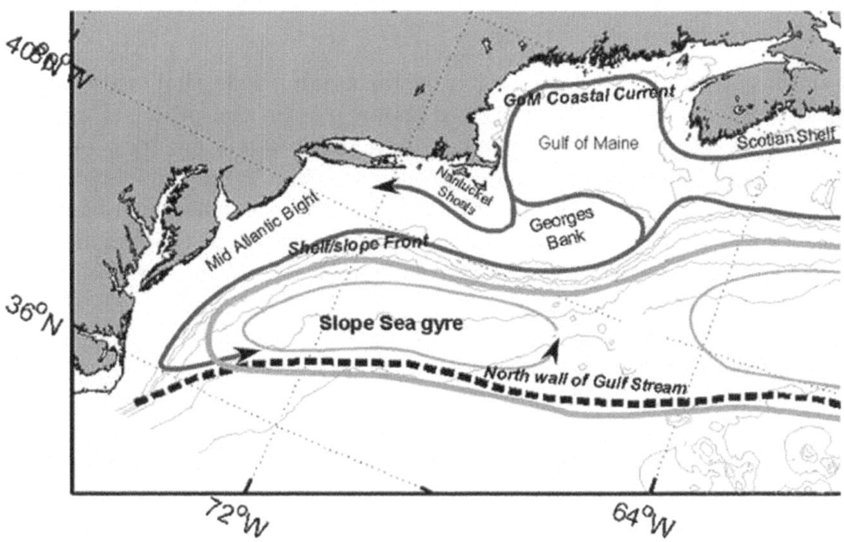

Fig. 24.4 Geography and schematic circulation in the Gulf of Maine–Middle Atlantic Bight. The dark lines on the shelf indicate the southwestward flow from the Scotian Shelf to the MAB. The light lines depict the currents that define the Slope Sea gyre, and the dashed line is the mean position of the Gulf Stream.

of the shelf break jet as well as upstream advective influences from the Scotian shelf (Flagg et al., 2006). As a result, this region has significant along- and cross-shelf fluxes of heat, freshwater, nutrients, and carbon that control the characteristics of water masses and the ecosystem both at the shelf break and in the neighbouring continental shelf and slope sea (Houghton and Visbeck, 1998; Ryan et al., 2001). Wind forcing is significant in much of the region and especially along the New Jersey coast, where local upwelling events on short timescales can intermittently overwhelm the buoyancy-driven coastal currents emanating from the Hudson and Delaware Rivers, and in Chesapeake Bay. This spectrum of forcing mechanisms that influence the shelf-break transition zone linking the circulation in the MAB and adjacent deep ocean makes the region a challenging laboratory for testing coastal ocean modelling skills.

The ROMS ESPreSSO (Experimental System for Predicting Shelf and Slope Optics) application covers the MAB Bight from the centre of Cape Cod southwards to Cape Hatters, from the coast to beyond the shelf break and shelf–slope front. In contrast to the EAC, tides are an important component of the observed variability and need to be included in the model. The current prototype system is a 5 km horizontal, 36 sigma-level vertical coordinate forced at the boundaries by the HYCOM NCODA forecast system (Cummings, 2005) and at the surface by the 12 km three-hourly forecast data from the North American Mesoscale (NAM) forecast system operated by NCEP. It also includes daily means of river discharge from USGS gauges and tidal boundary conditions are from the ADCIRC tidal model.

24.3.2 Preparing the model for assimilation

Model bias correction

Data assimilation methodologies are based on assumptions that are often violated by forward ocean models, and a very important step in the practical application of these data assimilation techniques is the elimination or minimization of such model deficiencies. One of the most common problems is that often models exhibit strong biases in the basic background state. It is important to note that such biases can have a negative impact that goes beyond 'interpolating' the bias to the regional domain in incremental systems such as IS4DVAR. Ocean mesoscale and sub-mesoscale variability tend to maximize in regions of strong gradients of water properties (both horizontal and vertical), and biases often affect the localization of such fronts and therefore the localization of the corrections or increments in the assimilation system. For instance, assimilation of altimeter data in a model with a too-shallow thermocline will tend to localize the increments at that depth and not at the depth where they are expected to occur, and the same could be expected with horizontal fronts such as the Gulf Stream and the shelf-break fronts in this application. As an example of the biases present in global models, Fig. 24.5 shows the subsurface temperature and salinity anomaly profiles (difference between observations and model predictions at the same location and time) of two global models that have data assimilation (HYCOM and Mercator). The profiles are sorted by water depth and the thick black line depicts the water depth of the first 1000 m. The biases in temperature for HYCOM and salinity for Mercator models are evident, especially in the deep ocean (water depths greater than 1000 m).

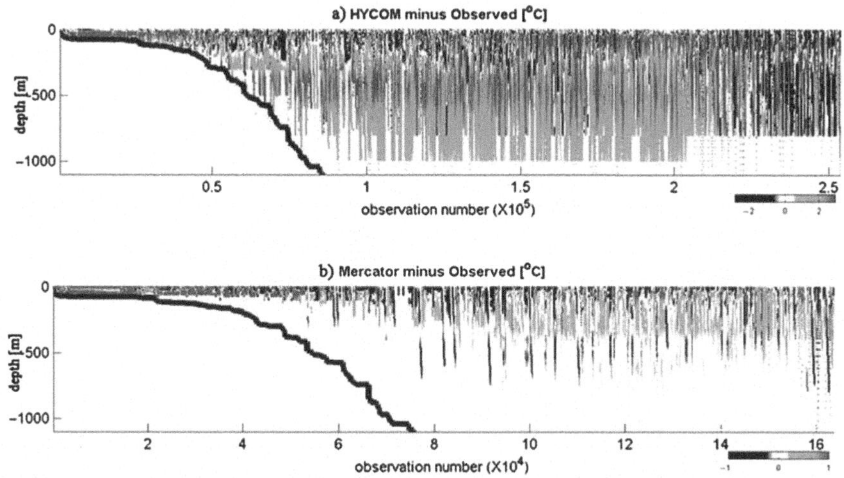

Fig. 24.5 Difference between observed and HYCOM model temperature (a) and difference between observed and MERCATOR model salinity (b) for the observations used for validation in this study. Individual profiles are sorted by water depth, and the spatial distribution of the profiles is presented in Fig. 24.9(b).

Regional models such as ROMS use boundary conditions from these models, and some corrections are needed to minimize such biases. In the ESPRESSO application, this bias is considerably reduced by replacing the seasonal cycle in temperature and salinity of HYCOM by that from a high-resolution climatology developed at Rutgers from a large set of historical data. For dynamical consistency, velocity and free surface heights are corrected by replacing the HYCOM mean with the results of a single assimilative ROMS climatological run.

Modelling of the background error covariances

An important component in the minimization problem is specification of the background covariance matrix **B**. The functions of **B** are to penalize the departures of the increments at time t_{i-1} from the original background and to act as a regularization term by spatially smoothing the updated initial conditions. For a detailed description of the formulation as implemented in ROMS, the reader is referred to Powell et al. (2008) and Moore et al. (2011). In the present configuration of ESPRESSO, cross-covariances between variables are introduced via the use of an adjoint model and the background error covariances are modeled via a block-diagonal diffusion operator that is scaled by a expected standard deviation of the model anomalies as in Weaver and Courtier (2001). For the MAB, the background seasonal state changes considerably during the year, affecting the localization and characteristics of the mesoscale and sub-mesoscale variability, and this strong seasonality needs to be reflected in the standard deviation file used to model **B**, especially in the shelf and shelf-break regions. During the winter, the strong atmospheric variability mixes all the water column in the

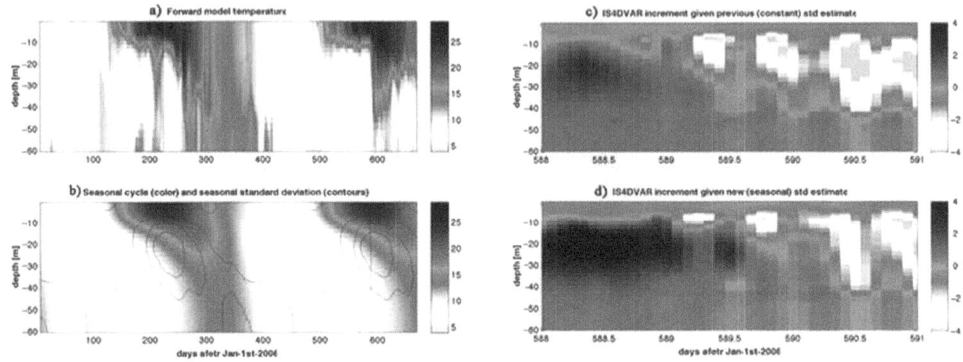

Fig. 24.6 (a) Seasonal variation in the MAB shelf subsurface temperature during two years of integration of the forward model. (b) Seasonal cycle after harmonic analysis of six years on model integration (shading) and seasonal standard deviation (contours) of the corresponding temperature anomalies. (c) increment of an analysis started at day 600 (the black line in (a)) using the annual mean of the standard deviation. (d) the same as (c), but using a seasonal standard deviation that better reflects the expected size and location of the anomalies.

shelf, whereas during the summer, strong stratification develops. Consequently, the mesoscale and sub-mesoscale variability tends to be distributed equally at all depths during the winter and concentrated in the thermocline during the summer. As an example, Fig. 24.6(a) shows the typical seasonal variation of temperature in the shelf during two years of integration of the ESPRESSO application. The erosion of the mixed layer during winter and strong stratification during the summer are evident. In order to compute a seasonally varying standard deviation, the model was integrated forward for six years and harmonic analysis was performed for each state variable at each model grid point. Then a three-month running standard deviation was used to compute a smooth version of the seasonally varying standard deviation, as shown in Fig. 24.6(b).

Figures 24.6(c) and (d) compare the inversion for a constant versus a seasonal standard deviation for one three-day assimilation window at the same grid point initialized at day 600 (black vertical line in Fig. 24.6b). The use of a seasonal standard deviation results in a larger and more localized anomaly around the seasonal thermocline, as should be expected at this time of the year.

24.3.3 Observation pre-processing

Since the focus of this chapter is on estimating how much subsurface information we can infer given surface information from satellites using IS4DVAR, we follow a similar approach to that used in the EAC: the satellite information is assimilated and the in situ hydrographic data is kept for evaluation of the resulting inversion. The observations used for validation are subsurface temperature and salinity from XBTs, CTDs, and autonomous gliders (see Fig. 24.9(b) for the spatial distribution), most of them collected during the summers of 2006 and 2007 during the Shallow Water Acoustics 2006 (SW06) field programme and during a prototype phase in 2007 for the upcoming Ocean Observatories Initiative (OOI) Pioneer Array (Todd et al., 2013). The assimilated observations and the pre-processing applied before they are used in the assimilation system is described next.

Along-track sea surface height anomalies

Until recently, altimeter data collected over the coastal ocean remained largely unexploited in the data archives because intrinsic difficulties in the corrections (especially the wet tropospheric component, the high-frequency atmospheric and oceanic signals, and the tides) and issues of land contamination in the footprint resulted in systematic flagging and rejection of these data. However, during the last few years, considerable research has been carried out into overcoming these problems, and these data have been re-processed in order to extend the observations of current and future altimeters as close as possible to the coast. For the specific case of the MAB, a large volume of data in the coastal zone ordinarily rejected by conventional quality control can be reclaimed by judicious application of the data error flags and revised wet tropospheric corrections (Feng and Vandemark, 2011), and this reprocessed data is used in the ESPRESSO assimilation system.

The observation assimilated in ROMS is total SSH defined as the sum of a mean dynamic topography (MDT) plus the SSH anomaly (measured by the satellite) plus tides. The MDT was determined by a single assimilation of the climatological mean of temperature and salinity forced at the boundaries by the bias-corrected climatological mean and at the surface by the long-term mean surface forcing. Unlike the EAC application presented in Section 24.2, the SSH data need also be modified to account for the tides that are also included in the MAB ROMS application (note that tides in regional models constitute a boundary problem and therefore should not be adjusted by the IS4DVAR algorithm used here). This step is very important, because variations in the time dimension during the analysis interval are fully accounted for by sampling the model at the precise time and location that the observations were made. (Thus, 4DVAR is sometimes characterized as a smoother or filter because the analysis at a given time is affected by observations after and/or before the analysis time.) Any mismatches at high frequency in the modelled and observed SSHA therefore enter into the cost function. This is potentially problematic in the MAB, since a large part of the observed variability is associated with tides, especially in the shelf region. In order to prevent the possibility that even small errors in the modeled and observed tide could dominate the IS4DVAR cost function the default corrections provided in the Radar Altimeter Database System (RADS; ⟨http://rads.tudelft.nl⟩) were applied to de-tide the along-track altimeter data, and then the model tides were added as follows. A one-year-long ROMS simulation without assimilation, forced by observed tidal harmonics at the boundaries, was run to compute the model's tidal harmonic sea level response at each grid point. The resulting harmonics were then used to predict the model tide at the times and positions of the along-track data and added to the de-tided SSH observations. Thus, the observations are adjusted to include the modelled tide and therefore will not enter into the IS4DVAR cost function; the model–data misfit in SSH is then dominated by processes other than the tides, such as mesoscale and sub-mesoscale variability, which is what we are trying to adjust via changes of the initial conditions.

Sea surface temperature

The SST data used are a combination of two sources. The main source is based on individual passes of AVHRR (about six per day) in the region, and it is complemented by a blended product (microwave plus infrared SST) when needed, as described next. The spatial resolution of AVHRR is higher than the model (1 km versus 5km), and therefore, when available, the variability of satellite SST is used to compute a model representation error as shown in Fig. 24.7. Large representation errors (up to 1°C) can occur in frontal zones such as the Gulf Stream edge and shelf–slope front (Fig. 24.7(b)), and they can be very small in highly homogeneous areas, such as the shelf. To avoid overfitting of the SST data, the representation error is constrained to not be smaller than the instrumental error (0.3°C in our case). Considering all the passes in one day (about six per day), an average of 70% of the ESPRESSO grid points are observed, with values as low as 40% coverage during winter months. Grid points not observed during one day by the six passes of the AVHRR radiometer are complemented with blended SST from remote sensing systems provided by NOAA Coastwatch. The through-cloud capabilities of microwave radiometers provide a valuable picture of global sea surface temperature, but at a resolution too low for the ESPRESSO application (about 5km). This gridded product is then 'blended' with one-day composites of SST derived from infrared measurements allowing higher spatial resolution and with SST near land. Therefore, every day, at least one observation of SST is provided to the assimilation system at each grid point (with the exceptions of grid points very close to the coast when no infrared observations are available at grid points not resolved by the 25 km microwave SST analysis).

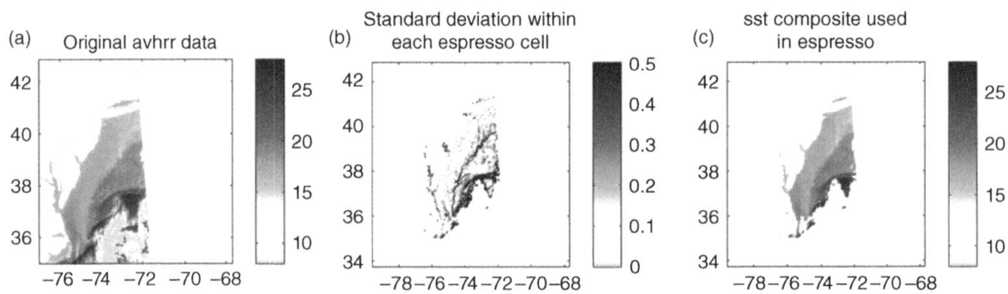

Fig. 24.7 (a) Example of an individual pass of AVHRR SST over the MAB region (light pixels are data flagged-out as clouds). (b) Standard deviation of all valid observations within a model grid cell. and (c) Mean SST within each model grid cell. The mean shown in (c) along with the spatial error shown in (b) are used in the assimilation system.

24.3.4 IS4DVAR data assimilation in the MAB

The assimilation is applied with an analysis window of three days during which the ROMS adjoint and tangent-linear models are iterated forwards and backwards in time to adjust the ocean state at the beginning of the analysis window so as to minimize the mismatch of model and data within the three-day interval. The state of the model at the end of each cycle is used as an initial guess for the next three-day assimilation cycle. On the continental shelf and slope, assimilation improves the correlation and decreases the RMS error.

Figure 24.8 depicts the pattern-centred correlation and RMS error between Jason along-track SSHA and our ROMS model without assimilation (a, c) and with assimilation (b, d) of data over a two-year period (2006–2007). This is therefore a representation of hind-cast skill, and since the model produces a full spatial analysis on the model grid, the assimilation solution is analogous to a gridded analysis sampled on the altimeter ground-tracks.

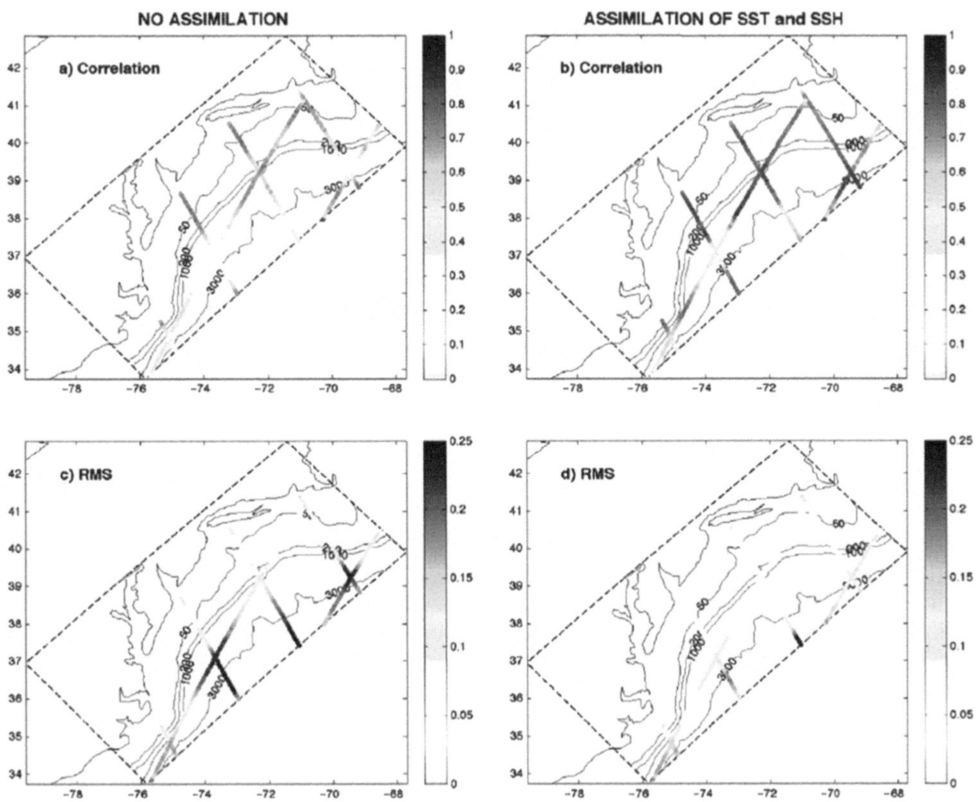

Fig. 24.8 Correlation (a, b) and RMS error (c, d) between Jason along-track SSHA and ROMS solutions. (a) and (c) without data assimilation; (b) and (d) with assimilation of SSH and SST.

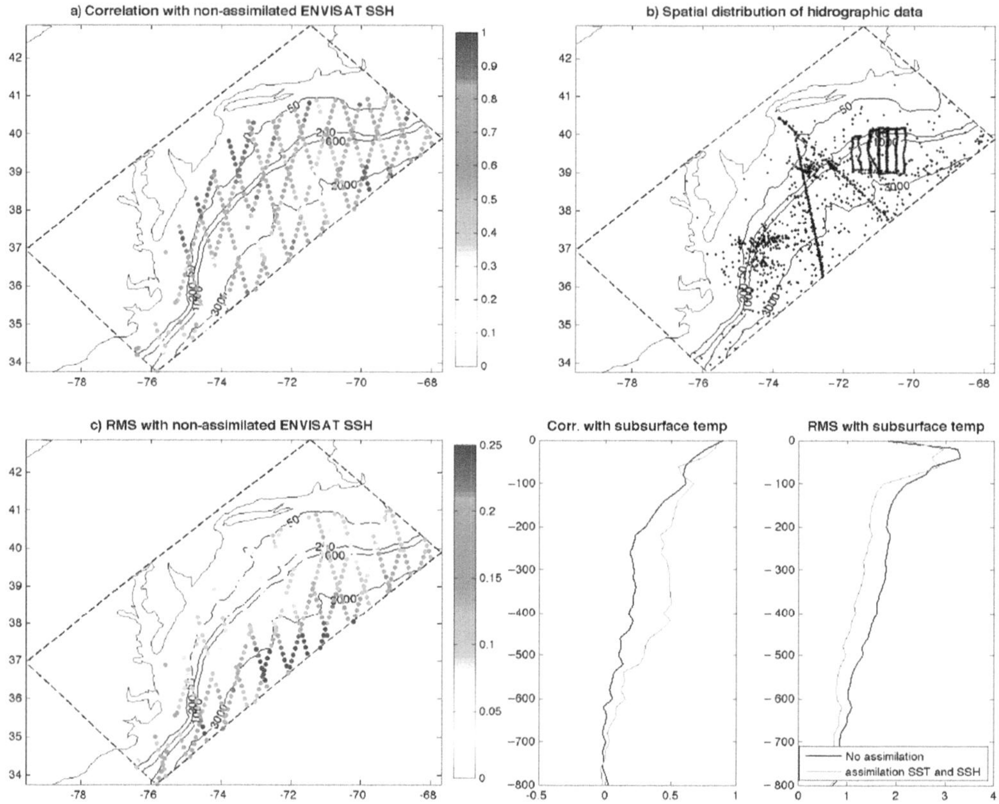

Fig. 24.9 Correlation (a) and RMS error (c) between ROMS with ENVISAT altimeter data over the two-year period 2006–2007, and correlation and RMS error between subsurface temperature and ROMS (d). The spatial distribution of the subsurface temperature data used in (d) is shown in (b).

Figure 24.9 shows the ability of the system to hindcast observations that are not assimilated. Correlation (Fig. 24.9(a)) and RMS error (Fig. 24.9(b)) between ROMS and ENVISAT altimeter data over the two-year period 2006–2007 (24 cycles) show similar patterns to those for the Jason data that were assimilated (Figs. 24.8(b,d)): high correlations and low RMS error in the shelf and Slope Sea. The unassimilated ENVISAT observations were made at different places and times from the Jason data, showing that the ROMS 4D-Var system is effective at interpolating nadir altimetry between ground-tracks in the coastal ocean. Figure 24.9(d) shows the skill that data assimilation adds to hind-cast subsurface temperature and salinity profiles observed at the points shown in Fig. 24.9(b). From 100 to 600 m depth, the assimilation of satellite data significantly improves the model vertical stratification, showing that variational methods are effective at propagating information to depths unobservable by satellite and to variables (here temperature) that are not present in the observation dataset assimilated. It is also hoped (work in progress) that the altimeter data could help to constrain the along- and across-shelf pressure gradients that are fundamental to the shelf circulation, the shelf–slope front, and the remotely generated forcing associated with the Gulf Stream.

We end this chapter by noting that while in situ observations in the area are available for the MAB and can be easily assimilated in ROMS, they are usually not available in real time, except for estimates of surface currents from a network of high-frequency coastal CODAR radars. A system similar to that validated here, with the addition of CODAR surface currents, has been operating in real time for the last couple of years and the analysis and forecast is made public daily at the ROMS website ⟨http://www.myroms.org/espresso⟩. In the future, it will assimilate subsurface temperature and salinity from ships, autonomous underwater vehicles, and/or profiling floats, as part of the Mid-Atlantic Regional Coastal Ocean Observing System (MARCOOS).

References

Bowen, M. M., Wilkin, J. L., and Emery, W. L. (2005). Variability and forcing of the East Australia Current. *J. Geophys. Res.*, **110**, C03019.

Chapman, D. C. (1985). Numerical treatment of cross-shelf open boundaries in a barotropic coastal ocean model. *J. Phys. Oceanogr.*, **15**, 1060–1075.

Chapman, D. C. and Lentz, S. J. (1994). Trapping of a coastal density front by the bottom boundary layer. *J. Phys. Oceanogr.*, **24**, 1464–1479.

Courtier, P., N., Thèpaut J., and Hollingsworth, A. (1994). A strategy for operational implementation of 4D-Var, using an incremental approach. *Q. J. R. Meteorol. Soc.*, **120**, 1367–1387.

Cummings, J. A. (2005). Operational multivariate ocean data assimilation. *Q. J. R. Meteorol. Soc.*, **131**, 3583–3604.

Feng, H. and Vandemark, D. (2011). Altimeter data evaluation in the Coastal Gulf of Maine and Mid-Atlantic Bight Regions. *Mar. Geodesy*, **34**, 340–363.

Flagg, C.N., Dunn, M., Wang, D.-P., Rossby, H. T., and Benway, R. L. (2006). A study of the currents of the outer shelf and upper slope from a decade of shipboard ADCP observations in the Middle Atlantic Bight. *J. Geophys. Res.*, **111**(C6), C06003.

Flather, R. A. (1976). A tidal model of the northwest European continental shelf. *Mem. Soc. R. Sci. Liege*, **6**(10), 141–164.

Haidvogel, D. B., Arango, H., Budgell, W. P., Cornuelle, B. D., Curchitser, E., Di Lorenzo, E., Fennel, K., Geyer, W. R., Hermann, A. J., Lanerolle, L., Levin, J., McWilliams, J. C., Miller, A. J., Moore, A. M., Powell, T. M., Shchepetkin, A. F., Sherwood, C. R., Signell, R. P., Warner, J. C., and Wilkin, J. L. (2008). Ocean forecasting in terrain-following coordinates: formulation and skill assessment of the Regional Ocean Modeling system. *J. Comput. Phys.*, **227**, 3595–3624.

Haidvogel, D. B., Arango, H. G., Hedstrom, K., Beckmann, A., Malanotte-Rizzoli, P., and Shchepetkin, A. F. (2000). Model evaluation experiments in the North Atlantic Basin: simulations in nonlinear terrain-following coordinates. *Dyn. Atmos. Oceans*, **32**, 239–281.

Houghton, R. and Visbeck, M. (1998). Upwelling and convergence in the Middle Atlantic Bight shelfbreak front. *Geophys. Res. Lett.*, **25**, 2765–2768.

Large, W. G., McWilliams, J. C., and Doney, S. C. (1994). Oceanic vertical mixing: A review and a model with a nonlocal boundary layer parameterization. *Rev. Geophys.*, **32**, 363–403.

Marchesiello, P. and Middleton, J. H. (2000). Modeling the East Australian Current in the western Tasman Sea. *J. Phys. Oceanogr.*, **30**, 2956–2971.

Mata, M. M., Tomczak, M., Wijffels, S. and Church, J. (2000). East Australian Current volume transports at 30°S: Estimates from the World Ocean Circulation Experiment hydrographic sections PR11/P6 and the PCM3 current meter array. *J. Geophys. Res.*, **105**, 28509–28526.

Moore, A. M., Arango, H. G., Broquet, G., Powell, B. S., Zavala-Garay, J. and Weaver, A. T. (2011). The Regional Ocean Modeling System (ROMS) 4-dimenensional variational data assimilation system, Part I: System overview and formulation. *Prog. Oceanogr.*, **91**, 34–49.

Powell, B. S., Arango, H. G., Moore, A. M., Di Lorenzo, E., Milliff, R. F. and Foley, D. (2008). 4DVAR data assimilation in the Intra-Americas Sea with the Regional Ocean Modeling System (ROMS). *Ocean Modelling*, **23**, 130–145.

Rasmond, T. E. (1992). The design and testing of the Navy Operational Global Atmospheric Prediction System. *Weather Forecasting*, **7**, 262–272.

Ridgway, K. R. and Godfrey, J. S. (1997). Seasonal cycle of the East Australia Current. *J. Geophys. Res.*, **102**, 22921–22936.

Ryan, J., Yoder, J., and Townsend, A. (2001). Influence of a Gulf Stream warm-core ring on water mass and chlorophyll distributions along the southern flank of Georges Bank. *Deep Sea Res.*, **48**, 159–178.

Sasaki, Y. (1970). Some basic formalisms on numerical variational analysis. *Mon. Weather Rev.*, **98**, 875–883.

Shchepetkin, A. F. and McWilliams, J. C. (2003). A method for computing horizontal pressure-gradient force in an oceanic model with a non-aligned vertical coordinate. *J. Geophys. Res.*, **108**, 3090.

Shchepetkin, A. F. and McWilliams, J. C. (2005). The Regional Ocean Modeling System (ROMS): a split-explicit, free-surface, topography-following coordinates ocean model. *Ocean Modelling*, **9**, 347–404.

Stammer, D. (1997). Global characteristics of ocean variability estimated from regional TOPEX/Poseidon altimeter measurements. *J. Phys. Oceanogr.*, **27**, 1743–1769.

Tiburg, E. C., Hulbert, H. E., O'Brien, J. J., and Shriver, J. F. (2001). The dynamics of the East Australia Current system: the Tasman Front, the East Auckland Current, and the East Cape Current. *J. Phys. Oceanogr.*, **31**, 2917–2943.

Todd, R. E., Gawarkiewicz, G. G., and Owens, W. B. (2013). Horizontal scales of variability over the Middle Atlantic Bight shelf-break and continental rise from finescale observations. *J. Phys. Oceanogr.* **43**, 222–230.

Weaver, A. T. and Courtier, P. (2001). Correlation modelling on the sphere using a generalized diffusion equation. *Q. J. R. Meteorol. Soc.*, **127**, 1815–1842.

Wilkin, J. L., Bowen, M. M. and Emery, W. J. (2002). Mapping mesoscale currents by optimal interpolation of satellite radiometer and altimeter data. *Ocean Dyn.*, **52**, 95–103.

Zavala-Garay, J., Wilkin, J. L., and Arango, H. G. (2012). Predictability of mesoscale variability in the East Australian Current given strong-constraint data assimilation. *J. Phys. Oceanogr.*, **42**, 1402–1420.

25
Data assimilation in glaciology

B. BONAN[1], M. NODET[1], O. OZENDA[1], and C. RITZ[2]

[1]Laboratoire Jean Kuntzmann
Université Grenoble Alpes, INRIA-CNRS
Grenoble, France

[2]Laboratoire de Glaciologie et Géophysique de l'Environnement
Université Grenoble Alpes, CNRS
Grenoble, France

Advanced Data Assimilation for Geosciences. First Edition.
Edited by É. Blayo, M. Bocquet, E. Cosme, and L. F. Cugliandolo.

Chapter Contents

25.1 Introduction

In this short chapter, we give one example of an inverse problem in glaciology. This problem is fairly simple to state: how to infer a climatic scenario (i.e. how to reconstruct past polar temperature) from ice volume records.

Ice-sheet dynamics is quite complex. Ice is a non-Newtonian fluid: its flow follows a highly nonlinear Stokes equation. One key factor of ice dynamics is surface mass balance, that is, the amount of ice accumulating at the surface (from snow precipitation) minus the amount of melting ice. This mass budget is particularly sensitive to temperature at the ice-sheet surface. The detailed modelling of ice-sheet dynamics is discussed in the textbooks by Greve and Blatter (2009) and Paterson (1981).

Ice volume observations are available from oceanic records, which give information about sea level and therefore about the amount of water stored in ice sheets, ice caps, and glaciers.

As far as we know, there has been little previous work on this subject. Bintanja et al. (2004) and Bintanja et al. (2005) used a very simple correction method, equivalent to the nudging method, with quite good results.

The idea of the present work is to explore the ability of the adjoint method to solve the inverse problem of reconstructing past temperatures given all available observations. We start here with a simplified ice-sheet model and perform twin experiments.

We present the ice-sheet model in Section 25.2, the adjoint method in Section 25.3, and numerical illustrations in Section 25.4.

25.2 Ice-sheet model

25.2.1 Ice dynamics processes

As the three-dimensional, full Stokes models are very costly to run over paleoclimatological time scales (thousands of years), large-scale ice-sheet models have been developed, using the so-called shallow-ice approximation (SIA). Such models are able to reproduce ice-sheet evolution over hundreds of thousands of years, for large ice sheets such as the Antarctica, Greenland, Laurentide, and Fennoscandian sheets. One such model, GRISLI, has been developed by Ritz et al. (2001). It is a two-dimensional thermomechanically coupled model, whose main physical processes are depicted in Fig. 25.1.

25.2.2 Model equations

As GRISLI is still a costly model, in order to study the paleoclimate inverse problem, we use a simplified, flowline, isothermal prototype of this model, on a flat bedrock topography and without floating shelves, which is faster to run. 'Flowline' means that the model is one-dimensional in space.

The first equation is the mass balance equation

$$\frac{\partial H}{\partial t} = \dot{b}_m - \frac{\partial \left(\overline{U} H\right)}{\partial x}, \quad H|_{t=0} = H_0,$$

where t is time, x is latitude, $H(t,x)$ is ice thickness, $H_0(x)$ is initial ice thickness, $\overline{U}(t,x)$ is the Eulerian velocity averaged over the ice thickness, and $\dot{b}_m(t,x)$ is the

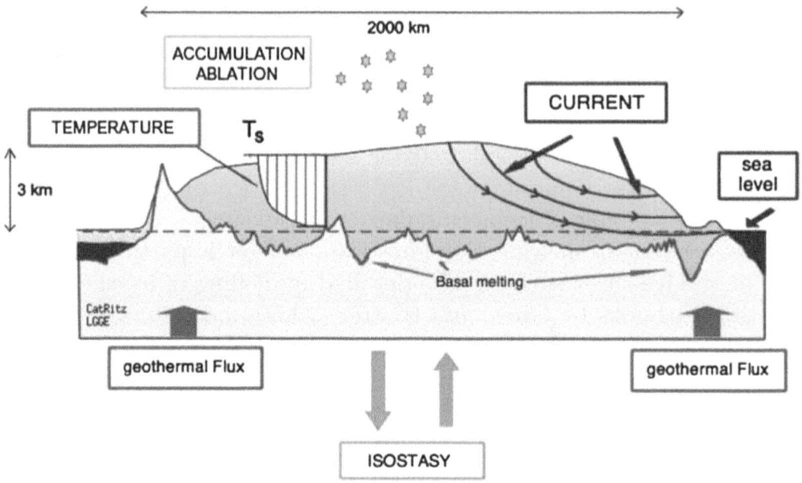

Fig. 25.1 Physical processes with an impact on ice-sheet dynamics.

surface mass balance rate. The vertically averaged ice velocity $\overline{U}(t)$ is a diagnostic variable, that is, it depends only on the geometrical variables (S, H) at time t, and is given by the following equation:

$$\begin{aligned}\overline{U} &= u_B - a_1 \frac{\partial S}{\partial x}\frac{H^2}{3} - a_2 \left(\frac{\partial S}{\partial x}\right)^3 \frac{H^4}{3},\\ S &= B + H,\\ H &\geq 0.\end{aligned}$$

where $u_B(t,x)$ is the basal velocity, $S(t,x)$ is the ice surface elevation, $B(t,x)$ is the ice base (bedrock elevation, taken to be constant in this study), and a_1 and a_2 are given coefficients.

25.2.3 Climatic scenario and surface mass balance

The surface mass balance is given as a function of the ice surface temperature:

$$\begin{aligned}\dot{b}_m &= \text{Acc} + \text{Abl},\\ \text{Acc} &= f(T_S),\\ \text{Abl} &= g(T_S),\end{aligned}$$

where $\text{Acc}(t,x)$ is the accumulation rate, $\text{Abl}(t,x)$ is the ablation rate, $T_S(t,x)$ is the surface temperature, and f and g are given functions. Figure 25.2 shows how accumulation and ablation depend on surface temperature.

Finally, the surface temperature itself depends on latitude x, on altitude S, and on the (poorly known) temperature at the northern point at sea level height, $T_{\text{clim}}(t)$:

$$T_S(t,x) = T_{\text{clim}}(t) + b\,x + c\,S(t,x),$$

with b and c given coefficients.

The aim of the problem is to infer $T_{\text{clim}}(t)$ from volume measurements.

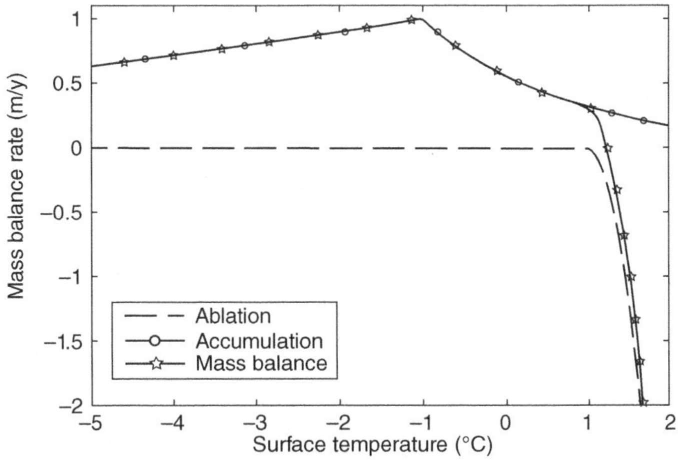

Fig. 25.2 Accumulation, ablation, and surface mass balance as functions of surface temperature.

25.3 Adjoint method and adjoint model

For further details we refer to the earlier chapter in the volume dealing with variational data assimilation and adjoint code construction.

25.3.1 Observations and control variables

We use ice volume observations, available from sea level records, an example of which is shown on Fig. 25.3 (Paillard and Parrenin, 2004). As we perform twin experiments, we generate the data from a reference 'true state' simulated using the model.

In the framework of four-dimensional variational assimilation, we write our inverse problem as a minimization problem: we want to find the 'best' temperature $T_{\text{clim}}(t)$ such that the misfit between modelled and observed volume is smallest. This means that our control variable is the climatic temperature $T_{\text{clim}}(t)$, and our cost function is written as follows:

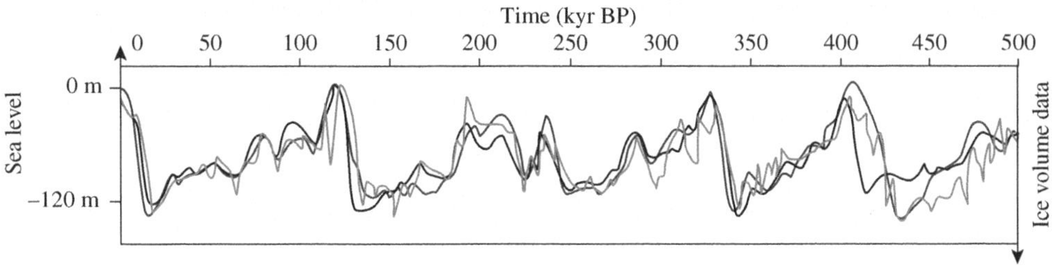

Fig. 25.3 Past ice volume records. The black curves shows: measured data from Raymo (1997); and the grey carves are models. (From Paillard and Parrenin (2004). Copyright © 2004, Elsevier.)

$$\mathcal{J}(T_{\text{clim}}) = \mathcal{J}^b + \int_{t_0}^{t_f} \|\text{Vol}(t) - \text{Vol}^{\text{obs}}(t)\|^2 \, \mathrm{d}t, \quad \text{Vol}(t) = \int_x H(t,x) \, \mathrm{d}x,$$

where the background term $\mathcal{J}^b$ is a classical regularizing term, involving a background temperature and background error covariance matrix:

$$\mathcal{J}^b(T_{\text{clim}}) = \|T_{\text{clim}} - T^b_{\text{clim}}\|^2_k.$$

The norm $\|.\|_k$ is classically given by

$$\|X\|^2_k = \int_{t=t_0}^{t_f} \int_{s=t_0}^{t_f} k(s,t)X(s)X(t) \, \mathrm{d}s \, \mathrm{d}t,$$

with k a regularizing kernel. In discrete form, this would be written as:

$$\|X\|^2_{B^{-1}} = X^{\mathrm{T}} B^{-1} X,$$

where B is the background error covariance matrix and X contains $T_{\text{clim}}(t_i) - T^b_{\text{clim}}(t_i)$ for all discrete control times t_i.

25.3.2 Adjoint code validation

We minimize the cost function using a quasi-Newton descent algorithm. This requires the gradient of $\mathcal{J}$, obtained using the adjoint model. As our flowline model is a simple model, we implemented the adjoint code by hand, using classical recipes for adjoint code construction, (see e.g. Giering and Kaminski, 1998). To validate the gradient, we perform first- and second-order tests; that is, we check that the first- and second-order Taylor approximation formula

$$\mathcal{J}(x + \alpha z) = \mathcal{J}(x) + \alpha \nabla \mathcal{J}(x) z + \frac{\alpha^2}{2} z^{\mathrm{T}} \nabla^2 \mathcal{J}(x) z + o(\alpha^2)$$

is valid. To do this, we form

$$\delta = \nabla \mathcal{J}(x) z, \quad \tau = \frac{\mathcal{J}(x + \alpha z) - \mathcal{J}(x)}{\alpha}, \quad \varepsilon = \frac{\tau(\alpha, z) - \delta(z)}{|\delta(z)|}, \quad r = \frac{\tau(\alpha, z) - \delta(z)}{\alpha},$$

and we check that ε tends to zero as α tends to zero and that r tends to a constant depending only on z. Figure 25.4 illustrates the gradient tests with our model.

25.4 Numerical results for twin experiments

In order to validate our method and ajoint model, we present a first illustration for twin experiments (Figs. 25.5 and 25.6). The principle of such experiments is as follows. First, we choose a reference climatic scenario T_{clim}, called the 'true' temperature forcing. Second, we perform a direct run of the model with the 'true' forcing and we generate synthetic observations (Fig. 25.5). Finally, we forget that we know the true temperature, we start with a constant first guess for T_{clim}, and we perform data assimilation using the adjoint method, in order to get an assimilated T_{clim}. We can then compare our result with the true T_{clim} (Fig. 25.6).

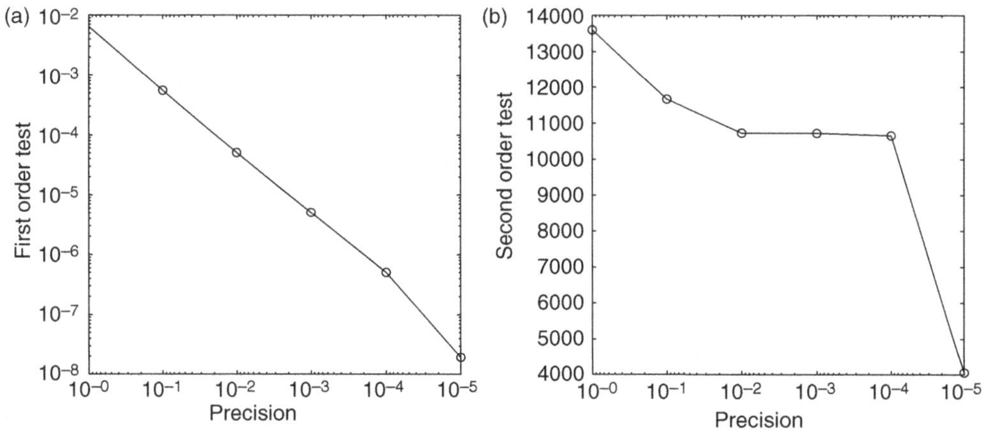

Fig. 25.4 (a) First-order and (b) second-order gradient tests.

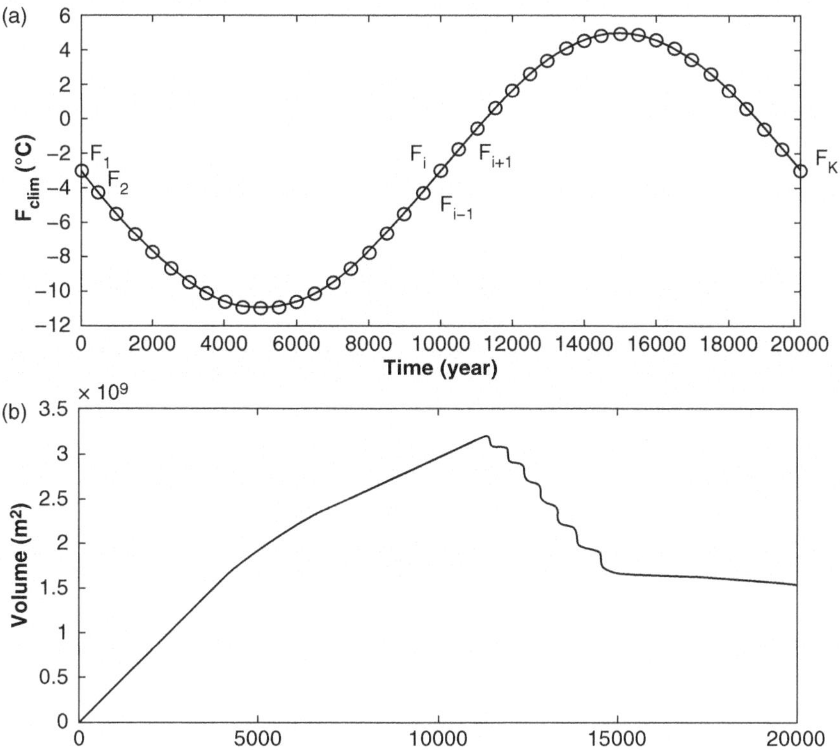

Fig. 25.5 Twin assimilation results: (a) true climatic scenario $T_{\rm clim}$; (b) generated volume data.

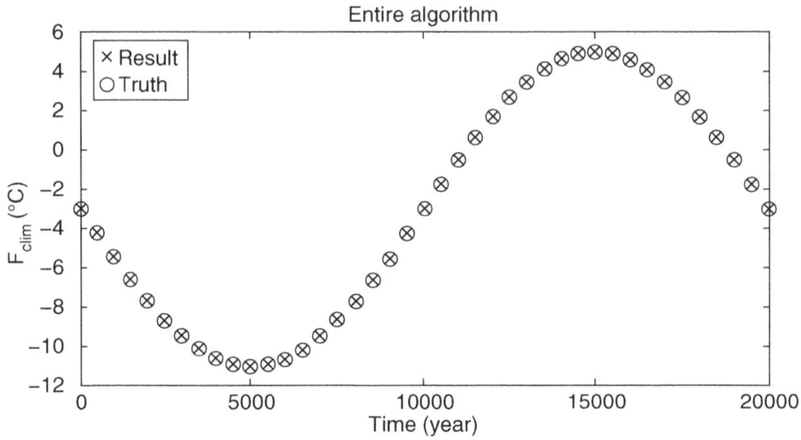

Fig. 25.6 Twin assimilation results: T_{clim} recovered as a result of data assimilation.

References

Bintanja, R., van de Wal, R. S. W., and Oerlemans, J. (2004). A new method to estimate ice age temperatures. *Climate Dyn*, **24**, 197–211.

Bintanja, R., van de Wal, R. S. W., and Oerlemans, J. (2005). Modelled atmospheric temperatures and global sea levels over the past million years. *Nature*, **437**, 125–128.

Giering, R. and Kaminski, T. (1998). Recipes for adjoint code construction. *ACM Trans. Math. Software (TOMS)*, **24**, 437–474.

Greve, R. and Blatter, H. (2009). *Dynamics of Ice-Sheets and Glaciers.* Springer, Berlin.

Paillard, D. and Parrenin, F. (2004). The Antarctic ice sheet and the triggering of deglaciations. *Earth Planet. Sci. Lett.*, **227**, 263–271.

Paterson, W. S. B. (1981). *The Physics Glaciers, 2nd edn.*. Pergamon Press, Oxford.

Raymo, M. E. (1997). The timing of major climate terminations. *Paleoceanography*, **12**, 577–585.

Ritz, C., Rommelaere, V., and Dumas, C. (2001). Modeling the evolution of Antarctic ice sheet over the last 420,000 years—Implications for altitude changes in the Vostok region. *J. Geophys. Res.*, **106**, 31.

The manufacturer's authorised representative in the EU for product safety is Oxford University Press España S.A. of El Parque Empresarial San Fernando de Henares, Avenida de Castilla, 2 - 28830 Madrid (www.oup.es/en or product.safety@oup.com). OUP España S.A. also acts as importer into Spain of products made by the manufacturer.
Printed and bound by CPI Group (UK) Ltd, Croydon, CR0 4YY
14/09/2026
02216290-0009